VOLUME 1

AUDIOPHILE VACUUM TUBE AMPLIFIERS

DESIGN, CONSTRUCTION, TESTING, REPAIRING & UPGRADING

IGOR S. POPOVICH

THIS PAGE WAS DELIBERATELY LEFT BLANK

To my father, Dr. Slobodan Popovich, with thanks for all his help and guidance

DISCLAIMER & COPYRIGHT NOTICE

The information contained in this book is to be taken in the context of general overview, not specific advice. You should not act on the information contained herein without seeking professional advice. Neither the author nor the publisher (or any other person involved in the publication, distribution or sale of this book) accepts any responsibility for the consequences that may arise from readers acting in accordance with the material given in the book. Professional advice about each particular case / instance should be sought.

Our choice of designs, parts, models and brands was based on their availability and educational value. We were not influenced or induced by anybody in our selection, and their use does not mean we actually endorse or recommend them. You should satisfy yourself that a particular device, component or method is suitable for your intended purpose.

Some circuit diagrams of commercial equipment discussed here were not published by the manufacturers, but were posted online by others, and we cannot confirm their accuracy or authenticity. They are used here for review and discussion purposes as "fair dealing", permitted by international copyright laws. Designs marked with a copyright symbol are intellectual property of their copyright holders and should not be used without their permission. They are discussed here from educational perspective only.

COMMERCIAL USE & INDEMNITY NOTICE

Our designs are available for construction but only for your personal, one-off, noncommercial use. Manufacturers must obtain a permission for commercial use.

The consequences of any modifications to and deviations from the featured designs are at your own risk, and no responsibility will be accepted by us.

Tube amplifiers involve lethal voltages, high temperatures and other hazards. By purchasing and reading this book you automatically agree to indemnify its author, publisher and retailer against any claims, of any nature, and for any reason!

Published by Career Professionals in Australia

P.O. Box 5668, Canning Vale South WA 6155, Australia

Revised edition, 2017

Bulk purchases

This book may be purchased in larger quantities for educational, business or promotional use. Please e-mail us at sales@careerprofessionals.com.au

National Library of Australia Cataloguing-in-Publication Data:

Popovich, Igor S.,

AUDIOPHILE VACUUM TUBE AMPLIFIERS - DESIGN, CONSTRUCTION, TESTING, REPAIRING & UPGRADING, Volume 1

ISBN: 978-0-9806223-2-4

1. Electrical engineering 2. Electronics 3. Hi-fi

I Igor S. Popovich II Title III Index

621.3

CONTENTS OF VOLUME 1

1. WHO WILL BENEFIT FROM THIS BOOK AND HOW Page 7
2. BASIC ELECTRONIC CIRCUIT THEORY Page 17
3. ELECTRONIC COMPONENTS Page 31
4. AUDIO FREQUENCY AMPLIFIERS Page 49
5. PHYSICAL FUNDAMENTALS OF VACUUM TUBE OPERATION Page 69
6. VOLTAGE AMPLIFICATION WITH TRIODES - THE COMMON CATHODE STAGE Page 85
7. OTHER VOLTAGE AMPLIFICATION STAGES WITH TRIODES Page 99
8. TETRODES AND PENTODES AS VOLTAGE AMPLIFIERS Page 123
9. FREQUENCY RESPONSE OF VACUUM TUBE AMPLIFIERS Page 137
10. IMPEDANCE-COUPLED STAGES AND INTERSTAGE TRANSFORMERS Page 153
11. NEGATIVE FEEDBACK Page 163
12. TONE CONTROLS, ACTIVE CROSSOVERS AND HEADPHONE AMPLIFIERS Page 179
13. PRACTICAL LINE-LEVEL PREAMPLIFIER DESIGNS Page 193
14. PHONO PREAMPLIFIERS Page 205
15. SINGLE-ENDED TRIODE OUTPUT STAGE Page 221
16. PRACTICAL SINGLE-ENDED TRIODE AMPLIFIER DESIGNS Page 231
17. PRACTICAL SINGLE-ENDED PSEUDO-TRIODE DESIGNS Page 255
18. SINGLE-ENDED PENTODE AND ULTRALINEAR OUTPUT STAGES Page 271
19. THE END MATTER OF VOLUME 1 Page 283

CONTENTS OF VOLUME 2

1. PRACTICAL SINGLE-ENDED PENTODE AND ULTRALINEAR DESIGNS Page 11
2. PUSH-PULL OUTPUT STAGES Page 21
3. PRACTICAL PUSH-PULL AMPLIFIER DESIGNS Page 45
4. BALANCED, BRIDGE AND OTL (OUTPUT TRANSFORMERLESS) AMPLIFIERS Page 67
5. THE DESIGN PROCESS Page 79
6. FUNDAMENTALS OF MAGNETIC CIRCUITS AND TRANSFORMERS Page 87
7. MAINS TRANSFORMERS AND FILTERING CHOKES Page 97
8. POWER SUPPLIES FOR TUBE AMPLIFIERS Page 115
9. AUDIO TRANSFORMERS Page 141
10. TROUBLESHOOTING AND REPAIRING TUBE AMPLIFIERS Page 169
11. UPGRADING & IMPROVING TUBE AMPLIFIERS Page 181
12. SOUND CONSTRUCTION PRACTICES Page 205
13. AUDIO TESTS & MEASUREMENTS Page 229
14. TESTING & MATCHING VACUUM TUBES Page 261
15. THE END MATTER OF VOLUME 2 Page 279

INTRODUCTION TO VOLUME 1

Humans, writers and DIY constructors included, usually overestimate their abilities and underestimate the difficulty or complexity of the tasks ahead of them. After having written eight nonfiction books in the last 20 years, my intuitive feeling was that this book would have 300 or so pages and that I would complete it in four to five months of full time writing. Ultimately, my intuition was proved wrong. It took me a year to finish almost 600 densely-packed pages! Nobody would buy a 600-page technical book, so it was decided to split the material in two volumes, to make it more user-friendly and more manageable.

The quantity and complexity of the material grew significantly from the project's inception to its completion. Instead of simply glossing over or not even mentioning certain aspects of tube technology, I made a faithful decision to try to cover (almost) everything of relevance.

Out of necessity, this tome, Volume 1, is more theoretical of the two. Circuit and tube fundamentals had to be covered in reasonable depth. Theory is intrinsically linked to models and mathematical analysis, and there is plenty of that here, not to prove my mathematical skills, but to give you the tools to perform similar analysis, should you be so inclined, in future. Volume 2 is much more practical, since most of the theory has been covered here.

You will notice a strong emphasis on systems thinking and systematic approach in this book. Sometimes we will study a circuit or an idea and widen the conclusions onto other similar circuits and designs (generalize). At other times we will study a class or family of designs or a general rule and then "zoom in" on specific issues or exceptions to such a rule. Without such a systematic and hierarchical approach the readers student (and many authors!) are in great danger of missing the wood for the trees. Understanding and mastering deeper principles is infinitely more important and beneficial than focusing on details. Details (specific designs, methods, components, figures, etc.) change, go out of production or fashion, but principles last forever.

However, that doesn't mean that the saying "The Devil is in the details" is not true. There is no room for any errors at high voltages and when using expensive vacuum tubes and transformers. A designer and DIY constructor must be able to deal with both macro and micro issues.

Finally, I want to thank you for buying this book and investing your time and energy in reading it. I hope our journey together will be illuminating, profitable and enjoyable for you.

Best wishes for your audiophile journey, see you on the road!

Igor S. Popovich

WHO WILL BENEFIT FROM THIS BOOK AND HOW

- ABOUT US
- WHO IS THIS BOOK FOR?
- THEORY, PRACTICE, MEASUREMENTS AND LISTENING IMPRESSIONS IN THIS BOOK
- HOW TO READ THIS BOOK
- GETTING IN TOUCH

"Miss a meal if you have to, but don't miss a book!"
James E. (Jim) Rohn, self-help philosopher

1

ABOUT US

Another book on vacuum tube amplifiers? Oh please ...

Welcome to your book! If you parted with your hard earned cash to buy it, you are most likely wondering now, have I just wasted my money? If you downloaded or photocopied this book, shame on you! No seriously, I bet you are still wandering, well I didn't pay anything for this, but is it worth investing my time and effort into reading it?

The main impetus behind our fateful decision to invest a few years of our lives into researching, writing, editing and publishing this book is our belief that we have the theoretical, practical and educational skills to accomplish this extremely ambitious task. Ultimately, you will be the judge of how realistic or delusional such a belief was.

Who are we to tell what to do and what not to

Before we start discussing ways of getting closer to that illusive goal of "high fidelity" (whatever that means) and musical reality (or, should we rather be seeking a musical illusion?), please allow me to introduce myself, and also the real guru behind all my knowledge, my technical mentor, Dr. Slobodan Popovich. That is why I use "we" in this book. I may be the guy who put it on paper, but without Dr Bob this book would not be in your hands. I still like to emphasize that I'm still an apprentice, still learning from the Master, and what a better way to learn than by writing a book!

ABOVE: Dr. Bob Popovich as a budding student of physics in 1958, repairing a tube amplifier as a means of financing his university studies.

Oh, yes, I forgot to mention one little detail, in the interest of full disclosure, Dr. Bob is also my dad! I used to call him The Old Man, but since I am well into my middle age and may seem old to many of you younger readers, I either need to start calling him a Very Old Man, which is a bit impractical, or stop it altogether, so from this point onward in this book I will call him Dr. Bob.

Dr. Bob, long retired from his academic career, was a physics professor by day and an electronics guru by night. He started his hands-on-training in tubes early, as a student in our native Yugoslavia. He went on to build receivers, hi-fi and guitar amplifiers for local bands, all using tube technology, of course.

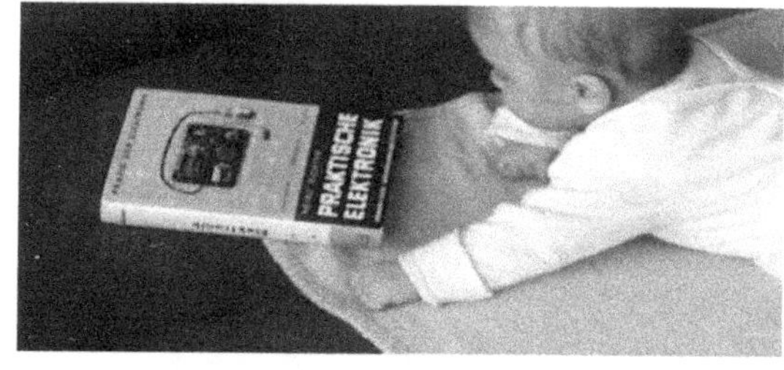

ABOVE: Be careful what kind of book you give your kids to "read", it may have a profound effect on their life. Here I am in 1963, "reading" my first book called "Practical Electronics", in German, no less.

In his spare time he even designed and built plastic welding machines that used RF (Radio Frequency) tube oscillators, fridge-size monsters with huge German tubes stamped "Wermacht".

So, for the first ten years of my life I thought Wermacht was a tube factory, just like Telefunken or Siemens. Bob picked these up for next to nothing from obsolete military stocks left behind by German occupiers after Tito's partisans kicked them out in 1945, at the end of WWII. Little I knew that 40 years later vacuum tubes would feature prominently in my life.

I started with theory first, completed my university degree in electronics, and then went on to learn the practicalities in various engineering positions in Australia. Unfortunately, tubes were out of the curriculum by the early eighties, so I had to educate myself all over again, with Bob's help, of course.

We have been experimenting designing and building tube (valve) amplifiers as a serious hobby, both guitar amps and those pejoratively called hi-fi or even "hi end". At one stage we thought of establishing a commercial operation, but decided against it. Manufacturing such a product in Australia, where labor costs are high and imported parts are even more expensive, would be a losing proposition.

By then (late 1990s) Australian retailers were starting to sell Chinese branded amplifiers which were cheaper than what only the parts would cost us.

Then, one sunny day I had an epiphany: Why don't we build an amplifier, film the whole process and make it into a video workshop so prospective DIY builders can see exactly how it's done? I had paid $1,700 for a Panasonic video camera a couple of years before and it was just sitting there, not being used much. My heart still aches when I remember, that was a huge sum at the time. So, we may as well get some of that investment back, I thought. We went ahead and started selling our know-how packaged as educational DIY DVD sets.

WHO IS THIS BOOK FOR?

Why design and build your own tube amplifier or preamplifier?

The love of good sound is the driving force behind every audiophile. It is not love-for-love's sake, although some audiophile components are so beautiful you can just look at them for hours. It is about emotions, love, and ultimately, making the listener feel better about him or herself, and the world around us. However, mot audiophiles have no technical knowledge, skills or even desire or inclination to build their own audiophile components (DACs, amplifiers, speakers, cables). If you are reading this book, you most likely posses such skills, and you are ready to raise to the challenge of constructing your own gear.

With all due respect to our less technically inclined audiophile friends, anybody can go into a hi-end shop and buy a tube amplifier. All you need is wads of money. Building your own gear can save you a small fortune, and bring you that smug feeling of satisfaction that your $2,000 DIY amplifier sounds as good or even better than $8,000 commercial models.

The feeling that you were able to conceptualize and manage such a project and achieve all or most of your design goals results in even higher satisfaction - it is one of the greatest feelings in the world, a source of pride and joy.

I would hazard a guess that to most constructors the equipment of their own making sounds better than commercial equivalents. The reason is simple - it is because they've made it!

The question of balance

The main issue in writing this book was deciding how deep to go into the physical properties of tubes and circuits, and, even more importantly, how deep to go into the mathematics that describes such physical behavior.

Inevitably, every book is a compromise, just as any amplifier design is. No matter where you draw the line, the material will be too highbrow for some, and too pedestrian for others. We tried to achieve both aims, to be as practical as possible, and, simultaneously, for those of you who don't mind relatively complex maths, to underpin the praxis with necessary theory. The theory explains the origin and appropriateness of empirical formulas and "rule-of-thumb" simplifications, their limits, the assumptions behind them and the necessary conditions for their use.

You may have read some books on tube amplifiers, in which their authors make unsubstantiated statements and sweeping generalizations. These leave you perplexed and/or annoyed, since the critical question "Why is that so?" isn't answered very often. For instance, something like "Some amplifier builders want to use tube rectifiers. There is nothing that a tube rectifier can do that a 5 cent silicon diode cannot do better."

If that was true, why would companies such as Messa Boogie include both types of rectifiers in their top models of guitar tube amplifiers, switch-selectable? Obviously, the two types of rectifiers behave differently and affect the sound of amplifier in various ways. Such flippant comments have no place in a serious book on vacuum tube technology.

On the other end of the range are books with 800+ pages whose authors go through millions (literally!) lines of formulas and calculations. Yet, after all that mathematics you are left none wiser, or, more likely, totally confused and stupidified. These guys miss the wood for the trees.

You may be thinking now, how dare this author be so arrogant to put all others down! I apologize if I came across as too critical or too cocky, that certainly wasn't my intention. I certainly don't know everything and always try to get to know my limitations, as you should do too. My aim here is simply to illustrate and justify the need for a book that will try to balance things out and present a unified view of the topic.

Amateurs, audiophiles and "professionals"

While some of you may have a technical or mathematical background, I suspect that most won't. You may be in search of a hobby, and amplifier designing and building is the greatest hobby of all. Or, you may have the basic knowledge of electronics and want to take things to a higher level. You may be an audiophile who is dissatisfied with constantly upgrading expensive "hi-end" gear that is long on promise but delivers ordinary results, and sick and tired of losing large wads of money to importers, dealers and other middlemen.

No matter what your situation is, it is reasonable to assume that you are not a professional in this field, i.e. that you are an amateur. Before you get offended, the word amateur comes from the Latin "amicus", meaning friend or companion. French "ami" and Italian "amico" derive from it too, as does the word "amore" or love. Thus, it means a person who does something out of love, and not for profit.

Unfortunately, the original meaning got corrupted over the millennia, and now amateur is used instead of "dilettante", a person who lacks knowledge and doesn't know what he or she is doing, compared to the so called "professionals", who are supposed to be all knowing, almost omnipotent.

We all know how overrated, even useless many "professionals" are in their "profession". Alas, we cannot rectify the crooked or correct the ingrained, and that is why instead of "amateur" we will use the term "audiophile" in this book.

Dictionaries define an audiophile as a person enthusiastic about high-fidelity sound reproduction. Since high-fidelity is another unfortunate term that is even harder to define, this definition leaves us none wiser. However, to bring this mildly philosophical pondering to some kind of conclusion, if you feel you would benefit from reading this book and if you are prepared to invest time, money and effort into reading it, you must be enthusiastic about vacuum tube amplifiers in particular and sound reproduction in general, and that definitely makes you an audiophile.

The great news for solid state "converts"

When Dr Bob said he could make a tube amplifier that would sound better than solid-state Marantz, Yamaha, Onkyo, Denon, or any of the then dominant mega-brands, I must admit I was apprehensive. I knew how capable he was, but I still wasn't a true believer. How can an amp with six tubes sound better than a solid-state amp designed by the best Japanese minds? Well, I still don't know how and why, but it did sound better. Much better in fact!

So, if you are a solid state "convert", here are the good news for you. Tube electronics is simpler than solid state electronics. The design and analysis of tube circuits is easier and faster, primarily because the input and output circuits are non-interactive, due to the almost infinite input impedance of most tube circuits. The base and collector currents in bipolar transistors are interactive, the input and output circuits cannot be analyzed separately and independently.

Secondly, if you understand FETs (Field Effect Transistors), you will understand vacuum tubes in no time. The internal workings cannot be more different, but the outward behavior is very similar! Pentodes and beam tubes are similar to bipolar transistors to the extent that both are current sources. A pentode is a voltage-controlled current source, a bipolar transistor is a current-controlled current source. More on that soon ...

Hi-fi amps compared to guitar (instrument) amps

I am sure one of the most common questions we'll get from prospective buyers will be "I am interested in guitar tube amps, does this book cover them?". The answer is yes and no. Yes, because the principles covered in the early chapters apply to any tube amplifier, be it for hi-fi, guitar, telecommunications or industrial use, and no, because the aims and guiding principles are different for hi-fi and for instrument amplifiers. The art of designing and building guitar amps revolves around achieving the required gain and the desired type of distortion, power sag and a distinctive character or voice of an amplifier, all of which are highly undesirable in hi-fi designs, which aim to be as neutral as possible and to reduce their own contribution to tonal balance.

Of course, there are always exceptions. Bass guitar and keyboard amps for instance, are closer to hi-fi amps, as are PA (public address) amps, since they need to be "clean", meaning they should not distort as lead guitar amps must be able to when the guitar player chooses to overdrive them.

In essence, a hi-fi tube amp is a reproduction device that amplifies the low voltage, low power input signal from the source (turntable, CD player, D/A converter, etc.) into a high current, high power electrical signal needed to drive the electromechanical transducer (loudspeaker). Both should be as "neutral" as possible, meaning they should add as little of their own sound coloration to the recorded sound. Real amps and especially loudspeakers are never neutral, they always add their "personality" or voicing to the ultimate result.

The basic job of a guitar amp is also to amplify the weak signal from a guitar pickup and drive the loudspeaker, but a guitar amp is also an instrument, not just a transducer, since its second function is to add its own voice to the overall result. In other words, a guitar amp should produce a tone or a variety of tones that guitar players are looking for, meaning they should distort the incoming guitar signal in certain desirable ways.

In addition to tonal and distortion controls, a guitar amp may include additional effects, the most common being tremolo (amplitude modulation of the signal) and reverb or echo.

In terms of functionality, modern hi-fi amps are very Spartan. Many don't even have volume control (these are called power amplifiers), since volume control is performed by the preamplifier. Tone controls are a thing of the past in hi-fi, as are presence and loudness controls. The prevalent contemporary approach in hi-fi circles is minimalist in essence, "Less is more!", meaning the less switches, potentiometers, cables, etc. the signal goes through, the better/cleaner/more pure the sound.

Guitar amps, being musical instruments, tend to include various controls, of which gain, master volume, tone (bass, middle and treble), "overdrive", "bright" and power attenuation are the most common. Then there are manufacturer-dependent tonal controls named "presence", "resonance", "texture", "contour", "crunch", "fuzz", "Tubescreamer, "Vaporizer" and heaps of others.

TUBE GUITAR AMPLIFIERS

- portable, sound adjustable by positioning and slanting of the amp
- distortion welcome, at least the desired kind
- the amplifier is part of the instrument, its distortion behavior and voicing are of paramount importance
- a wide range of inbuilt effects and controls (tone, presence, gain boost, reverb, tremolo, ...)
- a narrow range of tube types used (the top five being 12AX7, 6L6, 6V6, 6BQ5 and EL34)
- designs are very similar and relatively predictable
- tube guitar amps are relatively affordable, low powered ones priced under $500, high power ones $1,000-2,000, except "boutique" amps, which can be even more expensive

Hi-fi TUBE AMPLIFIERS AND SOUND REPRODUCTION

- stationary, setup critical for optimal results
- neutrality and distortion minimization are the main aims
- minimalist approach, volume control only, no tone controls
- a wide range of preamp and power tube types used
- designs & topologies vary widely (stereo amps, monoblocks, OTL amps, directly-coupled and transformer-coupled amps, etc.)
- good quality and brand names audiophile amps are expensive, $5-10,000 or much more

Hi-fi, hi-end and audiophile gear, is there a difference?

Hi-fi stands for "high-fidelity" and over the years has been attributed to amplifiers and systems deserving of this epithet, but also to some questionable audio components. The term itself is relative and therefore debatable.

Hi-end is equally problematic. The very premise of this book is to help you avoid forking out a small fortune on overpriced hi-end products by designing and making a better product yourself, for less than 20% of the cost! A pair of monoblocks, each with a single amplifying 6C33C-B (plus one more as a voltage regulator) selling for US$63,000? That is hi-end (price-wise) and is also preposterous.

The names and epithets are not the only problematic issue in hi-end audio. No other field suffers from so many myths and misconceptions, so much fluff and so many furfies. This is mostly due to the fact that very few designers, manufacturers, retailers and reviewers have any kind of engineering background, so ignorance and copycat attitude rule the field.

Of course, the audiophiles and the DIY community are not entirely without blame. We get many inquiries about our DIY DVD programs of the "I know nothing about electronics but want to build a tube amplifier. Will your DVD program make me able to do it?" type. This is akin of me saying " I know nothing about medicine or anatomy, but I want to perform cosmetic surgery on my wife!" Building an amplifier is a highly creative and highly demanding act, requiring not just practical, mechanical and soldering skills, but also a deep knowledge of tube electronics. The aim of this book is to get you closer to that level of mastery.

THEORY, PRACTICE, MEASUREMENTS AND LISTENING IMPRESSIONS IN THIS BOOK

From knowing WHY to knowing HOW (or the other way around?)

I still remember some of my fellow first-year electrical engineering students. Even at that tender age they were already building hi-fi amplifiers, phono stages and active crossovers, albeit of the solid state kind (that was in 1982, the solid state era). They would talk about MOSFETs, biasing, intermodulation distortion and heaps of other things that I had no clue about, as if they were talking about Spanish villages, as the saying goes. Yet, very few of those, let's call them young "practitioners", passed the mathematical analysis, calculus, solid state physics and other highly theoretical subjects in the first two academic years; almost none graduated four years later, as I did.

I must admit, even after graduating in electronics, I still had no clue about the things they talked about, and wasn't any closer to being able to design or build an amplifier. My degree in electronics was for all intents and purposes a degree in high-level mathematics, with very little practical applicability, almost useless from a practical point of view.

Referring now to the diagram on the next page, I knew a bit about WHY things were a certain way, but I had no clue about HOW to implement such theoretical models. The young DIYers knew the rules-of-thumb, or the shortcuts to practicality, but didn't understand why or how those shortcuts or rule-of-thumb were arrived at, by which process, what the limits of such simplifications were, when they applied and when they didn't.

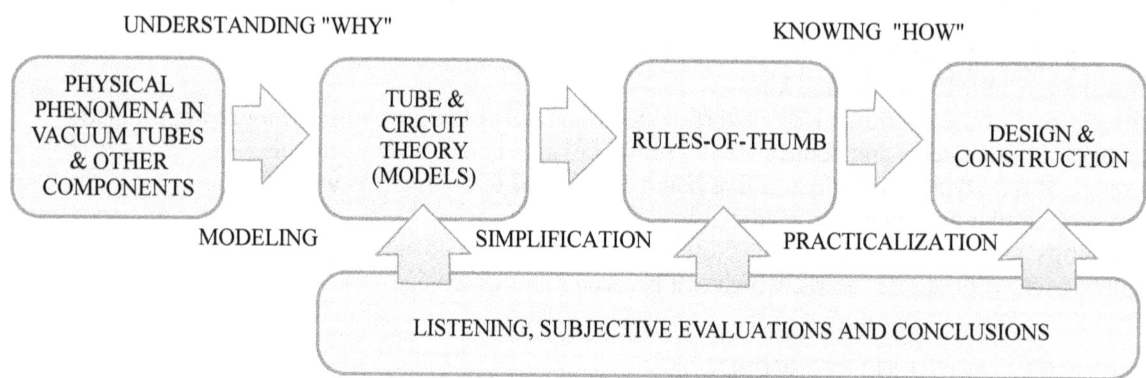

ABOVE: The model of audiophile and DIY understanding - how theory, practice and listening all fit nicely together

Strictly speaking, to build a vacuum tube amplifier of any kind, one does not need to understand the physical and chemical processes that take place inside vacuum tube or valves as Brits and Australians call them. However, it does help to at least have a basic idea. We will try to achieve that in this book, but won't dwell too long on it.

The same applies to other electronics components used in amplifiers, such as resistors, capacitors, wires and the class of components we call "magnetic components. These are filtering, anode and grid chokes, power transformers and audio transformers (output and interstage transformers).

So knowing exactly why things happen in a certain way is the domain of "professionals", a deeper level. Knowing what happens is the more basic or practical level.

In the model above, as we move from the leftmost box to the right, the building blocks of our understanding become more and more practical. They also become more simplified. We will often use simplified models of tubes and amplifier stages, so please understand that such models have their limitations. They don't apply to all situations and cases. For instance, one common method of analysis and design is to divide the frequency range of an amplifier in three frequency ranges or bands, bass, midrange and treble, and then proceed to analyze those using different models. This si a common practice in electronics engineering.

Rules-of-thumb are called *FaustFormeln* in German, most likely referring to Dr. Faust (or Faustus), a successful yet frustrated scholar searching for the true knowledge and the essence of life. Dr. Faust makes a pact with the Devil, exchanging his soul for unlimited knowledge, power and pleasure. FaustFormeln are simplified formulas and shortcuts through complex analysis or synthesis (design), which makes them easy to remember and fast to use in practice. They are not "carved in stone" and do not always apply, so should be considered flexible guidelines rather than strict laws.

RULE-OF-THUMB

A simplified formula or guide, valid in most but not all situations, considered a rule, but often broken by designers and constructors with deeper knowledge and more confidence.

Total beginners notwithstanding, I am convinced this book will have two major kinds of readers, those who understand maths and physics and want to master the practical aspects and get the "know-how", and those who are soldering-iron wizards but want to cross the knowing-doing gap from the opposite direction.

They may have a pretty good feel for what works and what doesn't, so now they want to systematize that knowledge and underpin it with some theory and deeper understanding of the substance.

Obviously, even after a lifetime devoted to this field you will not understand everything. For instance, some audiophiles or guitar players may claim that flat-plate Telefunken ECC83s sound better than their ribbed-plate cousins, made in the same factory and during the same period, or that mesh-plate 300Bs sound superior to their solid-anode brethren. Assuming their claims are correct (confirmed by hundreds of listeners on dozens of different amplifiers), does anybody really know why?

I doubt. We may each have a plausible explanation, but there can be no conclusive proof of the underlying cause! Some things simply are, and searching for the whys behind them is an exercise in futility.

However, don't let that discourage you. One reason many students abandon their studies is the paradox of getting deeply immersed in a subject. The more you learn, the faster you will realize the limits of your knowledge. You will find an answer to one question, but three new questions may be raised in the process.

You will always have more questions than answers, and that frustrates a certain people, the ones who claim that it's better to stay blissfully ignorant and happy, then to become frustrated in the ongoing quest for knowledge

What matters and what doesn't

Reviewers of audiophile equipment face a difficult balancing task. They have to reconcile the listening impressions with measurement results, the visual aspect of the reviewed products, their quality of manufacture or workmanship (is that sexist, should that be "workpersonship"?), the reliability and the quality of the after sales support and warranty. Probably the most important issue of all, the quality of the design, is an even bigger minefield. Design features such as the basic topology, only get mentioned in passing, if at all.

Obviously, reliability cannot be properly evaluated in a short time they spend with a piece of equipment (days and weeks instead of months or years), and neither can the quality and the inclusiveness of the warranty or the responsiveness during after sales support. Reliability can be inferred by a detailed analysis of the design, the currents and voltages and the component rating in the circuit, but no magazine that I know of pays any attention to that!

For instance, if one 300B SET amplifier biases the output tubes at 60mA and the other at 85mA, the one with higher idle current will chew through those tubes much faster and its long-term cost of ownership will be much higher (assuming of course that all other aspects of the two amplifiers are comparable and correctly engineered).

Some magazines, such as *Stereophile*, don't use any formulas or cumulative scores, others use a basic star system (up to 4 or 5 stars). One German magazine evaluates hi-fi components based on the 70% - 10% - 10% - 10% formula: 70% sound quality, 10% measurements, 10% construction quality, and 10% value for money. The cultural difference between the prudent American approach and the pedantic German attitude could not be starker.

Paradoxically, as soon as you start assigning percentages to different aspects of hi-fi gear, you raise more questions than you answer. For instance, assigning only 10% for "value-for-money", regardless of how you define this term (another minefield) is questionable!

We faced similar difficulties in writing this book. How deep should we go into theory, and how much should we talk about practice? Whatever balance we strike, we can annoy and alienate a great deal of readers, some of whom expect thorough analysis and calculations (greetings to my German friends!) , while others want to see practical circuits and wiring diagrams. An author should have a "typical" reader in his mind while writing a book, but a typical reader is an oxymoron, kind of like "military intelligence" or "practical theory". Anyway, no matter what kind of reader you are, I hope you will find something useful within these pages!

HOW TO READ THIS BOOK

One way to read a technical book like this one is to immediately go to a section or topic that interests you, and then to keep jumping back-and-forth to the related issues and chapters. This will, I suspect, be the way the more experienced designers and constructors will approach this book.

A more systematic approach is sequential, starting from the beginning and reading in order. This is what I would recommend. Although it seems more time consuming (since you will read about many issues you may already know a lot about), paradoxically, this approach is often faster. You will not miss anything, and you will not waste time flipping forward and backward trying to clarify an issue that you've overlooked and perhaps not fully understood.

Whatever you do, don't treat this book as Holy Scripture. Underline or highlight the important parts, write your thoughts and ideas on its margins, sketch diagrams and circuits in its blank spaces.

MEASURED RESULTS:

- BW: 15Hz - 35 kHz (-3dB, at $10V_{RMS}$ into 8Ω)
- V_{MAX}: $11V_{RMS}$
- $P_{MAX} = 15W$

Most of the circuit diagrams in this book have been tried in practice. When you see this type of frame you can rest assured that the design is either of a commercial amplifier or one that has been built and thoroughly tested by us.

RULE-OF-THUMB

Load impedance for triodes:

$Z_{AOPT} \approx 3r_I - 4r_I$

r_I = internal resistance of a tube

For amplifier builders who don't want to bother with high-level maths, models and similar highbrow concepts, these Rules-of-Thumb are simple shortcuts. Easy to memorize, they approximate and summarize much more complex formulas, methods and concepts.

Although each detailed circuit diagram in this book could be a DIY project by itself, small projects are framed and marked with this soldering-iron symbol.

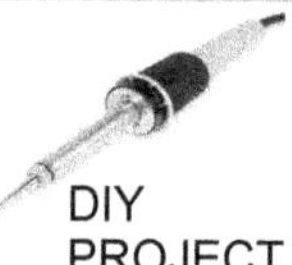

DIY PROJECT

TUBE PROFILE: XXXXX

Each of the tubes discussed or featured in the designs in this book will have its basic parameters and operational data summarized in box such as this one.

Commercial designers and manufacturers wish to protect their practical knowledge and "insider secrets". Framed boxes of this kind will emphasize lesser-known practical tips and tricks. Although a magician's hat and a magic wand are used as symbols, there is nothing magical about these trade secrets, all are underpinned by solid science and engineering principles.

TRADE TRICKS

DESIGN CALCULATION or ANALYSIS

The sharpened pencil symbolizes a numerical example or calculation.

IMPORTANT FORMULA

The calculator symbol indicates an important or often-used formula.

CRITICAL QUESTION

While there are no stupid questions (only stupid answers!), some questions are of far more importance than others. These are answered in frames with this symbol.

KEY FEATURES

The key aspects, strengths or interesting features of certain designs are summarized in a frame with a key.

A WARNING OR A VERY IMPORTANT POINT!

Some issues, myths and warnings are so important that they warrant being emphasized in a frame of this kind.

Currents, voltages and other markings on circuit diagrams

250V (boxed)	DC voltage in the marked node (quiescent state, no signal)
1V ~	AC signal voltage in the adjacent node (RMS or effective value)
→ 5mA	DC current through the adjacent branch (quiescent state, no signal)
A=55	Voltage amplification of the adjacent stage

Abbreviations used (in no particular order)

AC	Alternating current	MAX	Maximum	GND	Ground terminal
DC	Direct current	MIN	Minimum	COM	Common terminal
RIAA	Recording Industry Association of America	BNC	Bayonet Neill–Concelman, video & test equipment connector for coaxial cable	ESR, ESL	Equivalent series resistance and inductance (of a capacitor)
THD	Total harmonic distortion	IMD	Intermodulation distortion	ELCO	Electrolytic capacitor
PIO	Paper-in-oil (capacitor)	NOS	New Old Stock	CCS	Constant current source or sink
F&F	Film and foil (capacitor)	PP	Push-pull (amplifier)	CF	Cathode follower
MF	Metallized film (capacitor)	PP	Polypropylene (capacitor)	SPL	Sound pressure level
MM	Moving magnet (phono cartridge)	PP	Peak-to-peak (AC signal)	NFB	Negative feedback
MC	Moving coil (phono cartridge)	PPP	Parallel push-pull	PFB	Positive feedback
LOG	Logarithmic scale or taper (potentiometer)	NTC, PTC	Negative and positive temperature coefficient (of a resistor or other component)	GBW	Gain-bandwith product of a tube or amplifier
LIN	Linear scale or taper (potentiometer)	SET, PSET	Single-ended triode, parallel SET	EMF, CEMF	Electro-magnetive force, counter EMF
TR, VR	Turns or voltage ratio (of a transformer)	E/S	Electrostatic (field, interference or shield)	RCA	Unbalanced audio connector or Radio Corporation of America
IR	Impedance ratio (of a transformer)	SQ	Special quality (tube)	XLR	Balanced audio connector (3-pins)
RC	Resistor-capacitor coupling between stages	SRPP	Shunt-regulated push-pull (stage or amplifier)	SLO-BLO	Slow blowing fuse, delay fuse
LC	Inductive-capacitive coupling between stages	GOSS	Grain-oriented silicon steel (transformer lamination material)	FET, JFET, MOSFET	Field effect transistor, Junction FET, Metal Oxide FET
CG	Control grid (in a tube)	SG	Screen grid (in a tube)	SP	Suppressor grid
TPV	Turns-per-volt (of a transformer)	CT	Center tap (of a transformer)	DF	Damping factor
CRC	Capacitor-resistor-capacitor filter	CLC	Capacitor-inductor-capacitor filter	UL	Ultralinear (output stage)
AF	Audio frequency	RF	Radio frequency	RFI	Radio frequency interference
PSRR	Power Supply Rejection Ratio	CMRR	Common Mode Rejection Ratio	LCR	Inductance-capacitance-resistance meter or circuit
DHT	Directly heated triode	NFB	Negative feedback	RMS	Root-Mean-Square
KL	Kirchoff's Law	BW	Bandwidth	EM	Electromagnetic

GETTING IN TOUCH WITH US

If you've liked the book and benefited from it, the best way to repay a favor is to recommend it to your friends and colleagues, or to write an online review. If you spot an error or an omission or should you have any constructive criticism of the book, I'd really like to hear from you, so we can fix it together. Also, if you want to contribute diagrams or projects, or if you have any ideas on how to make the next edition better, please let me know.

My e-mail is igorpop@careerprofessionals.com.au

SYMBOLS USED

≈ APPROXIMATE

≡ EQUIVALENT

|| PARALLEL CONNECTION

RESISTORS

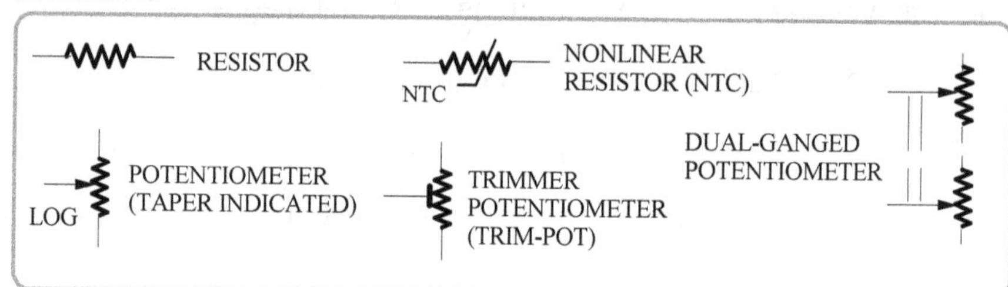

MAGNETIC COMPONENTS

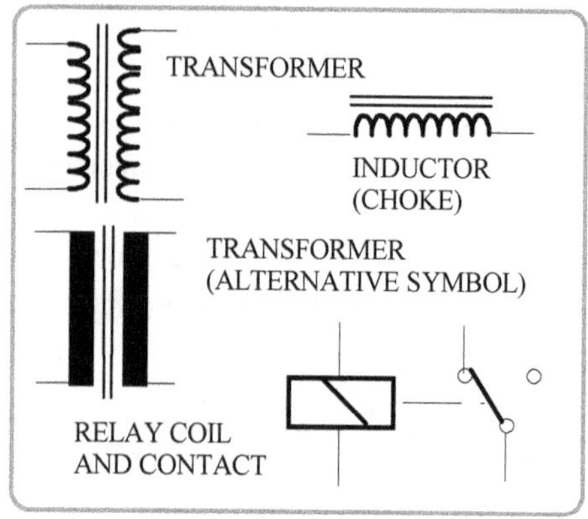

SEMICONDUCTORS

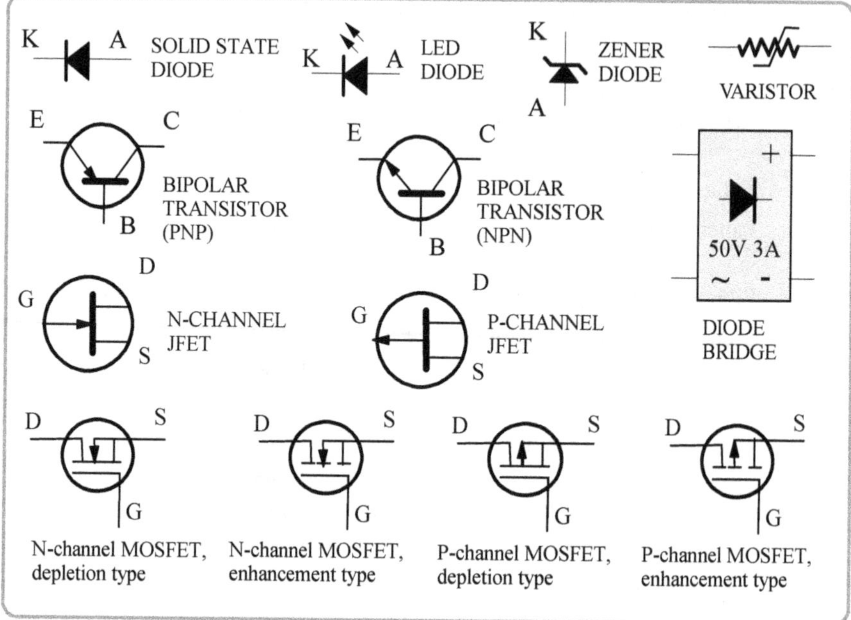

CAPACITORS

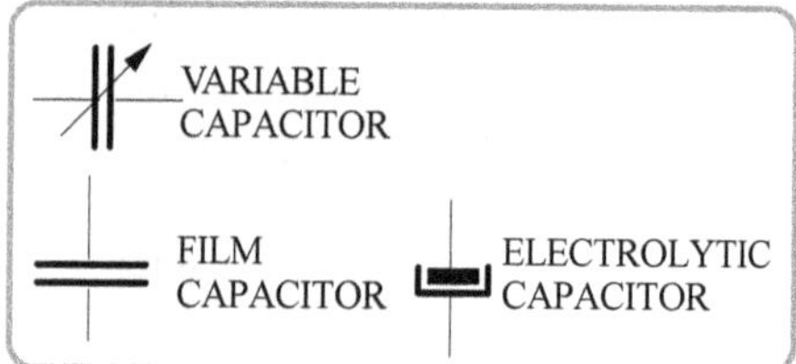

AC AND DC SOURCES

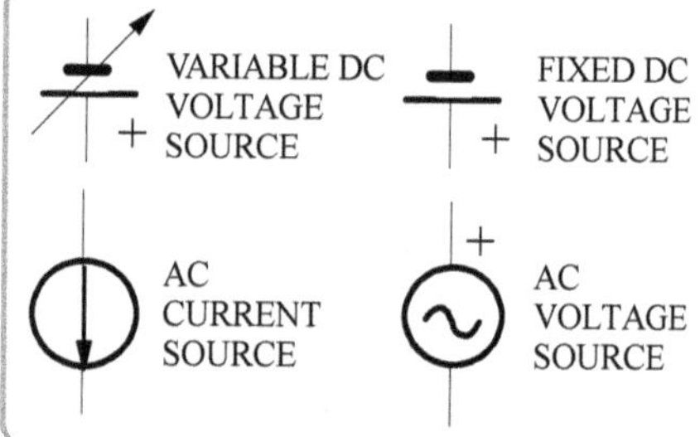

ELECTRON TUBES

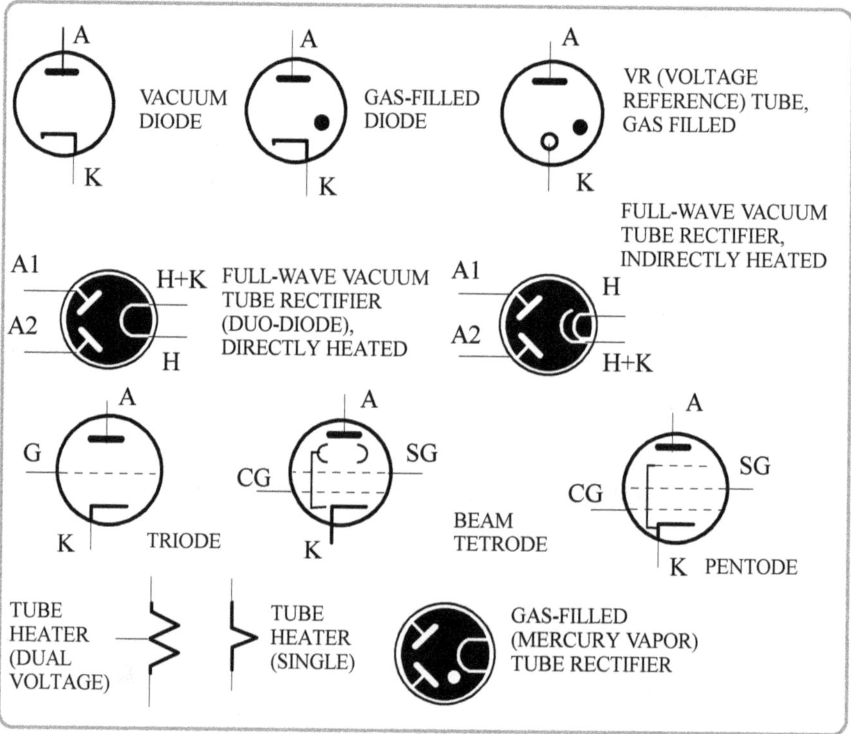

MISCELLANEOUS SYMBOLS

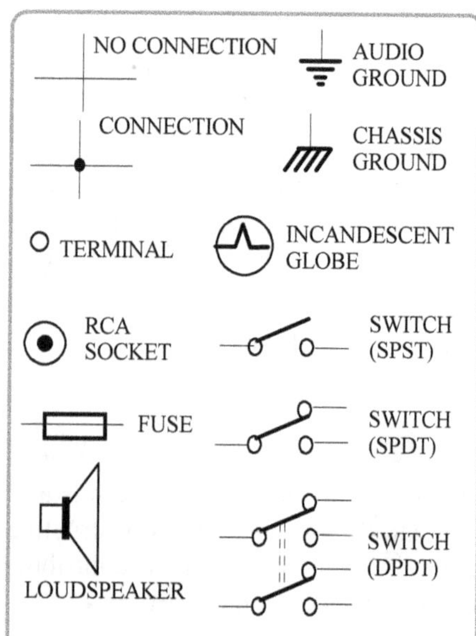

TEST INSTRUMENTS

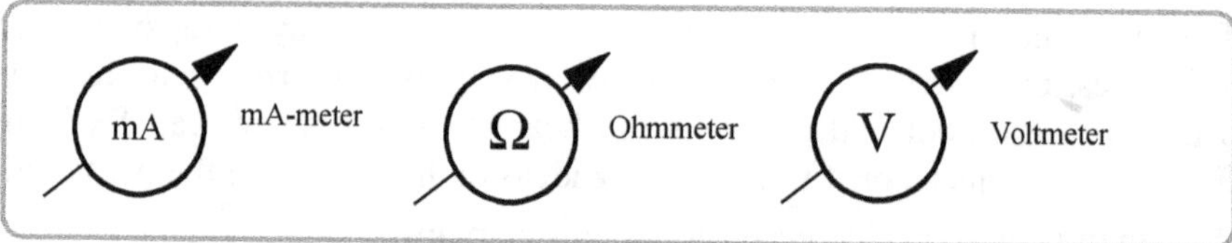

BASIC ELECTRONIC CIRCUIT THEORY

- FUNDAMENTAL LAWS OF ELECTRICAL ENGINEERING: DC circuits, Ohm's Law, Kirchoff's 1st or Voltage Law, Kirchoff's 2nd or Current Law
- CIRCUIT ANALYSIS TOOLS AND METHODS: Current loop method, Node voltage method, Superposition theorem, Reciprocity theorem
- AC VOLTAGES AND CURRENTS: Basics of vector algebra, Power in AC circuits, Signals in time- and frequency domains
- RESISTIVE, CAPACITIVE AND INDUCTIVE IMPEDANCE: Ideal resistors, Ideal capacitors, Ideal inductors, Real inductors, A brief introduction into electromagnetic induction, RCL circuits
- THE BLACK BOX CONCEPT: THEVENIN'S AND NORTON'S THEOREMS

"Well begun is half done."
Horace (Quintus Horatius Flaccus), Roman poet

2

FUNDAMENTAL LAWS OF ELECTRICAL ENGINEERING

Ohm's Law

There are three fundamental laws that underpin all other electrical and electronic engineering formulas and laws. Pretty much everything else is either based on these fundamentals or flows on directly from them. The first is Ohm's law, named after Georg Simon Ohm, German physicist and mathematician (1789–1854).

The simplest possible electrical circuit consists of one voltage source and one resistor. Without the load (resistor) connected, the battery terminals are open and no current flows - there is no "circuit". At such a low voltage (12 volts) an air around the battery is an insulator. Once there is a conductive path across the terminals (nodes) A and G, no matter how large or small such a resistance may be, the circuit is closed, and a current will flow.

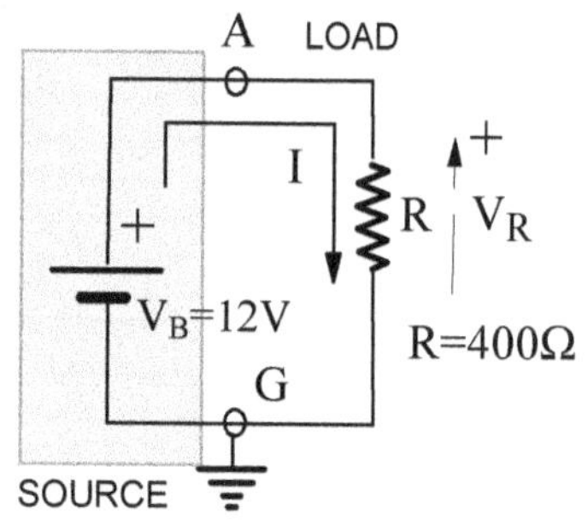

Resistors "resist" the flow of current and limit it in proportion to their resistance. The higher the resistance, the lower the current. A resistor is a physical component, a device, while resistance is its physical property.

A DC source could be a 1.5V AA battery or a 12 Volt car battery. Even though we use the battery symbol in circuit diagrams, this DC voltage source does not have to be of chemical nature (a battery), it could be a stand-alone fixed or adjustable mains-powered DC power supply on a test bench or a DC power supply inside a tube amplifier.

DC or "Direct Current" in voltage terms means that one terminal is always positive, and the other one always negative, with respect to each other. In electrical engineering positive and negative are relative designations.

One terminal is chosen as a referent point, in this case point G. Often (but not always), this point is grounded or "earthed", physically connected to the neutral or earth terminal of a mains supply, so a symbol G, GND or E is used. Likewise, COM or REF may be used to indicate that this point is common or referent point. Of course, point A could be grounded instead and be considered a referent point. Then point G would be at -12V with respect to ground.

Ohm's Law describes the relationship between three parameters in a circuit, voltage, resistance and current. Since there are three parameters, the same law can be written in three different ways: **V=RI, R=V/I or I=V/R**

These can also be expressed verbally in three ways. The voltage drop on a resistor is a product of its resistance R and the current I flowing through it. The resistance of a resistor is the ratio of the voltage across the resistor and the current flowing through it. The current through a resistor is the voltage drop across the resistor divided by the resistance of the load.

In this case the voltage drop V_R on the resistor equals the battery voltage (V_B=V_R), because there are no other resistors in the circuit. Notice the current flowing out of the positive end of the battery and into the positive end of the load (the resistor). Resistors don't have positive or negative ends, the + next to the resistor indicates the polarity of the voltage drop across it ! This is how you can tell the difference between sources and loads in a circuit. Currents come out of the + poles (or terminals) of sources and enter into the + poles of the loads.

Source, load and resistance

The term "load" is very important in circuit analysis. The battery is the source of electrical energy, the source of electrons, negatively charged particles whose flow is called "electrical current". The resistor is the load, meaning it uses that energy in some way. The "load" either dissipates energy by converting it into heat (an electrical heater, which is nothing but a powerful wirewound resistor in open air or in oil-filled tank) or it converts it into other forms of energy, such as mechanical energy, as electric motors do.

In amplifiers, resistors are used to produce DC voltage drops and thereby reduce voltage levels, or to convert the flow of alternating current into AC voltage signals. More on that soon. The unit for voltage (both AC and DC) is called Volt, in honor of Italian physicist Alessandro Volta (1745–1827).

Notice different ways resistors can be "marked" in a circuit. The symbol for resistance unit, also called Ohm, is a Greek capital letter Ω (omega). There are larger units, such as kiloohm (1kΩ = 1,000Ω) and MΩ (megaohm, 1 million ohms or 10^6Ω), and smaller units such as milliohm (0.001Ω).

In amplifier diagrams we usually write 1k2 instead of 1.2kΩ. Apart from avoiding awkward Greek letters (apologies to our Greek readers!), this has another advantage. Circuits get copied by hand or photocopied and the . (dot) between 1 and 2 in 1.2kΩ can easily get lost or omitted by mistake. Suddenly, 1.2kΩ becomes 12kΩ, which can have serious consequences, the finished amplifier may not work properly or even go up in smoke. Using 1k2 naming convention prevents this from happening!

CURRENT FLOW AND ELECTRON FLOW

In modern literature the positive flow of current is the opposite of the electron flow. Electrons are negatively charged particles, so in this circuit they will come out of the battery's negative terminal and flow upwards through resistor R towards the positive pole of the battery. Opposites attract in love and in electronics!

In some older and a few American books you may find the direction of electron flow as positive current flow, but there is no need to get confused - the positive current flow is nothing but a convention, you can use either standard providing you do it consistently.

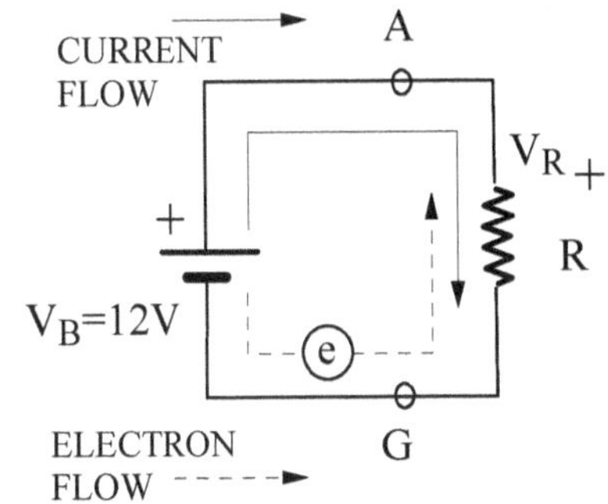

The "official" direction of current flow is the opposite of electron flow

Kirchoff's 1st or Voltage Law

The two laws were named after German physicist Gustav Robert Kirchoff (1824–1887). We'll illustrate them on a couple of examples. A simple DC circuit with one source and a load, two resistors in series, R_1 and R_2. The same current flows through them and the same current "develops" the voltage drops V_{R1} and V_{R2}. Kirchoff's First Law says: **The sum of voltage drops in a closed loop equals the sum of voltage sources.**

Here we introduce the notion of a "loop", which means going around a closed circuit and writing the voltage equation of such a loop. We can do it in two ways. We can put all voltage sources on one side of the equation and all voltage drops on the other, so again we get $V_B = V_{R1} + V_{R2}$

Alternatively, we can put all voltages on the same side of the equation. We can start at any point and go in any direction. Let's start in point A and go clockwise. The arrow will hit the + side of V_{R1} first, then the + side of V_{R2} and finally the - side of the battery V_B, so we write $V_{R1} + V_{R2} - V_B = 0$ Why zero? Because that's what 1st KL says, that the sum of voltages in a closed loop is always zero. When you look at the first equation, $V_B = V_{R1} + V_{R2}$, it is exactly the same equation, just written in a different way.

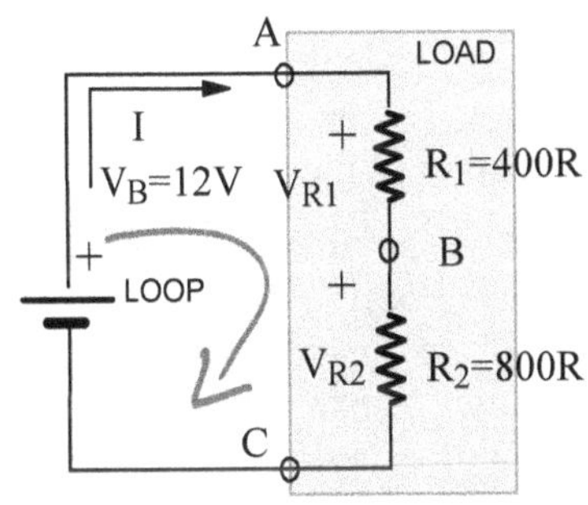

The 1st KL: The sum of voltages in a closed loop is zero.

Let's calculate the values for the current I and the two voltage drops V_{R1} and V_{R2}! Ohm's law says that $V_B = V_{R1} + V_{R2}$ and since $V_{R1} = IR_1$ and $V_{R2} = IR_2$ we have $V_B = IR_1 + IR_2 = I(R_1 + R_2)$

Finally $I = V_B/(R_1+R_2) = 12/(400+800) = 12/1{,}200 = 0.01$ A [Ampere]

Ampere is the international unit (abbreviated SI from French: Le Système International D'unités) for electric current, named after André-Marie Ampère (1775–1836), French mathematician and physicist. A smaller unit called mA (milliAmpere) is often used in electronics, 1mA= 0.001A or 10^{-3}A.

Now that we know the current flowing in this circuit, we can calculate the voltage drops using the Ohm's law: $V_{R1} = IR_1 = 0.01*400 = 4.0V$, $V_{R2} = IR_2 = 0.001*800 = 8.0V$

The two resistors "divided" the source voltage of $12V_{DC}$ (between points A and C) in proportion to their resistances, to 4 and 8 Volts. R_2's resistance is twice the resistance of R_1, so twice as much voltage is "dropped" across R_2 compared to R_1. We can write $V_{AC} = V_{AB}+V_{BC}$ or $V_{BC} = V_{AC}R_2/(R_1+R_2)$

This type of circuit, called a *voltage divider*, is extremely important and we will come back to it many times in this book. It applies to both AC and DC circuits. Remember this formula for voltage division well!

Kirchoff's 2nd or Current Law

While the 1st KL was about voltages in a loop, this 2nd law is about current entering or exiting a "node". A node is a point where three or more "branches" meet. **2nd KL: The sum of currents entering a node is equal to the sum of currents exiting that node.**

We know that $V_B=R_1I_1$ so $I_1 = V_B/R_1 = 12/400 = 0.03A$ Likewise, $V_B=R_2I_2$ so $I_2 = V_B/R_2 = 12/800 = 0.015$ A

Mr. Kirchoff would have us believe that $I = I_1 + I_2 = 0.03$ A+ 0.015 A = 0.045 A

Here we will introduce an important concept called "the equivalent circuit". We have two resistors connected in parallel and we can calculate or measure their "equivalent" resistance, that of a single resistor that could replace the two of them.

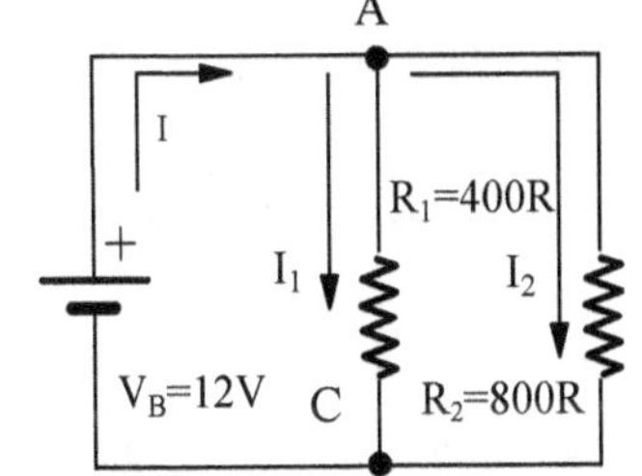

The 2nd KL: the sum of currents in any node of a network is zero!

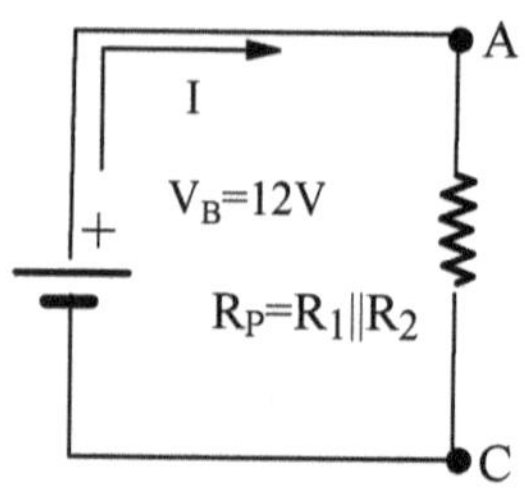

The equivalent circuit between points A and C as far as the battery is concerned

This is such a simple example that "the rest of the circuit" here means only the battery, but it will illustrate the principle regardless. The formula for the equivalent or parallel resistance R_P of N resistors connected in parallel is $\mathbf{1/R_P = 1/R_1 + 1/R_2 + ... + 1/R_N}$

In this case N=2 (2 resistors) so we have $1/R_P = 1/R_1 + 1/R_2$

Also, $1/R_P = (R_1+R_2)/(R_1*R_2)$ and finally $\mathbf{R_P=R_1*R_2/(R_1+R_2)}$

Instead of writing words "in parallel", a sign or symbol "||" is used to denote parallel connection, as in $R_P=R_1||R_2$!

In our case R_P=400*800/(400+800) = 320,000/1,200 = 266.67 Ω ≈ 267Ω. The symbol "≈" means "approximately". Now we can calculate the current supplied by the battery: $I = V_B/R_P$ = 12/266.67 = 0.045 A, the same result as before.

CIRCUIT ANALYSIS TOOLS AND METHODS

Current loop method

Consider this more complex circuit, with two voltage sources and three resistors. Again, it is a DC circuit, but this and all other methods equally apply to alternating current (AC) circuits. To determine voltages and currents in this circuit, the current loop method uses the 1st Kirchoff's law to write the loop equations for all the loops in the circuit. In our case we have three *branches*, two *nodes* (A and B) and two independent *loops*. The orientation of the loops does not matter. In our case it is clockwise, but you can go anti-clockwise if you prefer. However, once you choose the direction of the first loop, all other loops must be of the same direction.

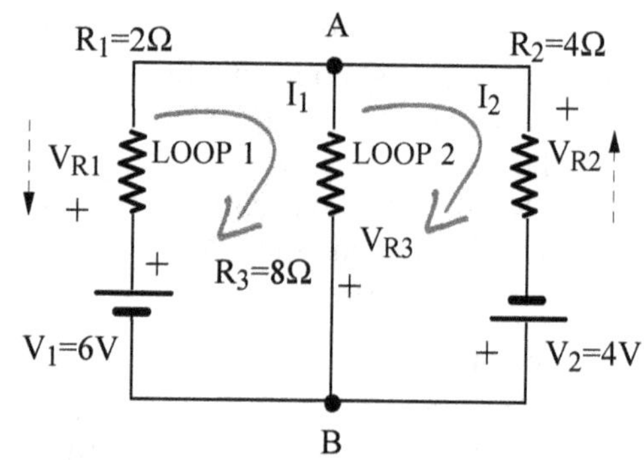

Voltage polarities are written as the arrow "enters" them. Let's follow Loop 1, or current I_1. Starting at B, I_1 flows upward and "hits" the voltage source's - or negative terminal, then it flows through resistor R_1. A current creates a positive potential at the point of entrance and negative potential at its exit, so the voltage drop across R_1 is V_{R1} with its plus as marked. Sometimes a dotted arrow is used as well (as illustrated), but that may be confused with current flow, so a + sign should be sufficient.

I_1 then reaches node A and flows through R_3 downwards into node B and thus completes its loop. Now we can write $-V_1+V_{R1}+V_{R3}=0$. We don't want voltages, so we substitute $V_{R1}=I_1R_1$ and $V_{R3}=(I_1-I_2)R_3$

Notice that the current through R_3 is I_1 flowing clockwise through it, in the "positive" direction, minus current I_2, flowing upwards through it or in the "negative" direction, so the current through R_3 is I_1-I_2!

Finally we have $-V_1+I_1R_1+I_1R_3-I_2R_3=0$

We write the second loop directly now starting at B. Again, it does not matter at which node you start from, providing you "complete" the loop: $I_2R_3-I_1R_3+I_2R_2-V_1=0$

Notice that in this loop the current I_2 through the common resistor R_3 is positive and I_1 flowing through the same resistor is now negative! We have two equations with two unknowns, I_1 and I_2. Once we solve this system of equations we can calculate any current or voltage in this circuit.

$-V_1+I_1R_1+I_1R_3-I_2R_3=0$ (Eq.1) and $I_2R_3-I_1R_3+I_2R_2-V_2=0$ (Eq.2)

If we substitute resistance values we get $-6+2I_1+8(I_1-I_2)=0$ and $8(I_2-I_1)+4I_2-4+0$ After solving the two equations by substitution, cancellation or matrix method, we get I_1=1.857A and I_2=1.571A

Since I_1 is larger than I_2, the resultant current I_1-I_2 =0.286A flows through R_3 downward (in the direction of the larger current, which is I_1). Now we know all currents in the circuit and can calculate any voltage we want.

Node voltage method

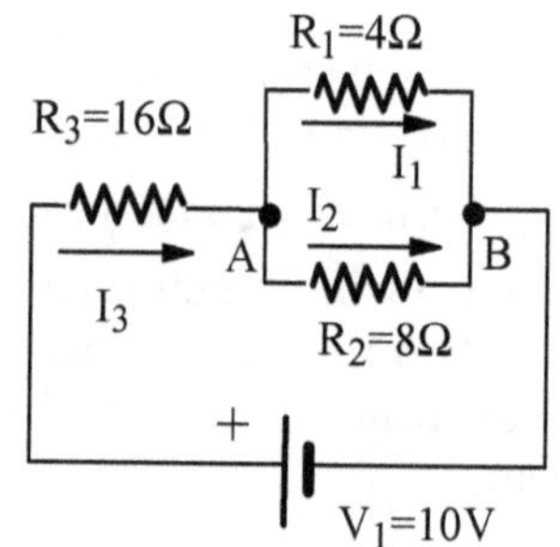

In simple circuits, instead of loops, we can write a single equation immediately. This method is one such shortcut. The 2nd KL for node A is $I_3=I_1+I_2$

Since $I_1*R_1 = I_2*R_2 = V_{AB}$ and the 1st KL says that $V_1 = V_{AB}+I_3*R_3$, so we can immediately write $I_3=I_1+I_2$ as $(V_1-V_{AB})/16 = V_{AB}/4 + V_{AB}/8$

Now we have one equation with one unknown, V_{AB}. Multiplying the whole equation by 16 we get $V_1-V_{AB} = 4V_{AB} + 2V_{AB}$ or $V_{AB}= V_1/5= 2V$!

Now that we know the two voltages we can calculate all the currents.

Superposition theorem

The superposition theorem is especially useful in the analysis of tube circuits. It simply says that in a linear network with two or more voltage or current sources, the current through any element (impedance or branch) is the sum of currents produced by each source separately, providing the internal resistances (or impedances for AC circuits) of these voltage sources are left in the circuit.

This means that we analyze the circuit twice if there are two sources (as in this example), three times if there are three sources, and so on. Here we have source V_1 with its internal resistance R_1 and source V_2 with its internal resistance R_2. It is the same circuit we analyzed in the section on loop currents, so the results must also be the same! Notice that two currents are flowing in the opposite directions, so $I_3 = I_{3A}-I_{3B} = 0.429-0.1429 = 0.286A$

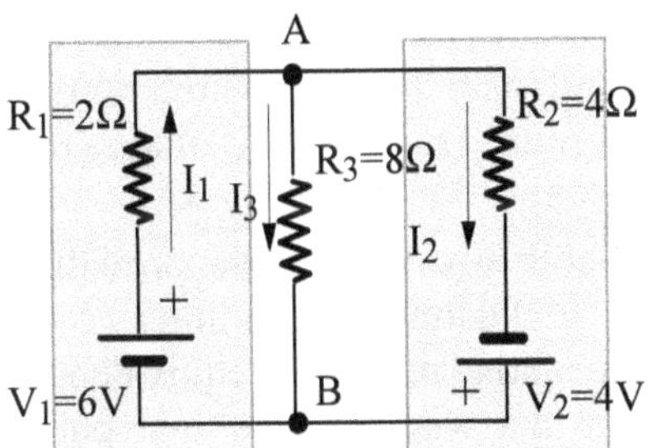

The original circuit to be analyzed, as for the current loop method on the previous page

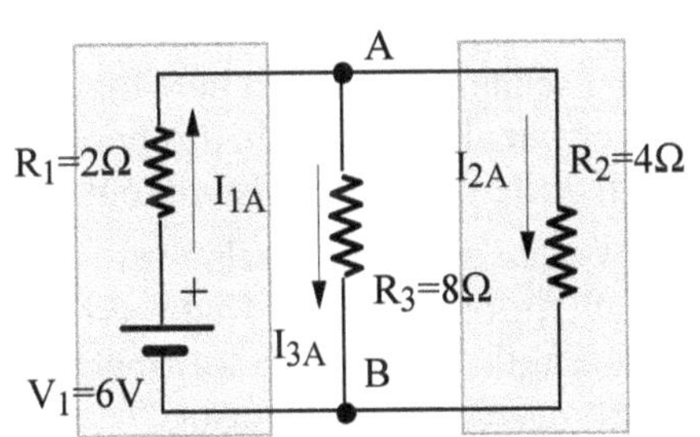

STEP 1: V_2 removed, analysis with V_1 only

$(V_1-V_{AB})/2 = V_{AB}/8 + V_{AB}/4$

$V_{AB} = 4V_1/7 = 3.428V$

$I_{3A}= V_{AB}/8= 0.429A$

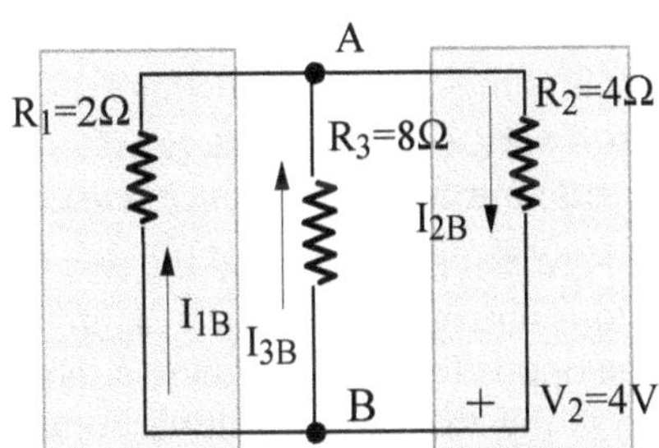

STEP 2: V_1 removed, analysis with V_2 only

$(V_2-V_{BA})/4 = V_{BA}/8 + V_{BA}/2$

$V_{BA} = 2V_2/7 = 1.143V$

$I_{3B}= V_{BA}/8= 0.1429A$

Reciprocity Theorem

T-circuit or T-network has three resistances (or for AC circuits, impedances), arranged in a shape of the letter T. X and B are input terminals and Y and B are output terminals. B is also called a common terminal, because it is common to both input and output. T-filters are very important in electronics, and are also used in the design of phono stages and preamplifiers, but here we will use DC analysis for simplicity reasons.

We have one voltage source V at the input of the T-network and are "measuring" the DC current at the output. A in a circle is a symbol for an ammeter. In an ideal case, its internal resistance is zero so it does not change the voltages or currents when inserted in the measured circuit.

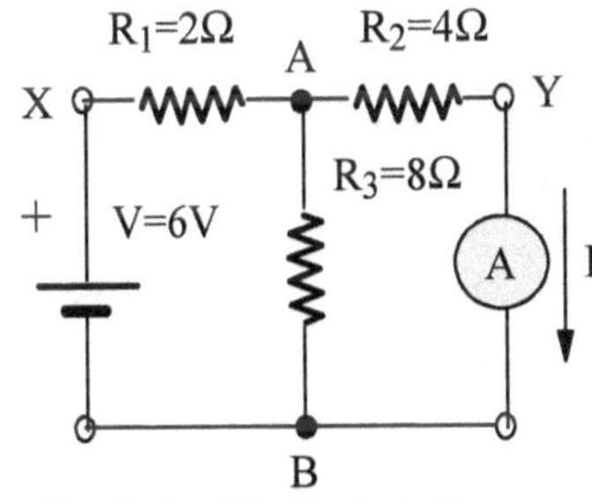

The T-circuit to illustrate the Reciprocity Theorem

The reciprocity theorem states that in any AC or DC network comprised of linear impedances (resistances), if a source voltage V is applied between any two nodes and the current I is measured or calculated in any branch, their ratio V/I will be equal to the ratio obtained if the positions of V and I are interchanged. The circuit is the same one we used in the superposition theorem discussions, so we already know that the current through R_2 is $I=V_{AB}/4 = 3.428/4 = 0.857A$! The voltage-current ratio or *transfer resistance* from the input to the output of the circuit is $V/I = 6/0.857 = 7\Omega$!

If the voltage source and the ammeter swap places, we analyze the circuit again: $(V-V_{AB})/4 = V_{AB}/2 + V_{AB}/8$ or $V_{AB}= 2*V/7 = 12/7 = 1.714V$! The current I through R_1 is now $I=V_{AB}/2=1.714/2= 0.857A$ The transfer resistance of the circuit is $V/I = 6/0.857 = 7\Omega$, exactly the same as in the previous case!

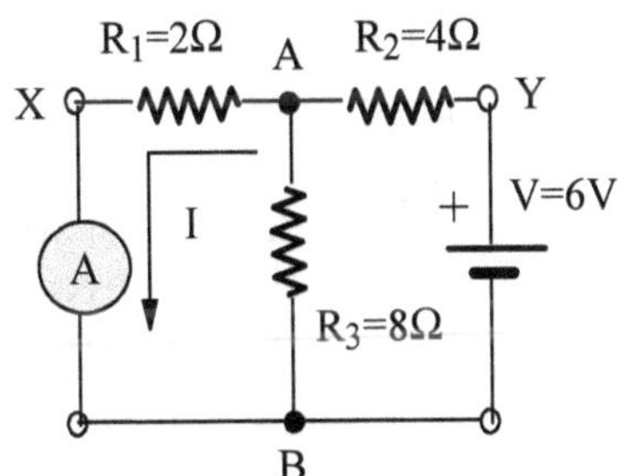

The positions of V and I interchanged

AC VOLTAGES AND CURRENTS

Basics of vector algebra

There are two types of physical quantities. Scalars (same root as the word "scale") can be fully described with only their magnitude, or numeric value. For instance, temperature and resistance are scalars. Vectors are described or fully defined by two parameters, the numeric value, or magnitude, and the direction. For instance, force is a vector. It isn't enough to know how much force you are using to push something away from you, but also the direction of such force. Vectors are symbolically represented by arrows. The length of the arrow is the vector's magnitude or amplitude, while the angle between such arrow and a referent line is the vector's "phase".

Vectors can be added and subtracted, but not by simply by adding their amplitudes. We have to do it in a two-dimensional space, and such addition, multiplication and other operations on vectors are called vector algebra.

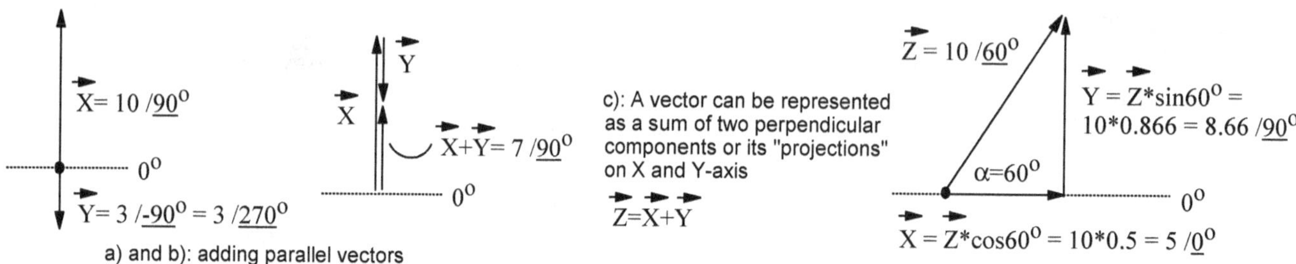

a) and b): adding parallel vectors

In the first example we see two parallel vectors, X and Y, but their directions are opposite. Vector X has a magnitude of 10 units (whatever that unit may be), vector Y's magnitude is 3 units. Notice how vectors have a little arrow on top of their names, that is to distinguish them from scalars. In our text here that is awkward to do due to the limitations of the software, so please understand that many physical quantities in this book will be vectors (such as AC voltages and currents) but for convenience reasons we will write their symbols without their arrows.

In this case it is easy to add the two vectors. We can move them relative to one another in any way we like, providing we don't change neither their magnitude (length) nor their angle. That was done in b), to illustrate that adding a vector of the opposite angle is akin to subtracting its length, in the same way that adding a negative number is equivalent to subtracting the same positive number. Rotating vectors or "phasors" were named that way because their phase changes constantly.

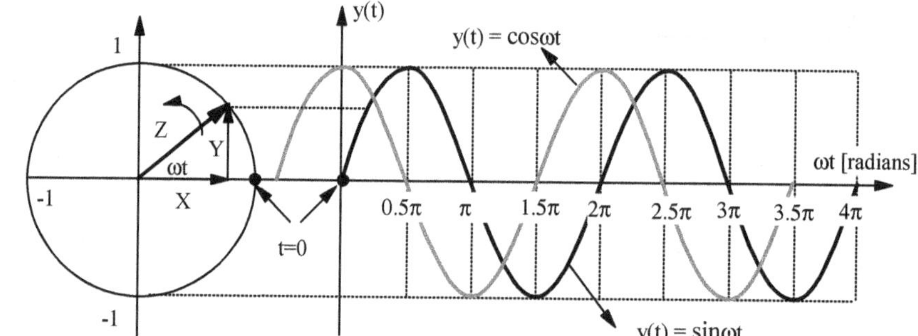

Phasors (rotating vectors) and sinusoidal signals

To truly understand the origin of sinusoidal waves, such as electrical voltages and currents, we need to study the behavior of a rotating vector Z. Phasor Z has an amplitude of 1, since the radius of the circle is also 1. The reason for this is that the maxima and minima of sine function in mathematics are +1 and -1!

Phasor Z rotates in the counter-clockwise direction, which is the positive direction in mathematics and physics. It starts rotating at a moment in time marked t=0. We are interested in the projections of vector Z on the X and Y axis.

ANGULAR AND NATURAL FREQUENCY OF A PERIODIC SIGNAL

$$\omega = 2*\pi/T = 2*\pi*f$$

At t=0 moment its projection on Y-axis is zero, marked with the dot on the waveform. At another point in time, as drawn, its projection on Y axis is also marked as Y. Again, we see that the two projections are also two perpendicular components of the vector Z. The sum of these two components always equals Z, or 1 in this case. If you trace the vertical projection Y in time you will get a sine wave described by a sine function: $y(t) = \sin\omega t$

The maximum is reached when the angle ωt=90 degrees or $\pi/2$ radians, then the projection goes down towards zero which is reached at ωt = 180 degrees or π radians. Then the same happens in the opposite direction. Eventually the phasor Z reaches the original position it had at t=0. Since the rotational speed and therefore the frequency of rotation is constant, the projected waveform repeats itself with the same amplitude and shape. That is why we call such signals periodic, because after each period they repeat themselves.

This period is marked capital T, to signify that it is a constant (a scalar) and to distinguish it from any moment in time "t". ω (lower case Greek letter "omega") is called angular or radian frequency and is related to the "natural" frequency by $\omega = 2\pi f$. Since f and T are reciprocal, $\omega = 2\pi/T$!

The same discussion applies to the horizontal projection. At t=0 such a projection as already at its maximum. This wave will look exactly the same as the sine wave or the vertical projection, but it will be shifted 90° or $\pi/2$ radians forward, it will "lead" the sine wave by 90°. This mathematical function is called a cosine function. The sine wave is described as $y(t) = \sin(\omega t)$, and the cosine wave is $x(t) = \cos(\omega t) = \sin(\omega t+\pi/2)$, so the cosine wave is simply a sine wave shifted in "phase by $\pi/2$ radians, 90 degrees or 1/4 of the period.

We will talk about "leading" and "lagging" phase very soon. These two sine and cosine waves can be represented by two perpendicular phasors (90° between them), of the same amplitude and rotating together with the same frequency, so the phase angle between them never changes. Notice that lower case symbols are used for instantaneous values of signals, while upper case letters are used for the peak, average and effective values.

To understand these three very important terms we need to move on from the mathematical definition of a sine or cosine wave to electrical signals such as voltages and currents.

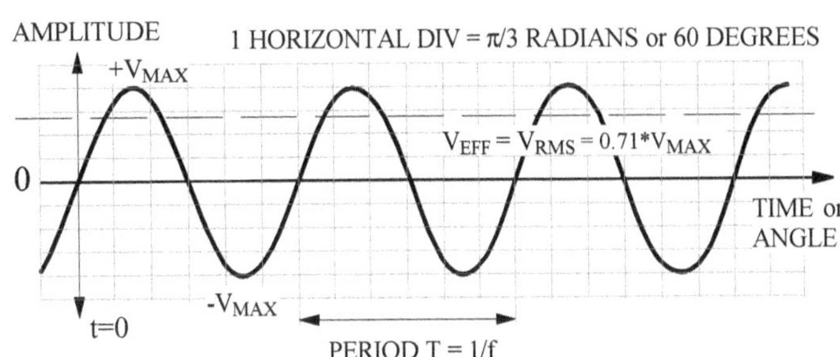

When an electrical conductor rotates in a magnetic field, an AC voltage is "induced" at its ends, and if a load is connected to it, an AC current will flow though the load. This is the operating principle of alternators which generate alternating (AC) voltages in power stations.

Generators are the opposite of electric motors. Mechanical energy from a steam turbine or a diesel engine drives the alternator, which converts such mechanical energy into electrical energy. Motors take electrical energy from the power grid and convert it into a rotational mechanical energy, to drive loads such as pumps, fans and compressors. Eventually, a "reduced amplitude" version of alternator's voltage ends up in your house as the "mains voltage". The frequency, the period and the waveform stay the same. The mains frequency in most countries is 50Hz, meaning their power generators rotate with a natural frequency of 50 times a second.

In some countries the mains frequency is 60Hz, meaning their power generators rotate faster, 60 times per second. The corresponding angular frequencies are $\omega=2\pi 50=314$ radians/second and $\omega=2\pi 60=377$ radians/sec. While the amplitude of a sine function was 1, the amplitude of sinusoidal voltages and currents can vary, so in general we write $v(t) = V_{MAX}\sin(\omega t)$, V_{MAX} is called the amplitude of the sinusoidal voltage and $T=1/f$ is its period.

Power in AC circuits

The effective or Root-Mean-Square value of an AC voltage is that value that produces the same heating effect (dissipates the same power) on a resistor as the equivalent DC voltage.

The RMS value of a sinusoidal signal is its amplitude divided by $\sqrt{2}$: $V_{EFF} = V_{MAX}/\sqrt{2} = V_{MAX}/1.41 = 0.71V_{MAX}$

It makes no intuitive sense that an AC voltage or current, which changes direction so many times a second, can produce any active power. However, just because the average value of a sine voltage is zero, does not mean that the power it delivers to a load is also zero!

Here we see one period of sinusoidal voltage $v(t) = V_{MAX}\sin(\omega t)$ and the AC power it dissipates across resistance R connected across it.

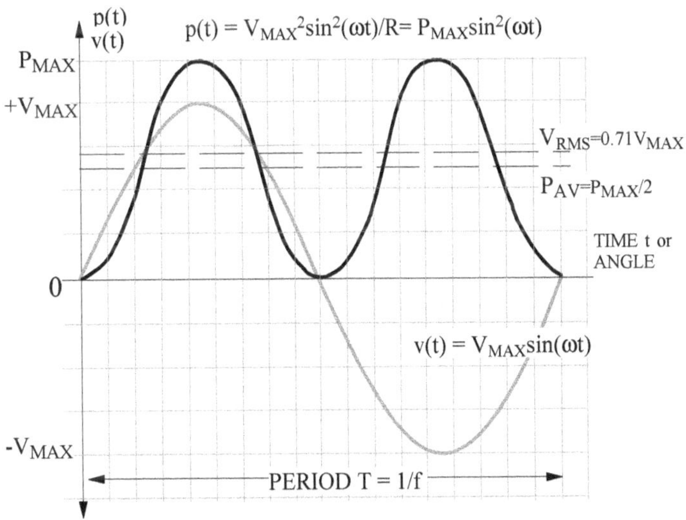

The power of a sine wave (voltage or current signal)

Remember, power is a square of the voltage divided by the resistance. Once squared, the power wave has the shape drawn in bold: $\mathbf{p(t)=v^2(t)/R}$. AC power also fluctuates periodically, but since a square of anything cannot be negative, the power on the load always has a positive average. It can be mathematically proven that the average of the power wave is exactly half of its peak value or $\mathbf{P_{AV}=P_{MAX}/2}$. Notice that at the precise moments when the power wave is at that value, the voltage wave is at its RMS value, or $0.71V_{MAX}$. This is not a coincidence, it is a very important relationship and we'll come back to it many times. Remember that when voltage drops from its value down to 71% of that value, the power on the load halves!

Signals in time- and frequency- domains

A sinusoidal signal can be expressed in two ways. The form we have been using, $v(t) = V_M\sin(\omega t)$ is called a time-domain form. The 18th century Swiss mathematician Euler defined the complex exponential $e^{j\theta}$ as a point in the complex plane, expressed as $e^{j\theta} = \cos\theta + j\sin\theta$ The "e" in this equation is the base of the natural logarithm, approx. 2.72. "j" is the operator called "i" in mathematics. It is defined as a square root of -1 or $i^2=-1$. This mathematical concept expands the real number system (where anything squared must be positive) to the complex number system.

To avoid confusion, in electrical engineering, which uses i(t) or just "i" to denote AC current, operator "j" is used instead of "i". ω is the angular frequency of the signal. The phase angle θ of the phasor is referenced to the cosine signal, so the two notations are:

TIME-DOMAIN: $v(t) = V_M\sin(\omega t)$

FREQUENCY-DOMAIN: $v(t) = V_M e^{j\theta} = V_M\angle\theta$

The notation $V_M\angle\theta$ is the expression in a polar form, where a phasor is defined by its amplitude V_M and its initial phase angle (in further text just "phase") θ.

Notice that in the definition circle illustrated, the amplitude is $V_M=1$, we are free to scale the phasors any way we like by multiplying (or "amplifying") them with V_M. We will use the complex plane a lot to depict the I-V characteristics of electronic components such as resistors, capacitors and inductors, and later on, when we study in more depth the amplifier behavior in the frequency domain.

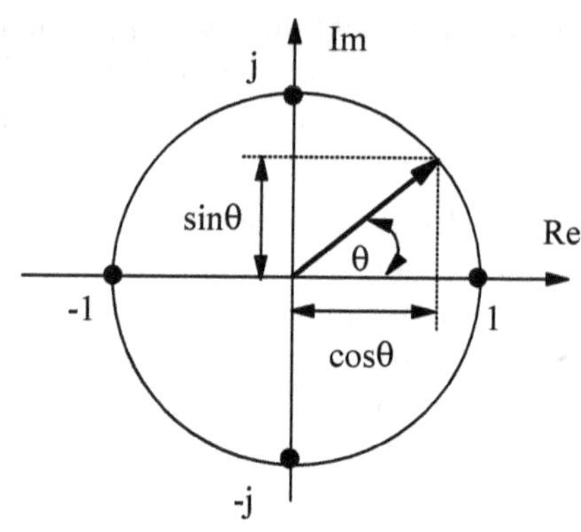

A phasor in a complex plane, with its projections onto the real (Re) and imaginary (Im) axis

RESISTIVE, CAPACITIVE AND INDUCTIVE IMPEDANCE

The equivalent of resistance in DC circuits is *impedance* (symbol Z) in AC circuits. While resistance is always a real number, impedance is a complex number, comprised of its real part (resistance) and its imaginary part or *reactance*: $Z(j\omega) = R(j\Omega) + jX(j\omega)$

Just as conductance (symbol G) is the reciprocal of resistance in DC circuits, the reciprocal of impedance is *admittance*, for which symbol Y is used. It is also a complex number, its real part being *conductance*, while its imaginary part is called *susceptance*: $Y(j\omega) = G(j\Omega) + jB(j\omega)$ Since $Y=1/Z$, its unit is $1/\Omega$, called Siemens [S]. Although Y is inverse of Z, generally speaking G is not an inverse of R and B is not an inverse of X!

The three basic passive components used in electronics, represent three fundamental physical properties, resistance, capacitance and inductance. As always, to make our journey smooth and gradual, we will start with ideal resistor, ideal capacitor and ideal inductor.

Ideal resistors

An ideal resistor has a constant resistance, meaning it does not change with temperature, and, more importantly, does not change with the frequency of the AC signal or with the magnitude of the current flowing through it.

The ideal resistor has no inductance and no capacitance. This means that it behaves in the same way in both DC and AC conditions. The vector diagram shows the current and voltage phasors "in phase" or parallel to one another. If the current flowing through an ideal resistor is a sine wave of a certain frequency f, the voltage drop such current will produce on the resistor will also be a sine wave, of the same frequency and the same phase.

The amplitude of the voltage drop can be determined by Ohms' Law, just as for DC conditions: $V_R = I_R{*}R$

An ideal resistor is a linear device, there is no amplitude distortion of either current or voltage, there is no phase shift between them, so no phase distortion. The higher the current through an ideal resistor, the higher the voltage drop across it, so the V-I graph is a perfectly straight line. In any operating point Q, the slope of this curve or the tangent of the angle between that line and X-axis is constant and equals the ratio of the voltage and the current in that point or the resistance R.

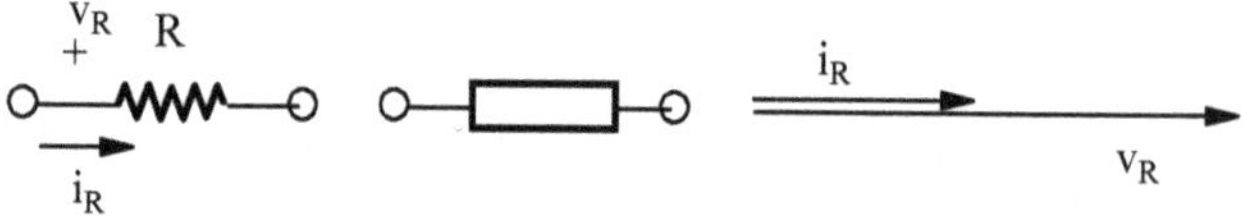

Ideal resistor: common symbols and the vector diagram (right)

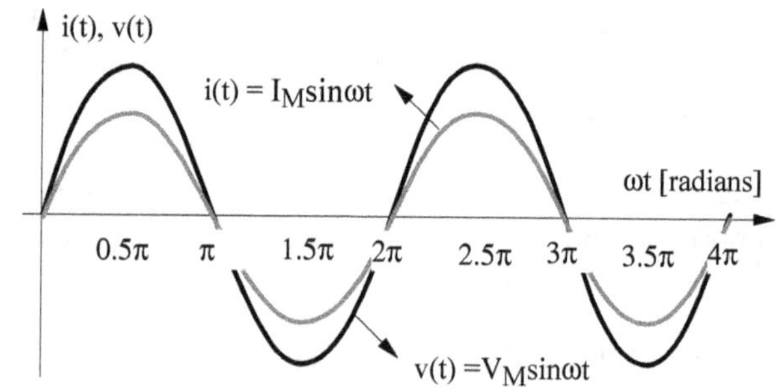

Ideal resistor: voltage and current are in phase (no phase shift)

BELOW: The impedance z_R of an ideal resistor is pure resistance, independent of frequency. The V-I relationship is linear. Its slope (tan α) is constant and equal to resistance R. The I-V relationship is more commonly used in tube characteristics. Its slope (tan β) is constant and equal to conductance (G=1/R).

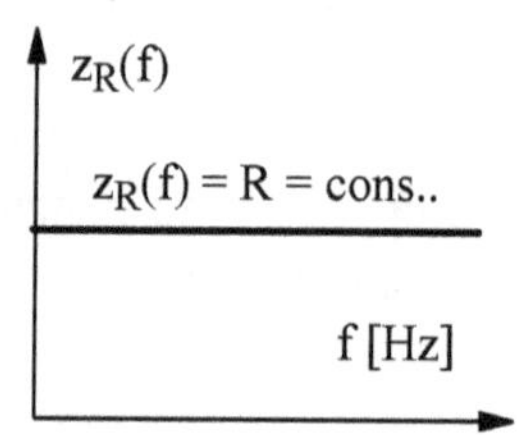

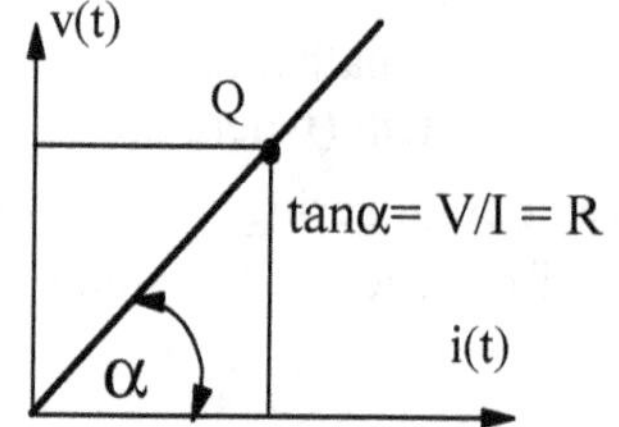

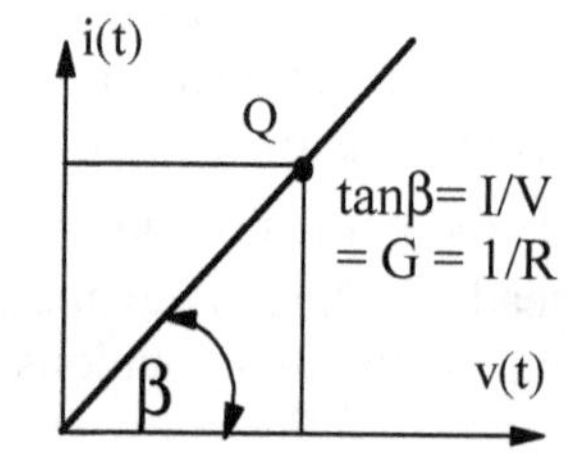

Ideal capacitors

An ideal capacitor would have an infinite resistance for DC current, so it would not allow any DC current to pass. The AC current through an ideal capacitor precedes the voltage by 90 degrees. If you cannot remember this fact, you can deduct it logically. Just remember that a discharged capacitor (no voltage across it and no charge held inside it) needs a current to start flowing first, in order to charge it. So, the current "leads", while the voltage across it develops slowly (raises), or "lags" behind the charging current. The impedance of an ideal capacitor is inversely proportional to frequency f and its capacitance C: $\mathbf{Z_C = 1/(2\pi fC)}$.

The higher the frequency of the signal, the lower the capacitor's impedance. This fact will have important repercussions in amplifier circuits, where unwanted or parasitic capacitances will form low pass filters with adjacent resistances and divert high frequency signals to ground, thereby limiting amplifiers' frequency range!

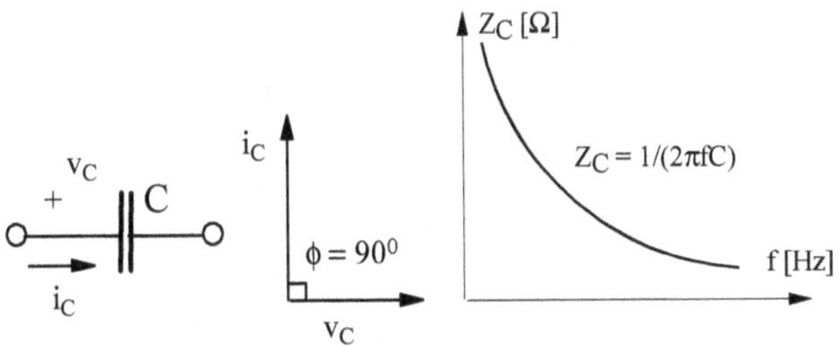

Ideal capacitor and its vector diagram: current leads voltage by π/2 radians or 90⁰

The impedance of an ideal capacitor is inversely proportional to frequency.

A 220μF electrolytic capacitor in a tube amp's power supply will have an impedance (reactance) of $Z_C = 1/(220*10^{-6}*2*3.14*100) = 7.2\Omega$ at 100 Hz, but only 0.072Ω at 10kHz. A 100 times higher frequency means 100 times lower impedance!

Capacitors are used for the following purposes:

1. To block the flow of DC currents, for instance between the high DC potential anode of an amplifier stage and the grid of the following stage which is at low DC potential.
2. To bypass AC signals around resistors, for example in the cathodes of amplification stages, where we don't want such signals to be attenuated by the cathode biasing resistors. The bypass capacitor forms a low-impedance path for audio signals, while the DC current cannot flow through the capacitor, only through the cathode resistor.
3. In various filters, such as tone controls and phono preamplifiers' RIAA filters.
4. As smoothing capacitors in amplifiers' power supplies, to reduce ripple (AC component on a DC power line) and to provide energy storage.

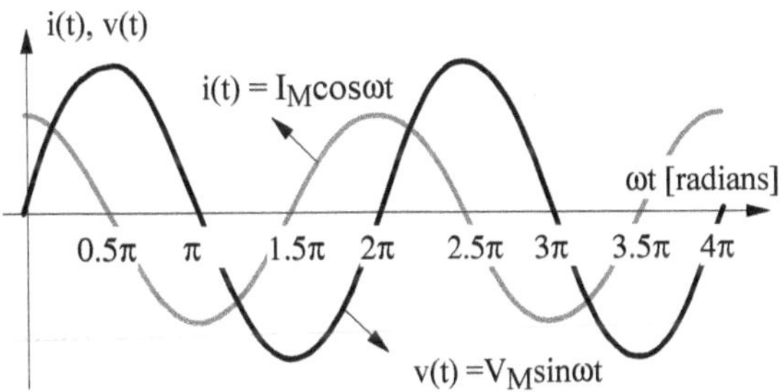

The phase relationship between the AC voltage across and current through an ideal capacitor.

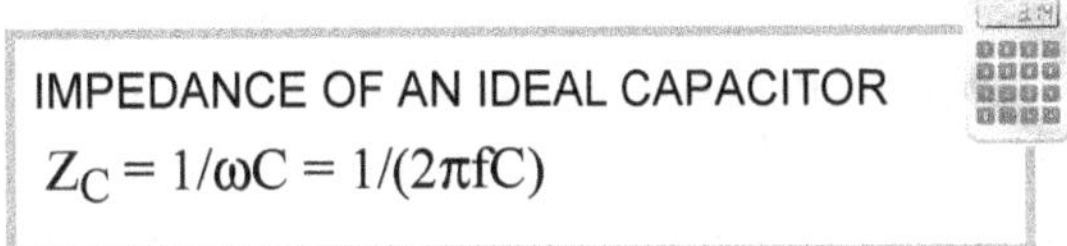

IMPEDANCE OF AN IDEAL CAPACITOR

$Z_C = 1/\omega C = 1/(2\pi fC)$

Ideal inductors

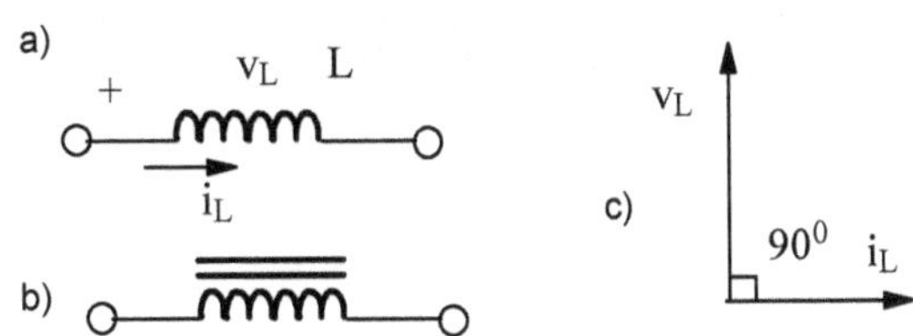

a) Ideal air inductor (no magnetic core)
b) Ideal inductor with magnetic core
c) Vector diagram shows voltage leading current by π/2 radians or 90⁰

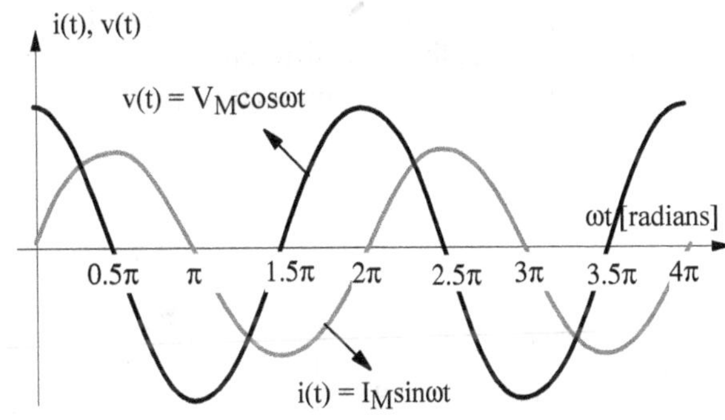

An ideal inductor (choke) has no resistance for DC current and no power losses, so there is no energy loss. The voltage on the choke precedes current, or you can say that the current lags voltage by 90 degrees, the opposite of a capacitor!

An inductor is named after a physical phenomenon called "electromagnetic induction, while a "choke" is a popular or pejorative name given to an inductor, due to its effect on alternating currents, which it suppresses or "chokes". In other words, an inductor or choke resists any change in the current that flows through it by inducing an opposite voltage which then tries to push the current in the opposite direction so as to neutralize the current that caused the induction in the first place. That mechanism is called *self-induction*! A choke has a low impedance for DC currents (ideally presents a zero DC resistance) and a high reactance to AC currents. This desirable combination of properties makes chokes useful in power supplies , as ripple filters, and in audio circuits as inductive loads.

The reactance of a choke is $X_L = \omega L = 2\pi fL$. Thus, the higher the frequency of an AC signal, the higher the reactance of a choke and the higher the choke's attenuation of the signal!

IMPEDANCE OF AN IDEAL INDUCTOR
$Z_L = \omega L = 2\pi fL$

A brief introduction into electromagnetic induction

Although we will study magnetic components such as transformers and chokes in detail later on, let's look at two practical examples, moving magnet and moving coil turntable cartridges. A moving coil (MC) cartridge has a diamond-tip needle sitting in a record's groove. As the groove's depth and width changes, the needle experiences mechanical forces which it transfers to a tiny coil or inductor inside the cartridge, which then moves in a stationary magnetic field produced by a permanent magnet. The movement of a conductor in a magnetic field induces electric current to flow in such a conductor. The mechanical force F of the movement, the density of the magnetic field's flux B and the induced current i(t) are related by $\mathbf{F(t)=Bi(t)}l$, where l is the length of the wire in the coil.

With a moving magnet (MM) cartridge, the situation is reversed. The magnet moves inside a stationary coil (inductor), so the magnetic field varies with music and induces current in the coil. Since stationary coils can be made larger and with more turns than those in moving coil cartridges, the induced voltages in MM case are much higher, 1-3 milliVolts, while MC cartridges produce weak outputs of around 0.1mV or 100μV!

It does not matter if a conductor moves in a stationary magnetic field or if a magnet moves inside a conductor shaped as a loop, any movement induces voltage and causes current to flow. Thus, any changing magnetic field causes an electric field, and vice versa. That is why this field of study is called *electromagnetism*!

In fact, chokes and transformers operate without any physical movement. It is enough for the current through a coil to vary in time, as AC current does, to induce an opposing voltage in the coil (inductor).

RCL circuits

If a series RCL circuit is connected to a source of AC signal, the same current flows through all three components. The voltage on the resistor is in phase with the current, but the voltages across the capacitor and the inductor are shifted in phase. One is lagging the current by 90 degrees (capacitor), while the other (inductor) is leading the current.

Since the voltages V_L and V_C are opposing each other, at a certain frequency f_R they will become equal and cancel each other out. Such a frequency is called a *resonant frequency* and the phenomenon is called *series resonance.*

$X_L=X_C$ or $\omega L=1/\omega C$ Solving the equation for ω, $\boldsymbol{\omega_R=\sqrt{(1/LC)}}$ **or** $\boldsymbol{f_R=\sqrt{(1/LC)}/2\pi}$.

At the resonant frequency f_R the overall impedance of the circuit is at its minimum, and equals the resistance R. At all other frequencies the resultant impedance is larger than R. Its amplitude can be calculated as $\boldsymbol{Z=\sqrt{[R^2 + (X_L-X_C)^2]}}$ and its phase angle is $\boldsymbol{\phi = \arctan [(X_L-X_C)/R]}$

There is another kind of a resonant circuit, and that is the parallel RCL circuit, where the resistor, capacitor and inductor are connected in parallel, so the same voltage is across each of them, and the currents through them vary with frequency. A typical example of such a system would be a dynamic loudspeaker that uses crossovers. A similar situation occurs there, the only difference being that at the parallel resonant frequency f_R the equivalent impedance of the whole circuit is at its maximum (R), and the impedance at all other frequencies is lower than R.

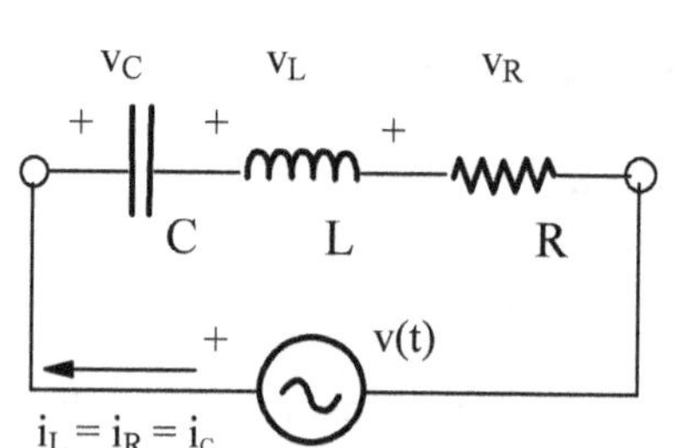

Series RCL or resonant circuit

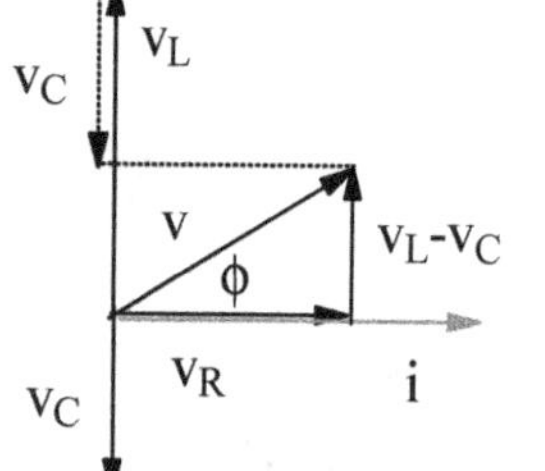

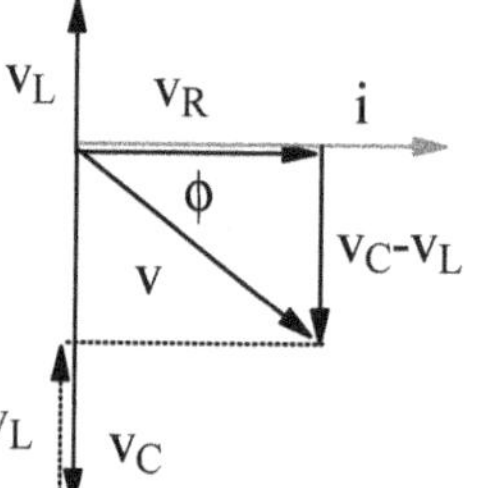

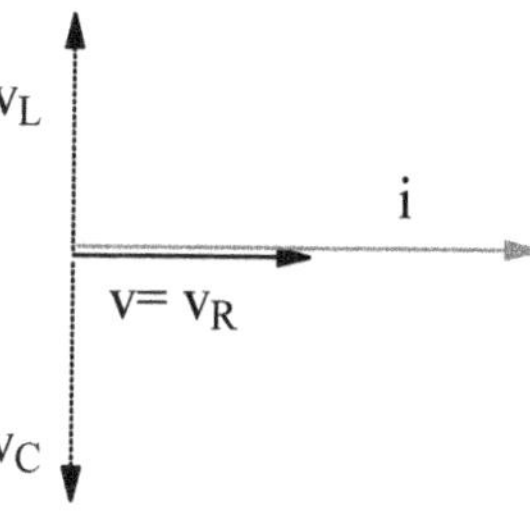

a) predominantly inductive (voltage leading current)
b) predominantly capacitive (voltage lagging current)
c) Resonance (the inductive reactance equals and cancels the capacitive reactance at the resonant frequency), the equivalent circuit is a pure resistance

Real inductors

Since inductors are made of many turns of wire on a bobbin (plastic frame), such a large length of relatively thin wire (typically 0.2-0.6 mm diameter) has a measurable resistance. We model real chokes as a series RL circuit, by adding an ideal resistor in series with the choke's inductance. The vector diagram for the real choke (next page) features angles δ and ϕ.

The same current flows through the resistance and the inductance, but there is a phase shift of 90 degrees or $\pi/2$ radians between the voltage drop on the resistance (which is in phase with the current) and the voltage drop on the inductance (which advances 90^0 ahead of the current).

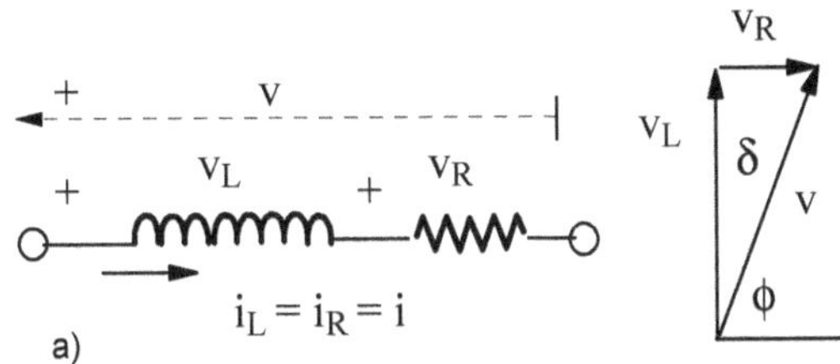

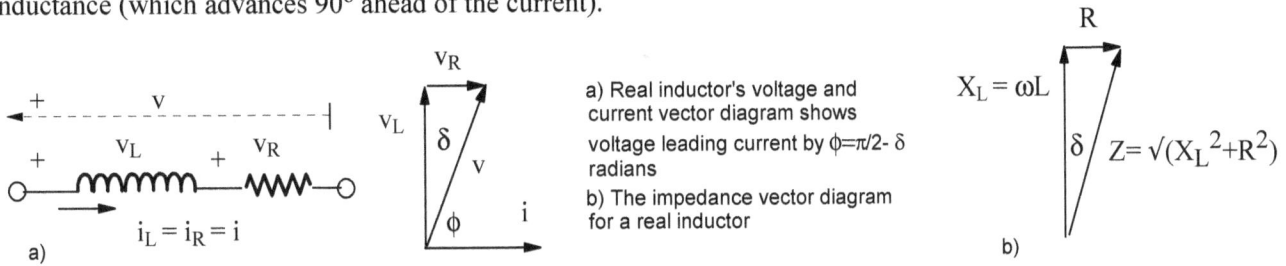

a) Real inductor's voltage and current vector diagram shows voltage leading current by $\phi=\pi/2-\delta$ radians

b) The impedance vector diagram for a real inductor

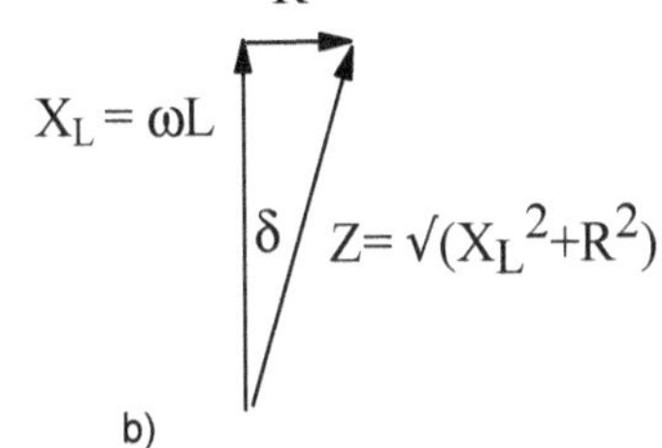

From the right angle voltage triangle we know that $\tan\delta = v_R/v_L$. Since $i=v_R/R= v_L/X_L$ and since $X_L = \omega L$, we get $v_R/R=v_L/\omega L$, $tg\delta = v_R/v_L = R/\omega L$

The quality factor or Q-factor for short, is the inverse of $\tan\delta$: $Q = 1/\tan\delta = \omega L/R$. The lower the tangent delta or the higher the Q-factor, the more the choke approaches the ideal inductor, meaning the lower its ohmic resistance (to DC current) and the lower the power losses. We will get deeper into the analysis and design of inductors later.

The losses and a vector diagram of a real capacitor

Real capacitors allow a small leakage DC current to pass through their dielectric. This behavior is modeled by adding a very large resistance in parallel to the ideal capacitor. The value of this leakage for a simple plate capacitor (to be discussed soon) would be $R_P= \rho d/A$. ρ (Greek letter "rho") is the specific resistance of the dielectric material, S is the surface area of the plates and d is the distance between them. Ideally ρ would be infinite (no conduction at all), meaning the parallel resistor would be an open circuit and the losses would be zero.

The vector diagram for the real capacitor (with leakage resistance) features angles δ (delta) and ϕ (phi), the sum of which is always 90^0 (degrees). Delta is zero for an ideal capacitor. From the right angle current triangle we know that $\tan\delta = i_R/i_C$. Since $v_C=v_R=Ri_R$ and the same voltage $v_C=v_R=X_Ci_C = 1/(\omega C)i_C$, it follows that $Ri_R = i_C/\omega C$

If we express $i_R/i_C = 1/(\omega RC)$ and take a tangent of the whole equation we get $\mathbf{\tan\delta = i_R/i_C = \tan[1/(\omega RC)] = D}$.

That parameter was named D for "dissipation factor" and is zero for an ideal capacitor. So, the lower the D-factor, the better the capacitor, the lower its dielectric losses. These losses are of course thermal, so they raise the temperature of the capacitor. This is most noticeable on vintage electrolytic capacitors which have high losses and are hot to touch. The top of the metal case often even bubbles up and such a bulge is a sure sign that the capacitor has reached the very end of its life and should be replaced immediately.

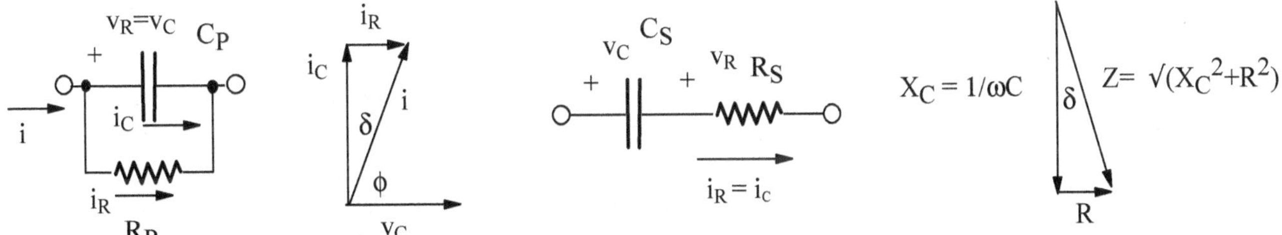

Parallel model of a capacitor and its vector diagram

Series equivalent circuit of a capacitor

The impedance diagram of a real capacitor

Eventually, the build up of gasses will cause the metal can to rupture and the capacitor will explode, spewing the gunk (electrolyte) all over the insides of the amplifier! Film capacitors do not use an electrochemical process as electrolytic capacitors do, so there is no gas buildup or possibility of explosion.

For both models the impedance diagram of a real capacitor is a right-angle triangle so Pythagora's Theorem applies again, helping us to calculate the modulus of impedance Z, the loss angle δ and the phase angle ϕ. The series resistance R_S in the second model is called ESR or Equivalent Series Resistance. It is an important parameter for electrolytic capacitors which should ideally be zero, but in reality can be quite high, up to 10Ω!

Series- and parallel-connected capacitors

In a parallel connection, each capacitor is connected to the same voltage, so its charge will depend on its capacitance - the higher the capacitance, the higher the stored charge. $Q = Q_1 + Q_2 + ... +Q_N = VC_1 + VC_2 + ... + VC_N = V(C_1 + C_2 + ... + C_N) = VC_{TOT}$.

Parallel connection is used in audio to increase the overall capacitance and to reduce the overall ESR. Assuming identical capacitors and a 6Ω ESR, by connecting four in parallel, the overall ESR will be reduced by the factor of 4, down to 6Ω/4 = 1.5Ω.

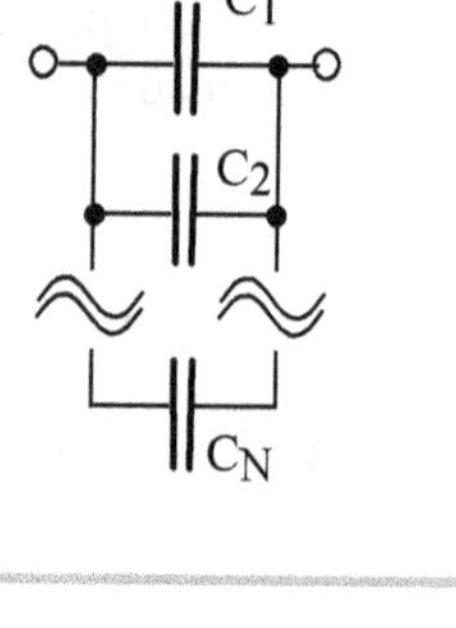

CAPACITORS IN PARALLEL
$C_{TOT}=C_1+C_2+ ... +C_N$

In a series connection, each capacitor will carry an equal charge Q. If potential drops across the capacitors are $V_1, V_2, ...V_N$, then $V_{TOT} = V_1 + V_2 + ... + V_N = Q/C_1 + Q/C_2 + ... + Q/C_N = Q(1/C_1 + 1/C_2 + ... + 1/C_N)$ The factor in brackets can be considered an equivalent capacitor, so $1/C_{TOT}=1/C_1 +1/C_2+ ... + 1/C_N$

Series connection is used in high voltage audio power supplies in cases where DC voltage of the power supply exceeds the voltage rating of available electrolytic capacitors, which is $450V_{DC}$ or $500V_{DC}$ at most. Two $450V_{DC}$ capacitors in series can then operate at voltages of up to 800V (a margin is always necessary).

However, since capacitance tolerances of these capacitors are wide, typically +/-20%, in the worst case scenario, one 470μF capacitor may have a capacitance 20% lower than this nominal value, or 376μF, while the other may have a 20% higher capacitance, or 564μF! Connected in series, they will have different voltage drops across them. The capacitor with larger capacitance will have a *smaller* voltage drop across it, while the cap with lower capacitance will be subjected to a *higher* voltage drop, which may exceed its rated voltage.

The rule is $V_1C_1=V_2C_2$, and since in our example $V_1+V_2 = V_{TOT} = 800V$, $C_1=376μF$ and $C_2=564μF$ we can calculate $V_1 = V_{TOT} * C_2/(C_1+C_2) = 0.6*V_{TOT} = 480V$ and $V_1 = 800-480 = 320V$!

Indeed, the voltage across C_1 is 480V, which has exceed its rated voltage of 450V. In this situation some capacitors would explode immediately, while others will work fine for a while but will suffer an early failure.

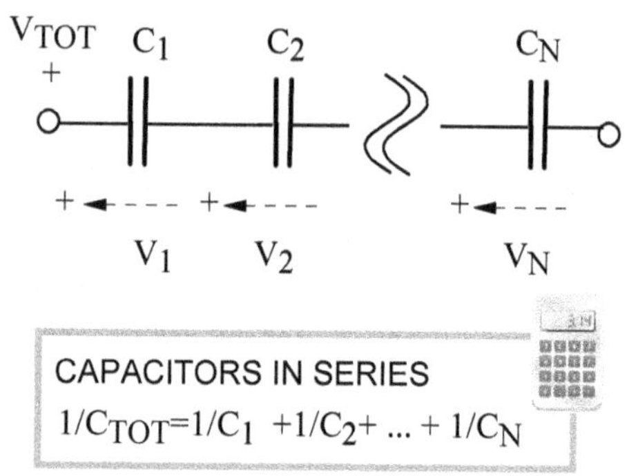

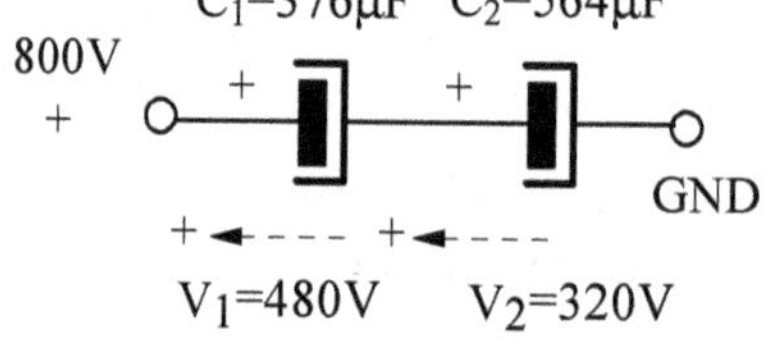

Two unequal electrolytic capacitors will divide voltages unequally!

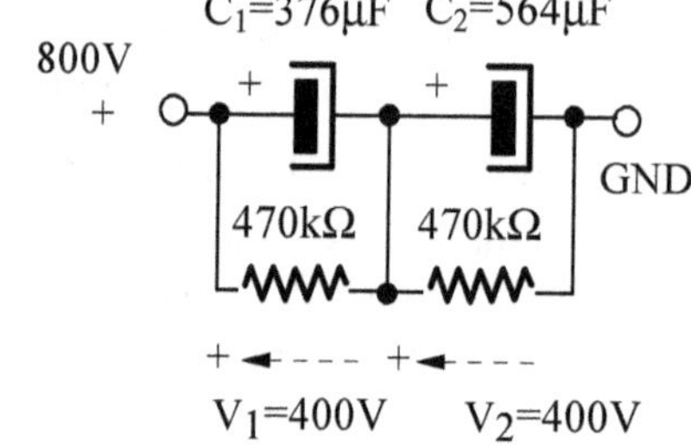

High resistance paralleled resistors will equalize the two voltage drops.

Series and parallel-connected resistors

By connecting resistors in series, their resistances add up, while the overall resistance of two or more paralleled resistors is lower than individual inductances. In parallel circuit the conductances add up, and since the conductances are inverse of resistances, we get the formula $1/R_{TOT}=1/R_1 +1/R_2+ ... + 1/R_N$!

The voltage across individual resistors will be proportional to their values, so the larger the resistance the larger the voltage drop across it (since the current through the lot is the same). Series resistors form a so called "voltage divider" network. Such resistive networks are often used in power supplies of tube amplifiers to provide required voltages from the same high voltage power source.

One advantage of paralleled resistors is the division of currents between them, so by using a few lower power rating resistors we can make up an equivalent resistor of a much higher power rating. This handy principle is often used to construct a dummy load for amplifier testing. Say you need an 8Ω load of at least 20W rating, but have only a bunch of 68Ω 3W resistors. By connecting eight of them in parallel you will get a total resistance of 68/8 = 8.5Ω (which for practical testing purposes is close enough to the nominal 8Ω), and a total power rating of 8*3 = 24 Watts!

R_1 R_2 R_N

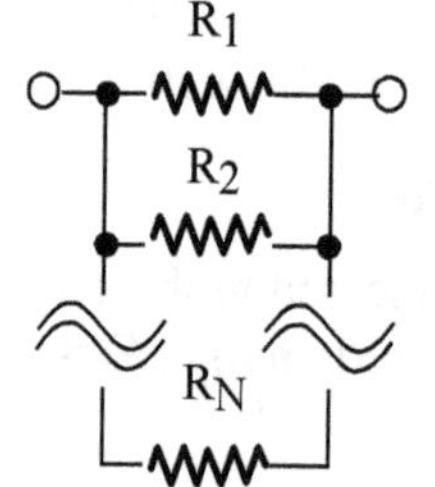

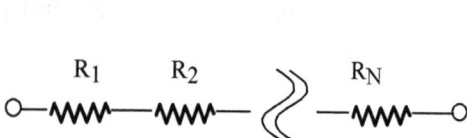

RESISTORS IN PARALLEL
$G_{TOT}=G_1+G_2+ ... +G_N$
$1/R_{TOT}=1/R_1 +1/R_2+ ... + 1/R_N$

RESISTORS IN SERIES
$R_{TOT}=R_1 +R_2+ ... +R_N$

The other advantage is that the overall tolerance is reduced. For instance, if you connect five identical 10% tolerance 10kΩ resistors, the parallel combination will have a resistance of 10/5=2kΩ, but the tolerance of such a network will be reduced by the same factor, 10%/5 = 2%. In such a case a careful selection of resistor values is usually not required!

Series and parallel-connected inductors

As with resistances, series-connected inductances add up, while the overall inductance of two or more paralleled inductors is lower than their individual inductances. Parallel connection of chokes is used in filtering circuits of tube amplifiers' power supplies if the current rating of one choke isn't sufficient, for instance if a total load of the amplifier's high voltage supply is 200mA, and 10H, 100 mA chokes are available. Two would be connected in parallel to satisfy the current demand, but the overall inductance would drop to 10H/2 = 5H!

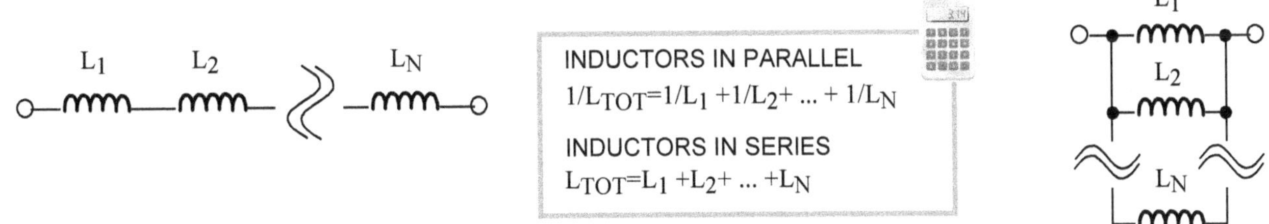

THE BLACK BOX CONCEPT: THEVENIN'S AND NORTON'S THEOREMS

We can evaluate the outward behavior of an electronic device without any knowledge of its internal details, its schematics or construction. By feeding suitable signals at its input and observing or measuring the signals at its output, we can determine what that system or device is (an amplifier, a filter, an oscillator, etc.) and how well it performs its intended function(s). This is the concept of a "black box". For instance, we use this principle when we get an unknown amplifier on our test bench and test its maximum power, distortion levels, frequency range, damping factor, signal-to-noise ratio, crosstalk (if it's a stereo amp) and other parameters of interest.

Two of the most important theoretical theorems in electrical engineering describe the behavior of any such "black box". Any linear electronic system or device can be represented as a black box with only two components, a signal source and an impedance.

Thevenin's equivalent circuit

From the point of output terminals A and B, any linear circuit can be replaced by a series combination of a voltage source equal to the open-circuit voltage at those terminals and an impedance looking back from the two terminals, once all *independent* voltage and current sources have been removed.

Norton's Theorem: From the point of output terminals A and B, any linear circuit can be replaced by a parallel combination of a current source equal to the short-circuit current flowing between these terminals and an impedance that would be measured at those terminals, after the removal of all *independent* voltage and current sources.

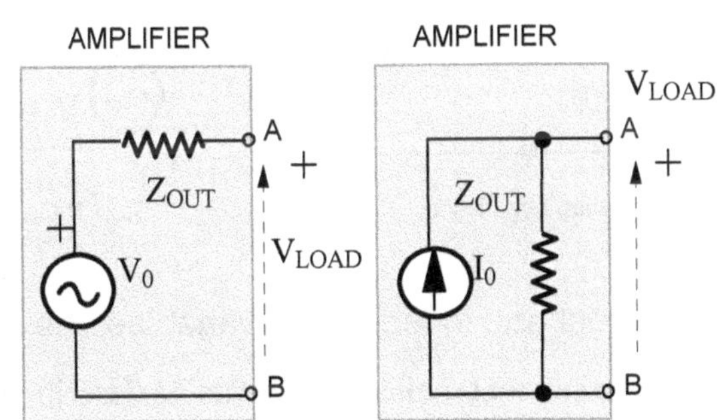

The Thevenin's (left) and Norton's (right) equivalent circuit of an audio amplifier

Let's use the simple circuit illustrated on the next page. We are interested in load terminals A and B, the 8Ω resistor connected between them is the load. In the first step we remove the load, leaving terminals A and B open and calculate the voltage between them. Since there is no current through the 2Ω resistor between terminals C and A, the voltage at A is the same as the voltage at C. The voltage source V_1 supplies two equal resistors (4Ω) in series, so the voltage between points C and B and thus A and B is half of V_1. So, our Thevenin's voltage $V_0(t)=0.5V_1(t)$.

Whatever the amplitude and frequency of this signal, the Thevenin's voltage will have the same frequency and shape (waveform), but its amplitude will be only half of V_1! For instance if $V_1(t)$ is the mains voltage in Australia with effective or RMS value of 240V, its amplitude V_{1MAX} will be 1.41*240V = 338V and its frequency will be 50Hz ($\omega=2\pi f=338$) so we can write $V_1(t)=338\sin(314t)$ and Thevenin's voltage will be $V_0(t)=169\sin(314t)$ [V]

In the second step we remove the ideal voltage source leaving a short circuit where it used to be (since it's internal impedance is zero). Now we look "back" to the left, into the circuit from terminals A and B and calculate the equivalent output impedance. We have two 4W resistors in parallel (equaling 2W) and then in series with another 2W resistor, for a total impedance of 4W! Now we can draw our Thevenin's equivalent circuit:

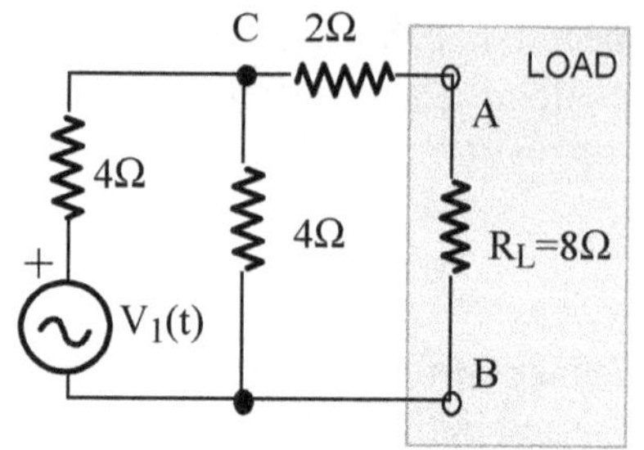

The original circuit

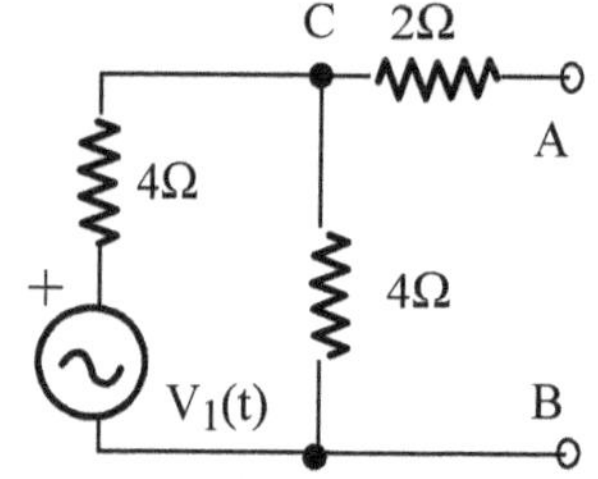

Step 1: Calculating Thevenin's voltage

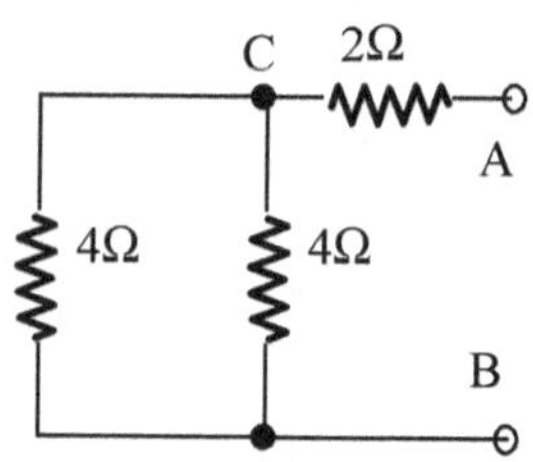

Step 2: Calculating the output impedance

A few important points:

1. The impedance of the load has no impact on the calculations.
2. More than one voltage source can be present, but the principle stays the same.
3. Dependent voltage sources are not removed in Step 1 calculations, only the independent sources.
4. The output impedance is usually complex, it has an active (resistive) and reactive (inductive or capacitive) component. We used pure resistance for the sake of simplicity.

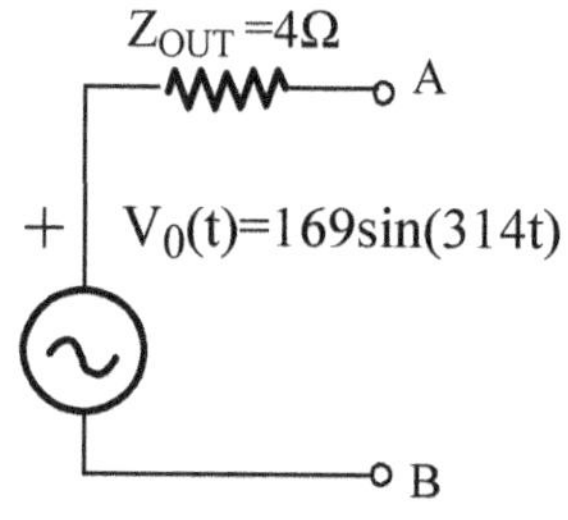

Thevenin's equivalent circuit for our example

Norton's equivalent circuit

Determining the output impedance of the Norton's model is identical to the method for Thevenin's model, so it won't be repeated here. To determine Norton's current we short-circuit terminals of interest (in this case A and B) and calculate the current through that branch.

$(V_1-V_{CB})/4 = V_{CB}/4 + V_{CB}/2$ so $V_{CB}=V_1/4$

Since $I_0 = V_CB/2$, $I_0 = V_1/8$, and again, assuming that $V_1(t)=338\sin(314t)$, we get $I_0(t)=42.25\sin(314t)$ [A]

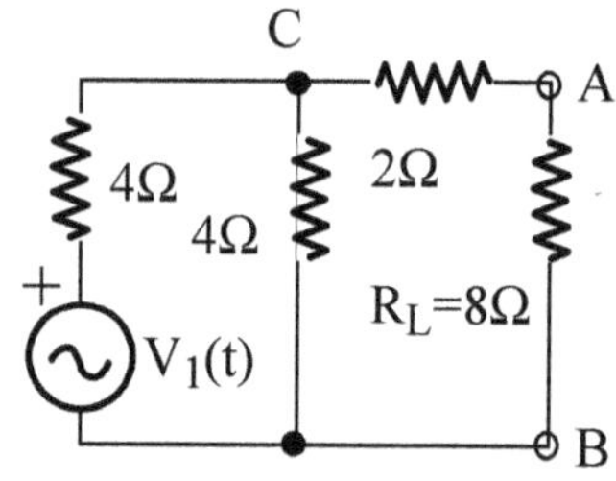

The original circuit

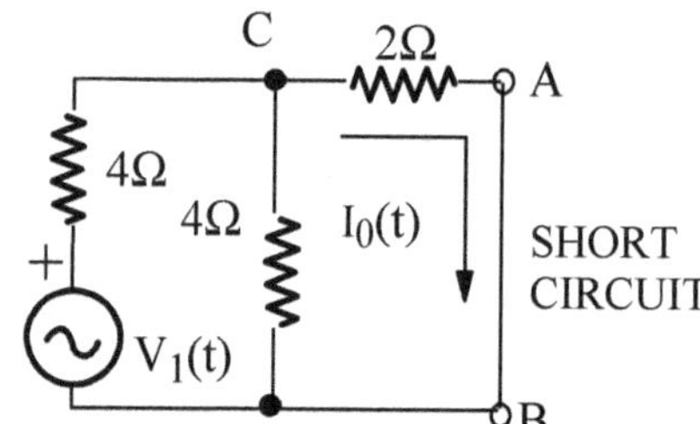

Step 1: Calculating short-circuit current

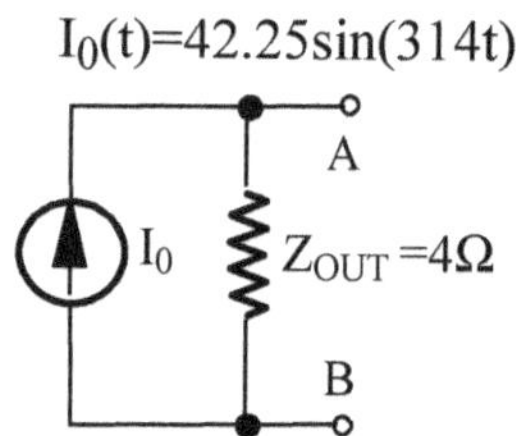

Norton's equivalent circuit

Dependent and independent variables, their units and notations

We have seen that mains voltage can be described by the formula $V(t)=V_{MAX}*\sin(\omega*t) = \sqrt{2}V_{EFF}*\sin(2*\pi*f*t)$ [V]! In this formula f (the mains frequency) is constant, it is either 50Hz or 60Hz (or "cycles per second"), depending on the country you live in. The peak V_{MAX} and the effective value V_{EFF} of the voltage are also nominally constant, although they fluctuate somewhat in real life, just as the frequency does, depending on the quality of power grid you are connected to. The factors 2π ($\pi\approx3.14$) and $\sqrt{2}$ are obviously fixed numbers, so the only variable is "t" or time. Thus, time is the *independent variable* while the amplitude of the mains AC voltage is the *dependent variable.*

The [V] at the end of the equation is the unit of the dependent variable, in this case the unit for voltage is a Volt. The square brackets will indicate a unit for the physical parameter described by the formula that precedes it.

Also, since writing the multiplication sign "*" or "x" many times during mathematical analysis can be tedious and slow us down, we will omit those symbols assuming that everyone understands that $V_{MAX}\sin(\omega t)$ means $V_{MAX}*\sin(\omega*t)$! Furthermore, symbol "x" for multiplication can be falsely interpreted as some kind of variable called x, so we better not use it. The bracket after sine, cosine, tangent or cotangent functions is also often omitted, so $V_{MAX}\sin\omega t$ is the ultimately simplified notation, but you must remember that in $\sin\omega t$, the sine function applies to the whole product ωt (the independent variable) and not just the first letter ω!

However, with the root symbol we should use brackets to indicate what it applies to, or what is "under" this symbol, otherwise errors are almost certain! For instance $\sqrt{2}V_{EFF}$ indicates that only number 2 is under the root, while $\sqrt{(2V_{EFF})}$ would mean that the whole product $2V_{EFF}$ is under the root symbol.

ELECTRONIC COMPONENTS

- FIXED RESISTORS: carbon composition, metal fim, carbon film, wirewound
- NONLINEAR RESISTORS: thermistors (NTC and PTC), ballast (current regulating) tubes
- VARIABLE RESISTORS: potentiometers and rheostats
- AUDIO ATTENUATORS
- OTHER COMPONENTS: fuses and switches
- CAPACITORS
- PASSIVE SEMICONDUCTORS: rectifier diodes, Zener diodes, varistors
- WIRES AND CABLES: interconnects, hookup wire, power cables

"Architecture starts when you carefully put two bricks together. There it begins."
Ludwig Mies van der Rohe, (1886–1969), avant-garde architect, famous for his 'less-is-more' approach to design

FIXED RESISTORS

Carbon composition resistors

The resistive element of carbon composition (CC) resistors is made of a finely ground mixture of powdered carbon and silica filler, with a small quantity of resin to act as a binding agent., compressed into a slug. The ends of this rod or pin are then sprayed with a thin coating of copper and the whole rod is cemented into a ceramic tube. Finally, spun brass caps, attached to tinned copper leads are pressed onto the ends. Alternatively, the insulating ceramic tube is omitted and the end connections are made by spinning copper wire around each end and dipping it in solder.

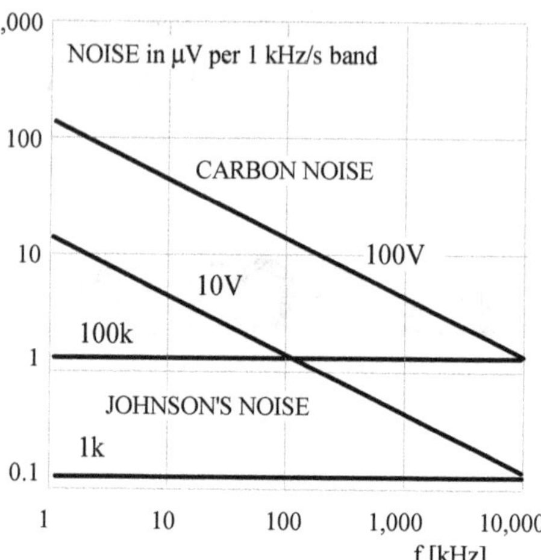

How two types of noise in carbon-composition (CC) resistors depend on frequency, voltage and resistance value.

CC resistors suffer from aging, their values drifting significantly over time. Their tolerances are the poorest of all resistor technologies. However, since they were the cheapest to produce in the 1950s and 1960s, they were widely used in vintage equipment.

There are two types of noise generated by the flow of current through carbon composition resistors, Johnson's noise, independent of frequency of the signal, and carbon noise, which increases with the voltage drop across the resistor and decreases with frequency.

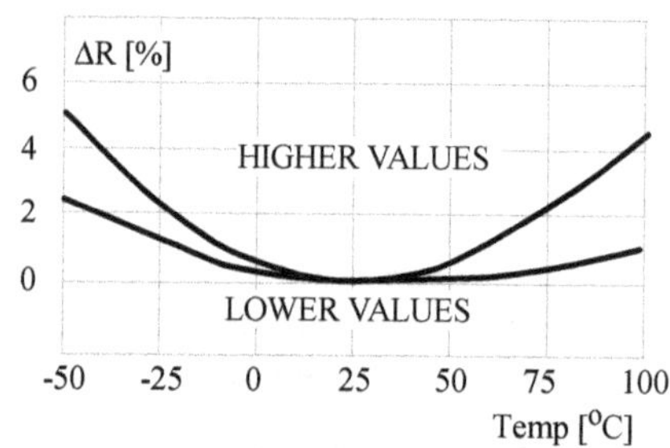

Relative temperature deviation of carbon composition resistors

As it often happens in audio, components and technologies get a new lease of life once they get a reputation of sounding good. The most illustrative examples are mercury-vapor rectifier tubes (and rectifier tubes in general), and carbon composition resistors. Generally, CC resistors sound "warmer" and more "musical" then the film types (especially compared to metal film), but also "grainier". Due to their high levels of noise, they should never be used as grid resistors, in phono stages or any other application involving low-level signals.

Carbon film, metal film and metal-oxide resistors

Carbon film resistors (also called carbon deposit type) are made by feeding a ceramic rod through a furnace with a controlled hydrocarbon atmosphere. The gas, either methane or benzene is cracked and carbonaceous coating is pyrolytically deposited onto the rod. The film can also be applied by spraying or dipping the rod into a carbon solution. In both cases a spiral groove or track is cut into the coating to increase the length of the resistance path. The pitch and the length of the groove are varied to adjust the resistance to the desired value.

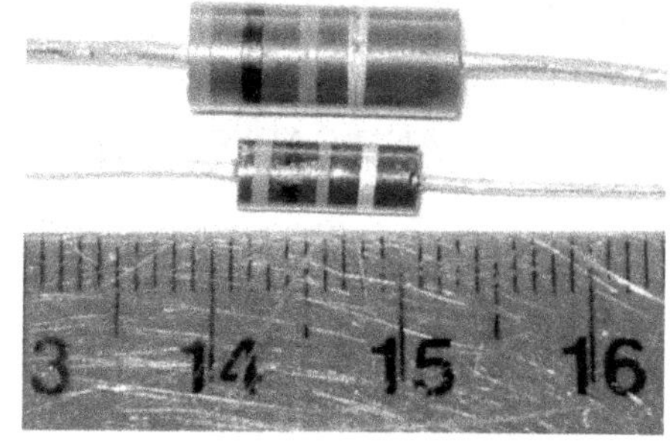

1/2W (lower) and 1W (upper) carbon composition resistors. The ruler is in cm.

Carbon film resistors have a higher negative temperature coefficient than carbon composition type.

The resistive material used in the manufacture of metal film resistors is usually nickel-chromium (NiCr), although other alloys containing gold, platinum, palladium, rhodium and tantalum nitride are also used. A cathodic deposition process in vacuum is followed by the groove incision. As with carbon film resistors, resistance and its temperature stability are strongly dependent on the thickness of the metal film, thicker films resulting in better stability and lower resistance.

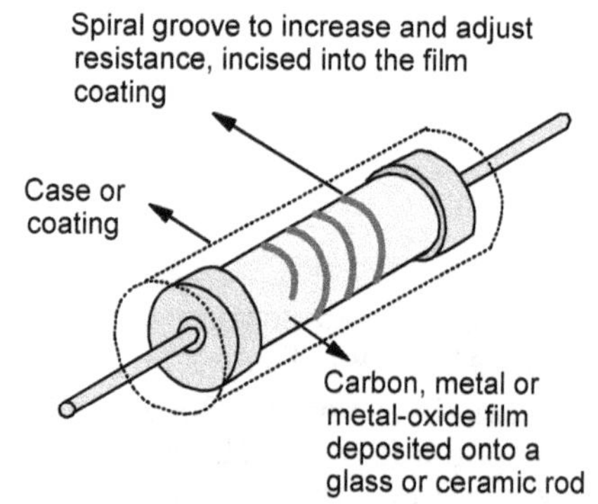

The internal construction of film resistors.

Metal-oxide film resistors are made by oxidizing reaction of a vapor or by spraying of a tin chloride solution on a heated glass or ceramic rod, so a thin film of tin oxide is fused onto the substrate. Antimony oxide may be added to increase the resistivity of the film. Metal oxide resistors can withstand higher operating temperatures than carbon or metal film resistors and are more stable and reliable.

Wirewound resistors

Wirewound resistors are made by winding a certain length of a special resistive wire around a ceramic rod. The whole thing is then dipped or coated in epoxy, or enclosed in an aluminium case which can be bolted onto a chassis, to act as a heatsink for better cooling.

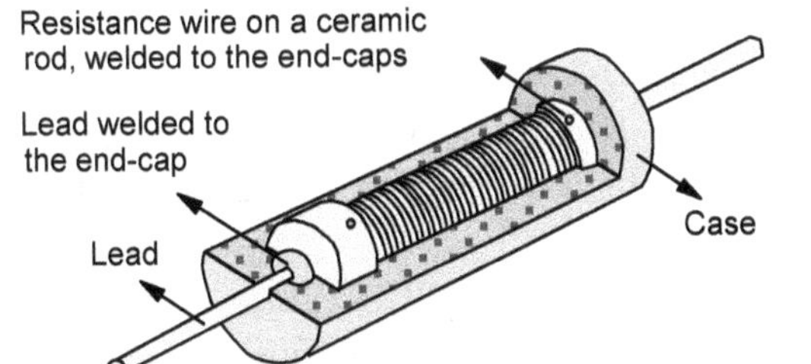

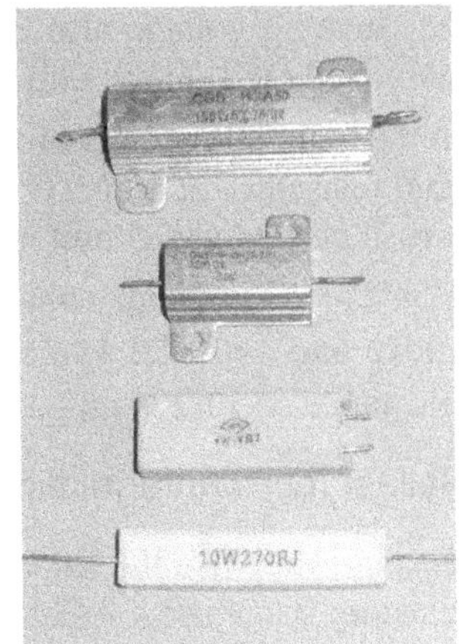

ABOVE LEFT: The cross-sectional view of a wirewound resistor.
ABOVE RIGHT: Axial resistors by Shallcross (7W) and Dale (5W), 5% tolerance.
RIGHT: A few more examples of wirewound resistors. Top two are in aluminium casing, meant to be bolted down to a heatsink or metal chassis. The bottom two are examples of radial (leads perpendicular to the body) and axial (leads parallel or in-line with the body) construction, 7 Watts and 10 Watts power dissipation respectively.

Since they are constructed in the same way as inductors are, compared to other resistor technologies the inductance of wirewound resistors is relatively high. There are non-inductive wirewound resistors made using special winding methods, such as the Ayrton-Perry winding.

Copper cannot be used in resistor making because its resistance varies a lot with temperature changes. Instead, materials such as constantan, nickel-chrome and manganin (copper-nickel-manganese alloy) are used because of their higher resistivity and low temperature coefficients.

Resistor color chart and standard values

Standard resistor values follow geometric progressions named E12, E24 and E48 series (10%, 5% and 2% tolerance). The values between 1 and 10 are listed below. All other values are obtained by multiplying those with 10, 100, 1000, and so on.

E12 series (10% tolerance): 1.00 1.20 1.50 1.80 2.20 2.70 3.30 3.90 4.70 5.60 6.80 8.20

E24 series (5% tolerance): 1.00 1.10 1.20 1.30 1.50 1.60 1.80 2.00 2.20 2.40 2.70 3.00 3.30 3.60 3.90 4.30 4.70 5.10 5.60 6.20 6.80 7.50 8.20 9.10

Example: Red - Violet - Red - Silver: 27 x 100 = 2k7 (2,700 ohms or 2.7 kohms), tolerance +/-10%

Say you need an 82k anode resistor and are rummaging through your resistor stash, which is not sorted.

Color	Band 1 1st digit	Band 2 2nd digit	Band 3 Multiplier	Band 4 Tolerance
Black	0	0	1	
Brown	1	1	10	+/- 1%
Red	2	2	100	+/- 2%
Orange	3	3	1000	
Yellow	4	4	10,000	
Green	5	5	100k	
Blue	6	6	1M	
Violet	7	7	10M	
Gray	8	8	100M	
White	9	9	1G	
Gold			0.1	+/- 5%
Silver			0.01	+/-10%
None				+/- 20%

8 is Gray, 2 is Red, the multiplier is 1,000 (to get from 82 to 82,000), which is Orange. So you need Gray-Red-Orange-Silver (from E12 series, with 10% tolerance) or Gray-Red-Orange-Gold (from E24 series, 5% tolerance) or Gray-Red-Orange-Red (from E48 series, 2% tolerance).

NONLINEAR RESISTORS

Thermistors (NTC and PTC)

The word "thermistor" comes from "thermal resistor". Thermistors are nonlinear resistors, meaning that the relationship between the voltage applied to them and the current flowing through them is not a straight line. Therefore, their resistance is not constant but varies with their temperature. PTC thermistors have a "positive thermal coefficient" while NTC means "negative thermal coefficient". NTC thermistors have their highest resistance when they are cold (no current flowing through them) and their resistance goes down rapidly (in an exponential fashion) as they heat up.

This makes NTC thermistors ideal as limiters for inrush currents. When a tube amp is switched on there is a current spike through the heaters of all tubes, which are cold (tube filaments have a positive thermal coefficient, their resistance goes up as they heat up).

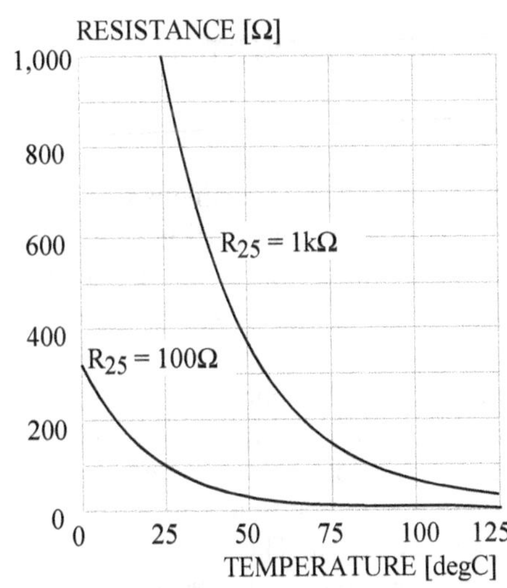

Typical resistance vs. temperature curves for two different NTC thermistors

The heating voltage also drops down due to poor regulation of power transformers (unless the heating voltage is stabilized DC). These current surges stress-out the filaments and shorten the life of vacuum tubes.

In the high voltage section, large capacity elcos are empty, and upon energizing, large charging currents flow. A high cold resistance of an NTC thermistor connected in series after the rectifier will limit the no-load voltage rise and the inrush current spike, and in turn reduce the stress on all components.

The "cold" resistance value of a thermistor is referenced at 25°C (abbreviated as R25). The R25 values are between 100Ω and 100 kΩ. Typically, thermistors for this application will have a cold resistance of 3,000Ω and a "hot" resistance of 30Ω, a ratio of 100:1!

Ballast (current regulating) tubes

Because they use the same bases and glass bulbs, ballast tubes may look like vacuum tubes, but are simply PTC resistors. They use a resistance wire with a positive temperature coefficient (PTC), sealed in a glass bulb containing hydrogen or helium gas. Threshold voltages range from 0.4 to 40 V while the voltage ranges vary between 3 and 70 volts. The rated currents are in the 50 mA to 6A range.

The operating principle is simple. Assume that the voltage across the tube increases. The current through the tube will increase as well, causing the filament wire to get hotter, which will increase its resistance and decrease the current. The tube acts as an automatic current regulator.

Unlike voltage regulator tubes, ballast tubes should not be operated in series, but tubes of the same rated voltage range can be paralleled. They can be used on AC, DC, or pulsating current.

The current-voltage curve of a ballast tube is illustrated below. V is the voltage across the tube and I/I_{NOM} is the ratio of the actual current through the tube and the nominal (rated) current. An approximate curve can be obtained by dividing or multiplying the current scales by the appropriate factor to get specific currents and voltages.

The regulation capabilities of any ballast tube can be estimated from the graph. As the voltage across the ballast tube increased from the threshold voltage V_T of 4V to the maximum voltage (11V in this case), which is an increase of 2.75 times or 275%, the current increased only 11-12%!

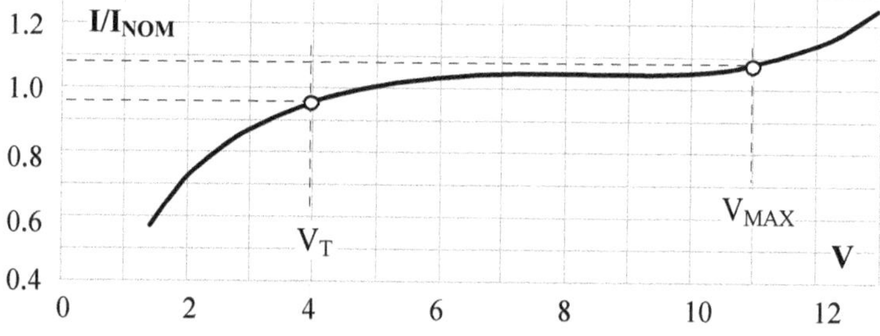

ABOVE: USA-made Amperite 10-4A ballast tube.
LEFT: The current-voltage curve of a typical ballast tube.

VARIABLE RESISTORS

Potentiometers and rheostats

Potentiometers are variable resistors, the ones used in audio are of a single-turn construction (270°), but precision 10-turn pots are also widely used in instrumentation and other applications where a more precise control of resistance is required.

Potentiometers that need to be adjusted frequently, such as those that control the volume or tone in amplifiers have a plastic or metal shaft onto which a knob is attached, either a push-on or screw-type. Potentiometers that are adjusted rarely, only during servicing or calibration, instead of a shaft and a knob have a slot for a screwdriver. These are usually mounted inside an amplifier or an instrument, so the user has no access to it, and are called trimmer-potentiometers or "trimpots".

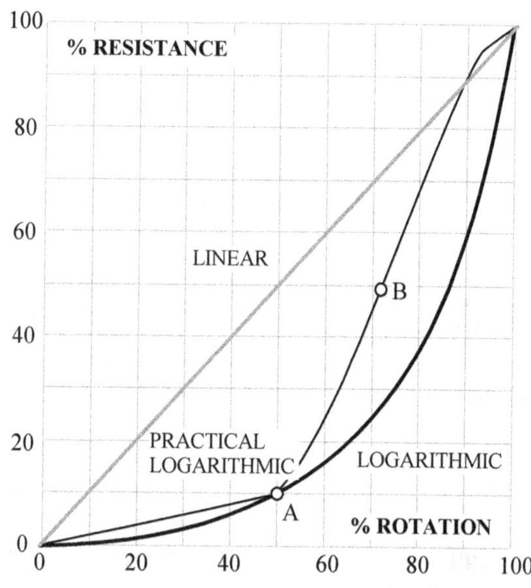

Percentage of resistance versus percentage of rotation curves for a linear potentiometer, nominal (ideal) logarithmic or audio potentiometer, and real (practical) logarithmic potentiometer

Film potentiometers are made by depositing a conductive film onto a plastic substrate. The conductive film is usually carbon or "cermet", a special conductive ceramic material. The wiper or sliding (rotating or linear) arm makes mechanical contact with the conductive film, and as such is the weakest link, since such contacts are abrasive and gauge the conductive layer, eventually making pots "scratchy" or "crackly", or even losing contact altogether.

The two main types of potentiometers are linear taper, usually marked "B" and logarithmic or audio taper, usually marked "A". The logarithmic potentiometers are used for volume controls, since human ears respond to sound pressure in a logarithmic manner.

Manufacturing true log tapers by carbon or plastic deposits is very difficult, so manufacturers usually resort to approximating the log curve with two linear segments.

As illustrated, up to 50% rotation, where amplifier's volume controls are set most of the time, the deviation from the real log curve) is quite small, and at higher volumes such error is not noticeable at all.

Rheostats are high-power potentiometers, usually used to control the flow of current in a circuit, or to control voltage if there is a significant load current, where ordinary, low power potentiometers would be overloaded and burn out.

Just like their fixed-resistance relatives, rheostats are wirewound components. The thinner the wire, the finer is the step-adjustment (the jump in resistance between adjacent turns), but the lower the current rating. The rheostat on the right, using a flat metal strip, has a much higher current rating but also a much lower resistance than the one on the left, which uses a round-profile wire. The higher powered rheostat has only around 36 steps (jumps), while the wirewound one has about three times more!

Two examples of power potentiometers or rheostats, bottom view.

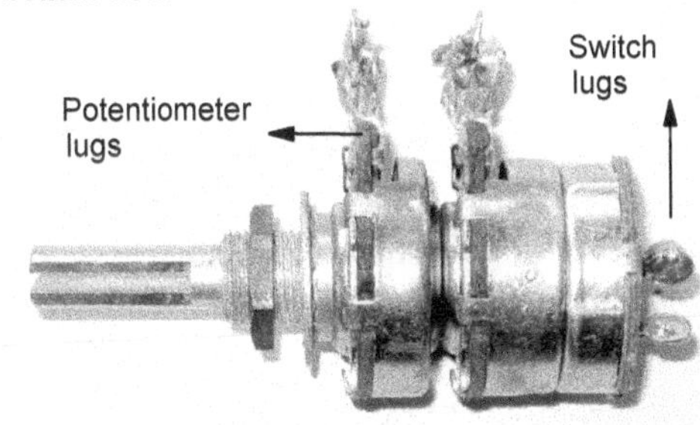

Dual-gang audio potentiometer with an on-off switch and a split-shaft

10-turn linear potentiometer (left) and a trimmer potentiometer (right)

These 50W 4Ω rheostats make great dummy loads. Since they are adjustable, you can ascertain how the power amp you are testing drives low impedance loads such as 1 or 2 Ω!

MAKING A DUMMY LOAD FOR AMPLIFIER TESTING

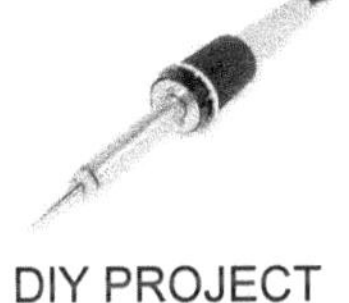

If you don't have 8Ω resistors rated at 20 or 50W, you can use resistors of lower power rating and/or lower or higher resistance. Combine them in parallel-series combinations to get the resistance and power rating you need. Two of many options are shown here.

You can mount the resistor on a tagboard or bolt them down to a metal heatsink of some sort.

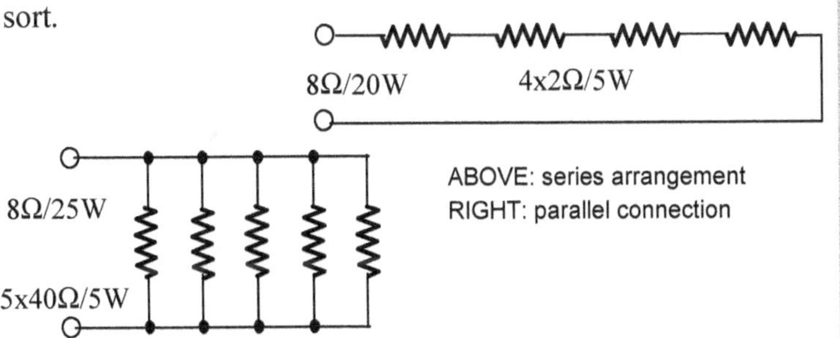

ABOVE: series arrangement
RIGHT: parallel connection

ALPS "Blue Velvet" potentiometer. Most sold as such are counterfeit! In late 1990s we contacted ALPS in Japan by fax and they advised us that they never produced such a pot and that the part number printed on the sticker is not recognized by them at all!

Testing potentiometers for tracking error

The most critical parameter of a dual-gang audio potentiometer used for volume control, is the tracking between the two gangs. A large difference in the resistance of the two gangs may manifest itself as one channel (with higher resistance between the pot's wiper and ground) of a stereo amplifier sounding louder than the other.

In this simple test any battery can be used, standard voltages of 1.5 or 9V are fine, or you can use a benchtop DC power supply. Since it works as a null-indicator in this bridge-circuit, the moving-coil instrument should have a center-zero, so it's able to deflect to either side.

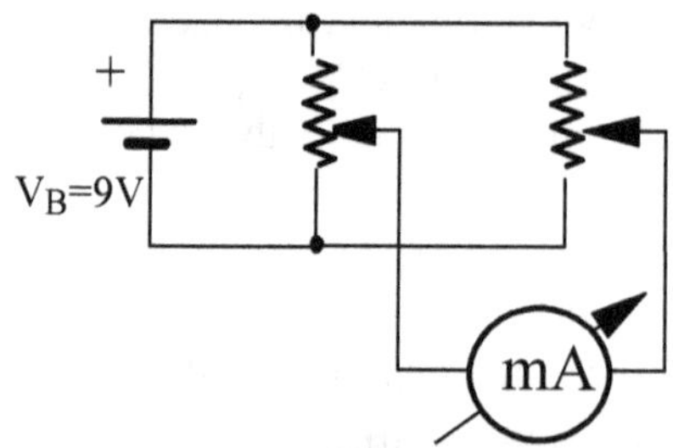

A simple setup to test potentiometers for tracking error requires only a DC battery and a zero-centered moving coil meter.

The four halves of the dual-ganged potentiometer form four legs of the bridge, the meter is connected in the bridge diagonal, between the two wipers. Ideally, the meter should stay dead center. The larger the deflection to either side, the bigger the discrepancy between the two gangs of the potentiometer! This test is only qualitative, to express the error quantitatively, an ohmmeter is needed.

A two ohmmeter method is also illustrated, but a single ohmmeter and a changeover switch can also be used, although in that case the test is not continuous but taken in a dozen or so points across the potentiometer's range, from 0 to 270 degrees.

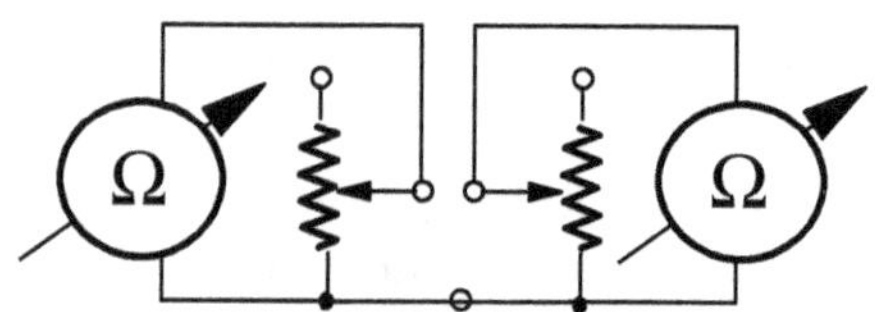

The alternative method of testing for balance requires the use of two analog ohmmeters (or multimeters on "ohms").

Audio attenuators

Both sonically and from reliability perspective (they become scratchy, lose contact and suffer from tracking errors), volume control pots are considered the weakest link in the signal chain. Audio attenuators are considered superior mainly because the thin carbon or conductive plastic film is replaced by a quality switch and a resistive divider. The minimum number of switched steps is 12, although most commercial units use at least 23 or 24 position switches. Obviously, the higher the number of discrete positions, the finer the resolution of volume control.

The steps or jumps in resistance between the switched positions are usually expressed in dB. They may be uniform, but need not be. At lower volumes finer steps may be used, say 2 dB, and at higher volumes larger jumps can be tolerated, 3 or 4 dB, for instance.

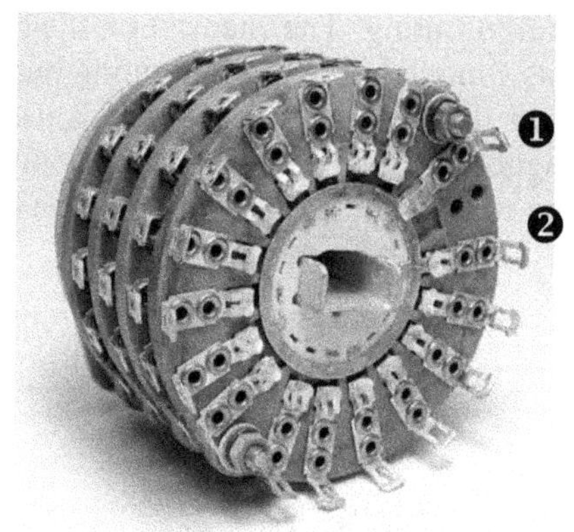

Rear view of a vintage 4-pole (four separate wafers), 18-position rotary switch needed for ladder and shunt-type attenuators. 1) is the common lug (always touching the ring), the switch is in position 18, the rotating pin connecting the last lug (2) in the anti-clockwise direction.

The simplest type of audio attenuator is the series type, where a string of resistors is connected between the input and common (ground). As the switch is cycled from the lowest (zero) volume (the bottom position on the drawing where $V_{OUT}=0$), more and more resistors are connected in series between the output and ground and less and less between the input and output. The input impedance of the series attenuator is constant.

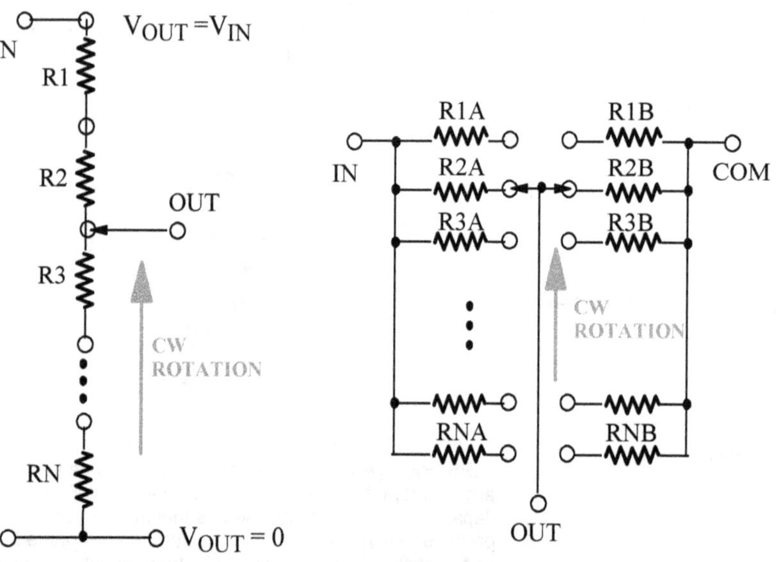

Series attenuator

Parallel or ladder-type attenuator

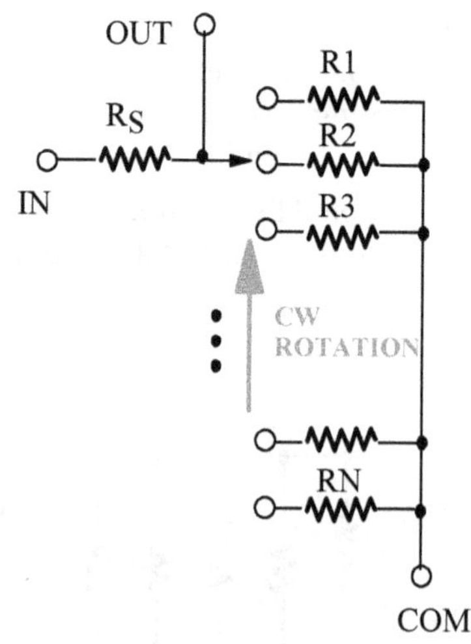

Shunt-type attenuator

However, no matter in which of the 24 positions the switch contact is, the signal goes through all 23 resistors (there are always N-1 resistors, if N is the number of switch positions). That is the biggest weakness of the series type: there are just too many solder joints and too many noisy resistors in the signal path. One wonders if any improvement over a good quality potentiometer has been achieved, and the answer is probably no!

The parallel or "ladder" type and the shunt attenuator both eliminate this problem by having only two resistors switched into the circuit at any time. The shunt type is simpler, since all positions share the same series resistor R_S, and only the shunt resistors R1 to RN are switched in and out. Since the series resistor is fixed and the shunt resistor's value gets progressively smaller and smaller as the switch is turned CCW (counterclockwise), the input impedance of a shunt attenuator is not constant. This could become an issue with some sources. The downside to the shunt type is that it presents a varying input resistance to the source driving the amp.

In the ladder type, each shunt resistor (R1B to RNB) has its own series resistor (R1A to RNA). This enables the designer to choose the values so a constant input impedance is achieved.

The switches used for audio attenuators are the "shorting" type (or "make before break" switches). The wiper will make contact with the next lug before breaking the contact with the previous lug. That way there are no "gaps" in the switching and no transients that may cause audible "bumps" or even damage sensitive speaker drivers.

Constructing your own attenuators is not for faint-hearted, manual dexterity and immense patience is required. If you value your time (hours can be spent sifting through resistors trying to find the required values) and sanity, relatively cheap ready-made units are available from China. Some are clunky and noisy, due to poor quality switches used, others are quite quiet and refined, so buy one first, to ascertain its quality and "feel". Of course, if you have money to burn, there are also many brand-name (read "expensive") audio attenuators.

OTHER COMPONENTS

Fuses

A fuse is a component designed to act as the weakest link in an electrical circuit and burn out ("blow") when the current through the circuit exceeds its rated value. In its application in audio it protects the primary winding of the mains transformer and the wiring inside the amplifier, so neither of these burn out before the fuse does. Thus, it is extremely important not to replace a fuse of a certain current rating with the one of a higher rating. If the current through the primary of a mains transformer is 1A and we replace a 1.5A fuse with a 3A fuse, should a fault develop (a short-circuit on the secondary side), the $150 transformer may burn out before the five cent fuse!

There are many types and standards of fuses. 3AG (American size) and M205 (European size) are the two miniature types most commonly used in electronics. Both are so called "cartridge type", of cylindrical shape. AG stands for "automotive glass", since these sizes (2AG, 3AG, etc.) were originally developed for use in cars. Paradoxically, cars today use blade-type fuses, which are easier to change in a hurry by simply unplugging them, while 3AG and M205 are almost always enclosed in a fuse holder whose top cap needs to be unscrewed for a fuse to be changed. 3AG fuses are 31.8mm long and 6.35mm in diameter, while metric-sized M205 fuses are 20mm long.

Two main types of fuses are used in electronics, fast and slow-acting (delayed-action fuses). Slow acting are often called SLO-BLO(W) and are designed to withstand higher inrush currents that flow when equipment is powered-up, but only for a short time. If the higher current persists the fuse melts.

The naming convention of 3AG fuses is best illustrated with an example: 3AGDA1.25 = 1.25A 250V Slow 3AG fuse. M205 fuses a include letter D for "Delayed action" in their model number: M205DA01.6R = 1.6A 250V Delayed M205 fuse. An example of a fast-acting M205 fuse would be M205002R (2A, 250V Fast M205 fuse)

The current draw of an amplifier should never exceed 75% of the nominal fuse current rating. Thus, a 1 Amp drawing amplifier should use a fuse of at least 1/0.75 = 1.35A rating!

However, ambient temperature also plays a role in fuse behavior. Although hi-fi amplifiers operate exclusively at room temperatures, depending on the topology and mechanical design of the chassis and element positioning, temperatures inside the chassis can reach higher levels, especially during hot summer days. A 30°C room temperature and 30°C temperature rise next to the mains transformer means a fuse would be at 60°C. According to the derating curve, a slow-blow fuse would need to be derated to around 80%. For instance, with a current of 1.5A, we would need a slow-blow fuse rated at 1.5/(0.75 x 0.80) = 2.5A!

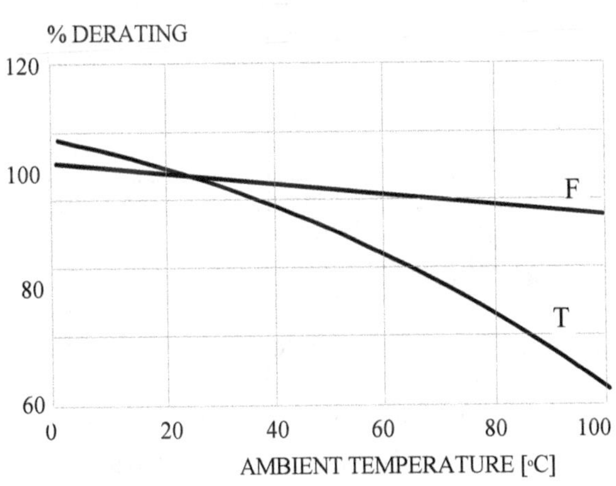

Temperature derating curves for FAST ACTING (F) and SLOW BLOW (T) fuses

Switches

Switches are classified according to the principle of operation and the configuration of contacts. The main types are toggle switches, slider switches and rotary switches.

The configurations are described in terms of the number poles and positions ("throws"). A single pole means a single switch, while double-pole describes two mechanically linked switches . "Single throw" means one active contact, double throw means one common contact alternatively connected to two active contacts. So, SPST = single pole single throw, SPDT = single pole double throw, DPDT = double pole double throw, 3PDT = triple pole double throw, and so on.

Most "universal" rotary switches come in the 12 contacts (soldering lugs) configuration. 12 is the product of poles and throws (1*12, 2*6, 4*3, 3*4 and 6*2). So it can be a SP12T switch, a single pole switch with 12 positions, DP6T, two-pole 6-position switch, 3P4T, three-pole 4-position switch, 4P3T, four-pole 3-position switch and 6PDT, six-pole 2-position switch.

Mains-rated rotary switches such as this $250V_{AC}$ rated Lorlin two-position on/off switch (below left) are relatively rare. The AC rating of its silver cadmium oxide contacts is 10A. The rated insulation resistance voltage is $500V_{DC}$. The body and the shaft are both made of plastic. Since most of our amplifiers have volume control, we use these as power on-off switches mounted at the front fascia in a symmetrical arrangement, with two identical knobs.

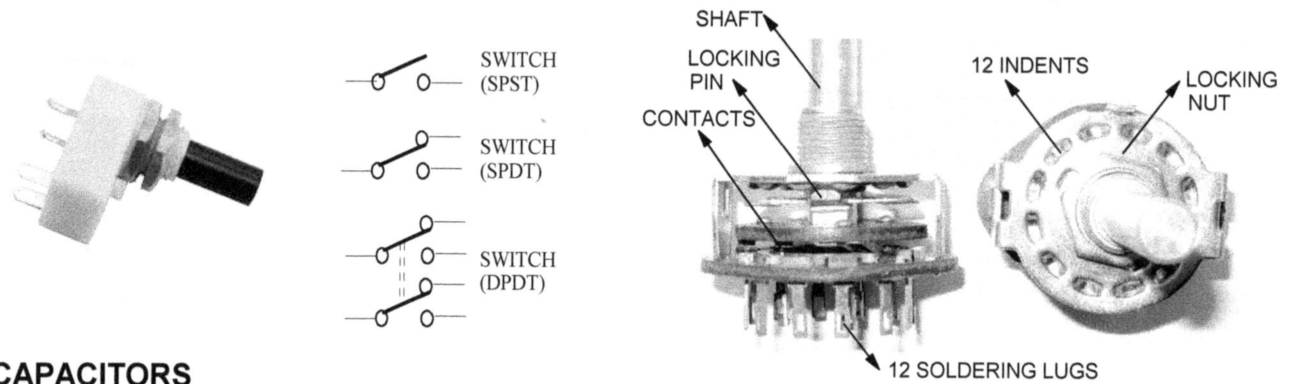

CAPACITORS

The physical nature of capacitance

The simplest capacitor is formed by two conductive plates with a dielectric in-between. The most common dielectric materials used are air (in variable or "tuning" capacitors), mica, oxide films (in electrolytic capacitors), plastic films or paper (in film capacitors) and ceramics. The capacitance of this simple structure can be expressed as $C_P=\varepsilon_0*k*A/d$ where ε_0 is the absolute dielectric constant (also called "permittivity") of free space, whose value is $\varepsilon_0=8{,}85*10^{-12}$ Farads/meter, and k is the relative dielectric constant of the dielectric. Typical values are listed in the table below.

DIELECTRIC	k	DIELECTRIC STRENGTH [kV/cm]
Air	1.0006	30
Polystyrene	2.5-2.7	197
Polyester	3.2	650
Polycarbonate	2.57	150
Polypropylene	2	236
Teflon	2	240
Impregnated paper	4-6	500
Mica	4.5-7.5	1,180
Aluminium Oxide	8.4	146
Tantalum Pentoxide	27	1,470
Ceramic (low loss)	6-20	80
Ceramic (high k)	100-1,000	350
Titanium Dioxide	80-120	6,000
Barium Titanate	200-16,000	300

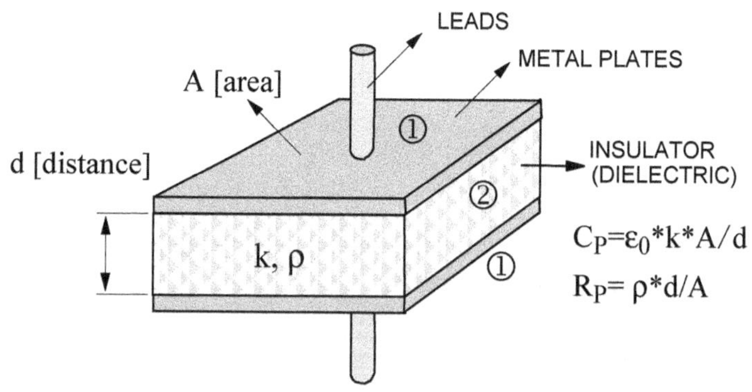

ABOVE: The simplest capacitor is formed by two conductive plates with a dielectric in-between (which could be air)

The term "capacitance" refers to the amount of electric charge a capacitor can store for each volt of potential difference between its plates: C= Q/V [Coulomb/ Volt] This unit is called a Farad, in honor of Michael Faraday, famous for his work in the fields of electromagnetism (Faraday's Equations) and electrochemistry.

When a discharged capacitor is connected to a DC voltage source (for instance an ordinary battery), a transient current will flow in this simple circuit, electrons will flow towards one electrode and accumulate there. The charge on the other plate will be of equal magnitude, but positive, which simply means it will be depleted of electrons.

An electric field now exists between the two plates, and a certain amount of electric energy is stored as electric "stress" in the dielectric material between the plates. Since the charging current has stopped flowing, in such an equilibrium or *steady-state*, the electric field is stationary, and is thus called *electrostatic* field.

The energy stored in a capacitor is $E= ½QV = ½CV^2$. For instance, in a typical tube amplifier's anode power supply, a 470µF electrolytic capacitor, charged to 400V will store energy of $E=½*470*10^{-6}*400^2= 37.6$ J (Joule).

A solid state amplifier working at 40V and using 10,000 µF of capacitance in its power supply will have a stored energy of only $E=½*10{,}000*10^{-6}*40^2 = 8$ J.

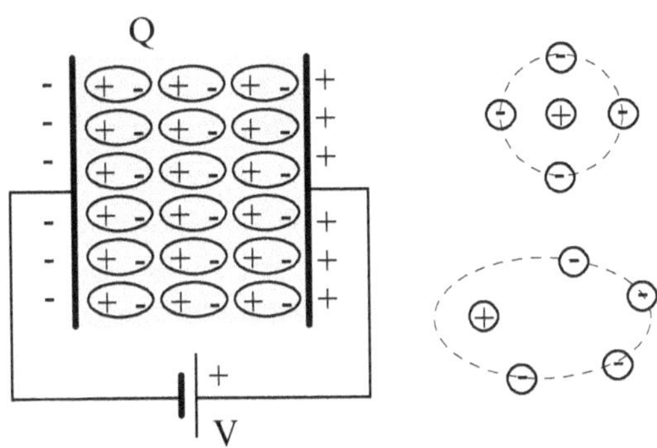

Without an electric field an atom has a positive nucleus and a cloud of negative charge made up of electrons rotating around it. Placing it inside an electric field "polarizes" the charges inside the atom and pulls the negatively charged cloud slightly away from the positive nucleus.

Film capacitors

Film capacitors are non-polarized, and are rated in terms of their capacitance, maximum working DC voltage and the dielectric material. Two main types are film&foil and metallized film capacitors. Film and foil technology uses two thin metal sheets or foils, acting as electrodes, and a plastic film in between them, acting as the dielectric. The three are then rolled together in a tubular shape. The thicker the dielectric, the higher the working voltage of a capacitor, and the bigger its physical size.

Metallized film technology does not use metal sheets or foils, the electrodes are metal layers that are deposited on both sides of the plastic film instead. Generally, for the same value and working voltage, metallized units are much smaller than the film&foil capacitors. Metallized capacitors have a higher voltage rating due to their self-healing property. An internal arc through the metallized paper or plastic film burns the very thin metal layer away, effectively clearing the short circuit.

Film & foil capacitors are made in two technologies. In extended foil capacitors foils and film are wound offset, with one metal foil protruding over the plastic film on one end and the other foil out the other end.

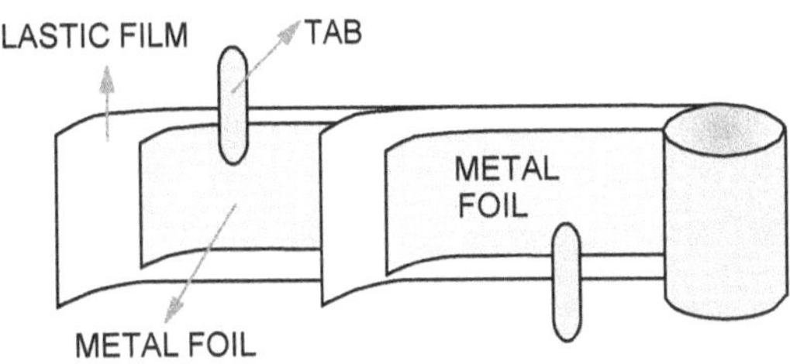

Film & Foil capacitor: Tab construction

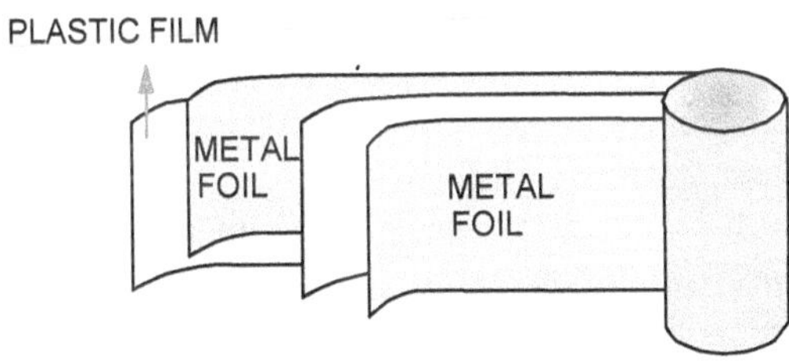

Film & foil capacitor of extended foil construction

Once the winding is finished the foils are crushed and metal lead are either soldered or welded to the crushed ends. This method effectively shorts the capacitor's parasitic inductance without affecting the capacitance.

The tab construction requires that plastic film be wider than the metal foil so metal tabs can be inserted in the roll, and the leads are then soldered onto the tabs. Once the leads are attached, the rolled capacitor is vacuum impregnated with a mineral vax, rigid resin, or synthetic or mineral oil (for paper capacitors).

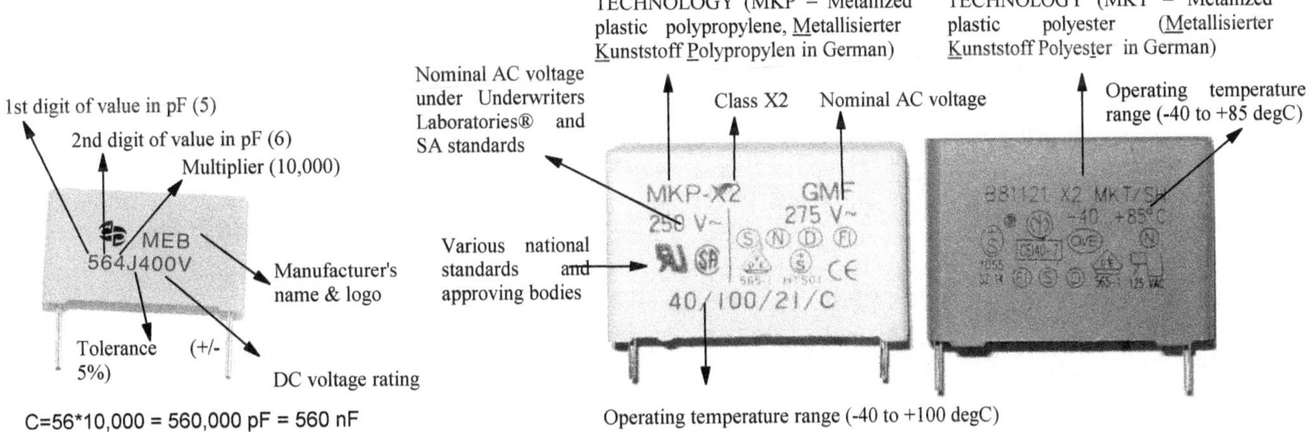

Two examples of X2-rated (for mains filtering applications) film capacitors, in metallized polypropylene film and metallized polyester film. These make great coupling capacitors as well.

A code imprinted on the case or body of a capacitor usually declares what the dielectric is. The codes are: KP - polypropylene film&foil, KS - Polystyrene film&foil, KT - Polyester film&foil, MKC metallized polycarbonate, MKP - metallized polypropylene, MKT - metallized polyester, MKY - metallized low-loss polypropylene, MKL or MKU - metallized lacquer (cellulose acetate).

Modern capacitors have their value or a code printed on their bodies. The code has three digits and one letter. The first two digits are the value, the third digit is the multiplier, and the letter is the tolerance. Some vintage units were marked using a color-coding system similar to that of resistors. The values are given in the tables.

NUMBER	Multiplier	LETTER	Tolerance
0	1	F	+/- 1%
1	10	G	+/- 2%
2	100	H	+/- 3%
3	1,000	J	+/- 5%
4	10,000	K	+/- 10%
5	100,000	M	+/- 20%
6	NOT USED	P	+100% -0%
7	NOT USED	Y	+50% -20%
8	0.01	Z	+80% -20%
9	0.1		

Color	Band 1	Band 2	Band 3	Band 4	Band 4	Band 5
	1st digit	2nd digit	Multiplier	Tolerance above 10pF	Tolerance below 10pF	DC voltage rating
Black	0	0	1	+/-20%	+/- 2pF	
Brown	1	1	10	+/- 1%	+/- 0.1pF	100V
Red	2	2	100	+/- 2%	+/- 0.25pF	250V
Orange	3	3	1000	+/- 3%		
Yellow	4	4	10,000	+/- 4%		400V
Green	5	5	100k	+/- 5%	+/- 0.5pF	
Blue	6	6	1M			600V
Violet	7	7				
Gray	8	8	0.01	+80% -20%		
White	9	9	0.1	+/-10%	+/- 0.1pF	
Gold			0.01	+/- 5%		
Silver			00.1	+/-10%		

The color-coding system for film capacitors.

Of all film capacitors polystyrene have the best electrical properties, very low temperature coefficient, almost linear, very small losses (low dissipation factor of approx. 0.01%) Their maximum temperature is only 85°C.

Polypropylene capacitors are almost as good, with a slightly higher temp. coefficient and dissipation factor. They can operate up to 105°C, and are cheaper than polystyrene equivalents!

Polyester capacitors are the least stable and have the highest losses and ass such are not recommended in hi-end audio.

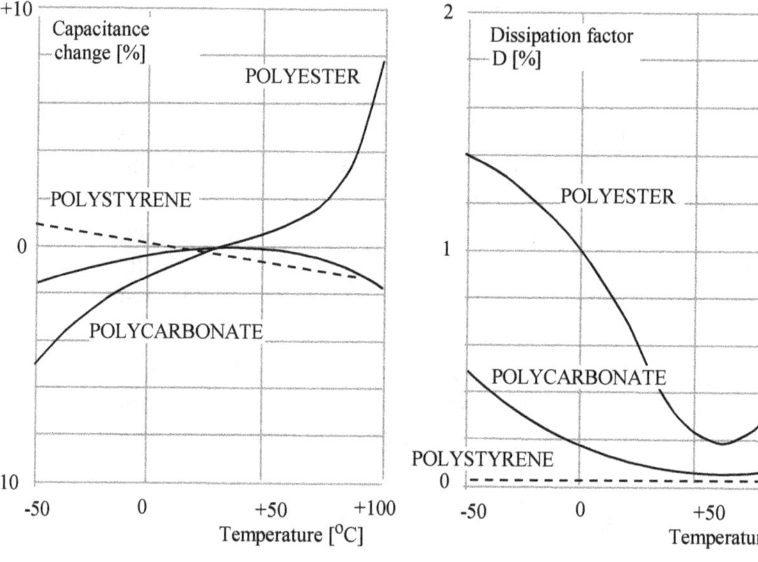

Capacitance and dissipation factor versus temperature for three types of film capacitors

Paper-in-oil capacitors

Paper-in-oil (PIO) is a vintage technology that precedes the use of plastics. Oil loses its insulating properties when subjected to high temperatures, and paper is inferior to plastics in every way, so these capacitors don't last nearly as long as the plastic film types. However, many audiophiles prefer the sound of PIO capacitors and this type of film capacitor is highly sought-after!

If you are opening a PIO capacitor to study its construction or have a leaking one, keep in mid that older units used PCB (polychlorinated biphenyl), a synthetic organic chemical compound used as a dielectric fluid ("oil"). PCBs are cancerogenic and were banned in USA in 1979. Notice the printing on the vintage Mitsubishi capacitor (probably dating back to early 1980s), "No PCBs". Some of the trade names for PCB oil are Pyranol, Pyrenol, Chlorinol, Aroclor, Asbestol, Askarel, Phenoclor, Pyralène and Clophen.

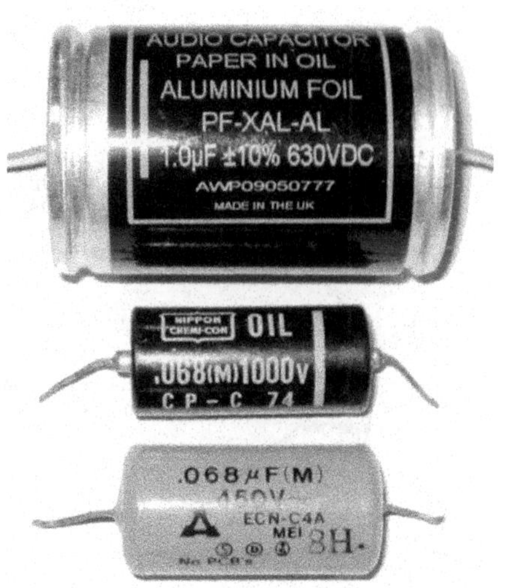

RIGHT (Top-to-bottom): Aluminum foil PIO capacitor (1μF), and two Japanese PIO caps from 1970s, by Nippon Chemi-con and Mitsubishi (both 68 nF). The Mitsubishi capacitor is rated at $450V_{AC}$, which is roughly 450*2.8 = 1,260 V_{DC}!

Ceramic capacitors

Ceramic capacitors are non-polarized types manufactured in a myriad of shapes, sizes and technologies. The main ones are rod- (or tubular), disc- and monolithic type. Tubular type can have radial or axial leads, and can be molded (resembling carbon composition resistors!), enamel coated or dipped phenolic. Monolithic ceramic caps can also look like low-wattage resistors (axial), these are glass-sealed type. The other variant is dipped polymer coated, usually rectangular in shape.

There are two main classes of ceramic capacitors. Class I temperature-compensating ceramic capacitors have a linear temperature coefficient of capacitance (TCC), expressed in parts-per-million-per-degree-Celsius (ppm/°C). Below 100Mhz their voltage and frequency coefficients are negligible, as are their capacitance changes caused by temperature changes and aging. This makes them widely used in applications where accurate and stable capacitance values are required, such as tuned circuits and RC filters and networks. These low-loss types have a very high leakage resistance (1,000 MΩ) and can be used instead of mica types in many applications.

The ordinary kind is called a "Class II high-K" type. K refers to their high dielectric constant, enabling them to pack more capacitance into a smaller size then Class I, but their tolerances are wider and stability inferior. Their capacitance varies widely with temperature, as does their dissipation factor (losses) and insulation resistance.

Type II ceramic capacitors use dielectrics based on barium titanate to which modifiers such as niobium, tantalum pentoxide or bismuth stannate are added. These materials are ferroelectric, meaning they exhibit a hysteresis effect similar to that found in magnetic materials, but in this case we are not referring to a magnetic but an electrical field. This ferroelectric phenomenon results in wide variations of capacitance and D-factor with voltage and frequency changes.

Mica capacitors

Mica is a naturally occurring mineral whose layered structure makes it easy to cleave or split into sheets of uniform thickness, down to $5*10^{-6}$ mm! It has a high dielectric constant (6.85) and high insulation voltage (3,000 volts per mil) and it is chemically inert. This makes it into an ideal dielectric and mica capacitors do not age or change their properties over time and are very stable under temperature and electric stress.

Alternate layers of metal foil and mica sheets are sandwiched together, or silver is deposited directly onto mica sheets instead of foil. The finished stack is clamped together and leads are spot welded or soldered on. The assembly is then treated (usually with silicone) to prevent moisture absorption and sealed in a molded case, or dipped in epoxy resin. Synthetic mica was also developed, a mixture of aluminium and magnesium oxides, sand and cryolite, which is then heated to 1,300 °C and gradually cooled over a few days. It has a similar layered structure to natural mica but is more difficult to cleave.

Mica capacitors are made in values from 1pF to 100nF. The relatively low upper limit is due to the fact that mica sheets are not flexible and cannot be rolled into tubular structures as film capacitors can. Very low leakage currents and dissipation factors, together with small tolerances and high temperature stability make mice capacitors invaluable in high precision and high frequency applications. 1kV voltage ratings are common, with some capacitors rated even up to 10kV!

Trimmer capacitors

You may have seen large multi-plate air tuning capacitors inside old radio sets. Disc ceramic trimmer capacitors work on the same principle. The rotor is a thin disc made of Class I or Class II ceramic dielectric. A silver layer is deposited, screened onto the top surface of the rotor, and a similar electrode is applied to the stator, which also serves as the outer shell of the capacitor. Turning the rotor changes the effective plate area.

While disc ceramic trimmers go from minimum to maximum capacitance value in half-a-turn, tubular trimmers cover their capacitance range in many turns, and are used for more precise adjustments of small capacitance values (kinda like single-turn and ten-turn potentiometers).

The other type of trimmer capacitor are compression mica type, where a thin layer of mica is sandwiched between two spring metal conduction plates. An insulated screw through the center of this assembly changes the compression on the plates and varies the distance between them, thus varying the capacitance. Usually it is not just the mica sheet acting as dielectric but a layer of air as well, in series with the mica.

Mica trimmers are nonlinear and not particularly stable, their value drifts with time. The capacitance values range from 1 pF to 3 nF.

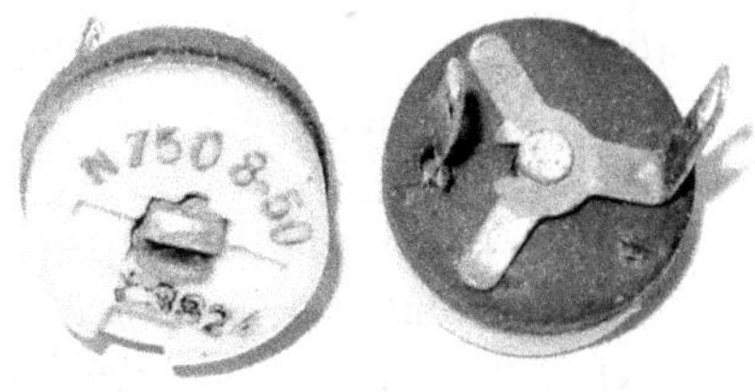

A typical disc ceramic trimmer capacitor.

Electrolytic capacitors

No other electronic component has as many symbols in use as elcos. Symbol c) is mostly used in vintage American literature, while a) and b) are more common in Europe. To add to the confusion, many American books and diagrams also use symbol c) for non-polarized capacitors (film, mica, etc.) as well. The difference in shape between the two electrodes has a significant meaning, with which those authors seem unfamiliar with!

We will use symbol a) in this book. The bottom electrode is shaped to symbolize the can into which the anode or positive electrode is immersed.

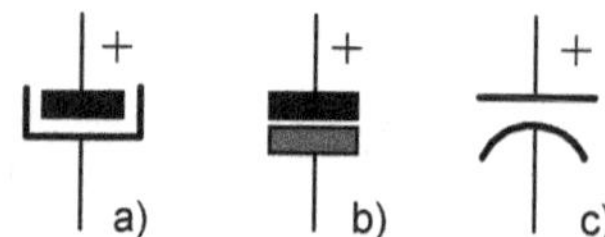

Three common symbols for electrolytic capacitors. The + sign is usually omitted, since it can be determined from the orientation of the symbol. In a) and c) the upper electrode is positive.

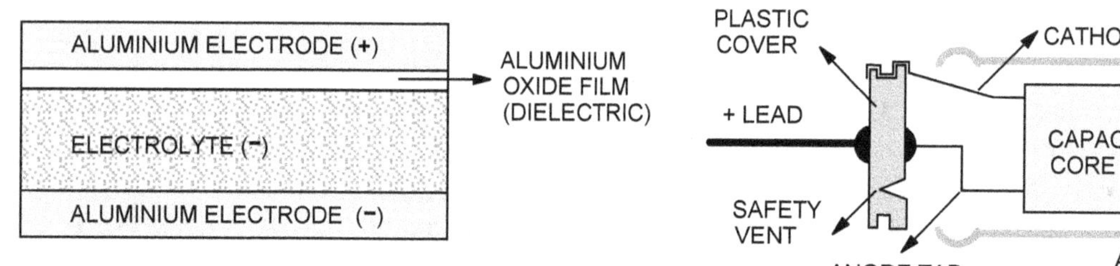

Cross-section through an axial electrolytic capacitor

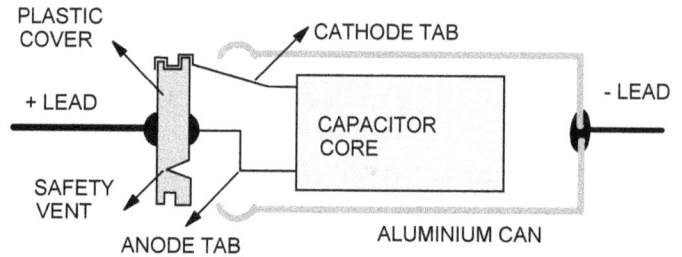

The mechanical construction of an axial elco

During manufacturing, a thin oxide film is formed on the anode (positive metal electrode). This film acts as a dielectric. It is surrounded by electrolyte, into which the negative electrode is also immersed, whose job is only to make the contact with the outer surface of the dielectric via the electrolyte. This contact is made by electrolytic conduction, and that is why this class of capacitors are called electrolytic.

The thickness of the oxide film is proportional to the forming DC voltage, but the thinner the film, the higher the capacitance. That is why capacitors rated at 450 or 500V seldom go higher than 680 μF in value!

The oxide film deteriorates without the forming voltage, that is why electrolytic capacitors deteriorate more on the shelf than working inside an amplifier. It is also why NOS capacitors that have been on a shelf for years or even decades should not simply be installed into an amplifier. They should be subjected to the reforming process, connected to lower DC voltage initially (say 100V instead of 400V) and then the voltage is gradually increased to their rated voltage and about 10% higher, which a healthy capacitor should handle without any ill effects (500V instead of 450V for instance).

The anode must stay positive. If voltage polarity is reversed, the capacitor will be destroyed and most likely will explode, spewing the electrolyte all over the amplifier. The same happens when their rated voltage is exceeded. This is why it is critical to observe the marked polarity and voltage rating of these capacitors.

Despite its shortcomings, the electrolytic capacitor technology is constantly improving, and their ever shrinking sizes are a testament to that. The smaller capacitor on the left (240μF/450V_{DC} by Hitachi Japan) and the one on the right (220μF/450V_{DC} by Nippon Chemi-con Japan) are recent units, made in the last 10 years. In the middle is 25+ years old 220μF/400V_{DC} capacitor (also by Nippon Chemi-con), which is almost double in size (volume).

The older capacitor is also inferior in two other aspects. It has a lower voltage rating (400V compared to 450V), and a lower maximum allowable temperature of 85 degC, compared to 105 degC for the two modern caps.

Some manufacturers use proprietary technologies, such as Elna in their Cerafine™ range, which uses super fine ceramic particles and sound better than ordinary elcos, most likely due to improved charge and discharge speeds and reduced distortion.

In a similar fashion, the out-of-production and thus super-expensive Black Gate™ series used fine graphite particles as the separator between its aluminium oxide anode and cathode. They even invented a name behind it all, calling it "The Transcendent Electron Transfer Theory". They used to be manufactured under license by Rubycon in Japan.

Three electrolytic capacitors of similar values and DC voltage ratings, but different physical sizes.

The life of an electrolytic capacitor

In their data sheets for long-life 350-450V_{DC} aluminium electrolytic capacitors Vishay specifies the following:

Useful life at 85°C: 10,000 h, Useful life at 40°C, 1.4 x I_R applied: 400,000 h and Shelf life at 0 V, 85°C: 1,000 h

Notice that an increase in operating temperature from 40°C to 85°C reduces life by the factor of 40! (400,000/10,000 = 40). Also, notice that the shelf life is only 1/10 of the operating life (working in equipment), 1,000 hours compared to 10,000 hours. Electrolytic capacitors last longer when energized in equipment than sitting on the shelf. The power constantly forms the oxide film, which quickly and irreversibly deteriorates on the shelf.

Vintage capacitor bridges test capacitors at high voltage, typically up to 600V_{DC}. They don't just measure capacitance, but can measure the leakage currents at high voltages and can also be used to "reform" old elcos. These invaluable instruments are covered in Volume 2 of this book. Digital LCR meters also measure losses (D-factor) but at very low voltages. They are more precise and faster to use than vintage bridges, so get both instruments.

Be very careful who you buy your elcos from. As with sealed lead acid batteries (which also deteriorate quickly sitting on a shelf) ask for capacitors' age and for a money-back-guarantee in case the seller was not truthful - most will not be, or will have no clue about how long their wares have been sitting on the shelf.

Improving the high frequency behavior of electrolytic capacitors

Film caps behave in the same fashion across the whole audio frequency spectrum, unlike electrolytic caps, whose capacitance drops and whose inductance and resistance rapidly rise with the increase in frequency of the signal they have to handle. Elcos lose their capacitance at higher frequencies where their resistive and inductive components become dominant. A qualitative frequency curve (right) shows these three regions.

The equivalent circuit of a capacitor shows the leakage resistance R_L, which models a resistance that allows a very small but always present DC leakage current to flow. The other three components are a series LCR circuit. L is the ES or Equivalent Series Inductance, and R is ESR, or the Equivalent Series Resistance.

If you still want to use ELCOs, here is a trick that will make your preamp or amp sound noticeably better. Bypass all of them with a small film capacitor of the best quality you can get. Usually one bypass is enough (1μF film in our example), but if you are really fanatical, bypass it again and again with progressively smaller value film caps. Some audiophiles and amp builders claim that this type of bypassing is detrimental to sound, as always, the opinions are divided.

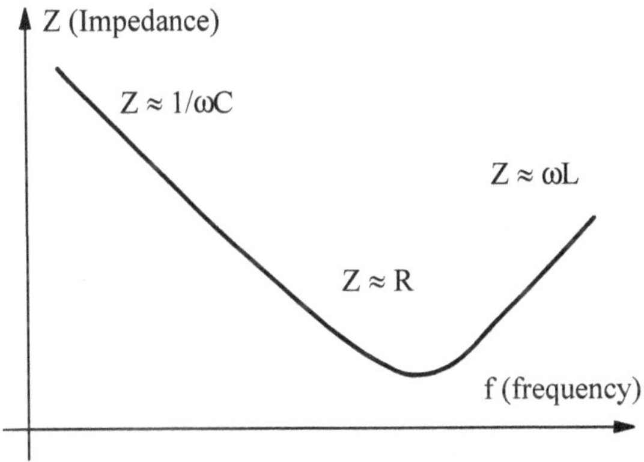

ABOVE: The impedance versus frequency curve for electrolytic capacitors. Above a certain frequency the capacitor turns into an inductor!

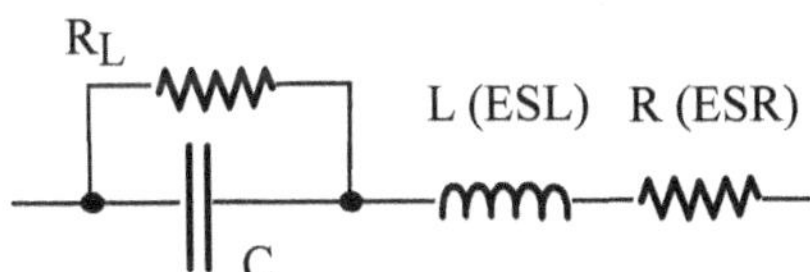

ABOVE: A simplified model of an electrolytic capacitor.

Using film instead of electrolytic capacitors

Motor start (MS) capacitors can be found at surplus sales for a dollar or two, but retail prices in Australia are incredibly high, $30-80 each. Most are self-healing polypropylene film capacitors, highly durable and reliable, and in an amp, used as power supply filtering caps, they will last forever! They will not age, leak, overheat or explode as ELCOs can, and they will sound noticeably better.

Physically, these film capacitors are very large, so fitting them in small amplifier chassis may not be possible.

Don't be fooled by their lowish AC voltage rating. Remember, AC is a sine wave, it has a positive and negative crest. So, if an MS capacitor is rated at say 450 V_{AC}, that is the effective value. The peak value is 41% higher, then we double it to get peak-to-peak value which is their DC rating!

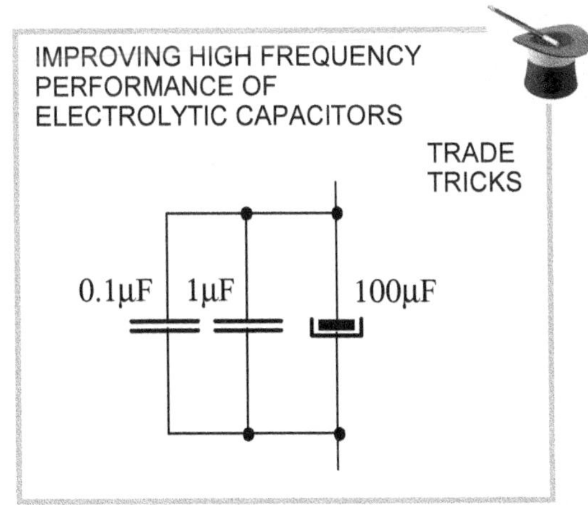

$V_{EFF} = V_{RMS} = V_{AC}$

Peak voltage $V_P = 1.41V_{RMS}$

DC voltage equals Peak-to-Peak voltage: $V_{DC}=V_{PP}=2V_P = 2.82V_{RMS}$

Rating example: $250V_{AC}$ written on the capacitor

$V_{DC} = 2.82*250V_{AC} = 700$ V

Even a modestly rated 250 V_{AC} capacitor has a DC rating of 700 Volts! A 450 V_{AC} rated capacitor would have a DC rating of 1,269 Volts!

DC VOLTAGE RATING OF MOTOR-START & OTHER AC-RATED FILM CAPACITORS

$V_{DC} \approx 3*V_{AC}$

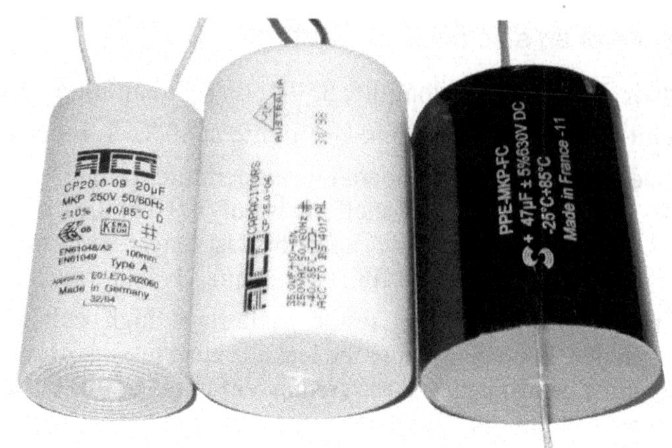

ABOVE: Two types of MKP motor-start capacitors made by ATCO (20µF and 35µF), rated at $250V_{AC}$ (up to 700 V_{DC}), compared in size with a similar audio capacitor made by Solen (47µF), rated at 630 V_{DC}

Typical parasitic capacitances

Not all capacitance is wanted or desired. Every time you have two metal parts or two conductors adjacent to each other, you have an unwanted capacitor. Such parasitic capacitance between components and within components themselves, such as the inter-electrode capacitance of tubes, affect circuit properties and its operation, especially at higher frequencies. Here are a few typical values of parasitic capacitances encountered in audio work:

- Ceramic octal tube socket - between pins and case (skirt): 1.5 pF
- 1W carbon-film resistor - 2.0 pF
- 100kΩ potentiometer, case to element: 9.8 pF
- Single-core shielded cable: 65 pF/m
- Two-core shielded cable: 44 pF/m between conductors, 85pF/m between each conductor and shield
- Audiophile quality speaker cables: 300 pF/m
- Teflon-insulated silver-plated copper interconnect cables (twisted pair): 91 pF/m
- Coaxial copper interconnect cables: 80 pF/m

PASSIVE SEMICONDUCTORS

Rectifier diodes

A silicon diode is a PN-junction, which conducts current when forward biased (P-side or anode is positive with respect to the cathode) and does not conduct when reverse-biased (anode is negative with respect to cathode). Vacuum tube diodes behave in a similar manner, but work on a totally different physical principle, as we'll see very soon.

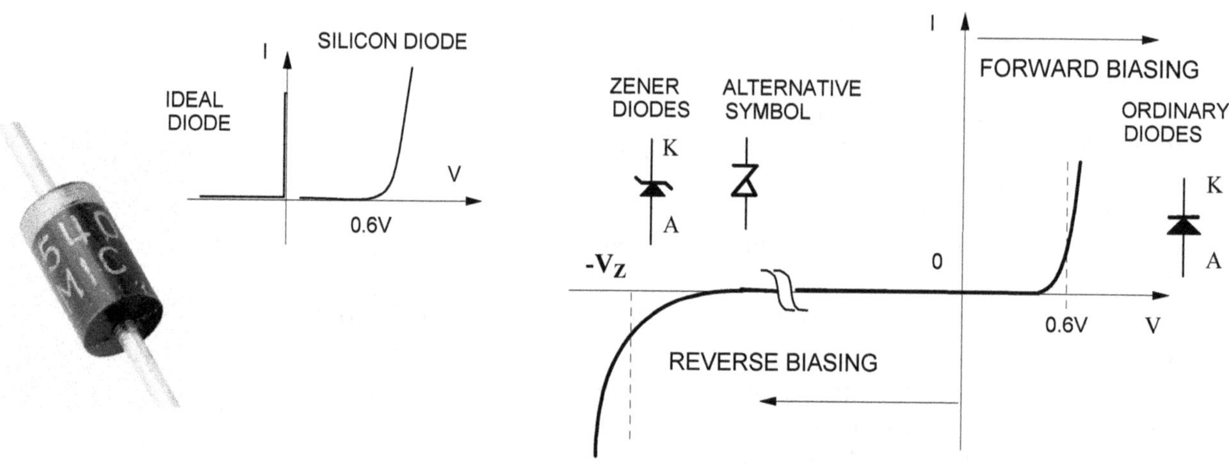

While ordinary silicon diodes work in the forward biasing regime, Zener diodes are reversely biased. The inverted SS symbol indicates that a range (section) has been omitted, that the graph is not to scale, as here with reverse (negative) voltage.

1N4004 is a general purpose 1A rectifier, with max. reverse voltage rating of 400V. The V_{RRM} (Peak Repetitive Reverse Voltage) ratings for other diodes in the same series are 50V (1N4001), 100V (1N4002), 200V (1N4003), 400V (1N4004), 600V (1N4005), 800V (1N4006) and 1,000V (1N4007)! As you can see, there is some method to the naming madness, but logically one would think that 4007 meant 4A, 700V, but it does NOT, it means 1A, 1,000V! 1N5404 (pictured on the previous page) is a general purpose 400V rectifier diode rated at 3A.

Zener diodes

While ordinary silicon diodes work in the forward biasing regime, Zener diodes are reversely biased (negative DC voltages on the anode). As the reverse bias on the diode increases, the diode does not conduct and its current is zero. Once the reverse voltage reaches the value of Zener or *avalanche* voltage $-V_Z$, the Zener diode starts conducting abruptly and its current increases rapidly, while the voltage stays at the $-V_Z$ level.

The curves are exaggerated for educational purposes, in reality the curve around $-V_Z$ is much sharper and more vertical! Zener diodes are available in a range of voltages. Zener diodes of different V_Z voltages can be connected in series if a higher Zener voltage is needed, or diodes with identical V_Z can be in paralleled if a higher current capacity is required. More on Zener diodes and their use as voltage regulators in Volume 2 of this book.

Duo-diodes and rectifier bridges

Duo-diodes are used extensively in switch-mode power supplies and DC-DC converters. Their ratings range from low voltage Schottky diodes, such as MBR3045PT (45V, 30A Schottky duo-diode, common cathode) to high reverse voltage super-fast passivated rectifiers such as STPR1660CT. The numbers mean duo-diode (CT=Centre Tap) rated at 16A (first two digits is the current rating), 600V max. reverse voltage (last two digits 60).

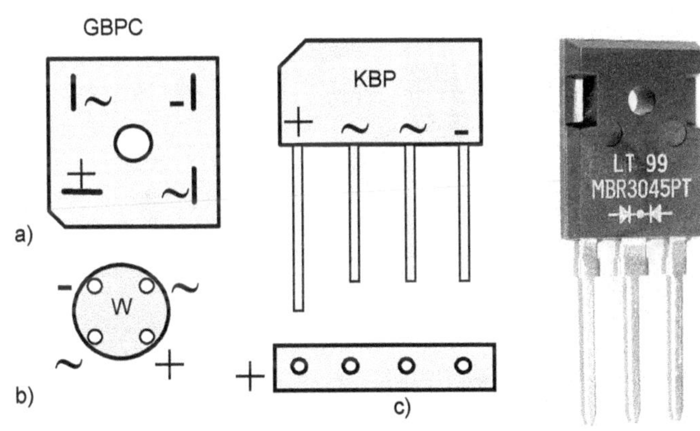

The most common modern types of rectifier bridges:
a) high power bridge, lug-side (bottom) view, GBPC type
b) low to medium power bridge, bottom view, W-type
c) medium power bridge, side and bottom view, KPB type

Like the one pictured, most have the two cathodes joined. There are also duo-diodes with common anodes, but they much less common. Rectifier bridges come in various sizes, ratings and packaging. The naming is not uniform, so if you are trying to decipher voltage and current rating from the number code, that is not always straightforward. BR810 is probably the easiest example, meaning 1,000V, 8A bridge rectifier. KBP04M is a 400V, 1.5A bridge while GBPC5006 is a 600V 50A bridge.

Although semiconductor components are cheap, it pays to keep a stash of salvaged diodes and bridges. UPS, battery charges, PC power supplies and similar equipment with DC-DC converters and switch-mode power supplies are good sources of quality diodes and bridges.

Varistors

Although varistors are classified as semiconductors, they have no PN-junctions. Vintage varistors were made from silicon carbide mixed with a ceramic binder. The mix was then pressed or extruded into a desired shape (disc, rod, washer) and sintered, resulting in a hard ceramic-like material.

More modern zinc oxide varistors have sharper I-A characteristic. Polycrystalline zinc oxide is mixed with molten bismuth oxide and sintered. The bismuth oxide forms a rigid coating around the zinc oxide grains, so the varistor acts as an open circuit at low applied voltages, but when the voltage across it exceeds the value of its "clamping voltage" V_C the varistor suddenly changes its properties from a very high to a very low resistance and conducts.

This makes them the simplest, cheapest and also most effective overvoltage protective device and transient suppressor.

Due to their symmetrical I-V characteristic, varistors work in both DC and AC circuits, in parallel with the load.

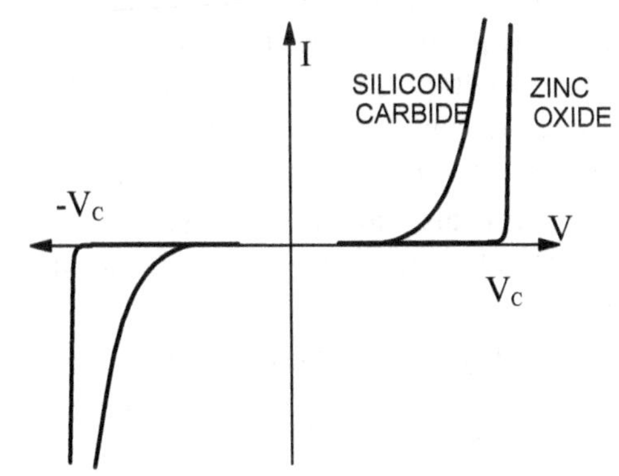

The current-voltage characteristics of silicon carbide and zinc oxide varistors

The voltage ratings range from 12 to 1,000V, while the peak current-handling ratings range from 10 to 2,000A! 275V-rated varistors are most often used on 220-240V mains circuits.

The energy of the voltage spike or transient has to be dissipated within the varistor, so varistors are also rated in terms of their energy-absorption capabilities, up to 160J (Joules).

TYPE D 82 - value, 1 - multiplier ($10^1 = 10$), 820V

14 D 82 1 K

DIAMETER in mm VOLTAGE RATING Tolerance (K=+10%)

Explanation of varistor part numbers

WIRES AND CABLES

Interconnects

Since they transmit the lowest level signals in the audio chain, and since their sonic signature (a nice euphemism for all kinds of distortion) is amplified through the rest of the system, interconnect cables are arguably the most important of all. Their capacitance and resistance cause signal loss that will impact high frequency reproduction.

We couldn't carry out extensive tests on thousands of available products and cable types (coaxial, shielded, braided, twisted, etc.) so we took a few cables we had lying around and performed a basic LCR-meter test, the capacitance between the active pin and shield or common conductor (since one cable was unshielded) and the D-factor, an indicator of losses. The lower the either figure the better the cable, but not necessarily better sounding.

Notice that cheaper cables, Ultralink CS1 (c) and the standard garden variety (d) had the lowest losses, while the most expensive of the four, Neotech's "Element" (a) had the highest D-factor.

Since it is of braided construction, it was expected that it would have the lowest capacitance.

Likewise, the cheapest generic budget cable had by far the lowest inductance, followed by the Ultralink CS1 (Contractor Series). It seems that the more expensive the cable, the higher the inductance!

The same lessons as with buying any other audio component or DIY part apply:

1. Caveat Emptor (Buyer Beware)! Don't fall for glitzy advertising, empty marketing hype and visuals designed to impress you.
2. Listen to expensive cables *with your system* before you part with lots of money.
3. A more expensive product does not necessarily sound better.

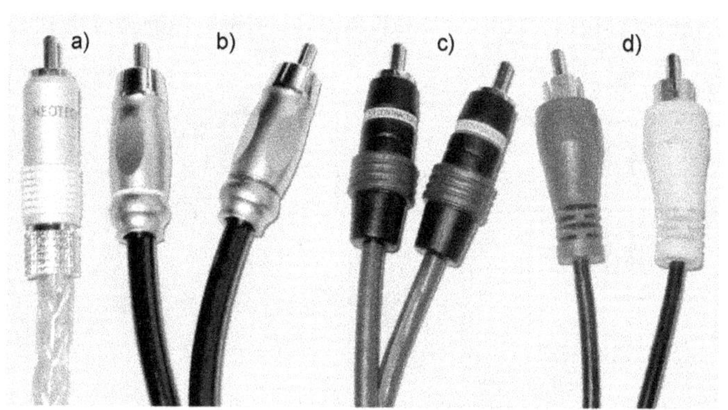

INTERCONNECT CABLE (L-R)	C @1kHz [pF/m]	D	L [μH]	Q
Neotech "Element"	42.5	0.017	1.2	0.089
Neotech "Origin"	70.0	0.008	1.7	0.074
Ultralink CS1	87.0	0.003	0.9	0.030
Budget	83.0	0.004	0.3	0.008

Hookup wire

To paraphrase the quote from Woody Allen's movie *Midnight in Paris* ("Knowing that Paris exists and anyone could choose to live anywhere else in the world will always be a mystery to me"), knowing that Teflon®, Tefzel® and Kynar® wires exists and anyone would choose to buy PVC-insulated, low-grade copper hookup wires at high prices in electronic hobby shops will always be a mystery to me. We did that once in a local electronic store in Perth, Australia, and threw the wire away. Just touching its end with a soldering iron would cause its insulation to shrink 5-6mm! For less than half of its price we got silver-plated avionics Teflon®-insulated wire from an ebay seller in USA, also great for making your own interconnects, and, when paralleled, even speaker cables.

Teflon® is DuPont's trade name for PTFE, short for Polytetrafluoroethylene. PTFE-insulated wires are resistant to most chemicals, thermal aging, soldering iron damage, they are flame- and moisture-proof. Apart from its excellent mechanical, thermal and chemical characteristics, PTFE has superior electrical properties such as low power loss. It comes in AWG sizes 14, 16, 18, 20 and 24.

Teflon®-insulated wire is always silver-plated. It comes in various voltage ratings, something to be mindful of before ordering, typically 250, 600 and 1,000V_{AC}. Unfortunately, Teflon®-insulated wires are hard to work with, they are difficult to strip, but solder well.

An alternative is to use a quality bare copper or silver-plated copper wire and feed it through translucent Teflon® tubing.

Tefzel® is another DuPont trade name, this time for ethylene tetrafluoroethylene or ETFE. Tefzel®-insulated wire wrapping wire is an unorthodox option that some constructors swear by. Available in a range of colors, it comes in 26 a.w.g and 30 a.w.g sizes. Tefzel® insulated wire is the aviation industry standard (M22759/16) due to its high abrasion resistance and lightweight nature. Rated at 600V, its temperature rating is -55°C to 150°C. The conductor is usually stranded tinned copper with extruded ETFE insulation.

Polyvinylidene difluoride (PVDF) is a fluoropolymer better known under one of its trade names, Kynar®. Kynar®-insulated wirewrap wire is made of solid silver-plated Oxygen Free High Conductivity (OFHC) copper conductor with Kynar® insulation. Its UL temperature rating is 105°C, but is otherwise rated at 125°C. This makes it suitable for top-cap connections as the insulation will not melt in contact with a hot vacuum tube glass bulb. It comes in AWG size 24, 26, 28 and 30 and it usually has a voltage rating of $300V_{DC}$, but it takes higher voltages without any problems. There is also a multi-strand variety, with one, two and more conductors in one cable, most commonly with 19 strands each.

Power cables

Technically, or theoretically, since they are so removed from the main audio circuit, power cables should have no or very little impact on the sound of an amplifier or any other audio component. In reality, no matter how incredible that claim may sound, from personal experience, good quality power cables can have more sonic impact than interconnects or speaker cables.

Quality power cables come in all sizes, materials and types of construction. As a first example, Monster Powerline 400 Signature Series features double Mylar insulation and aluminium sheet wrap in between, together with a copper drain wire. The 95% copper braid shields the three round conductors using multi-bundled wire technology.

Monster Powerline 400 Signature Series power cable

The drain wire and the braid should be connected together with the green earth wire to the earth pin at the mains connector plug. As with all shields, the other ends of the shield and drain wire are left unconnected at the IEC plug.

For some strange reason the Audioquest AC-12 cable is not shielded, and that is its main drawback. Its cross-section shows a symmetrical arrangement of eight connectors (four for the phase and four for the neutral) around the central earth conductor. The earth is a bundle of 19 wires (AWG #25 size), while each "live" connector is a solid AWG #18 wire made of long grain copper. Both the primary and secondary insulation is PVC with a thin internal Mylar jacket, the overall outside diameter is 11mm.

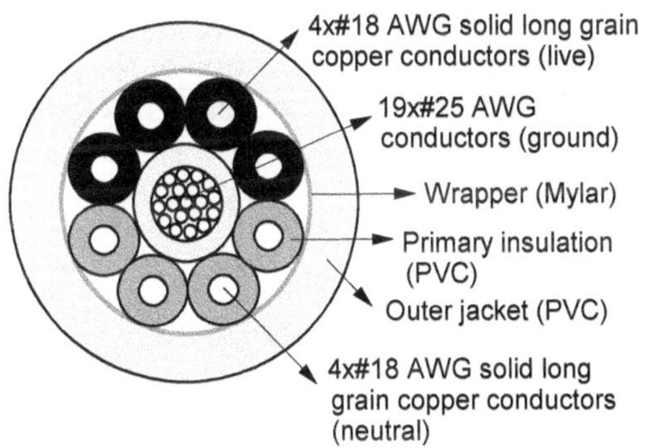

Internal construction of Audioquest AC-12 power cable

In Volume 2 of this book we will talk in more depth about capacitive and inductive coupling between wires and cables, and various ways to avoid interference, ground loops and other related problems.

Fakes and originals

Be they interconnects, speaker or power cables, audiophile cables are expensive, some very expensive, so it should come as no surprise that there is a whole industry in China and other countries producing fakes that to untrained eye are indistinguishable from the originals. As with all fakes (watches, jeans, etc.) there is a sure way to tell, a lower, or *much* lower price.

If the original costs $50 per foot and an ebay seller in China is selling it for $10 per foot, you know something's not quite right. Even in a high markup hi-fi industry a seller cannot stay in business on five times lower prices.

However, while the purity and crystalline structure may not be as good as the original's, the fake can still be miles better than your garden variety cable. Most of these are made from inferior materials, made to look good (purple sheath, gold plated connectors, etc.) and cleverly packaged to impress the gullible and sold at extortionist prices in retail shops.

Contacts - the invisible components

When two or more wires or component leads are joined together, a contact is formed. For best audio performance a contact must be clean and tight. Clean means that the wires and leads must be free from dirt, dust, grime and oxidization, i.e. that they must be cleaned before they are joined together.

The best electrical contact is mechanically sound. When wires are crimped together or wire-wrapped onto a terminal, the shape of the metal is deformed, forming a "cold-weld", an airtight connection, meaning the contact will not oxidize over time since air cannot get into it. Thus, if you can, twist wires together first, or wrap them around terminals before soldering them together.

Solder is a necessary evil. It is made of at least two, often three of four dissimilar metals, and then used on copper or tinned copper leads. When dissimilar metals are joined a parasitic diode is formed, which acts as a micro signal rectifier and an RF (radio) signal detector. Of course, these are minute effects, but their cumulative effect may be noticeable. Thus, it is wise to minimize them by reducing the number of soldered joints in an amplifier.

Copper, silver, gold, nickel and beryllium: a crash-course in audio metallurgy

As for the pins, spades and lugs used on speaker cables and RCA plugs and sockets for interconnects, the aesthetics often interfere with the sonics. Almost all are gold plated, and to make the gold shiny, nickel plating was done first underneath the gold layer. Sonically, that nickel layer is the worst offender. Its distortion is irritating and causes listener fatigue. Gold also distorts, and, despite a common perception, just because it is noble metal does not automatically mean that it is a good electrical conductor, not as good as silver, anyway, and not even as good as copper.

Technically, gold is only used to prevent contact corrosion, but since such a thin layer very quickly rubs off by frequent plugging and unplugging, I suspect the main reason is of marketing nature - anything plated in gold seems more expensive and "hi-end".

When added to copper, aluminium, iron and nickel, beryllium improves their physical properties. Many audio and power connectors use gold plated brass for the body and gold plated beryllium copper for the pins. 2.0% beryllium copper alloy is nonmagnetic, does not suffer from metal fatigue, retains the high electrical and thermal conductivity of copper, and is much stronger and harder than copper alone, which is a soft metal. Some manufacturers such as Oyaide use a beryllium copper base for their mains plugs, first platinum-plated, then plated with palladium.

However, many audiophiles don't like the sonic signature of beryllium copper alloys either, preferring pure copper. One may argue that a concoction of four metals (copper, beryllium, platinum and palladium) violates the Simplicity Rule. On the other hand, copper connectors oxidize quickly, so some sort of plating is required unless you tighten the hell out of your speaker's copper spades (to create an airtight contact) and never touch them again.

AUDIO FREQUENCY AMPLIFIERS

- PROPERTIES OF IDEAL AMPLIFIERS
- DECIBELS AND LOGARITHMIC SCALES
- REAL AMPLIFIERS
- DISTORTION IN AUDIO AMPLIFIERS
- AMPLIFIER IN THE SIGNAL CHAIN
- WHAT DETERMINES THE SOUND OF AMPLIFIER?

"I don't regard this amplifier as a hi-fi product at all. It is actually a tone control, and an unpredictable one at that."

Stereophile's John Atkinson in his review of Cary CAD-300 SEI amplifier, Sep. 1995

PROPERTIES OF IDEAL AMPLIFIERS

As the very word suggests, an amplifier amplifies some kind of signal, from a sensor or transducer, or from a music source such as a phono cartridge or a CD player. Before we get into real amplifiers, let's talk about the ideals we are trying to achieve, or to come as close to as possible.

Voltage-controlled voltage amplifier

The ideal voltage amplifier is a source of AC output or load voltage v_L, which is controlled by and proportional to the input voltage v_{IN}. μ (Greek letter "mju") is the amplification factor, which is constant. The + markings refer to one particular moment in time, since the AC voltage changes polarity f times a second, where f is the frequency of the signal. Sinusoidal waveform will be assumed in the further study, unless noted otherwise, hence the sine waveform inside the voltage source.

The i_L-v_L lines are perfectly straight, vertical and equidistant (equally spaced). This is a linear amplifier without any distortion or power losses. Its input resistance is infinite (there is no input current flowing), so the power input is zero. The output or internal resistance is zero (there are no thermal losses within the amplifier).

The amplifier is oblivious to the resistance of the load, no matter how high or low it may be, its voltage will stay strictly proportional to the input voltage. Whatever voltage comes in, no matter what its frequency or shape may be, it will be multiplied μ times and imposed onto the load. Since the input power into the amplifier is zero, its power gain (the ratio of output to input power) is infinitely high. An ideal vacuum triode is one such amplifier.

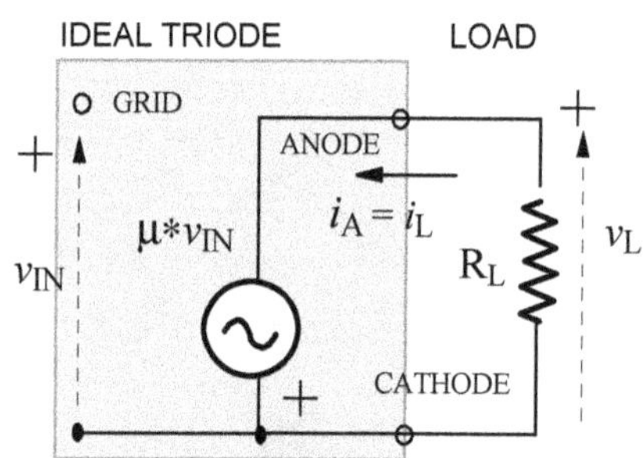

LEFT: The equivalent circuit of an ideal voltage-controlled voltage amplifier. Triode parameters are used, which will be explained soon.

RIGHT: The i_L-v_L characteristic of an ideal voltage-controlled voltage amplifier.

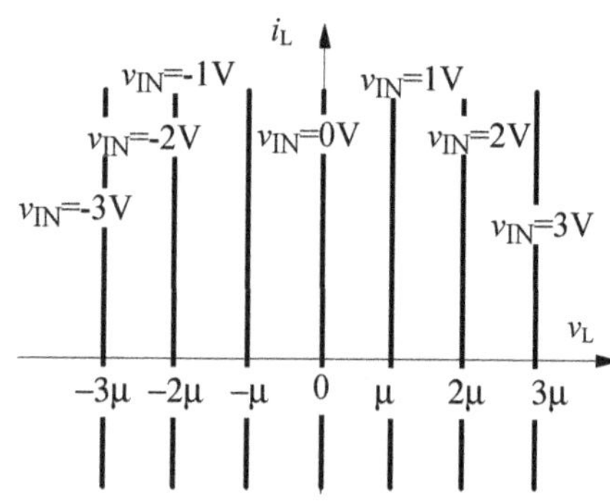

Current-controlled current amplifier

The ideal current amplifier is a source of AC load current i_L, which is controlled by and proportional to the input current i_{IN}, or $i_{OUT} = \alpha i_{IN}$. α (alpha) is the current amplification factor. The i_L-v_L lines are perfectly straight, horizontal and equidistant (equally spaced). This is a linear amplifier without any distortion or power losses. The input voltage is zero, so the input power is zero, and the power gain is again infinite.

Bipolar transistors are current-controlled current sources, although far from ideal. In the common emitter circuit, whose model is pictured, the input current is base current, and the output current is collector current, the load is connected between the collector and the common terminal, in this case emitter.

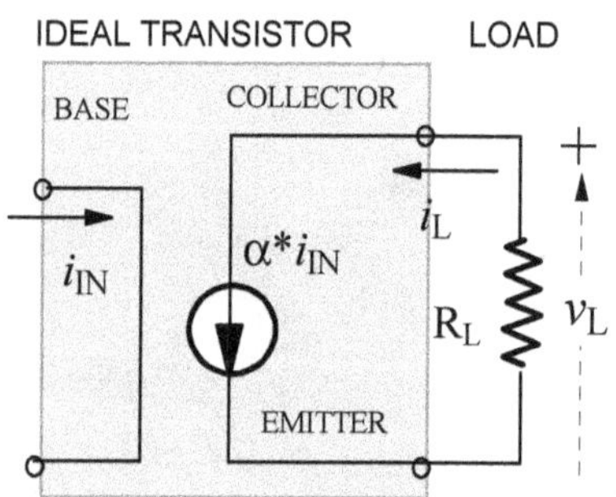

LEFT: The equivalent circuit of an ideal current-controlled current amplifier. Bipolar transistor parameters are used.

RIGHT: The i_L-v_L characteristic of an ideal current-controlled current amplifier. Horizontal "curves" mean the output currents are constant, regardless of the load resistance or load voltage v_L.

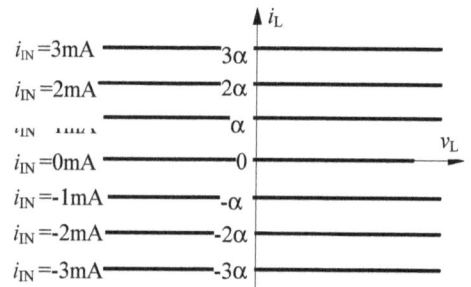

i_L
i_{IN}=3mA 3α
i_{IN}=2mA 2α
α
v_L
i_{IN}=0mA 0
i_{IN}=-1mA -α
i_{IN}=-2mA -2α
i_{IN}=-3mA -3α

Voltage-controlled current amplifier

Ideal amplifiers are classified by two parameters, voltage and current, and since these amps have one input and one output, there may be four types of amplifiers: voltage-controlled voltage amps, current-controlled current amps, voltage-controlled current amps, and current-controlled voltage amps.

Voltage controlled current amps supply AC load current that is proportional to the AC signal voltage: $i_L = gm * v_{IN}$. The gm factor is called transconductance or mutual conductance (between the output and input), its unit being mA/V.

Conductance is the inverse of resistance, with a dimension 1/Ω. Since Ω=V/A, the unit for conductance has a dimension of current divided by voltage: I/V and in the International System of Units (SI) is called a Siemens, with a symbol S, named after Ernst Werner Siemens (1816–1892), German inventor and industrialist. Siemens AG is a huge multinational industrial concern, tubes made by Siemens in 1950s and 60s are still in circulation.

In American literature you will come across a unit called micromho. Micro means 10^{-6} or one millionth-part of "mho", which is, believe it or not, ohm spelled backwards. Ah, those weird Yanks. So 1 Siemens = 1A/V = 1,000 mA/V = 1,000,000 micromhos. European tube testers are calibrated in mA/V or mS, which is 1,000 micromhos. So, if EF86 pentode in a particular operating point has a gm of 1,800 micromhos that mutual conductance can be expressed as 1.8 mA/V or 0.0018 S!

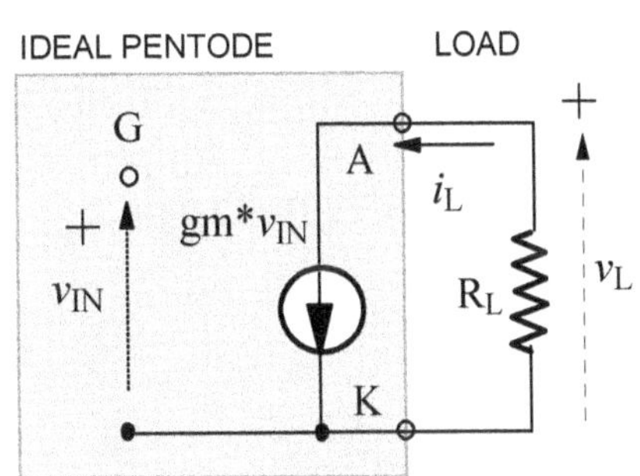

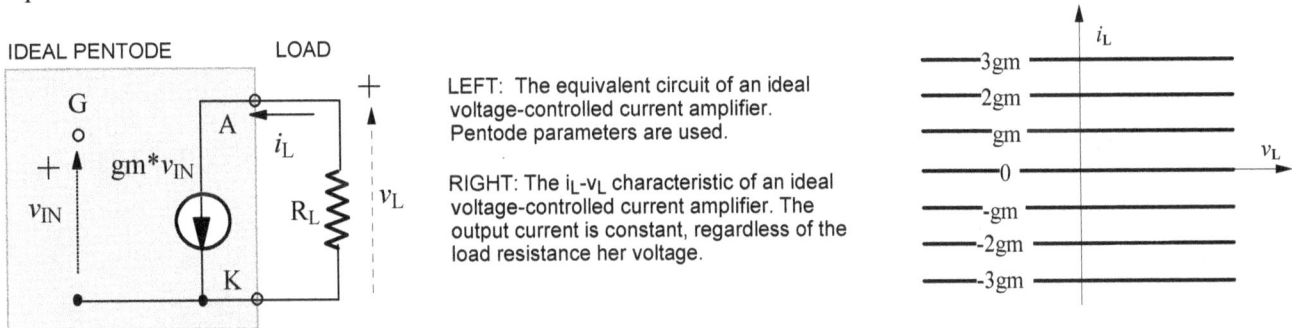

LEFT: The equivalent circuit of an ideal voltage-controlled current amplifier. Pentode parameters are used.

RIGHT: The i_L-v_L characteristic of an ideal voltage-controlled current amplifier. The output current is constant, regardless of the load resistance her voltage.

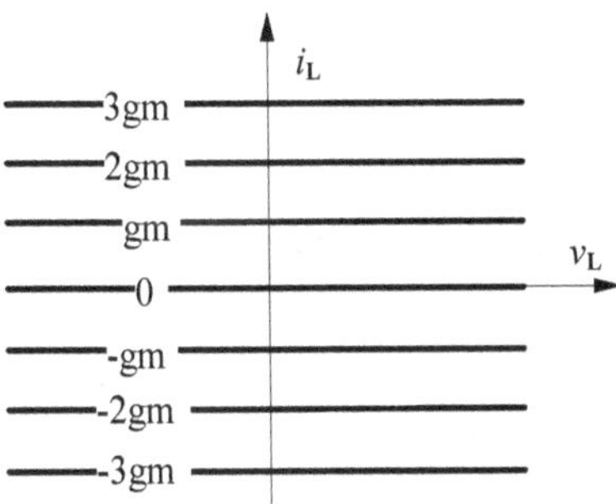

A pentode is indeed a real-life example of a voltage-controlled current amplifier, albeit not an ideal one, just as a triode is a real-life (meaning "imperfect") voltage-controlled voltage source.

Understanding ideal amplifiers is a vital first step in building a long-lasting knowledge, and an intuitive feeling for vacuum tube electronics. We will evaluate various tube types and circuit configurations mainly on the basis of how close they come the these ideal amplifiers. There are also current-controlled voltage amplifiers, but we will not talk about them in this book.

DECIBELS AND LOGARITHMIC SCALES

The proper and improper use of decibels

Before we move on to the properties of real or imperfect amplifiers, due to their importance, we must digress a bit and talk about decibels and logarithmic scales. deciBel or "dB" is a unit for a power ratio or gain: $\mathbf{P_G = 10*log(P/P_0)}$ **[dB]** where P is the power produced by a system or a device, P_0 is the referent power chosen for a particular purpose, comparison or measurement. Various referent power levels have been used over the decades, and that created a confusion among users. We will talk about that in a minute.

The power gain of an amplifier is $P=V^2/R$ we get $P=10*\log(P_{OUT}/P_{IN}) = 10\log[(V_{OUT}^2/R_{OUT})/(V_{IN}^2/R_{IN})] = 10\log[(V_{OUT}/V_{IN})^2(R_{OUT}/R_{IN})] = 20\log(V_{OUT}/V_{IN}) + 10\log(R_{IN}/R_{OUT})$ [dB]

If, and only if $R_{OUT}=R_{IN}$, we get $\mathbf{P = 20log(V_{OUT}/V_{IN})}$

A tube amplifier has an input impedance of 47kΩ and supplies an 8Ω load. Its voltage gain is 15 times (A=15). What is the voltage gain in dB?

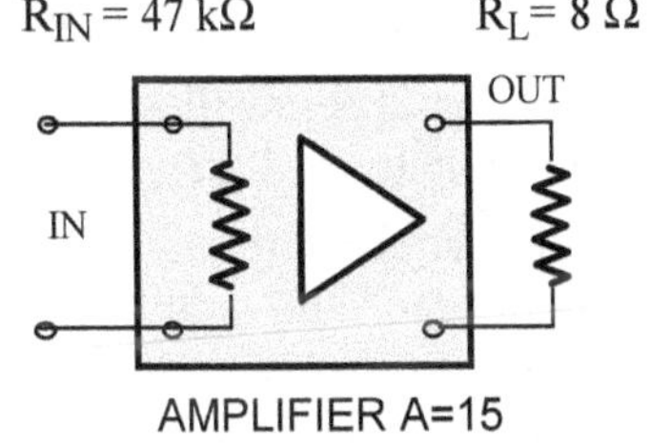

AMPLIFIER A=15

These days we simply calculate $P = 20\log(V_{OUT}/V_{IN}) = 20\log15 = 23.5$ dB! That is certainly convenient, but, strictly speaking, it is not correct, because this simplification is only correct when $R_{IN}=R_{OUT}$, and in our case the input and output resistances are very different!

The correct answer is $P=20\log(V_{OUT}/V_{IN}) + 10\log(R_{IN}/R_{OUT}) = 20\log15 + 10\log(47{,}000/8) = 23.5+37.7 = 61.2$dB

Since the last factor $10\log(R_{IN}/R_{OUT})$ almost always gets omitted (its not even mentioned in nine out of ten books!) we will adopt this simplification too. Although improper from the strict definition point-of-view, such usage of dB works, providing it is used consistently, meaning as long as everyone (mis)uses it in the same way!

Different "types" of decibels and dB scales

If you thought the confusion surrounding the use of dB is over, you were wrong. Humans can and do overcomplicate things on a grand scale, and electronic engineering is a good example of such tendencies. There are two different dB units, dBm and dBV. dBm (also called dBmW, only compounding the confusion) has a communication origin, referenced to the 1 mW of power that a sine source of 0.775 V_{RMS} would dissipate on a 600Ω resistive load. dBV is the unit we have been talking about so far, used for voltage ratios such as amplification and attenuation factors.

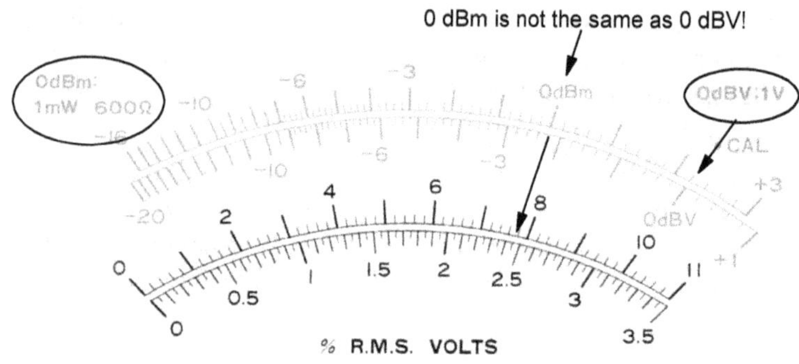

See how the two dB scales have a different zero on this analog meter face, from a harmonic distortion analyzer: 0 dBm is at 0.775 V_{RMS} (7.75 on the lower scale) and 0 dBV is at 1.0 V_{RMS} (10 on the lower scale)!

The reference power level for dBm is P_0=0.001 W, so P [dBm] = 10log(P/0.001)

To convert dBV to dBm we need the same referent impedance (R), in this case 600Ω. Remember, $P=V^2/R$!

Taking the log of both sides of that equation log(P) = 2*log(V) - log(R) or dBm - 30 = dBV - 10log(R) Finally, dBV= dBm - 30 + 10log(600) = dBm -30 +27.78 so **dBV= dBm-2.218**

The logarithmic power curve

Let's return to the half-power frequencies, or, as they are more commonly called, -3dB frequencies. To understand the logarithmic nature of the amplifier's power drop at those frequency limits we need to study the logarithmic power curve, the logarithmic equation $\mathbf{P = 10log(P_2/P_1)}$ **[dB]** in a visual form. The ratio P_2/P_1 is on the horizontal axis (the abscissa) and the decibels of that ratio are on the vertical axis (the ordinate).

When $P_1=P_2$ their ratio is 1, and log1=0 (because 10^0 = 1) Any number to the power of zero equals one (Point A).

As P_2 becomes larger than P_1, the log curve is in the positive territory, and as P_2/P_1drops under 1 the curve goes into the negative values.

In point E, P_2 is twice as large as P_1, which is +3dB. In point D, P_2 is half as large as P_1, which is -3dB. These are half-power points.

In point C, P_2/P_1=10, and log10=1, so P_2/P_1=10dB. Similarly in point B, P_2/P_1 = 0.1 so P_2/P_1=-10 dB.

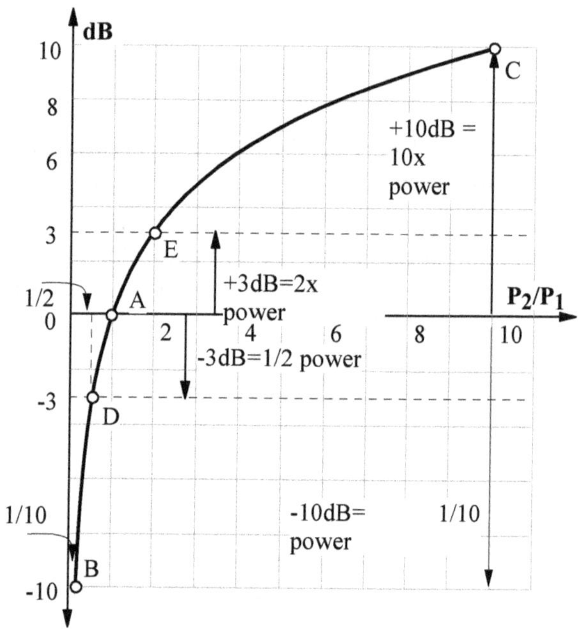

LIN and LOG scales

Four types of X-Y scales are used. The independent variable (X-axis) and the dependent variable (Y-axis), can be presented using a linear (LIN) or logarithmic (LOG) scale, resulting in four possibilities: LIN-LIN, LIN-LOG, LOG-LIN and LOG-LOG. LOG scale is used when the range of the variable depicted is wide, as in audio frequency range, which is studied from 1Hz to 100,000 Hz, or 5 decades. Each decade is a factor of 10:1.

The example below illustrates amplifier's gain, where a linear horizontal scale would be impractical, since it would be impossible to read the gain from the curve at the frequency extremes, where most of our interest lies.

In some books you will see the frequency curve of a 1st order system (a simple RC filter or an amplifier stage) with straight "end" and in others with a curved end. The straight end means the vertical scale is logarithmic (LOG) and it thus has no end, there is no vertical zero "0" marked! A curved end means the vertical scale is linear (LIN) with a definite end, which is zero level at the horizontal axis!

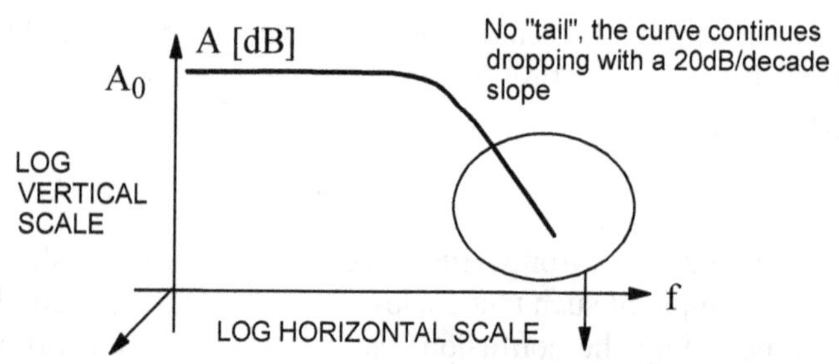

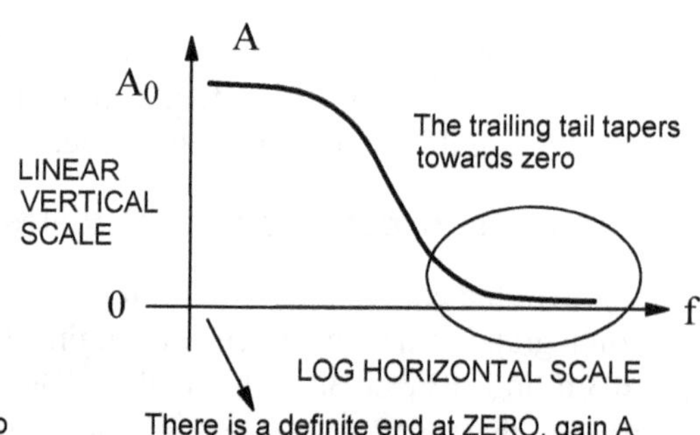

REAL AMPLIFIERS

Harmonics and spectra

Before we start analyzing the spectra of real audio amplifiers, we need to understand spectra of signals. A square wave is not just the most illustrative example, but is also of the most practical importance, as you will see in the chapter on audio tests and measurements in Volume 2 of this book.

A square wave is comprised of an infinite number of sine waves, but only odd harmonics, 1st, 3rd, 5th, 7th, etc. are present, there are no even harmonics. The 1st harmonic is a sine wave of the same frequency as the square wave, the frequency of the 3rd harmonic is three times the frequency of the 1st, and so on.

Notice how the amplitude of the harmonics drops off rapidly in the exponential fashion. Although the illustration only shows spectral components up to the 11th harmonic, the harmonics continue indefinitely, there is the 13th, the 15th, ..., the 127th, ..., the 367th, etc., but because their amplitudes are so infinitesimally small we usually neglect all other harmonics above the certain value.

For a square wave the amplitudes of harmonics are $A_N = A/N$, so the amplitude of the third harmonic is 1/3 of the fundamental's amplitude, the fifth harmonic has a 1/5 of the 1st harmonic amplitude, and so on.

This spectral distribution applies to any square wave. For instance, if f_1 was 800 Hz sine wave of the amplitude A_1=10V, the third harmonic would have a frequency three times higher or 2,400 kHz and an amplitude three times smaller, or $A_3=A_1/3 = 10/3 = 3.33$V. The fifth harmonic will be $f_5=5f_1=5*800= 4,000$ Hz, its amplitude will be $A_5=A_1/5 = 10/5 = 2.0$V and so on.

This illustration shows the contribution of the first three harmonics, the 1st, the 3rd and the 5th. Notice how such a sum already resembles a square wave. As the harmonics are added, the waveform will more and more resemble a square wave.

Although amplifiers are modeled, analyzed and tested mostly using sine waves of a fixed frequency, there are no such pure tones in nature. Sounds have complex waveforms, comprised of sine waves of various amplitudes and frequencies.

A spectrum of a signal is depicted in an X-Y manner with frequency on the X-scale and the amplitude of various harmonics on the dB vertical scale. Two examples of musical instrument spectra are depicted here, violin's and flute's. The first or the lowest frequency is called the "fundamental" or first harmonic. The second harmonic has twice the frequency of the first, the frequency of the third harmonic is triple the fundamental frequency and so on.

Notice that the amplitudes of some of the harmonics are actually higher than the fundamental tone. Also, not all harmonics are depicted in the violin's case, only those up to 1,500 Hz.

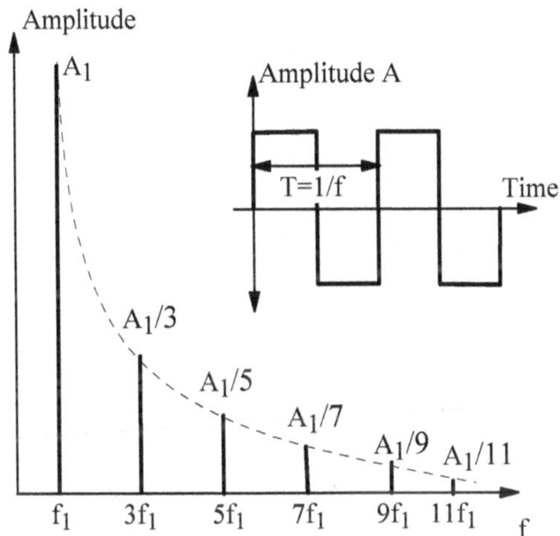

ABOVE: The spectrum of a square wave
BELOW: The fundamental and first two harmonics of a square wave added together.

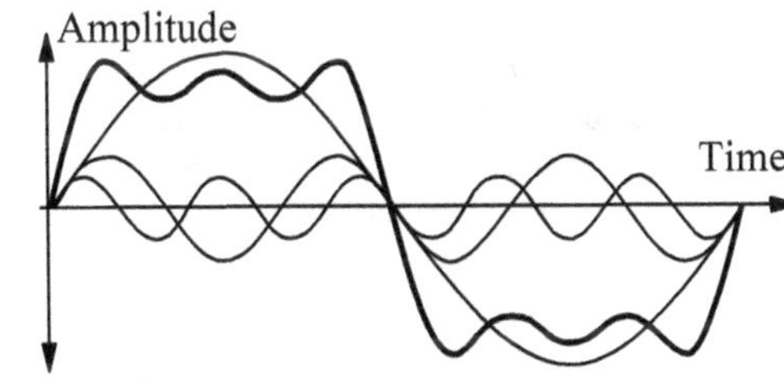

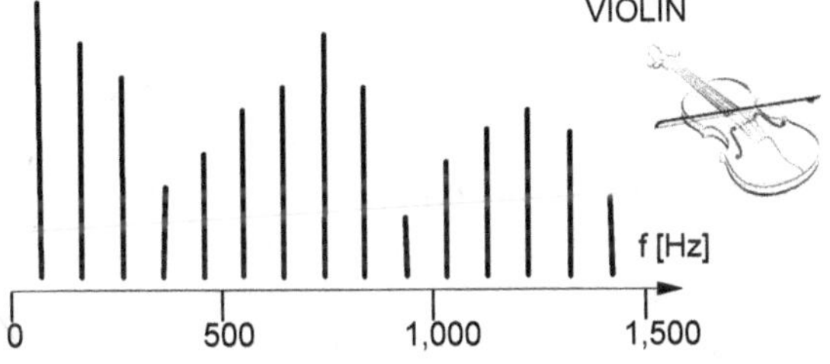

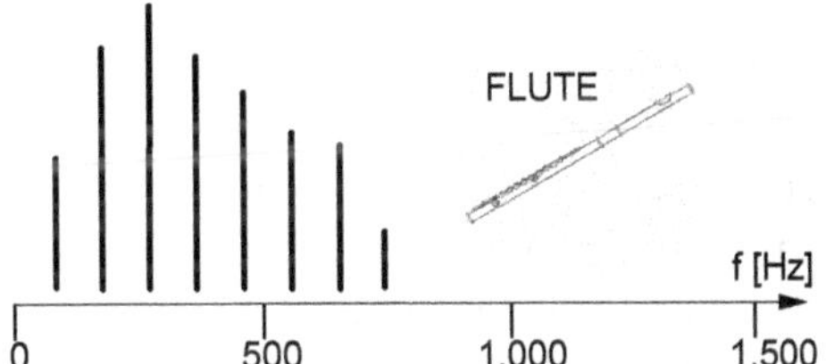

The amplitude and phase characteristics of real amplifiers

An ideal amplifier would have an infinite frequency range, meaning it would amplify all frequencies, from DC (f=0) to infinity. Real amplifiers and preamplifiers behave like bandpass filters. They are unable to amplify very low and very high frequencies as well as they can amplify the midrange band of frequencies.

At a certain low frequency f_L, usually in the region of 10-20 Hz, the amplification factor drops to 0.707 or 71% of its midrange value A_0 (usually specified at 1kHz). As the frequency is increased, at a certain frequency f_U the amplification factor again drops to 71% of its midrange value. These are -3dB points, also called "half-power" or corner frequencies.

How does output power drop by half if the voltage drops to 71% or $1/\sqrt{(2)}$ of its midband value? It's all a matter of mathematics: The power on the load R is $P=V^2/R$, so the midband power is $P_0 = V_0^2/R$ and the half-power is $(V_0*0.707)^2/R = (V_0/\sqrt{2})^2/R = V_0^2/2/R = V_0^2/2R = P_0/2$

The frequency characteristic of a typical amplifying stage is illustrated here using a linear vertical scale, so it is not in dB but in absolute numbers. The midrange gain is 30 and the phase shift is 180 degrees, meaning the output signal lags behind the input signal by 180°. This amplifying stage inverts the phase of the signal. In popular speak, the input and output signals are "out-of-phase".

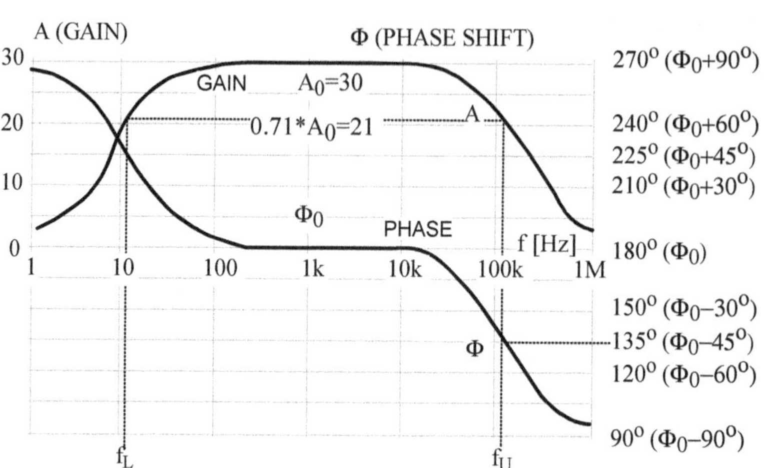

Typical amplitude versus frequency and phase versus frequency characteristics of a real amplifier and the meaning of -3dB points

The phase characteristic of an ideal amplifier would be a straight line - there would be no phase shift between the input and output signals, or such a phase shift would be constant, i.e. the same for all frequencies.

As with signal's amplitude, the phase relationships also change at frequency extremes. At low frequencies, the phase lag is increased, and at high frequencies the lag is decreased. Notice that at our half-power or -3dB points, frequencies f_L and f_U, the phase changes +/-45° from the midrange phase. At f_L the phase is 180°+45° = 225°, and at f_U the phase is 180+45 = 225°.

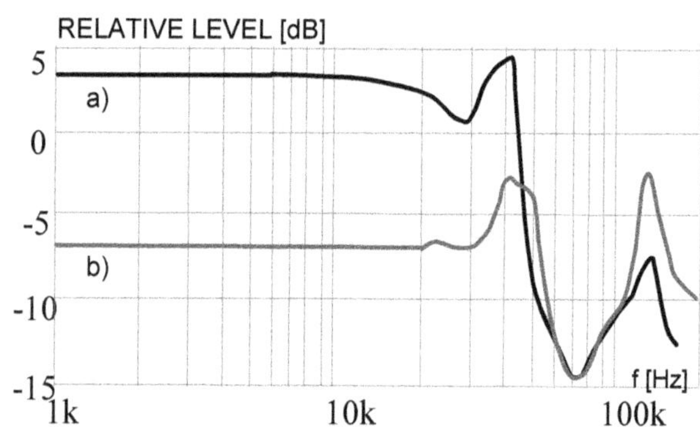

Gain vs. frequency of Cary CAD-805 amplifier
a) minimum feedback, b) maximum feedback
Source: *Audio*, July 1995 review

The "typical" gain vs. frequency curve shown above is actually anything but typical. Typical is the word text book writers and university professors use to illustrate a point or present a simplified argument.

If you get such a curve from your DIY amplifier, you'd be doing really well! Many commercial amplifiers have a much more malign A-f characteristics, and I am not talking just about sub-$1,000 Chinese-made designs.

Look at the A-f curves of Cary CAD-805 amplifier, redrawn from "Audio" magazine's review. Negative feedback is adjustable by a potentiometer, but no matter how much NFB is used, the amplifier exhibits two amplitude peaks and a serious dip at around 70 kHz. The peaks are due to the output transformer resonances at around 40kHz and 130kHz, where leakage inductances react with distributed parasitic capacitances.

What frequency range should we aim for in our designs?

But, you must be protesting now, these peaks and dips are above the upper audible limit of 20 kHz, we cannot hear such ultrasonic frequencies anyway! Well, we can't hear them directly, but indirectly we hear their "artifacts", their interactions with the audible frequencies, so we should care about ultrasonic behavior of amplifiers!

Although 20Hz - 20kHz is the "official" audio range, audiophile components sound much better if their frequency response extends at least 2X higher for power amps and 5X higher for the preamps. A similar (and even more important) requirement applies to the low frequency range. A superior power amp should have its lower half-power or -3dB frequency f_L as low as possible, 10Hz would be great, although that is very difficult to achieve due to the limitation imposed by the output transformers. Preamplifiers can be made to go down even lower, to 4 or 5 Hz, since most don't use output transformers and the power levels are low, because they only provide voltage amplification.

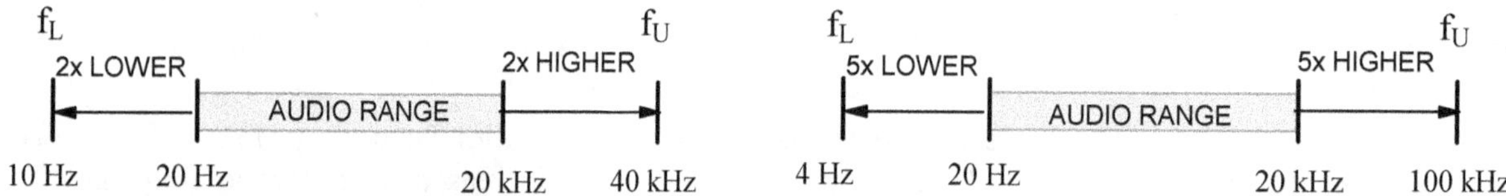

Rules-of-thumb for the target frequency range of a quality power amplifier (LEFT) and a line-level preamplifier (RIGHT).

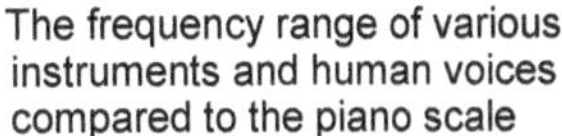
The frequency range of various instruments and human voices compared to the piano scale

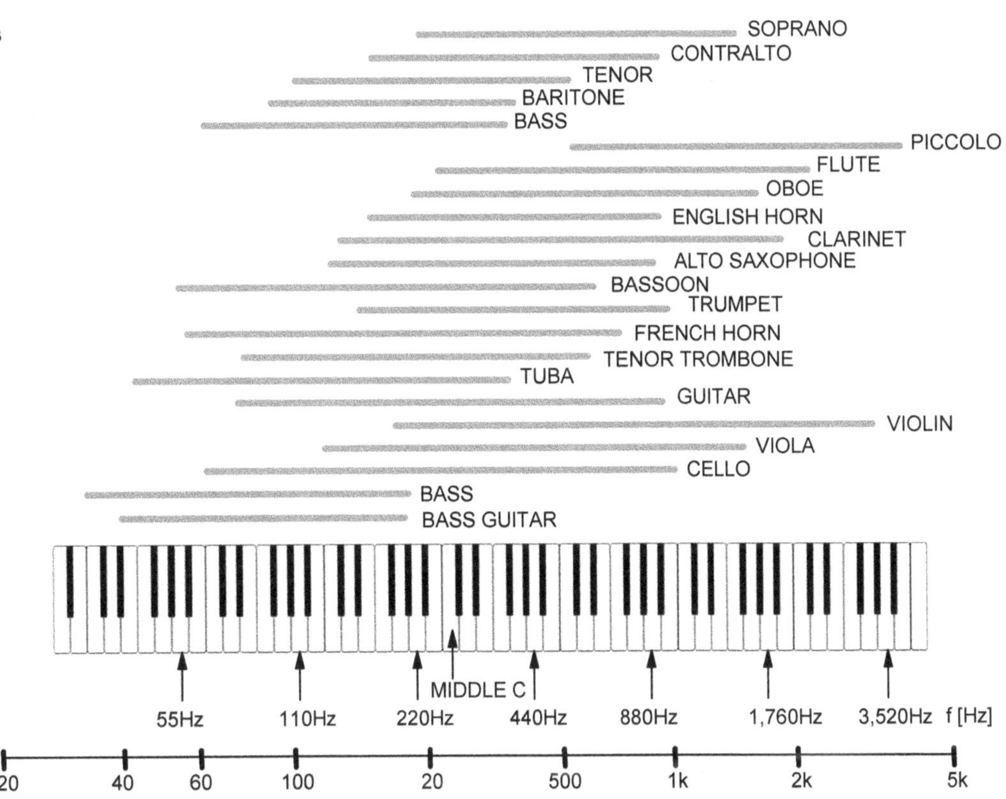

This may seem strange for at least four reasons. Firstly, at the source, if you look at the frequency range of musical instruments, none of their fundamental harmonics exceed 4 kHz. Secondly, at the end of the musical chain and at the bottom of the frequency spectrum, no loudspeaker that I know of can reproduce 20 Hz, let alone 5 Hz. The higher frequencies can be easily reproduced, tweeters can and do go much higher than 20kHz, just ask your dog or cat how the music you listen to sounds to them.

Thirdly, the recorded material, be it an LP, a CD or a digital file, is limited in its own frequency range, most cannot reproduce anything over 20kHz due to sampling limitations in CD players and similar confines for other media. Finally, even if those sources and transducers could reproduce a wider audio spectrum, no human can hear frequencies below 20Hz or above 20kHz, so why would we impose such difficult design goals on ourselves?

The first argument is misleading, it mentions only the fundamental tone. While the highest notes played on a violin or piano are only around 3kHz and 4kHz respectively, their harmonics extend well into the ultrasonic range (above 20kHz)! It is the harmonics and their relative amplitudes and interactions that define specific instrument sounds.

Ultimately, the short answer is that "wideband" amps and preamps generally sound better than those whose frequency ranges are narrower. Although you cannot *directly* hear those sub-harmonics and high frequency overtones, you will hear a different tonal presentation with them present and absent.

The "wideband versus limited high frequency extension" debate

However, the wideband school of audio design is not universally accepted. Some audio designers choose to limit the upper frequency extension of their designs, claiming that such amplifiers sound softer and more musical. There is some truth in their claim that significantly extending the bandwidth only gives prominence to higher harmonics, of which the odd ones (5th 7th, etc.) are particularly objectionable (harsh sounding and irritating).

Plus, it is not just the higher harmonics by themselves, but also the intermodulation distortion products that lie in the 20-60 kHz band, which limited bandwidth amplifiers would cutoff or at least attenuate significantly, but wider bandwidth amplifiers would reproduce at full levels!

HF extension needs to be limited if high levels of negative feedback are used, which can become positive at such high frequencies and make amplifiers unstable. So, the truth is that designs that produce low levels of THD and IM distortion and use low levels of negative feedback, will probably benefit from a wide bandwidth, while poor designs (distorting and oscillating) will suffer from it.

Power bandwidth

The -3dB frequencies of power amplifiers depend greatly on the output power level. It is customary to perform these measurements at two power levels, usually at 1 Watt (2.83V_{RMS} into 8Ω load) and at the rated power output.

The frequency range at the rated power output will always be narrower than the -3dB range at 1 Watt. In tube amplifiers this is primarily due to the imperfections of the output transformer, whose performance is limited at both low and high frequencies. In the example illustrated here, the single-ended triode amplifier rated at P_R=15W was tested and the half-power frequencies (also called -3dB frequencies) were found to be f_{LR}= 10Hz and f_{UR} = 49 kHz. "R" stands for "rated" power.

However, when the same amplifier was tested at a lower power output level of 1 Watt, the -3dB frequencies were f_{L0}=5Hz and f_{U0}=57kHz, meaning the frequency range was much wider. A tube amplifier's frequency range will be wider at low and narrower at high power levels. That is why quoting a frequency range without specifying the power level and the load impedance is meaningless.

The narrower frequency range (or bandwidth) of an amplifier at higher power levels is due to the limited primary impedance and the saturation of the output transformer's magnetic core at low frequencies and parasitic capacitive effects at high frequencies.

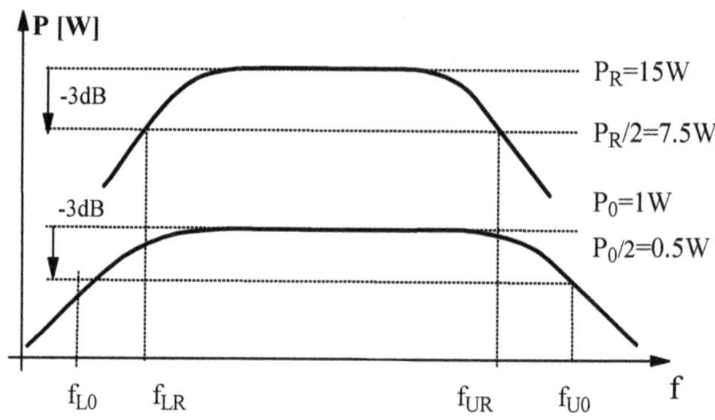

ABOVE: How bandwidth of a certain tube amplifier varies with output power level, at 1W output and the rated 15W output

DISTORTION IN AUDIO AMPLIFIERS

While distortion may be a desirable feature of guitar amplifiers, it is generally avoided in high fidelity ones. I say generally, because not all distortion is bad or unpleasant sounding, and also because certain measures that reduce distortion may also negatively impact the other aspects of sound quality. A typical example is negative feedback.

Harmonic distortion (THD)

Assuming a simple sine wave signal at the input of a real amplifier, the output will also be a sine wave, but slightly distorted. Such a distortion cannot normally be detected visually, by observing the waveform on an oscilloscope, for instance, except in the cases of severe distortion. Nevertheless, should such a signal be brought to the input of a distortion or spectrum analyzer, the presence of new harmonics will be detected.

Real amplifiers generate harmonics due to their nonlinear input-output or "transfer" characteristics. We will study these in great detail, so we can understand the various measures that can be taken to "straighten" or "linearize" the operation of nonlinear devices such as vacuum tubes. By the way, for the transistor fanatics in our midst, bipolar transistors are even less linear than vacuum tubes. The most linear of all amplifying devices is a triode!

A, A_1, f, f_1 — A, A_1, A_2, A_3, A_4, f, f_1 $2f_1$ $3f_1$ $4f_1$

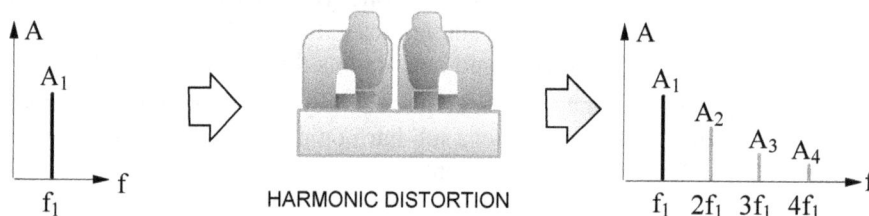

HARMONIC DISTORTION

Usually the second harmonic (twice the signal's frequency) has the highest amplitude, followed by the third, the fourth and so on. The relative amplitudes of various harmonics depend on the type of amplifying device (triodes distort differently from pentodes and beam tubes), the power level at which the measurement is taken, the design of the circuit (single-ended or push-pull, and many other factors). More on different spectral "signatures" very soon.

Intermodulation (IM) distortion

To explain this type of distortion, let's look at the way it's measured. Two pure sine wave signals are mixed and fed into an amplifier. One is of a lower frequency f_1 (usually the mains frequency, 50 or 60Hz), the other of fifty times higher frequency f_2. Assuming f_1=50Hz, f_2=2,500Hz. Various commercial analyzers use different frequencies but the principle behind their operation is the same. The amplitude of the lower frequency signal is adjusted so that it's four times higher than the amplitude of the higher frequency signal ($A_1=4A_2$).

An ideal amplifier would amplify both signals equally (say 20 times), but a real amplifier will also generate two unwanted signals (or "side-bands) of the higher frequency signal. One will be of f_2-f_1 frequency (or 2,500-50 = 2,450Hz in our case), the other will have a frequency f_2+f_1 (or 2,500+50 = 2,550Hz)!

In severe cases of distortion, second side-bands will be also be generated, f_2-$2f_1$ and f_2+$2f_1$. The situation is analogous to AM radio, although frequencies in question aren't in the radio but in the audio range.

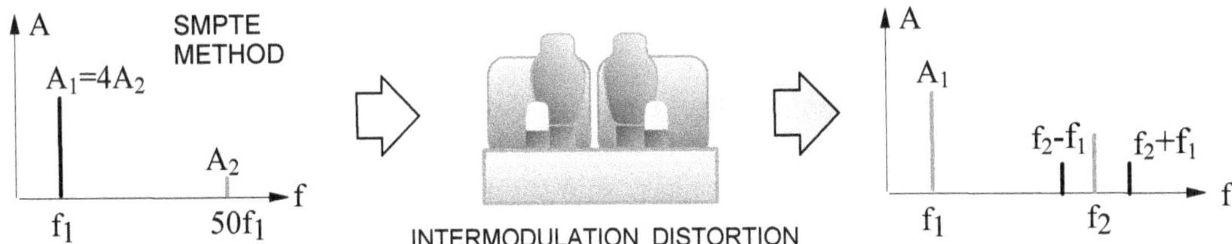

The output signal will be an amplitude-modulated carrier, the f_2 signal is the high frequency (HF) carrier, while the low frequency (LF) signal (f_1) modulates the LF signal's amplitude.

IM distortion is more unpleasant to human ear than harmonic distortion and thus its reduction should be even higher on the priority list of an amplifier designer than the reduction of harmonic distortion.

Going back to the desirability of a wide bandwidth debate, a poor quality amplifier may generate distortion tones at say 28kHz and 31kHz. Due to IM distortion these will then produce sidebands, one of which, 31-28=3 kHz, will fall in the audible range, dead smack in the midrange where human ear is the most sensitive!

Attenuation distortion

The illustrated situation shows a complex audio signal at amplifier's input (thicker trace), comprising of two sine waves, a fundamental and its 3rd harmonic (thinner traces). If the third harmonic lies in the frequency region where the amplification factor A of the amplifier starts dropping (around and above the f_U frequency), the 3rd harmonic will be amplified less than the fundamental tone, and the waveform of their sum, the output signal, will differ from the waveform of the input signal. This kind of distortion is called attenuation distortion, because higher harmonics are attenuated (or not amplified as much) compared to lower harmonics.

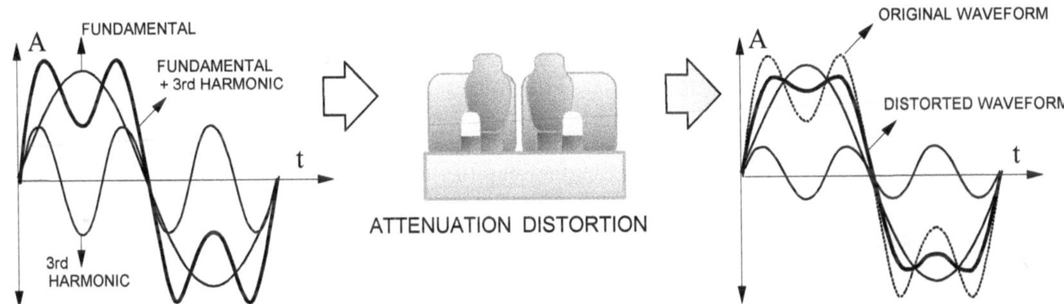

For instance, assuming the fundamental tone has frequency f_1=15kHz, the third harmonic will be at f_3=3*15=45kHz. An amplifier with A=20 and f_U=45 kHz (numbers chosen for the sake of simplicity) will amplify the fundamental 20 times, but the 3rd harmonic will only be amplified only 20*0.71 = 14.14 times!

Delay distortion

The phase angle between the input and output signal stays constant through the midrange frequencies, but changes significantly at frequency extremes. Just as in the last example, the illustration shows a complex input audio signal comprising of the a fundamental and its 3rd harmonic. Again, if the third harmonic lies in the frequency region where the phase angle changes from its midrange value, the 3rd harmonic will be shifted in phase by the angle θ.

Their sum, the output signal, will differ from the waveform of the input signal. This kind of distortion is called delay distortion. Looking at how different the shape of the resultant output voltage is from the input waveform, you'd think that this kind of distortion would be the most serious and malign of all. The truth is that our ears do not object to this kind of distortion. In fact, they don't even detect it! Exactly why that happens (or rather why doesn't it happen), is beyond the scope of this book, but boy, aren't we glad, one less issue to worry about.

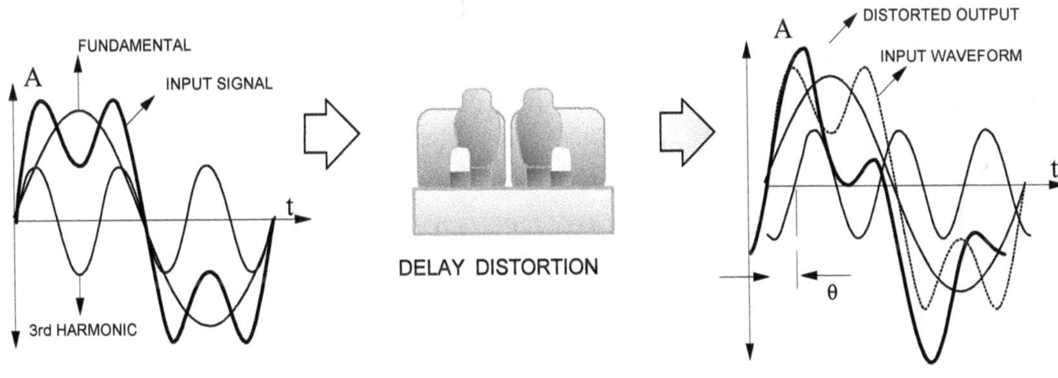

Amplifier behavior in time- and frequency domains

Now that we understand the concept of harmonics in a two-dimensional domain or time domain , we can study it in a 3D space, and introduce the frequency domain. The amplitude-time graph shows a distorted periodic waveform which can be broken down into two components, the fundamental harmonic and the second harmonic of twice the frequency of the fundamental ($2f_1$). There is also a phase shift α (alpha) between the fundamental and the second harmonic. This is "time domain" in which we see the amplitude and phase relationships of these three signals. Test instrument that displays signal waveforms in time domain is called an oscilloscope.

If we depict these harmonics in a three-dimensional space and add frequency as the 3rd dimension, it would look like this. The projection on the A-f plane will result in the spectrum, as illustrated. This is the frequency domain. We don't see the waveforms any more (we know they are sinusoidal signals, anyway), but we see the frequencies, the absolute and relative amplitudes of all harmonics, something we don't see in the time domain. Test instrument that displays amplitudes of signal harmonics in frequency domain is called a spectrum analyzer.

The two depictions illustrate different aspects of the same signal. In this case we have only the fundamental (1st harmonic) and the second harmonic, but for generally there'll be higher-order harmonics present as well.

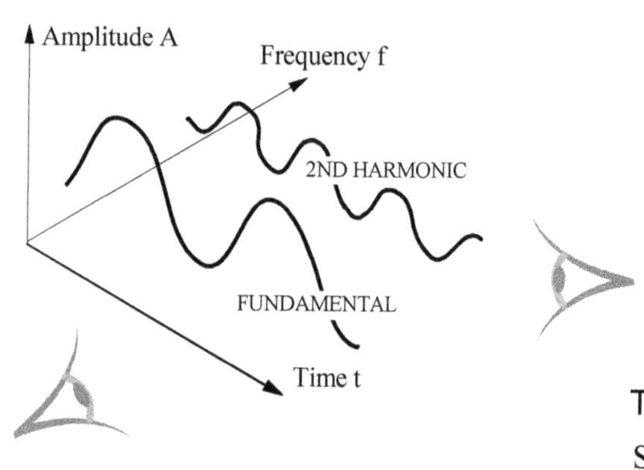

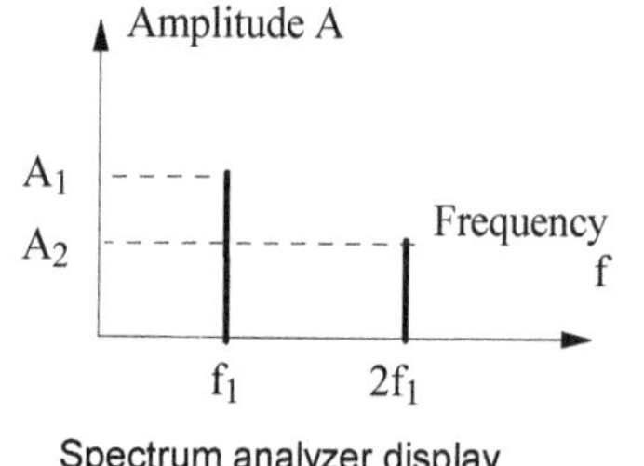

Spectrum analyzer display

LEFT: The relationship between the time and frequency domains in a 3D space

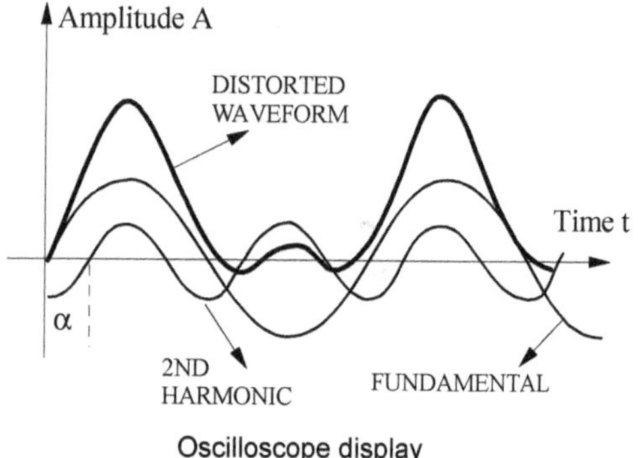

Oscilloscope display

The tale of two spectra

Spectral diagrams answer the common question in audio: Why do triode amplifiers usually sound better then those using pentodes? The spectra of 2A3 single-ended amplifier at half-a-watt and three watt power outputs show a dominant second harmonic.

At lower power levels that is the only harmonic measurable. The third and fourth only appear when the amp approaches its maximum power of 3.5W. The harsh sounding 3rd harmonic is "masked" by the harmonically pleasing second harmonic.

The pentode's sonic signature is very different. Even at low power levels the third and the fifth harmonics are present and higher order ones are measurable. At higher power levels those odd harmonics increase rapidly (from -53 dB to -25 dB, in case of the 3rd harmonic) and that is what make smost pentode amps sound shrill and harsh compared to triodes.

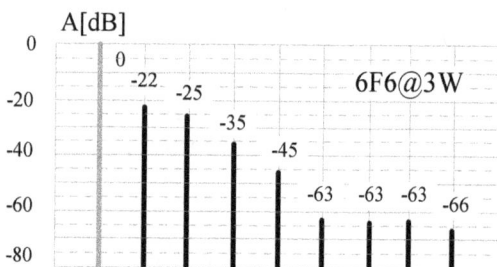

A[dB]
0
-20
-40
-60
-80
6F6@3W
0 -22 -25 -35 -45 -63 -63 -63 -66

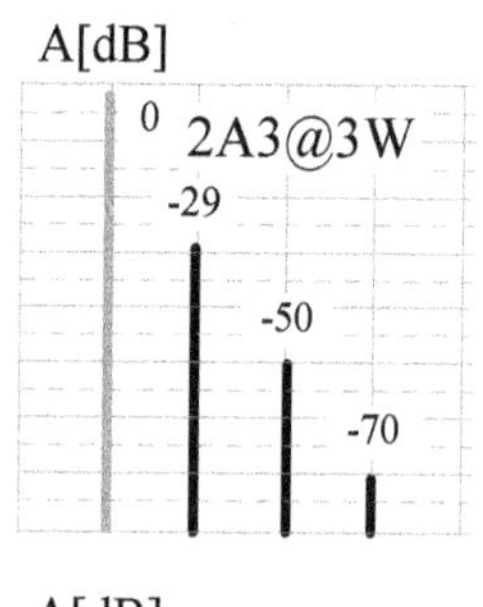

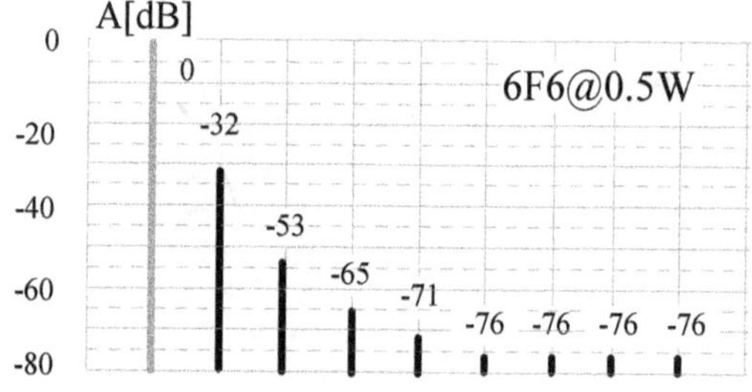

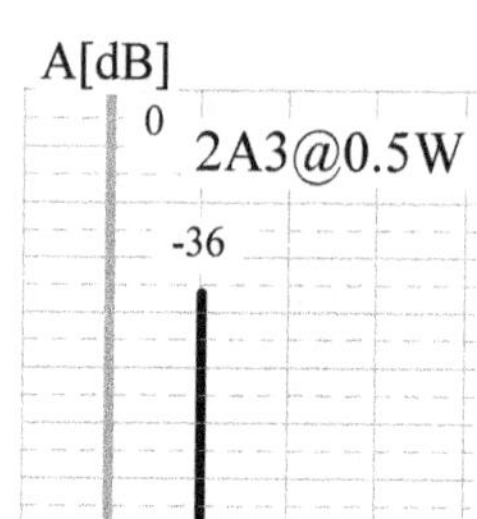

Harmonic distortion spectra of two low powered SE amplifiers, 6F6 pentode and 2A3 triode, for two power levels.

Calculating distortion figures from the harmonic distortion spectrum

How do we determine harmonic distortion in % from the spectral figures in dB, such as those just illustrated? The fundamental formula is dB = 20log(x), so "x" would be our harmonic distortion coefficient H (but not in % yet!) Remember, if the base of the logarithm is not specified it is assumed to be 10, so log(x) really means $\log_{10}(x)$. Therefore x=10(db/20)

Let's say the 2nd harmonic H_2 is 40 dB below the zero level of the 1st harmonic H_1. We have $H_2 = 10^{(db/20)} = 10^{(-40/20)} = 10^{(-2)} = 0.01$ To convert that figure into percentages, multiply it by 100 and get $H_2 = 1.0$ %

One way to remember this is that -20 dB means 10% distortion, -40dB equals 1% distortion, -60dB corresponds to 0.1% distortion, or, in other words, for every 20 dB drop the distortion reduces by the factor of 10!

What if, for some reason, the amplitude of the 1st harmonic is not 0dB but some other figure, such as -4 dB? Well, this is where the beauty of decibels comes into play. Since the dB is a relative unit, to get a relative difference between the two harmonics, simply subtract one figure from the other! Let's say the 2nd harmonic is 40 dB below the -4 dB level of the 1st harmonic H_1. Since all harmonics must be referenced to the fundamental's level we have $H_2 = H_2 - H_1 = -40-(-4) = -40+4 = -36$ dB

Now $H_2 = 10^{(db/20)}*100$ [%] $= 10^{(-36/20)}*100$ [%] = 1.585 %

Calculating THD from individual harmonics

If the amplitudes of the individual harmonics are known (in Volts), the overall THD figure can be calculated as **THD [%] =** $\mathbf{\sqrt{(H_2^2+H_3^2+ ...+H_N^2)}/H_1 *100\%}$

THD

THD [%] = $\sqrt{(H_2^2+H_3^2+ ...+H_N^2)}/H_1 *100\%$

Let's calculate THD figure for the 2A3 SET amplifier at the output level of 3 Watts (spectrum on the previous page). The fundamental H_1 is at 0dB, the 2nd harmonic H_2 is at -29dB, the 3rd H_3 at -50 and the 4th H_4 at -70 dB.

Since $-29dB=20\log(H_2/H_1)$ we get $H_2/H_1=0.03548$. Since $-50=20\log(H_3/H_1)$ we get $H_3/H_1=0.003162$, and $H_4/H_1=0.00031623$. Our harmonics are already expressed as percentages of H_1, so the square root of H_1^2 in the formula cancels out H_1 in the denominator and we have

THD= $\sqrt{(H_2^2+H_3^2+...+H_N^2)}*100\%$ if H_2 to H_N are in % of H_1. In our case THD [%] = $\sqrt{(0.03548^2+ 0.003162^2+ 0.00031623^2)}*100\% = \sqrt{(3.548^2+0.3162^2+0.031623^2)}*100\% = 3.562\%$

Why do tube and solid-state amplifiers sound different?

If you are reading this book you most likely don't need to be convinced of the sonic beauty of tube amplifiers. Making wide-sweeping statements is risky and often counterproductive, since it borders on generalization. That is why I used "different" in the subtitle instead of "better". Even that statement will be challenged by many readers. I still remember an interview with Tim de Paravicini, where he claimed that he could design & build a tube amplifier to sound like a great solid-state amp and vice versa, and I entirely agree with him. Two examples come to mind.

In my teenage years I owned a few different mid-fi systems by Grundig and Siemens, which were 100% solid state but sounded very warm and tube-like. Then, during our ValveMark years, we designed and built a 300B amplifier with active tube HV regulation that sounded great, but didn't sound like a typical 300B amp at all.

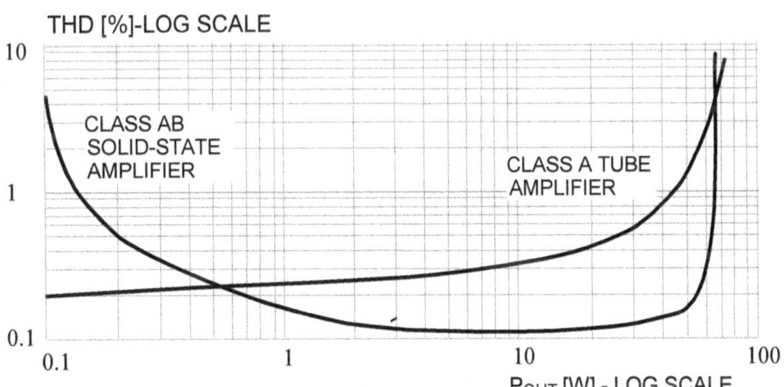

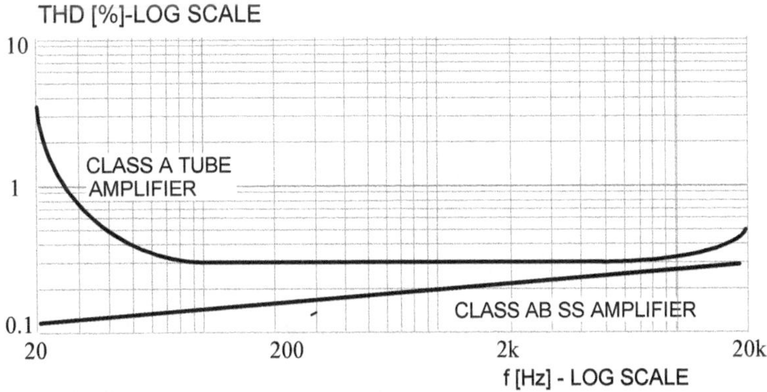

THD versus output power characteristics of a typical Class AB solid state and Class A tube amplifier (above) and THD versus frequency characteristics of a typical Class AB solid state and Class A tube amplifier

It had many qualities of a very good solid-state Class A amplifier, but it still retained some of that 300B magic, without that syrupy over-the-top coloration some 300B designs suffer from.

Although they have numerical scales, don't get stuck on the figures, the graphs on the previous page are meant to qualitatively illustrate two of many reasons why tube and SS amps sound different.

Firstly, notice that SS amplifiers' distortion at low power levels is relatively high, and then it comes down with an increase in output levels, only to jump suddenly once the clipping levels are reached. Distortion of a typical tube amplifier is low at low power levels and gradually creeps up with increasing output levels. Clipping is gradual and soft, not sudden, unless high levels of feedback are used, in which case it resembles solid-state clipping.

Perhaps that explains why tube amps sound better at low power levels and why their distortion at high power levels often goes unnoticed, due to its gradual nature. When SS amps reach clipping levels one would need to be totally deaf not to notice such unpleasant harshness.

The second graph illustrates that THD (Total Harmonic Distortion) in SS amps increases steadily with increasing frequencies. Tube amps distort more at very low frequencies, especially at high power levels, due to saturation of the output transformers' core, but then the harmonic distortion is more or less constant through the midband, increasing only in the upper frequency band above 10kHz.

AMPLIFIER IN THE SIGNAL CHAIN

Voltage, current and power amplification through the audio chain

For a 2A3 triode amplifier supplying $5V_{RMS}$ to an 8Ω load, the output power is $P_{OUT} = V^2/R = 25/8 = 3.125W$. The AC (signal) current through 8Ω load is $I_L = V_L/R = 5/8 = 0.625A$

For a high power tube amp supplying $40V_{RMS}$ to an 8Ω load, the output power is $P_{OUT} = V^2/R = 40^2/8 = 200W$, while the current through the load is $I_L=V_L/R = 40/8 = 5A$

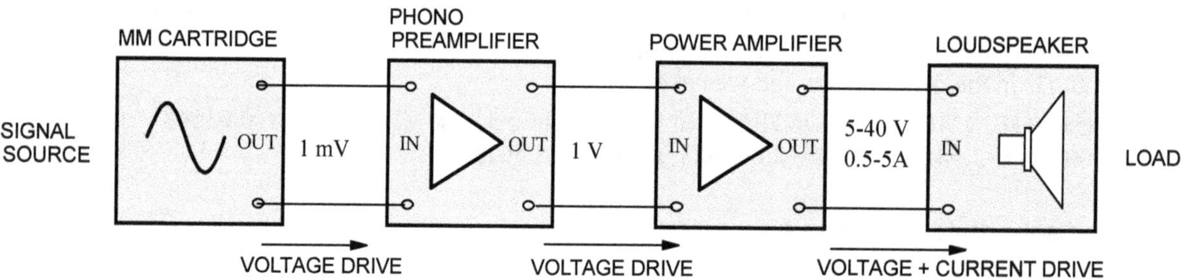

All amplification devices upstream of the power amplifier provide only voltage amplification, while the power amplifier usually amplifies both the voltage and current.

We can now proceed with the analysis of a multistage amplification chain. We have an audiophile system comprising of a signal source, a turntable with a moving coil cartridge (not shown), an MC step-up transformer, a MM (moving magnet) phono preamplifier, and a power amplifier. The amplification factors are given for each component. What is the overall amplification factor in dB?

There are two ways to do this. You can calculate the gain of each stage or device in dB and then simply add them up: A_{TOTAL} [dB] $=A_1+A_2+A_3 = 103.5$ dB , or, you can calculate the overall gain of the system $A_{TOTAL}=A_1A_2A_3 =$ 15,000 and then convert it into dB: A_{TOTAL} [dB] $= 20\log A_{TOTAL} = 20*\log 150{,}000 = 20*5.176 = 103.5$ dB

A_1=20 dB, A_2=60 dB, A_3=23.5 dB

0.1 mV → A_1=10 → 1 mV → A_2=1000 → 1 V → A_3=15 → 15V

MC STEP-UP TRANSFORMER | PHONO PREAMPLIFIER | POWER AMPLIFIER

$A_{TOTAL}=A_1A_2A_3 = 150{,}000$

A_{TOTAL} [dB]$=A_1+A_2+A_3 = 103.5$ dB

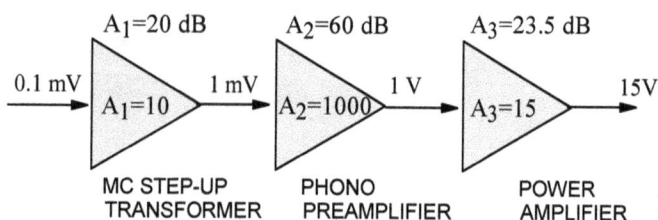

The importance of speakers' sensitivity

Consider two audiophile setups, a low power triode amplifier driving efficient dynamic speakers of 93 dB/W sensitivity, and a high power parallel push-pull tube amplifier driving hybrid electrostatic speakers with a dynamic woofer of a low 83dB/W sensitivity . Which one will sound louder in the same room?

If we express the ratio of power outputs of the two amps as a ratio we get $P_2/P_1 = 100/10 = 10$ or in dB terms $P2/P1 = 10\log P_2/P_1 = 10\log 10 = 10$ dB

I deliberately chose nice & easy numbers (but very realistic). We could have figured that result from the dB power graph a few pages back. Ten times higher power of the push-pull amp is +10 dB above the output level of the low powered amp.

However, its speakers have 10 dB/W lower sensitivity, so to get the same SPL (sound pressure level) as the low powered amp with high efficiency speakers, the amp must have 10 dB higher amplification. Therefore, both setups will sound equally loud! What we gained in amplifying power we lost by using very inefficient speakers!

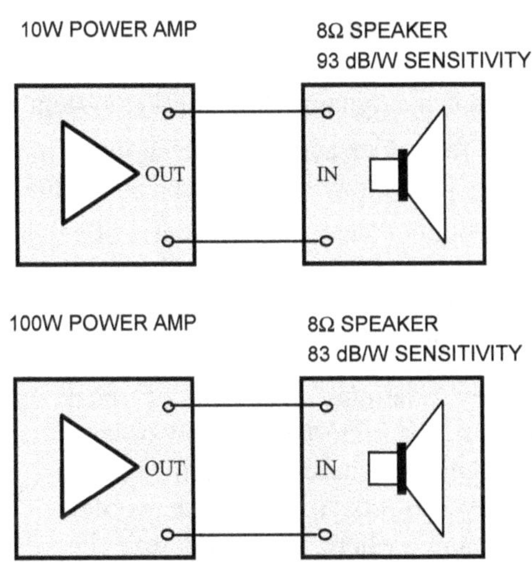

right: Since we don't listen to amplifiers alone, but through loudspeakers, the maximum power of an amplifier does not matter on its own, the loudspeaker sensitivity also has to be taken into account. The two systems illustrated produce the same SPL!

The importance of amplifiers' low output impedance

The following discussion is somewhat simplified. For instance, it assumes that amplifier's output impedance is constant throughout the whole frequency range, which isn't strictly true. It also assumes that in all other respects the amplifier is ideal, that it can provide any current demanded by the load, and that is only true for a limited range of load impedances. Nevertheless, for this purpose it's close enough to reality.

This scenario illustrates issues arising from high output impedance of an amplifier (especially single-ended amplifiers without negative feedback), and from its interaction with varying load impedance (the voltage divider effect). At frequencies where speaker's impedance is at its minimum (say 2Ω), $V_L/V_0 = R_L/(R_L+Z_{OUT}) = 2/(2+3) = 0.4$ or 40% At resonant frequencies where speaker's impedance is at its peak (say 22Ω), $V_L/V_0 = R_L/(R_L+Z_{OUT}) = 22/(22+3) = 22/25 = 0.88$ or 88%

The output voltage will vary $20\log(0.88/0.40) = 20\log 2.2 = 6.85$ dB!

Assuming an ideal amplifier with $Z_{OUT}=0$, the output voltage of a 300B amplifier capable of producing 8 Watts on an 8Ω load would be $V=\sqrt{(PR_L)} = \sqrt{(8*8)} = 8$ V

A real amplifier, driving a speaker whose impedance dropped to 2Ω, would have its output voltage divided between the output impedance of the amplifier and the load, so the voltage on the load would be $V_L = V_0R_L/(R_L+Z_{OUT}) = 8*2/(2+3) = 8*2/5 = 3.2$V

The power fed to the load would be $P=V^2/R_L = 3.2^2/2 = 5.12$W, not 8W as it should be!

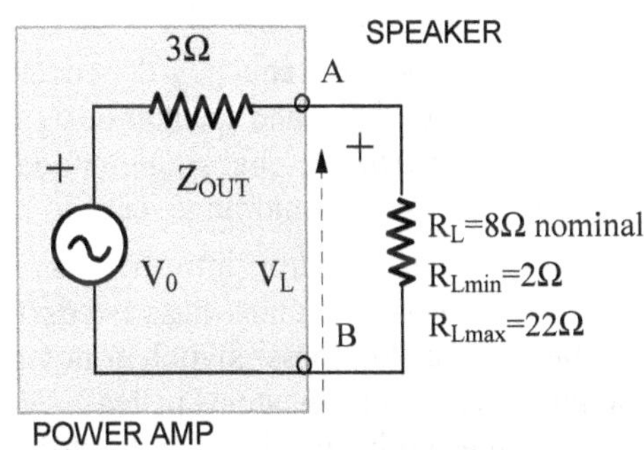

ABOVE: The output of an amplifier behaves as a frequency- dependent voltage divider

Output impedance versus frequency

Damping factor (DF) is a ratio of load impedance and amplifier's internal or output impedance: $DF= R_L/Z_{OUT}$ It helps us predict how well a particular amp would control the speaker cone.

Due to output transformer's leakage inductance that increases with frequency, the output impedance of a tube amplifier rises with frequency, especially in the absence of negative feedback. The same amplifier even with a relatively mild feedback exhibits a much lower rise in the output impedance compared to the rise when NFB is removed. This could be one of the possible factors that have made us conclude (after extensive listening evaluations) that a mild feedback is usually beneficial and preferable to no NFB at all.

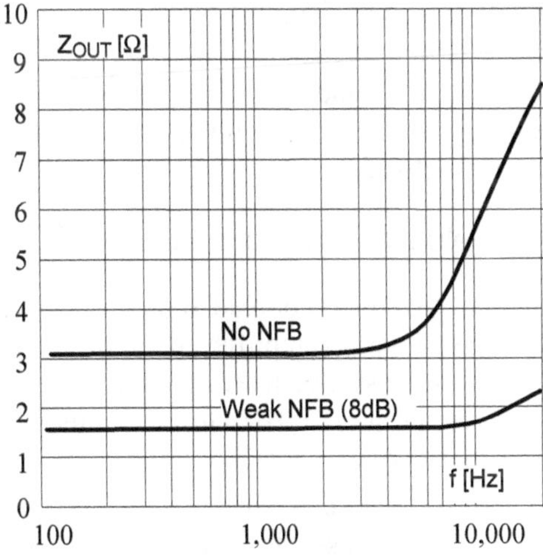

RIGHT: How the output impedance of a typical 300B single-ended triode amplifier (8Ω output) varies with frequency with and without negative feedback.

This significant rise of Z_{OUT} at high frequencies may seem intolerable, but luckily, there are mitigating factors. The most significant rise is in the 10-20 kHz band, and very little power is needed in that frequency band to drive a typical tweeter in a dynamic loudspeaker system.

Very few older audiophiles can hear frequencies above 15 kHz, so even a significant drop in the upper frequency power levels may go unnoticed and actually "mellow" the amplifier's sound!

Last, but not least, the impedance of a typical dynamic loudspeaker also rises at these frequencies, so while the output impedance of a typical amplifier Z_{OUT} rises in the absolute sense, the rise in the load impedance tends to offset it to some degree. Remember, DF is measured with a fixed resistive load, as in the example below, not with real loudspeakers!

Finally, damping factor is of most interest and importance at low or bass frequencies, where an amplifier must supply lots of current to the woofer coil and where damping of the woofer's cone is of paramount importance for the speed, definition and prominence of the bass.

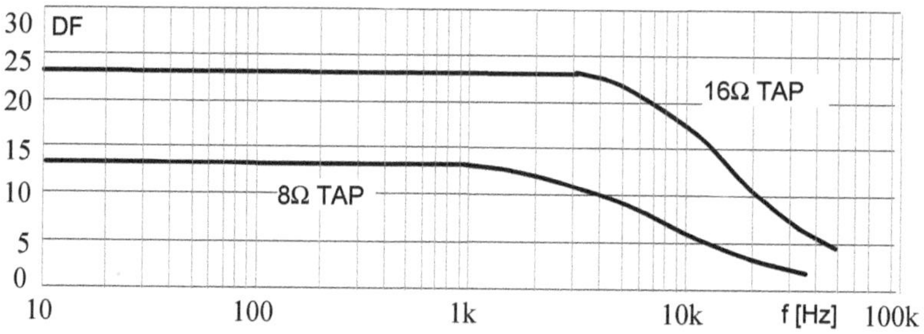

Damping factor of a typical push-pull tube amplifier versus frequency

WHAT DETERMINES THE SOUND OF AMPLIFIER?

The short answer to this question is *everything (to some extent)*! The long answer is this whole book. All the issues we are going to cover, from those we will mention only in passing, to those that we will analyze in great detail, will impact the sound of an amplifier or preamplifier to some extent.

Why do amplifiers sound different at lower volume levels?

Volume control potentiometers in amplifiers do not use linear taper (the relationship between output resistance of a potentiometer and the angle of rotation), since human ears are not linear sensors, but react to sound pressure in a logarithmic manner. That is why the logarithmic taper is commonly called an audio taper.

Human hearing is also frequency-dependent, and since our ears are most sensitive in the frequency range between 500 Hz and 5kHz, amplifier and speaker designers should strive to minimize the distortion of their amplifiers and speakers in the midrange. The fact that single-ended triodes amps have a relatively large following could be explained by their "magical" and "warm" midrange, caused by pleasant-sounding 2nd harmonic distortion.

As signal frequencies drop into the bass region, higher SPL levels are needed to achieve the same perception of loudness. At lower volumes bass needs to be boosted in order to restore the original balance, so many vintage amplifiers had a "loudness" switch or potentiometer. This simple filter did not (could not!) boost low frequencies but attenuated the high frequencies instead. Due to their minimalist approach, no modern audiophile amplifier has such a feature. Just as with capacitors, audiophiles "blacklisted" potentiometers as detrimental to sound, so that was the nail in the coffin for the loudness control in particular and tone controls in general.

The same conclusion applies to higher frequencies, although to a lesser extent. The nature of our hearing correlates quite well with the frequency-dependent nature of amplification, so one of the most common methods used in audio engineering is to study the three frequency bands separately.

Our modeling of vacuum tube amplifiers and audio transformers will also be done that way, with separate models for the low-, mid- and high- frequency ranges. The limits of these frequency bands aren't strictly defined, the low frequency range extends up to 100 or 200 Hz, the high frequencies are above, say, 10kHz, and the midrange is the range in between.

A-B evaluations and loudness levels

You may have heard or read about A-B listening tests, and about the oft-repeated rule that no matter what component is being evaluated (amplifier, preamplifier, CD player, phono stage, etc.), the listening tests must be done at the same level of loudness. So, in serious A-B tests, it isn't enough just to swap amplifiers and roughly adjust the volume, a tone generator, a test record or a test CD must be used (with calibrated tones), together with a SPL (Sound Pressure Level) meter to ascertain that is really the case.

Assuming we are comparing amplifiers, the conventional wisdom says that the louder amplifier will sound better. Is that really the case? It will probably sound more "confident" (How do you measure confidence in a piece of equipment?) and even more dynamic, but what about distortion?

About 250 years ago, Italian composer and violinist Giuseppe Tartini discovered an important psychoacoustic phenomenon: when two notes of different frequencies are played simultaneously, a listener can "hear' additional tones whose frequencies are the sum and difference of the two frequencies. We now call this type of distortion intermodulation distortion, which seems to be caused by the non-linearity of the inner ear.

A hundred years after Tartini, the famous Hermann Helmholtz in his book "On the Sensations of Tone as a Physiological Basis for the Theory of Music", first published in 1862, showed by experimental and mathematical analysis the existence and nature of these combination tones, and that they increase steeply with rising intensity of the fundamental tones.

So, we could hypothesize that increasing volume levels would sharply increase the IM distortion in our ears and make us perceive the louder amplifiers as harsher or more irritating, at least on a subliminal level. This could be compared to listening to solid-state amplifiers, which sound nice initially, but soon listener's fatigue sets in and the longer you listen, the more "uneasy" you feel. Instead of being drawn deeper and deeper into the musical pathos, as it happens with the best tube amplifiers, you are compelled to stop listening altogether.

Why tubes of the same type sound different?

Most audiophiles will agree that different tubes (of the same type) make the same amplifier sound different. Tube "rolling" makes it easy to compare the sound of various tubes, by simply swapping them around in an amplifier and listening critically. If you take a dozen or so tubes of the same type from various manufacturers, say 12AX7 (ECC83), you will notice that anodes or plates are of different shapes and sizes. So are the cathodes and the grids, but these usually cannot be seen from the outside.

Some anodes are twice as large than the others (JJ has the smallest anode here, Ei the largest). Some anodes are "boxed", others flat, some are ribbed, some smooth and, even more importantly, they are made of very different materials, some shiny silver in color (nickel?), such as Ei, some matte gray (JJ and RCA), others black (Arcturus).

Other parts of these tubes will also differ, the material used for cathodes, heaters, grids and other electrodes, the chemical composition of the glass will vary, the thickness and construction geometry of electrodes will be different, and so on. These variations are not obvious, tubes would need to be destroyed and taken apart for further inspection, and only chemical analysis would detect metallurgical differences.

All marked 12AX7, but are they all the same tube?
L-R: JJ ECC83S (Slovakia), RCA (USA), Ei (Serbia), Arcturus brand (unknown origin)

It is indeed a miracle, or rather a long stretch of imagination to call all these very different tubes "12AX7". Sure, they all have a similar amplification factor μ and the other two basic tube parameters (transconductance and internal impedance), but may not behave in the same manner in other respects. No wander they sound different in the same amplifier!

By the same token, other parts will also impart their character onto the final sound of an amplifier.

Passive components (resistors, capacitors, inductors and transformers) are a significant factor in the final voice of an amplifier, but not the only one. Hookup wire, shielded cables, tube sockets, input connectors and speaker binding posts all impact the sound. Even fuses and power cords, far removed from the signal chain, contribute sonically!

In the early days of my audio journey, the engineer in me dismissed the claims about power cords as ridiculous, until I tried a few and noticed significant differences. Some improved the bass immensely, others improved the "speed" and the dynamic of an amplifier, many changed its tonal balance. Power cables had more impact on the sound than interconnect or speaker cables. Since that makes no technical or theoretical sense whatsoever, I realized that neat models and simple measurements are one thing, but the messy reality is something else.

Why amplifiers that measure well don't necessarily sound better?

The first issue is the lack of standard load which would emulate a typical loudspeaker. What is a typical loudspeaker is the ultimate question, and the answer is that there is no such thing! Thus, we test amplifiers using a resistive dummy load, whose impedance is independent of frequency and is pure resistance. Loudspeakers are complex loads whose impedance is highly frequency dependent, either inductive (dynamic speakers) or capacitive (electrostatic speakers).

They also generate counter electromotive force (EMF), which is fed back into the output stage of an amplifier and interacts with it in mostly unpredictable ways.

Many tests (frequency range, input impedance, output impedance and damping factor, harmonic distortion, maximum power) use only a single test frequency of a fixed amplitude, while music signals are complex waveforms comprised of dozens of harmonics of rapidly changing amplitudes and phase.

THD measurements do not make a distinction between pleasant and rich sounding even harmonics and harsh and irritating odd harmonics. It is not so much the amplitudes of the harmonics that determine the sound of an amplifier, but its whole spectral "signature". Spectral analysis gives us the relative amplitudes of all harmonics, so such a test is more meaningful than simple THD measurements, but despite knowing the number of dB each harmonic is below the fundamental, we still cannot predict how the overall spectral signature would sound to our ears.

Finally, we measure things that are easy and convenient to measure, and not necessarily those that have the greatest impact on the sound fidelity! The most important aspects such as rhythm and pace, the sense of "presence", "musicality", "microdynamics" or "transparency" cannot be correlated with measurements at all.

The dielectric theory of sound

Audiophiles tend to hate capacitors with passion (especially the electrolytic kind!), claiming they are detrimental to sound quality. This disdain is probably the main reason behind the resurgence of directly-coupled and transformer-coupled designs. Sure, transformers sound different from capacitors, just as pentodes sound different from triodes, or tube amps from transistor amps. Perhaps their sonic signature (read "distortion") is more pleasing to the human ear. However, as we will see soon, replacing coupling capacitors with direct coupling or with interstage transformers solves one problem but creates two or more new ones!

Capacitors are inevitable; they are everywhere. As a matter of fact, you are sitting inside a huge capacitor right now. The wiring in the roof of your house is one conductive plate, the earth or ground is another, and you and everything around you (including the air) is the dielectric between the two plates.

Even if you remove all the coupling capacitors in an amplifier, there will still be dozens of capacitances left. These are parasitic capacitances between the cables, components (resistors, capacitors, inductors, tube sockets) and chassis. Another issue are tubes' internal capacitances, and there is sweet nothing you can do about them.

All other components, from potentiometers to filtering chokes also have parasitic capacitances and all those capacitances will have an impact on the sound of an amplifier. You cannot see them or touch them, they are not *discrete* capacitors, but parasitic (unwanted) capacitances *distributed* across cables and components.

The sound of a capacitor, or any other component for that matter, is determined by two main factors. One is its conductive parts, the cathode, the grid and the anode of the tube, the conductors in a cable, the metal plates and leads of a capacitor, the winding wire of a transformer. The other factor is the dielectric between those conductors.

The same cable, say silver plated OFC (oxygen-free copper) interconnects, will sound different with a Teflon® (PTFE) jacket and with a PVC jacket. The same output transformer will sound different if impregnated paper is used for insulation between winding sections instead of plastic film such a Mylar® or Kapton®.

The sound of audio transformers will also depend on the properties of the core or laminations used. A transformer with GOSS (Grain-Oriented Silicon Steel) core will sound different from an otherwise identical transformer wound on an ordinary 3% silicon steel core, or on a 49% nickel core such as Permalloy®.

The amplifier-speaker interface

One of the greatest paradoxes in audio is that we spend so much time, effort and money to make tube amplifiers as close to perfection as possible, and then connect them to (apart from a turntable) the worst link in any audio system - the imperfect loudspeaker.

A loudspeaker is a frequency-dependent complex-impedance load, so it is impossible to predict how a certain amplifier-speaker combination will sound. The only way to ascertain that is to try the two together.

Even with a resistive dummy load the square wave response of Audio Research D-70 amplifier (RIGHT) is very poor. Add 2μF of capacitance in parallel (a rough way to simulate an electrostatic speaker) and you get a hump that does not even resemble the original square wave.

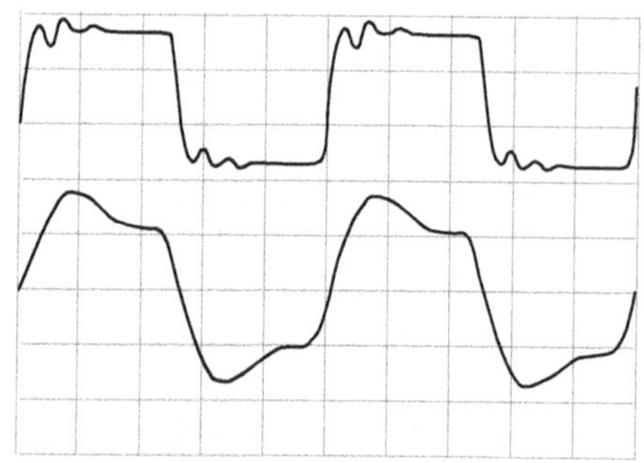

10 kHz square wave response of Audio Research D-70 amplifier a) into 16Ω load b) 16Ω paralleled by 2μF (Source: *Audio*, June 1984 review)

The final transducer - the listening room

Most listening rooms either sound "too harsh, "cold" and "clinical", or they sound dull and "dead". The first indicates that there are too many solid surfaces in the room, for instance a tiled floor, bare walls or exposed windows (without curtains or blinds), objects or surfaces that reflect sound waves.

The second indicates the opposite situation, there are too many soft surfaces, furnishings, carpets, curtains and other plush objects that absorb sound waves. Therefore, your first task in any room is to strike an optimal balance between the reflective and absorptive surfaces. That does not have to involve the installation of expensive professional acoustic panels and bass traps or ugly egg cartons, a typical budget solution.

For instance, in our demonstration room we have a floating timber floor, and initially the sound was too harsh for my liking. Once we added a $200 (3 x 2 meters) wool rug in front of the speakers all was well. The sound became soft and warm, the rug completely tamed down the offending reflections of the floor.

An automotive analogy is in order. An amplifier is akin to the engine of a car. The speakers can be likened to the transmission, wheels and tires. The room, as the ultimate acoustic transducer, is the equivalent of the road on which the car travels. You may have the best engine, but if the transmission isn't optimized, the car will not drive well. Likewise, even the best car and the best tires will lose their grip on a lousy road. I call it "The Weakest Link" syndrome:

THE WEAKEST LINK SYNDROME:

An amplifier or a whole audiophile system is only as good as its weakest link.

Optimize the dimensions of your listening room and speaker positioning

There are three candidates for optimal proportions of a listening room. The simplest has proportions 2:1:1 (length:width:height). These were often used in Egyptian and ancient Greek temples, also later on in Romanesque and gothic churches. For a 6 m long listening room the width and the ceiling height would be 3m.

The second set of numbers is 2:1:1.118, used for the shape of "King's Chamber" in the Great Pyramid of Giza. For a 6 m long listening room, the width would be 3m and the ceiling height would need to be 3.354m.

Finally, named "Golden Solid" or "Golden Cuboid", these proportions were used for Egyptian tombs and medieval churches. Φ is the "golden ratio". The dimensions progress in a 1 - Φ - Φ^2 order, also called the Fibonacci sequence. For a 6 m long listening room the width would be 3.71m and the ceiling height needs to be 2.29m. This is the closest to the often used games room size in Australia of 6 x 4 x 2.2m.

The same ratios and sequences have successfully been applied to the dimensioning of musical instruments, loudspeaker boxes, and, of most interest to us here, the placement of the speakers in a room.

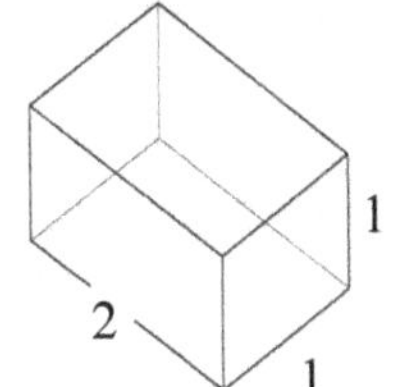

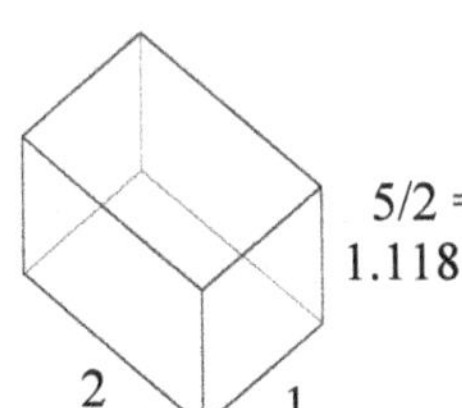

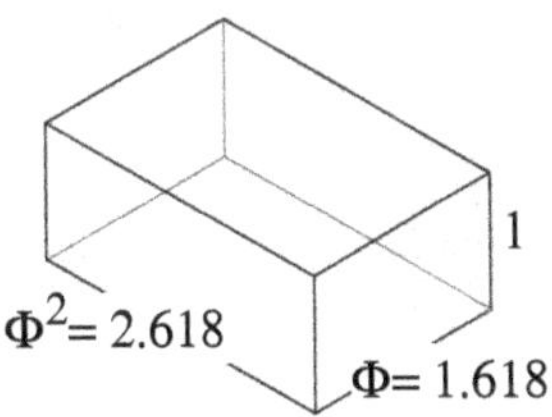

LEFT: Three chamber proportions commonly used throughout history

BELOW: Speaker positioning based on the Golden Ratio

We are assuming an equilateral triangle between the speakers and the "sweet spot" (all three sides A meters in length). This "A" corresponds to "1" unit in the Golden Ratio. The distance between the speakers and the further wall is then Φ times longer, or 1.618*A. Obviously, the distance from the closer side wall is then 0.816*A.

Firstly, measure the width of your listening room (W). Since W=2.236*A, divide W with 2.236. That is your "A". Say your room is 3.71 m wide. Then A = 3.71/2.236 = 1.66 m, meaning your speakers should be 1.66m from the front wall and the sweet spot should be 0.866*A = 0.866*1.66 = 1.44 m from the front line of the speakers (X-X).

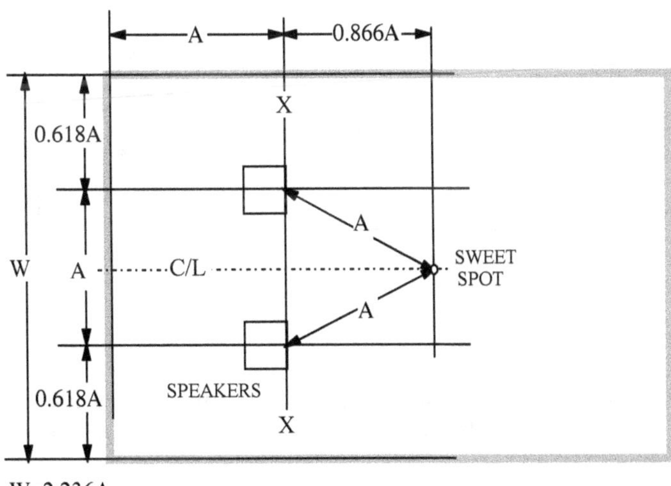

Position all hi-fi components as far away from your speakers as possible!

Monoblock amplifiers are usually considered the pinnacle of the hi-end game. The channels are completely independent (they have their own power supplies), so there is no crosstalk within the amplification stage. Also, they generally store more energy in their power supplies, so have more dynamic reserve.

From the aesthetic point-of-view, monoblock amps also look great when placed on the floor, each next (usually slightly towards the back) to the speaker they are driving. We often find such setups at prospective buyers' listening rooms, but very rarely at dealers' showrooms. And there is a good reason for that.

Firstly, loudspeakers vibrate, and these vibrations are transmitted to the floor, and from there to anything on the floor, including equipment racks and amplifiers sitting on the floor. This is especially an issue with timber floors, which vibrate more than the concrete ones. Audio components should not vibrate mechanically (except the moving parts of a phono cartridges) so they should not be sitting on the floor.

Secondly, especially with the rear-firing bass-reflex speakers, there may be significant acoustical feedback from the speaker to the tube amplifier. All tubes are microphonic to some extent, some very much so, and if they are exposed to sound waves, that will result in unwanted reproduction of the fed-back sound.

Have somebody gently tap a tube in your amp or preamp with their finger nails while you listen with your ear on the speaker. You will usually hear a sound, sometimes it will be faint, in other cases very loud. Very few tubes aren't microphonic at all, the mighty ECC40 duo-triode comes to mind as one such tube.

Before you even start listening to a tube component in dealer's demo room or when buying privately, perform this simple microphony test. If the tube amp or preamp fails it, don't even listen to it, move on.

Fixing microphonic amplifiers is always difficult, and removing the microphony completely is usually impossible (especially if printed circuit boards are used), so it simply isn't worth your time and money. Some China-made amplifiers we analyzed over the years were incredibly microphonic, beyond any help.

The same acoustic phenomenon will happen with acoustic feedback from the speakers to the amp or preamp, it's just that it will be masked by the main signal of the amplifier. Although you will not be able to isolate it acoustically or electrically in order to listen to it by itself, it is there and it will cause distortion, smearing, loss of focus, it will impact the sound stage, accuracy and transparency of the whole system.

Finally, the rule of thumb says that the lower the signal amplitude, the shorter the signal-carrying cables should be. Thus, turntable interconnects should be as short as possible, the phono stage should be very close to the cartridge. Likewise, line-level signal interconnects should be shorter than speaker cables. Positioning monoblocks behind speakers shortens the speaker cables considerably, which is a nice benefit, but it lengthens interconnect cables between the source and the amps significantly, typically from 0.5m to 2-3m, and that is bad, bad, bad!

The illustration shows a proper (LEFT) and improper way (RIGHT) of equipment positioning. The source(s) and amplifiers should be located as far back from the speakers as possible, and nothing else should be placed behind, around or in front of the speakers - no coffee tables, no CD racks, shelves, plants, nothing.

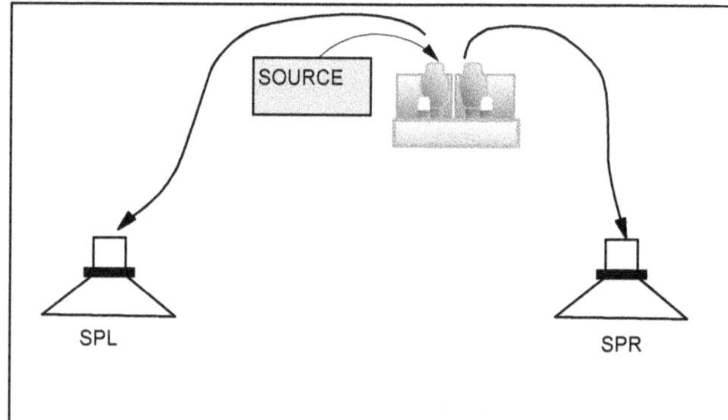

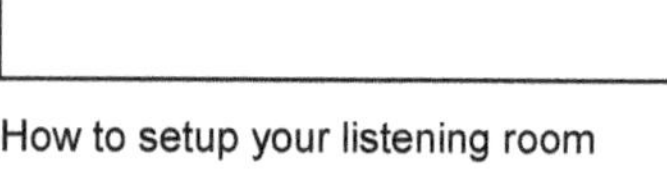
How to setup your listening room

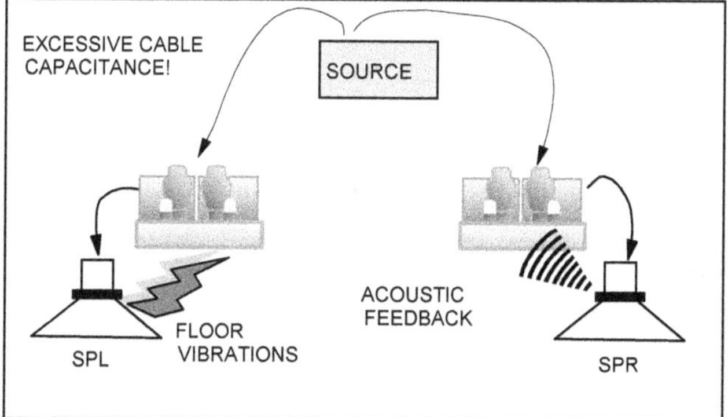

Amplifiers sitting on a floor, close to speakers and far from the source are a NO-NO!

The Setup Rule

You may be considering selling a wonderful piece of secondhand gear which does not go well with the rest of your system, and be fooled into thinking that it isn't such a great piece of gear after all, and risk selling it way below its true value.

Or, in a much more common scenario, the shrewd seller (either a dealer or a private seller) has, through trial and error and lots of experimentation, found components that match the component he is selling extremely well.

Now your perception of the quality of this component is likely to be much higher than it really deserves, for its weaknesses have been cleverly and carefully masked or compensated for by the other components in the system.

This is what smart hi-fi dealers (retailers) do. Unless you are buying the whole system from them (even then it may sound much worse in your home due to the issues with your listening room), you are unlikely to get such synergy once you bring the component home. So, never buy straight away, always ask to borrow the unit and evaluate it with your system, in your room. Most shortcomings in audiophile systems are in the mismatch category. The components simply don't go well together, or, even more often, the biggest culprit usually isn't any of these components, but the listening room itself. The system just doesn't sound right in that particular room.

THE "YOU HAVE BEEN SET UP" RULE:

Never buy anything based only on how it sounds in a dealer's showroom! Ask to borrow the gear and try it in your room with the rest of your system. If a retailer isn't happy to do so, be sure that many of his (more desperate) competitors will be.

Constant vigilance: nothing beats a critical eye and some basic knowledge

Apart from cleverly matched setups by shrewd dealers and hi-fi retailers, the biggest pitfall in buying audio components surely must be magazine reviews. As in any profession, there are decent and honest hi-fi reviewers, and there are those who have no technical knowledge or any real expertise, and therefore should not be trusted. Many are biased towards certain technologies or brands. All have their preferences and all think they know what things should sound like. The only problem is, it is not what they think that matters, it is what you think!

Sure, read the reviews, but read them critically. Read at least three reviews of the same piece of gear, never base your purchasing decisions on only one piece of information, no matter how much you respect or trust its source.

Magazines must make money, and they make it from advertising, not from magazine sales, or, heaven forbid, from subscriptions. It takes a very brave editor to publish a negative review of an XYZ brand amplifier when that same XYZ corporation spends $150,000 per year on advertising with the said magazine!

I have nothing against hi-fi reviewers, they have a job to do and they do it based on their values and beliefs, so I am also warning you against listening to your friends' opinions. Sure, their intentions are most likely good, but you are not your friends. Your idea of a good sound may be different from theirs. Plus, what sounds good in their room, may not sound good in yours.

Arrested development: There is nothing truly new in tube audio

We have now reached the end of the introductory chapters, whose aim was to put a few issues into some kind of perspective, before we get into the more technical aspects of amplifier design and construction. After some raving and ranting, a few philosophical issues, and a quick summary of the whys and hows of good sound, a bit of tube history should complete this chapter nicely.

This super-simplified timeline is a brief overview of audiophile amplification trends and milestones. As you can see, the most significant developments and most of the key or seminal products (tube types and amplifier models) happened before 1960.

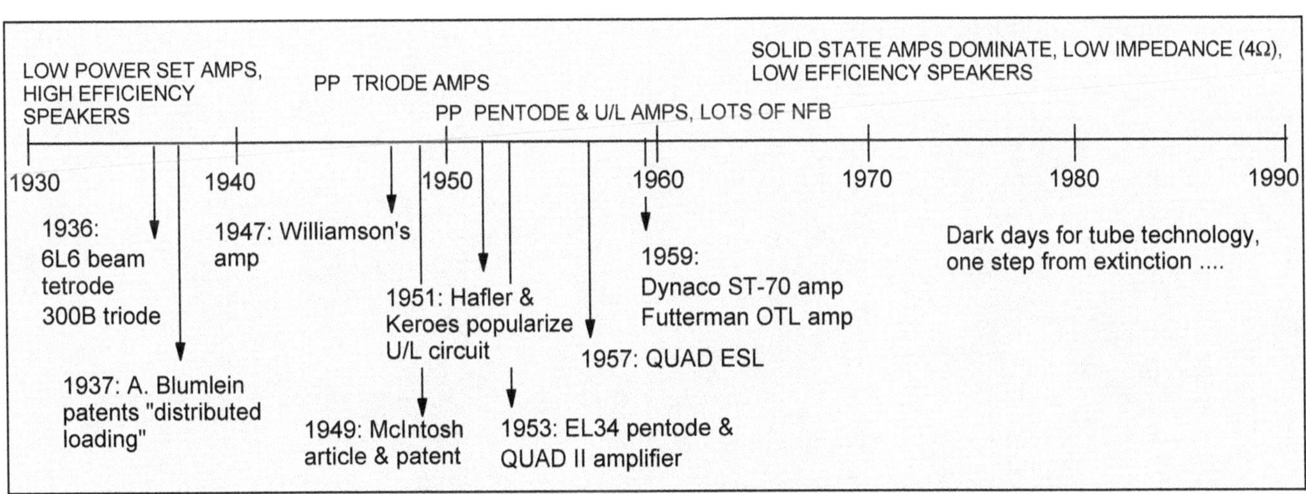

The three decades that followed (the 1960-1990) were a dark and uneventful age, dominated by transistors, during which tubes suffered a near-death experience.

Luckily, a new dawn in the early 1990s started tubes' revival, an era that has lasted to this very day. However, very few (if any!) truly novel designs or developments took place since 1990. Sadly, many so called "designers" simply rehashed the old circuits from tube manuals or copied commercial designs from the 1950s and 60s. Luminaries such as Tim de Paravicini, the late Kondo-san of Audio Note fame and a few other contemporary innovators are notable exceptions.

PHYSICAL FUNDAMENTALS OF VACUUM TUBE OPERATION

- ELECTRON EMISSION AND VACUUM DIODE
- VACUUM DIODE IN DC AND AC CIRCUITS
- ELECTRONIC TUBES - NAMING CONVENTIONS, BASES AND PIN NUMBERING
- CONSTRUCTION & MANUFACTURE OF AN ELECTRON TUBE
- TRIODE PARAMETERS AND CHARACTERISTICS

"If you want to find the secrets of the universe, think in terms of energy, frequency and vibration."
Nikola Tesla, Serbian inventor

5

ELECTRON EMISSION AND VACUUM DIODE

Thermionic emission

In any material its atoms or molecules are at rest only at absolute temperature of -273°C. At any other temperature they are in a continual state of motion. Higher temperatures make this movement more vigorous and increase the velocity of electrons in the material. This "liberation" of electrons from a metallic electrode by the virtue of its increased temperature is known as *thermionic emission* and the emitter electrode is called a *cathode.*

There are other ways of supplying energy to electrons so they break away from the surface of the metal. When certain materials are exposed to light energy they emit electrons through photoelectric emission. Likewise, a mere presence of a strong electrostatic filed (high voltage gradient) may be enough to liberate electrons through *cold or field emission.*

Finally, when metal electrode is bombarded by ions or electrons of sufficient energy, some of that energy is absorbed by surface electrons which may be enough to exceed the work function of the metal and liberate the electrons. This phenomenon is called a *secondary emission*, meaning the released electrons are secondary electrons while the energy is provided by the "primary" electrons.

Work function E_W is the minimum energy that must be imparted to an electron so it can break free. It is measured in eV (electronVolts) and ranges from 1 to 6 eV. ElectronVolt (eV) is a unit of energy and is simply 1Volt multiplied by the charge of one electron $q_e = 1.602*10^{-19}$ [coulomb]. The lower the work function the lower the temperature required for an emitter to release electrons. Thus, from the table below, oxide coated cathodes can supply the same quantity of electrons as tungsten filaments but at much lower temperatures (1.00 eV compared to 4.52 eV)!

Tungsten is a one metal suitable for use in vacuum tubes as a heater/cathode, due to its high melting point at around 3,600°C but a useful emission of electrons is achieved at a much lower temperature of 2,200°C. Its proper name is Wolfram (symbol W in the Periodic Table of Elements, atomic number 74).

The addition of a small quantity of thorium to tungsten lowers the work function to around 2.6 and thus enables this type of filament to operate at a much lower temperature of around 1,600°C. A heat treatment manufacturing process brings thorium to the surface where it forms a thin molecular layer. Thoriated tungsten filaments are used in tubes operating at anode voltages above 1kV (1,000V) such as 211, 845 and many others. Their white glow is very bright.

RIGHT: The work function, emission efficiency and operating temperature range for various materials

MATERIAL	$A_0[A/m^{2o}K^2]$	E_W [eV]	Emission efficiency [mA/W]	Operating temp. [°K]
Thorium	$60*10^4$	3.4	-	-
Tungsten	$60*10^4$	4.52	4 - 20	2,500-2,600
Thoriated tungsten	$3*10^4$	2.63	50 - 100	1,900-2,000
Oxide coated	$0.01*10^4$	1.00	100-1,000	1,000-1,200

Richardson-Dushman equation

In 1901 British physicist Owen W. Richardson published the results of his experiments in thermionic emission. 22 years later Saul Dushman derived a modified equation also describing the electron emission from metals as an exponential function of temperature of the emitter.

Although strictly speaking they are two different equations, many textbooks treat them as one, using Dushman's form for the thermionic current density: $J= A_0T^2e^{-b/T}$ [A/cm²] where A is an experimentally determined constant (approx. 60.2 A/m²°K²), T-emitter's temperature in °K (°K=°C+273), e=2.7183 (the base of the system of natural logarithms), $b=11,600*E_W$ is the temperature equivalent of work function and is actually $b = E_Wq_e/k$, q_e is electron charge in coulombs and k is Boltzman's constant (k=1.371 Joule/°K).

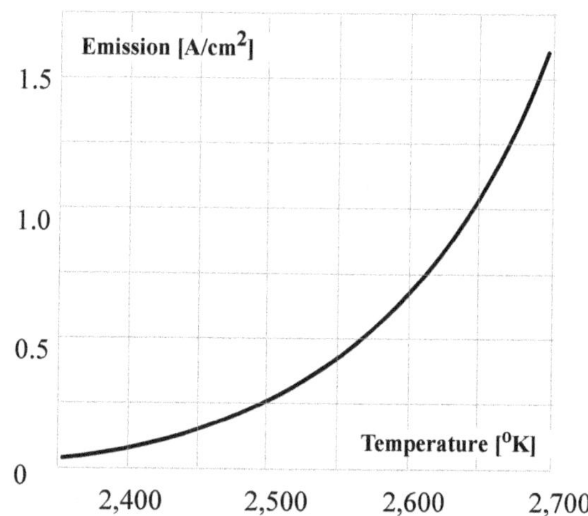

ABOVE: Emission current density versus temperature for a tungsten cathode

Directly- and indirectly-heated tubes

Early vacuum tubes were directly-heated or filamentary type. Typical examples still in use today are 2A3, 300B and 211 power triodes. The heater filament acts as a cathode as well, i.e. it emits electrons itself. In indirectly-heated tubes an insulated heater wire is placed inside a cathode, a metal sleeve that gets energy as heat from the heater (conducted through a ceramic insulator) and the cathode then emits electrons.

Oxide-coated indirectly-heated cathodes and their advantages over the filamentary type

Most vacuum tubes operating below 1kV use oxide-coated cathodes. In indirectly-heated tubes a separate heater heats a cathode coated with a mixture of barium and strontium oxides. The cathode is made of nickel, nickel with a few percent of cobalt or silicon, platinum or Konal (alloy made of nickel, cobalt, iron and titanium).

The heater is a tungsten wire loop coated with a heat-resistant insulating material such as aluminium or beryllium oxide, placed inside the cathode sleeve. This refractory coating conducts the heat to the cathode and insulates it electrically from the heater wire. The oxides have a very low work function and provide copious electrons at temperatures as low as 750°C (dull-red heat).

Oxide-coated cathodes have five important advantages over filamentary ones. Their efficiency is much higher. Secondly, they have a longer life of several thousand hours. SQ (Special Quality) tubes are guaranteed for 10,000 hours. At slightly reduced filament voltages some tubes last 100-200,000 hours!

The third advantage is easier manufacture. The more rigid structure makes smaller mechanical tolerances between electrodes possible, resulting in improved performance.

Fourthly, filamentary tubes are sensitive to vibration and shock, due to delicate suspended filament, so indirectly-heated tubes are much more robust and reliable.

The fifth advantage is the result of the fact that with indirectly-heated cathodes all points on the cathode are at the same electrical potential. This potential remains constant with AC heating and due to the greater bulk and its thermal inertia, there is no appreciable variation in cathode temperature throughout of the AC cycle. That is not the case with filamentary type where there is a voltage drop between one end of the filament and the other.

ABOVE: The internal views of a directly-heated triode (300B). The anode is the dark gray plate on the left photo. Once the anode was removed, the heater (double-M vertical wire) and control grid (horizontally- wound wire) could be seen. Notice the two round getter dishes between the ceramic base and the metal structure.

The space charge or "electron cloud"

A thermionic diode is the simplest vacuum tube with only two electrodes, a cathode and anode. Assuming a filamentary diode, where the heater acts as a cathode as well, once the heater voltage is applied the filament heats up and starts emitting electrons. If there is no positive voltage on the anode, the emitted electrons disperse into the surrounding space and form a cloud or "space charge" around the cathode. The cloud is negatively charged and repels newly-arrived electrons back to the cathode. Soon an equilibrium is established, meaning that as some excited electrons leave the cathode, as many others are forced back to it.

The density of the cloud is the greatest in the immediate vicinity of the filament and reduces with the increased distance from it. Although the parallel plate structure is not used in practical diodes, due to its simplicity it is useful as a teaching tool to explain the behavior of electrons in a vacuum tube (diagram on the next page).

If anode is now connected to a source of positive voltage, it will start attracting negative electrons from the cloud. Those closest to the anode will be swept to it first. As electrons are reaching the anode, others come from the cathode and take their place in the electron cloud. The cloud is not affected by the anode, which is unable to disperse it.

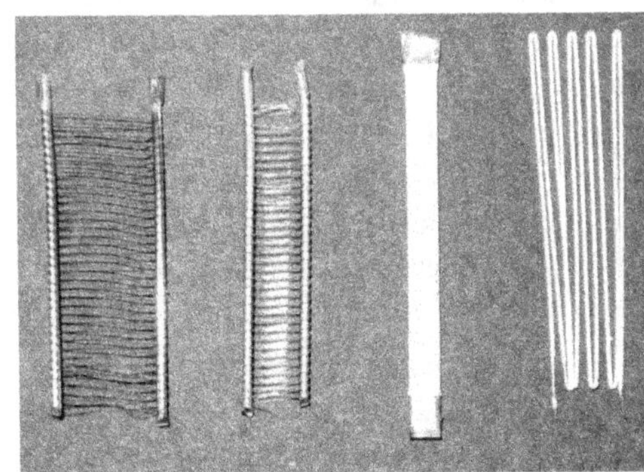

ABOVE:(L-R): The screen grid and the gold control grid with their vertical mechanical support rods. Next is the cathode sleeve and the folded heater filament pulled out of the cathode sleeve of 5881 beam tetrode.

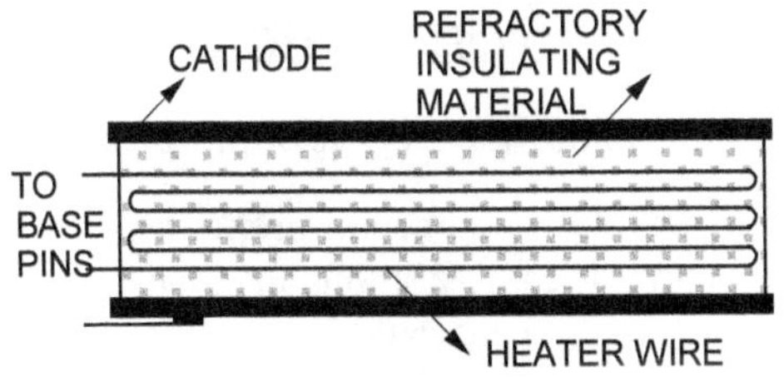

ABOVE: Construction of an indirectly-heated cathode-heater assembly

The electron cloud (space charge) inside a vacuum diode reduces the inter-electrode potential from the linear distribution (dotted line) and causes its nonlinear character. The potential in front of the cathode is actually lower (more negative) then at the cathode surface itself. This potential minimum can be considered a virtual cathode, which retards electrons emitted by the cathode and allows only the faster, more "energetic" ones to reach the anode. This flow of electrons is an electric current through the vacuum, and is said to be space-charge limited.

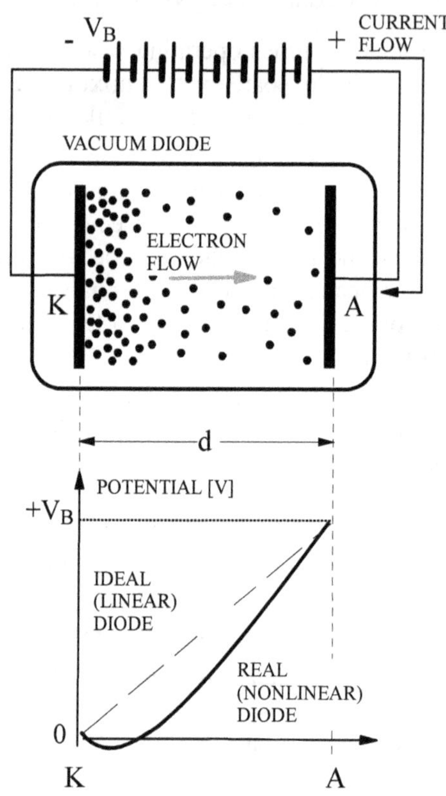

ABOVE: The electron cloud (space charge) inside a vacuum diode reduces the inter-electrode potential from the linear distribution (dotted line) and causes its nonlinear character.

The Child-Langmuir equation or "Three-halves power law"

The physics and the mathematics that describe the behavior of electrons inside a tube are extremely complex and as such are beyond the scope and DIY character of this book, so it will suffice to say that the anode current density J varies as the 3/2 power of the anode voltage. This formula was experimentally derived by both Child and Langmuir, working separately, and is thus called the Child-Langmuir equation or the "three-halves power law":

$\mathbf{J=2.33x10^{-6}\,V_B^{3/2}/d^2}$ [A/m^2], where V_B is the anode voltage and d is the distance between the anode and cathode.

In cylindrical tubes the current is also in the form $\mathbf{I_A=KV_B^{3/2}}$. The constant K depends on the geometry of a specific tube and is called *perveance*. High perveance tubes are characterized by high anode currents at low anode voltages and have a reputation for sounding better than the low perveance ones!

Anode characteristic of a vacuum diode

If we connect a variable DC voltage source across a diode (+ on the anode and -on the cathode) and hook up a mA-meter in series to measure the DC current, with increased V_{AK} the current will increase exponentially. The anode is drawing electrons from the space charge and these electrons are replenished by the cathode. The anode current is limited only by the space-charge and the cathode can produce more electrons than anode can attract.

However, above a certain voltage V_C the rise will become slower, the current has reached the saturation level. The anode has depleted all the electrons from the space charge and the cathode cannot produce enough electrons to reestablish the space-cloud. The anode current is said to be temperature limited. Normally tubes operate in the space charge region.

Theoretically, in saturation, the current should stay constant with increasing anode voltage. The continuing rise, most pronounced with oxide-coated emitters, happens because of the so called Schottky effect. The electric field lowers the work function of the filament and thus additional electrons are released.

In both regions the current is affected by the temperature of the emitter, T2 in the illustration is a higher temperature than T1.

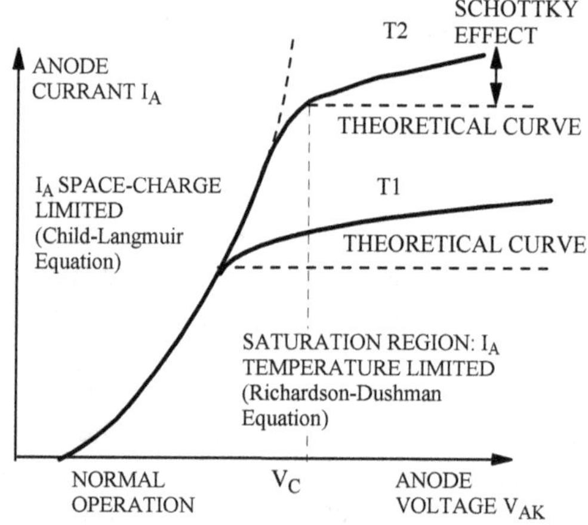

ABOVE: The anode characteristic of a vacuum diode for two different cathode temperatures (T2 > T1).

VACUUM DIODE IN DC AND AC CIRCUITS

Static resistance of a vacuum diode

To illustrate fundamental properties of a vacuum diode, let's connect it to a source of relatively high DC voltage, 200V in this case. Without any load connected, the internal resistance of the diode will limit the current in this circuit and all the power would be dissipated into heat inside the diode.

However, the DC current through the diode may exceed the maximum value declared by the manufacturer and the device may be damaged or destroyed, so we need to connect a load in series with the diode. It does not really matter where we place it, between the anode and the DC source (as illustrated), or between the cathode and the source.

The basic concept of Ohm's law applies here, but Ohm's law assumes linear resistances and here we have a nonlinear diode, whose resistance we don't know.

Moreover, the diode's resistance changes with the magnitude of the current flowing through it. Likewise, the voltage drop across the diode is not linearly proportional to the current. How would we analyze this kind of circuit?

One option is to use the I-V curves for the specific diode, given by its manufacturer, and to find the solution graphically. The other is to simplify things in a prudent way and "linearize" the diode. We will do that soon while performing the AC analysis.

The circuit is a simple voltage divider. Depending on the static resistance of the tube and the load, the DC source's voltage will be split in some proportion between the two: $V_B = V_L + V_A$ ("A" for anode-cathode voltage).

The I-V curve for the diode is given, and on it we draw a "load line" for the resistor R_L. We only need two points to define a line. One is the battery voltage V_B (200V). We mark that point on the voltage axis (point A). The second point can be chosen arbitrarily. Simply choose a voltage and calculate the corresponding current.

For instance, if we move 200V to the left of point A (right on the Y- or vertical-axis), the current in that point B would be I_B=200/25,000 = 8 mA. We mark that value and have our point B. Now we draw the load line for R_L=25kΩ.

Once the R_L line is drawn, the intersection with tube's I-V curve is our operating or "quiescent" point Q. All we need to do now is draw a horizontal line towards the vertical or current axis and read the value of the current in point Q as I_Q = 3.3 mA. The current through the diode tube and the load is the same current $I_Q = I_L = I_A$, so we can call it simply I.

Now we draw a vertical line down to the voltage axis and read the V_A=120V. That is the voltage drop on the diode tube. The rest of the 200V battery voltage is dropped across the load resistor (V_L=200-120 = 80V). We have "solved" our circuit, we know all currents and voltages.

The "static" or DC resistance of the tube R_I in point Q is the voltage drop across it divided by the current"

$\mathbf{R_I = V_A/I_L}$ = 120V/0.0033A = 36,364Ω or approx. 36kΩ!

For each value of the load, the operating point Q will be different. Two additional cases are illustrated, for the load resistor of twice the previous value or 50kΩ and half the previous value or 12.5kΩ (steeper curve).

For increased load resistance the current decreases and the operating point moves to the left, meaning less and less voltage is dropped across the tube and more and more voltage is across the load. As an exercise, calculate the internal DC resistance of this tube in Q1 and Q2.

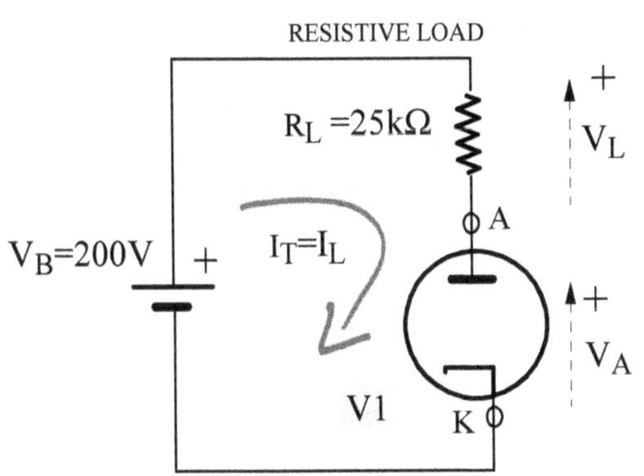

ABOVE: Vacuum diode in a simple series DC circuit with a resistive load

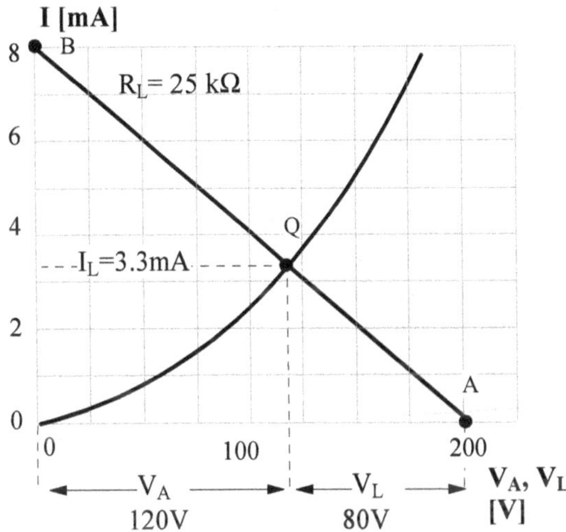

ABOVE: Graphical analysis of the series DC circuit with vacuum diode

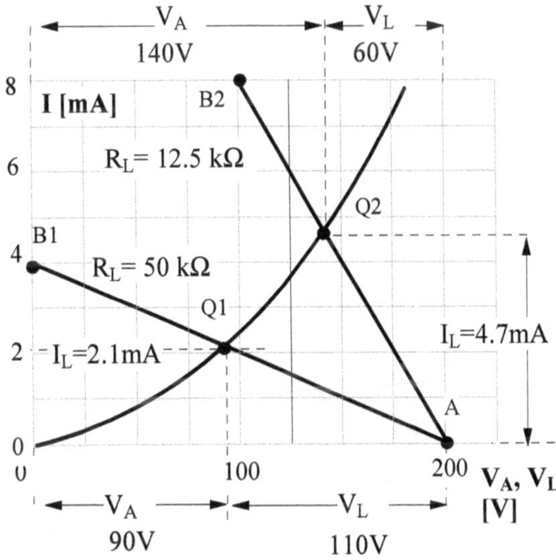

ABOVE: Graphical analysis of the series diode DC circuit for two different load values

Instead of "plate" resistance we will use *internal resistance*. This term may be confusing to some. It is not the resistance of the material the anode or "plate" is made of, but the parameter used to model the behavior of a tube as a variable resistance, when viewed from the outside, from the circuit perspective, as with Thevenin's model.

Linearizing diode's anode characteristics

Since the load resistance is usually much higher than the R_I of a diode tube, linearizing their I_A-V_A curves does not introduce a significant error, as illustrated for 6AX5-GT rectifier. The slope is R_I=120V/300 mA = 400Ω With a power supply of 400V_{DC} and a load current of 200mA the load is R_L=400V/0.2A = 2,000Ω, which is five times higher than the 400Ω internal resistance of the 6AX5 diode.

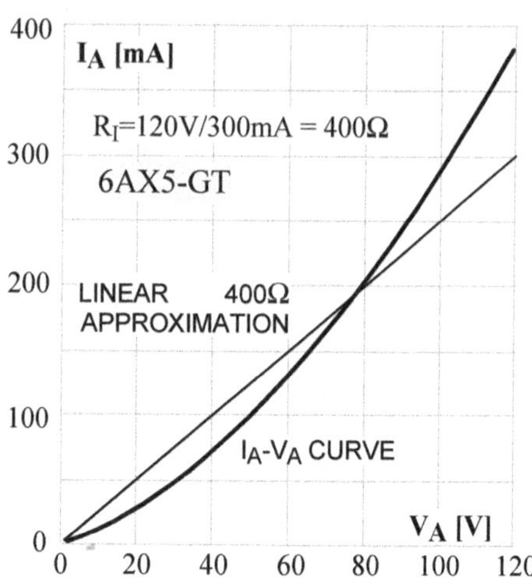

Linearizing a diode curve does not introduce a significant error and can be used for quick approximations in practice. The same applies to triodes (coming soon).

Although diodes can be found in radio and communication circuits as detectors and modulators, and are invaluable in pulse and digital electronics, in audio amplifiers diodes are used mostly as rectifiers, so we will limit ourselves to that application. Let's see how a vacuum diode behaves in alternating current circuits.

Vacuum diode in AC circuits

A vacuum diode V1 is connected across AC voltage source (electro-motive force "e"), and load resistor R_L is connected between its cathode and ground. What will be the waveform and amplitude of the AC voltage on the load resistor?

Again, we can solve this circuit graphically, in a tedious, point-by-point fashion. For each point on the sine wave of the input voltage we would find a projected point, reflected of the diode's I-V curve.

The positive halves of the sine voltage are passed on by the vacuum diode. The valve is fully open and the AC current flows. For negative peaks the diode does not conduct, because the AC signal makes its anode negative with reference to its cathode. The valve is closed and no current flows through it.

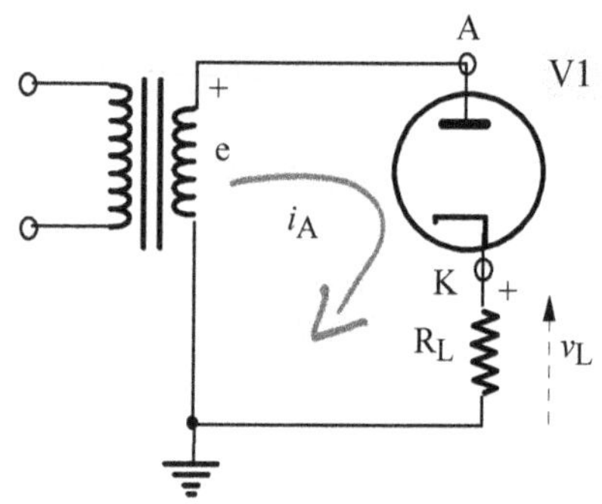

ABOVE: Vacuum diode and its load connected to an AC voltage source.

LEFT: The graphical point-by-point method of constructing the output current's waveform

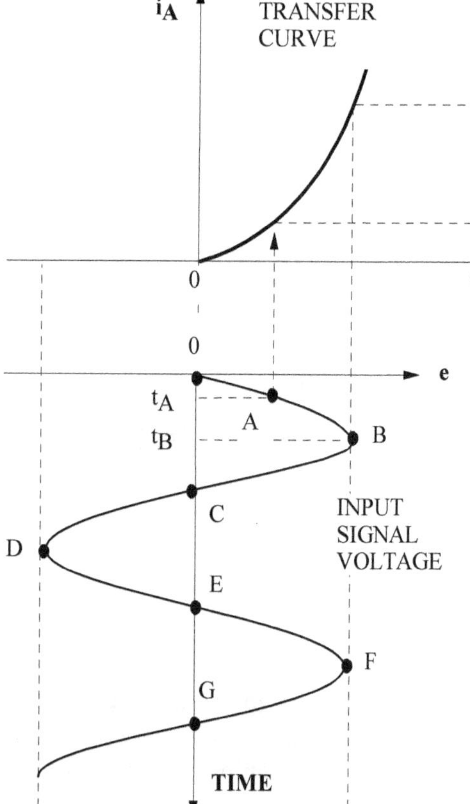

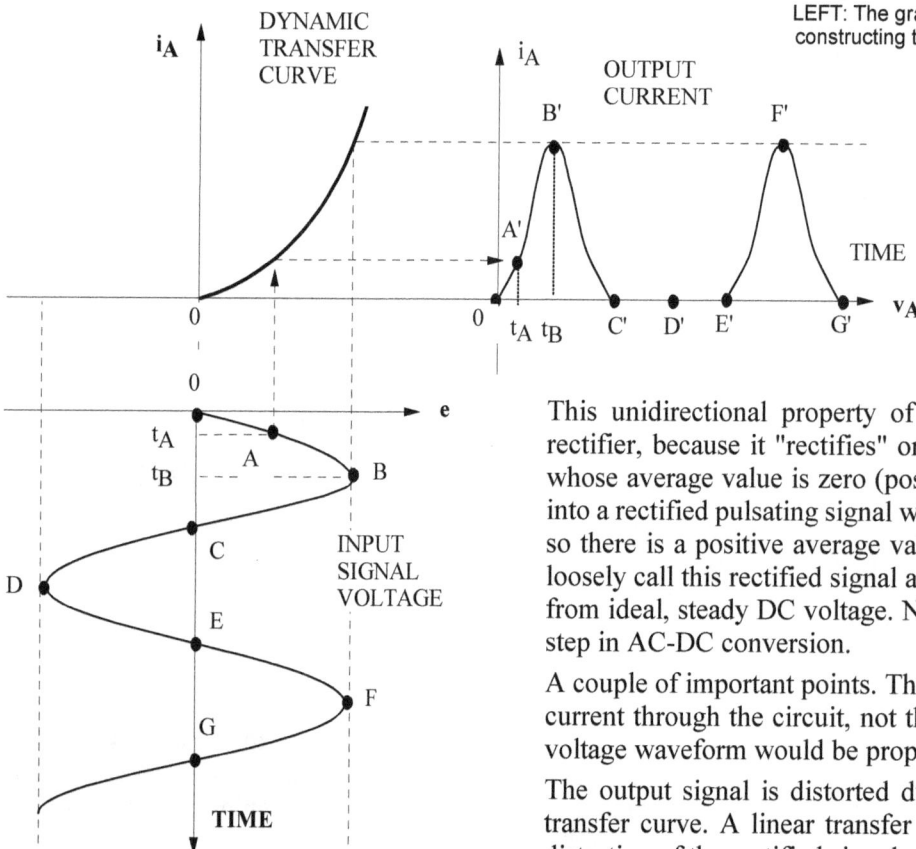

This unidirectional property of the diode makes it behave as a rectifier, because it "rectifies" or "straightens" the input AC signal whose average value is zero (positive peaks cancel negative peaks) into a rectified pulsating signal where there are no negative currents, so there is a positive average value of the rectified signal. We can loosely call this rectified signal a DC current or voltage, but it is far from ideal, steady DC voltage. Nevertheless, it is an important first step in AC-DC conversion.

A couple of important points. The waveform thus constructed is the current through the circuit, not the voltage. For a resistive load the voltage waveform would be proportional to this current.

The output signal is distorted due to the curvature of the diode's transfer curve. A linear transfer curve (a line) would result in no distortion of the rectified signal.

Dynamic resistance of a vacuum diode

A diode's resistance to AC currents, or its impedance, is different from its resistance to DC currents. The dynamic resistance in any point along the I-V curve is the slope of the tangent in that point. Once we draw the tangent, we arbitrarily chose ΔV and ΔI and read their values of the two axis. Greek capital letter "delta", symbol Δ, stands for "difference". We have ΔV = 150-120 = 30V and ΔI = 4.9-3.3 = 1.6 mA

The dynamic internal resistance of the tube is $r_I = \Delta V/\Delta I$ = 30/0.0016 = 18,750 Ω or 18.75kΩ

The symbol for dynamic or AC resistance is a lower case letter r. In this case, the AC internal resistance of the diode is about half its DC resistance (18kΩ versus 36kΩ).

Again, for estimation purposes a nonlinear diode curve can be approximated with a line, which would mean a constant AC and DC resistance across its operating range.

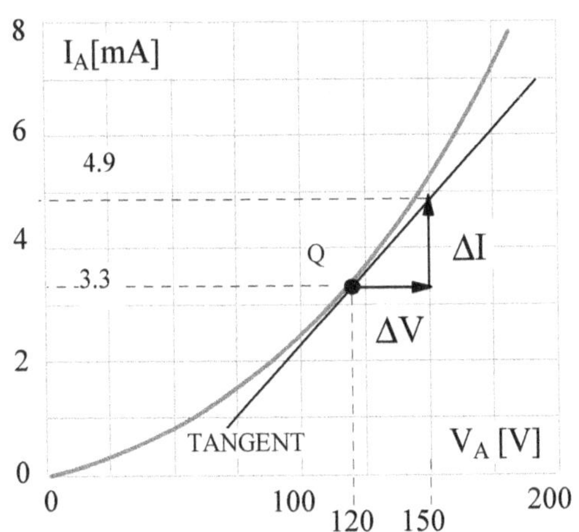

Determining the internal dynamic resistance of a diode from its I-V curve

Constructing a dynamic transfer characteristic

Static curves, published by tube manufacturers, are only valid for DC conditions. To analyze AC circuits we need to convert them into dynamic ones. There is only one static curve for a particular diode, but each load resistance will result in a different dynamic situation and a different curve! This is why manufacturers don't publish dynamic curves. Here is how to do it:

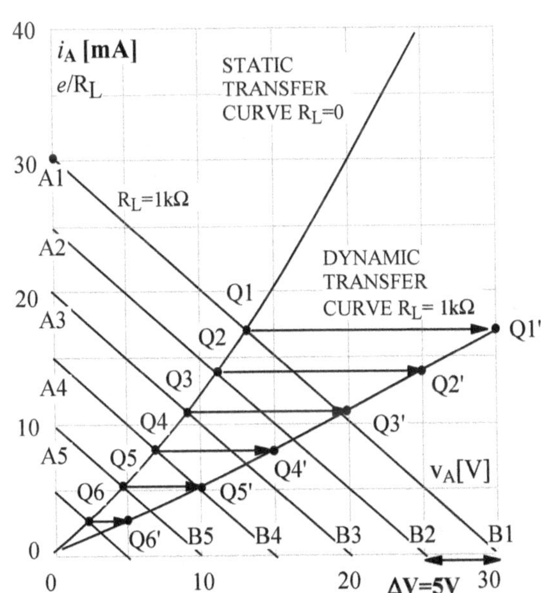

ABOVE: Construction of a dynamic transfer curve

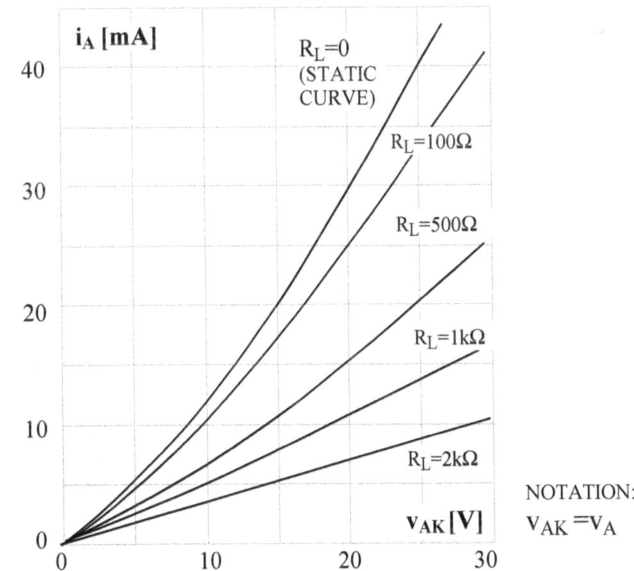

ABOVE: Dynamic transfer curve for various load resistances

1. Draw the load line AB starting with point B1 (V_A=30V) and mark operating point Q1.
2. Find the horizontal projection of the point Q1 onto the vertical line through B1, and mark it Q1'.
3. From B1 move the load line to the left a fixed voltage step ΔV, in this case 5V, towards the lower current/voltage area, to get a new point B2. Find the new intersection with the static curve, new operating point Q2.
4. Find the horizontal projection of the point Q2 onto the vertical line through B2 and mark it Q2'.
5. Repeat the process a few more times to get points Q3' - Q6'.
6. Connect these projected points and you have a dynamic load line for the specified load resistance (1kΩ load resistor in this example).

The curve you will get will not be a perfect straight line, it will still have some curvature. The illustration on the right shows four such dynamic curves for four different load resistances. With increasing load resistance they get more and more horizontal and more linear. This is a very important concept. The diode is not becoming more linear, it is still the same nonlinear diode, but the circuit it is in "behaves" more and more like a linear circuit. Large load resistances "linearize" the circuit so the non-linearity of the diode is less and less of an issue.

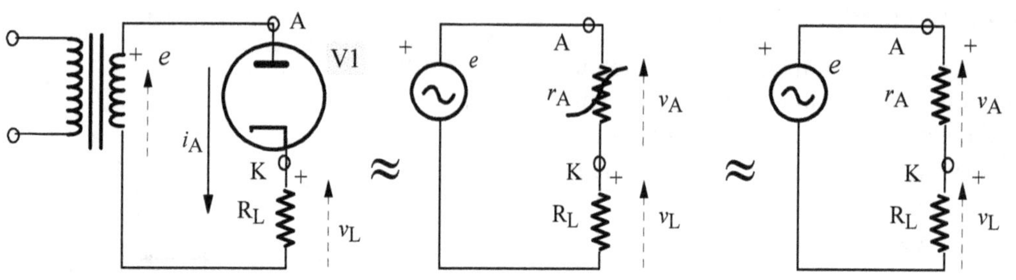

LEFT: Vacuum diode with a load resistor: a voltage divider with one nonlinear resistor (diode) and a linear load, further approximated by a linear voltage divider. The mains transformer was also approximated as an ideal voltage source.

Ideal and real diode

As we will do with triodes, we distinguish between two idealized concepts, the ideal diode and the linear diode. Ideal diode has no internal resistance, it is completely linear, and behaves in a way best illustrated by its I-V characteristic. For all AC voltages at its input, the ideal diode cuts-off their negative half-waves and passes the positive half-waves unchanged. There is no voltage drop V_A across the ideal diode, so its I-V characteristic is a flat line I_A=0 for negative voltages and a vertical line V_A=0 for all positive signal values.

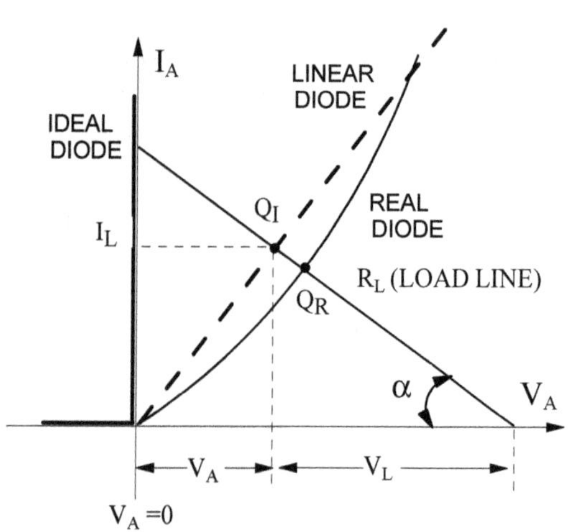

ABOVE: The comparison of I-V characteristics of an ideal diode, linear diode and real diode

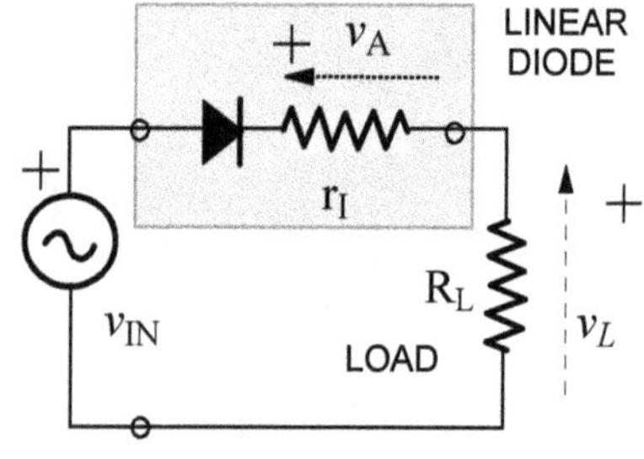

ABOVE: The equivalent model of a linear diode with a load (half-wave rectifier circuit)

This assumption of no internal resistance and no voltage drop across a diode is unrealistic for design use, so we introduce a concept of a linear diode, which is simply an ideal diode in series with its internal resistance r_I.

Now, with a load connected at the output of such a simple circuit, the shape of the positive waveform is still the same, but its value or amplitude is reduced by the voltage drop on the diode (V_A). The angle of the load line with X- or voltage-axis is $\alpha = \Delta I_A/\Delta V_A = 1/R_L$.

The current flowing in this circuit is $I_L = V_{IN}/(r_I + R_L)$.

Linearizing circuits introduces an error, the real operating point Q_R is not the same as the idealized operating point Q_I. The real values of the currents and voltages are also different from the linearized results. This doesn't mean that models, simplifications and approximations don't work and are therefore useless, but that a designer needs to understand their limitations and decide how appropriate such manipulations are for each situation.

Things should be simplified up to a certain point, but no further. Where that point lies will depend on many factors, objective and subjective. Different designers will simplify and model the same real-life circuit differently and get different results.

One graph that is missing below is the output voltage of a real diode. Since the difference between the real and linearized diode I-V characteristic isn't great, it would resemble the output voltage of the linearized diode, but with some distortion of the waveform, which may not be noticeable by an untrained eye and thus would be difficult to depict on such a small size drawing.

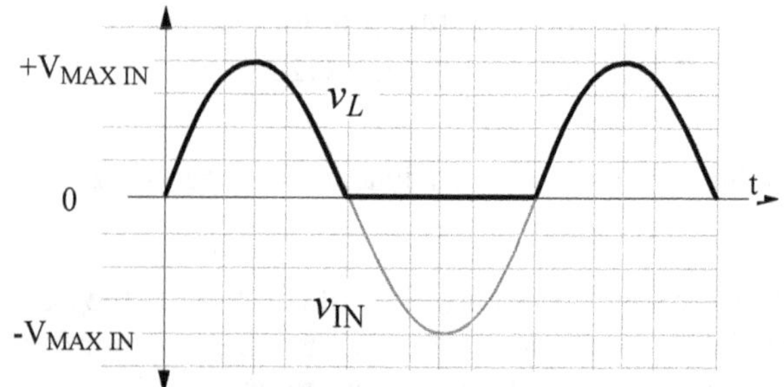

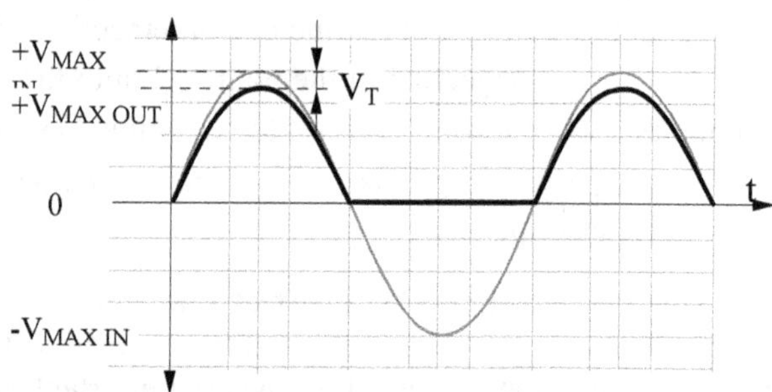

ABOVE: The input sine voltage and the output voltage (half-wave rectified waveform) of an ideal diode (LEFT) and a linear diode (RIGHT).

Anode dissipation

One of fundamental principles in classical physics is the Law of Conservation of Energy, which states that energy cannot be created or destroyed, only transformed from one kind to another.

The analogy principle links mechanical, hydraulic and electrical phenomena. For instance, a body with a mass m, resting at an elevated height h above ground, possesses potential energy $E_P=mgh$, where g is the gravitational constant or acceleration. If allowed to fall, the body gains velocity and just before it hits the ground it reaches a speed of $v=\sqrt{(2gh)}$. This can be derived by making $E_P=E_K$ and expressing v, since all potential energy has been converted into kinetic (motional) energy (conservation principle).

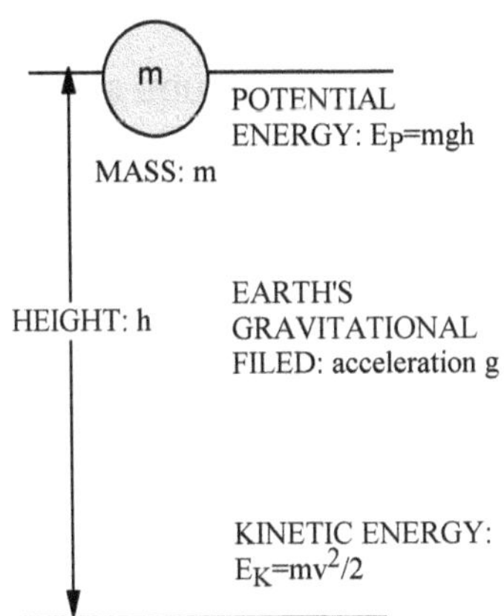

ABOVE: The electromechanical analogy: a body falling from a height h in Earth's gravitational field is akin to an electron traveling from cathode to anode in tube's electrostatic field.

Once the body hits the ground, all the motional energy is converted into heat. This analogy is useful in explaining the energy conversion inside a vacuum tube. Soon we will analyze the anode efficiency of a triode.

The strength of a uniform electric field between two parallel plates (anode and cathode) as illustrated recently, is $E_F=V_B/d$. The kinetic energy of an electron $mv^2/2$ equals the mechanical work done by its transit from cathode to anode, and equals the electrical work done, or $E=E_F*e$, where m is mass of an electron ($9.11x10^{-31}$ kg) and e is electron charge ($1.6x10^{-19}$ coulomb). From here the speed of an electron can be calculated once the accelerating voltage V_B is known: $v=0.593x10^6\sqrt{V_B}$.

This kinetic energy of electrons traversing from cathode to anode and hitting the anode's surface at such a high speed is converted into heat at the anode. Of course, the heat cannot be allowed to accumulate, for that would quickly raise the temperature of the anode's material and eventually melt it away, destroying the tube. So, the heat has to be dissipated from the anode into the surrounding vacuum, through the glass bulb and into the ambient air.

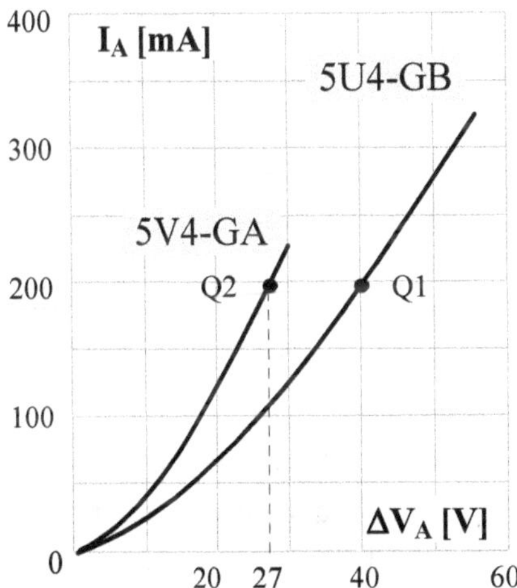

ABOVE: I_A-V_A curves for two vacuum tube rectifiers (duo-diodes), 5V4-GA and 5U4-GB.

For both rectifier diodes and amplifying tubes in steady-state operation, the rate of heat generation equals the heat dissipation away from the anode, so its temperature remains constant and within allowable limits.

In a quiescent point Q (determined by I_A and V_A), the dissipated heat is equal to electrical power and that is $P=I_AV_A$. For instance, directly heated duo-diode 5U4-GB in Q1 dissipates 40V*0.2A=8W.

Indirectly heated 5V4G for the same current (Q2) dissipates 27V*0.2A= 5.4W and is thus a *more efficient* rectifier, since it wastes less electrical energy into heat.

Generally, the larger the area of the anode, the higher the tube's power dissipation rating, which makes intuitive sense, since the speed of cooling is proportional to this area. However, the material the anode is made of also plays a part. Although 300B's anode is larger than the graphite anode of Svetlana's SV572-10 triode, it is only rated at 40 Watts, compared to 125 Watts for SV572-10.

Tube overheating is a serious condition. It releases the occluded gasses from the anode and the glass bulb (mainly hydrogen, nitrogen, carbon monoxide and dioxide) into the inter-electrode space. This gas is ionized by high speed electrons, the tube arcs over and a heavy current of those positive ions starts flowing through sustained gas discharge, causing damage and ultimate destruction of the tube.

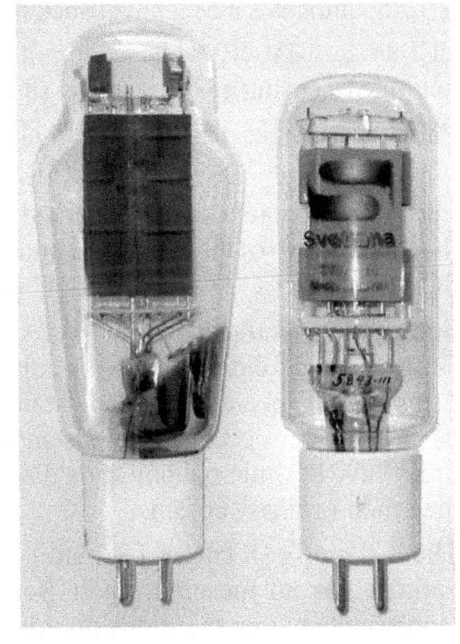

RIGHT: Despite its 20% larger anode surface area, the power dissipation of 300B (on the left) is only 40 Watts, compared to 125 Watts for SV572-10 (tube on the right)! The secret is in the graphite anode of SV572-10. Despite what many think or claim, size does not always matter!

ELECTRON TUBES - NAMING CONVENTIONS, BASES AND PIN NUMBERING

The language of tube electronics

Vacuum tubes, as they are called in United States, are also called "electronic tubes" or "thermionic valves". Strictly speaking, the "high level" name is electron tube, which can then be divided into vacuum tubes and gas-filled or gas tubes. Thus, electron tube is a device in which conduction by electrons takes place in a vacuum or in a gaseous medium within an airtight envelope.

The term "valve" is used mostly in UK and Australian English, and in other languages, like "valvola" in Italian or "válvula" in Spanish. In some countries the term for electronic tube is "lamp", such as "lampa" in Serbian and "lampe" in French. Germans call it "Röhre".

While the term electron "tube" probably comes form the tubular shape of the glass bulb, "valve" most likely originated in the mechanical-electrical analogy that explains the operating principle of the triode. Just as turning the handle on a mechanical valve controls the flow of water (or any gas or liquid) through it, changing the voltage (or bias) on triode's grid controls the flow of electrons through it. The water pressure is analogous to anode voltage, and the water flow corresponds to the flow of the current. Americans use the colloquial term "plate" instead of the proper term "anode". In this book we will use "vacuum tube" and "anode".

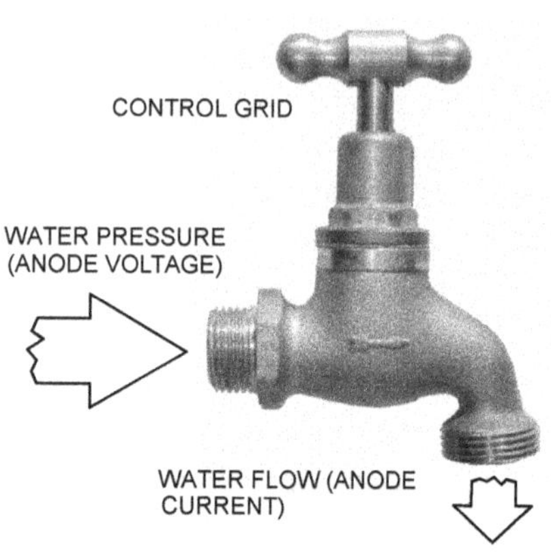

The water tap analogy of a triode explains why vacuum tubes are also called "valves"

Tube naming "conventions"

European tube manufacturers at least tried to follow some kind of naming system. The same cannot be said for their American competitors. Standardization, globalization and similar concepts were not in dictionaries during those halcyon days of electron tube manufacture. European system uses 4 letters followed by 2 or 3 digits. The meanings are in the tables below.

EXAMPLES:

AD1 was a vintage power triode (D) using 4.0V heating (A)

ECC82 means a double triode (CC) with 6.3V heating (E) and using a Noval socket (8).

ECC40 is a double triode (CC) with 6.3V heating (E) and using a RimLock socket (4).

PL519 means a TV output pentode (L) with 300mA heater current (P), "5" indicates a Magnoval socket.

ECF86 is 6.3V (E) voltage amplifier triode (C) and preamplifier pentode (F) using a Noval socket (8).

PY88 is a single diode (Y for "half-wave rectifier"), with 300mA heater (P) using a Noval socket (8).

Special quality tubes where marked SQ and/or named by reversing the order of two of the letters and numbers. For instance E82CC is the SQ version of ECC82, while E86C is an SQ version of EC86.

Some tube manufacturers (and also OEMs such as Hewlett- Packard, Beckman and Tektronix) used a color-coding system, where tips (or "nipples") of specially selected preamp and power tubes would be marked with special heat-resistant red, yellow or blue paint. These tubes have low microphony and noise levels and usually sound better than their ordinary cousins.

The KT prefix (KT66, KT88, etc.) means "Kinkless Tetrode". The number has no meaning, apart from the fact that the higher the number, the higher the anode dissipation of the tube.

The American naming convention is less consistent and not at all self-explanatory. The first number indicates heating voltage, so 12AU7 is a tube with a 12.6V heater, while 6L6 has a 6.3V heater.

1st number	
3	Octal
4	Rimlock
5	Magnoval (large 9-pin)
8	Noval (small 9-pin)
9	Miniature 7-pin
1st letter	
A	4.0V heater
B	180mA heater
C	200mA heater
D	Battery-powered tubes, 1.25-1.4V heater
E	6.3V heater voltage
F	13V heater
G	5.0V heater
P	300mA heater current
U	100mA heater current
O, Z	Cold cathode tubes
2nd, 3rd & 4th letter	
A	Single diode
B	Duo-diode
C	Triode (voltage amplifier)
D	Power triode
E	Tetrode
F	Preamplifier pentode
H	Hexode
L	Output (power) pentode
M	Indicator tube
N	Thyratron
X, W	Gas-filled tube
Y	High voltage half-wave rectifier
Z	High voltage full-wave rectifier

European tube naming system

The last digit indicates the number of electrodes + heater, so 2A3 triode has three elements, one electrode is a heater as well (directly-heated cathode). 6L6 has five electrodes + a separate heater, for a total of six elements. 12AU7 has six electrodes (two triodes) and a common heater, thus 7 in total.

Then there are "numerically-named" tubes such as 6922, 5687 and many others, which are usually the "industrial" versions of their consumer-aimed equivalents. These numbers have no meaning and that is why I titled this section naming "conventions" instead of naming "standards". As different tube manufacturers developed new tubes, they simply named them at their convenience without any regard for standardization.

Not to be outdone, the USA military established their own naming, where numbers are usually preceded by the acronym JAN, which stands for "Joint Army Navy".

Tube bases and pin numbering

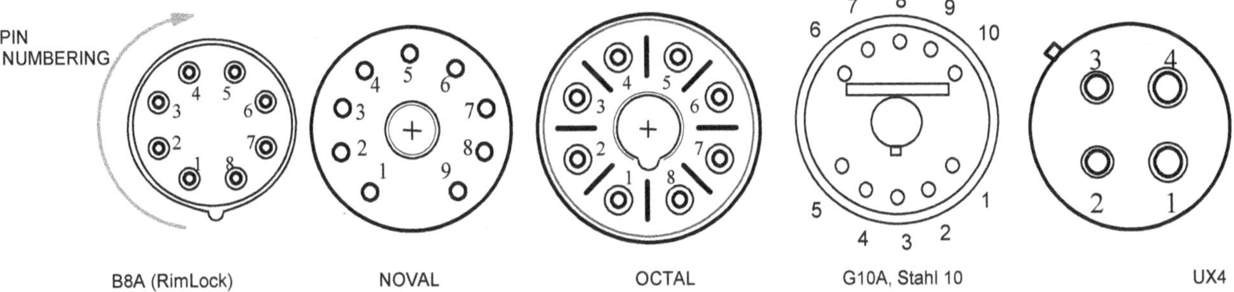

No matter what base a tube uses, the pins are always numbered in a clockwise fashion if viewed from underneath (pins facing you). The same view is of their sockets if viewed from the lug side (where components are soldered).

CONSTRUCTION & MANUFACTURING OF AN ELECTRON TUBE

The physical construction of a directly-heated triode

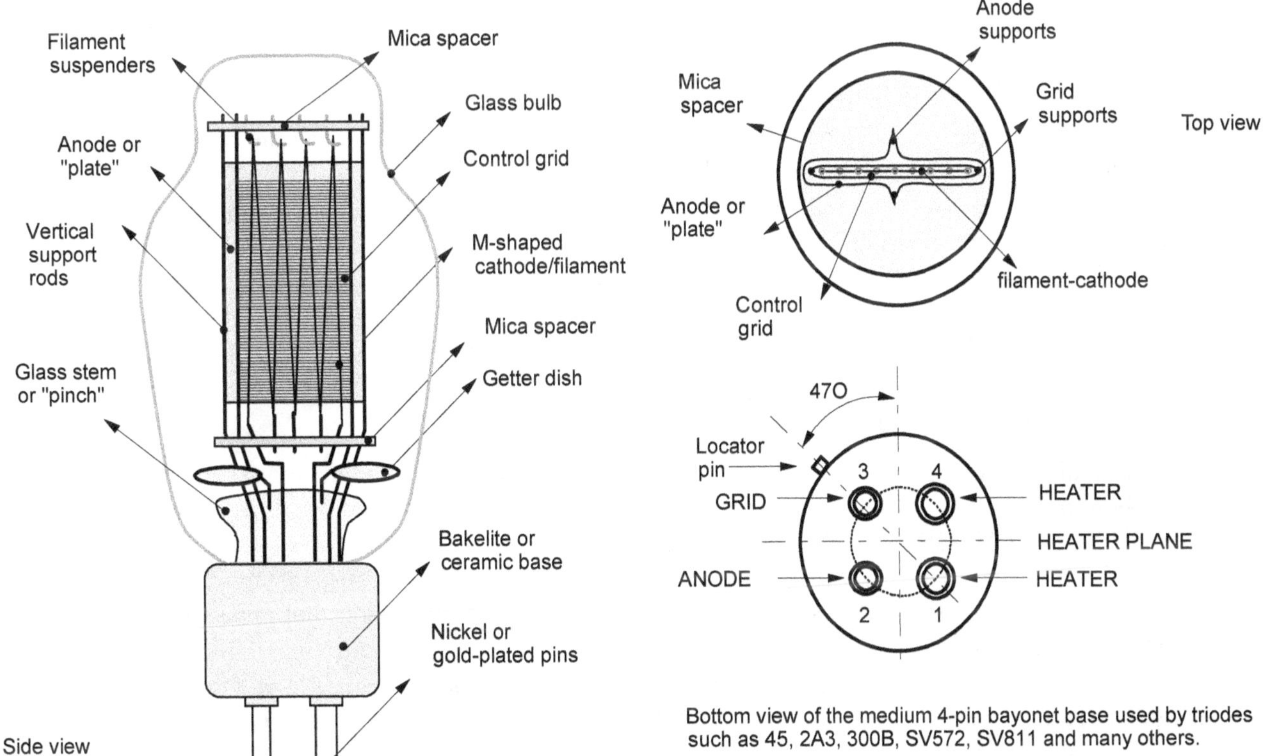

Bottom view of the medium 4-pin bayonet base used by triodes such as 45, 2A3, 300B, SV572, SV811 and many others.

The evacuation process and flashing the getter

Once a tube is put together and mounted inside a glass bulb, air is exhausted from the tube using a vacuum pump. The glass and the metal parts are outgassed by baking the tube in an oven. This is not done for individual tubes, but a whole batch at a time. The metallic structure is further heated in a high-frequency induction furnace. A strong magnetic field induces eddy currents in the metal parts and such currents heat the metal parts quickly.

As the bulb is being evacuated, a small amount of barium compound (barium with aluminum or magnesium) sits in a little dish or two. As the tube is sealed off from the vacuum pump, an inductive heater "activates" the getter. The strong magnetic field "flashes" the getter, which spreads out of its dish and forms a silvery film deposit on the inside of the glass.

During this process the getter bonds or chemically combines with molecules of various gases such as oxygen, hydrogen, nitrogen, carbon monoxide and carbon dioxide. The getter stays active and maintains the vacuum throughout the life of the tube. It is not effective against inert gases such as argon or helium.

RIGHT: Although of the same vintage and taken out of the same push-pull amplifier, these two Philips 6550 tubes show different getter condition. The milky-gray getter on the left has almost disappeared, while the one on the right, although of much smaller diameter compared to its original size, is still black and has more life left in it.

Activating the cathode

Once a batch of tubes is finished, their oxide-coated cathodes still have low emission and must be "activated". This process involves operating tubes' heaters at higher voltages for several minutes, to achieve cathode temperatures way above normal. In the second step the heater voltage is lowered and anodes are connected to anode voltage source so anode currents are flowing. During this stage anode currents will suddenly rapidly rise to a higher value, indicating increased emission from the oxide layer and the end of the activation process.

The activation process for filamentary-type tubes requires three steps. The filament is heated to about 2,500°C for a few minutes. This high temperature cleans the tungsten surface and reduces some of the thoria inside tungsten into metallic thorium. The filament temperature is then maintained at around 2,100°C for 30 minutes. The thorium is still diffusing to the surface at that temperature until a one molecule thick adsorbed layer of thorium covers the surface of the filament. In the final step the filament temperature is further reduced to its normal operating range, between 1,500°C and 1,700°C.

Two main causes of reduced emission - cathode poisoning and high resistance interface layer

The quality of vacuum inside a thermionic tube is the principal factor affecting its life and performance level. Residual gas left in tubes with oxide-coated cathodes attacks the oxide layer by oxidizing the barium and causing a serious reduction in emission. This phenomenon is known as *cathode poisoning.*

Cathode's emission in those tubes is also reduced by the development of a high resistance interface layer between the cathode's metal substrate and the barium-strontium oxides on its surface. The main cause of this layer's formation is the presence of silicon and other impurities in the cathode's material. This fault is prevented by making cathode out of platinum or 4% tungsten-nickel alloy with low levels of impurities. Leaving tube equipment powered up without any signal seems to cause or at least accelerate the development of this interface layer.

WHY IS THERE A BLUISH GLOW INSIDE SOME VACUUM TUBES?

There are three main types of bluish glow. Fluorescence is usually located close to the inner surface of the glass bulb, and is most noticeable in power tubes. It is believed to be caused by stray electrons striking the glass coated with getter material.

The ionization of gas residues also produces a similar glow, but located within the metal structure and not around the glass surface, so the two should not be confused. Fluorescence does not impact tube's performance, but the gas ionization may.

Finally, a blue-violet glow in mercury vapor rectifier tubes and gas-filled tubes such as Thyratron, voltage regulator and voltage reference tubes is normal. Voltage regulator (VR) tubes using neon instead of argon glow a pink-orange color.

TRIODE PARAMETERS AND CHARACTERISTICS

Triode in 3D

Since there are three variables of interest, grid voltage, anode voltage and anode current, triode's behavior is a contour in 3-dimensional space. Dealing with a 3D curves on 2D paper is awkward, it is much easier to draw two variables in two-dimensions, with the third variable as a parameter.

There are three possible triode graphs. The I_A vs. V_A graph with V_G as a parameter is also called anode or plate characteristics. The I_A vs. V_G graph with V_A as a parameter depicts one or more transfer curves, and V_A vs. V_G graph with I_A as a parameter is called "constant current characteristics" (since I_A is kept constant, which is the meaning of "parameter")!

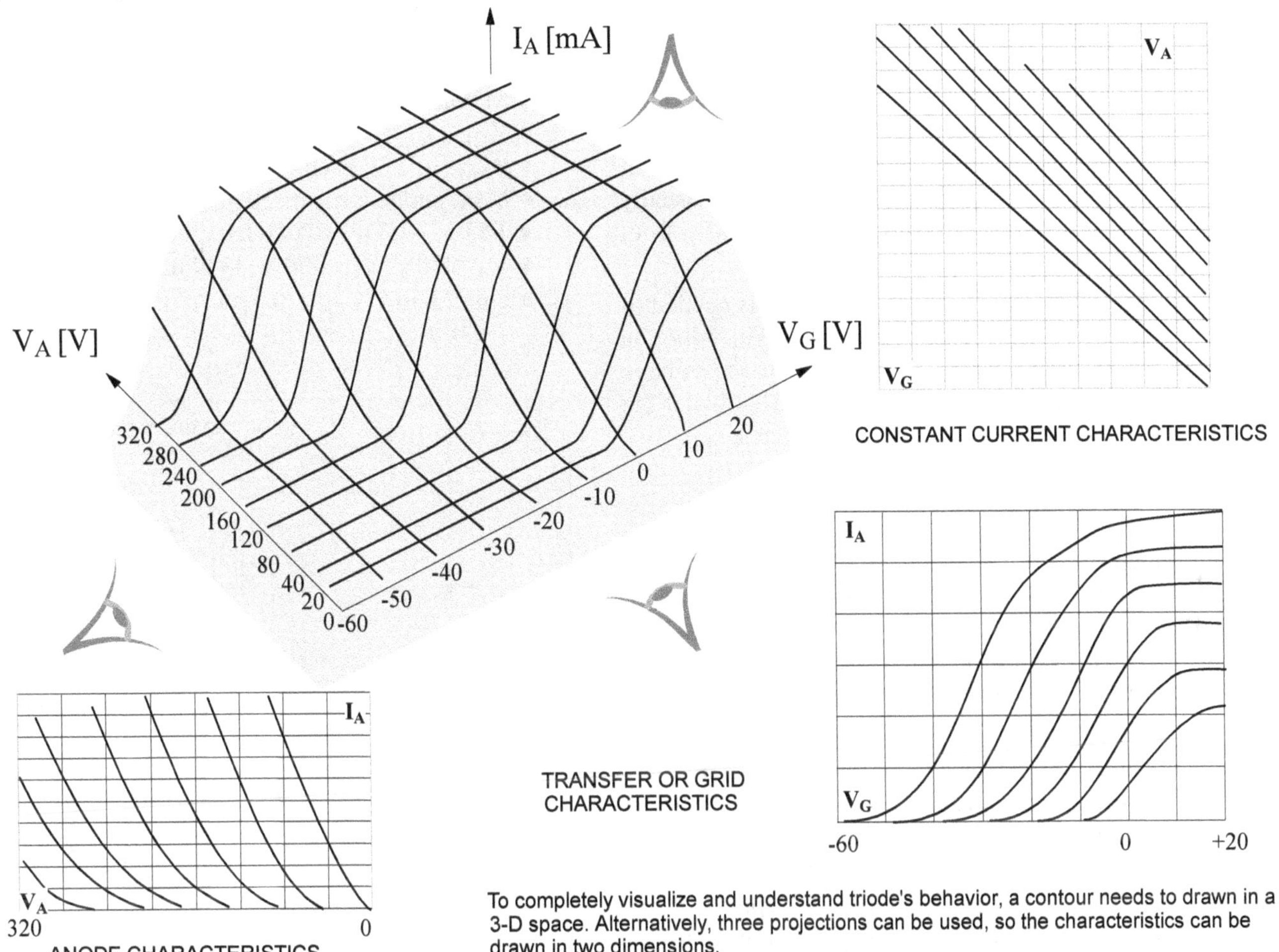

To completely visualize and understand triode's behavior, a contour needs to drawn in a 3-D space. Alternatively, three projections can be used, so the characteristics can be drawn in two dimensions.

Current control by triode's grid, amplification factor and "reachthrough"

We have seen that the current through a vacuum diode is a function of one variable, the anode voltage: $I_A = KV_A^{3/2}$. In triodes, apart from depending on the anode voltage, the anode current is a function of another variable as well, the voltage on the control grid: $I_K = K(V_G + DV_A)^{3/2}$

Since $I_K = I_G + I_A$ (cathode current is a sum of grid and anode currents), and in most circuits (but not all!) the grid current is negligibly small compared to anode current and can be considered zero, this equation is usually written as $I_A = K(V_G + DV_A)^{3/2}$, but that is only valid if $I_G = 0$!

The constant K is called perveance, as in a diode, and the constant D comes from German noun *Durchgrief* meaning "reachthrough", since the anode voltage and its electric field has to reach through the grid to "get" electrons from the cathode. It is a measure of the degree of anode's control over the current, its "effectiveness".

The concept is not used in English literature at all, Anglo Saxons for some reason exclusively use the inverse concept called voltage amplification factor, for which Greek letter μ (pronounced mju) is used. Since **$\mu = 1/D$**, the Anglo literature writes this equation as $I_A = K(V_G + V_A/\mu)^{3/2}$, or, more often, they use the alternative way of expressing anode current, $I_A = A(\mu V_G + V_A)^{3/2}$, where constant A loses its meaning, it is not perveance any more.

For instance, the D-factor for LS50 power pentode is 20% or 0.02, meaning its amplification factor is 1/0.02 = 5. For 12AX7 triode, with μ=100, D is 1/100 = 0.01 or 1%. These differences can be interpreted in a sense that anode current in 12AX7 is much more sensitive to grid voltage changes (20 times more sensitive, since its μ is 20 time higher!) than for the LS50 pentode.

Using the reachthrough figures for interpretation, in LS50 with D=20%, 20% of anode current changes are due to the anode voltage influence and 80% as a result of grid voltage changes, while in 12AX7 triode, only 1% of anode current changes come from changes in the anode voltage and 99% result from the grid action, meaning that the grid is 20 times more effective in the 12AX7's case.

The dame of voltage amplification, 6SN7 duo-triode

6SN7 duo-triode is highly regarded by audiophiles. Let's use it as an example to explain three most import triode parameters and three types of triode characteristics.

Transfer characteristics show how anode current I_A changes with a DC bias on the control grid (V_G), with anode voltage V_A as a parameter (kept constant). There is a different curve for each value of V_A, so we are really talking about an infinite number or a family of curves. Manufacturers usually publish at least two or three of them. We have six here, from 50V to 300V.

As V_A is increased, a higher negative grid bias is needed to keep the anode current I_A at the same level (draw a horizontal line through all six curves). The anode voltage has a large impact on triode's anode current. That is not the case for tetrodes and pentodes as we will see later.

TUBE PROFILE: 6SN7GT - 12SN7GT

- Indirectly-heated medium μ dual triode
- Octal socket
- Heater: 6.3V, 0.6 A - 12.6V, 0.3A
- Maximum anode voltage 450 V_{DC}
- $V_{HKMAX} = 100\ V_{DC}$
- P_{AMAX}: 5 W each triode, 7.5 W both
- Max. anode current: 20 mA
- TYPICAL OPERATION:
- V_A=250V, V_G=-8.0 V, I_0=9 mA
- gm=2.6 mA/V, μ=20, r_I = 7.7 kΩ

Transconductance or mutual conductance

From introductory calculus, a derivative "d" of a function is the slope of a tangent to that function in a particular point. Most DIY constructors would rather have a root-canal job than anything to do with calculus, so in the first approximation the derivative "d" can be replaced by a small difference, for which Greek capital letter Delta is used (Δ).

Transconductance or mutual conductance (gm in English and S in many other languages, from Steilheit in German) is defined as a change in I_A (anode current) caused by a small change in V_G (DC voltage on control gird or "bias" voltage): **$gm=\Delta I_A/\Delta V_G$** That would be the slope or steepness of the transfer curves in any chosen point, thus justifying the use of letter S.

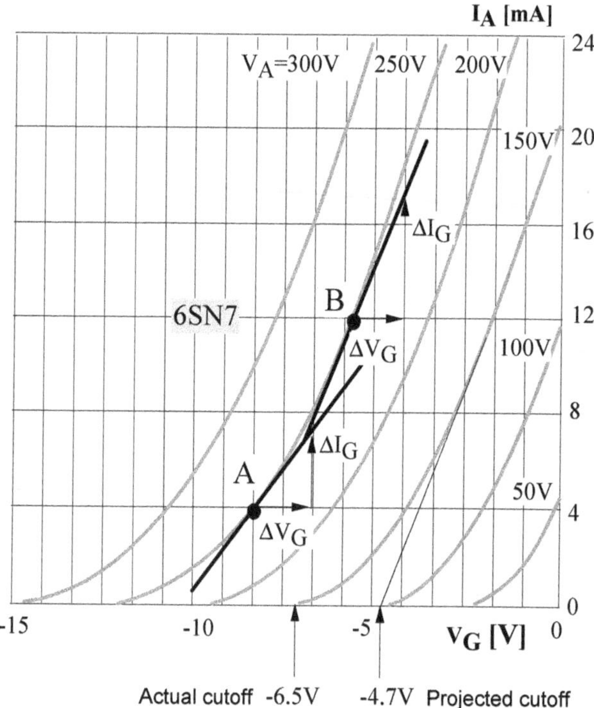

RIGHT: The transfer curves for 6SN7 triode: determining mutual conductance gm. The definitions of the actual and projected cutoff points are also illustrated.

Notice how transfer characteristics aren't straight but curved, especially at low anode currents. This means that the slope of a tangent will vary with anode current, which then implies that gm will not be constant, but will vary from low values at small anode currents, to high values at large I_A! This causes a great deal of angst and confusion amongst the users of tube testers. Each model uses a different anode and a different bias voltage during tube tests, meaning they test the same tube in different points along the transfer curve or in totally different transfer curves! Thus, their test results vary widely! More on that issue in Volume 2 of this book.

Let's calculate mutual conductance in two operating points, A and B on the same curve, V_A=250V. In A, we move 2 divisions to the right (a change of ΔV_G = 2 V in bias, from -10 to -8V) and then go upwards until we hit the tangent drawn in point A. We read ΔI_A on the Y-axis of around 3.2 mA (a jump from 3.8 to 7.0 mA). Now we have gm = $\Delta I_A/\Delta V_G$ = 3.2/2 = 1.6 mA/V. In point B let's move 1.7 V to the right and get ΔI_A of 5 mA (a jump from 12 to 17 mA), so gm= $\Delta I_A/\Delta V_G$ =5/1.7 = 2.94 mA/V!

Since our increment ΔV_G was large, this is not a very precise way of determining gm, but is quite adequate for quick estimation and even for design purposes.

Mutual conductance of any active or amplifying device is proportional to its output current. For a bipolar silicon transistor gm is proportional to the collector current, $gm \propto I_C$, for a FET it is proportional to a square root of drain current ($gm \propto I_D^{1/2}$), while in a triode it is proportional to a third root of the anode current ($gm \propto I_A^{1/3}$).

The bipolar transistor has the highest gain of all three, but its maximum input voltage is very low, in the order of 10mV. FETs and triodes can take input signals of 1-10V or more!

Actual and projected (extended) cutoff

If we connect a triode to a fixed positive anode voltage (in this case +150V) and vary the negative DC voltage on its grid, we will get a transfer curve for that voltage. At some value of grid voltage called *cutoff grid voltage* the anode current will cease or be cut off completely (-6.5V on the graph).

In some applications, for instance when biasing push-pull output stages of power amplifiers operating in Class B, it is helpful to use projected or extended cutoff voltage value (in this case -4.7V). This value is obtained by extending the linear part of the transfer curve down until it reaches the x-axis or abscissa ($-V_G$ axis).

Since $I_A = K(V_G + DV_A)^{3/2}$, for that current to be zero the term in the brackets must be zero. Thus, $-V_G$ must equal DV_A or $-V_G=V_A/\mu$! In the illustrated case $V_A/\mu = 150/\mu = 4.7$, so $\mu = 150/4.7 = 32$ This means we can estimate the amplification factor μ from the projected cutoff value!

Triode's amplification factor μ and the extremely important Barkhousen's equation

The tube amplification factor is defined as the change of anode voltage divided by the change of grid voltage that caused it: $\mu=\Delta V_A/\Delta V_G$ The two identical units (volts) cancel one another, so μ is dimensionless factor.

In point C (I_A-V_A graph below), to estimate μ we choose a jump in grid bias V_G and red the corresponding difference in anode voltage (ΔV_A). In this case $\mu = \Delta V_A/\Delta V_G = (140\text{-}100)/2 = 40/2 = 20$

Now that we know how to read the tube parameters from the graphs, the good news is that usually we don't have to bother doing it all. In most cases, at least for tubes intended for audio applications, manufacturers had published the μ-gm-r_I graphs.

Notice how μ (amplification factor) is constant for almost the whole range of anode currents, except the very low current levels. In that low current region the internal resistance of the tube shoots up, while gm drops rapidly off as the anode current drops. Normally, a designer would avoid using a tube in that region. Even from the relationships of the three curves, we can conclude that gm multiplied by r_I equals μ or **$\mu=gm*r_I$**

This equation is called the Barkhousen's equation. Remember it well, we will use it very often!

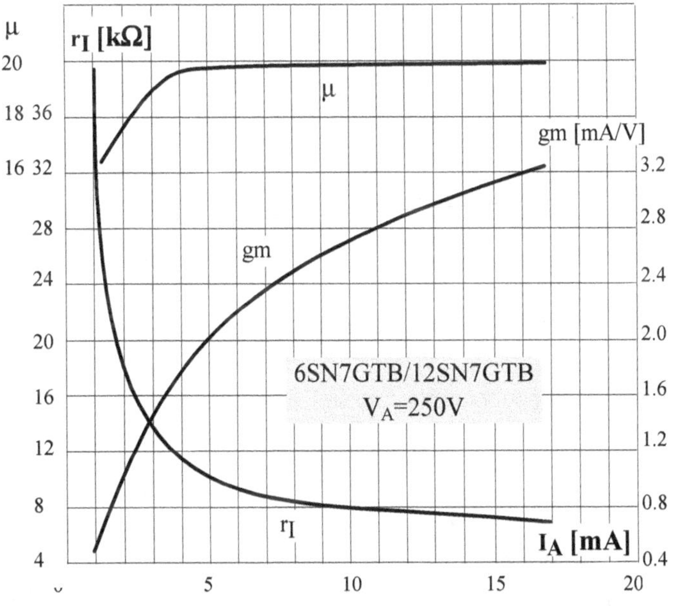

ABOVE: How the three tube parameters vary for 6SN7 triode

Internal resistance of a triode

6SN7 anode curves describe the relationship between anode current and the voltage drop across the tube (between its anode and cathode). The slope of the curves is inversely proportional to the internal resistance of the tube, and we can determine such resistance graphically from the curves, just as we have done for the mutual conductance.

The slope is low at low anode currents, meaning the internal resistance is high, and as anode current goes up, the slope rises and the internal resistance falls. Let's choose a couple of points at the same anode voltage as before (250V).

In A: $r_I = \Delta V_A/\Delta I_A = 50/4.2 = 11{,}900\ \Omega$

In B: $r_I = \Delta V_A/\Delta I_A = 25/4.1 = 6{,}100\ \Omega$

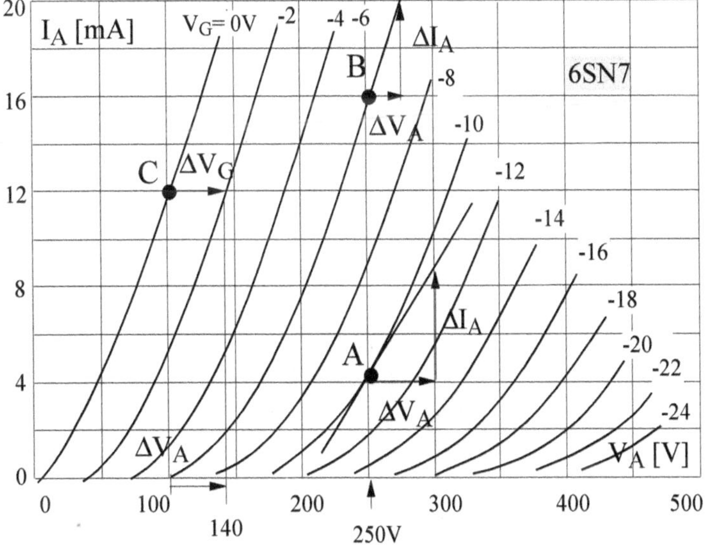

RIGHT: Estimating 6SN7 triode's internal resistance r_I in points A and B, and its amplification factor in point C, using its anode characteristics.

Constant current curves

These aren't commonly published or as often used as the other two types of characteristics. Three parameters mean two degrees of freedom (N-1), so choosing two automatically determines the third and designers prefer to use the other two graphs.

For instance, if we want to position the quiescent point of a 6SN7 triode stage at 200V anode voltage and want 8mA of DC anode current, from 6SN7 CC curves we go across from 200V to the left until we hit the 8mA curve. That is our quiescent point Q. Then we go perpendicularly down until we hit the V_G axis , on which we read the required grid bias voltage, in this case -6V!

CC curves are ideally equidistant straight lines, and real curves come pretty close to that. Their slope is the amplification factor ($\Delta V_A/\Delta V_G = \mu$), which remains constant except for the region of very low anode currents.

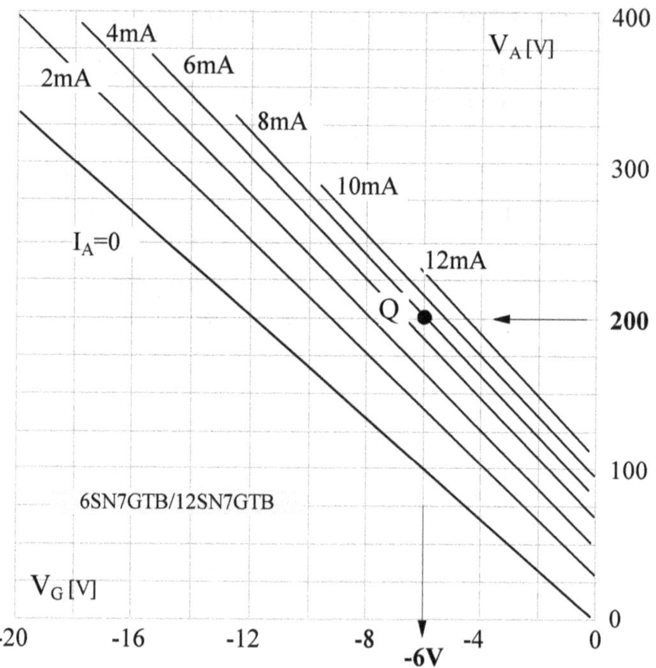

ABOVE RIGHT: Constant current characteristics of 6SN7 triode

Symbols and examples of real triodes

A directly heated triode is the simplest amplifying vacuum tube. The same element acts as a heater and a cathode, only one grid and a metal plate (anode).

When naming tubes only "active" elements are considered , so a triode has a cathode, a control grid and an anode, while for instance a separate heater in indirectly-heated tubes is not counted (considered an auxiliary element). A tetrode would have 4 active elements, a pentode five, a hexode six, a heptode seven, etc.

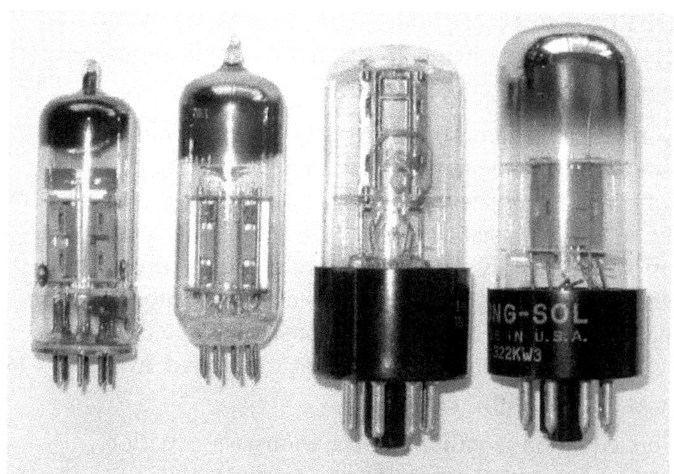

ABOVE: Four good sounding duo-triodes (L-R): ECC40 (Rimlock base), 6N6P (Noval base), 12SN7GT and 12SL7GT (Octal base). The first three have higher anode dissipation ratings but lower μ, 12SL7 has a higher μ but a lower anode dissipation rating. Notice the differences in anode (plate) sizes!

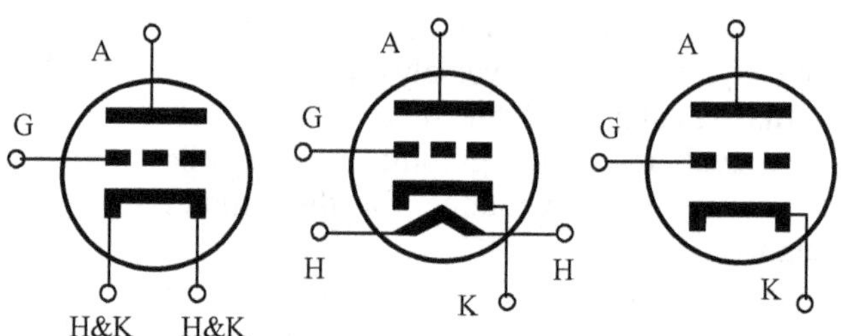

ABOVE: Symbols for a directly-heated (filamentary-type) triode (left) and indirectly-heated triode (middle). The heater of indirectly-heated tubes is usually omitted from the symbols used in circuit diagrams (right) for clarity, it is assumed that it is connected to a suitable power source.

VOLTAGE AMPLIFICATION WITH TRIODES: THE COMMON CATHODE STAGE

- TRIODE AS AN AMPLIFIER
- COMMON CATHODE AMPLIFYING STAGE - GRAPHICAL ANALYSIS
- COMMON CATHODE AMPLIFYING STAGE - ANALYTICAL METHODS
- POWER RELATIONSHIPS AND EFFICIENCY OF THE COMMON CATHODE STAGE
- REACTIVE LOADS AND THE LOAD "LINE"
- AC AND DC INTERSTAGE COUPLING
- DIRECTLY-COUPLED AMPLIFIERS

"Any sufficiently advanced technology is indistinguishable from magic."
Arthur C. Clarke, English science fiction writer

6

TRIODE AS AN AMPLIFIER

Three possible ways to use a triode as an amplifier

Since a triode has three active electrodes, it can be connected in three different ways. One of the electrodes is common to both the input and output circuits. Notice a pair of terminals at the input circuit and the output circuit, with a common or shared terminal. The common terminal is grounded for AC signals, but it can be at any DC voltage. This statement is profound and very important for understanding of tube circuits!

The cathode bypass capacitor C_K in grounded cathode amplifier is a short circuit for AC signals, thus grounding the cathode K (zero AC potential). The DC voltage on the cathode is used as bias voltage, so for DC currents and voltages ("conditions") cathode K is NOT at ground potential! Suitable DC voltages on all electrodes are necessary for proper operation of a triode as an amplifier of AC signals. Thus DC conditions have to be designed, analyzed and optimized first!

The coupling capacitor C_C (called *coupling* since it connects or couples this stage to the following stage or external load) is also a short circuit for AC signals, as is the power supply $+V_{BB}$. It is assumed that the internal resistance of that power supply is zero (an ideal power supply) and "zero resistance" is another way of saying "short circuit"!

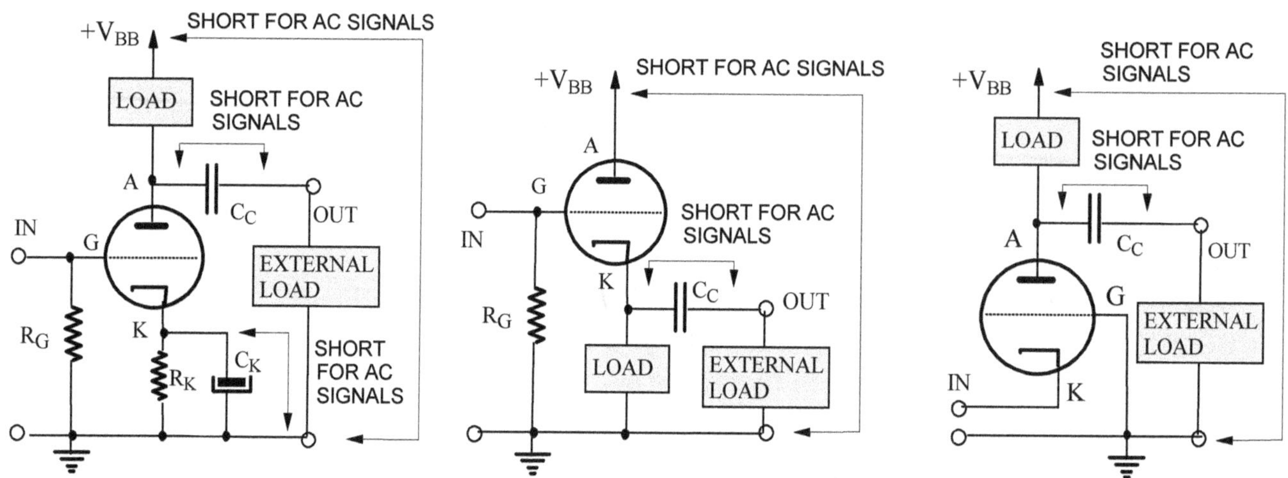

ABOVE (L-R): Common cathode (CC) stage, common anode stage, a.k.a. cathode follower (CF) and common grid stage (CG)

The meaning of input- and output-circuits and the "load" concept

There are two circuits in all amplifiers, the input and the output circuit. In the CC circuit, the input is at the grid, so the input circuit is the grid-cathode circuit. The output is taken from the anode, so the output circuit is the anode-cathode circuit. In the CF circuit, the input is at the grid, so the input circuit is the grid-anode circuit. The output is taken from the cathode, so the output circuit is the cathode-anode circuit. In the CG circuit, the input is at the cathode, so the input circuit is the cathode-grid circuit. The output is taken from the anode, so the output circuit is the anode-grid circuit.

In any case, the "load" is *always* connected in the output circuit! In audio tube amplifiers the load can be a resistor, an inductor (choke), a constant current source or sink, a transformer or a loudspeaker. Notice that "loads" are always present, they are part of the amplifying stage or an amplifier, while external loads may or may not be present. For instance, a preamplifier without any amplifier connected as a load will still operate, it will still amplify signals from your CD player for instance, so an external load is not necessary for its operation.

In CC and CG amplifiers the load is *only* in the output circuit, but in the case of cathode follower the load is in both the input and output circuits. This leads to a specific kind of interaction between the input and output circuits in a cathode follower, called negative feedback from the output to the input, which we will investigate very soon.

Transistor or "TRANSfer resISTOR"

The name transistor is a short for TRANSfer resISTOR. It is also a 3-terminal device (emitter, base and collector), so all concepts discussed so far apply to transistor circuits as well. The "transfer" of resistance is a mildly awkward way of saying that a transistor amplifying stage can have a low input and high output impedance, or a high input and low output impedance, or anything in between. A triode is also a "transfer resistor. CF has a high input and low output impedance, while CG circuit has a low input and high output impedance. CC stage has a medium-to-high input and low-to-medium output impedance. So, impedance "transfer" or change is an important quality of tubes as well, sometimes utilized as a benefit, at other times dealt with as an undesirable but unavoidable reality.

For instance in power amplifiers the high output impedance of tubes forces us to use awkward, bulky and expensive audio output transformers to match low impedance loads (loudspeakers) to high impedance drivers (tubes).

Now that we understand the basic concepts of the three basic amplifying stages, let's start with the most common one, the CC stage.

COMMON CATHODE AMPLIFYING STAGE - GRAPHICAL ANALYSIS

The biasing concept

With no DC voltage difference between its grid and cathode, a common cathode stage would behave as a rectifier - half of the input signal would simply be cutoff and the output signal waveform would be materially different from the input waveform (as on p74).

For this amplifying stage to behave as a linear amplifier the grid potential needs to be negative with respect to the cathode. This negative voltage between the grid and cathode is called the bias voltage, and it can be implemented in many ways. The simplest to comprehend is the use of a dedicated battery or DC voltage source V_{CC}, with + pole connected to the cathode and its - pole at the grid.

The amplifying principle is illustrated with help of the dynamic transfer characteristic (as for diodes, see p75). The input sinewave voltage $v_G(t)$ is applied to the grid, the input of the stage, centered around the negative bias voltage (-6.0V in this case). Anode current $i_A(t)$ is the output signal, in phase with the grid voltage. Soon we will explain how dynamic transfer curve is constructed for triodes .

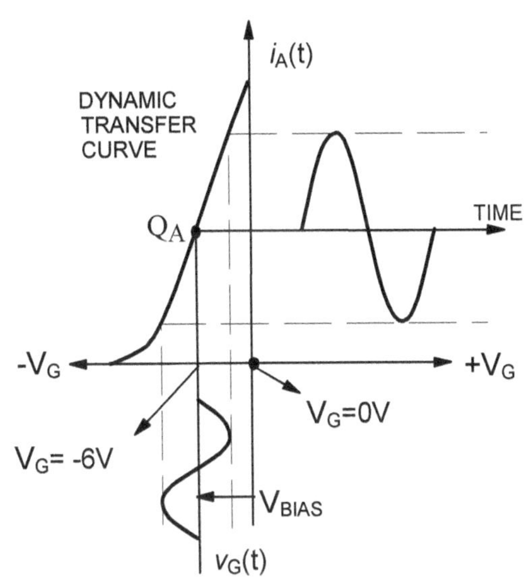

ABOVE: For linear operation, a single-ended output tube needs to have its grid negatively biased

Voltages and currents in the common cathode triode stage

The name common-cathode signifies the fact that cathode is the common or reference terminal, shared by both the input and the output circuits. The total grid voltage we will call v_C, a sum of the AC component v_G and DC bias V_G. Likewise, the anode voltage $v_B=V_A+v_A$ is the sum of the AC signal voltage v_A (which in this case is also the voltage across the load $v_A=v_L$) and DC component V_A, which in the quiescent point (without any signal) equals V_0. The total anode current is $i_B = I_A+i_A$, again a sum of an AC and DC component.

Without any input signal ($v_S=0$) the anode current is a DC current, steady at $I_A=I_0$ level. The anode voltage V_A is also steady, at its quiescent $V_A=V_0$ level. As the grid voltage starts oscillating around the bias voltage $V_G=-V_{CC}$ in a sinusoidal fashion, described by the formula $v_G=v_S+V_{CC}$ the anode current starts changing too, its waveform is in phase with the grid voltage. v_G is instantaneous total grid voltage while v_S is the AC signal on the grid.

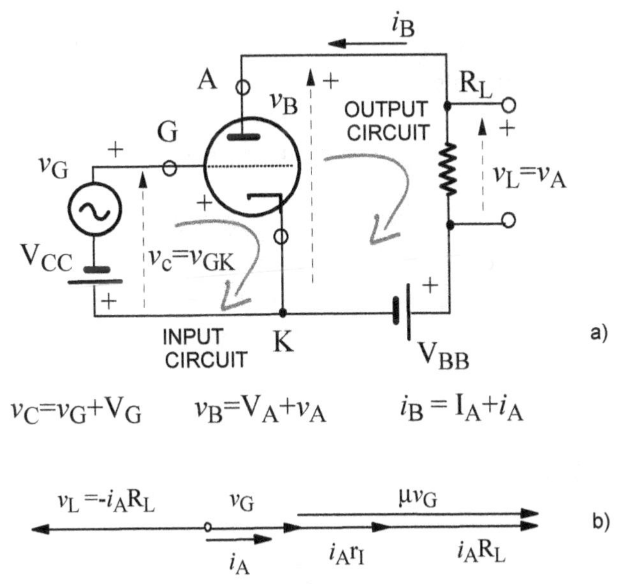

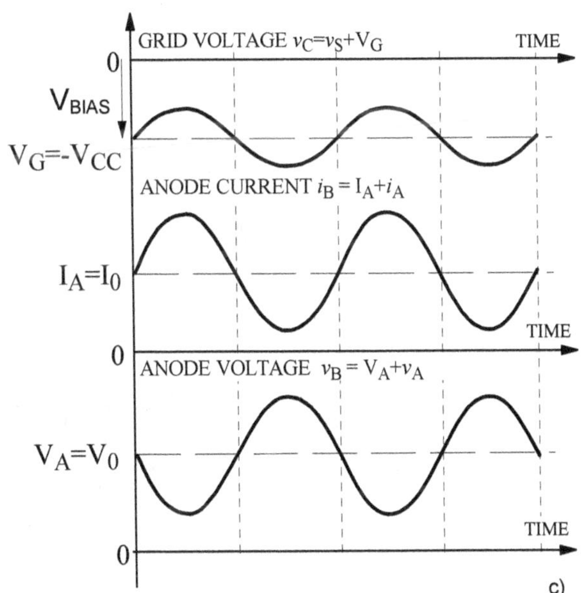

a) Common cathode triode stage with resistive load b) Vector diagram
c) Current and voltage phase relationships

The anode current flows through the load R_L and causes a voltage drop on this resistor ($i_A R_L$) to vary in a sinusoidal fashion, too. The variations of the load voltage are much larger than the variations of the grid voltage and thus voltage amplification is achieved! However, due to the way the load is connected, the output voltage is of the opposite or "inverted" phase compared to the input (grid) signal!

The modulation aspect of anode current and voltage

Just as the total anode current is the sum of DC current and AC (signal) current, so the anode voltage is the sum of the DC voltage and the AC or signal voltage. The "zero" line of the sine wave is not zero volts, but the DC voltage in the operating point Q, and that in this case (ECC40 stage that will be analyzed soon) is $V_0=235V_{DC}$!

For emphasis the instantaneous values of anode voltage are shaded, and those values goes from the zero volts line to the AC component of the voltage. We could say that the AC signal "modulates" the DC voltage between the anode and cathode. In an ideal case (no distortion), the dips and the peaks around the V_0 level are equal, and the average voltage is V_0, since the positive and negative dips cancel each others out. In reality, as illustrated for ECC40 CC stage, the negative "dips" are larger than the positive "peaks", meaning the output signal is distorted.

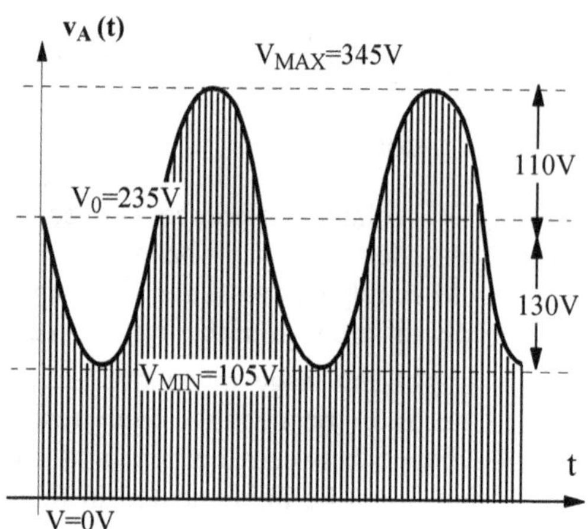

ABOVE: The output (anode) voltage waveform for the ECC40 common cathode stage that we will use as our main example in further deliberations. The drawing emphasizes its "modulation" nature.

Real instead of "typical" triodes

The framed tube profiles in this book are snapshots of each tube's power ratings, DC conditions required for its proper operation, and its essential parameters such as gm, μ and r_I, and as such are not a substitute for tube data sheets which go into more detail.

Instead of talking about a "typical" triode, as many books do, (after all, what is a "typical triode?), let's introduce another dame of voltage amplification. Not nearly as famous or common as 6SN7, yet, in no way inferior, ECC40 was developed in Philips labs in Holland in the early 1950s.

TUBE PROFILE: ECC40

- Medium μ dual triode
- RimLock socket, 6.3V, 0.6 A heater
- V_{AMAX}= 300V, V_{HKMAX}= 175V_{DC}
- P_{AMAX}= 1.5 W
- TYPICAL OPERATION:
- V_A=250V, V_G=-5.5V, I_0=6 mA
- gm=2.9 mA/V, μ= 32, r_I = 11 kΩ

Solidly built, not microphonic at all, great looking and excellent sounding duo-triode, ECC40 is perhaps the most underrated valve or all, certainly one of the mighty Philips' greatest achievements.

6SN7 is hailed as a linear (relatively speaking) and good sounding triode, but I believe ECC40 sounds better, cleaner and smoother (less grainy), yet still enchanting. Let's compare the static transfer characteristics of the two, for the same anode voltage, of course, say 250V! That will give us the first comparison of their linearity.

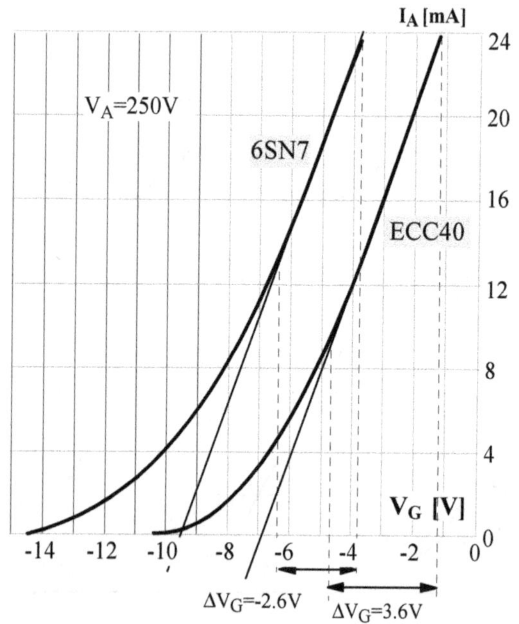

We see an uncanny similarity, the same slope (meaning the same transconductance). It is also noticeable that ECC40 needs a lower bias, meaning it cannot take as large input voltage as 6SN7 can. The cutoff for ECC40 is at -10V on the grid, compared to -14.5V for 6SN7.

RIGHT: This broken-in-transport ECC40 shows the very sturdy internal construction. The nipple on the metal skirt clicks into a groove in the RimLock socket which locks the tube in place.

SIDE-BY-SIDE: Transfer characteristics of 6SN7 compared to ECC40 for 250V anode voltage (LEFT)

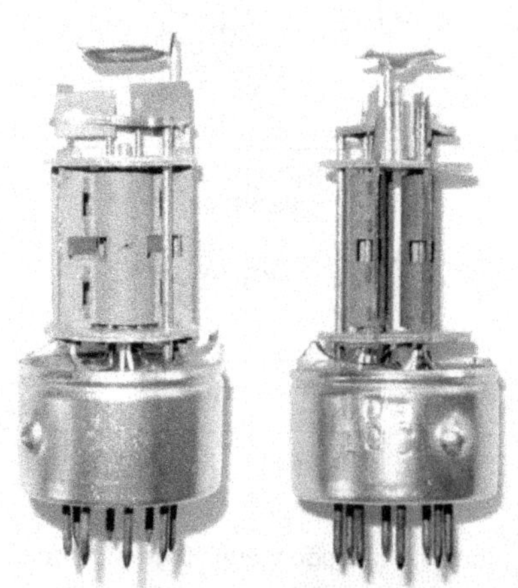

Constructing a dynamic transfer curve for a triode

While tube manufacturers publish at lest two or three static transfer curves, they never publish dynamic transfer curves, the ones we need to analyze the performance of a tube amplification stage. They don't do it because dynamic transfer curves depend not just on the internal parameters of the tube, but also on the load resistance and the anode supply voltage, meaning there is an infinite number of combinations of the two external parameters, and an infinite number of such curves. The good news is that it is relatively easy to construct such a curve yourself. The procedure for triodes (illustrated below for 6SN7 triode) is essentially the same as the one for diodes.

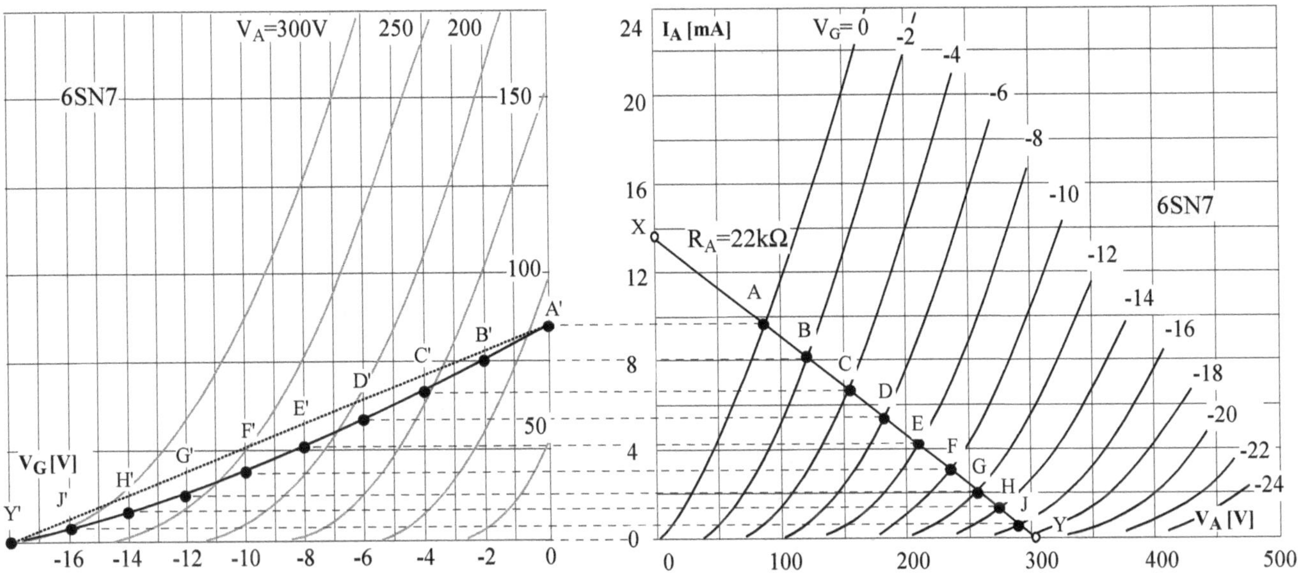

ABOVE: How to construct a dynamic transfer curve for a particular load line

Let's say you've selected the load value of R_A=22k and you have available anode voltage supply of 300V_{DC}. The first point is for I_A=0, that is your +V_B = 300V (point Y). You need one more point, and since 300V/22k = 13.64 mA, that is your point X (I_A=13.6mA). Now you can draw the load line between them.

The load line will intersect anode curves in points A, B C, D, etc. To construct the dynamic transfer characteristic for this tube and this load, for each of these points draw a horizontal line to the transfer characteristic until you reach the vertical line for its V_G.

For instance, from point A we draw a horizontal line to V_G=0V and mark that point A'. For point B, whose V_G=-2V, we draw a horizontal line until we hit V_G=-2V vertical line on the TX graph, and mark that point B'. Once all the points A' to J' are marked, we draw a curve through them.

This dynamic transfer curve will look almost linear, but if you draw a straight line between the end points it will become obvious that it is far from straight. However, it is much more linear than any of the static transfer curves. The presence of the load straightens or "linearizes" the transfer function.

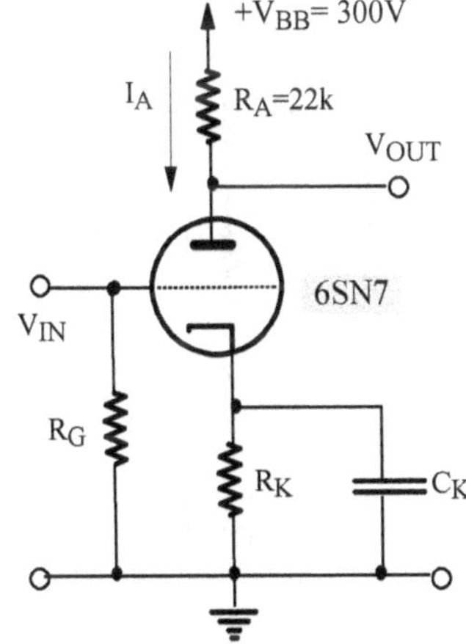

ABOVE: Common cathode triode stage whose dynamic transfer curve is illustrated above

22 kΩ is a relatively low anode load for 6SN7. If you repeat the same exercise for a higher anode load, say 47kΩ or 82 kΩ, you will notice that the higher the anode load, the more linear the transfer curve. This means that higher anode loads will result in less distortion!

Graphical analysis of the ECC40 common cathode stage

We have already mentioned that a small signal model is one tool a designer has available. The other, equally important tool is graphical analysis, based on tube characteristics, either those published by tube manufacturers or our own graphs, drawn as the result of our own measurements.

The AC voltage signal at the input (grid), centered at the quiescent DC voltage V_{G0} (the bias voltage), causes the operating point to move up and down the load line. The variations in grid voltage cause changes in anode current ΔI_A, which are of the opposite polarity (inverted phase) The variations ΔI_A cause in-phase variations of the plate voltage, due to the voltage drop on the anode resistor. In this case the anode resistor R_A is the load resistor R_L, since nothing is coupled at the output of the stage!

The graph illustrates a large grid voltage signal of $10V_{PP}$ (peak-to-peak). To calculate the voltage gain, read the peak-to-peak swing of the anode voltage (105 to 345V) divide the two and you get $A = \Delta V_A/\Delta V_G = (345\text{-}105)/(-10\text{-}(-1)) = 240/(-10) = -24$ The negative sign signifies that the two voltages are of the opposite phase (180° phase shift)!

This amplifying stage was not chosen for any particular reason. It hasn't been optimized, it simply illustrates the operational principles, characteristics and shortcomings of triode amplifiers.

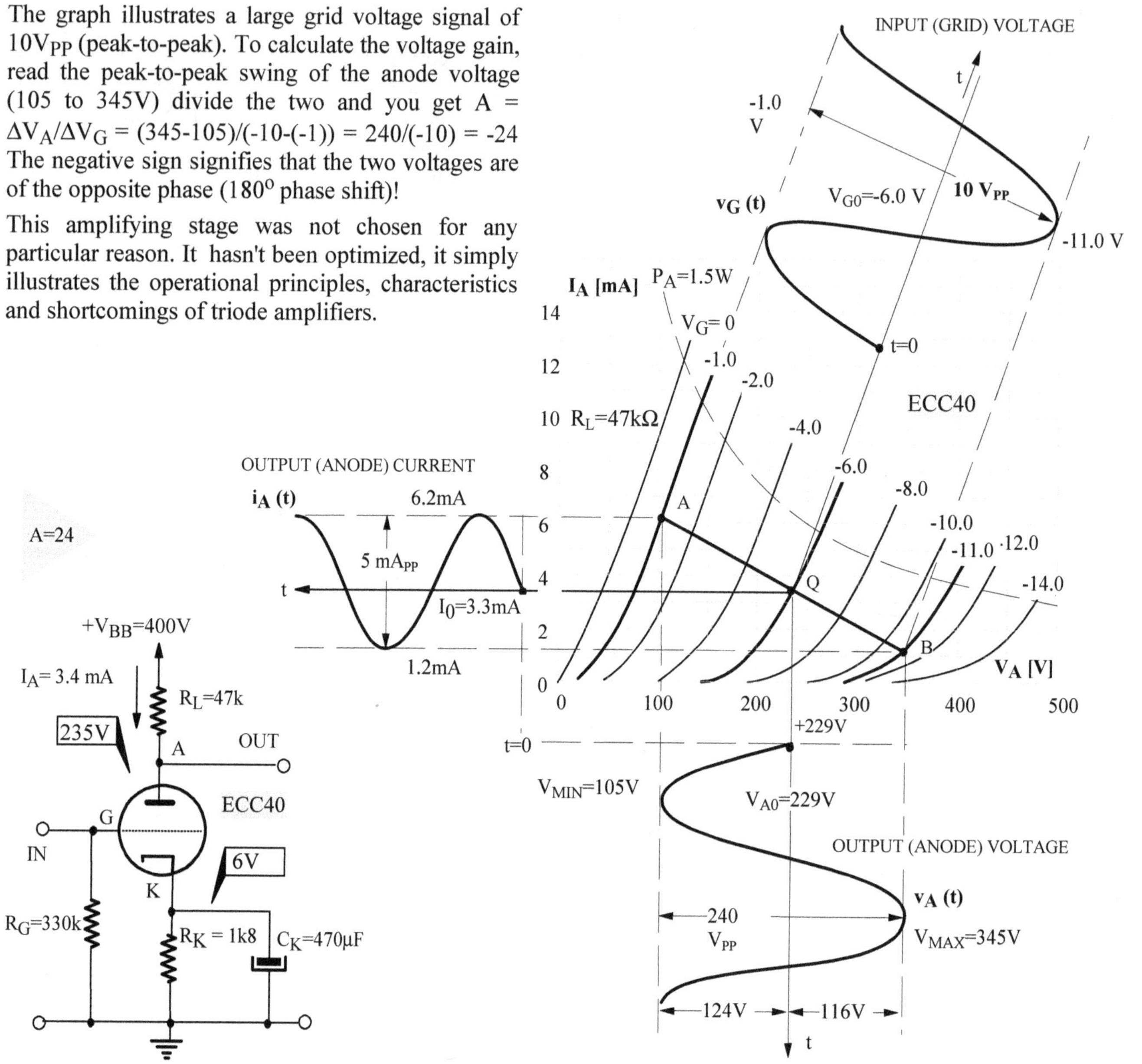

ABOVE: The common cathode stage we will use in our analysis of a triode as an amplifier, with DC voltages in main points (DC currents do not have to be written, since they can be calculated from voltages and resistances). The voltage amplification of the whole stage at midrange frequencies is specified in a triangle. If the voltage between the control grid CG and ground is not marked, it means it is at ground potential.

The boundaries of the operating region

To minimize distortion, ensure reliable operation and a long tube life, five regions need to be avoided:

1. Positive grid voltages, where grid current starts flowing and input resistance drops significantly. Power tubes specifically designed to operate with grid currents are an exception.
2. Above the maximum anode currrent (I_{AMAX})
3. Very low anode currents, where the characteristics are very curved and distortion is very high (I_{AMIN})
4. Above the P_{MAX} parabola, where the power dissipation of the anode is exceeded
5. Above the specified maximum anode voltage allowed for a particular tube (V_{AMAX}). Consult tube's data sheets.

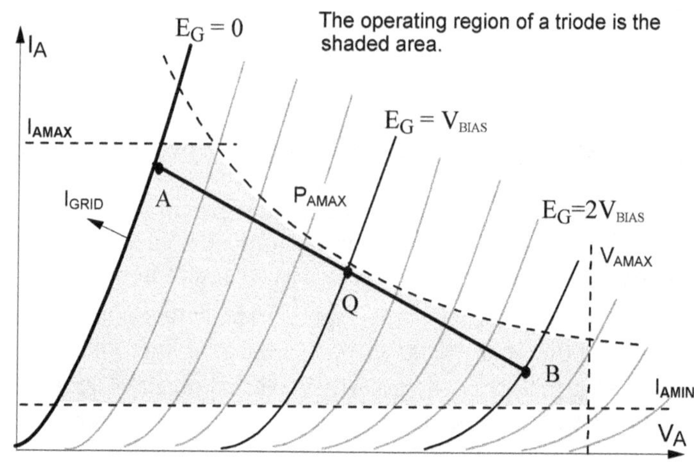

The operating region of a triode is the shaded area.

Choosing the load resistance and the quiescent point

Assuming we limit ourselves to the safe operating region as discussed, we can position the operating point right on the maximum power dissipation parabola (Q1). That would result in the maximum output power (large current swing multiplied by a large voltage swing), which is one of our goals in power or output stages. However, in preamp stages we are not concerned about power, we want maximum voltage gain with minimum distortion.

Operation with a higher anode (load) impedance would give us a higher voltage gain. We will soon see how to make this load line almost horizontal, for even low distortion and even higher gain.

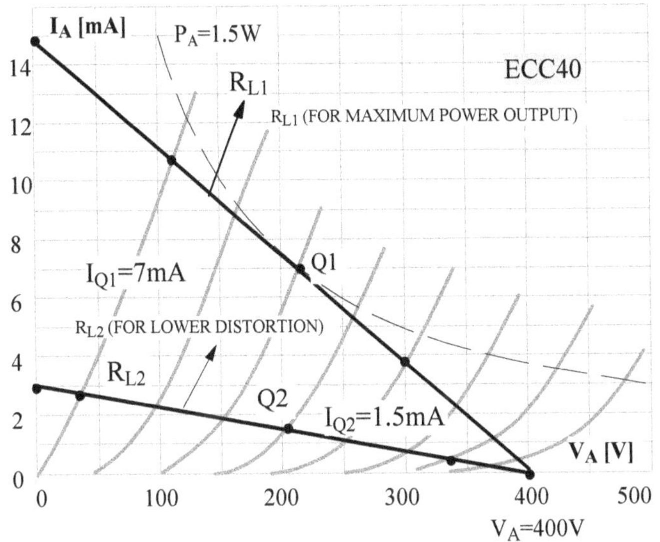

RIGHT: The choice of load resistance is a choice between the maximum output power (lower load resistance) and maximum linearity-minimum distortion (higher load resistance)

Biasing options

For proper operation, the grid of the amplifying tube must be negatively biased in respect to the cathode. The obvious way to do so is to connect the grid leak resistor R_G instead to ground, to a source of required negative DC voltage $-V_G$. This type of bias is called a fixed or external bias. The fixed bias name is actually an oxymoron, since it is actually adjustable by the designer or even often by the user. Perhaps it is called "fixed" because the tube has no "say" in it, it is a fixed external voltage imposed on the tube.

Lithium-iron or NiMh (Nickel-MetalHydrate) batteries will provide a stable grid voltage for thousands of hours, since the grid current is next to zero. Both fixed and battery biasing require the grid to be capacitively-coupled to the previous stage so that the voltage offset in the input signal does not change the bias or that the battery does not discharge through the output resistance of the previous stage.

Making the grid negative is equivalent to making the cathode positive, and the following four biasing methods use that approach. The DC cathode current I_K flows through the cathode resistor R_K and creates a voltage drop across it (V_K) making the cathode positive against the grid. That kind of bias is called self-bias, cathode-bias or auto-bias. The tube biases itself. The value of R_K depends on the required operating point Q and steady-state cathode current I_K, plus on the desired value of the bias voltage. Voltage V_K is the bias voltage.

Obviously, if the DC current through the tube varies, the bias will change, too, and that is usually to be avoided. The following three methods keep the cathode voltage V_K much less dependent on the cathode current I_K.

Again, rechargeable miniature batteries have a very low internal resistance and can also be inserted in the cathode as in d), without creating cathode degeneration (negative feedback) because of un-bypassed cathode resistance.

The internal resistance of a typical small-signal silicon diode is only around 5-6Ω, while red or infrared light-emitting diodes have a slightly lower r_I of 4-5Ω. Green and blue LEDs have a higher r_I, in the 30Ω region.

Zener diodes are ideal to keep the cathode voltage constant. They can be connected in series to get the exact bias voltage required, or a programmable integrated circuit can be used, where the Zener voltage is "programmed" by two resistors in a voltage divider arrangement. Diodes, especially Zener diodes, are noisy and should be bypassed by a good quality ceramic or film capacitor, as illustrated below.

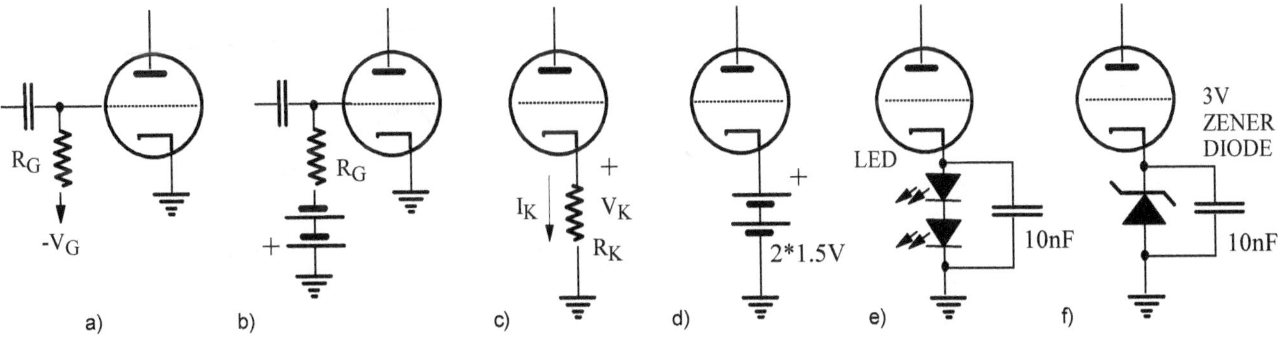

ABOVE: Six ways of providing a DC grid bias: a) fixed external negative grid bias voltage from a power supply b) negative grid bias voltage from a battery c) cathode-, self- or automatic-bias via the cathode resistor voltage drop d) cathode bias via batteries in the cathode e) cathode-bias via the voltage drop on silicon or LED diodes f) Zener diode or programmable voltage reference integrated circuit

Cathode- or automatic bias: choosing R_K and C_K

The flow of 3.4mA of DC current through a 1k8 cathode resistor creates voltage drop $V_K=I_KR_K$, and since in a triode cathode current is equal to anode current, V_K=3.4mA*1k8 = 6V

R_K is bypassed for AC signals by the cathode capacitor C_K. The value of this capacitor is chosen so that its reactance is much smaller than R_K at the lowest frequency the amplifier must reproduce. The rule-of-thumb is that $X_C=1/(\omega C) = R_K/10$ or smaller! That is why R_K does not feature in our small-signal AC model (next topic). It assumes that C_K behaves as a short-circuit and makes R_K irrelevant for AC signal. Let's see if that was a correct assumption!

Most evaluations and practical measurements start in the midrange of frequencies, with f=1kHz being a convenient choice. At this frequency the reactance (reactive impedance) of C_K is $X_C=1/(\omega C)=1/(2*\pi*1{,}000*470*10^{-6}) = 1/(2*\pi*0.47) = 0.34\Omega$ That can indeed be considered a short circuit, it is 1,800/0.34 = 5,315 times lower than R_K!

Assuming we want this amplification stage to reproduce frequencies down to at least 10Hz, the bypass capacitor's reactance at that frequency will be $X_C=1/(\omega C) = 1/(2*\pi*10*470*10^{-6}) = 1/(2*\pi*0.0047) = 1/2.95= 34\Omega$

While much higher than at 1kHz, even that impedance can be considered a short circuit, it is 1,800/34 = 53 times lower than R_K! Since 10 times is the accepted minimum, it seems we have chosen our C_K very well!

COMMON CATHODE AMPLIFYING STAGE - ANALYTICAL METHODS

Small signal linear model (equivalent circuit) of a triode

Just as we have done with diodes, in most deliberations from now on we will use the linear triode model. Just like any model, it is an approximation. The smaller the amplitude of the signal around the quiescent operating point Q, the better this small-signal model fits reality.

The main reason we use simplified (not entirely acccurate) linear models is that we know how to deal with and solve linear equations, but we don't know how to analyze nonlinear systems, which are described by nonlinear equations. Luckily, for most purposes in electrical engineering such models are perfectly adequate and useful.

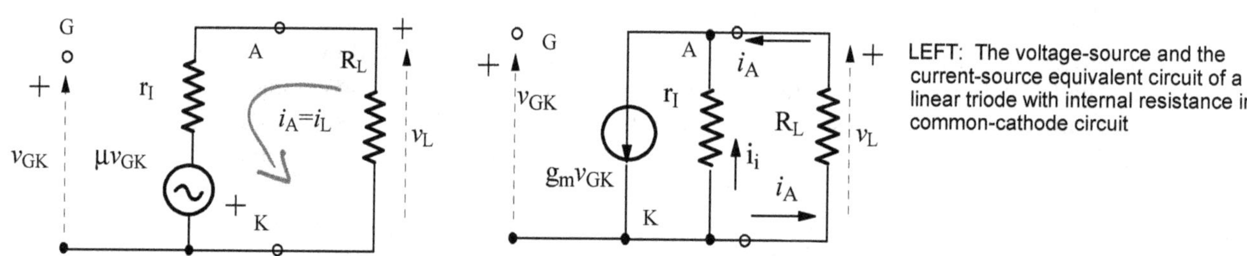

LEFT: The voltage-source and the current-source equivalent circuit of a linear triode with internal resistance in common-cathode circuit

Anode characteristics have anode current I_A as a dependent variable, anode voltage V_A as independent variable and grid bias voltage E_G (or V_G) as a parameter. If only one grid bias voltage was possible, there would be only one I_A-V_A curve, as in the case of a diode. With triodes, the grid can be at any bias voltage the designer chooses, so instead of one curve, we have an infinite number of curves. However, for practical reasons we only draw half-a dozen to a dozen such curves, usually for uniformly selected grid voltages.

The voltage-source model of a linear triode had its internal resistance r_I in series with the voltage source μv_{GK}.

Thevenin's theorem says that this is equivalent to a current source $g_m v_{GK}$ in parallel with the same resistance r_I (Norton's theorem). While the voltage source model makes certain analysis easier, the current source model is more convenient in other calculations, so we will use both.

Although a triode is not a very good current source (it's behavior is much closer to a voltage source), we can still use the current source model if it makes number crunching easier. While the anode current is in phase with the grid signal, the voltage drop that such current creates on the anode resistor (load) is of the opposite or "inverted" phase, meaning the phase shift between the input grid voltage and output voltage on the anode (or load) is shifted by 180°. The common cathode stage is a phase inverting amplifier.

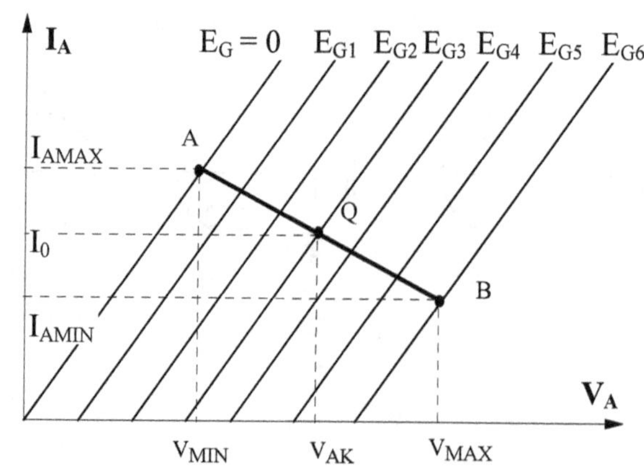

ABOVE: Anode curves for a linear triode: equidistant straight lines, so there is no distortion of any kind

The analysis of the common cathode stage using a small signal model

We estimated the amplification of our ECC40 CC stage using graphical means, let's see what results our new tool, the small signal equivalent circuit, will provide, and compare the two. Normally, you won't have to use both, one estimation is sufficient. However, this is an educational book, and just as we should get to know the soldering iron, the multimeter and the oscilloscope we use in amplifiers' construction and testing, we must understand the tools we use in their design!

Using a voltage model of the CC stage, the equation for the current loop is $i_A = -\mu v_{GK}/(r_I + R_L)$ and since $v_{OUT} = v_L = i_A R_L$ we have $v_{OUT} = -\mu v_{GK} R_L/(r_I + R_L)$. The voltage amplification is

$\mathbf{A = v_{OUT} / v_{IN} = -\mu^* R_L/(r_I + R_L)}$ For ECC40 stage with R_L=47k, r_I=11kΩ and μ=32, A= -32*47/(11+47) = -32*0.81= -25.9

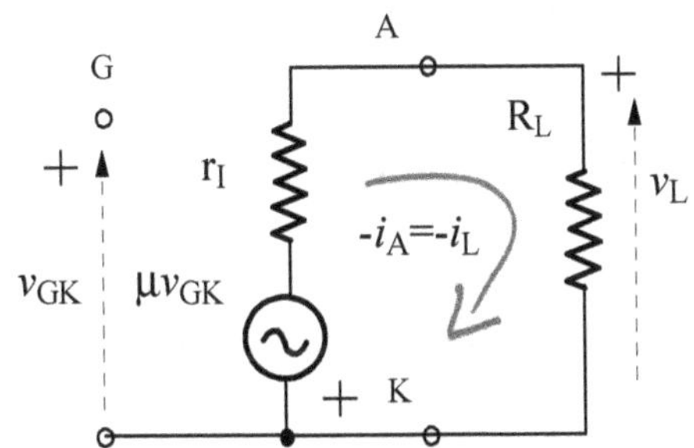

ABOVE: The small signal voltage model of a common cathode stage with a bypassed cathode resistor

From the graphs we estimated the gain as A=-24 and now we calculated A=-26 by using the linear model. Why this discrepancy of 7.7%?

Firstly, the curves are approximate and our drawing isn't very precise. Secondly, the linear model is a simplification and as such assumes that I_A-V_A curves are straight lines, which they are not.

And finally, considering that parameters of real tubes vary widelz from the published "bogey" or average figures, for design purposes the accuracy under 10% is acceptable, and anything under 5% is excellent!

We have already explained how C_K bypasses the signal frequencies to ground and how the cathode resistor R_K does not appear in the Ac model. Its only purpose is to provide the negative DC bias for the grid, making cathode positive compared to the grid, which is equivalent to making the grid negative with respect to the cathode.

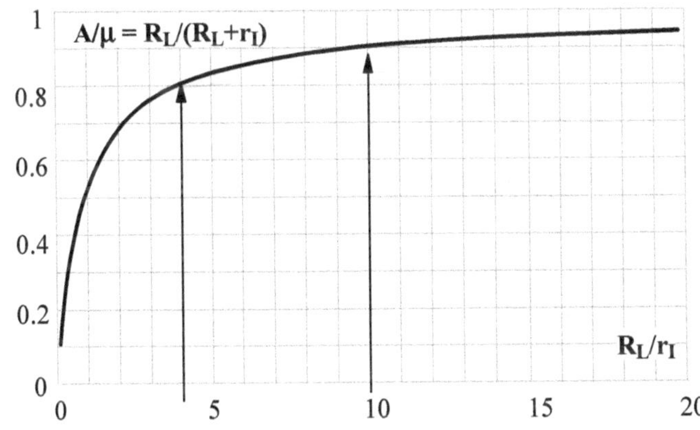

ABOVE: How the gain of common cathode stage (referenced to μ) varies with the size of the anode load R_L (referenced to r_I)

You may ask yourself how come the anode DC power supply voltage V_{BB}=+400V does not come into play either? Well, assuming the internal impedance (resistance) of this power supply is zero (another simplification), such a DC voltage source does not exist at all for the AC signal.

For AC signal, point V_{BB} (the other side of the anode or load resistor) is at GND (COM) potential, as is the cathode K! This simplifies our circuit immensely.

It is interesting to see how quickly the amplification factor of the common cathode stage (A) approaches the μ of a triode. For $R_L/(R_L+r_I)$=4, we get 80% of μ, for $R_L/(R_L+r_I)$=10 we get A=0.9μ, and further increase does not yield any significant increase, which limits $R_L/(R_L+r_I)$ to the 4-10 range.

Dynamic mutual conductance of a common cathode stage

The mutual conductance of a triode (gm) defined as a change in anode current caused by the change of grid voltage is the *static* transconductance, of a tube itself, in static conditions (no AC signal applied). These are figures quoted by tube manufacturers in data sheets and manuals, specified in a certain quiescent or operating point, the slope of the static transfer curve in that particular point.

Now that we have our dynamic transfer curve, what is the significance of its slope? Isn't that some kind of transconductance? Yes it is!

Since $v_{OUT} = -\mu v_{GK} R_L/(r_I + R_L)$ and if we substitute $\mu = gmr_I$ and $v_{OUT} = i_{OUT} R_L$ into that formula we get $i_{OUT} R_L = -gmr_I v_{GK} R_L/(r_I + R_L)$ The R_L terms on each side cancel one another, and if we move v_{GK} to the left side, we get $i_{OUT}/v_{GK} = -gmR_L/(r_I + R_L)$

Differentiating the equation gives us $\mathbf{di_{OUT}/dv_{GK} = S_{DYN}} = \mathbf{gmR_L/(r_I + R_L)}$ The amplification of the common cathode stage can also be written as $\mathbf{A = S_{DYN} R_L}$

S_{DYN} is called *dynamic transconductance*, or transconductance of the whole common cathode amplification stage. There is no symbol for it in English-language literature, Europeans use capital S for static and lower-case s for dynamic transconductance (or "steepness"), so we'll call it S_{DYN}.

Due to the presence of R_L in the denominator of the fraction, the dynamic transconductance is always lower than the static one. Theoretically, if we could make $R_L=0$, the two would be equal. S_{DYN} is useful for calculations where the input signal is very small, and accurate graphical methods (projections onto the dynamic curve) are not practical.

Since the dynamic TX curve is almost a straight line (nearly constant slope), that means that the dynamic transconductance does not vary much at all, unlike the static gm!

POWER RELATIONSHIPS AND EFFICIENCY OF THE COMMON CATHODE STAGE

Efficiency of ideal triode stages

Firstly, let's consider an ideal triode with a resistive load R_L in the anode circuit. The quiescent point is Q, and the load line is AB. $I_{AMAX} = 2I_0 = E_{BB}/R_L$, the anode voltage in the quiescent point is half of the high voltage battery voltage V_{BB}. The efficiency of the whole stage in percentages is defined as η= (AC output power)/(DC input power)*100%
DC input power is $P_{IN}=V_{BB}I_0$

AC output power is $P_{OUT}= V_{RMS}I_{RMS} = (V_P/\sqrt{2})(I_P/\sqrt{2}) = V_PI_P/2$ Since the peak values are $V_P = (V_{BB}-V_{AMIN})/2$ and $I_P=(I_{AMAX} -I_{AMIN})/2$ we get

$$\mathbf{P_{OUT}= V_PI_P/2 = (V_{BB}-V_{AMIN})(I_{AMAX} -I_{AMIN})/8}$$

$$\boldsymbol{\eta} \mathbf{= (V_{BB}-V_{AMIN})(I_{AMAX} -I_{AMIN})/8(V_{BB}*I_0)*100\%}$$

The maximal swing or peak value I_P for the output current is $I_{AMAX}-I_0$ in the positive direction and $I_0-0=I_0$ in the negative direction. For an ideal triode these are equal since I_0 is in the middle of the range ($I_0=I_{AMAX}/2$). Since $I_{AMAX} = 2I_0$, $I_{AMIN}=0$, $V_{MIN}=0$ the maximum efficiency of this amplifying stage is $\eta= V_{BB}2I_0/8V_{BB}I_0)*100\% = 1/4*100\% = 25\%$

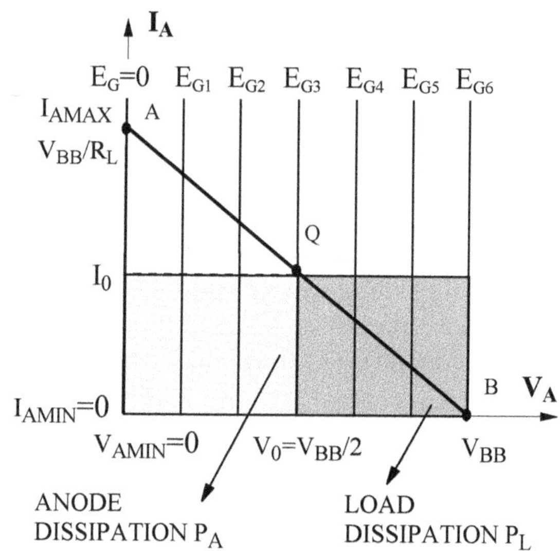

ABOVE: Ideal anode curves and load line for the resistive load

The DC anode current I_A flows through the load and dissipates power in it, which is one reason for low efficiency. In steady-state, without any AC signal, the DC power dissipated in the anode is $P_A=V_0*I_0$, (the area of the light gray-shaded rectangle). The power dissipated as heat on the load is $P_L= (V_{BB}-V_0)I_0$ (the area of the ldark gray-shaded rectangle). The total DC power provided by the power supply is $P=P_A+P_L= V_{BB}*I_0$ (the area of both rectangles together).

Again, in this ideal case, since Q is in the middle of V_{BB}, $P_A=P_L$, at idle (no signal) 50% of the power is wasted as heat on the tube and the other half on the load!

For real triodes the range A-B is much smaller and V_0 is closer to V_{BB}. Thus, due to such position of the quiescent point Q, the dissipation on the tube is higher than the dissipation on the load and the overall efficiency is lower than the 25% theoretical maximum. Let's see what we get for our practical ECC40 CC stage!

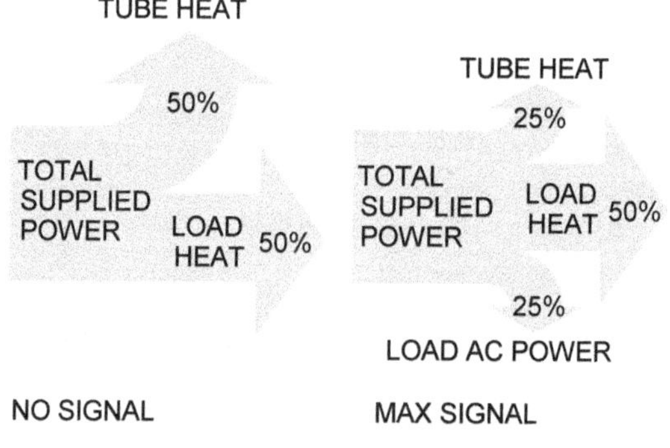

ABOVE: With or without the input signal, half of the supplied power is wasted on the load as heat.

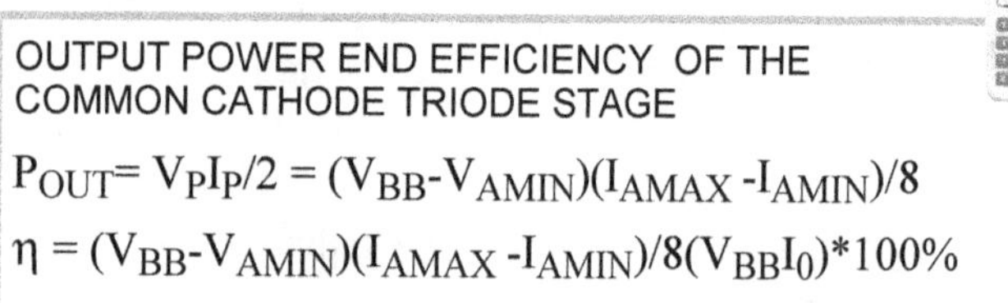

OUTPUT POWER END EFFICIENCY OF THE COMMON CATHODE TRIODE STAGE

$$P_{OUT}= V_PI_P/2 = (V_{BB}-V_{AMIN})(I_{AMAX} -I_{AMIN})/8$$

$$\eta = (V_{BB}-V_{AMIN})(I_{AMAX} -I_{AMIN})/8(V_{BB}I_0)*100\%$$

Efficiency of a real common cathode stage

One way to estimate the output signal power is $P_{OUT}= (\Delta V_A \Delta I_A)/8 = 240*0.005/8 = 0.15W$ Another way is to calculate the power of the positive and negative halves of the signal range and add them up: $P_{OUT+} = V_{AQ}I_{AQ}/2 = (229-105)*(0.0062-0.0033)/2 = 124*0.0029/2 = 0.1798$ W and $P_{OUT-} = V_{QB}I_{QB}/2 = (345-229)*(0.0033-0.0012)/2 = 116*0.0021/2 = 0.1218$ W

$P_{OUT}= (P_{OUT+} + P_{OUT-})/2 = (0.18+0.12)/2 = 0.15W$ Notice that P_{OUT+} and P_{OUT-} are not equal, which means that distortion is present.

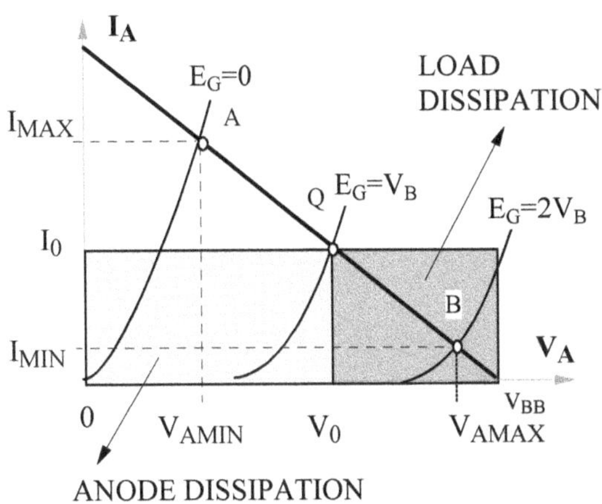

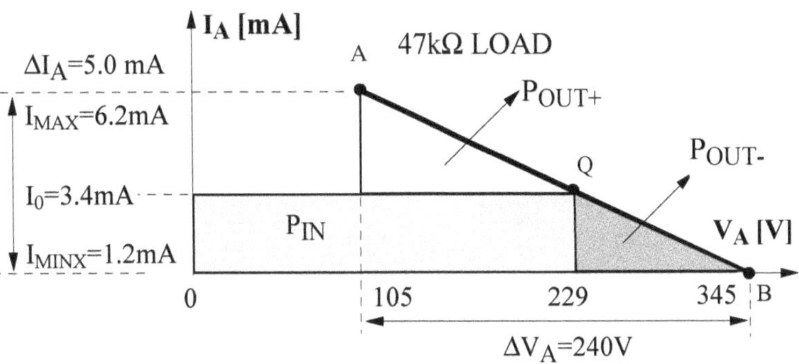

The anode and load dissipation for a real triode stage (LEFT) and for the specific ECC40 stage being discussed (ABOVE)

We can look at the efficiency of amplifying stages from two perspectives. One is the overall efficiency, comparing the AC signal power at the output to the DC power that the power supply must supply. The power supply provides $P=P_A+P_L= V_{BB}I_0 = 345*0.0034 = 1.17W$

The efficiency of the whole stage is $\eta = P_{OUT} /P*100$ [%] = $0.15/1.17*100 = 12.8\%$

The other type of efficiency of interest is anode efficiency, again, comparing the AC signal power at the output to the DC power supplied to the anode only, disregarding the losses on the load. The anode dissipation without the signal is $P_A=P_{DC} =V_0I_0 = 229V*3.4mA=0.78W$

The anode efficiency is $\eta_A=P_{OUT}/P_A*100[\%]$ or $\eta_A=0.15/0.78*100 = 19.2\%$

Whichever way you look at it, 12.8% or 19.2%, the CC stage with a load in the anode circuit is a lousy converter of DC into AC power. However, we should not be concerned at all, we are not power utilities, where each 0.1% of reduced efficiency (thermal loss) results in millions of dollars of financial loss.

To understand the power relationships in CC stage fully, study the graph on the right. DC power from the power supply (area within the rectangle divided by the period T) is $P_{DC}=V_0I_0$

Anode dissipation of the tube (gray-shaded area/T) is P_A, as we have seen before. Without the AC signal that area would be equal to $P_{DC}=V_0I_0$ (the area of the rectangle).

The AC output power is $P_{AC} = P_{DC}- P_A$ = (area within the rectangle - diagonally-hatched area)/T = crosshatched area/T

Notice that in the first half of the period T P_A exceeds P_{DC} (two "humps"), which need to be subtracted from P_{DC} in the second half-period to get P_{AC}.

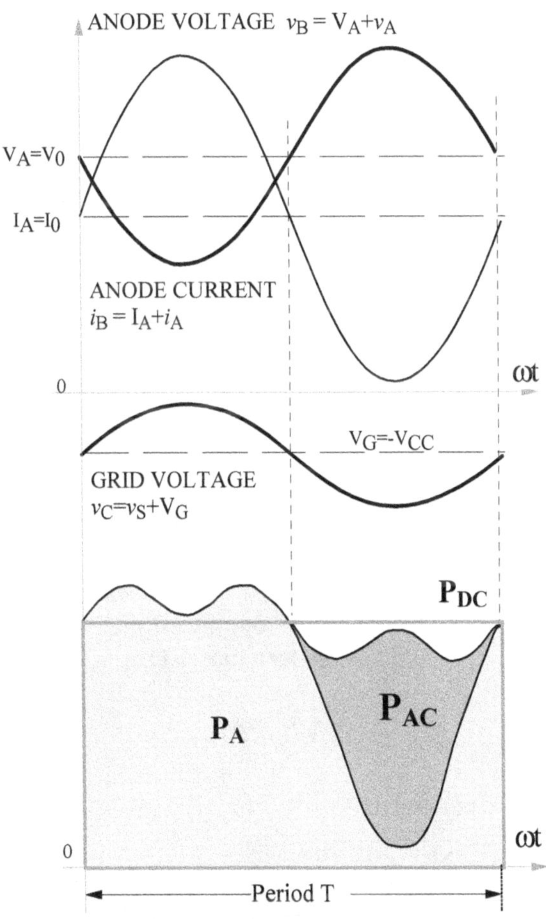

Voltages, currents and power relationships in a CC stage

Voltage transfer characteristic of a common cathode stage

The voltage transfer curve of any device visually illustrates the relationship between the output AC (signal) voltage (dependent variable on Y-axis) and input AC voltage (independent variable on X-axis). For the common cathode amplifier the output voltage is anode voltage and the input is grid voltage.

The transfer curve of an ideal amplifier is a straight line, again meaning there is no dynamic "compression" or distortion. Due to the curvature of triode characteristics, which are not equidistant, and due to the bundling together of the curves in the region of low anode voltages (gray shading on the graph), the transfer curve of the CC stage shows a pronounced curvature (points F-G-H) in the region of high negative grid voltages and low anode currents.

Constructing a transfer curve for a resistive load line is very easy. For each intersection of the load line with anode curves (points A to H) draw a vertical line down to the V_A axis and read the value of V_A. These are your output voltages, which you then mark on the V_A-V_G graph on the left, join the dots and voila!

BELOW: Point-by-point construction of the voltage transfer characteristic of our ECC40 CC stage. The nonlinear region is shaded.

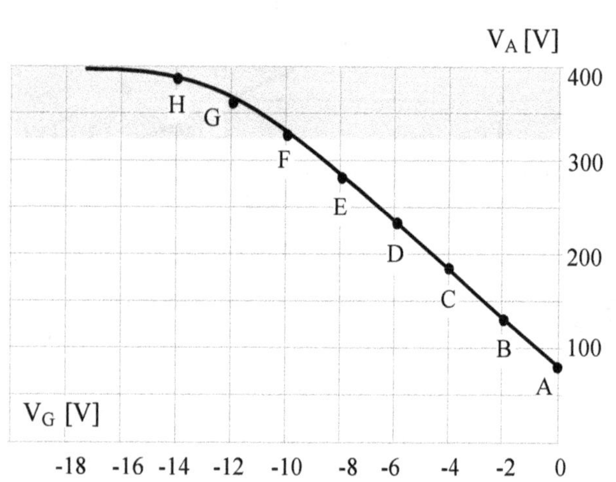

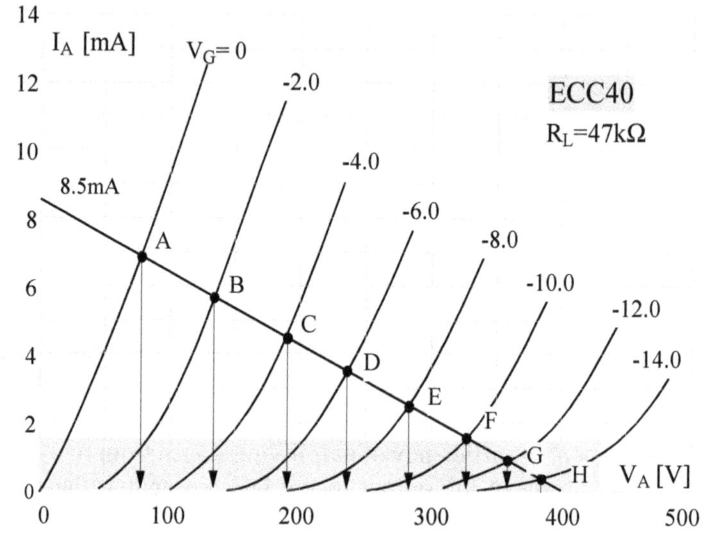

Common cathode stage: estimating harmonic distortion

An ideal dynamic transfer characteristic would be a straight line, and the grid signal would be "reflected" into the anode circuit without any distortion.

Compared to the ideal (linear) transfer characteristic, triodes' parabolic curve results in an asymmetric waveform of the output (anode) current. The negative (bottom) half of the sine wave is widened and flattened, with the slight narrowing of the positive half.

This type of amplitude distortion results in even harmonics, with the second harmonic being dominant, resulting in that warm and harmonically pleasing triode sound.

For triode stages, distortion is predominately 2nd harmonic and can be roughly estimated by the current distortion formula (CDF) which says:

$D_2 = [(I_{AMAX}+I_{AMIN})-2I_0]/2(I_{AMAX}-I_{AMIN})*100\%$

This is equivalent to

$D_2 = (\Delta I_+ - \Delta I_-)/2\Delta I*100\%$

For our ECC40 stage we have $D_2 = [(6.2 + 1.2) - 2*3.4]/2(6.2-1.2)*100\% = 0.6/10*100 = 6.0\%$

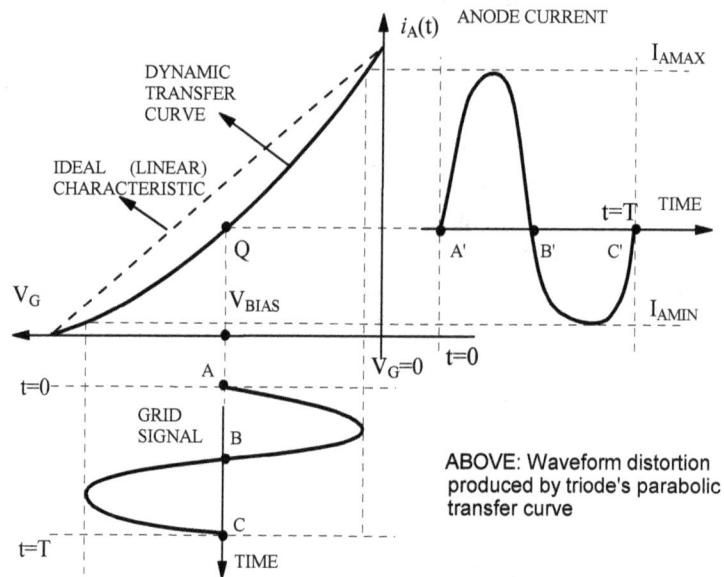

ABOVE: Waveform distortion produced by triode's parabolic transfer curve

TRIODE DISTORTION ESTIMATION FORMULAS

$D_2 = (\Delta I_+ - \Delta I_-)/2\Delta I*100$ [%]

$D_2 = (\Delta V_+ - \Delta V_-)/2\Delta V*100$ [%]

This seems a high figure, but keep in mind this is the worst case scenario, for the maximum voltage swing, with $10V_{PP}$ input signal at the grids (or $3.55V_{RMS}$) and $240V_{PP}$ ($85.2V_{RMS}$) at the anode (output signal).

Some contemporary authors, such as Morgan Jones in his book "Valve Amplifiers", advocate the use of the voltage-based formula to estimate the 2nd harmonic distortion, although they don't explain how they arrived at such formula: $D_2 = [V_0 - (V_{MAX}+V_{MIN})/2]/(V_{MAX}-V_{MIN})*100$ [%]

It is not mentioned in the vintage tube books or any literature, which only gives the current distortion formula. The voltage distortion formula can also be written in a shorter form: $D_2 = (\Delta V_+ - \Delta V_-)/2\Delta V*100$ [%]

However, the two formulas usually yield totally different results! Substituting our ECC40 stage's figures of $\Delta V_- = 345-229 = 116V$, $\Delta V_+ = 229-105 = 124V$ and $\Delta V = 345-105 = 240V$ we get $D_2 = (124-116)/2*240*100 = 1.67\%$

Clearly, both results cannot possibly be simultaneously correct, so take your pick.

AC AND DC INTERSTAGE COUPLING

RC coupling

In practical amplifiers, the common cathode stage is either directly or capacitively coupled to the grid of the next stage. The value of the coupling capacitor C is chosen so it represents a short circuit at audio frequencies (see p145-146). Also, the DC power supply is assumed to have a zero internal resistance, therefore the $+V_{BB}$ point is at ground potential for AC signals. For AC signals R_A is in parallel with the grid resistor R_G of the following stage, forming a parallel load resistance of $R_L = 47||220 = 38.7k\Omega$ Voltage amplification is $A = -\mu R_L/(r_I+R_L) = -32*38.7/(11+38.7) = -32*0.78 = -24.9$ Without the coupled second stage, we had $A=-25.9$, so for practical estimation purposes the gain reduction (due to grid resistor) is negligible.

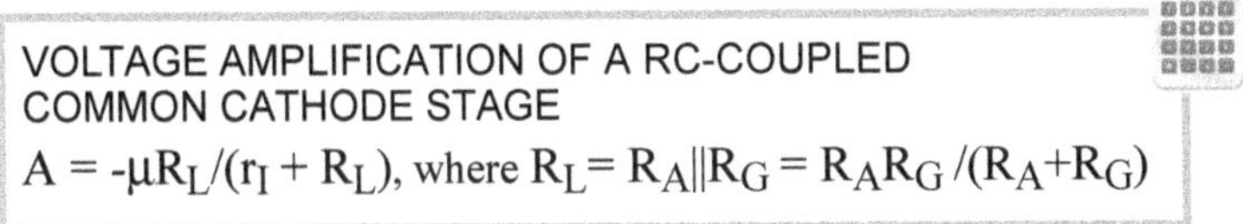

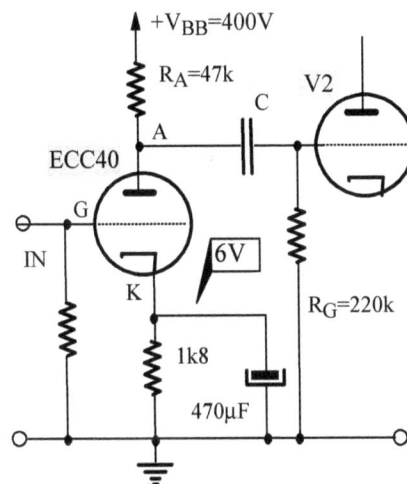

VOLTAGE AMPLIFICATION OF A RC-COUPLED COMMON CATHODE STAGE

$A = -\mu R_L/(r_I + R_L)$, where $R_L = R_A||R_G = R_A R_G/(R_A+R_G)$

RIGHT: The small signal model of capacitive coupled stages at midrange frequencies

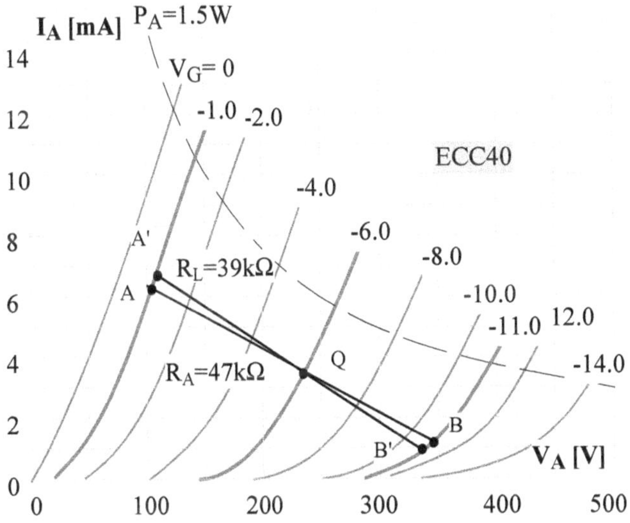

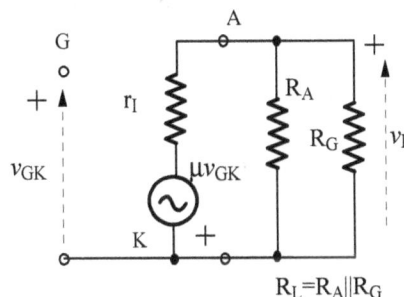

ABOVE: The AC and DC load lines for RC-coupled ECC40 stage

DC- and AC-load lines

The AC load line (determined by $R_L=R_A||R_G$) still goes through the quiescent point Q, but the intersection with $V_G=-1$ line is higher (A'), while the intersection with the $V_G=-11V$ is lower (B').

The anode voltage V_A swing is a bit narrower, and thus the amplification is a bit lower, as we have just calculated using the small signal model.

The AC load line in this case is only slightly different from the DC loadline, because the grid resistor of the next stage R_G is large compared to the anode resistor R_A (220kΩ versus 47kΩ), so their parallel combination R_L is only slightly smaller than R_A (38.7kΩ versus 47kΩ).

Directly-coupled amplifier stages

The main cause of loss of gain at low frequencies is the coupling capacitor C (see p145-146). Since capacitors cannot pass DC signals, this type of coupling is called AC coupling. The idea to eliminate the coupling capacitor is as old as the tube technology. In directly-coupled stages there is DC connection between the anode of the first tube and the grid of the following stage. Two possibilities are illustrated on the next page.

The circuit on the left is essentially an RC-coupled amplifier, but the coupling capacitor between stages has been replaced by the bias battery V_{CC2} and the grid-leak resistor of V2 has been eliminated. Since the DC voltage in point A1 is relatively high (100V or more) the bias battery V_{CC2} is needed to bias the second tube properly by subtracting that voltage from the anode voltage in point A1. For instance if $V_{A1}=100V$ and the bias for V2 needs to be -15V (the voltage between G1 and it cathode at ground potential), than a 115V battery is needed.

Batteries are not user friendly. They get depleted, they age and lose capacity and have to be replaced. Plus, high voltage batteries (in essence dozens of low voltage "cells" strung together) are bulky, complex and expensive. The most common vintage DC design, the Loftin-White amplifier, eliminates the bias battery and two separate anode batteries (or power supplies) and instead uses a single high voltage power supply. Proper biasing is achieved through precision resistive voltage dividers R1-R2-R3 and R4-R5.

To provide low-impedance paths for AC signals, and to help maintain the polarizing potentials constant, four bypass capacitors are also needed (not shown for clarity).

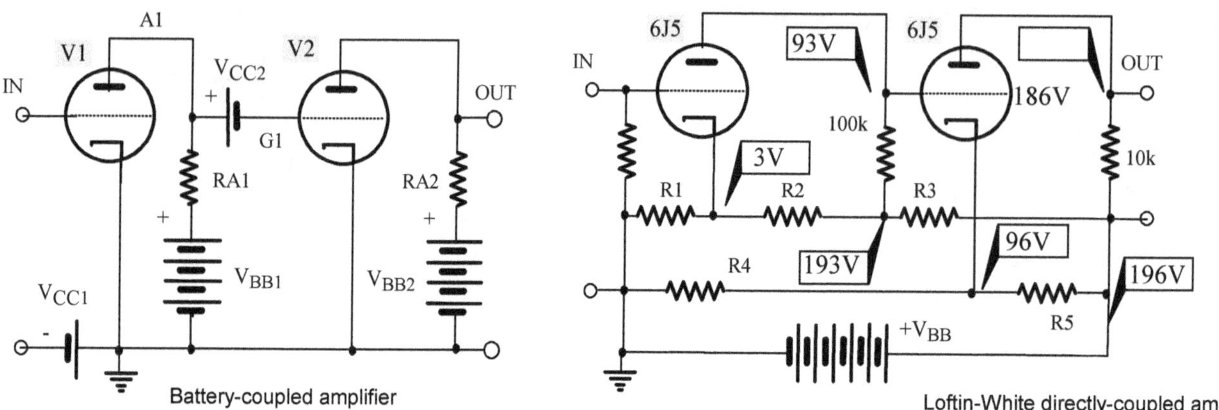

Battery-coupled amplifier

Loftin-White directly-coupled amplifier

Notice that cathodes of the two tubes are at very different potentials. If directly-heated tubes are used, each must have its own heater supply. Even with indirectly-heated tubes there is a danger that the maximum heater-cathode voltage will be exceeded. For most duo-triodes that limit is only around 90-100V. A few notable exceptions (12AX7 comes to mind) have a higher limit of 200V. In this case one cathode is at 3V and the other at 96V, and for 6J5 triode V_{HK} should not be higher than 90V. Thus, if a common heater supply is used for the two tubes, it should not be grounded but elevated (referenced) to somewhere around $+45V_{DC}$! More on this issue in Vol 2!

Since a small change in a DC grid voltage produces a relatively large change in tube's anode current and corresponding voltage drop across the load or its anode resistor, it can be said that such DC voltage changes are "amplified". However, you should notice by now the fundamental problem with DC amplifiers, and that is that they cannot distinguish between a slow change in the DC level of the input signal and any other change that can cause the change in DC voltages around the circuit.

For instance, resistors age and their value changes, thus changing voltage drops across them. Tubes age and their parameters drift, power supplies cannot keep DC voltages absolutely constant. Temperature changes and changes in electron emission from a spot in tube's cathode also affect the operation of a DC amplifier. Thus, highly regulated power supplies MUST be used, and even then the results are depressing.

DC amplifiers violate The Simplicity Rule and create too many problems just to eliminate one or two coupling capacitors. Ultimately, while amplifying signals of a few Hz or even down to a DC level may be necessary in instrumentation and special industrial applications, audio is not one of them. No speaker can go below say 20Hz anyway, so using DC amplifiers in audio would be sheer madness.

DC-coupled preamp stages

Just because a completely DC-coupled (from input to output) audio amplifier isn't viable, it does not mean that we couldn't directly-couple two stages. However, this is not a DC amplifier, since it uses cathode bypass capacitors and an output coupling capacitor.

The first stage is designed in the usual way, with a proviso that neither the anode resistor nor the anode current are too small, for that would result in a high anode DC voltage. In this case a standard 47kΩ load is used, and the current is lowish 2.3mA, so the anode voltage is around 106V.

Assuming we want to bias the second tube at -6V, its cathode has to be at 106+6 = 112V potential. With the chosen current of 2.4mA, that means it needs a cathode resistor $R_K=V_K/I_K= 112/0.0024 = 47k\Omega$! This pretty much concludes the design.

The first stage amplifies 15 times, the second 11 times, for a total of 15*11= 165 or 20*log165 = 44.4dB.

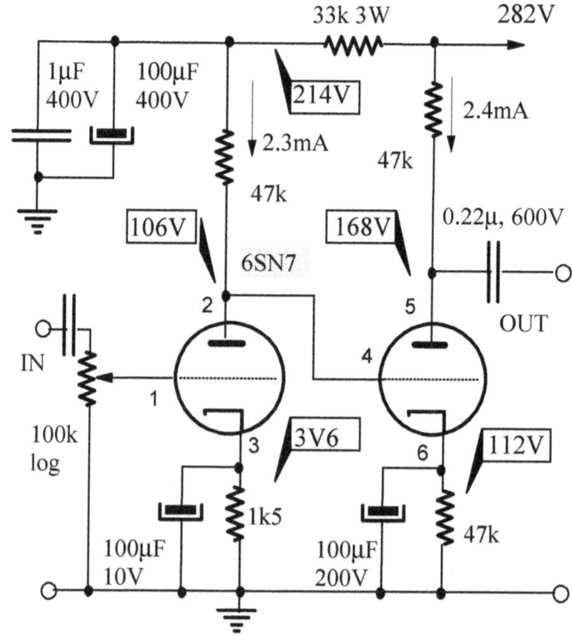

A true 2-stage DC amplifier would look identical except that those three capacitors would not be used. That means both stages would have local negative feedback in the cathode circuits, and, due to a very high value of its R_K, the second stage's amplification factor would be quite small.

We haven't yet discussed the impact of un-bypassed cathode resistors, that will be done in the next chapter, so, after reading it, come back here and calculate the voltage gain of this design with un-bypassed cathode resistors.

OTHER VOLTAGE AMPLIFICATION STAGES WITH TRIODES

- COMMON CATHODE STAGE WITH UN-BYPASSED CATHODE RESISTOR
- CATHODE FOLLOWER
- THE CATHODE FOLLOWER QUARTER-BRIDGE DRIVER STAGE
- COMMON GRID STAGE
- CASCODE
- SRPP
- THE WHITE FOLLOWER
- DIFFERENTIAL AMPLIFIER
- ACTIVE LOADS: CONSTANT CURRENT SOURCES AND SINKS

"As a culture we seem to have trouble distinguishing science from pseudoscience, history from pseudohistory, and sense from nonsense."
Michael Shermer, "Why People Believe Weird Things"

7

COMMON CATHODE STAGE WITH UN-BYPASSED CATHODE RESISTOR

This simple circuit is the most important of all amplification stages. At least a dozen or so various circuits have their origins here, so despite the tedious and heavy algebra, we need to analyze it inside and out. You will see why very soon. If you understand this circuit, you will understand all amplification stages and all phase inverters as well!

There are two power supplies, the positive anode supply $+E_{BB}$ and the negative cathode supply $-E_{CC}$. E_{BB} is always needed, but in many circuits $-E_{CC}$ is not included, so that end of the cathode resistor R_K is grounded.

The circuit also has two possible outputs, V_A from the anode and V_K from the cathode. Again, some circuits derived from this one use V_A output only, others use V_K output only, but there are also circuits that use both.

Once we draw the equivalent circuit and write the equations, depending on what we want to express (amplification factors, output impedances, input impedance, etc.), we manipulate those two equations in various ways.

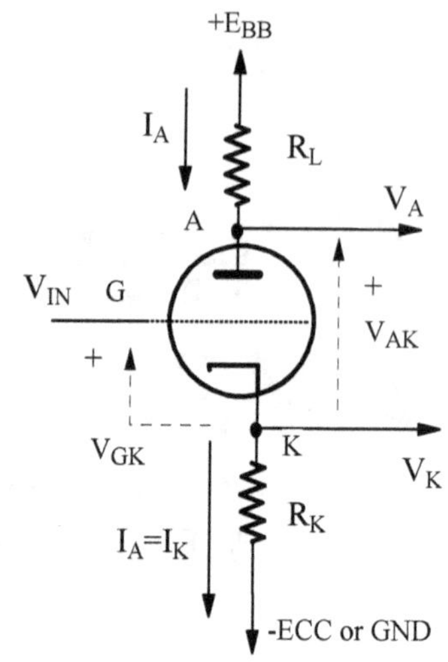

ABOVE: Currents and voltages in common cathode stage with un-bypassed cathode resistor

BELOW: Its smalll signal model

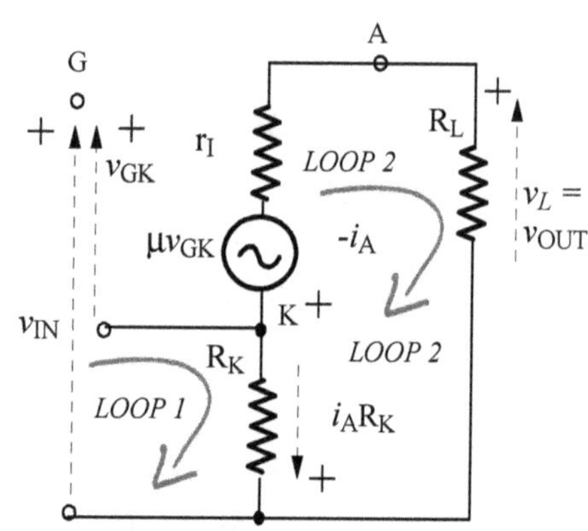

Amplification factor of the anode output

The voltage equations for the input and output loop are

$V_{IN} - V_{GK} + I_A R_K = 0$ and $-\mu V_{GK} - I_A r_I - I_A R_L - I_A R_K = 0$

Since $V_{OUT} = I_A R_L$, to eliminate V_{GK} we write $V_{GK} = V_{IN} + I_A R_K$ from the first loop and substitute it instead of V_{GK} into the second equation:

$-\mu(V_{IN} + I_A R_K) - IA(rI+RL+RK) = 0$ or $-\mu V_{IN} = I_A(r_I+R_L+R_K+\mu R_K)$

$I_A = -\mu V_{IN} /(r_I+R_L+R_K+\mu R_K)$

Now we substitute the right side of this equation into the third equation for V_{OUT} and get $V_{OUT} = I_A R_L = -\mu V_{IN} R_L/[r_I+R_L+(1+\mu)R_K]$

We finally get the anode amplification factor $A_A = V_{OUT}/V_{IN}$ or

$$\mathbf{A_A = -\mu R_L/[r_I+R_L+(1+\mu)R_K]}$$

CHECK: If R_K was zero (for AC signal) as when bypassed to ground by a capacitor, the gain would be $\mathbf{A_A = -\mu R_L/(r_I+R_L)}$, just as we had before. So, the CC stage with a bypassed R_K is just a special case of this circuit!

ANODE & CATHODE VOLTAGE AMPLIFICATION OF COMMON CATHODE STAGE

$A_A = -\mu R_L/[r_I+R_L+(1+\mu)R_K]$

$A_K = \mu R_K/[(1+\mu)R_K + r_I + R_L]$

Local NFB through un-bypassed cathode resistor

Going back to the CC stage with ECC40, only this time leaving R_K un-bypassed, the gain is now $A_A = -\mu R_L/[r_I+R_L+(1+\mu)R_K] = -32*47,000/[11,000+47,000+(1+32)1,800] = -32*47,000/117,400 = 12.8$

The gain was 25.9 before, now its only 12.8. Why?

Assume the signal V_{IN} is rising, making the grid more positive with respect to cathode. This reduction in the negative bias will cause the anode current to rise, and because it also flows through the cathode resistor, the voltage drop on the cathode resistor will also increase, making the cathode more positive, equivalent to making the grid more negative vis-a-vis the cathode. The two voltages are opposing each other, the tube behaves as if the rise in the input signal voltage was smaller, resulting in smaller output voltage, so the amplification of the stage is lower.

The cathode resistor provides "degeneration" or negative feedback. Since this feedback AC voltage on the cathode resistor is proportional to load current, it's a current feedback, and because it is added to the input AC signal in series (input loop), it is a series-applied feedback. The input and output circuit share the cathode resistor R_K, so the input to the stage is not independent of the output any more (as it was when R_K was bypassed).

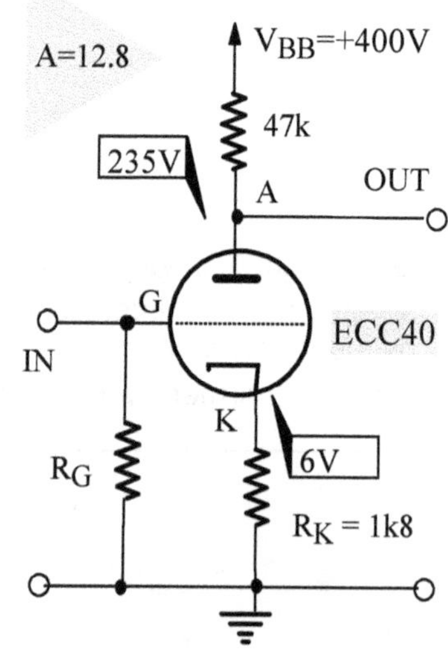

RIGHT: Practical common cathode ECC40 triode stage with un-bypassed cathode resistor

As we will see soon in the chapter about negative feedback, gain with feedback, or "closed loop" gain is $A_F = A/(1+A\beta)$, where A is the gain of the stage without feedback, and β is the "feedback ratio", or the fraction of the output that is fed-back to the input. The term $(1+A\beta)$ is called "feedback factor".

In this case, since $A=-\mu R_L/(r_I+R_L)$ and $A_F = -\mu R_L/[r_I+R_L+(1+\mu)R_K]$, the ratio of feedback voltage to output voltage is the same as the resistance ratio, so the feedback ratio is $\beta = R_K/(R_L+R_K)$ x100 [%].

Whenever you find or derive a formula and you don't have full confidence in it (its source or your own mathematical manipulation skills) check two things. Firstly, carry out dimensional analysis. In this case all three factors are resistances, so upper resistance units cancel the bottom ones, leaving β as dimensionless quantity, which is what it should be. Only by multiplying it with 100 do we get percentages, so it passes the first test.

Secondly, test the equation on simple or special cases. For instance, in this case, if R_K was zero in this AC analysis (bypassed to ground by a capacitor) β would be $0/(R_L+0)$ x100 or 0% and there would be no feedback. So, this formula passes the second test as well. Of course, these two simple tests do not guarantee that a formula is correct, they are "necessary but not sufficient" conditions. Remember that phrase well, it is extremely important in electronics and in life in general. However, if a formula fails either of these tests, it's definitely wrong.

Amplification factor of the cathode output

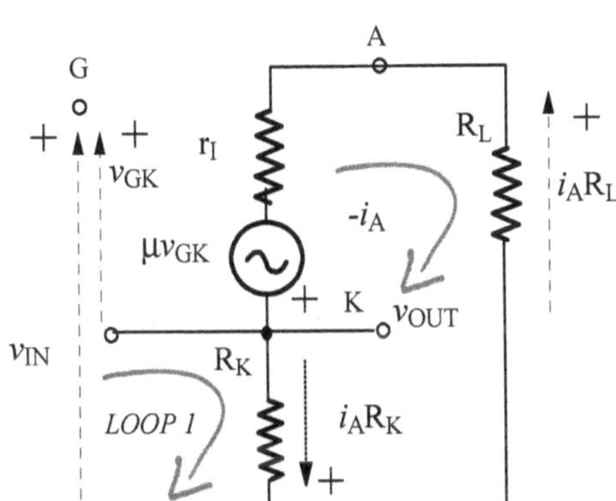

Going back to the same two loop equations: $V_{IN} - V_{GK} + I_A R_K = 0$ and $-\mu V_{GK} - I_A r_I - I_A R_L - I_A R_K = 0$

We also know that $V_{OUT} = -I_A R_K$

From the 1st equation.: $V_{GK}= I_AR_K+V_{IN}$ and substitute it into the second: $-\mu(I_AR_K +V_{IN}) - I_A r_I - I_A R_L - I_AR_K = 0$

$- I_A(\mu R_K + r_I+ R_L+R_K) -\mu V_{IN} = 0$

Since $I_A = -V_{OUT} /R_K$, we can substitute that into the eq. above and move $-\mu V_{IN}$ to the other side: $V_{OUT}(\mu R_K + r_I+ R_L+R_K)/R_K = \mu V_{IN}$

The amplification factor $A_K = V_{OUT}/V_{IN}$ or

$$\mathbf{A_K = \mu R_K/[(1+\mu)R_K + r_I+ R_L]}$$

Output impedance of the anode output

The procedure for calculating the output impedance of an amplifier (or any other 4-pole network) is to short circuit the input terminals and connect a voltage source instead of the load resistor, across the output terminals. The we calculate the current flowing in the output circuit and the test voltage divided by such current is the output impedance Z_{OUT}.

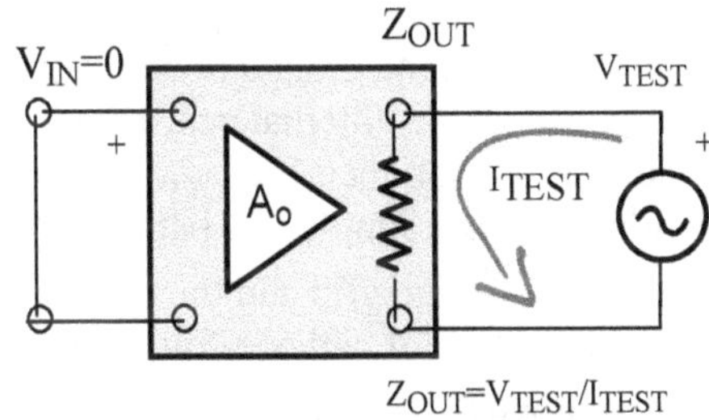

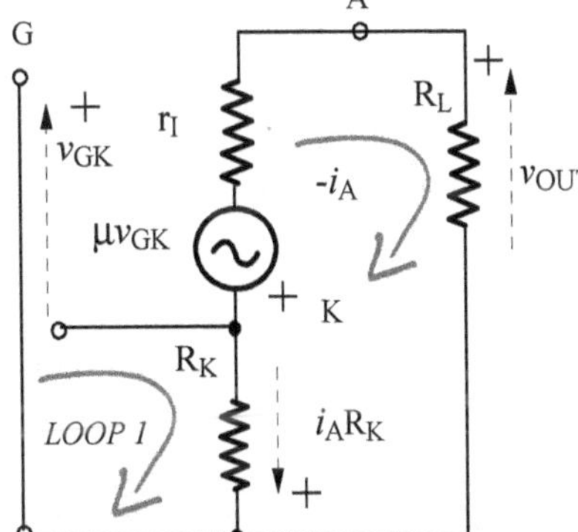

The voltage equations for the input and output loop are:

$- V_{GK} + I_AR_K = 0$ or $V_{GK} = I_AR_K$ and $-\mu V_{GK} - I_A r_I + V_{OUT} - I_AR_K = 0$

Eliminating V_{GK} from the second equation: $\mu I_AR_K + I_A r_I + I_AR_K = V_{OUT}$

$I_A[(1+\mu)R_K+ r_I] = V_{OUT}$ Since $V_{OUT}/I_A=R_{OUT}$ we get $R_{OUT}=r_I+ (1+\mu)R_K$

To do this the easy way we used a trick and considered R_L an external load resistor, but in reality, that resistor is part of the amplification stage and is connected in parallel with the R_{OUT} just calculated so the output impedance of the anode output is $Z_O=[r_I + (1+\mu)R_K] \parallel R_L$ or

$$\mathbf{Z_{OUTA} = R_L[r_I + (1+\mu)R_K]/[R_L+ r_I + (1+\mu)R_K]}$$

Output impedance of the cathode output

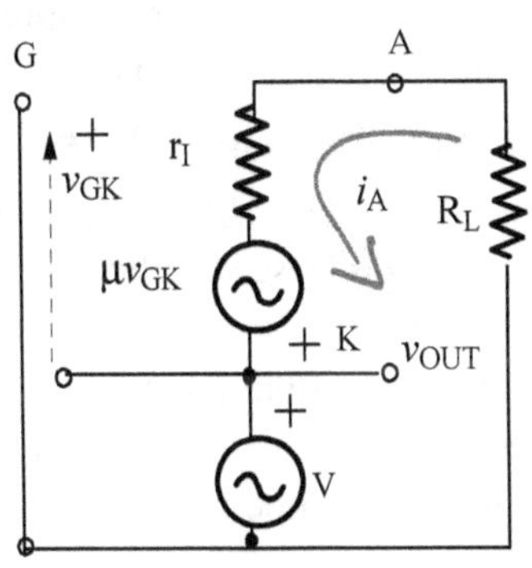

Again, to find the output impedance at the cathode output, we replace the cathode resistor with a voltage generator and short the input terminals. The input loop is now $-V_{GK} - V = 0$ or $V_{GK} = -V$ and the output loop:

$V -\mu V_{GK} - I_A r_I - I_A R_L= 0$

Eliminating V_{GK}: $V+\mu V-I_A(r_I +R_L)=0$ or $V(1+\mu)=I_A(r_I +R_L)$

The impedance at the cathode K: $Z_K = V/I_A$ so $Z_K = (r_I +R_L)/(1+\mu)$

Again, just as with the anode output impedance, we temporarily disconnected the load (which in this case is the cathode resistor R_K) to calculate the output impedance, so now we have to "connect" it back in parallel with the result we got $Z_{OUTK} = Z_K \| R_K = [(r_I + R_L)/(1+\mu)] \| R_K$

$Z_{OUTA} = R_L[r_I + (1+\mu)R_K]/[R_L + r_I + (1+\mu)R_K]$
$Z_{OUTK} = Z_K \| R_K = [(r_I + R_L)/(1+\mu)] \| R_K$

Un-bypassed cathode resistor and the load line

Load line's intersection points with X- and Y-axis are found in the same way as before. Instead of E_{BB} we now have $E_{BB}+E_{CC}$ (point A), and instead of E_{BB}/R_L in point B we now have $(E_{BB}+E_{CC})/(R_L+R_K)$. So far so good.

However, the bias voltages are a problem. Those given in anode graphs are between the grid and cathode, assuming a grounded cathode. In this case, however, the cathode is NOT at the ground potential!

We have to modify grid voltages for all curves. Assuming that the first curve is for V_G=0V, we will now have $V_1 = V_{G1} + I_{A1}R_K - E_{CC} = 0 + I_{A1}R_K - E_{CC}$. Likewise, as per the illustrated example, instead of V_{G2} we have $V_2 = V_{G2} + I_{A2}R_K - E_{CC}$. A practical example will be worked out soon for a cathode follower, the only difference being that R_L in that case will be zero.

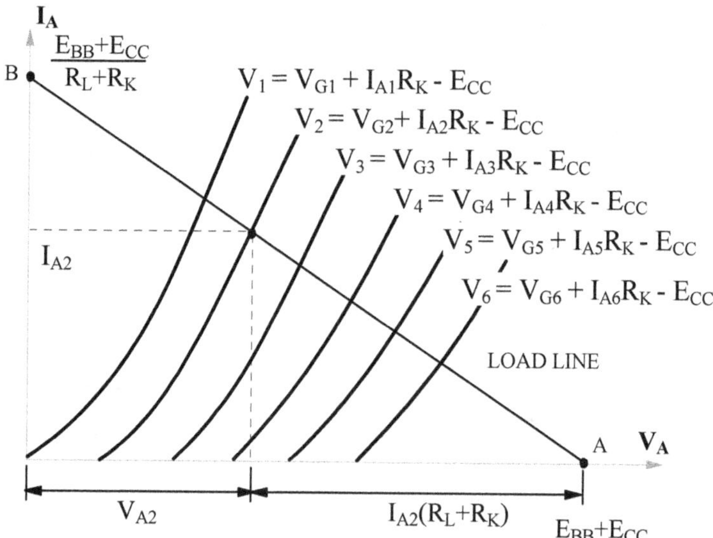

CATHODE FOLLOWER (COMMON ANODE STAGE)

Cathode follower (CF) is a special case of the circuit analyzed above when R_L=0. The anode is directly connected to a positive DC supply voltage E_{BB}, therefore it is at the ground or common potential for AC signals. This, of course, assumes that the internal resistance of the E_{BB} power supply is zero. This condition is not met in most practical designs, except those using highly regulated power supplies or battery power.

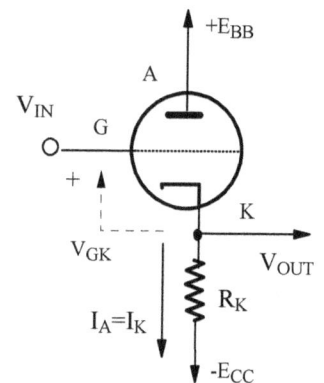

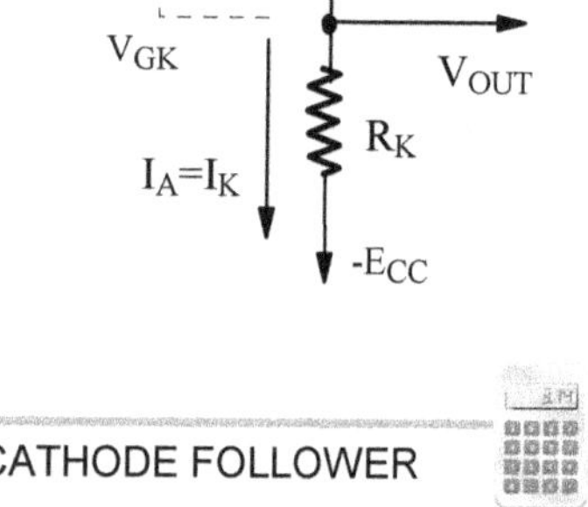

The input is at the grid and the output is taken from the cathode, that is why the "official" name for cathode follower is common or "grounded" anode stage.

If we remove R_L from the amplification factor formula $A_K = \mu R_K/[(1+\mu)R_K + r_I + R_L]$, we get the voltage amplification for cathode follower:

$\mathbf{A_K = \mu R_K/[(1+\mu)R_K + r_I]}$

To get a better feel lets go back again to our circuit with ECC40, only this time removing R_L and taking the output from the cathode.

CATHODE FOLLOWER
$A_K = \mu R_K/[(1+\mu)R_K + r_I]$
$Z_O = r_I/(1+\mu) \| R_K$

As before R_K =1k8, r_I=11k and μ=32, so the gain is now $A_K = \mu R_K/[(1+\mu)R_K + r_I] = 32*1{,}800/[(1+32)1{,}800+11{,}000] = 32*1{,}800/70{,}400 = 0.82$

We have lost all the gain! This "amplification" stage is now acting as an attenuator, the output voltage is lower than the input voltage (only 82% of it)! Why are we even mentioning such a seemingly useless circuit?

Just as with people, first impressions of circuits are often wrong. Remember the output impedance of the cathode output, $Z_{OUT} = (r_I + R_L)/(1+\mu) \| R_K$? For cathode follower R_L=0 so its output impedance is $\mathbf{Z_O = r_I/(1+\mu) \| R_K}$

For our ECC40 circuit $Z_O = 11{,}000/(1+32) \| 1{,}800 = 333 \| 1{,}800 = 281\ \Omega$

Anything below 1kΩ is a very low output impedance. For comparison, the output impedance of a common cathode stage with bypassed cathode resistor is $Z_{OUT} = r_I \| R_L = R_L r_I/(R_L + r_I)$, which for our amplification stage with ECC40 would be $Z_{OUT} = 11k \| 47k = 8k9$!

Ideally, the output impedance of any amplification stage should be zero (ideal voltage source), or as low as possible. Almost 9kΩ from a common cathode stage is not that good, around 280Ω from a cathode follower is much better. As for the input impedance, it is determined by the grid resistor R_G, which can be in the order of 1 MΩ or higher.

Now you see that CF is not really an amplifier, but an impedance "transformer", its high input impedance means it will not load the previous stage, and its very low output impedance means it will not suffer from the loading effect of the following stage. No loading means no distortion and no loss of signal, so CF is an ideal output stage for preamplifiers that have to drive long interconnect cables and also an ideal driver stage for output stages that operate in Class A_2 or AB_2 . Grid current flows in those classes of operation and the input impedance of the power tubes drops very low due to such grid current. More on that soon.

Practical cathode follower with ECC40

The equivalent circuit of a cathode follower

It may prove useful to draw an equivalent circuit diagram for CF just as we did for the CC stage. You can see that the schematic looks identical, except that the values of tube's parameters are different. Instead of the voltage source $v_{GK}*\mu$ we now have $v_{GK}*\mu/(\mu+1)$. The higher the μ, the closer this factor gets to 1, and the closer the output voltage gets to the value of the input voltage. The cathode or output voltage closely "follows" the input or grid voltage, and that's how cathode follower got its name. Notice that the output voltage is in phase with the input, so CF does not invert phase.

Secondly, notice the value for the equivalent internal impedance of the tube in cathode follower connection: it is the internal impedance of the tube r_I divided by the factor $(1+\mu)$. It is that low equivalent internal impedance that gives CF its main strength, low output impedance!

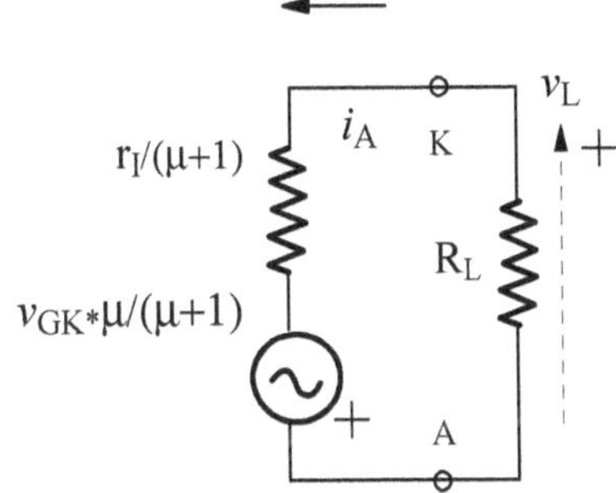

The equivalent circuit

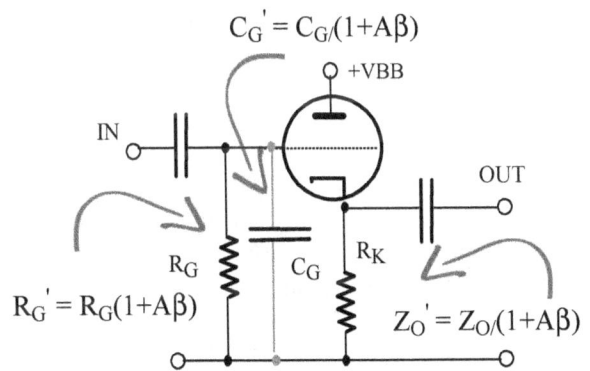

The feedback ratio and feedback factor of a cathode follower

To summarize and simplify the CF analysis, consider the in-principle diagram below. Since the whole output signal is fed back into the input circuit, the feedback ratio is β=-1 (100% negative feedback). This makes the feedback factor (1+Aβ)= (1-A) All input impedances are multiplied by this feedback factor (1+Aβ) or (1-A)! The grid resistor's new effective value is now $R_G(1+A\beta)$! Since A=0.82 in our ECC40 case and 1+Aβ= 1–0.82 = 0.18, instead of the 1M resistance of the grid resistor, the input resistance of the cathode follower is 1M/0.18 =1M *5.55 = 5.55MΩ!

Since the capacitive impedance between the grid and cathode is also multiplied by (1+Aβ), this is equivalent to the input capacitance being much lower (since $X_C=1/\omega C$!)

The output impedance of the stage would be the internal impedance of the tube r_I=11k in parallel with R_K=1k8 or 1.55k. NFB reduces the output impedance by the factor (1+Aβ) or 5.55 times to arrive at Z_O' = 1,550/5.5 = 281 Ω, exactly the result we got earlier.

Cathode follower biasing arrangements

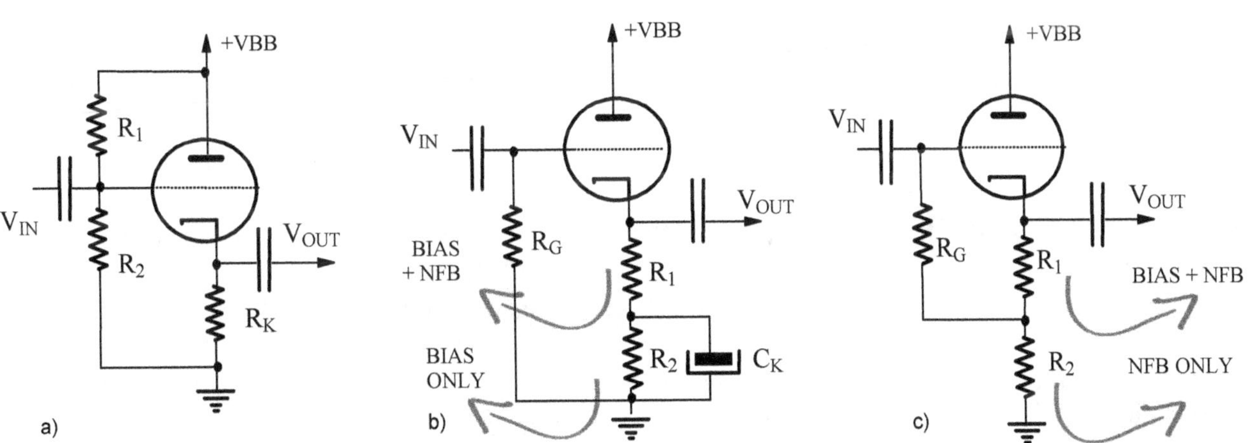

Three biasing methods for cathode followers and other circuits with un-bypassed R_K

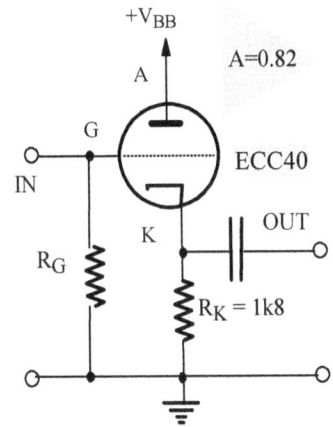

A CF stage can be biased in various ways. The arrangement (a) is the easiest to comprehend. The grid is biased by the voltage divider formed by R_1 and R_2.

Once the operating point is positioned on the anode characteristics, the cathode current and the V_{GK} is known. The value of the R_K is calculated based on the desired cathode current. The parallel combination of R_1 and R_2 should not exceed the maximum allowed grid resistance specified for the particular tube used. This arrangement is never used in practice, the other two biasing methods are superior because they offer a much higher input impedance.

In (b) R_1 and R_2 determine the cathode voltage and thus the grid bias voltage, while the un-bypassed resistor R_1 provides the cathode feedback. A larger output voltage swing is achievable compared to a), since the cathode can be at a much higher potential.

Since circuit (c) offers the best performance, the (b) circuit isn't common either.

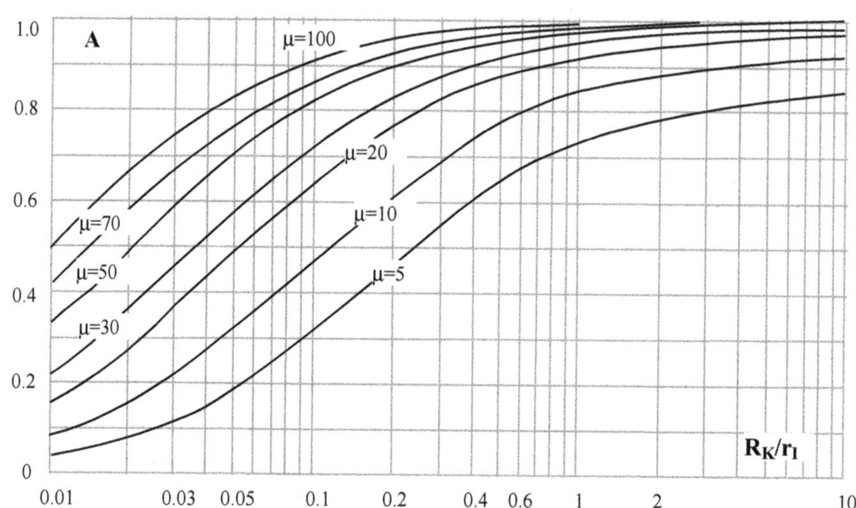

How the gain of cathode follower varies with μ of the tube and the ratio R_K/r_I

The "Bootstrapping" concept

Let's consider a practical CF circuit using another great tube, 12AU7 duo-triode, which also goes by the name ECC82 and many, many equivalent or near-equivalent designations: E82CC, 6189, 5814, B329, CV491, E2163, E812CC, ...

12AU7 has low internal impedance and high anode dissipation (2.75W), so it can operate at relatively high anode currents of 10-20mA, and high current stages generally sound better than low current ones!

TUBE PROFILE: 12AU7 (ECC82)

- Indirectly-heated medium μ dual triode
- Noval socket
- Heater: 6.3V/300mA or 12.6V/150mA
- V_{AMAX}=300V, V_{HKMAX}= 100V
- P_{AMAX}=2.75 W, I_{AMAX}=20 mA
- TYPICAL OPERATION:
- V_A=250V, V_G=-8.5V, I_0=10.5 mA
- gm=2.2 mA/V, μ=17, r_I = 7.7 kΩ

I_K=120/17,157=7 mA
V_{BIAS}=I_K*857= -6V
so V_G=120-6=114V

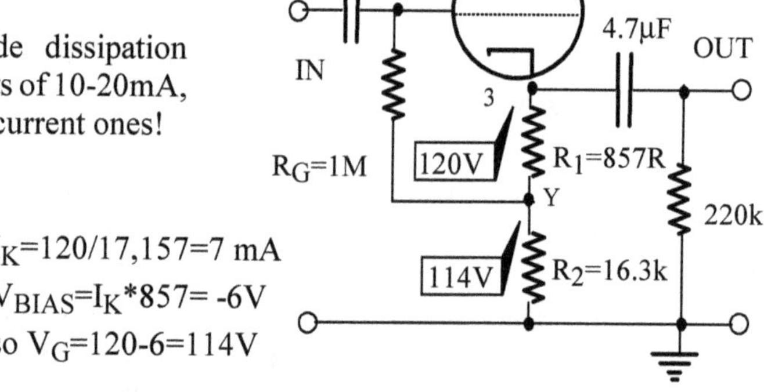

ABOVE: Practical cathode follower with ECC82. Add it to the output of your CD-player or any line or phono stage. Use it for decoupling stages in a power amplifier. Make it into a stand-alone buffer. This little circuit performs well and sounds good.

Bootstrapping is an American term that likens circuit action to a person pulling themselves up by their own bootstraps or shoelaces. While I don't think that is physically possible, the concept works in electronics. I've tried to come up with a better linguistic alternative but failed, so for now we have no choice but to use it.

In this case, the grid resistor is bootstrapped, from point Y back to the input (point X), which is equivalent to an increase of the input impedance Z_{IN} from approximately R_G to a higher value $Z_{IN}= R_G/[1-AR_2/(R_1+R_2)]$

The higher R_2 and the smaller R_1, the higher the bootstrapping effect and the higher the input impedance. With R_K =17k16, r_I=7k7 and μ=17, the gain is $A_K=\mu R_K/[(1+\mu)R_K + r_I]$ = 17*17,160/[(1+17)17,160+7,700] = 0.90

The 857R and 16k3 resistors form a voltage divider, so the voltage in point Y is 16,300/(857+16,300)*V_{OUT} = 0.95V_{OUT}, where V_{OUT} is 0.9V_{IN} in this case, so V_Y=0.95*0.9*V_{IN}= 0.855*V_{IN}

That means that the input signal "sees" a much reduced voltage across the grid-leak resistor, only 0.145V_{IN} in this case, which is equivalent to the input impedance of $Z_{IN} = R_G/0.145$ = 6.9 MΩ, an increase of almost 7 times!

The output impedance of this CF is $Z_{OUT}=r_I/(1+\mu)||R_K=r_I/(\mu+1+r_I/R_K)$ =7,700/(17+1+7,700/17,157) = 417Ω

Voltage transfer characteristic of a cathode follower

The voltage transfer curve (also called the dynamic response curve) is a plot of the output voltage versus the input voltage. It enables the designer to chose the most suitable grid bias voltage, the maximum input voltage that can be accepted by the stage and the voltage gain.

Once the desired load line is positioned on the anode characteristics, anode currents in a few points (5-6 is enough) are read from the intersections of the load line with anode curves. In step two, the output voltage (cathode voltage) is calculated by multiplying the anode (cathode) currents by the cathode resistance. Finally, in step 3, the input voltages in selected points are calculated by using the formula $V_{IN}=V_G+I_AR_K$.

Notice how this transfer curve is actually a straight line (compared to that of a CC stage on p95-96), confirming the fact that cathode follower is a very linear stage with extremely low distortion!

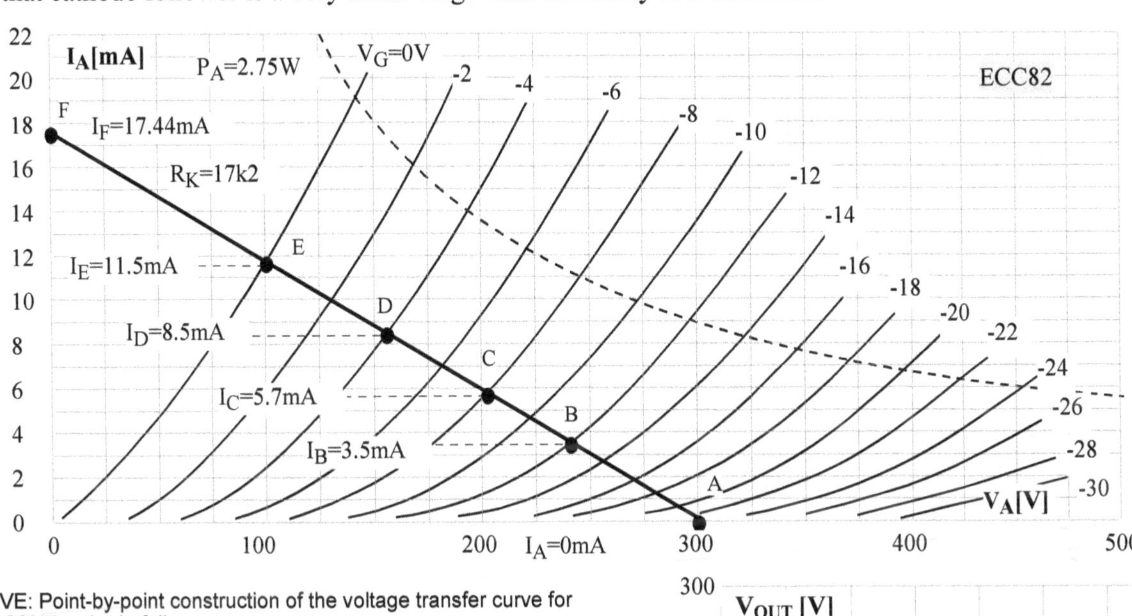

ABOVE: Point-by-point construction of the voltage transfer curve for the 12AU7 cathode follower

POINT	I_A [mA]	V_{OUT} [V]	V_{IN} [V]
A	0	0	-26
B	3.5	60.2	-12+60.2=48.2
C	5.7	98	-8+98=90
D	8.5	146	-4+146=142
E	11.5	198	0+198=198

RIGHT: The voltage transfer curve of the 12AU7 cathode follower

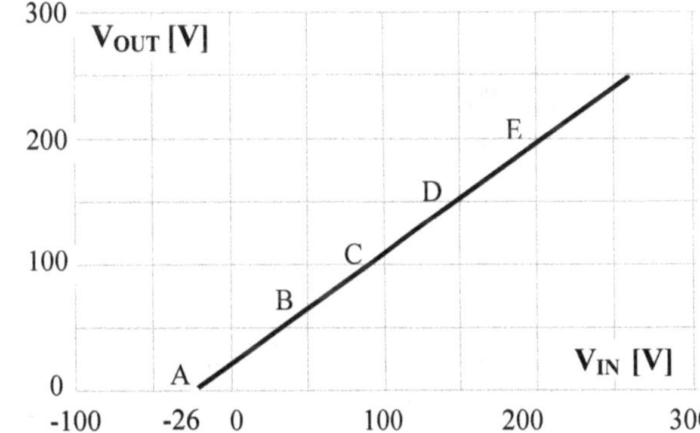

THE CATHODE FOLLOWER QUARTER-BRIDGE DRIVER STAGE

5687 duo-triode

5687 is a high perveance, high emission duo-triode. Each triode anode dissipation is rated at 4.2 Watts or 7.2 Watts together, so it is a high power, high anode current triode. There is a strong correlation between the anode current and the quality of the sound of both power and preamp tubes, such as 5687: the higher the current, the better the stage sounds.

Even the grid of 5687 can take 7 mA, which is higher than the anode of 12AX7! While μ is an par with 12AU7, gm is high and r_I is 2.5 times lower than 12AU7, around 3 kΩ. A high amplification factor is not needed in the cathode follower application anyway, but low internal impedance is crucial for driving the output tubes into class A_2 operation.

By paralleling two triodes we can get an internal resistance of 1k5 and gm of 10.8 mA/V. That translates into very dynamic sound, minimal internal losses, and low distortion! All in all, this is one awesome driver tube.

TUBE PROFILE: 5687

- Indirectly-heated medium μ dual triode
- Noval socket
- Heater: 6.3V, 0.95 A or 12.6V, 0.45A
- $V_{AMAX}=300V_{DC}$, $V_{HKMAX}=90V_{DC}$
- P_{AMAX}= 4.2 W, I_{GMAX}= 5 mA
- TYPICAL OPERATION:
- $V_A=250V$, $V_G=-12.5V$, $I_0=12$ mA
- gm=5.4 mA/V, μ=16, r_I = 3 kΩ

PRACTICAL DESIGN: Quarter-bridge driver stage with a symmetrical power supply

This cathode follower driver stage with symmetrical (+/-150V to +/-200V) DC power supply and direct coupling to the grid of the output tube acts as an impedance transformer can be used with a variety of driver and power tubes. The best choices are high current, low impedance driver tubes such as 6CG7, 12BH7, 6SN7 and 5687! It decouples the preamp stages so the output stage does not load them, thus minimizing distortion. It also enables direct coupling to the output stage, eliminating one capacitor from the signal path and its associated time constant. This widens the frequency range and makes global negative feedback more stable at high frequencies. The symmetrical power supply maximizes voltage swing and headroom, again, minimizing distortion.

Due to its low output impedance and high current capability, it can drive the power stage deep into class A_2. Since capacitive coupling cannot be used with class A_2, the only other option is to use an interstage transformer, and we will soon see what headaches that involves. Anyway, why use a chunky, expensive, distorting, sound-coloring piece of iron when you can use this elegant, cheap and relatively simple "electronic" interstage transformer?

Another strong point of this marvelous circuit is that its negative power supply makes a fixed bias arrangement very easy to implement. Instead of biasing the power tube directly, you bias the driver tube in the cathode follower, adjusting its cathode voltage to suit the required grid bias of the power tube.

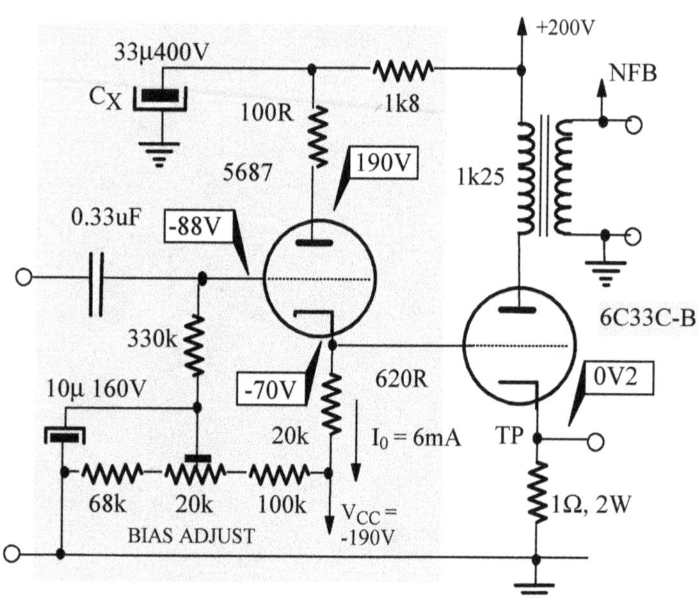

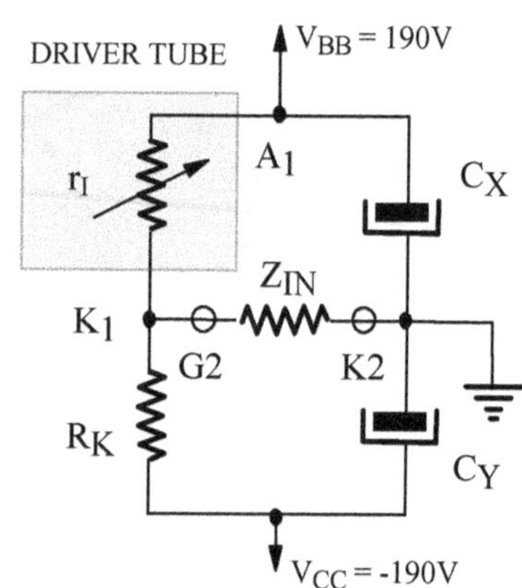

ABOVE: The simplified model reveals the quarter-bridge topology that may not be obvious from the circuit diagram.

LEFT: The schematics of the universal cathode follower driver stage, in this case driving 6C33C-B triode

The circuit is actually a quarter-bridge. The internal impedance r_I of the driver tube is the active element (the only one that changes out of four arms, that's why it's called a quarter bridge), R_K (the cathode resistor) is in the bottom branch or arm of the bridge, while two power supply filtering capacitors C_X and C_Y (not shown on the diagram, located in the power supply section) are in the 3rd and 4th arm of the bridge.

Finally, it provides a safe startup for directly-heated output tubes. When the amplifier is switched on, a directly-heated tube such as 300B or 2A3 will heat up much faster than the indirectly-heated 5687 driver. Since there is no cathode current flowing through R_K of the 5687 driver, its internal resistance r_I is very high and the full -190V is on the grid of the output tube, ensuring that the tube is in cutoff (no anode current flows at all). Once the 5687 warms up, its r_I decreases, its cathode current starts flowing and the bias on the output tube's grid slowly raises to the quiescent state of around -70V, by coincidence, the same bias as for the 6C33C-B triode in the circuit above.

Positioning the operating point

As with all cathode follower, we have to convert grid voltages from the graph into the actual grid voltages needed to analyze the circuit. Three points are enough, the quiescent point Q and two maximum swing points X and Y.

$V_{GX}' = V_{GX} + I_{AX}R_K + V_{CC} = 0 + 17*20 - 190 = 340-190 = 150V$

$V_{GQ}' = V_{GQ} + I_{AQ}R_K + V_{CC} = -14 + 6*20 - 190 = 106-190 = -84V$

$V_{GY}' = V_{GY} + I_{AY}R_K + V_{CC} = -26 + 0*20 - 190 = -26-190 = -216$

$\Delta V_{G+} = 150-(-84) = 150+84 = 234V \quad \Delta V_{G-} = 216-84 = 132V$

The positive signal of up to 234 V_{PEAK} can be accommodated, while the negative peak can only be 132V. So, it seems that we haven't positioned the quiescent point Q properly. However, firstly, the whole signal can be 264 V_{PP} or 94 V_{RMS} without clipping, and that is far in excess needed for triodes such as 300B, 211, 845 or 6C33C-B.

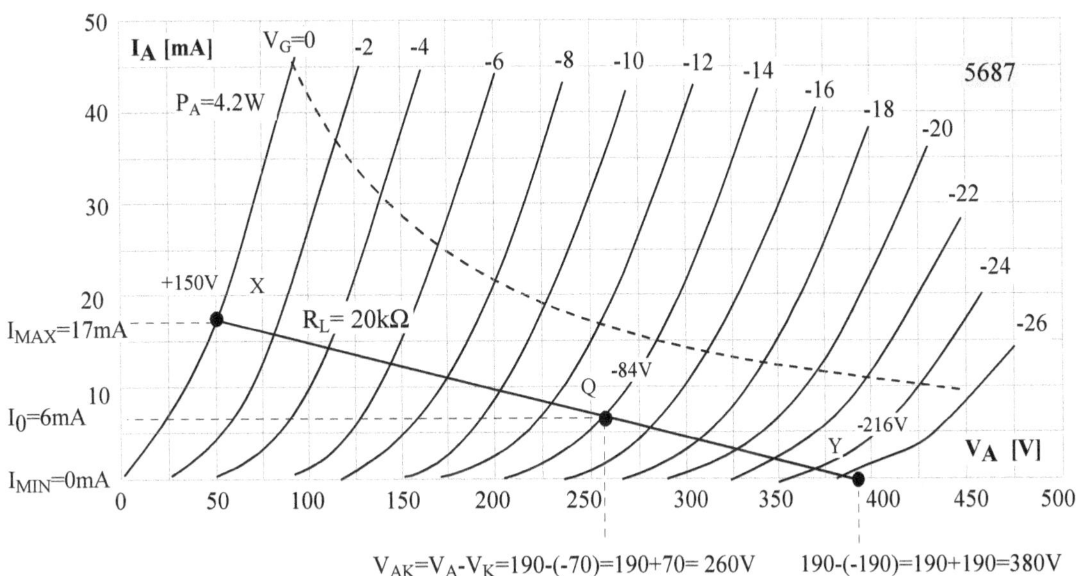

Secondly, we determined the bias voltages experimentally, not graphically, and the final voltages allowed for the maximum output power of the amplifiers this circuit was used in, meaning the positive and negative swings were equal! One possible explanation is that the plate curves published and reproduced here are not identical to the anode characteristics of the 5687 tubes we used.

Instead of Q being at the intersection of V_{AK}=260V and V_G=-14V, it would be enough if all the tube curves were shifted 2V to the right, and such intersection was with the V_G=-12V curve instead! That would have balanced or equalized the maximum positive and negative swings, so the maximum peak-to-peak swing of 150+216 = 366V_{PP} would be available or 130V_{RMS}. Another example of how blind reliance on published figures and graphs can actually lead to sub-optimal designs.

The 6SN7 - 12SN7 version

If you are one of those 6SN7 addicts and cannot live without the beautiful coloration (but coloration nevertheless!) of this old dame, you can use it instead of 5687. Here are the main figures:

V_G = -90+80 = -10V

V_{RK} = 200-80=120V, R_K=20kΩ

$I_K=V_{RK}/R_K$= 120/20 = 6mA

V_{AK}=200-17+80 = 263V

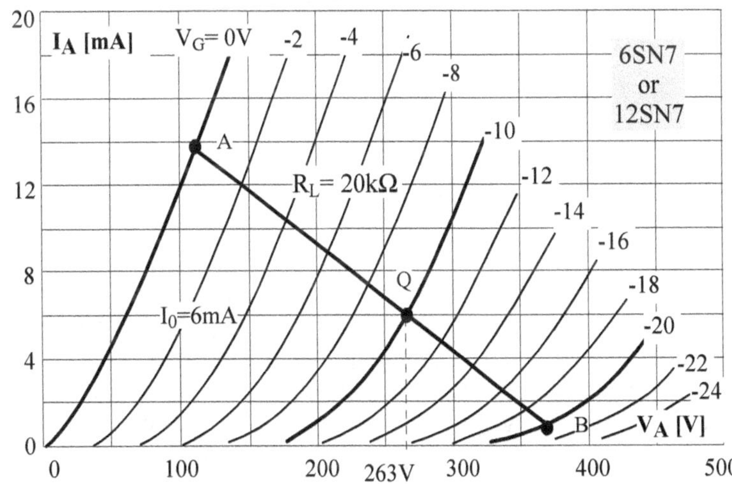

Common-cathode stage with local anode-to-grid negative feedback

A stage with a local negative feedback from its anode to its grid has an amplification factor:

$$A=\frac{1}{\frac{R_{IN}}{R_F}+\frac{1}{A_0}\left(\frac{R_{IN}}{R_G}+\frac{R_{IN}}{R_F}+1\right)}$$

By substituting $\alpha=R_F/R_{IN}$ and rearranging:

$$A=\frac{\beta A_0}{\frac{R_F}{R_G}+A_0+\alpha+1}$$

In this case $A_0 = -\mu{*}R_L/(r_I + R_L)$

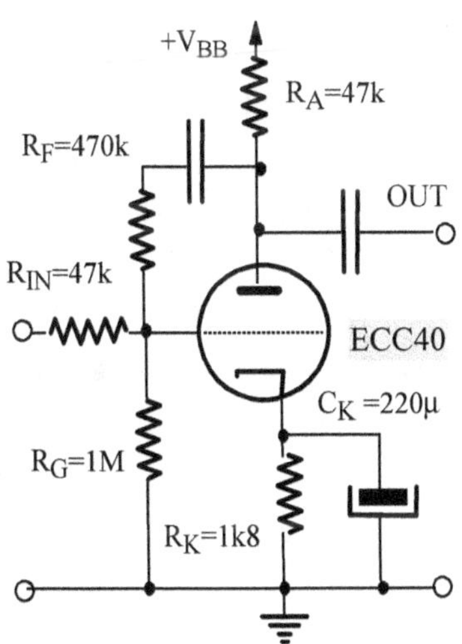

For our ECC40 stage with R_L =47k, r_I=11 kΩ and µ=32, we have

A = -32*47/(11+47)= -32 *0.81 = -26

$\alpha=R_F/R_{IN}$ = 10 so A= 10*26/(0.47/1+26+10=1)= 260/37.47 = 6.94

This assumes that the output impedance of the preceding stage is zero, if it is not, it simply needs to be added to R_{IN}. To produce the same output voltage as before the NFB was applied, the input signal needs to be 26/7 = 3.7 times larger, the input sensitivity has been reduced by this factor, named feedback factor $(1+\beta A_0)$. Expressed in dB, in this case, 20log 3.7 = 11.4dB. Distortion and noise are also reduced by the same factor as gain, or 11.4dB.

Anode follower

If you are a solid-state buff or a recent convert to tubology, this stage may seem familiar to you. Once you replace the tube and its anode and cathode elements with an operational amplifier, you should recognize this circuit! We have a special case where $R_F=R_{IN}$ and $\alpha=R_F/R_{IN}=1$

R_G is on the other side of R_{IN} so it does not feature in the gain equation, which has now simplified down to $A=\alpha A_0/(A_0+\alpha+1) = 1*26/(26+1+1) = 26/28 = 0.93$

The anode resistance has been reduced down to $R_{ANEW} = R_A/(A_0/A) = 47/(26/0.93) = 47/27.96 = 1.68k\Omega$ The output impedance is $Z_{OUT} = R_A||R_{ANEW} = (47*1.68)/(47+1.68) = 1.62k\Omega$

The distortion and noise have been reduced 26/0.93 = 28 times (or 29dB). Since these figures and behavior reminded someone of the cathode follower, this circuit was named anode follower (since the output is taken from the anode). The analogy isn't complete though, because in cathode follower the cathode or output voltage really follows the grid or input voltage, and the two voltages are in phase. In anode follower the output voltage is of the opposite phase (inverted).

Benefits? As always, NFB increased the upper -3dB frequency, increased the input headroom, reduced distortion and noise, and made the performance of the circuit much less dependent on the variation in tube parameters.

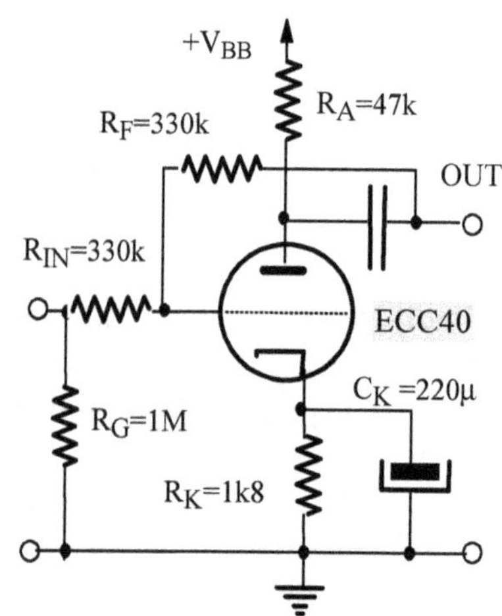

When $R_F=R_{IN}$ the amplitude of the output signal is approximately equal to that of the input signal, but the phase is reversed

THE COMMON GRID STAGE

The grounded-grid or common-grid circuit is useful in interfacing a source of low impedance, for instance a transistor preamplifier, to a load of high impedance such as tube output stage. It is also used instead of step-up transformers as a pre-preamplifier for MC (moving coil) phono cartridges.

Moving coil (MC) cartridges generate a very low voltage output, in the order of 0.1 mV or 100 µV. One way to bring those miniuscule voltages to a level needed to drive moving magnet (MM) phono preamplifiers is to use step-up transformers. Since MC transformers are expensive ($500 - $5,000+), some manufacturers use an additional amplification stage for MC cartridges. The gain does not have to be very high, 10-20 is all that is needed and almost any triode will achieve that (even lowish gain triodes such as 12AU7). However, more importantly, a very low input impedance is needed.

The stage has a similar gain to common-cathode stage, but it does not invert phase. The grid acts as an electrostatic screen between the cathode and anode, and since it is grounded for AC signals, the GG stage does not suffer from the dreaded Miller effect (more on that soon!). This gives it a very wide bandwidth and low noise, which is important when dealing with 0.1-0.5 mV moving coil signals.

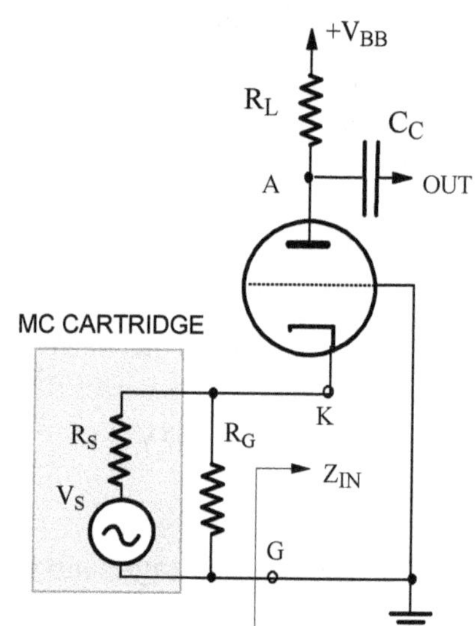

Common grid stage as a phono preamplifier

Ignoring the source resistance R_S for now, the loop equation is $\mu v_{GK} - v_S = i_A(r_I + R_S + R_L)$ (*) Since $v_{GK} = -v_S$ (**) we can substitute it into equation (*) and write $v_S(\mu+1) = i_A(r_I + R_L)$

Since $v_{OUT}=i_AR_L$ we get $A = (v_{OUT}/v_S) = i_AR_L/v_S = R_L/[(r_I + R_L)/(\mu+1)]$

Finally, the voltage amplification factor A is $\mathbf{A=(\mu + 1)R_L/[r_I + R_L]}$

The analysis is a bit more complex should we include the internal resistance R_S of the voltage source v_S (such as MC cartridge), but the end result is

$$\mathbf{A = [(\mu+1)R_L]/[r_I+(\mu+1)R_S + R_L]}$$

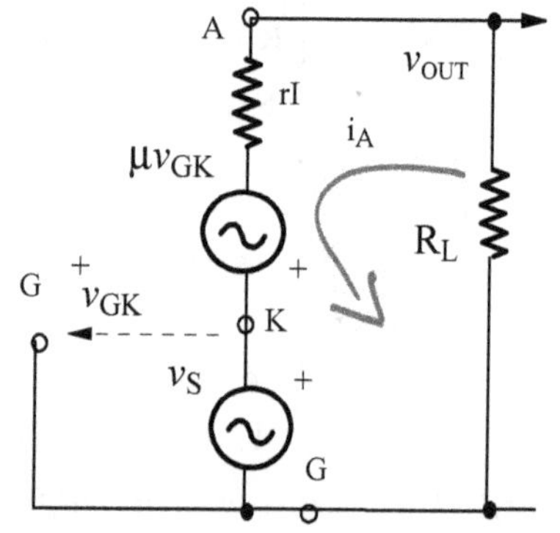

The equivalent circuit without the source resistance R_S

For example, Ortofon MC-1 cartridge has Rs=5Ω, so the gain with a 12AX7 triode (r_I=65kΩ, μ=100 and R_L=100kΩ) would be A = [(100+1)*100,000]/[65,000 + (101)*5 + 100,000] = 101*0.6 = 60

The output impedance is $R_{OUT} = r_I+(\mu+1)R_S$, and the input impedance (terminals K and G) is $R_{IN} = (V_G/I_P)=(rI+R_L)/(\mu+1)$

For the same circuit with 12AX7 tube, R_{IN} = (65+100)/(100+1) = 1.63kΩ.

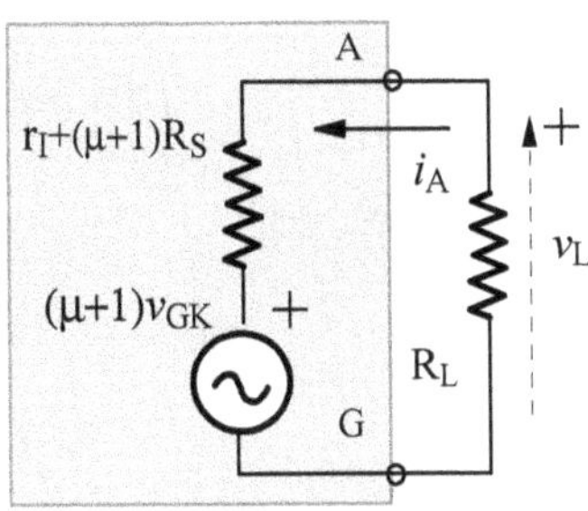

ABOVE: The final equivalent circuit of the GG stage

THE CASCODE

ECC88 and its equivalents

ECC88 (6DJ8, 6922) is a duo-triode developed for cascode circuits used in vintage TV tuners, tube oscilloscopes and other instrumentation. The special quality version was named E88CC and usually has gold-plated pins.

PCC88, the 7V heater version also works well on 6.3V. Not all ebay sellers know this so they price it lower, making it a cheaper alternative. PCC88 also has a much higher maximum heater to cathode voltage, which may be crucial in some circuits. Also, by underheating a tube, its life is prolonged and noise is reduced!

The internal shield connected to pin 9 needs to be grounded.

The ECC88 anode curves are linear and equidistant, so we can anticipate low distortion. ECC88 works well at very low anode voltages (12-48V), and achieves gains of 14-22 per stage. With only two stages an overall gain at 1 kHz of 40 (or 32 dB) can be achieved. This we will use soon in our battery phono stage designs.

TUBE PROFILE: ECC88 (6DJ8, 6922)

- Indirectly-heated duo-triode
- Noval socket, 6.3V, 365 mA heater
- Maximum anode voltage $130V_{DC}$
- $V_{HKMAX}=150V_{DC}$
- P_{AMAX}=1.8 W, I_{AMAX}=25 mA
- TYPICAL OPERATION:
- V_A=90V, V_G=-1.3V, I_0=15 mA
- gm = 12.5 mA/V, μ=33, r_I = 2.6 kΩ

The philosophy behind the cascode

Just like a pentode, a cascode has high gain, high output impedance and very low input capacitance, meaning the Miller effect does not affect it that much, so it can operate successfully at very high frequencies. Cascode does not suffer from pentodes' drawbacks such as noise and microphonics.

The bottom triode works in common-cathode mode. Its output feeds the cathode of the upper tube, which has its grid at the fixed DC potential and zero AC potential, meaning it works in the common grid mode.

For the sake of simplicity, since transfer curves are given for V_A=90V and 150V, we will design the cascode so that the bottom triode has V_{AK}=90V and the upper one V_{AK}=150V!

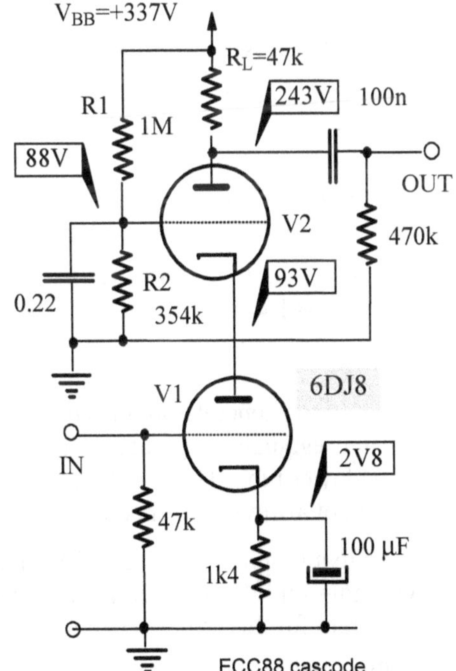

ECC88 cascode

Topology

The upper triode's grid is bypassed to ground for AC signals, which means that there is no feedback from plate to grid, and the grid acts as a shield to prevent feedback to the cathode, eliminating the dreaded Miller Effect (to be discussed soon).

The lower half of the stage contributes a relatively small amount of gain, and is not subject to the feedback effect since it is operating as a current amplifier (there is very little voltage swing on the plate, so there is little or no signal to feed back).

Although in the block diagram the two tubes were drawn horizontally, as in the *cascade* arrangement, to prevent confusion and save space a *cascode* is always drawn in a vertical arrangement, as are many other totem pole arrangements (SRPP, μ-follower), when one tube is in the anode or cathode circuit of the other.

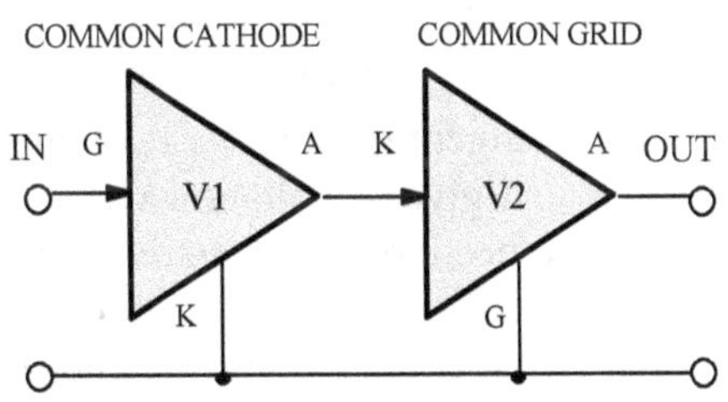

ABOVE: The block diagram of the cascode

CASCODE GAIN

$A \approx gm_1 * R_L$

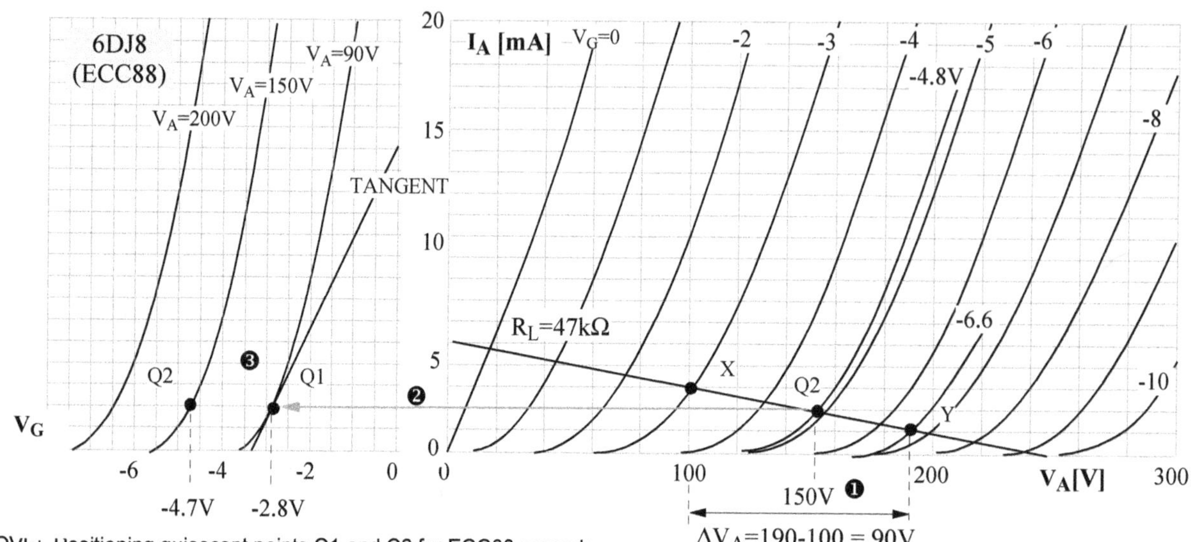

ABOVE: Positioning quiescent points Q1 and Q2 for ECC88 cascode

DC currents and voltages

Starting with the upper triode, we have chosen V_{AK2}=150V and 2mA of current (1). That determines our quiescent point Q2. The grid bias required is -4.8V. Going across (2) to the transfer curve for V_A=90V (for the lower triode), we get its required bias of -2.8V. That is Q1.

V1's anode is 90V above its cathode voltage, which is equal to its bias voltage, so V_{K1}=2.8V and V_{A1}=92.8V. Also, since V_{AK2}=150V and V2 cathode is at 92.8V, the anode of the upper tube must be at 92.8+150 = 242.8V!

The grid of V2 must be 4.8V below its cathode voltage, which is 92.8V, so V_{G2}=92.8-4.7 = 88.1V

Now we need to choose the anode resistor. Let's use 47kΩ, a standard value. 2mA of current will cause 94V voltage drop on this resistor. Thus, our power supply must provide V_{BB}=94+242.8= 337 V! We will use a voltage divider to bias the upper tube: $V_{G2}= R_2/(R_1+R_2)V_{BB}$ so we need to chose one of the two resistors. The maximum grid-leak resistor for ECC88 is 1MΩ, and that resistance will be the parallel combination of R_1 and R_2, so let's chose R_1=1M.

$V_{G2}R_1=(V_{BB}-V_{G2})R_2$, so $R_2=R_1V_{G2}/(V_{BB}-V_{G2})$=88(336.8-88)= 354kΩ Finally, the cathode resistor of V1 needs to be 2.8V/2mA= 1.4kΩ

AC gain

The lower triode generates all the current swing, its gm determines the gm of the whole cascode, and this allows us to immediately estimate the gain: $A \approx gm_1R_L$ We know the load resistance of the upper tube (47k) but need to determine gm of the lower tube. That can be done by drawing a tangent on the 90V transfer curve (3) or, by using the diagram (often but not always published in the data sheet) showing how three parameters vary with anode current.

Although that graph is for V_{AK}=150V and our V_{AK}=90V, we can estimate gm=3mA/V. Transconductance is low since our chosen anode current for this high current tube is very low (2mA).

Choosing a higher anode current would cause a huge voltage drop on the 47k load resistor and the needed V_{BB} would very quickly exceed 500-600V,which is not an "easy", practical voltage! In our case $A \approx gm_1R_L$ = 3mA/V*47,000 V/A= 3*47 = 141

Using the tangent method we read gm1=14mA/3.4V = 4.1mA/V, which would give us voltage gain of $A \approx gm_1R_L$ = 4.1*47 = 193! There is no need to lose sleep over such a large disparity, both are rough estimates anyway, but illustrate an important point, that cascode's gain is in the same ballpark as pentode's (100-200)!

The anode characteristics of two 6SN7 triodes in a cascode connection (RIGHT) indeed reminds us of pentode curves, although the characteristics are less steep in this case, indicating lower internal resistance of the composite tube, which can be modeled with the following parameters:

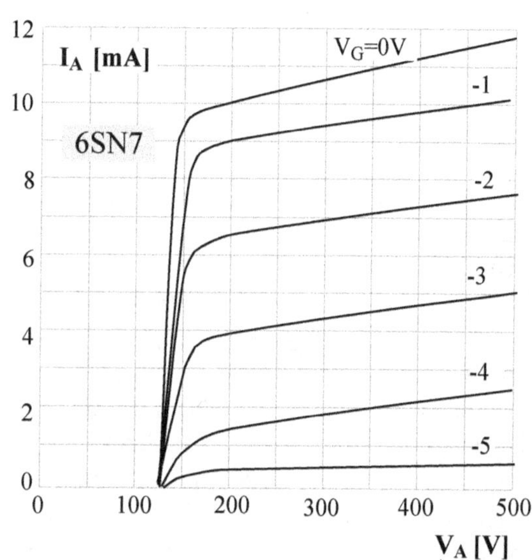

ABOVE: Composite anode curves for 6SN7 cascode

$\mu_C = \mu(\mu+1) \approx \mu^2$, $r_{IC} = (\mu+2)r_I$, and $gm_C = gm(\mu+1)/(\mu+2) \approx gm$, or in words, the same transconductance but squared amplification factor!

For 6SN7 cascode, $\mu_C=\mu^2=20^2=400$, $r_{IC}=(\mu+2)r_I= 2*7k\Omega=150k\Omega$ and $gm_C=gm(\mu+1)/(\mu+2)=2.8$ mA/V

Considering that 6SN7 has the amplification factor μ of only 20, that is an extremely high gain. The exact gain is always lower than the quick approximation: $A=-1/\{1/gmR_A +[(r_I+R_A)/R_A]/[\mu(\mu+1)]\}$ r_I is the internal resistance of the tube, around 6 kΩ in this case.

Going back to the ECC88 example, finding the gain of the upper tube is easy from the anode characteristics. Going from Q 1.8V to the left (V_G=-3V) we get point X, and moving 1.8V to the right we get point Y with V_G=-6.6V! So ΔV_G=3.6V, and drawing vertical lines from X and Y gives us ΔV_A=190-100 = 90V, so A=90/3.6 = 25

Remember, the upper stage (common grid) does not invert the phase. The gain of the lower stage is now somewhere between 141/25 =5.6 and 193/25=7.7 !

SRPP (SHUNT-REGULATED PUSH-PULL or SERIES-BALANCED AMPLIFIER)

Shunt-Regulated Push-Pull stage or SRPP has a high Z_{IN} and moderate Z_{OUT}, much higher than a cathode follower. Its amplification is higher than that of a single-tube common cathode stage, but much lower than can be achieved with two common cathode stages in cascade. None of these are its advantages. The main strengths of the SRPP stage are wide bandwidth (excellent transients), low distortion and a low component count (simple topology). Tubes V1 and V2 don't have to be identical, but usually are, since one duo-triode can be used for both.

The heater supply of the upper tube (V2) maz need to be elevated to a higher DC level, because most preamp tubes have maximum heater-cathode voltage around 90-100V, so if the cathode of V2 as at 130V and the cathode of V1 is close to zero potential (a few volts), the heater should be elevated to somewhere in between, say 60 Volts or so.

The bottom triode is operating in the common cathode mode. Its anode AC current creates a voltage drop across the anode resistor R_2, which functions as a simple current-to-voltage converter.

This voltage is fed directly into the grid of V2, so the upper triode works as cathode follower. Indeed, its anode is connected straight to the high voltage $+V_B$ and is at ground potential for AC signals. Now we understand a bit about SRPP, but we are still none wiser: why is this circuit called "push-pull"?

If $R_1=R_2$, the anode voltage of V1 will be half of the supply voltage or $+V_B/2$, which is also the biasing voltage for the grid of the upper triode. If we "dissect" or de-couple the upper triode form the bottom one, we get a circuit depicted here.

Without a signal, the same anode/cathode current flows through both tubes.

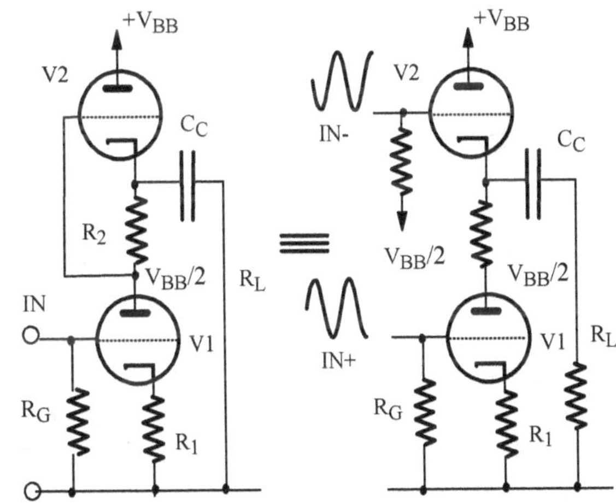

ABOVE: The dissection of SRPP stage reveals its hidden push-pull character!

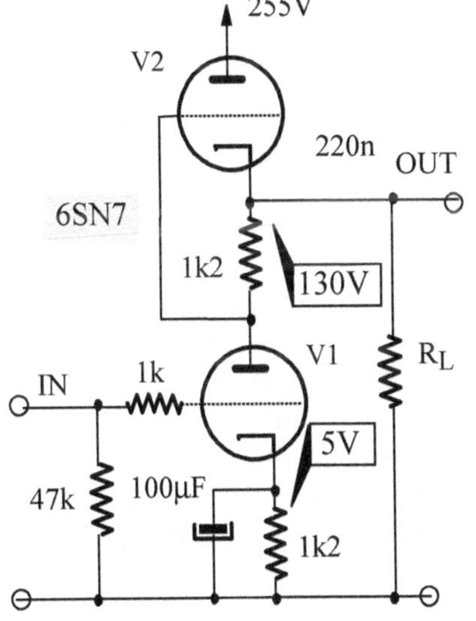

ABOVE: PRACTICAL DESIGN of SRPP with a 6SN7 duo-triode

The positive-going AC signal on the grid of the bottom triode is amplified, inverted and as a negative-going signal it's fed into the upper tube's grid.

Instead of bringing those symmetrical signals in from a preceding phase splitter, the SRPP circuit does that job "internally". Therefore, it is a phase-splitter, a voltage amplifier and an output stage all in one!

If the bottom tube stops conducting, so will the upper tube, so the circuit can only operate in class A! As V1 conducts more and pulls the output voltage down, V2 conducts less, and as V2 pushes the output voltage up by conducting more, V1 conducts less and pulls its anode voltage down, therefore a true push-pull action is achieved.

When it was originally conceived and patented in 1940 (U.S. patent no. 2,310,342) SRPP was called a "balanced direct and alternating current amplifier" and "series amplifier" because as far as the signal was concerned, the two tubes were connected in series with one another, just as they would be in a horizontally-drawn cascade arrangement.

Most SRPP stages don't work as such at all!

SRPP would have to be the least understood and most "misnamed" amplifying stage of all. In fact, 90+% of circuits named SRPP do not operate as such! The secret, or the critical difference is in the load. Just like any output stage (and SRPP surely is an output stage), SRPP needs to be able to push current through its load. In the original topology the circuit formed two branches of a bridge, with resistors forming the other two branches, and the load was connected across the bridge's output.

In schematics where SRPP drives a finite resistance or an impedance such as a tone-stack, a RIAA filter or an interstage transformer, the load current flows and SRPP operates as such.

However, most applications of this stage have it driving an almost infinite impedance of the next stage's grid, which draws practically no current. Except a small current charging the parasitic and Miller capacitance of the grid circuit, SRPP does not push any significant current through such infinite load and does not work as SRPP at all. The stage has "degenerated" into a common cathode triode stage (V1) with an active anode load (V2).

The top tube does not even work as a constant current source as is often claimed. It is merely an "active resistor" whose value is $R_{A1}=r_{I2}+(\mu_2+1)R_2$! For the ECC88 SRPP design below, this anode resistance would be $R_{A1}=2.6+(33+1)0.68=2.6+23=25.7k\Omega$. A 27kΩ resistor could do the same job!

For the 6SN7 version, $R_{A1}=7.7+(20+1)1.2=33k\Omega$. A=31.6 (30dB), THD=0.09%, $Z_{OUT}=3k\Omega$, frequency range (-3dB): 19Hz - 48 kHz

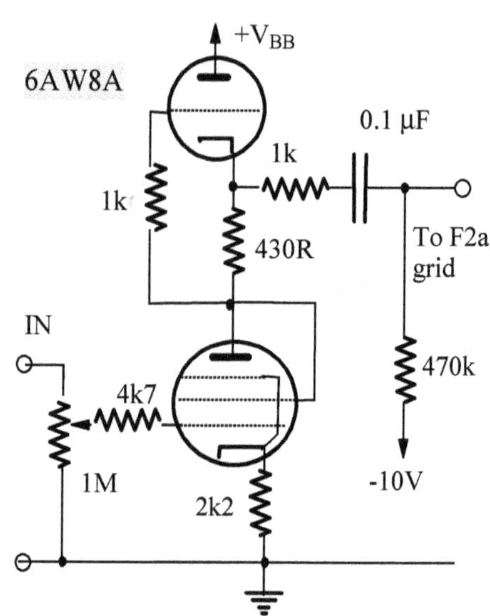

ABOVE: "SRPP" using dissimilar tubes, a triode-pentode 6AW8A and different cathode resistances, (Shindo Cortese amplifier)

Project: SRPP as a CD player output stage

Most CD player output stages use cheap bipolar transistors. The biggest improvement you can make to their sound is to pull those out like rotten teeth and replace them with a tube output stage.

This can be a simple cathode follower, if you don't need to amplify the signal. Most D/A converters produce $2V_{RMS}$ so a CF would be fine. The low impedance of the added CF will also make your CD player much better able to drive the long or high-capacitance interconnect cables.

If you want to boost the signal's amplitude as well, with 15-20 Volts at the output of SRPP, you could even drive a single-ended pentode or U/L power stage directly, totally obviating the need for a preamp or the input stages of your power amp! Extreme cool!

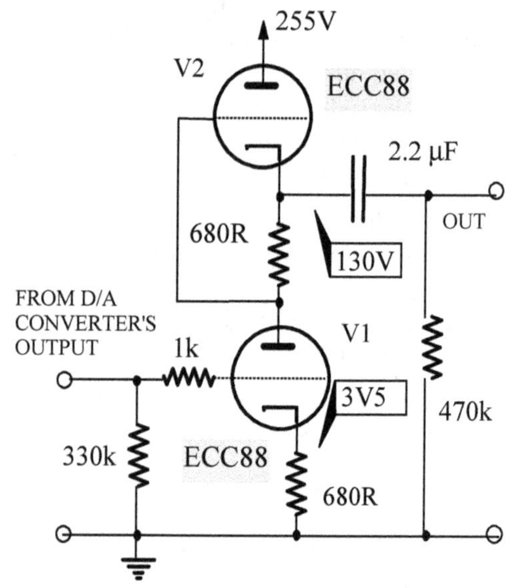

ABOVE: SRPP as a CD-player's output stage

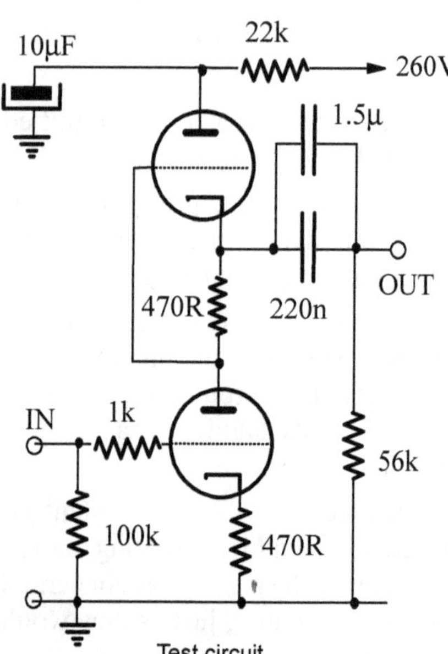

Test circuit

Practical experiment: SRPP shoot-out

In this simple experiment we took half-a-dozen Noval tubes of different types (but the same pinout!) and plugged them into the same circuit. The voltage amplification and the output impedance of the stage were then measured, plus the frequency range (-3dB bandwidth). The lower -3dB frequency was not affected by the change in tubes because it was set by the circuit components and the limits of our function generator. Even the range of upper -3dB frequencies was not that wide, from 94 to 154 kHz, with 123 kHz being the average.

The amplification depended a lot on the tube used, ranging from only 5 to 17, which was to be expected, since µ varied a lot among this bunch. The output impedance was either around 1.5kΩ or around 3.3kΩ.

TUBE	ORIGIN	BW (-3dB)	A	Z_{OUT}
6H1П	Russia	8Hz-100kHz	17	3k5
6N11	China	8Hz-154kHz	11	1k5
6CG7	Lafayette, USA	8Hz-94kHz	5	3k3
6H6П	Russia	8Hz-120kHz	8	1k5
6BQ7A	GE, USA	8Hz-145kHz	12	3k2
E88CC	Zaerix, UK	8Hz-127kHz	13	1k5

The sound character of SRPP stages can be described with a catchy CCC acronym: Clear, Controlled, but often Clinical -sounding. There is some truth in this, but it is a gross simplification nevertheless. When working as a true PP amplifier, the 2nd harmonic and all even harmonic distortion is canceled, or, at least, significantly reduced in the SRPP stage.

In most cases, where SRPP works as an ordinary CC stage, there is no cancellation of even harmonic distortion components, so the sound of the SRPP stage is warmer, with more single-ended coloration.

TUBE PROFILE: ECC83 (12AX7)

- Indirectly-heated duo-triode
- Noval socket
- Heater: 6.3V/300mA or 12.6V/150mA
- V_A= 300V_{DC}, V_{HKMAX}=180V_{DC}
- P_{AMAX}=1W, I_{KMAX}=8 mA
- TYPICAL OPERATION:
- V_A=250V, V_G=-2.0V, I_0=1.2 mA
- gm = 1.6 mA/V, μ=100, r_I = 62.5 kΩ

THE WHITE FOLLOWER

MC-7R is Chinese-made line-level preamp. The first stage, despite its extremely large anode resistance (270kΩ!), due to very strong negative feedback, does not amplify at all, the gain is around 0.9! Plus, for line level voltages of around 1V at the input of its grid the stage starts to distort, so it seems poorly designed.

All the amplification comes from the second stage. With 100 mV at the input, we get 90 mV at the first anode, 60 mV on the second tube's grid and 2 Volts at the output of the preamp, which means the overall gain is around 20.

The output stage uses ECC82 and is based on the second stage of the circuit patented in 1953 in USA by V. J. Cooper et al (British Marconi employees) titled "Stabilized Thermionic Amplifier", US Patent number 2,661,398. This configuration was patented in UK in 1940 by E. L. C. White (patent number GB564250) and in USA in 1944 (US patent number 2,358,428), and became known as the White Cathode Follower.

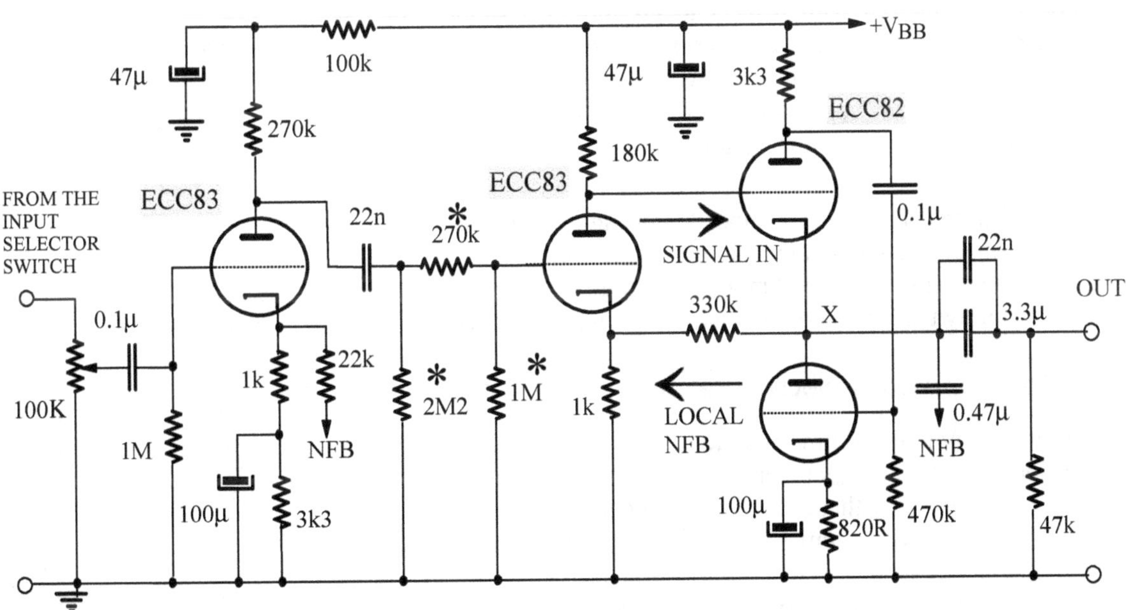

The upper triode (V3) is driven by the input signal, while the lower triode's (V4) amplification factor and resistor R_2 determine the reduction in Z_{OUT}. Output impedances under 10Ω are possible. The stage is effectively a push-pull amplifier where the upper triode provides the load current when the input signal is positive, and the lower tube drives the load during the negative periods of the voltage on the upper grid. Higher linearity (lower distortion) than ordinary CF is also claimed.

As in most totem-pole circuits the cathode of upper triodes are on approximately 1/2 of +V_{BB} voltage, so if +V_{BB} is 300V, there would be 150V in point X, meaning that for most tubes (except 12AX7), to avoid the breakdown in H-K insulation, the heaters of these tubes must be referenced to a higher DC voltage point, 60-80V.

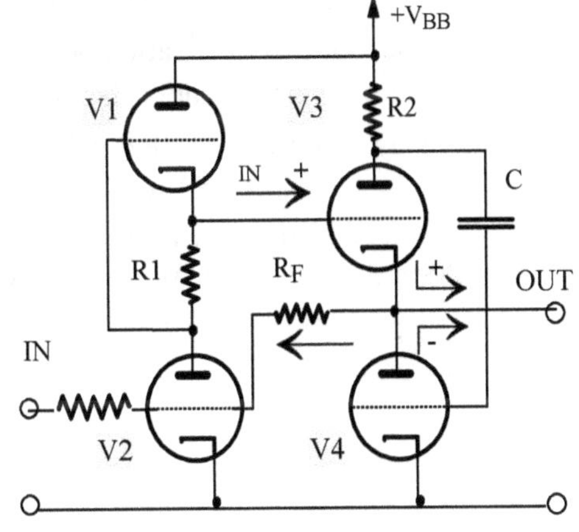

I must admit that we've never used the White Follower, the "ordinary" cathode followers with output impedances of 200-400Ω are more than adequate for most audio purposes! As if there was anything ordinary about this extraordinary performer called cathode follower! As many over-the-top circuit concoctions do, White Follower violates one of the cardinal rules of audio design, the Simplicity Rule.

THE SIMPLICITY RULE: Between a simple circuit that performs well and a complex circuit that performs marginally better, chose the simpler option. It is faster and easier to build, and to troubleshoot later on.

THE DIFFERENTIAL AMPLIFIER

A differential or difference amplifier has two inputs and two outputs. In the ideal case, it amplifies only the difference in the voltage levels between the two inputs, while any signal common to both inputs is not amplified. In reality, of course, there is some amplification of the common-mode signals.

Obviously differential amps lend themselves to balanced topologies and phase inverters since the output signals are identical and out-of-phase. The output resistance is identical to that of the grounded cathode stage, $r_I||R_A$ (internal resistance of the valve in parallel with its anode resistor).

The gain of a differential amp is the same as for the common cathode stage: $V_{OUT}=-\mu R_A/[r_I + R_A](V_{IN1}-V_{IN2})$

The cathode resistor R_K must not be bypassed, but the gain does not depend on its value. This assumes that the two halves of the circuit are identical. Any difference in component values or tube mismatch will affect the operation of the circuit and diminish its benefits. So, matched resistors and matched tubes are a must here!

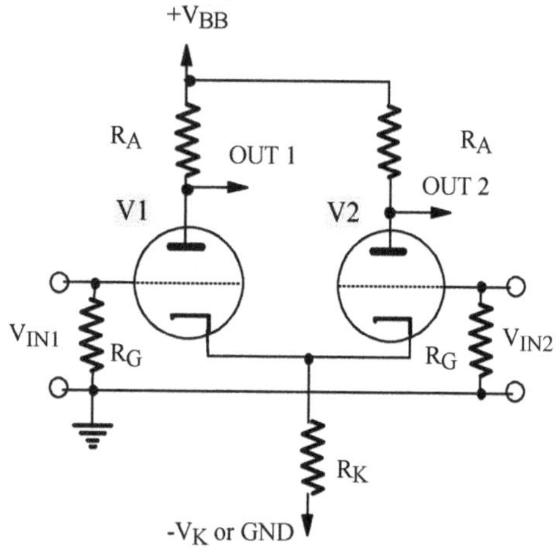

ABOVE: Differential amplifier with triodes

The outputs are balanced with respect to ground. If a single-ended output is needed, one of the anode resistors is made zero, trurning V1 into a cathode follower, its output feeding the cathode of V2 (cathode coupling). In this case $V_{OUT}=-\mu R_A/[2r_I+ R_A](V_{IN1}-V_{IN2})$ so the amplification is lower due to the factor 2 which multiplies r_I in the denominator.

Four topologies are obtainable: balanced input and balanced output, balanced input and single-ended output, single-ended input and balanced output, and single-ended input with single-ended output. The last option is rarely used, no benefit is gained from such a circuit.

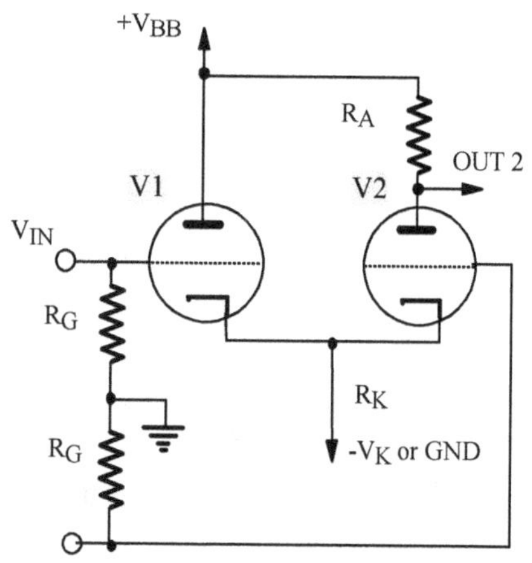

ABOVE: Differential amplifier with balanced input and single-ended output

CMRR (Common Mode Rejection Ratio)

CMRR is an acronym for Common Mode Rejection Ratio and describes an amplifier's ability to reject common mode signals at its input. Common mode signals are those that appear on both input terminals, for instance interference, hum and other induced nasties. $CMRR=A_{DIF}/A_{COM}$ is a ratio of amplification factor for differential signals and amplification factor for common-mode signals.

Since $\mathbf{A_{COM}= -gmR_A/(1+2gmR_K)}$ and $\mathbf{A_{DIF}=-gmR_A/2}$ we get $CMRR = (1+2gmR_K)/2 \approx gmR_K = \mu R_K/r_I$

The best tubes for this application, as is the case in most audio applications, are the one with high transconductance gm, meaning high amplification factor μ and low internal resistance r_I!

The other way to increase CMRR is to make R_K as large as possible. However, due to a large voltage drop on R_K that would require a very high $+V_{BB}$ voltage, a very high $-V_K$ voltage supply or both.

Ideally, we need a component that has a very low DC resistance, (so the $+V_{BB}$ and $-V_K$ supply voltages don't have to very high) and a very high AC impedance. One such device is a CCS (Constant Current Sink). More on constant current sources and sinks very soon.

DIFFERENTIAL AMPLIFIER

$A_V=-\mu R_A/(r_I + R_A)$

$CMRR = (1+2gmR_K)/2 \approx gmR_K = \mu R_K/r_I$

In single-ended (unbalanced) applications the second input of the differential amplifier does not need to be grounded. It can be used to bring back a negative feedback signal through a voltage divider $R_2/(R_1+R_2)$. More on negative feedback soon.

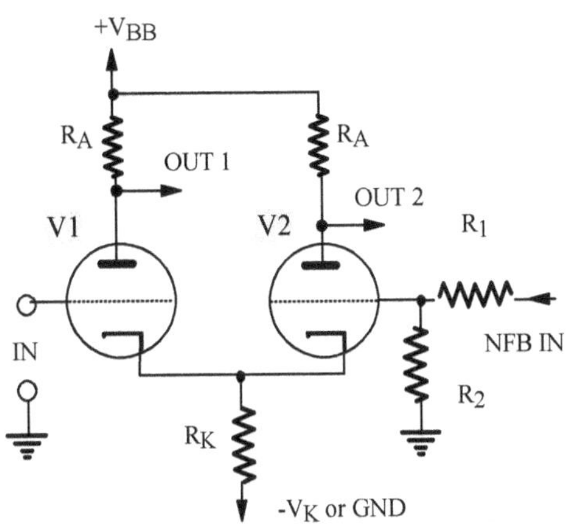

ABOVE: With a single-ended input into V1, the input into V2 can be used for negative feedback signal!

Cascode differential amplifier (Hedge circuit)

In his June 1956 article in Wireless World, L.B.Hedge, Ph.D. introduced a "Long-tailed Cascode-Pair" as a combined preamplifier and phase splitter stage, driving 1625 output tubes (a 12V heater version of 807), with a switchable triode- U/L-pentode output.

He investigated two choices for V1 and V2, 7N7 (1/2 of 6SN7) and 7F7 triodes (1/2 of 6SL7), and in the final version of the design settled on 7F7, presumably to get more gain and thus a higher input sensitivity. However, even the low μ tube such as 6SN7 in this circuit would produce gains of 75-120, depending on the anode resistor and the anode voltage used, and that is more than enough to drive the output stage, even after the application of negative feedback.

He didn't use a constant-current source for the differential amplifier, but did include a negative supply of -75V for the 50kΩ cathode resistor.

Rogers RD Senior MkII amplifier featured a Hedge front end driving EL34 push-pull stage directly. Four different negative feedback arrangements were used, making this one of the most feedback-rich designs!

Firstly, a local NFB through 3M3 resistor from the anode to the grid of the input tube (12AX7). Two 1M resistors form a voltage divider from the anodes of two upper triodes, and from that point NFB is taken to the grid of the second upper triode.

NFB #3 is taken before the coupling capacitor to the lower output tube back to the grid of the lower triode (3M3 resistor). Finally, the usual global feedback from the output is taken to the 330Ω resistor in the cathode circuit of the Hedge stage.

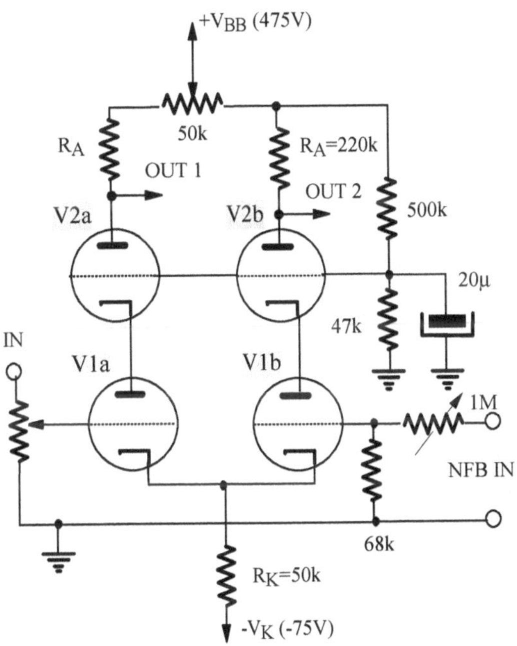

ABOVE RIGHT: The original Hedge circuit

BELOW: Rogers RD Senior MkII amplifier featured a Hedge front end driving EL34 PP stage directly. Notice quadruple NFB!

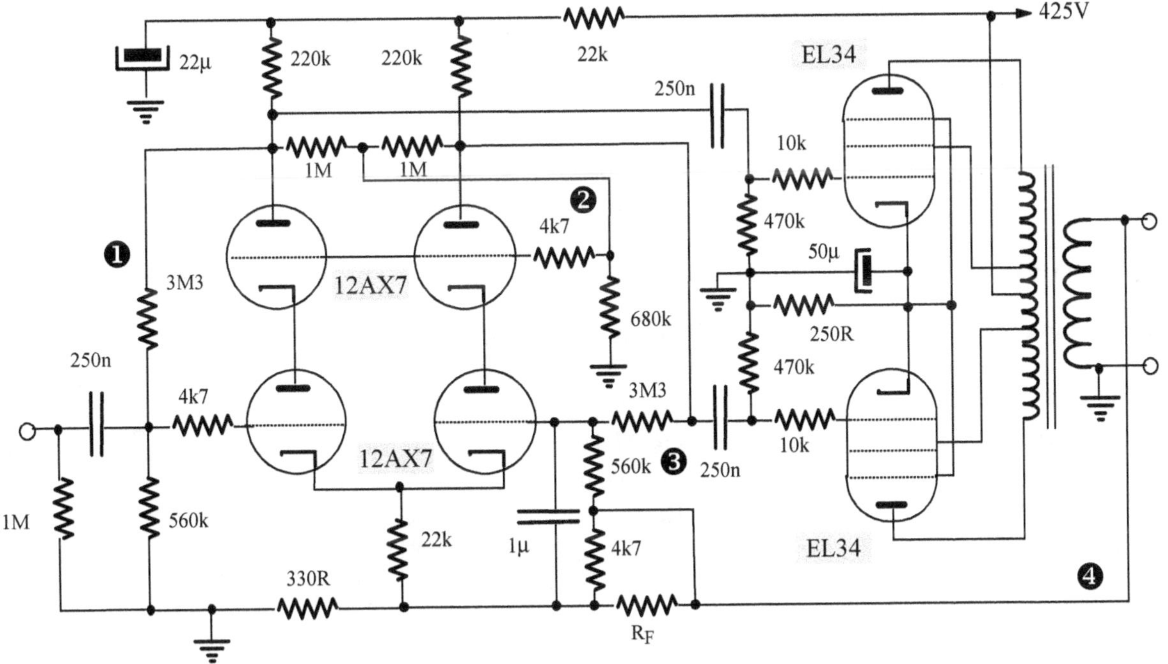

ACTIVE LOADS: CONSTANT CURRENT SOURCES AND SINKS

Consider a simple common cathode stage with ECC88 triode and 27kΩ anode resistor. The quiescent point Q is at I_A=6.5mA and V_0=145V. With the maximum input (grid) voltage swing between 0V and -8V, the operating point moves between point A and B. The anode voltage swings between 35V and 235V, for a total of 200V. The voltage amplification is thus A_{AB}=-200/8 = -25

With a constant current source the XY load line is horizontal (infinite AC load resistance and constant anode current). The anode voltage swings between 20V and 255V, a total of 235V. The voltage amplification is A_{XY}=235/8 = 30 The increased voltage gain is a nice bonus, the primary aim was to improve linearity and reduce distortion.

With a 27kΩ load, the anode voltage swing QA is 145-35=110V, while QB=235-145= 90V, the total voltage swing is AB=200V. Using the voltage formula the 2nd harmonic is D_{2AB} = (QA-QB)/2AB = (110-90)/(2*200) = 20V/400V= 0.05 = 5%

With CCS the anode voltage swing QX is 145-20=125V, while QY=255-145= 110V, so the 2nd harmonic is now D_{2XY} = (QX-QY)/2XY = (125-110)/(2*235) = 15V/470V= 0.032 = 3.2%

In this case, the use of CCS reduced harmonic distortion by 36%. With less linear tubes and different operating points and load resistances, the distortion reduction would be even larger.

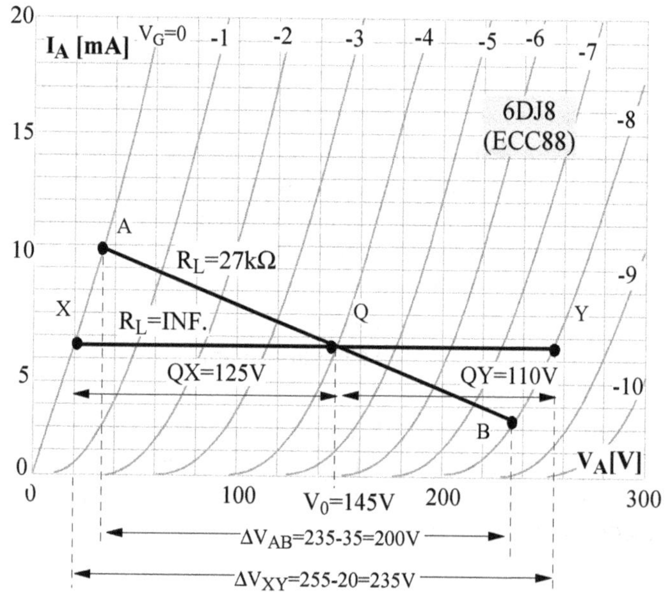

Resistive loads have the same resistance for DC and AC signal. Active loads, constant current sources or sinks (CCS), have similar properties to chokes, a very high AC impedance and a very low DC resistance. That makes them desirable in situations where a designer wants to maximize the gain of a stage, without having to use high voltages (due to large DC voltage drops on load resistors) or very low tube currents.

If used in the cathode of the output stage, current sinks obviate the need for external bias voltage or cathode resistors. The current sink acts as an automatic cathode bias regulator. No matter what the voltage on the cathode is, the current sink behaves as a variable resistor, maintaining constant cathode current through the output tube(s).

Pentodes are closer to ideal current sources than triodes but CCS can be realized with solid state active components as well, JFETs, MOSFETs, bipolar transistors, even integrated circuits can be used. Three terminal voltage regulators make excellent CCS with an addition of only one external resistor.

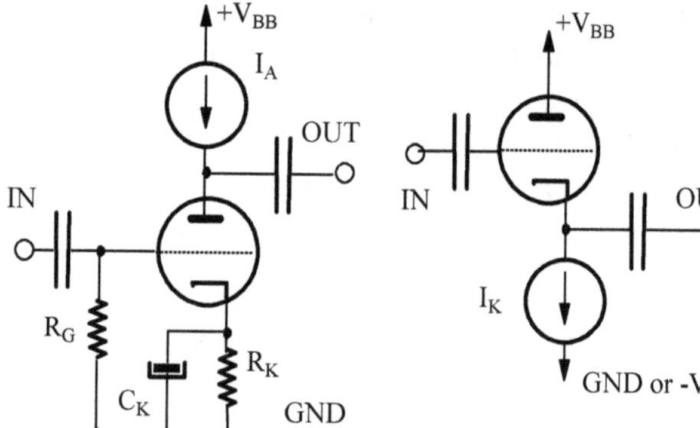

Constant current source as an active anode load

Constant current sink providing bias for a cathode follower

RIGHT: Constant current source as an active anode load of the bottom triode

Constant current sources and sinks with triodes

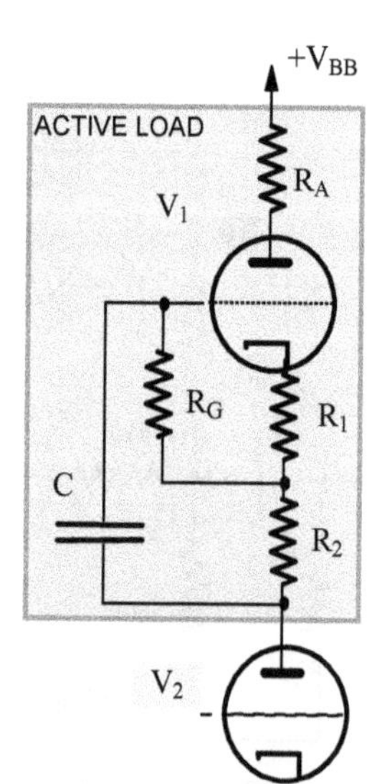

A constant current source is a resistance amplifier. Used as an anode load for tube V2, its resistors are sized the same way as for the self-biased cathode follower. R_1 and R_G determine the bias for the upper triode (V1), and R_1 and R_2 form its cathode load.

V_2 sees the anode resistance $R_A' = r_I + (\mu+1)(R_1+R_2)$! r_I is the internal resistance of V1 and μ is its amplification factor. The two tubes don't have to be the same, of course, but often are since usually both come in the same duo-triode. For ECC81 tube the values could be R_1=60R, R_2=1k, which at μ=66 and r_I=11kΩ would give us R_A' = 81kΩ!

This result is nothing to write home about, an 82kΩ resistor would do the job, but it would not keep the anode current constant, as this active load does.

Pentode as a CCS

As a voltage source, a triode is not an ideal choice of a constant current source or sink, because its internal resistance is very low, in the order of 5-15kΩ. A pentode, which is inherently a constant current device, does that job much better! We haven't talked about pentodes yet, so may wish to jump forward to the next chapter and study their behavior, but for now let's have a cursory glance at their anode characteristics.

The internal resistance of a typical small-signal pentode (6SJ7) can be roughly estimated from its anode curves. The positioning of the quiescent point Q does not matter one bit, for instance it is here at approx. V_A=200V and I_A=6.1 mA.

If we move 100V down to V_A=100V and then 100V up to V_A=300V, the anode voltage changes 200V, but look what happens to anode current: I_A changes from 6.05 to 6.15mA or 0.1mA! Actually it changes much less, but let's round it up, since we cannot estimate this properly from such an imprecise, hand-drawn graph anyway.

The internal resistance of 6SJ7 pentode in this operating point is $r_I=\Delta V_A/I_A$=200V/0.1mA=2MΩ! Compared to the previous example, a triode with r_I=11kΩ, this is the real deal.

The pentode is biased using cathode bias, the suppressor grid SUP and the control grid CG are connected to ground (in this application we don't want any AC signal on the control grid).

The screen grid SG is biased to the desired screen DC voltage through the resistor R_S and capacitor C_S. The graph here is for V_S=100V so you may as well design the CCS around that screen voltage. Of course, any V_S can be used, although it will have a effect on the anode current. In pentodes it is the screen voltage that determines the anode current, not the anode voltage, as we have just ascertained.

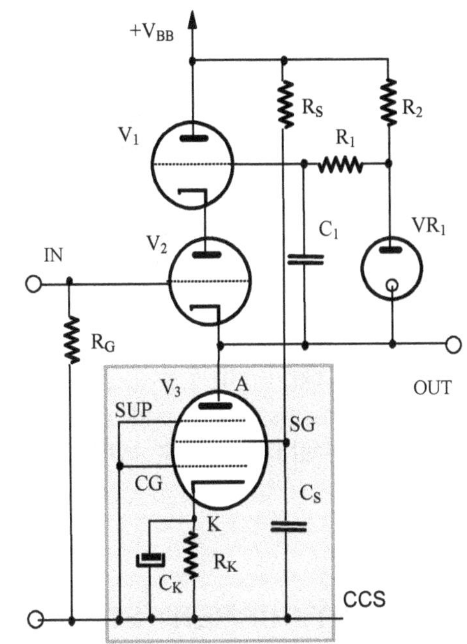

ABOVE: Low distortion cathode follower with a pentode CCS.

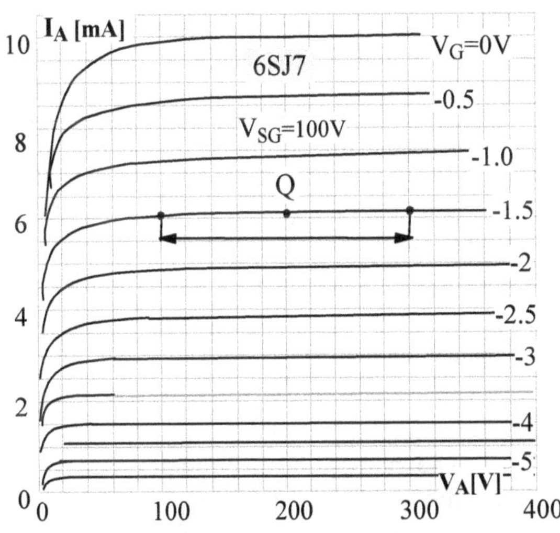

ABOVE: Pentode's almost horizontal output characteristics mean extremely high internal impedance!

PRACTICAL DESIGN: 12AU7 cathode follower

For the standard cathode follower we have I_K= 120/17,157 = 7.0 mA The cathode bias voltage is $V_{BIAS}=I_K$*857 = 7*0.857 = -6V, so V_G = 120-6 = 114V

For the CF with the current sink (BELOW RIGHT), from the anode characteristics graph on the next page, we can see that at V_{A2} = 114V and I_A = 7mA, the bias voltage is -2.2V.

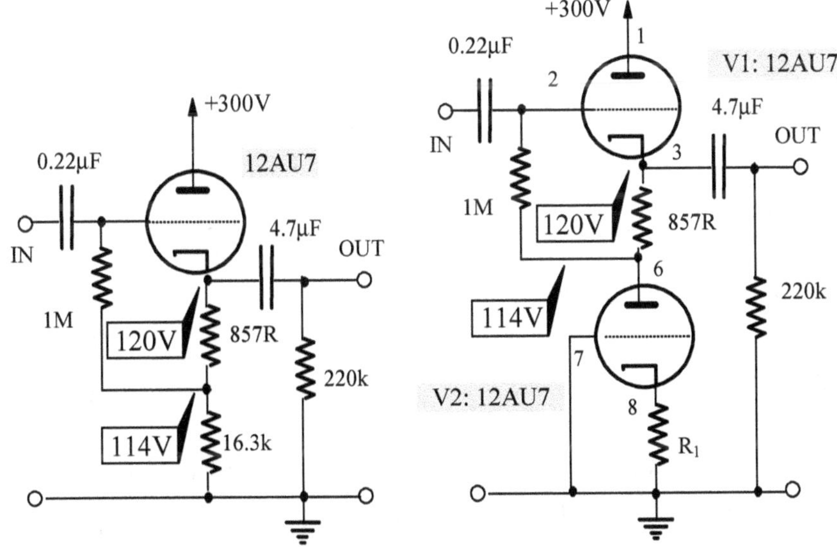

Now we can calculate the bias resistor for the CC sink R_1= 2.2/0.007 = 314Ω

We need to estimate the internal resistance of the 12AU7 tube in the operating point Q2, which is around 8,500Ω.

The CC sink is now the load for V1 and its value for AC signals is $r_K=r_I+(\mu+1)R_1$ = 8,500+21*330 = 8k5+6k9 = 15.4kΩ

Instead of the operating point for the upper valve V1 swinging between points X and Y, the new load line is horizontal (because the current is constant), and with the grid signal between 0 and -12 Volts, the operating point swings along the horizontal A-B load line.

The negative anode voltage swing X-Q_1 was 180-100=80V and the positive swing Q_1-Y was much smaller (240-180 = 60V). Now the positive swing is 110V and the negative is 100V, a much better balance.

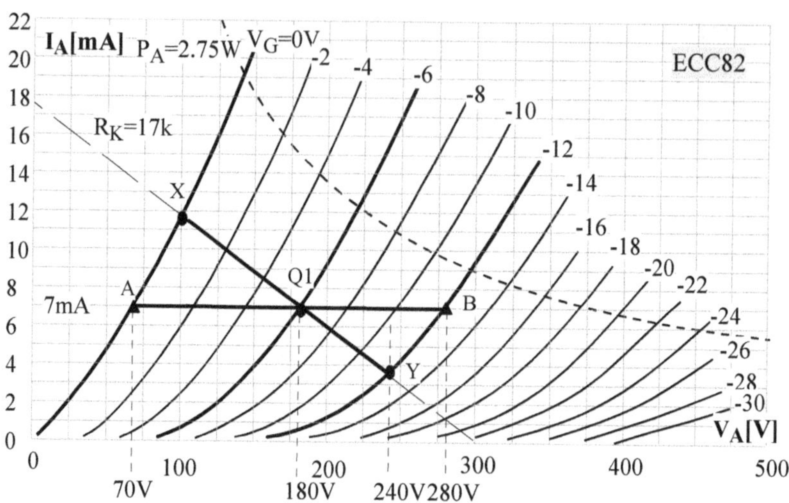

ABOVE: The load lines and operating paths for a 12AU7 cathode follower with a 17kΩ resistive load (XY) and with a pentode CCS (AB).

CONSTANT CURRENT SOURCES/SINKS WITH SEMICONDUCTOR COMPONENTS

JFET - Junction Field-Effect Transistor

JFETs were developed after bipolar transistors, and before their MOSFET brethren, following the spectacular advances in epitaxial silicon technology. We mention FETs here for a few reasons. From the outside-in perspective, as amplifying devices, (disregarding their semiconductor nature), they behave in a way very similar to vacuum tubes, more precisely pentodes. Both are voltage-controlled current sources, with very high input resistance and a very high output resistance.

Unlike bipolar transistors, where there is interaction between inputs and output circuits, unipolar FETs make interfacing with vacuum tubes very easy, and as such are often used in hybrid circuits.

FETs have a few advantages over bipolar transistors which makes them superior amplifying devices, not just in terms of higher reliability and easy circuit design, but also because of their pleasing sonic signatures. FETs are also often used in constant current sources and sinks, so we need to understand their behavior first.

In one operating region (triode region) JFET works as a voltage-controlled variable resistor. The narrow semiconductor channel (in this case N- or negative type) provides the conductive path between source S and drain D (next page). When the gate G is negatively biased (just like a grid in a vacuum tube), the width of the top and bottom P-regions is increased and the width of the N-channel is reduced, increasing the resistance of the channel.

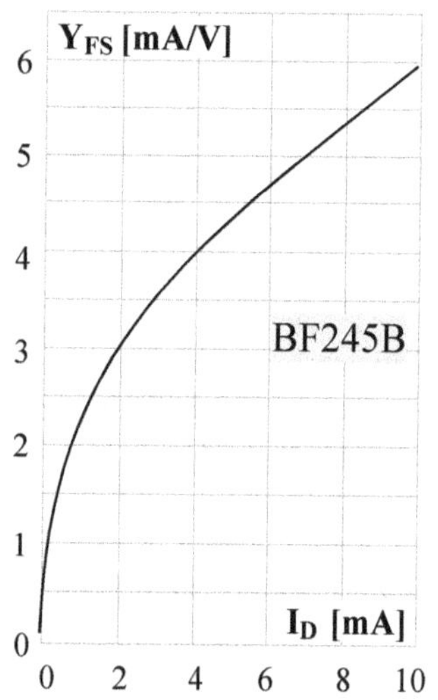

ABOVE: Forward transfer admittance versus drain current for BF245B JFET

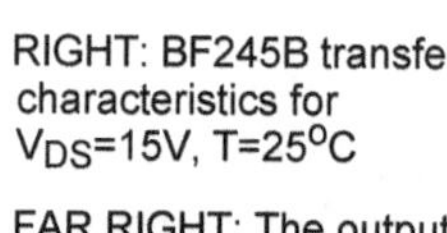

RIGHT: BF245B transfer characteristics for V_{DS}=15V, T=25°C

FAR RIGHT: The output characteristics (I_D vs. V_{DS}) for BF245B JFET

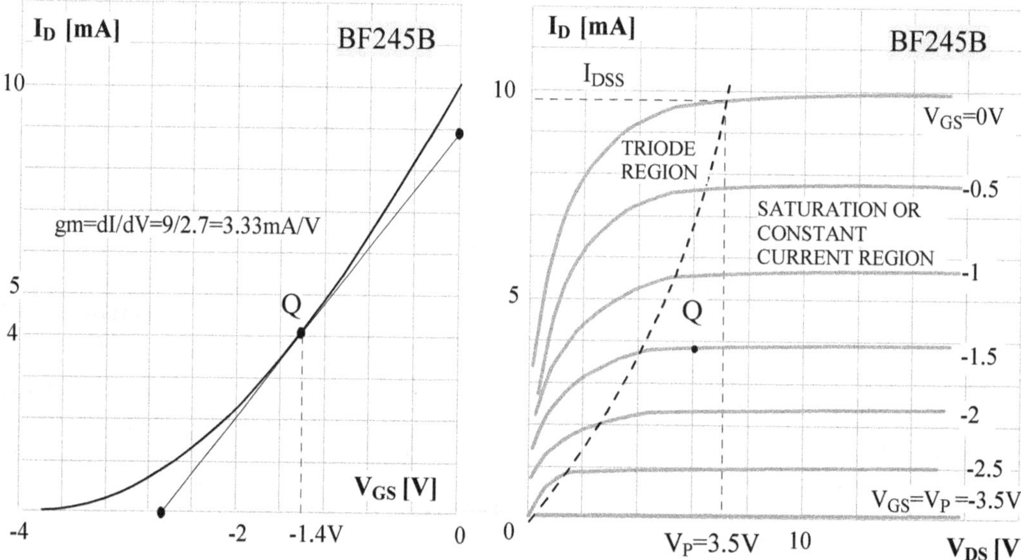

The grid-source voltage needed to reduce the channel thickness to zero is known as the pinch-off voltage V_P. The current through the channel does not cease when the drain voltage V_{DS} reaches the V_P value because a voltage equal to V_P exists between the pinch-off point and the source S, so the resulting electric field along the channel causes the current to flow.

As drain voltage is increased above the V_P value, the drain current increases only slightly, remaining practically independent of the drain voltage. In this saturation region FET behaves as a voltage-controlled current source (just like a pentode). It is this region that we are interested in.

The drain current that flows when V_{GS}=0V and V_{DS}=V_P is called I_{DSS}, or saturated drain current with shorted input (gate-source). FETs are semiconductors and their manufacturing tolerances are very wide (compared to vacuum tubes), so the data sheet for BF245B JFET, as a typical example of a small signal JFET says that I_{DSS} is 6-15mA.

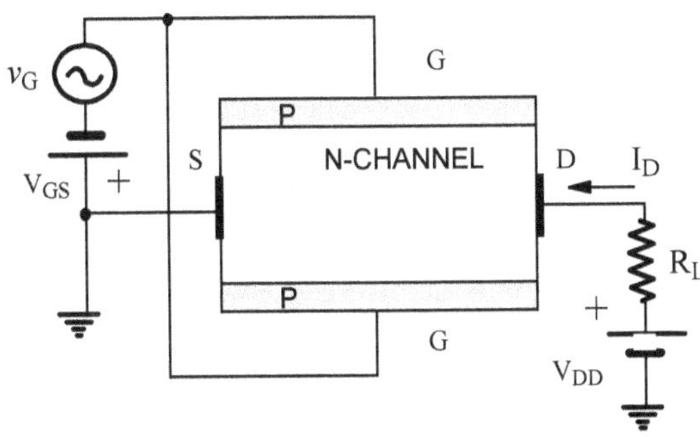

ABOVE: The physical operation of N-channel junction FET

CCS with field-effect transistors

When used in a constant current mode, FETs don't not need a Zener diode for voltage reference. Also, in comparison with bipolar transistors, FETs have a much higher temperature stability due to their negative gain-temperature coefficient, which protects them from thermal runaway. The source-drain current is primarily determined by the input voltage (V_{GS}) and it stays constant with an increasing load demand, which makes a FET an ideal current source.

In the simplest CCS illustrated in a) V_{GS}=0 and the output conductance g_O is equal to g_{OS} (g_{OSS} is also used), which is the small signal common-source short-circuit *output conductance*, specified in data sheets.

The constant current is the I_{DSS} current of the chosen FET, which is the drain current when the gate is connected to the source.

As g_{OS} ranges from around 1µS to 50µS or more, depending on the FET type, the output impedance of this simple CCS will vary from only 20kΩ to over 1MΩ.

The addition of a resistor R_S in the source circuit allows for the adjustment of V_{GS}, by which the mentioned zero coefficient of temperature drift can be achieved.

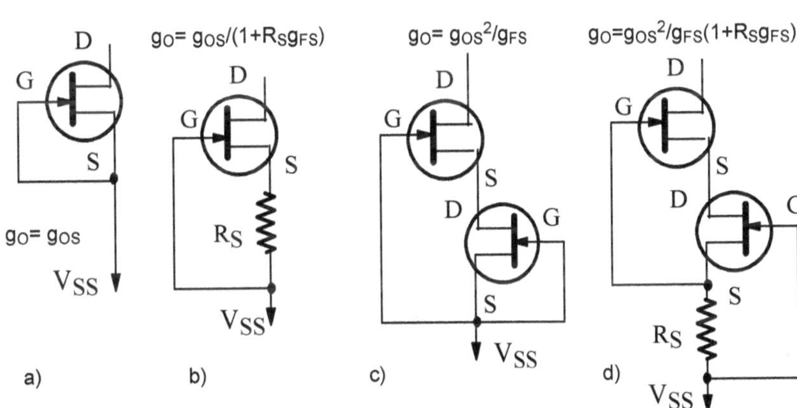

ABOVE: The simplest CCS with field-effect transistors

The freedom of adjusting V_{GS} allows for any constant current to be had, according to the basic FET formula $\mathbf{I_{DS}= I_{DSS}(1-V_{GS}/V_P)^2}$, where V_P is the pinch voltage, again, specified for every FET in its data sheet. The output conductance is reduced in this case by the factor $1+R_Sg_{FS}$, where g_{FS} is the *small signal common-source short-circuit forward transfer conductance* (isn't that a mouthful?), equivalent to gm in a vacuum tube.

By cascading two FETs, as in c) and d), much lower output conductance values can be achieved. For proper operation both FETs must be operated with a sufficient drain-gate voltage, V_{DG}, which should never drop below the value $V_{DG} = 2V_{GS(off)}$, otherwise g_{OS} will be significantly increased, and so will g_O! $V_{GS(off)}$ is the cutoff voltage.

FETs with long gates and very low g_{OS} values make the best current sources. Currents of 10 mA or more can be achieved, with internal impedances of 1MΩ or more.

PRACTICAL DESIGN: Differential amplifier with FET and bipolar CCS

Since differential amplifiers have high CMRR and can be used with a floating signal source, they are ideal as input stages of phono preamplifiers. To minimize noise and improve overall performance, let's use a FET constant current sink instead of the cathode resistor. A simple voltage regulator with a bipolar transistor provides a stable source of -6V for the CCS. The differential amplifier needs 2mA for each triode.

This simple CCS is a self-biased JFET. The source resistor provides a gate-source or bias voltage, which is proportional to the source current. The smaller the resistor, the higher the current.

Since we already have the FET selected, the design involves only a selection of the source resistor R_S, which must be adjusted so that at the required current the suitable gate-source voltage is achieved. Since our V_{SS} voltage is -6V, we start by positioning the operating point Q at the same value V_{DS}, so V_{DS} =6V.

Let's say we need 4 mA for our differential amplifier (2mA each tube), so I_D=4mA. This determines our Q. From the graph we read V_{GS}=-1.4V, and since $V_{GS} = I_D R_S$, we get the required value of the source resistor $R_S=V_G/I_D$= 1.4V/4mA= 350Ω. We used a 1k or 2k trimpot so we could precisely adjust the current to the required value.

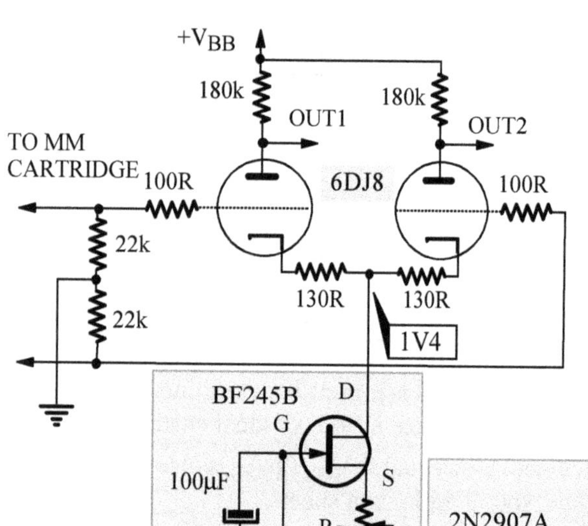

The equivalent internal resistance of the CCS is $R_I = R_{DS}(1 + gmR_S)$ or in terms of conductance $g_O= g_{OS}/(1+R_S g_{FS})$ = $40\mu S/(1+350*4*10^{-3})$ = $40\mu S/2.4$ = $16.7\mu S$, so R_I = $1/g_O$= 60kΩ

This particular type of FET was not selected for its low g_{OS}, actually its g_{OS} is quite high, that is why only 60kΩ was achieved. This was simply one FET we had in our stash. There are FETs with much lower internal conductance (higher internal resistance), but this example is meant to illustrate typical results that can be achieved.

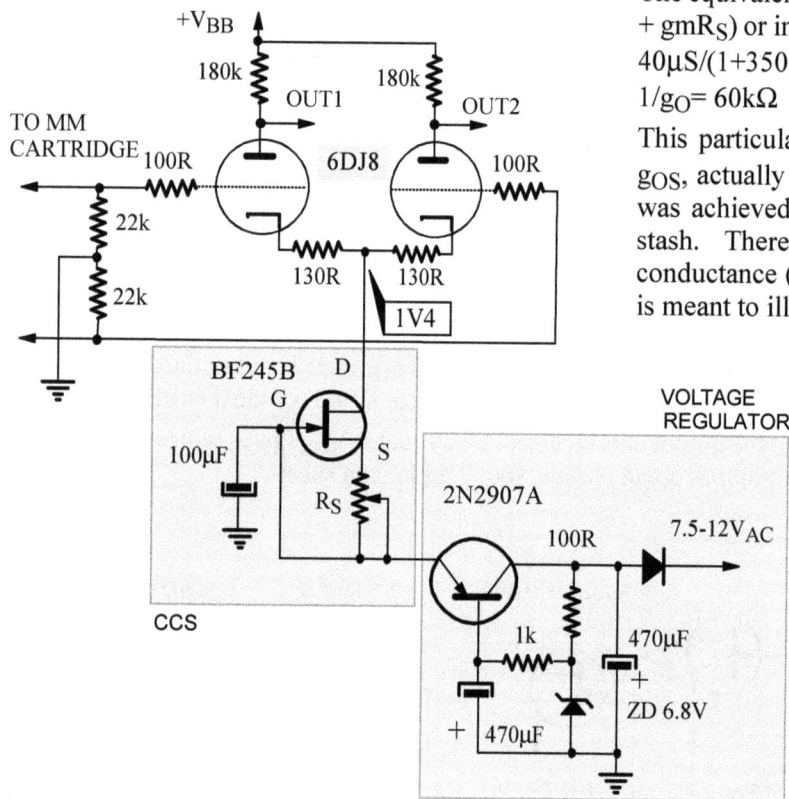

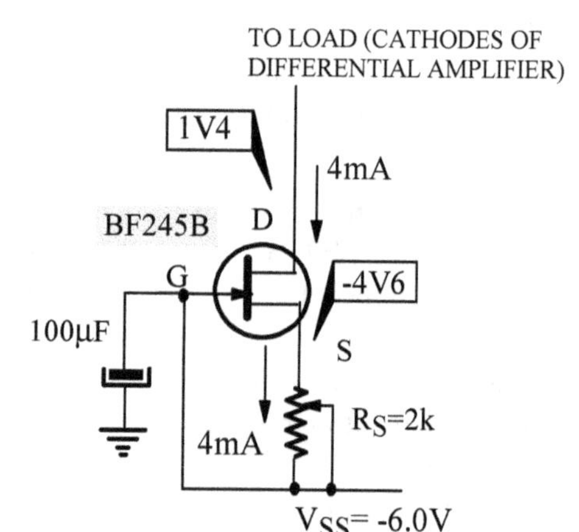

Instead of a CCS we could easily use a 60kΩ resistor, but firstly, we would then need a source of negative voltage equal to the DC voltage drop on such a resistor, or $V_K=R_S I_K$=60*4= -240V! Secondly, and more importantly, the current regulation would be way inferior to the one achieved with a FET.

PRACTICAL DESIGN: CCS with bipolar transistors

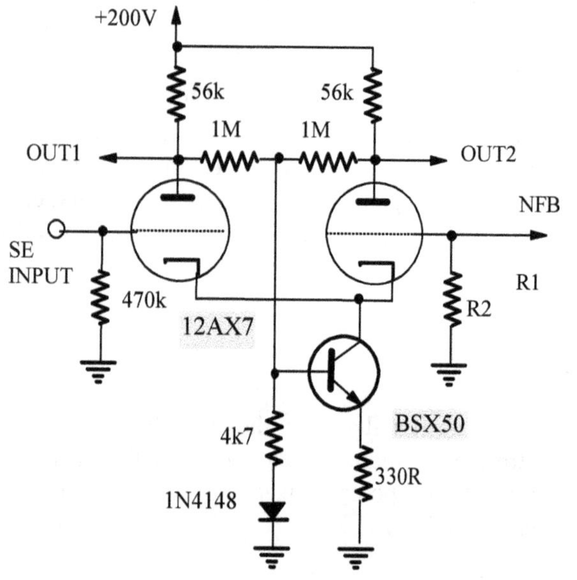

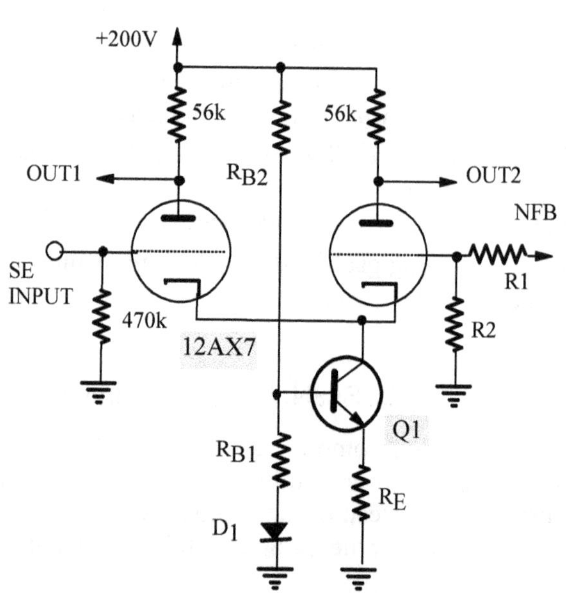

Two differential amplifiers using bipolar CCS are illustrated, using different anode connection points. The sum of cathode currents of V1 and V2 flows through the collector and is approximately equal to the emitter current of the transistor. Diode D1 in the stabilized bias circuit compensates for the temperature variations of the transistor's base-emitter PN junction voltage. The base is biased either from the $+V_{BB}$ of the differential tube pair (the upper connection of 56k anode resistors, as in the circuit below right) or from the reference point obtained by summing the DC voltages of the two anodes through 1M resistors (circuit below left).

1N4750 is a 27V 1W planar silicon Zener diode, and it keeps the base of the BSX50 transistor at around -27V (plus the voltage drop on the 1N4148 diode). The collector current is adjusted by changing the resistance of the 5k trimmer potentiometer in the emitter circuit. The 10µF elco bypasses the transistor's base to ground for AC signals, while 56k resistor is the series resistor need for the Zener diode to operate.

As the base-emitter junction voltage varies with temperature so does the 1N4148 diode's PN junction, thus acting as a temperature compensating circuit.

The other circuit (far right) works as a self-regulating loop. The voltage drop across the top part of the trimmer pot is the base-emitter voltage of T2, and is constant (0.7V). Should the current I_0 want to rise, so would the current through that resistor. The voltage drop across that resistor would increase and thus turn on T2 harder, lowering the voltage at the base of T1.

Its emitter current would decrease, thus counteracting any possible rise in the load current I_0! Both transistors are BC639.

There is no need for a Zener reference or for a temperature-compensating diode since the two transistors' PN junctions compensate one another.

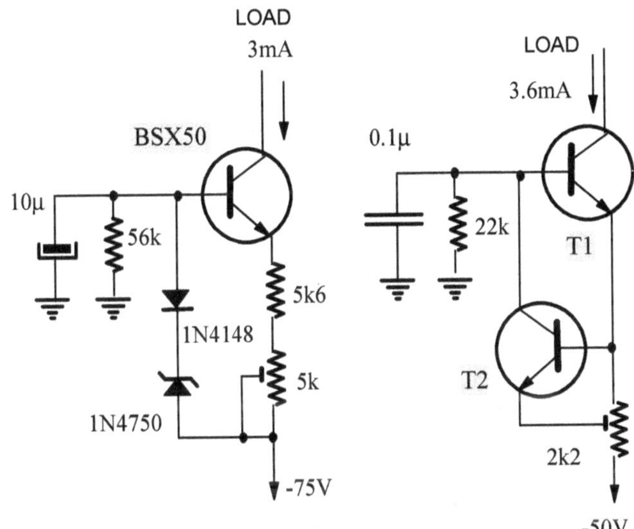

LEFT: CCS with a single bipolar transistor
RIGHT: CCS for the input stage with two bipolar transistors

Voltage regulator ICs as CCS

LM317 is a floating 3-terminal voltage regulator that can be operated as a simple CC source or sink. When operating in the current mode the internal reference voltage of 1.25V appears between terminals OUT and ADJ, so a simple resistor between them will set the desired current: $I = 1.25V/R_X$.

For proper operation about 2.5V are needed across the IC, plus 1.25V ref. voltage, for a minimum total of around 3.75V. The current should be at least 10mA, meaning R_X should not be larger than 125Ω!

Just like all semiconductors, the voltage regulator must be properly mounted if cooling is required via a heatsink. If TO-220 version is used, the top metal lug is connected to the V_{OUT} terminal. The metal body must not be bolted directly onto the metal chassis (to act as a heatsink), it must be insulated. TO-220 insulating washers are available. The contact area must be smooth and clean.

Once the IC is installed, test the resistance between its body (and the mounting bolt) and the chassis. An ohmmeter should show infinite resistance (no galvanic contact).

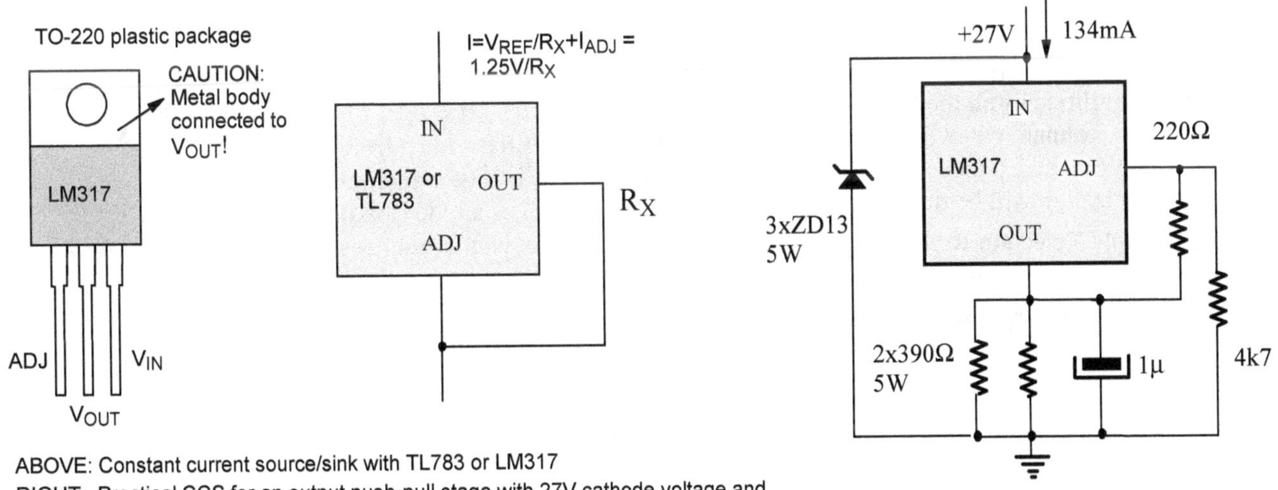

ABOVE: Constant current source/sink with TL783 or LM317
RIGHT: Practical CCS for an output push-pull stage with 27V cathode voltage and 134mA current (two tubes)

Power supply ripple rejection

One advantage of using constant current sources and sinks in amplifying stages is an improved rejection of the hum-causing ripple on top of the DC power supply lines. Ripple is a complex waveform that depends on the type of rectifier and filter used (discussed in Volume 2 of this book), but in the first approximation can be considered a sine wave superimposed on the DC supply. Indeed, no matter what its waveform, its first harmonic is always a sine wave of the double mains frequency, 100Hz or 120Hz.

For our CC stage with ECC40 we can model this situation with a source of AC signal v_R in series with a DC source V_{BB}. In the small signal equivalent model the DC source V_{BB} is a short circuit, and since there is no signal at the grid, the μv_G is also be shorted out.

We now have a simple series circuit with the ripple source and three resistances, the internal resistance of the tube r_I, the internal resistance of the source (power supply) R_S and the load resistance R_L. The ripple output is from the anode, just as with the audio signal, and the circuit is a simple resistive voltage divider: $\mathbf{v_{OUT}=v_R[r_I/(r_I+R_L+R_S)]}$.

Ideally, v_{OUT} should be as small as possible. There are four factors in this equation, but the tube's internal resistance r_I is in both nominator and denominator, so we won't find much luck there.

We could make the load resistor R_L (which is in the denominator) as high as possible. That would increase the gain of the stage and reduce the DC current through it, which may not be desirable from the sonic perspective. Stages with high anode DC current usually sound better than the "starved" or low-current designs.

Of course, we may use anode choke instead of a resistor, the choke would have a very low DC resistance but very high AC reactance and that would increase the rejection of ripple. Likewise, as we have just seen, we could use an active load, a constant current source, instead of the simple anode resistor, and that would significantly improve PSRR.

The internal resistance of the power supply R_S is in the denominator, so a high power supply internal impedance or resistance would reduce ripple. However, this is not a practical way of achieving our aim, since high power supply impedance is undesirable from all other aspects of design (regulation, efficiency, sag at higher power levels).

Ultimately, we are left with the only realistic way to reduce the amplification of the ripple through the subsequent stages, and that is to go back to the source and reduce the ripple in the power supply itself, thus reducing the factor v_R!

The ripple situation is similar to the noise issue, in that the most critical stage is the first stage, since any noise or ripple at its output is amplified by all the following stages! Towards the end of this volume we will talk about preamp-tube choices and in Volume 2 topics such as topologies, layouts and noise minimization will be discussed.

The Power Supply Rejection Ratio or PSRR can be defined as $PSRR=v_R/v_{OUT}=(r_I+R_L+R_S)/r_I$ or in dB:

$PSRR= \log\{(r_I+R_L+R_S)/r_I\}$ [dB]

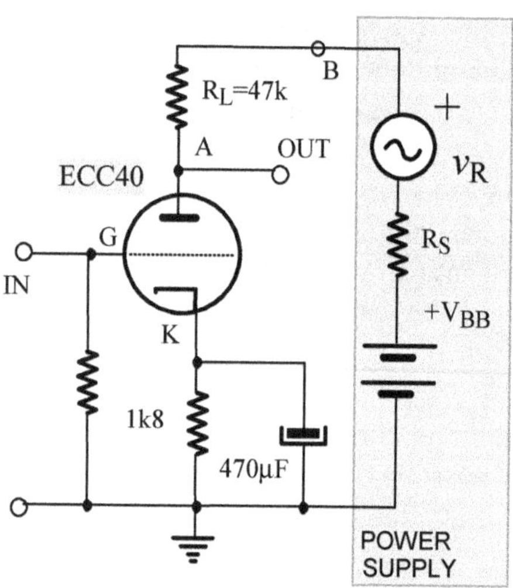

ABOVE: Real DC power supplies always contain some ripple (AC component superimposed on top of the DC voltage) and an internal resistance R_S

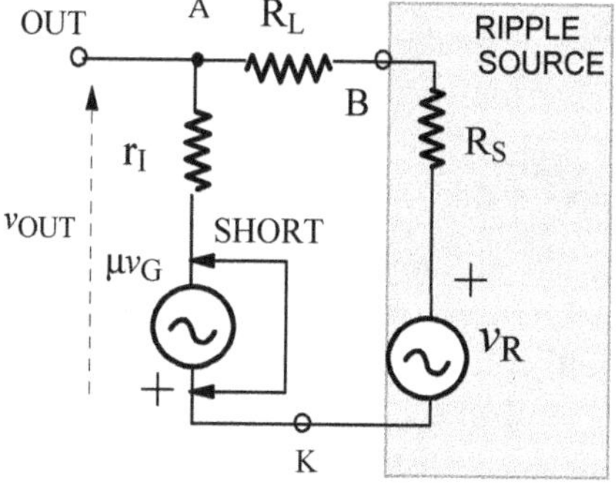

ABOVE: The small signal AC model of a common cathode stage for ripple voltage

POWER SUPPLY REJECTION RATIO

$PSRR= \log\{(r_I+R_L+R_S)/r_I\}$ [dB]

r_I is tube's internal resistance

R_L is anode (load) resistance

R_S is internal resistance of the power supply

TETRODES, PENTODES AND BEAM-POWER TUBES

- TETRODES
- PENTODES
- BEAM POWER TUBES
- 6K7 PENTODE DRIVER STAGE
- LINE STAGE WITH TRIODE-CONNECTED EF86 PENTODE
- THE μ FOLLOWER

"Science is like sex: sometimes something useful comes out, but that is not the reason we are doing it."
Richard P. Feynman, American theoretical physicist

TETRODES

Early experiments revealed that the insertion of a second grid between the control grid and anode of a triode provides electrostatic shielding between them to such extent that the input capacitance is reduced by the factor of 1,000 or more! Thus, tetrodes performed much better than triodes at high frequencies, so in 1928 the first commercial tetrode was introduced.

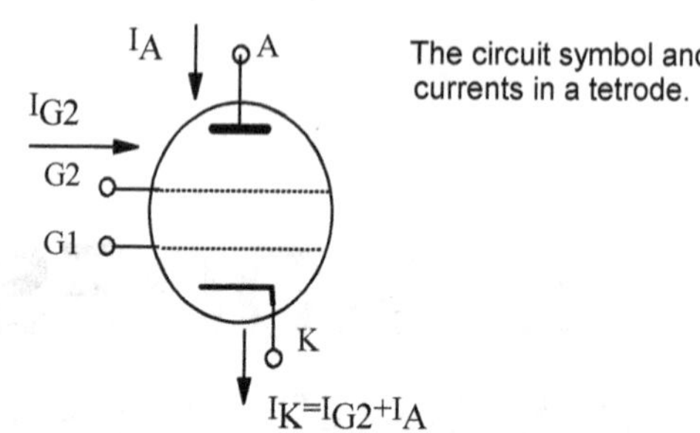

The circuit symbol and currents in a tetrode.

This addition of G2 or screen grid means there are now three element voltages and two currents, the anode and screen current, so the behavior of a tetrode is much more complex than that of a triode. However, the screen is usually held at a constant voltage and does not feature in AC models, so the same models as those for triodes can be used. The screen must be positive vis-a-vis cathode, otherwise there would be no electron flow.

A positive screen takes over the electron accelerating function of the anode, and, since it is physically much closer to the control grid G1 it has much more effect on the current flow than the anode itself, almost as much control over anode current as the control grid. Depending on the spacing between the control and screen grids and on the pitch of the grid wires (how closely they are wound), the degree of screen's control can be varied.

The finer the pitch and the closer the spacing between the screen grid's turns, the more control the screen grid will have over anode current. In both situations the screen grid will intercept more electrons (allow less of them to reach the anode), and therefore in both cases the screen current will increase.

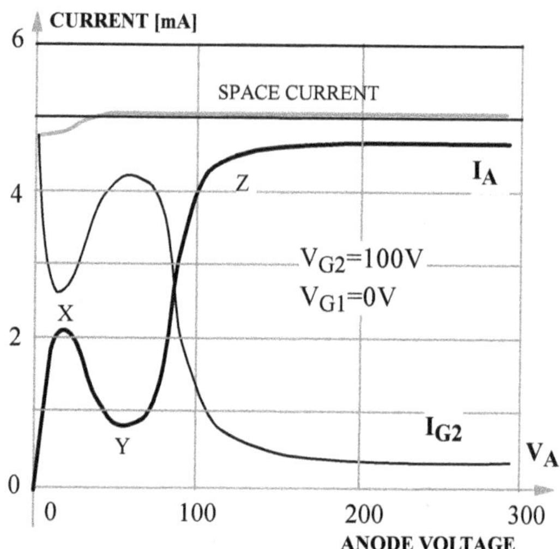

ABOVE: Screen and anode currents in a tetrode. The tube used was actually a pentode, 6J7, but connected as a tetrode (suppressor grid tied to the screen). This was done so we can directly compare its tetrode and pentode curves (NEXT PAGE)!

The reduction in input capacitance

The insertion of a screen grid seems to simply divide the anode-grid capacitance C_{AG} into two, which would halve it, since the two capacitances are connected in series. It still doesn't explain the significant (1,000-fold) reduction in C_{AG}! A model will help, as always.

The fact that screen S is at cathode (K) potential is the reason the capacitance is reduced. The dreaded Miller effect (feedback from anode to grid) cannot occur and thus the capacitance is not multiplied by the amplification factor of the stage because no voltage appears across the connection of the two capacitances!

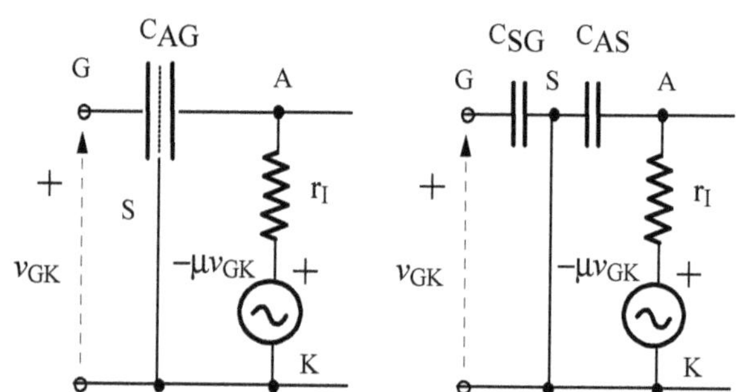

ABOVE: The AC model of a tetrode to help explain the reduction in input capacitance.

Secondary emission

The anode voltage in a tetrode may not have much effect on the anode current, but it does determine the way the total space current is distributed between the anode and the screen. Since the total space current is constant (except at very low anode voltages), the screen and anode currents are inverse or mirror images of one another!

The anode current curve has a very nonlinear shape. As the anode voltage is increased, the anode current peaks and then starts dropping, only to resume its rise before reaching the plateau where it's practically constant. A tetrode exhibits a negative internal impedance in the region where the current drops with the rising voltage X-Y, which is sometimes used in its applications as an oscillator, but is unwelcome in its use as an audio amplifier.

The kink is caused by a physical phenomenon called secondary emission, which begins at point X. While the primary emission is that of electrons from cathode, secondary emission occurs when high energy, high velocity primary electrons strike the anode and transfer enough energy to anode's electrons, so these leave the anode and flow back attracted by the positive screen grid. It is also possible for the screen grid to emit secondary electrons.

To avoid this nonlinear region, the lowest anode voltage in an amplifier would have to be above point Z, which is 100V in this case, so a tetrode amplifier would not achieve a large anode voltage swing and thus its voltage and power amplification would be limited. This is the reason that tetrodes were short lived and were replaced by pentodes only a year or two after their invention.

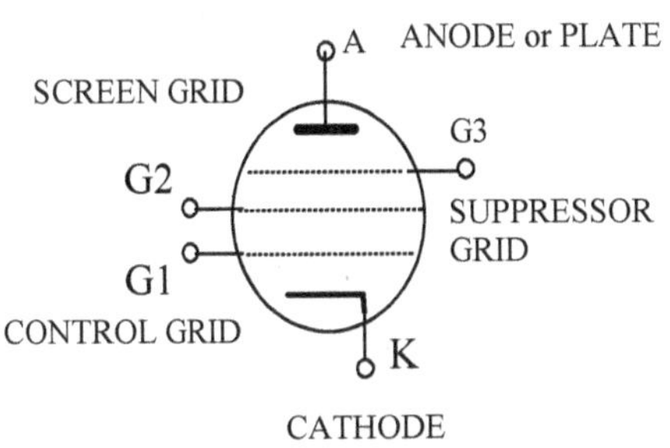

PENTODE CONSTRUCTION AND OPERATION

To overcome tetrode's deficiencies, primarily the kink in its anode characteristics, a pentode was introduced only a year later, in 1929. An additional grid called suppressor grid or G3 eliminated the effects of secondary emission and the associated kink, further improved the shielding and increased internal resistance to very high levels, thus increasing the amplification factors as well.

The suppressor grid is usually internally connected to the cathode, although some pentodes have it brought out to its own pin, most notable being EL34.

The suppressor is at a negative potential with regard to the screen, so it provides a retarding force that prevents secondary electrons emitted by the screen from flowing to the anode. More importantly, providing its potential is lower (more negative) than that of the anode, it also constrains the secondary electrons from the anode and returns them back to the anode.

In some pentodes with aligned grids the possibility of screen or suppressor grid intercepting electrons before they reach the anode is minimized and screen grid current is reduced. Not all pentodes use such alignment, but beam power tubes do.

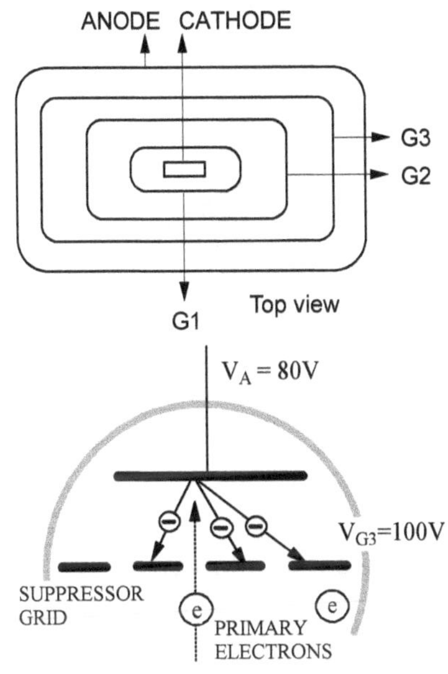

Space current = anode current + screen grid current

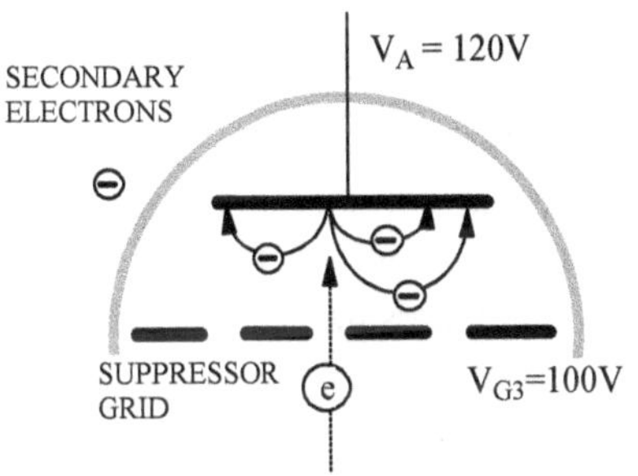

ABOVE: The secondary electrons are returned to the anode providing the anode is more positive than the suppressor grid. High speed primary electrons are not affected.

The magnitude of anode voltage plays only a minor part in the operational behavior of a pentode. The current such a pentode pushes through the load depends on gm, the pentode's transconductance, and on v_{GK}, the AC signal between the control grid and the cathode. However, there is one thing we cannot see from this graph, and that is the importance of the screen grid (or G2) for the operation of a pentode. What anode is for a triode, the screen grid is for a pentode, so the screen grid DC voltage determines its behavior.

Real pentodes don't have an infinitely high internal impedance or resistance. The internal resistance of a pentode can be read or calculated from its anode curves: $r_I=\Delta V/\Delta I$ The presence of pentode's finite internal resistance means that in our model not all of the current source's current (gmv_{GK}) will reach the load, so I_A will be lower than for an ideal pentode.

For instance, for EL84 power pentode $\Delta V=80V$ and $\Delta I=2$ mA, so $r_I=80/0.002 = 40{,}000\ \Omega$. This is a much higher internal resistance than in power triode's case (typically 200 -1,000Ω)!

The slope of the curves also means that the plate voltage does have some impact on the operation of a real pentode, as V_A increases, the plate current also increases, albeit very slightly.

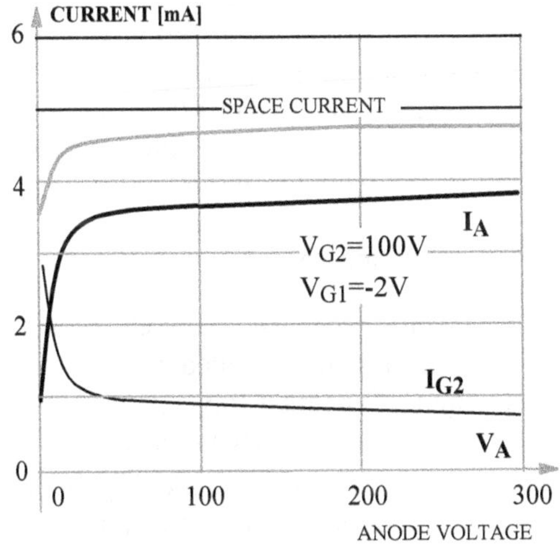

RIGHT: Screen and anode currents in 6J7 pentode. The kink in tetrode's anode curves is gone and the knee is at a much lower anode voltage, meaning much larger anode voltage swings are possible.

Small signal models of an ideal and linear pentode

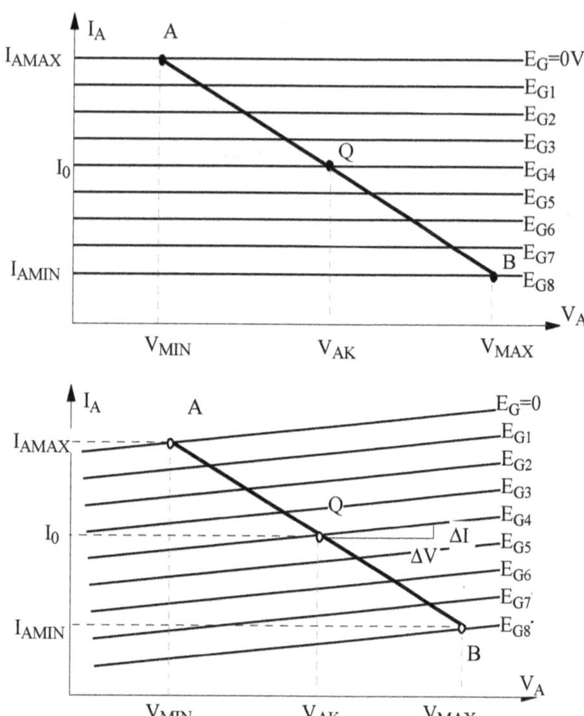

Ideal pentode's anode curves are equidistant horizontal lines, meaning its anode current does not depend at all on the anode voltage. It behaves as an ideal current source (infinitely high internal resistance).

For the linear pentode, anode characteristics are still equidistant, but they aren't horizontal any more. The internal resistance is not infinite but has some very high value r_I, modeled as a parallel resistance.

RIGHT: Linear model of a real pentode. The current is proportional to the grid signal v_{GK}: $i = gmv_{GK}$.

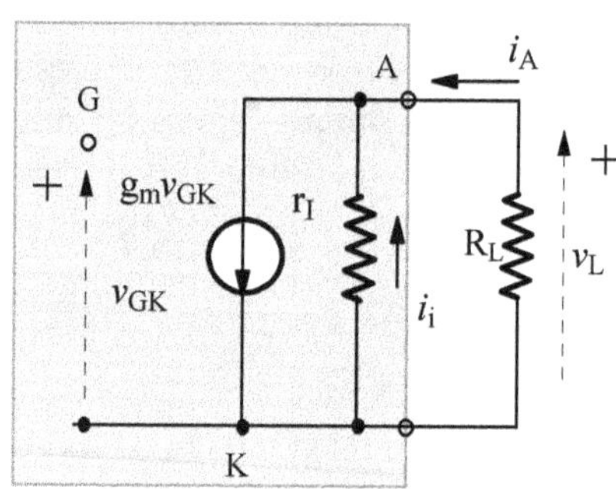

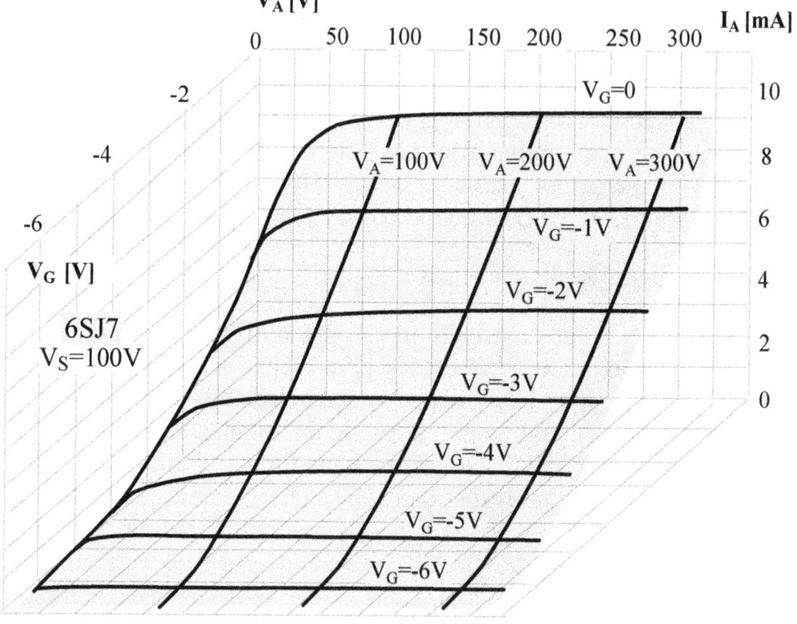

ABOVE: 3D characteristics of 6SJ7 small signal pentode with screen voltage of $100V_{DC}$

Pentode's operation in 3D

Since three electrodes impact the anode current in pentode, i_A is a function of three variables, grid voltage, screen voltage and anode voltage:

$$i_A = gm(v_{G1K} + v_{SK}/\mu_S + v_{AK}/\mu)$$

The screen amplification factor μ_{G1G2} (between the control grid and screen grid) is sometimes named μ_S!

Dealing with three variables is too complicated, so to simplify analysis, parameters gm, μ and μ_S are considered constant. However, they do depend on the operating point of the pentode.

The error introduced by this simplification is not significant, since other aspects of analysis and modeling introduce similar errors, as do tube data sheets. Going into tedious detail in one aspect of analysis while other factors vary widely would be insane.

The operating region

The operating region of a pentode is determined in the same way as for triodes. It is limited by the V_G=0V anode curve, the maximum dissipation P_{AMAX} curve and the maximum anode voltage V_{AMAX}, declared in data sheets.

The illustrated way of positioning the load line (for maximum power output) applies only to output pentodes, the small signal pentodes used as voltage amplifiers are designed using different criteria, as will be explained soon.

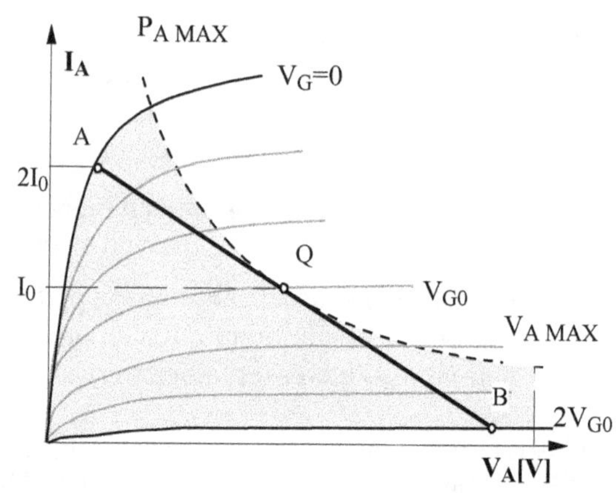

ABOVE: The boundaries of the operating region and the load line for a power pentode

Beam power tubes

The advances in tube design did not stop in 1930 with pentodes. Just as adding a third grid to a tetrode has done in pentodes, the addition of electron beam-forming plates also eliminated the dreadful kink, but also brought a few additional benefits that made beam power tetrodes superior to power pentodes. In beam power tubes or "kinkless tetrodes" the beam-forming plates restrict the operation of the tube to the regions where the space-charge is effective. The cloud of primary electrons acts as a virtual suppressor and prevents the flow of secondary electrons from anode to the screen. This phenomenon requires that the space charge cloud be at a much lower potential than the anode, requiring a large spacing between the screen and anode.

To limit the excessive increase in screen current, tube designers placed the screen grid windings in the shadow of the control grid. This deflects electrons away from the screen and allows more of them to pass through the screen.

The region of negative internal resistance and the "kink" in the anode curves is thus eliminated, prompting the whole series of European output tubes to be named kinkless tetrodes: KT66, KT77, KT88, KT90, KT100, KT120, KT150. Of course, Americans were not to be outdone, with their masterpieces such as 6L6 (plus all its variants including the industrial version 5881), 6550, 7027 and others.

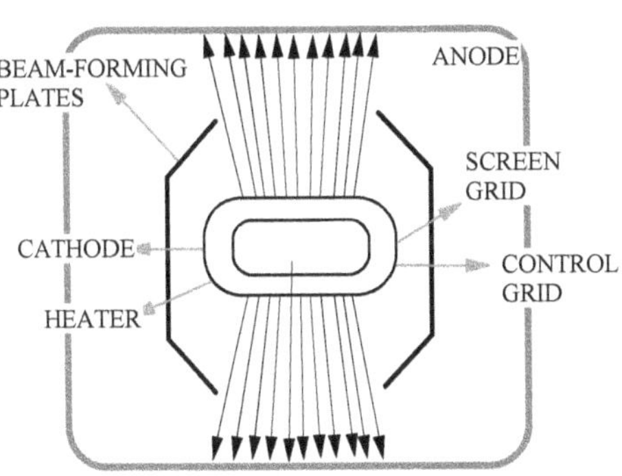

ABOVE: Horizontal cross-section of a beam power tube. The electron flow is indicated by arrows.

The rounding of the pentodes' characteristics limits the lowest anode voltage achievable under dynamic conditions and causes distortion. This rounding is the result of the inefficiency of suppressor grids at lower anode voltages, when some secondary electrons evade the pickup by the suppressor grid.

Beam power tubes have steeper (faster rising) curves in the low anode voltage region. V_{MINA} is closer to zero than V_{MINB} for pentodes, which results in a wider anode voltage swing for beam power tubes, higher power outputs and improved efficiency. Their curves are also more horizontal in the higher V_A range, leading to KT tubes' better linearity and lower distortion.

For the sake of simplicity and linguistic brevity in further text we will use the term pentode for both true pentodes and for beam power tubes (which, strictly speaking, are tetrodes). This is justified because for practical purposes their in-circuit behavior is all but identical.

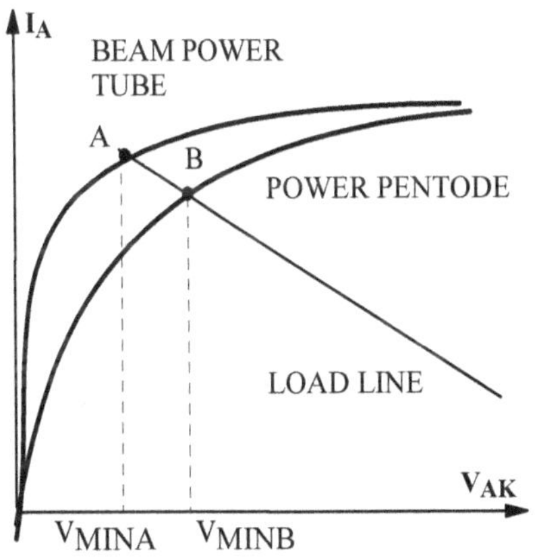

ABOVE: Beam power tubes versus pentodes

Physical construction of a beam power tube

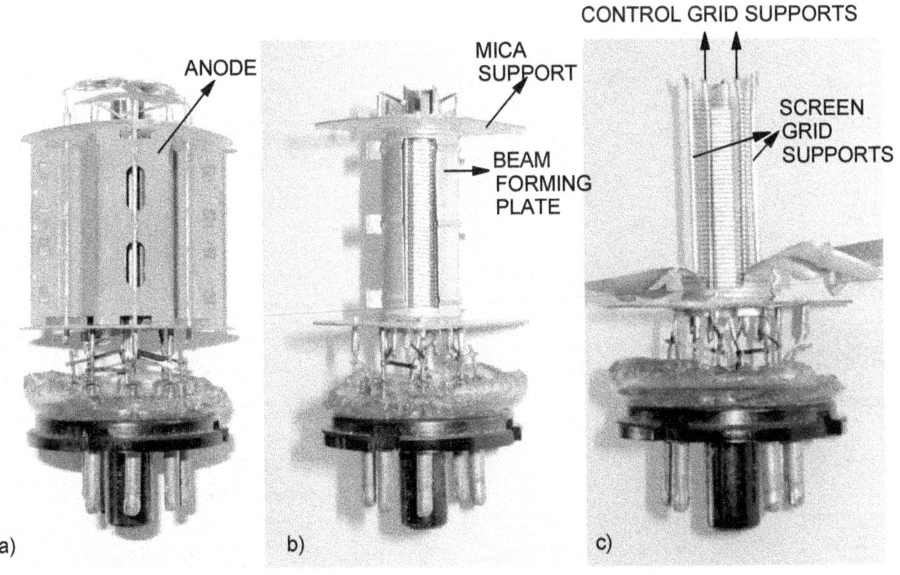

LEFT: a) The internals of 5881 beam power tube with glass removed

b) Beam forming plates became visible (the cylinder with the window in the middle) once the anode was removed

c) Two grids (screen and control grid), with the white-colored cathode (from the oxide coating) visible at the center.

Measuring and plotting pentode's characteristics

Since there is an additional control electrode (the screen grid), the test circuit for tetrodes and pentodes requires an additional DC voltage source and test instruments for the screen circuit, a mA-meter and a voltmeter.

Should you wish to investigate the effect of the suppressor grid as well, another variable voltage source is needed. Although seldom used as a control electrode (it is usually kept at a constant DC potential), a suppressor grid tied to a negative voltage source (vis-a-vis the cathode of course) can be used to control the anode current as well!

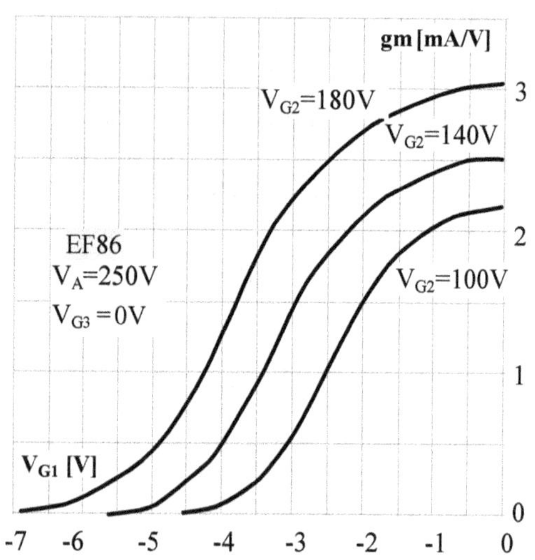

LEFT: How mutual conductance of a pentode varies with control grid and screen grid voltage (EF86)

BELOW: The circuit for recording transfer, anode and constant current curves of tetrodes and pentodes. Heater connection and its power supply is not shown.

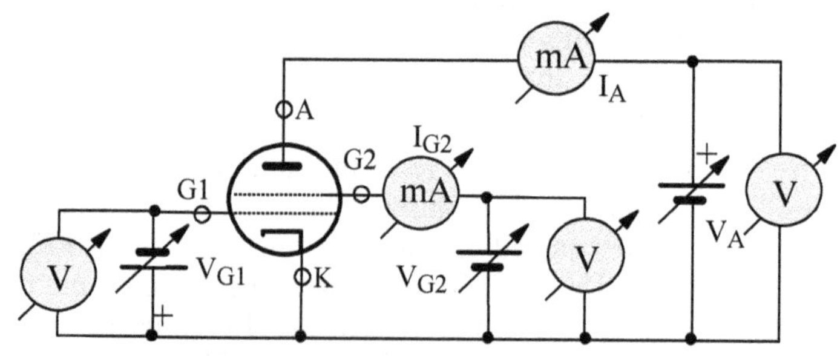

Pentode's distortion "signature"

The cubic or S-shaped dynamic transfer curve of pentodes and beam power tubes produces a large 3rd and a smaller 2nd harmonic, resulting in a symmetrically distorted waveform, with flattened peaks of the sinewave. The dominant 3rd harmonic, which is discordant (unpleasant sounding) is the primary reason pentodes have never been loved by audiophiles, even in 1960s, when they were widely used in mid-fi amplifiers of Dynaco or Eico class.

The second and third harmonic distortion in pentode stages can be estimated by the "5-point method".

$D_2 = [(I_{AMAX} + I_{AMIN}) - 2I_0]/[I_{AMAX} - I_{AMIN} + 1.41(I_X - I_Y)]*100\%$

$D_3 = [I_{AMAX} - I_{AMIN} - 1.41(I_X - I_Y)]/(I_{AMAX} - I_{AMIN} + 1.41(I_X - I_Y))*100\%$

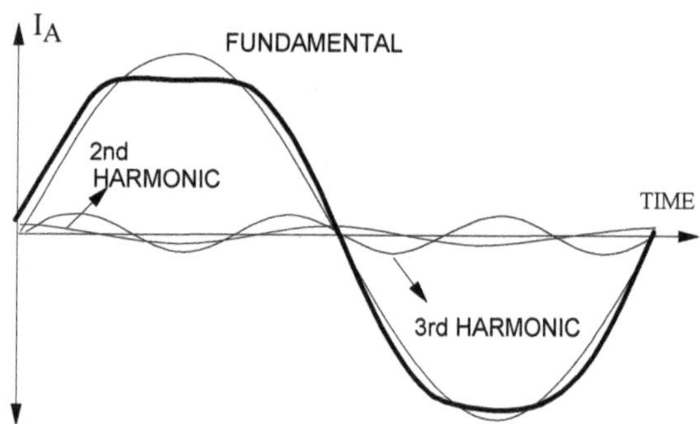

ABOVE: Waveform distortion produced by typical pentode's cubic transfer curve

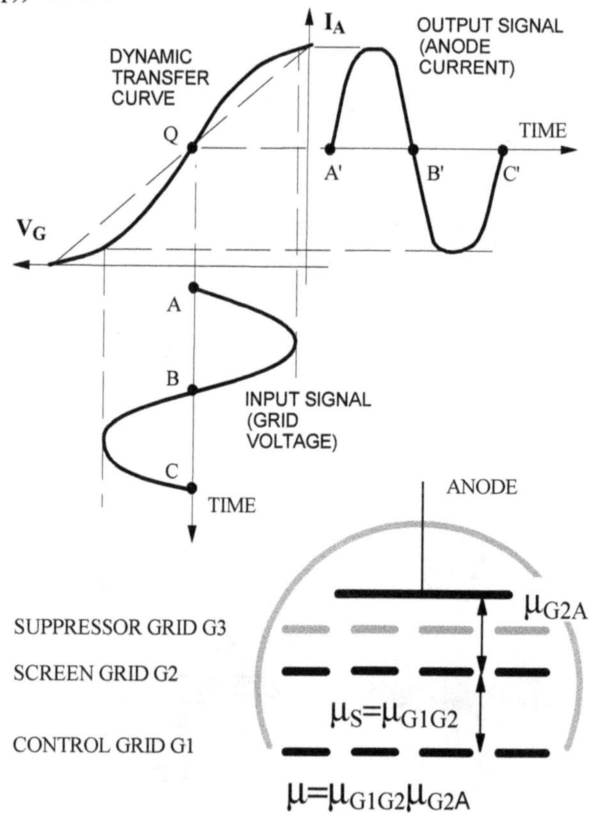

The amplification factor of a pentode

The overall amplification factor of a pentode can be expressed as a product of the amplification factors between the control grid and screen grid, and between screen grid and anode: $\mu = \mu_{G1G2}\mu_{G2A}$ The alternative symbol for screen amplification factor μ_{G1G2} is μ_S.

For example, for EF86 the figures are $\mu_{G1G2}=38$ and $\mu_{G2A}=122$, so $\mu = 38*122 = 4{,}625$

For power pentodes the figures are lower. EL34 has $\mu_{G1G2}=15$ and $\mu_{G2A}=11$, so $\mu = 165$

When a pentode is strapped as a triode, the amplification factor is approximately equal to that of the screen, so $\mu_{TRIODE} \approx \mu_S$ Screen transconductance gm_S and anode transconductance gm are proportional to the ratio of screen and anode currents: $gm_S = gm_{(pentode)} I_S / I_A$

$\mu = \mu_S \mu_{G2A}$ and $\mu_{TRIODE} \approx \mu_S$
SCREEN TRANSCONDUCTANCE:
$gm_S = gm_{(pentode)} I_S / I_A$

Providing screen voltage

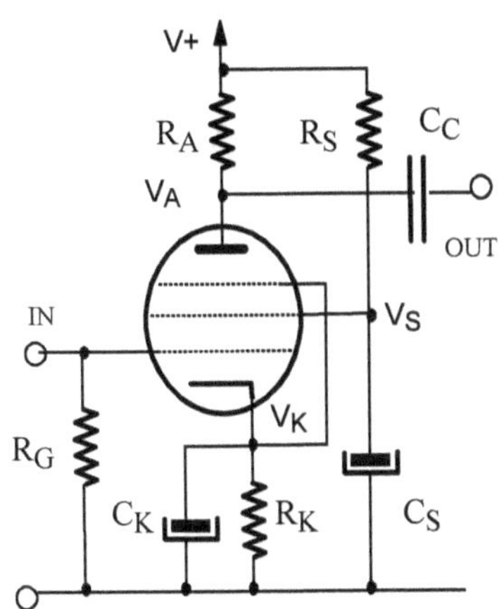

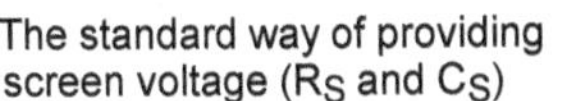

The standard way of providing screen voltage (R_S and C_S)

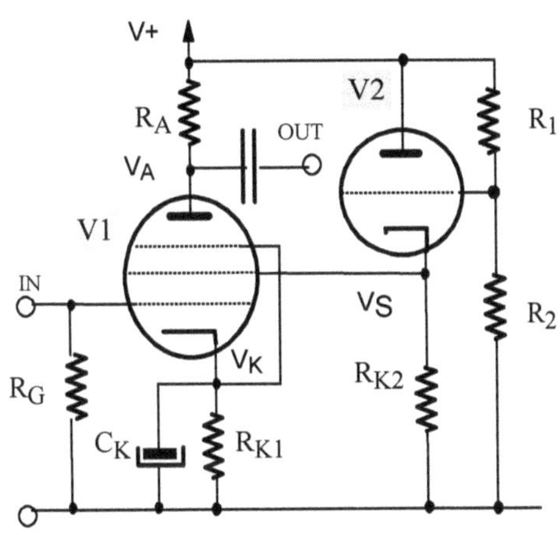

Triode cathode follower providing a stable screen DC voltage, so R_S and C_S are not needed.

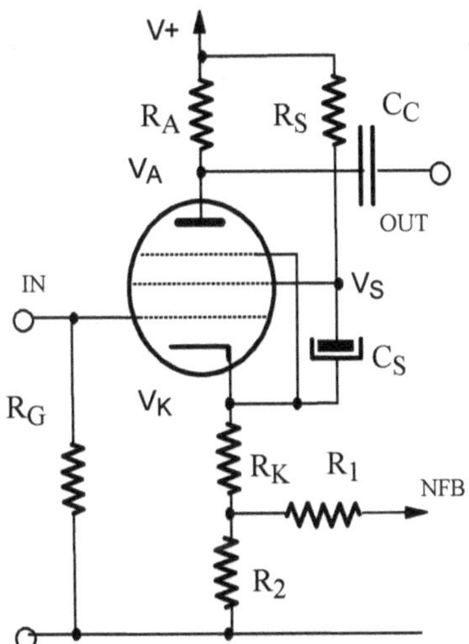

When un-bypassed cathode resistor is used, the screen capacitor C_S should be returned to the cathode and not the ground!

Since the maximum allowed screen voltage in small signal pentodes is much lower than the allowable anode voltage, it is usually supplied from the anode supply through a voltage dropping screen resistor R_S. This is akin to biasing a triode by a cathode resistor through the voltage drop the cathode current makes on that resistor. However, that means the screen DC voltage will vary as the screen current changes and that is clearly undesirable due to its high effect on operating conditions. The screen capacitor C_S needs to bypass any signal voltage from the grid down to ground, so there is no AC signal across it and the negative feedback does not develop!

For best results, the screen should be supplied from a fully regulated DC voltage source, or at least stabilized to some extent by the use of cathode follower, a Zener diode or a voltage regulator tube, as illustrated.

In output stages, audio power pentodes were specifically designed to have the same maximum allowed voltage on anode and the screen, so the two are simply connected together to a high voltage source V_{BB}. Many tubes not designed for audio use have much lower V_{G2MAX} and require a dedicated lower screen supply voltage source, which needs to be better filtered and better regulated than the anode supply.

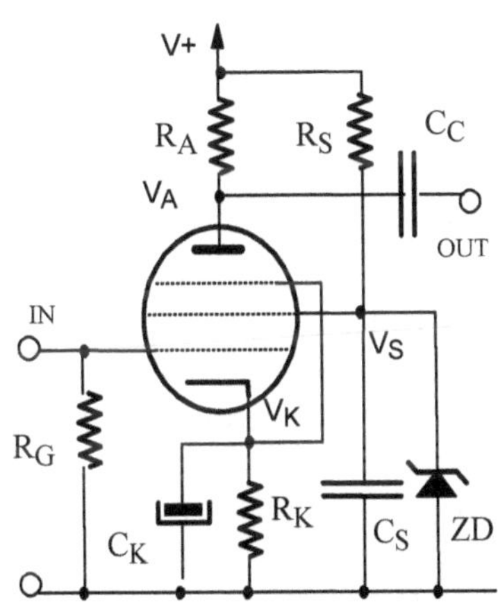

ABOVE: A Zener diode or VR tube can keep the screen voltage V_S steady, but must be bypassed to ground by a film capacitor.

BELOW: Anode characteristics of 6SJ7 pentode

The real pentode example: 6SJ7

The anode characteristics for pentodes include not just I_A-V_A curves, but also I_{G2}-V_A curves, the screen grid current curves.

Two operating points are given in the table (from data sheets), for two different anode voltages (100V and 250V).

Notice how the anode voltage does not impact the operation of a pentode, providing the screen voltage is the same (as it is here), the anode current change (2.9 to 3.0mA) and the screen current change (0.9 to 0.8mA) are minimal and can be neglected!

The ratio of the anode and screen currents is relatively constant (providing the operating point is above the "knee", that the anode voltage is higher than 50V for 6SJ7).

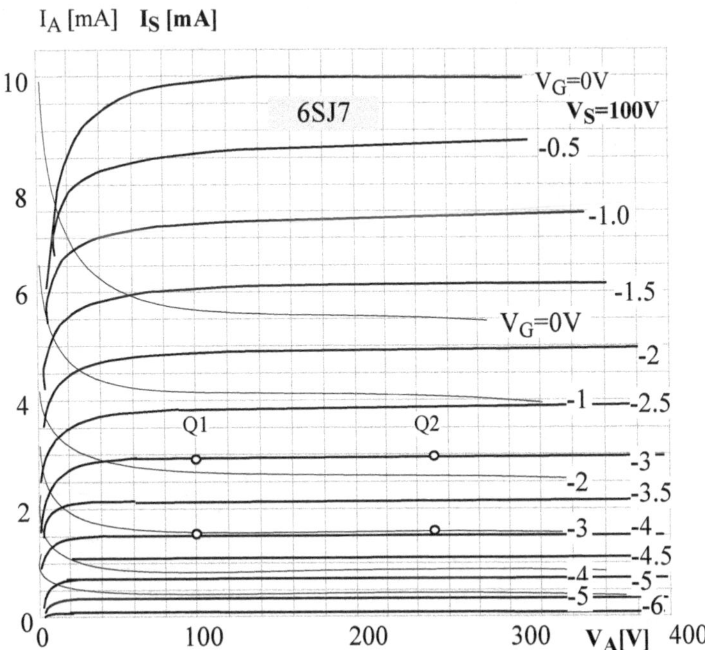

In this case 2.9/0.9 =3.22 or 3/0.8 = 3.75 so for anode voltages between 100 and 250V we can use a ratio of 3.5 to estimate screen currents. However, notice a discrepancy between the table and the graphs, redrawn from the same GE data sheet! The graphs show that screen current at both operating points is around 1.6mA, not 0.8-0.9mA. In our analysis we will assume that graphs are correct.

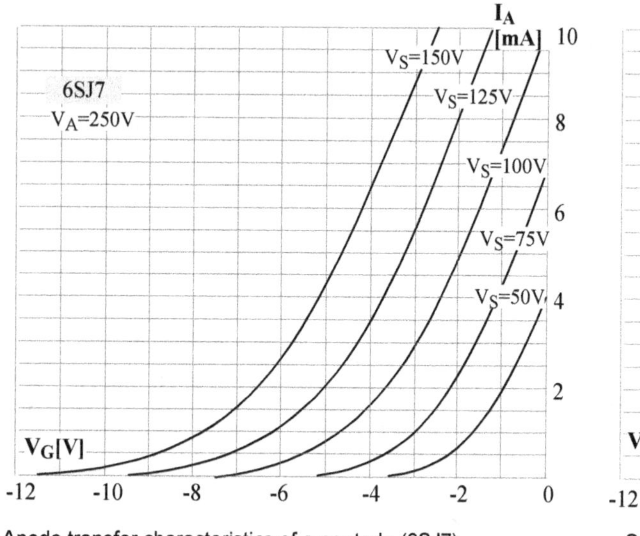

Anode transfer characteristics of a pentode (6SJ7)

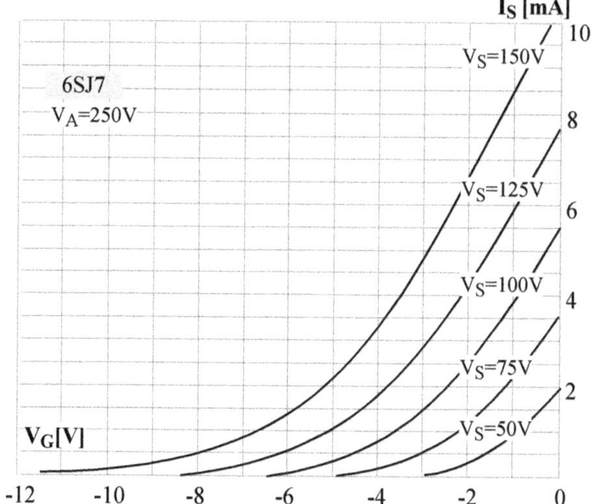

Screen grid transfer characteristics of a pentode (6SJ7)

TUBE PROFILE: 6SJ7

- Indirectly-heated sharp cutoff pentode
- Noval socket, 6.3V/825 mA heater
- Maximum heater-cathode voltage: $90V_{DC}$
- Anode/Screen dissipation: 2.5W/0.7W
- Max. anode/screen voltage: 300V/125V
- R_{GMAX}=1MΩ

6SJ7	Q1	Q2
Anode volts [V]	100	250
Control grid bias [V]	-3	-3
V_{G2} [V]	100	100
V_{G3} [V]	0	0
I_A [mA]	2.9	3
I_S [mA]	0.9	0.8
r_I [kΩ]	700	1,000
gm [mA/V]	1.58	1.65

Operating parameters for 6SJ7 pentode at two anode voltages.

DESIGN OF A PENTODE STAGE

Constructing approximate anode characteristics for any screen voltage

Pentode data sheets only publish anode characteristics for a few screen voltages, often only one, better ones up to three or four. However, you can construct approximate anode curves for any screen voltage, providing you have a static transfer curve for that voltage. There are five transfer curves published for 6SJ7 pentode, for screen voltages of 50, 75, 100, 125 and 150V. Even if there is no published curve for your desired screen voltage, say 60V, you can firstly estimate the transfer curve, and then construct anode curves from it. The method is illustrated below.

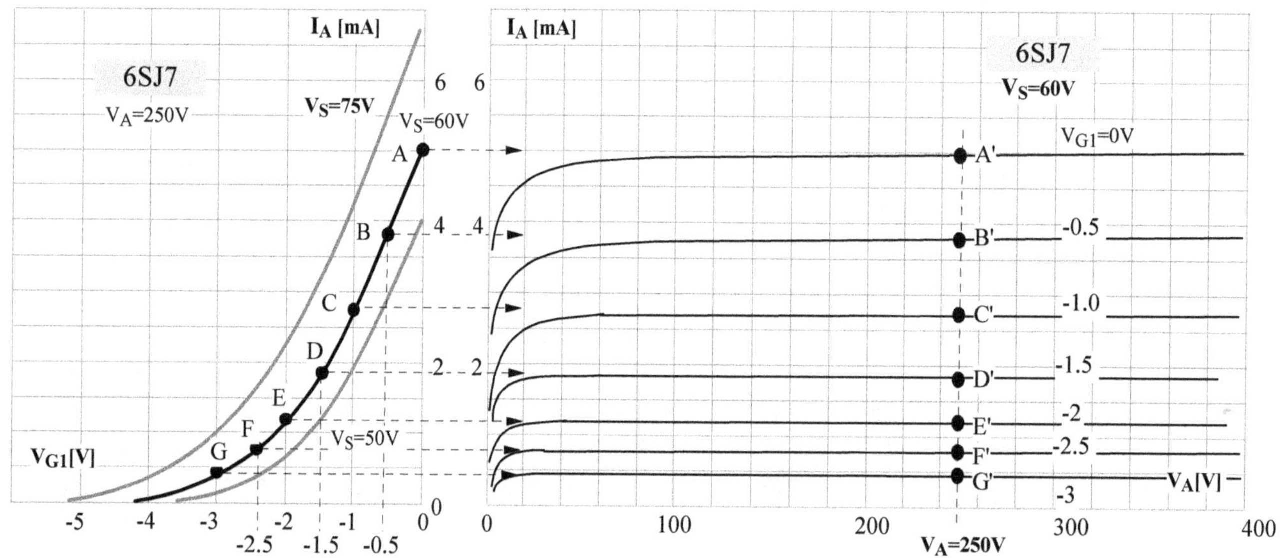

Say we need anode curves for 60V screen voltage. Starting with V_G=0V estimate point A and transfer it horizontally onto the I_A-V_A chart. Then, draw a vertical line through V_G=-0.5V from point B. Again, transfer it horizontally onto the I_A-V_A chart and draw the second curve. Repeat the process for control grid voltage steps: V_G=-1V, -1.5V, -2V, etc. The transferred points A',B',C', etc. must lie on the anode voltage specified on the transfer graph, in this case V_A=250V!

Choosing the load resistance

Let's look at three possible choices for anode resistor. With R_L=100kΩ, Q needs to sit in the middle of the grid swing at V_{G1}=-1.5V. With the grid voltage swing ΔV_G=1V, the anode voltage swing is ΔV_A= 180-40=140V, so the amplification of the stage is $A=\Delta V_A/\Delta V_G$=140/1= 140

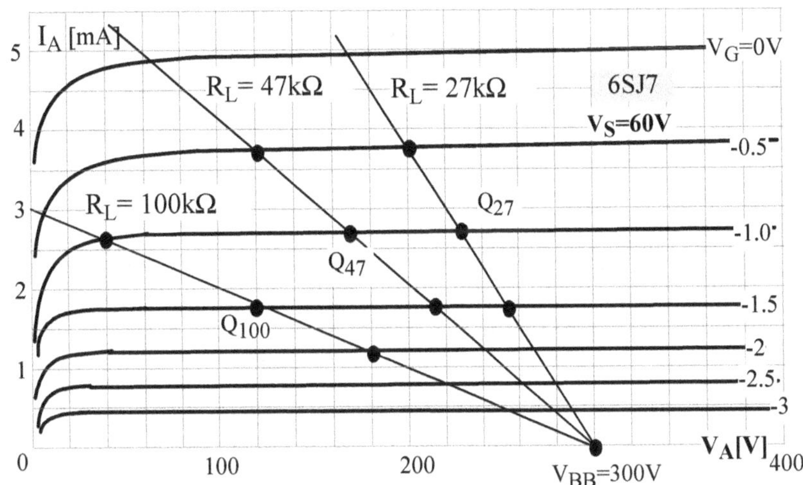

Reducing R_L to 47kΩ, Q sits at -1V. With the grid voltage swing ΔV_G=1V, the anode voltage swing is ΔV_A= 215-120=95V, so the amplification of the stage is lowered to A=95!

With an even smaller R_L of 27kΩ, Q sits again at -1, and with the same grid voltage swing ΔV_G=1V, the anode voltage swing is ΔV_A= 255-200=55V, so the amplification of the stage is A=55!

The gain is highest with 100kΩ load, but so will also be the output impedance of the stage. The compressed curves near the top will result in clipping and distortion. This compression will be less pronounced with a lower load of 47kΩ. The gain has dropped from 140 to 95, but the distortion and the output impedance will be lower.

The loadline of the 27kΩ load is fully above the knee in the curve, so the characteristics are not bunched at the top end any more. Their spacing is similar to triode's and the sound will move towards the "triode tone". The third and other odd harmonics are reduced, while the 2nd and even harmonics dominate. There is no rapid increase in screen current any more on positive signal peaks.

The stage gain of 55 may seem low for a pentode, but that is still higher than we usually get from a 12AX7 common cathode stage, so still higher than with any triode!

The load line and the quiescent point

Designing a pentode stage is not much different from that of the triode amplifier. Let's use datasheet curves for the screen voltage of 100V and the anode resistance value of 47kΩ.

Assuming you can choose V_{BB} freely, draw the load line so that point A does not sit in the curved part of the characteristics. To get a decent input swing (from -2V to -4V) we need to have V_{BB} of around 300V. Thus, Q needs to sit in the middle of the grid swing, or at V_{G1}=-3V. This has determined the quiescent anode voltage as V_0=160V.

The maximum grid voltage swing is ΔV_G=4-2=2V, the maximum anode voltage swing is ΔV_A= 230-70=106V, so the amplification of the stage is $A=\Delta V_A/\Delta V_G$=160/2= 80

Even higher amplification factors are possible with larger anode loads and higher V_{BB}.

To determine the values of screen and cathode resistors we need to know the screen current in Q.

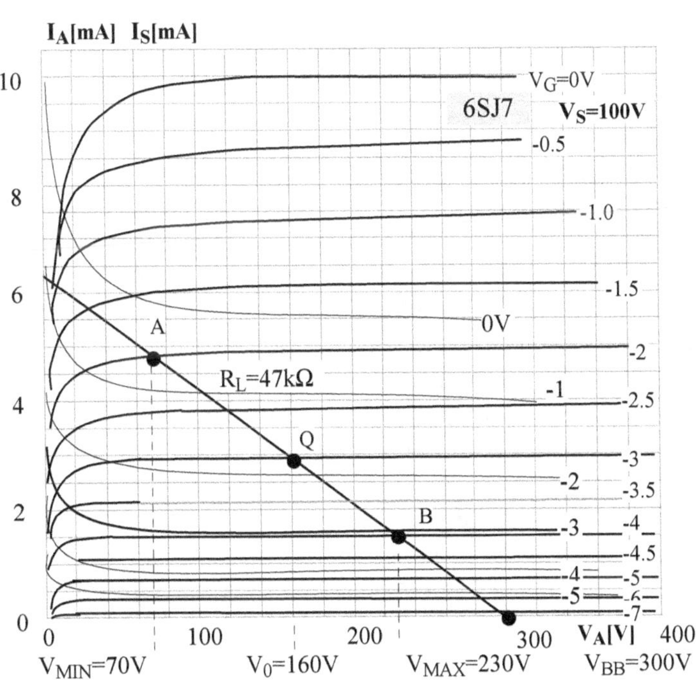

Designing the screen circuit

The graphs show that screen current at the quiescent point Q is around 1.6mA. If an RC circuit is used for the screen voltage, we know that V_{BB}=300V, that I_S=1.6mA and that V_S needs to be 100V. Thus the value of the resistor R_S must be $R_S=(V_{BB}-V_S)/I_S = (300-100)/1.6 = 200/1.6 = 125k\Omega$

The first rule-of-thumb for sizing C_S (the screen bypass capacitor) is that its reactance at the lowest desired operating frequency (-3dB) should be ten times lower than R_S. Let's say that is f_L=5Hz, then $\omega_L=2\pi f_L=10\pi=31.42$ So $X_C=1/\omega_L C_S= R_S/10 = 125{,}000/10 = 12{,}500$, or $C_S=1/X_C\omega_L=1/(12{,}500*31.42) = 2.55\mu F$

Another rule-of-thumb some designers use is the time constant of the screen circuit $\tau=R_SC_S$. From here (again, for f_L=5Hz) we have $C_S=1/(2\pi R_S f_L)$ = 0.26μF This is ten times lower value than that obtained by the first method, and is the absolute minimum we should use.

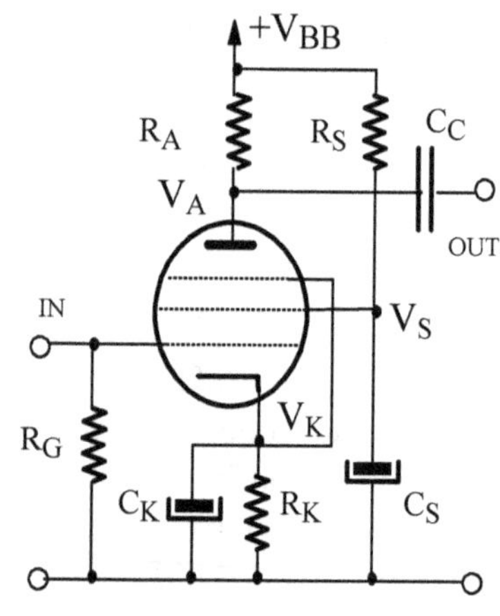

ABOVE: RC screen bias

If a voltage divider is used for the screen voltage, we know that V_{BB}=300V, I_S=1.6mA and that V_S must be 100V. We can choose the value of R_2 so the total current drawn by the screen biasing circuit does not exceed a certain value. This will depend on how much spare current capacity we have in our power supply for the preamplifier circuit. Let's say we want a total of I_T=10mA. Thus, 10mA will flow through R_1 and I_T-I_S= 10-1.6 = 8.4mA through R_2. Thus $R_1=(V_{BB}-V_S)/I_T = (300-100)/10 = 200/10 = 20k\Omega$

$R_2=V_S/(I_T-I_S) = 100/8.4 = 11.9k\Omega$

These are not exact formulas and there is a degree of error. The total resistance is R_1+R_2 =31.9kΩ, so the total current will be 300V/31.9kΩ = 9.4mA, while we assumed 10mA. This is close enough. Since 20kΩ and 11.9kΩ aren't standard resistor values anyway, we will choose 22k and 12k, so the total current will be 300/34k = 8.8mA

Now 22k*8.8mA=193V, so our screen voltage may end up being 300-193=107V instead of 100V, which is acceptable. Alternatively, increase R_1 to 25k and you will get down to around 100V on the screen.

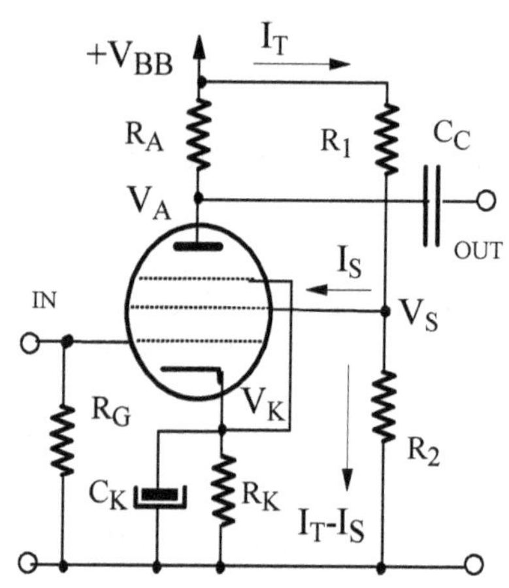

ABOVE: Screen bias through a resistive voltage divider

Cathode circuit

The cathode current is the sum of screen and anode currents $I_K=I_S+I_A$= 1.6+3 = 4.6mA The control grid must be at -3V with respect to the cathode, so the cathode must be at +3V! Since $V_K=I_K*R_K$, the required cathode resistor is $R_K=V_K/I_K$= 3/4.6 = 0.652kΩ. The standard value is 680Ω.

The cathode bypass capacitor is sized in the same way as the screen capacitor. For f_L=5Hz we have $C_K=1/(2\pi R_K f_L)$ = 46.8μF Use standard value of 47μF, or, for lower f_L, use 100 or 220μF.

The effect of screen impedance on pentode's amplification

An AC signal on pentode's control grid causes changes not only in anode current but also in screen current. These current changes will result in a voltage drop across any impedance in the screen circuit and thus also between screen and cathode. Assuming that the usual way of biasing the screen is used, a voltage dropping resistor R_S and a capacitor C_S, the impedance in the screen circuit is the internal dynamic screen resistance r_S in series with the parallel combination of the screen resistor and capacitor, as depicted in the equivalent circuit.

This voltage drop across the screen impedance reduces the gain of the pentode stage, acting as a kind of negative feedback. In an ideal case with Z_S=0 (the reactance X_C of the capacitor is so low that it practically represents a short across R_S) only the internal resistance r_S affects gain, so the ratio of this maximum amplification A_{MAX} and the actual amplification factor A is $\mathbf{A/A_{MAX}=r_S/(r_S+R_S)}$

Normally R_S is much larger than r_S, so their parallel combination is only slightly lower than r_S, so for complete bypassing the capacitor C_S must have a reactance at least ten times lower than r_S!

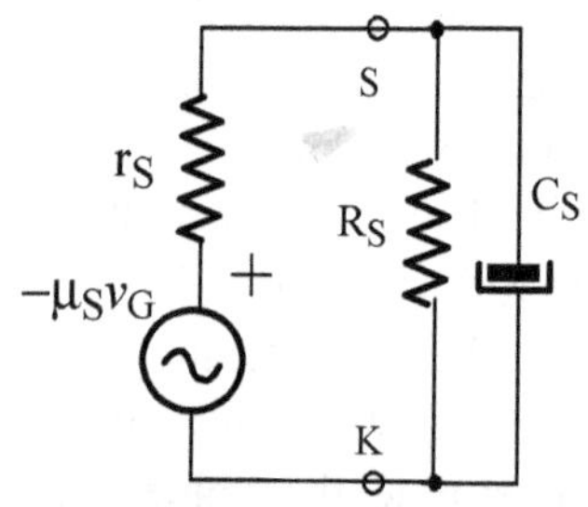

ABOVE: The equivalent circuit of the screen grid

Side-by-side: Triodes versus pentodes

Finally, let's compare 12AU7 triode and EF86 pentode! Typically, triodes have a higher anode power rating and work with higher anode currents (10mA vs. 3mA here).

The transconductance gm values are similar, but since pentode's internal resistance is 10-20 times higher, its amplification factor μ is also 10-20 times higher, making pentodes much better voltage amplifiers than triodes.

However, pentodes require very small control grid bias, meaning they cannot take as large input signals as triodes can.

	12AU7	EF86
P_{MAX} [W]	2.75	1
V_A[V]	250	250
Bias [V]	-8.5	-2.0
V_{G2} [V]	N/A	140
V_{G3} [V]	N/A	0
I_A [mA]	10.5	3
r_I [kΩ]	7.7	100
μ	17	190
gm [mA/V]	2.2	2.0

ABOVE: Side-by-side comparison of typical operating conditions of a voltage amplifying triode (12AU7) and a pentode (EF86)

PRACTICAL DESIGN: 6K7 DRIVER STAGE

First introduced in 1936, the metal 6K7 and the glass version 6K7G were used in RF- and IF-stages of radio receivers with AVC (Automatic Voltage Control). 6K7GT also goes under the aliases CV1942 and VT-86B. The top cap is the control grid. Not the prettiest looking tube by any stretch of imagination, but good sounding and cheap to buy.

Since the voltage drop on the 180k screen resistor is 435-54=381V, the screen current is I_S= 381/180 = 2.12mA Likewise, the voltage drop on the 120k anode resistor is 435-260 = 175V, so the current through it is I_A= 175/120 = 1.5mA!

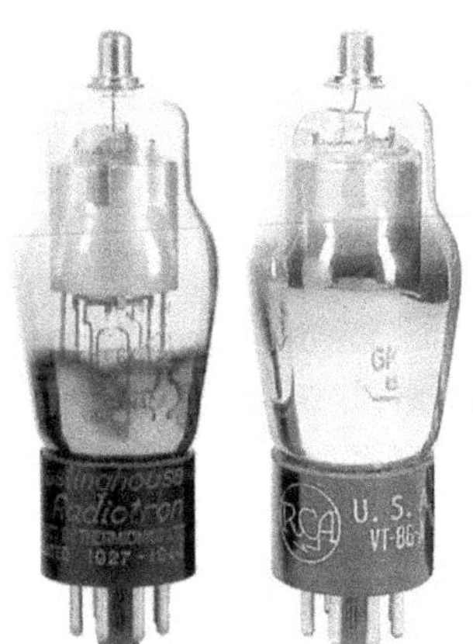

TUBE PROFILE: 6K7

- Indirectly-heated remote-cutoff variable μ pentode
- Octal socket, 6.3V, 300 mA heater
- V_{AMAX}/V_{AMAX} 300/125 V_{DC}
- P_{AMAX}= 2.75 W, I_{AMAX}=20 mA
- TYPICAL OPERATION PENTODE:
- V_A=250V, V_S=100V, V_G=-3V
- I_{A0}=7 mA, I_S=1.7mA
- gm =1.45 mA/V, μ=1,160, r_I=800 kΩ

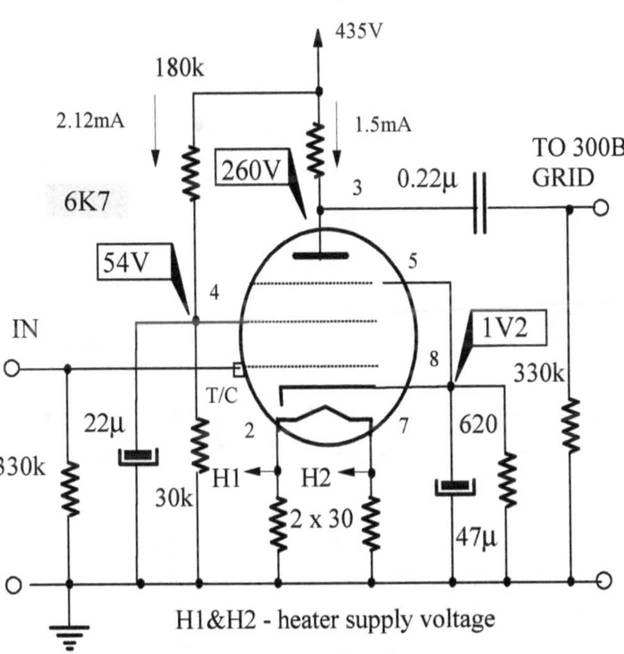

ABOVE: The input/driver stage of TAD-30 amplifier

ABOVE: Chinese-made TAD-30 amplifier can use either 2A3 or 300B output tubes (manually switchable heater voltage).

PRACTICAL DESIGN: LINE STAGE WITH TRIODE-CONNECTED EF86 PENTODE

EF86 pentode connected as a triode has very straight and equidistant anode curves (I_A - V_A graph below). The 6AU6 and E280F pentodes also make great sounding and very linear triodes. There are two possible ways to connect pentodes as triodes. Connecting both screen and suppressor grids to the anode creates a medium μ triode whose amplification factor is approximately equal to that of the screen ($\mu \approx \mu_S$). Connecting the screen grid to control grid results in a high μ triode, as per the illustration below.

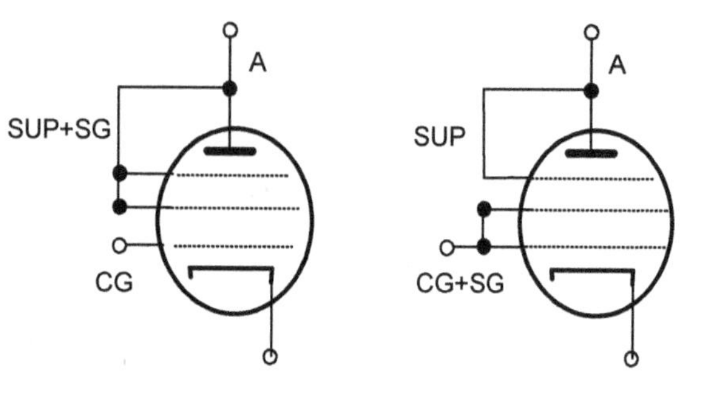

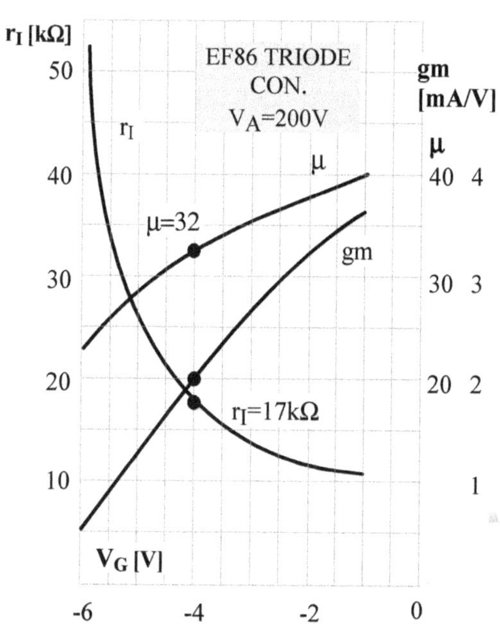

TUBE PROFILE: EF86

- Low noise voltage amplifying pentode
- Noval socket, heater: 6.3V, 200 mA
- Maximum anode/screen voltage 300/200 V_{DC}
- P_{AMAX}=1 W, I_{KMAX}=6 mA
- TYPICAL OPERATION PENTODE:
- V_A=250V, V_S=140V, V_G=-2.0V, I_A=3mA, I_S=0.6mA
- gm =1.8 mA/V, μ=4,500, r_I=2.5MΩ

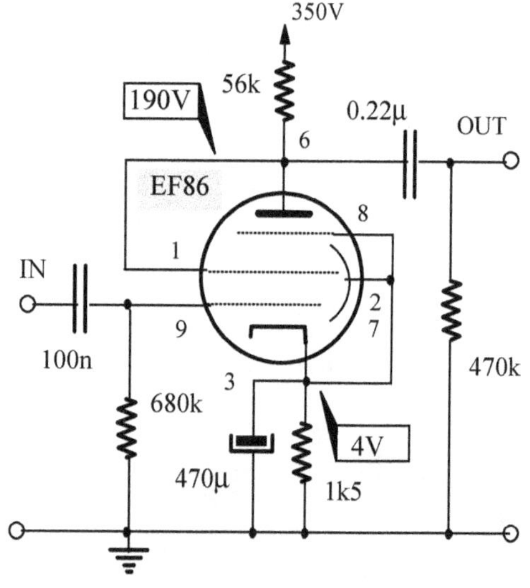

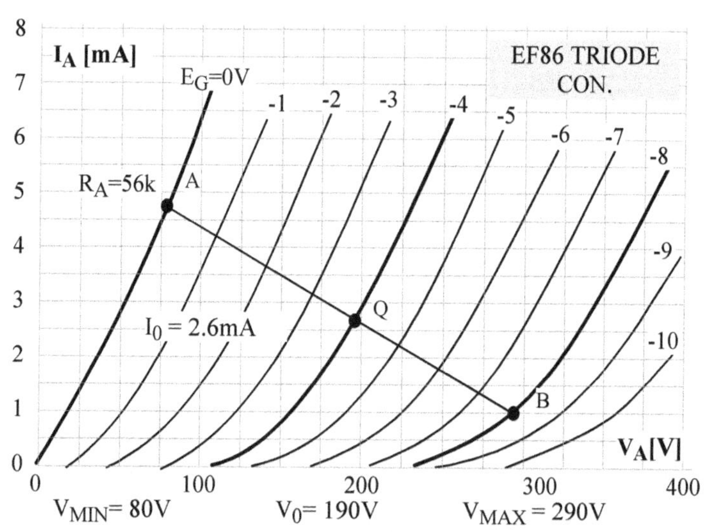

If we limit the input signal to $4V_{PEAK}$ or $8V_{PP}$, we can bias the stage around V_G = -4V This should leave us plenty of headroom, considering that a typical CD player's output is 1-2 V_{RMS}. We read the anode current in the operating point Q as I_0 = 2.6mA and the plate voltage as V_0 = 190V.

We know I_0=2.6 mA at V_K=4V, so the required cathode resistor needs to be $R_K = V_K/I_0 = 4/0.0026 = 1{,}538\ \Omega$ The closest E12 standard value is 1.5 kΩ, perfect. Voltage gain from the graph: $A = -\Delta v_A/\Delta v_G = -(290-80)/8 = -26.25$

Using a small-signal model: $A=-\mu R_A/(r_I+R_A)=-32*56/(17+56) = -24.5$

The largest signal will be distorted the most, so in the worst case scenario $D_2 =(\Delta I_+ - \Delta I_-)/2\Delta I = (2.1-1.6)/2*3.7 =$ 6.8%, which seems high but is actually a very good result. In a line-stage application we will never use $210V_{PP}$ or $74V_{RMS}$ at the output. At $10V_{RMS}$ output or less, the distortion will be under 0.1% !

The 0.22μF output capacitor, together with the 470k resistor forms a high pass filter, the -3dB cutoff frequency is 1.5 Hz in this case. The same applies to the CR filter at the input, whose -3dB frequency f_L is 2.3Hz.

The time constant of the cathode RC network is $\tau = 470\mu * 1k5 = 0.71$ seconds, so its f_L frequency is 0.2 Hz.

THE μ FOLLOWER

If the anode load of a triode has a low DC resistance but very high (ideally infinite) AC impedance, the voltage amplification factor of such a stage would be $A= v_{OUT}/v_{GK}= -\mu r_A/(r_I+r_A)=-\mu 1/(1+ r_I/r_A)$ If $r_A \rightarrow \propto$ then $r_I/r_A \rightarrow 0$ and $A \approx -\mu$ Thus, the amplification of the stage "follows" the μ of the tube, giving the stage its popular name, the μ-follower.

The design of the μ-follower is a concurrent design of the two stages in a "totem-pole" or series arrangement.

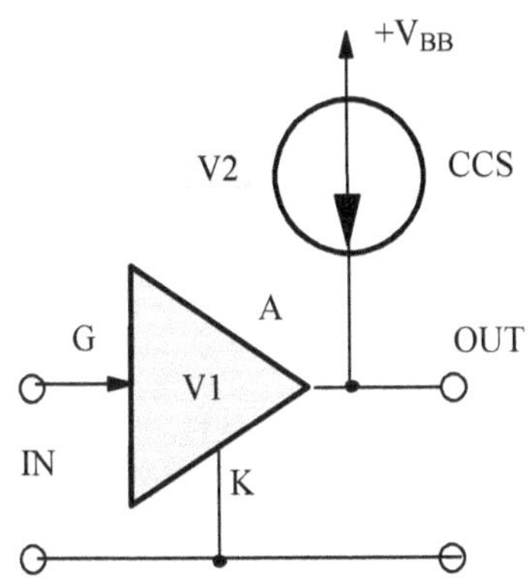

The block diagram of μ follower

Triode μ follower

A closer inspection of the detailed diagram (next page) reveals that AC signal is fed from the anode of V1 through capacitor C_{G1} to the grid of V2, while the output of V2 and the whole stage is taken from the cathode of V2. Its anode is at $+V_{BB}$ DC potential (no anode resistor) or at zero AC level. Thus, as you should recognize by now, the upper tube works as a cathode follower (CF).

So, the μ-follower is a "vertical" or cascode equivalent of a "horizontal" arrangement (cascade) of a common-cathode stage followed by a CF! A cascade is much easier to design because the two stages are independent, which also gives us total flexibility in choosing anode currents and positioning the operating points.

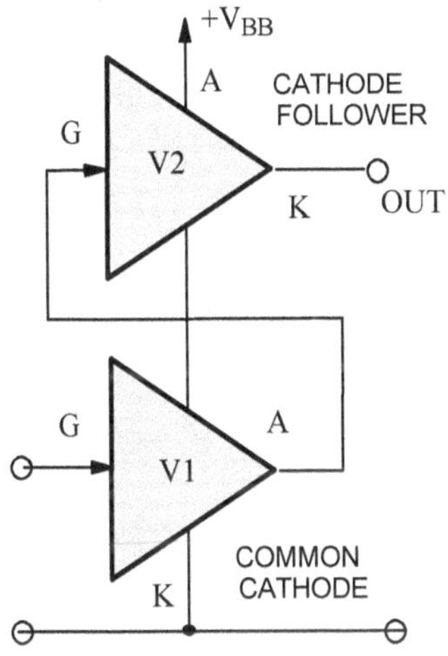

Alternative view

Now we start with the bottom tube, V1. Let's choose 2 mA again and 90V between anode and cathode. From the graph the bias is -2.8V, so our cathode resistor needs to be the same value as in our calculations for cascode, 2.8/2=1k4!

Being lazy, let's choose the same operating point for the upper triode and the same "load" resistor R_2=47kΩ. The -2.8V grid bias for V2 is created by the voltage drop across R_1, and with 2mA of current R_1=2.8/0.002 = 1k4 (again). The total load for the upper tube is 48.4kΩ!

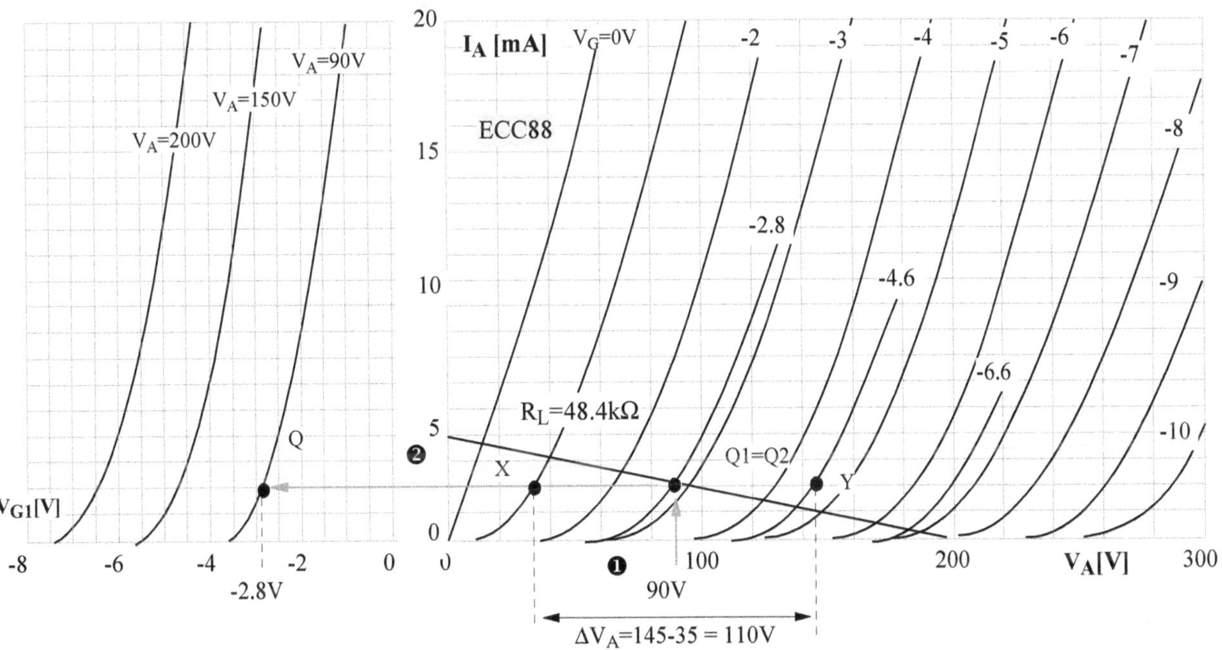

The voltage difference between K2 and A1 is 48.4*2 = 96.8V, meaning that K2 is at 96.8V+90V+2.8V= 189.6V above ground level (190V). Adding 90V for V2, we get the required power supply voltage of 189.6+90 = 280V.

The gain of the cathode follower is $A_K= \mu R_K/[(1+\mu)R_K + r_I]$, and since r_I=10kΩ and μ=28, we get A_K= 0.96

The dynamic load that the bottom tube (V1) "sees" is $r_A=R_L/(1-A_K)$ = 48.4/(1-0.96) = 48.4kΩ*25 = 1.2MΩ! This means that the loadline for V1 is practically horizontal. We can estimate the gain of the lower tube from the graph: ΔV_G=3.6V, ΔV_A=110V, so A=110/3.6 = 30.6

The input resistance of the cathode follower is $r_{IN} = R_{G2}/[1-A_V R_2/(R_2+R_1)]$ which for R_{G2}=470k equals r_{IN}= 470/(1-0.96*0.971) = 6.94MΩ ≈ 7MΩ

The value of C_G is determined based on the desired -3dB half-power frequency $f_L=1/(2\pi r_{IN} C_G)$. Choosing say C_G=22nF we get $f_L=1/(2\pi*7*10^6*22*10^{-9})$ = 1/0.968 ≈ 1 Hz, which is perfect.

Pentode μ follower

A current source has a very high internal AC resistance, and a pentode is by its inherent behavior a natural current source. However, the price to be paid is a much higher complexity, due to the presence of the screen grid and the associated circuitry.

A cascade is much easier to design than the pentode μ-follower because the two stages are independent. In this case the anode current of V1 and the cathode current of V2 must be carefully chosen, because their difference biases the upper tube through its voltage drop across R_1!

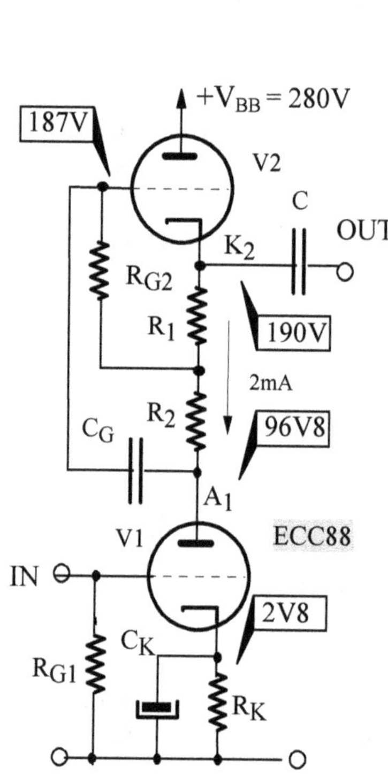

Triode μ–follower with important DC voltages and currents marked

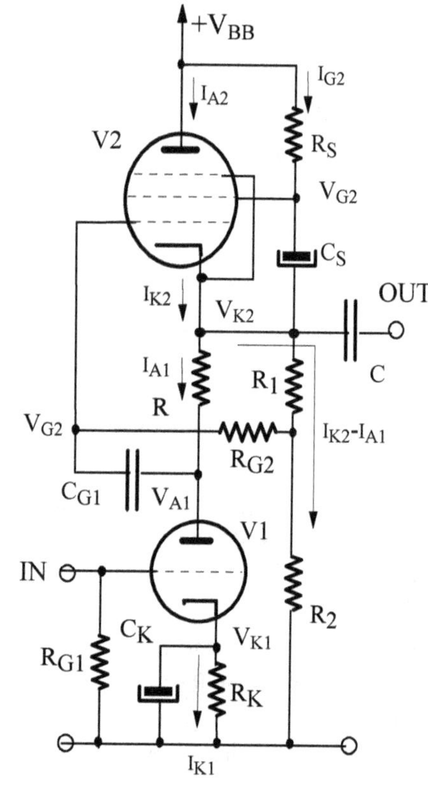

μ -follower with upper pentode

FREQUENCY RESPONSE OF TUBE AMPLIFIERS

- LINEAR SYSTEMS, FILTERS AND THEIR TRANSFER FUNCTIONS
- THE COMMON CATHODE AMPLIFIER STAGE AT LOW FREQUENCIES
- THE COMMON CATHODE AMPLIFIER STAGE AT HIGH FREQUENCIES
- THE AMPLITUDE AND PHASE CHARACTERISTICS OF RC-COUPLED STAGES
- FREQUENCY RESPONSE AND PULSE RESPONSE OF CASCADED STAGES

"I think I understand what *Stereophile* readers look for from the measurements: a sense of any technical flaws inherent in the design. For example, you see some products that have a very difficult time reproducing a semblance of a square wave. You would tend to think that an amplifier should do that credibly. If it does not, then you begin to question the fundamental design."

Steve Bednarski from BAT, in a *Stereophile* interview, Dec. 1995

LINEAR SYSTEMS, FILTERS AND THEIR TRANSFER FUNCTIONS

To fully understand the inner workings and outward behavior of tube amplifiers, we must first review the basics of the linear systems theory. Unfortunately, we cannot study this topic without some pretty heavy mathematics. If you get lost, skim through this section and review its main conclusions and their practical consequences. Remember the rules-of-thumb and you should be able to get by without fully understanding the "whys" behind them.

Zeroes and poles of a transfer function

The linear system or network (a filter or amplifier as typical examples) does something to the input signal. The "transfer function" of the system H(s) determines how the system changes the input signal, and can be expressed as ratio of two polynomials: $H(s)= Z(s)/P(s)$

Polynomials with real coefficients can be factorized so we have

$H(s)= H[(s-z_1)(s-z_2) \ldots (s-z_M)]/[(s-p_1)(s-p_2) \ldots (s-p_N)]$

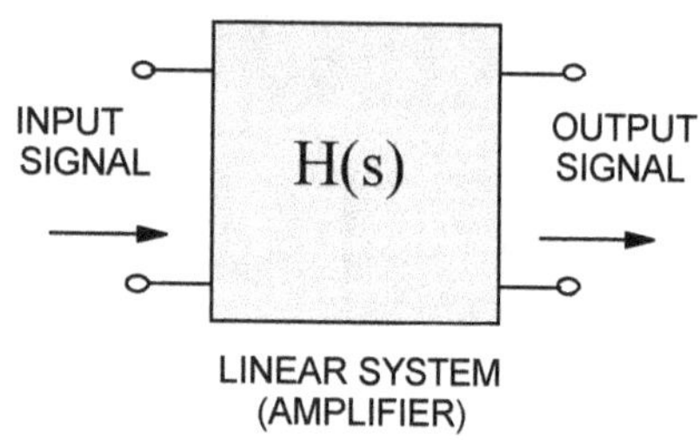

The operator "s" is an independent variable, a complex number of the form $s=\sigma+j\Omega$, which can be represented graphically in a complex plane with σ on the X-axis and ω on the vertical or imaginary axis. Since we are mostly interested in system response to sine wave signals, in that case we will replace the complex operator "s" by "$j\omega$".

The numerator is a polynomial of the M-th degree, the denominator is a polynomial of the N-th degree. The zeroes of the numerator, z_1 to z_M, have a dimension of frequency, just as do zeroes of the denominator, p_1-p_N. To distinguish between the two sets of frequencies, we call the zeroes of the upper polynomial zeroes of the transfer function (z_1 to z_M) and zeroes of the denominator we call "poles of the transfer function" (p_1 to p_N). These can be located on the complex s plane, small circles are used for zeroes and crosses for the poles.

The transfer function is a complex number, it has a modulus and phase, which are often studied separately.

Since there are two possible input and two possible output signals, there are four possible transfer functions, outlined in the table. Of most interest in tube amplifies is the voltage amplification factor $A_V(s)=v_{OUT}/v_{IN}$ but everything we say about $A_V(s)$ applies to the other three types of transfer functions. To simplify writing, instead of A_V we will simply write A, so keep in mind that this is the voltage amplification.

The four types of transfer functions

INPUT SIGNAL	OUTPUT SIGNAL	H(s)	NAME
v_{IN}	v_{OUT}	$A_V(s)=v_{OUT}/v_{IN}$	VOLTAGE AMPLIFICATION FACTOR
i_{IN}	i_{OUT}	$A_I(s)=i_{OUT}/i_{IN}$	CURRENT AMPLIFICATION FACTOR
v_{IN}	i_{OUT}	$G(s)=i_{OUT}/v_{IN}$	TRANSFER ADMITTANCE
i_{IN}	v_{OUT}	$Z(s)=v_{OUT}/i_{IN}$	TRANSFER IMPEDANCE

Bode plots

Henrik Bode was one of the pioneers of linear system analysis. The amplitude-frequency and phase-frequency plots using asymptotes are named in his honor. A simple tool to use, they will enable you to quickly determine the behavior of any system, providing you know its transfer function. Constructing the asymptotes in Bode's frequency plots is very easy. The rules are simple:

A constant (frequency-independent term) is represented by a horizontal line. A pole introduces a downward slope at -20dB/decade, starting at its angular frequency ω_P. A zero introduces a upward slope at +20dB/decade, starting at its angular frequency ω_Z. And finally, a double pole or double zero have a double slope, +/-40 dB/decade.

Where the pole and zero asymptotes overlap, the resultant slope is their difference. In the filter above, with increasing frequency (left to right) the asymptote of the pole (1) is active first and the resultant characteristic is equal to its slope. At ω_Z the zero introduces the upward slope (2) as a result the two slopes cancel each other and the resultant (3) is a horizontal line from that frequency onward!

The best way to ease into this relatively complex mathematical treatise is through the analysis of basic filters, the building blocks of all circuits.

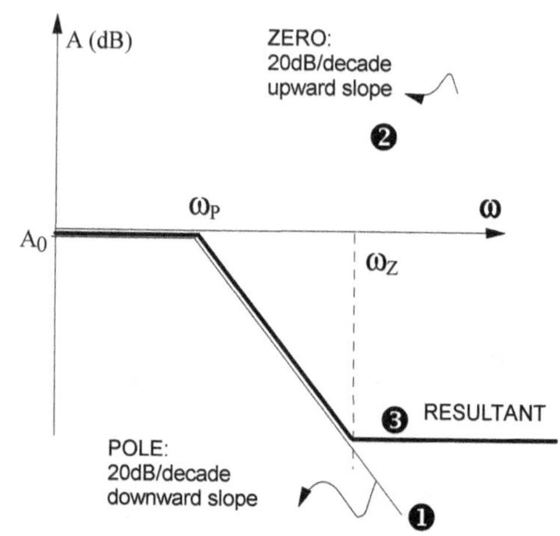

How the overall Bode plot is constructed from individual zero and pole characteristics

1st-order low pass RC filter

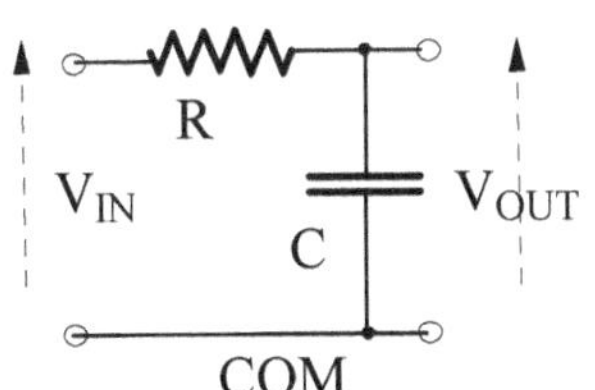

This filter is a resistive-capacitive voltage divider. For low frequencies the reactance of the capacitor shunting the output (load) is high and all input signal passes through the series resistor R and reaches the load.

As the signal frequency increases, the reactance $X_C=1/\omega C$ gets smaller and smaller, meaning more and more of the input signal is shunted (diverted) to ground (COM).

The voltage transfer function is $A_V= V_{OUT}/V_{IN} = A(s) = X_C(s)/[R+X_C(s)] = (1/sC)/(R+1/sC) = 1/(1+RCs)$ We divide both the numerator and denominator by RC and get $A(s)=(1/RC)/(s+1/RC)$

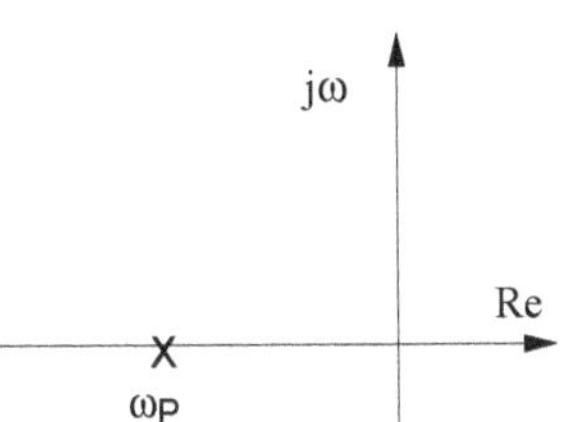

The pole-zero diagram for 1st order RC network

For sinusoidal signals $s=j\omega$ so we have $A(j\omega) =1/(1+j\omega RC) = 1/(1+j\omega/\omega p)$, where the pole frequency is $\omega_P=1/RC$, this is also the upper -3dB frequency f_U where the output drops -3dB! This frequency is also called a "corner", "break", "turnover" or "half-power" frequency.

To draw the Bode diagram and its asymptotes, first we assume that $\omega \ll \omega_P$. In that case A=1, or in dB A=20log1= 20*0 = 0 dB this represents a horizontal line ❶.

If $\omega \gg \omega_P$, $A = \omega_P/\omega$ or in dB $A = 20\log(\omega_P/\omega) = 20\log\omega_P - 20\log\omega$ For $\omega=\omega_P$ A= 0 dB This represents a downward sloping asymptote with a slope of -6 dB per octave or -20dB per decade ❷

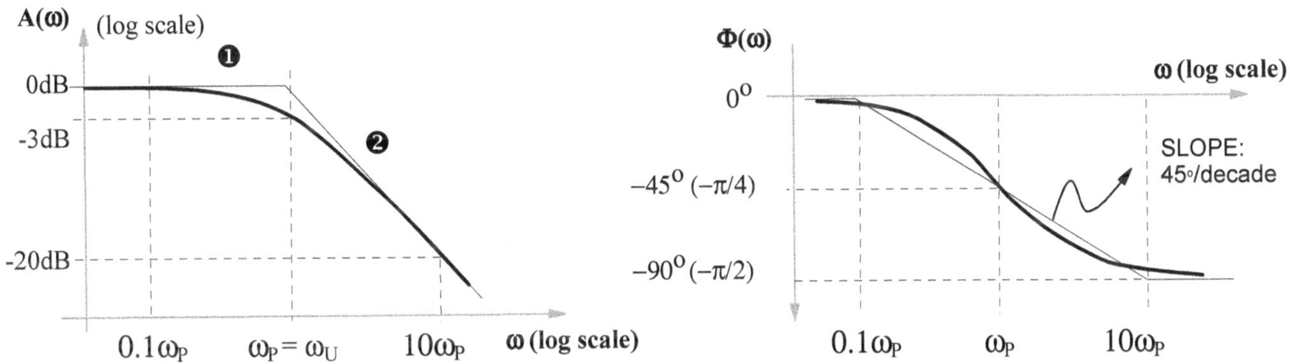

ABOVE: Bode diagrams for the modulus $A(\omega)$ and phase $\Phi(\omega)$ of the A_V transfer function
BELOW: The RC network's response to square wave signals of various frequencies with reference to the upper -3dB frequency f_U

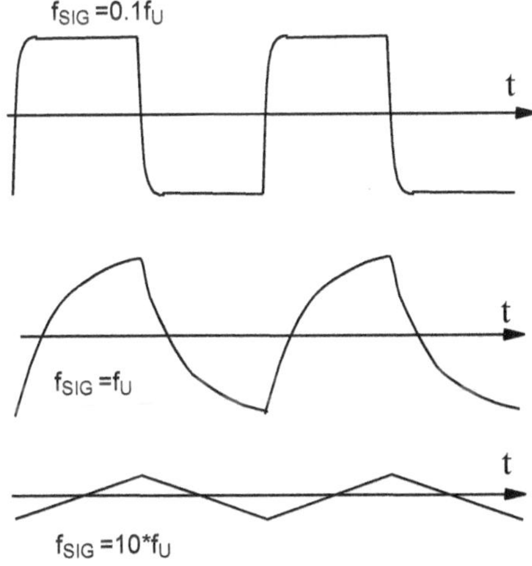

Since the output voltage lags behind the input voltage this circuit is also called a phase lag circuit.

With a square wave signal at its input, the output of the low pass filter will depend on the frequency of the signal. For $f_{SIG} \ll f_U$, the output voltage will have a small rounding of the corners, as illustrated. As the frequency of the input signal is increased, the closer it gets to the upper -3dB frequency f_U, the more rounded the output signal will get, until it resembles a sawtooth shape. Because the output signal is an integral of the input signal, this circuit is also called the *integrating circuit.*

For $f_{SIG} \gg f_U$ (where f_{SIG} is more than ten times f_U), the output voltage degenerates into a practically linear sawtooth shape, as illustrated.

This filter and its behavior is extremely important in audio since it models the behavior of an audio amplifier at high frequencies.

The meaning of the time constant

The time constant τ (Greek letter "tau") of a circuit is the reciprocal value of the angular frequency ω of its pole (Greek letter "omega"). If a DC voltage or a square wave signal is applied to a fully discharged capacitor, the time constant is the time required for the voltage on the capacitor to reach 63% of its final value. In the same point the voltage drop across the resistor is 100-63 = 37%

As a "charging" and "discharging" parameter it depends on the value of the capacitance and resistance. For the CR filter just discussed, $\omega_P = 1/RC$, so $\tau = 1/\omega_P = RC$!

Say a DC voltage V_{IN} ("step-function") is applied to this circuit at t=0. The capacitor is fully discharged. At t=0 the full V_{IN} is across the resistor and the output voltage is zero. The capacitor starts charging at a charging rate determined by τ! The larger the time constant the slower the charging rate! After $t=\tau$ the output voltage (across the capacitor) is at 63% of the DC voltage at the input.

The voltage across the resistor also follows an exponential curve and falls at the same rate, so after $t=\tau$ it is 1-0.63 or 37% of V_{IN}.

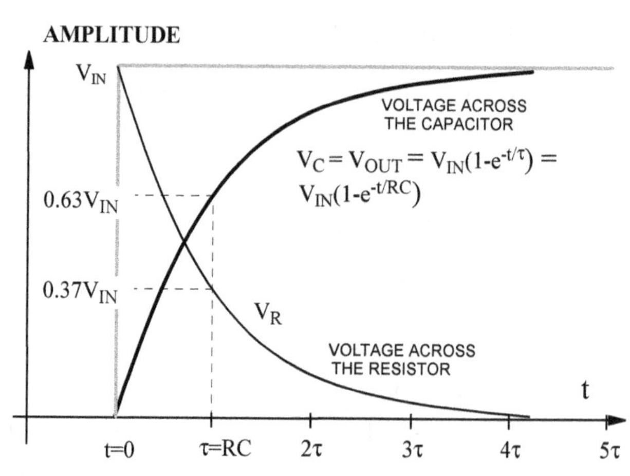

1st-order high pass RC filter

$A_V(s)=R/[R+Z_C(s)] = R/(R+1/sC) = RCs/(1+RCs)$

Divide both the numerator and denominator by RC and get $A(s)=s/(s+1/RC)$ The transfer function has a form $A(s)=s/(s+\omega_P)$ There is one zero at $\omega_Z=0$ and one pole at $\omega_P=1/RC$, this is also the lower -3dB frequency f_L where the output drops -3dB!

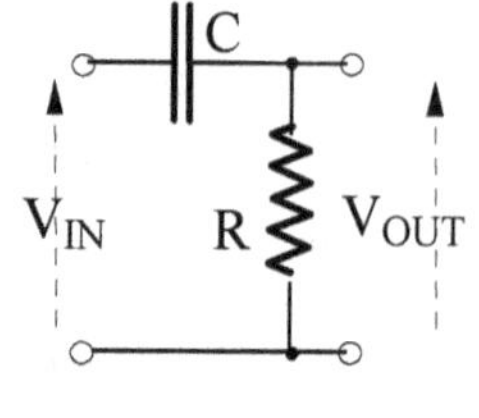

For sinusoidal signals $s=j\omega$ so we have $A(j\omega)=j\omega RC/(1+j\omega RC) = j\omega/\omega_P/(1+j\omega/\omega_P)$, where the pole frequency is $\omega_P=1/RC$

To draw the Bode diagram and its asymptotes, first we assume that $\omega \gg \omega_P$. In that case A=1, or in dB A=20log1= 20*0 = 0 dB This represents a horizontal line ❶.

For $\omega \ll \omega_P$, $A = \omega_P/\omega$ or in dB $A = 20\log(\omega_P/\omega) = 20\log\omega_P - 20\log\omega$

For $\omega=\omega_P$ A= 0 This represents a upward sloping asymptote with a slope of +6 dB per octave or +20dB per decade ❷

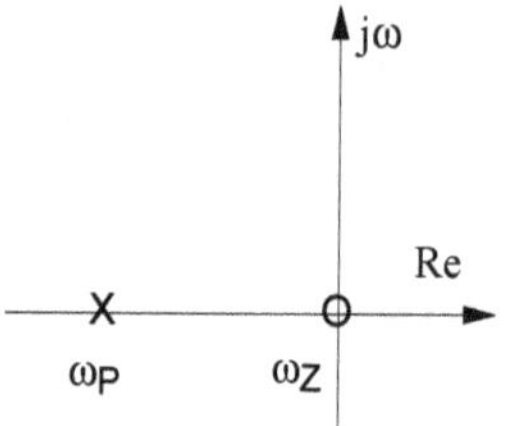

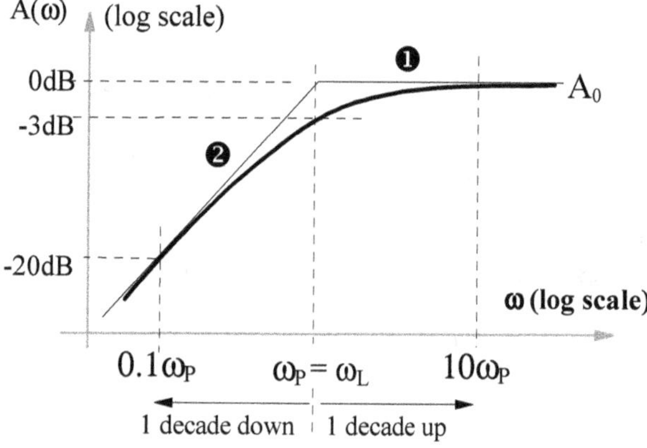

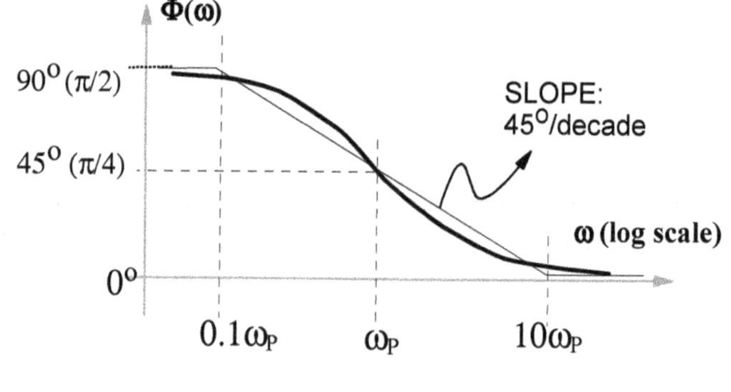

ABOVE: Bode diagrams for the 1st-order high pass RC filter
RIGHT: The response of the CR network or high pass filter to square wave signals of various frequencies

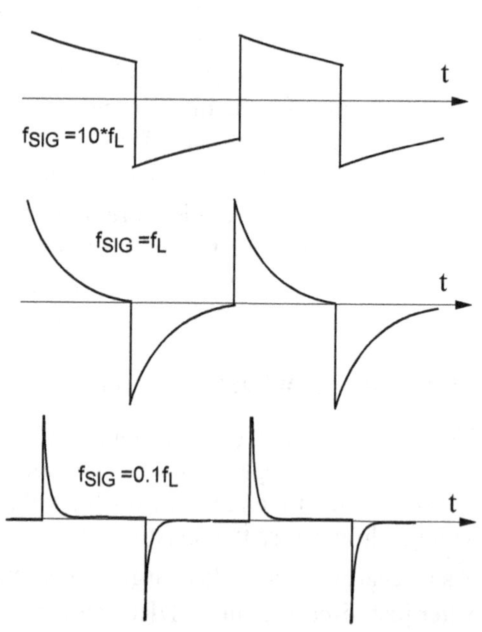

Since the output voltage leads the input voltage this circuit is also called a phase lead circuit.

With a square wave signal at its input, the output of the low pass filter will depend on the frequency of the signal. For high frequencies, where $f_{SIG} \gg f_U$, the output voltage will have a small, almost linear droop of the upper edge, as illustrated.

As the frequency of the input signal is lowered, the closer it gets to the lower -3dB frequency f_L (the pole frequency), the more pronounced the droop. Because the output signal is a derivative of the input signal, this circuit is also called the differentiating circuit.

For $f_{SIG} \ll f_U$ (where f_{SIG} is more than ten times f_U), the output voltage degenerates into sharp pulses, as illustrated.

This filter and its behavior is also extremely important in audio since it models the behavior of an audio amplifier at low frequencies.

Limited phase lag network

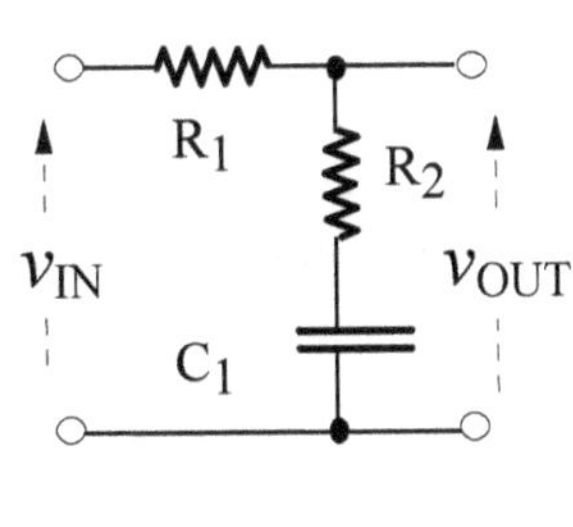

This important network is the basic building block of most RIAA filters used in phono stages (preamplifiers). The transfer function has one pole and one zero at frequencies ω_P and ω_Z: $H(j\omega) = V_{OUT}/V_{IN} = (1+j\omega C_1R_2)/[1+j\omega C_1(R_1+R_2)] = (1+j\omega\tau_Z)/[1+j\omega\tau_P)]$

At the frequency of the "zero", the capacitive impedance Zc or Xc is equal to the resistance R_2 or $1/(\omega_ZC_1) = R_2$ The time constant of the zero $\tau_Z = 1/\omega_Z$ so $\tau_Z = R_2C_1$! At the frequency of the "pole", the capacitive impedance Zc or Xc is equal to the total series resistance R_1+R_2, or $1/(\omega_PC_1)=R_1+R_2$ so $\omega_P = 1/[(R_1+R_2)C_1]$ and $\tau_P = (R_1+R_2)C_1$

The final output level A_X is equal to the ratio $\omega_P/\omega_Z = R_2/(R_1+R_2)$ or, if logarithmic scale is used, $A_X = 20\log[R_2/(R_1+R_2)]$

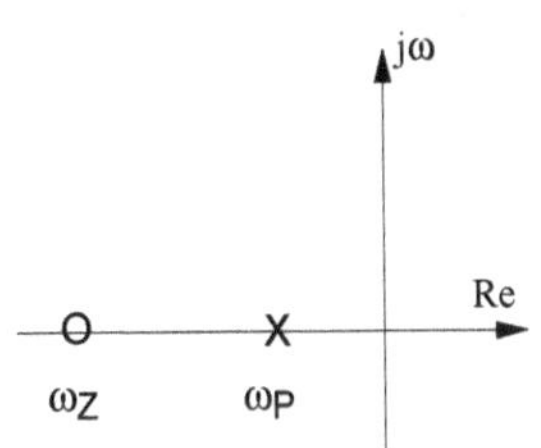

The pole-zero diagram for the limited phase lag network

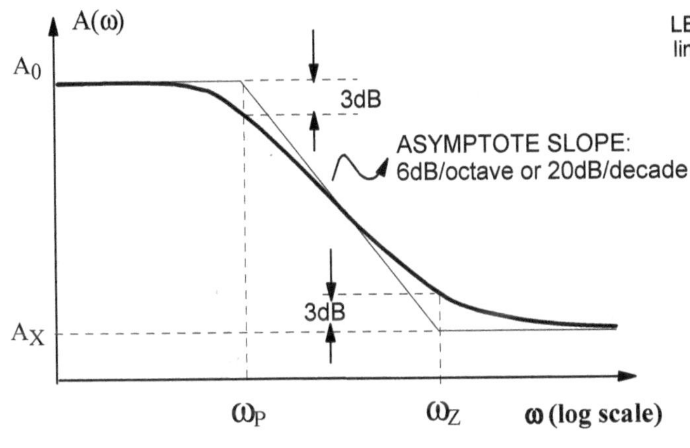

LEFT: Bode plot for the limited phase lag network

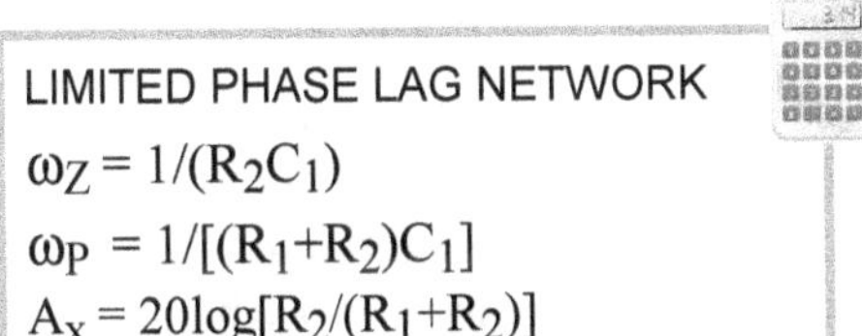

LIMITED PHASE LAG NETWORK

$\omega_Z = 1/(R_2C_1)$

$\omega_P = 1/[(R_1+R_2)C_1]$

$A_X = 20\log[R_2/(R_1+R_2)]$

Limited phase lead network

This network also has one pole and one zero at frequencies ω_P and ω_Z and is important since it models the behavior of a common cathode amplifying stage at low frequencies: $H(j\omega)=V_{OUT}/V_{IN}= R_2(1+j\omega\tau_Z)/[(R_1+R_2)(1+j\omega\tau_P)] = R_2(1+j\omega C_1R_1)/[(R_1+R_2)(1+j\omega C_1(R_1||R_2))]$

At the frequency of the zero, the capacitive impedance Z_C or X_C is equal to the resistance R_1 or $1/(\omega_ZC_1) = R_1$

The time constant of the zero $\tau_Z = 1/\omega_Z$ and $\omega_Z = 1/(R_1C_1)$, so $\tau_Z=R_1C_1$ and at the frequency of the pole, the capacitive reactance X_C is equal to $R_P=R_1||R_2$, the parallel combination of $R_1||R_2 = R_1R_2/(R_1+R_2)$, so $\omega_P=1/[(R_PC_1]$ and $\tau_P=R_PC_1$

The final output level A_X is equal to the ratio $\omega_Z/\omega_P = (R_1+R_2)/R_2$, or, if logarithmic scale is used, $A_X = 20\log[(R_1+R_2)/R_2]$

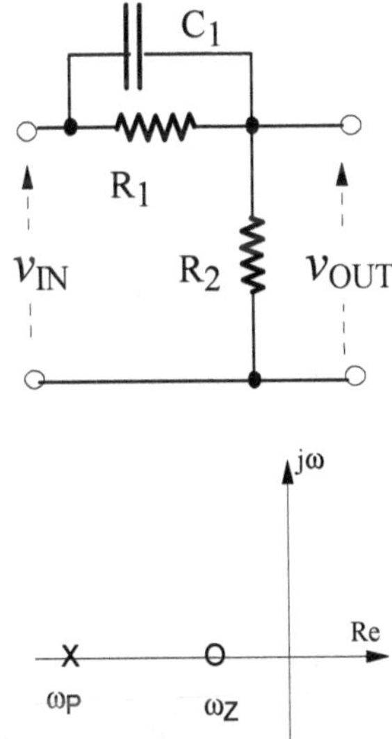

The pole-zero diagram for the limited phase lead network

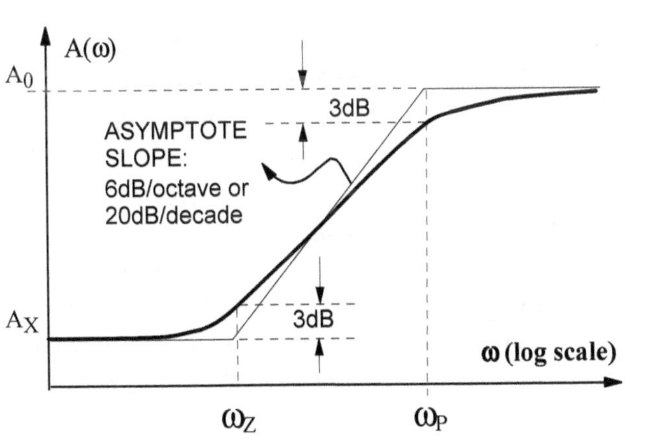

LEFT: Bode plot for the limited phase lead network

LIMITED PHASE LEAD NETWORK

$\omega_Z = 1/(R_1C_1)$

$\omega_P = 1/[(R_PC_1]$

$R_P= R_1R2/(R_1+R_2)$

$A_X = 20\log[R_2/(R_1+R_2)]$

2nd-order low pass filter

This type of circuit is extremely important in audio, since it models the behavior of audio transformers at higher frequencies and tube amplifiers driving an electrostatic speakers at mid-to-high frequencies. The resistance R is the sum of R_S (the secondary winding's resistance) and primary resistance reflected onto the secondary side R_P/IR where IR is the impedance ratio of the transformer. C is the total parasitic capacitance of the secondary winding and of the primary winding reflected onto the secondary side. If the transformer drives a grid of the following tube (interstage transformer), the grid-to-ground capacitance of that tube must also be added to capacitance C (since they are in parallel). In that case the load resistance (across the output) is considered infinite, providing the grid does not draw current, as in class A_1 and AB_1 operation.

$A(s) = V_{OUT}/V_{IN} = Z_C(s)/[R+Z_L(s)+Z_C(s)] = (1/sC)/(R+sL+1/sC) = 1/(LCs^2+RCs+1)$

$$A(s) = \frac{1/(LC)}{s^2 + sR/L + 1/(LC)}$$

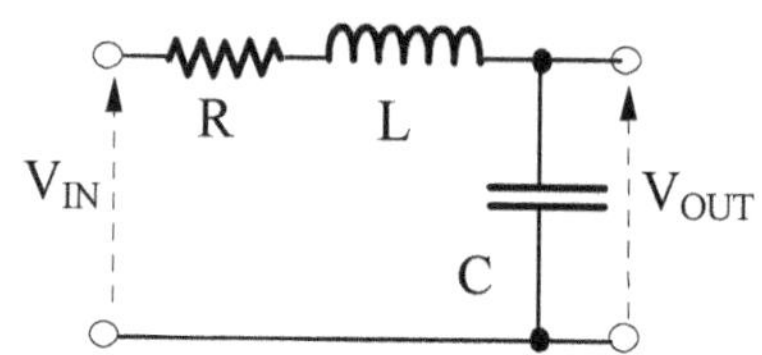

This transfer function has a form $A(s) = \omega_N^2/(s^2+2\xi\omega_N s+\omega_N^2)$ or $A(s) = \omega_N^2/(s^2+\omega_N/Qs+\omega_N^2)$ where $\boldsymbol{\omega_N=\sqrt{(1/LC)}}$ is the natural frequency of the circuit (don't confuse it with the resonant frequency!), ξ (ksi) is the damping factor $\boldsymbol{\xi=0.5R\sqrt{(LC)}}$ and Q is the quality factor, which is related to the damping factor $\boldsymbol{Q=1/(2\xi)}$**, or** $\boldsymbol{Q=\sqrt{(LC)}/R}$

The locations of poles and the transient response of linear systems

Poles of a transfer function are more important than zeroes, since only the nature and the location of the poles determines the transient response and stability of a system such as an amplifier. The root-locus method is a powerful tool that tracks the position of the poles in the complex plane as one parameter is varied. That factor is usually the amplification factor, useful when the stability of amplifiers with negative feedback is studied. As the amplification factor changes, the roots of the same amplifier change their positions, relative and absolute. At some point the poles may reach the imaginary axis and even cross it, making negative feedback turn into a positive one, thus causing oscillations!

An example is illustrated below for a 2nd order system with two different real poles, p1 and p2. As the values for L and C are changed and the damping factor is increased, these poles converge toward one another, until they become equal (one double pole).

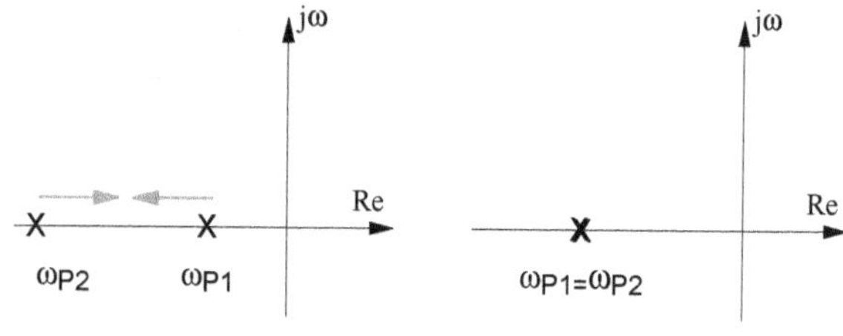

jω

ωP1 X

Re

ωP2 X

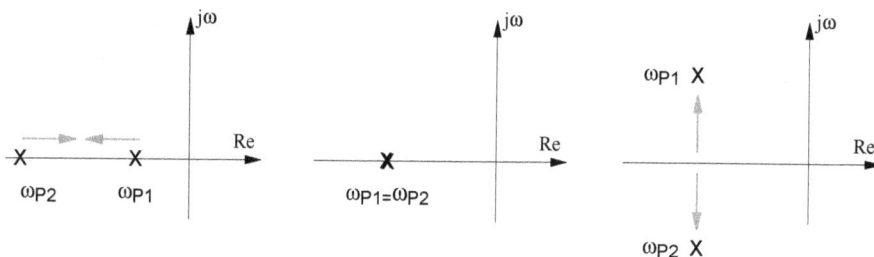

The pole-zero diagram for the 2-order RLC network (L-R): as damping ξ decreases, two real poles get closer and converge into one double pole for ξ=1, then they diverge vertically and become conjugated complex pairs for ξ<1 !

ξ>1 or Q<0.5, overdamped response, two real poles

ξ=1 or Q=0.5, critical damping, one real double-pole

ξ<1 or Q>0.5, underdamped response, two conjugated-complex poles

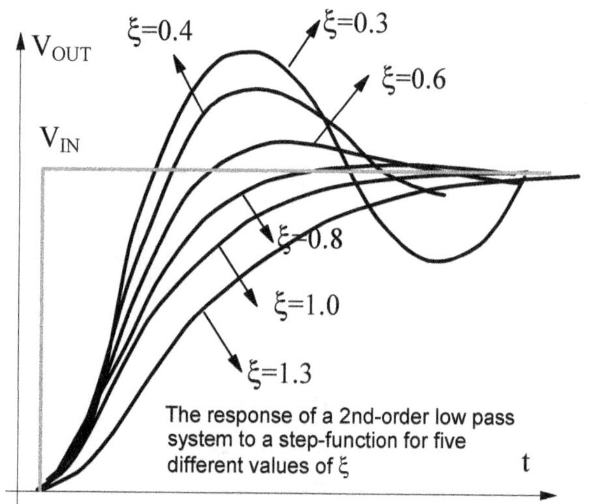

The response of a 2nd-order low pass system to a step-function for five different values of ξ

Then, further decrease in the damping factor splits the poles again and turns them into a pair of complex poles, and complex poles result in oscillatory response!

The response of the system to a step-function or a square wave input depends again on the nature of its poles.

For ξ>1 (two real poles) the response is overdamped and too sluggish. It takes too long to respond to fast changes or to reach a new steady state. For ξ=1 or Q=0.5 the system is critically damped, the two real poles are equal. For ξ<1 there is overshoot which gets progressively bigger as ξ becomes smaller. Typically, in audio amplifier response we aim for ξ between 0.4 and 0.8, with ξ=0.8 being optimal! Again, it is a compromise between fast rise (but high overshoot and long settling time) and no overshoot but very slow rise.

The amplitude vs. frequency characteristic of the 2nd order system has a peak at a resonant frequency for underdamped cases ($\xi<1$ or $Q>0.5$) The lower the damping the higher the resonant peak. The resonant frequency increases slightly as the damping is decreased, so the resonant frequency for Q=1.6 for instance is higher than for Q=1.0! The resonant frequency and the natural frequency ω_N are linked by the formula $\omega_R = \omega_N\sqrt{(1-2\xi^2)}$

Another important aspect of this circuit is that the amplification curve drops at a rate of 40dB/decade or twice the slope of the first order systems!

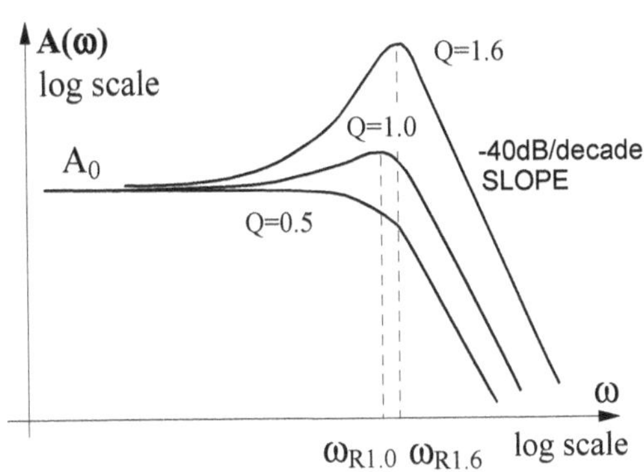

RIGHT:The A-f curve of a 2nd order low pass system for three values of Q

Unstable systems

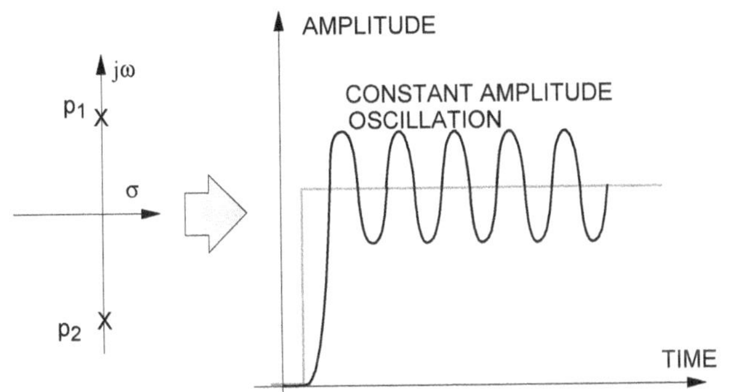

Poles on the imaginary axis - constant amplitude oscillation

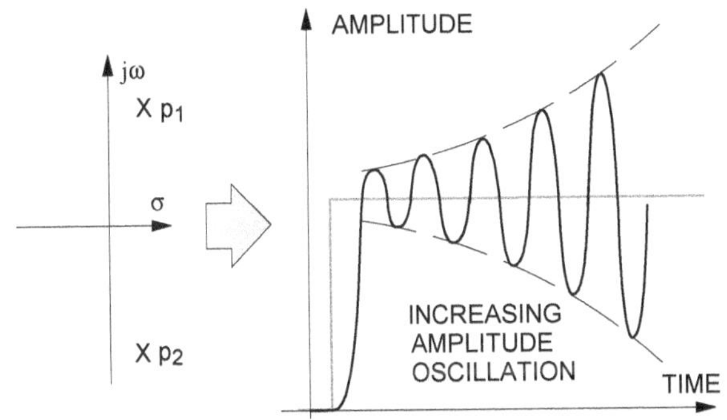

Poles in the right half of the complex plane - oscillation with an exponentially increasing amplitude

Once the poles of its transfer function reach the jω axis, the amplifier become unstable and turns into an oscillator. Its response never settles down. If the poles sit right on the jω axis, the amplitude of oscillation is constant, and for the poles with positive σ, located in the positive half of the complex plane, the oscillations would increase in amplitude until some restriction is reached, typically until the amplifying tube, transistor or output transformer saturate, or the signal amplitude reaches the power supply level and cannot increase any more. Soon, after we cover negative feedback, we will study this instability issue a bit deeper, and see what causes it.

THE COMMON CATHODE AMPLIFIER STAGE AT LOW FREQUENCIES

Sources of phase shift and frequency-dependent attenuation in a tube amplifier

It is surprising and somewhat disheartening that there are so many unwanted filters in a typical tube amplifier. The illustration shows the simplest possible layout, only one input stage, capacitively coupled to the output stage.

The two coupling capacitors and the following grid resistors form high-pass filters, marked (1). If not designed properly they will limit the lower -3dB frequency and weaken the bass. The same applies to the cathode bypass capacitors, marked with (2). These are also high-pass filters and will also attenuate low frequencies.

At low frequencies the output transformer (next page) behaves as a high-pass filter, marked with (3). The output voltage's phase leads the input voltage, again attenuating low frequencies and reducing bass power.

At high frequencies the output transformer can be modeled as a 2nd order low-pass LC filter. The output voltage lags the input voltage, but being a 2nd order system the attenuation is faster, 12dB/octave or 40dB/decade. All other filters here are 1st order, with lower slope of 6 dB/octave or 20dB/decade.

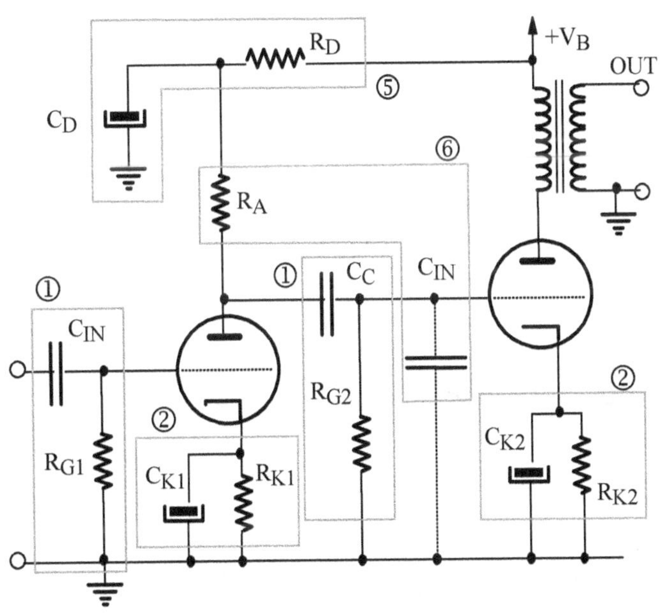

The decoupling filter (5) is a low-pass filter and as such its output voltage lags the input voltage at low frequencies. The AC voltage in this case is the unwanted component of audio signal propagating between stages through the power supply and as such this filter (if properly designed) does not affect the frequency range of the main signal path.

Finally, No. 6: the anode resistor R_A and the parasitic grid-cathode capacitance (input capacitance of the power tube) form a low-pass filter. Its output voltage lags the input voltage at high frequencies and this interaction limits the upper -3dB frequency of the amplifier.

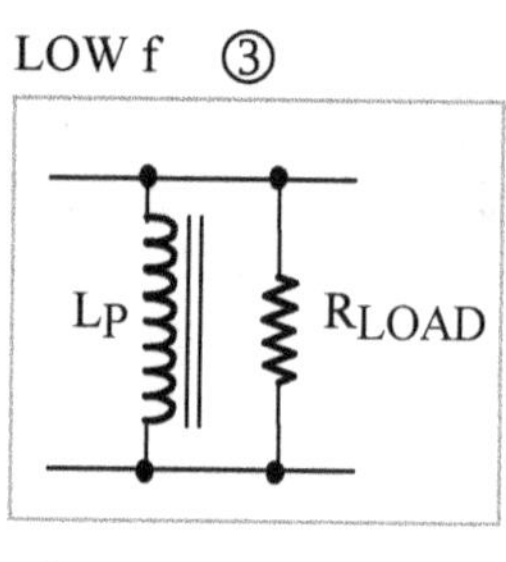

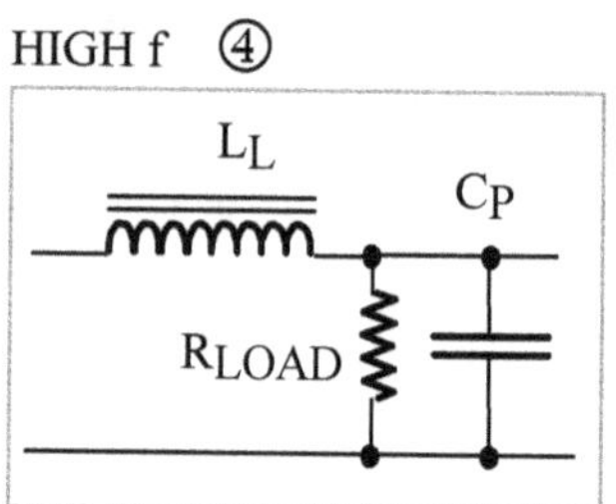

Simplified models of an output transformer at low and high frequencies.

The effect of the cathode-bypass capacitor at low frequencies

So far we have been studying the ECC40 triode stage at midrange frequencies, assuming that the cathode resistance has been bypassed for AC signals by the low reactance of the bypass capacitor C_K. Since $X_C=1/\omega C_K$, as the signal frequency drops into the low frequency range (below 100 Hz), this reactance becomes progressively bigger and cannot be considered a short circuit across R_K.

In the equivalent circuit as we have done with the completely un-bypassed cathode resistor, we can remove the reference to v_{GK} and simplify the circuit as in the second, equivalent model with v_{IN} as the input voltage and scaled cathode impedance. Both R_K and X_C (which in parallel form the cathode impedance Z_K) are multiplied by $(1+\mu)$, which means that C_K is actually divided by $(1+\mu)$. Now we can write

$$A(j\omega) = \frac{-\mu R_L}{r_I + R_L + (1+\mu)Z_K} = \frac{-\mu R_L}{r_I + R_L + \dfrac{(1+\mu)R_K}{(1+j\omega)R_K C_K}}$$

After factoring r_I+R_L out and multiplying both numerator and denominator by $j\omega R_K C_K$ we get

$$A(j\omega) = \frac{-\mu R_L}{r_I + R_L}\ \frac{(1+j\omega)R_K C_K}{1+\dfrac{(1+\mu)R_K}{r_I + R_L} + j\omega R_K C_K}$$

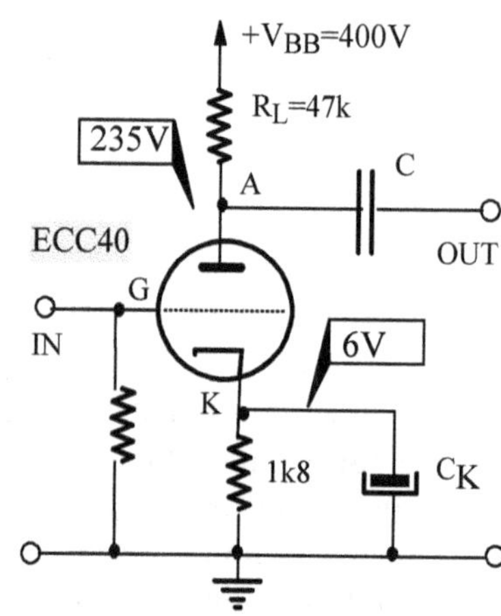

The common cathode stage whose frequency characteristic we will study in this section.

The first factor is the midrange gain, and the second factor modifies this gain at low frequencies. We introduce two new terms: $\omega_0=1/R_K C_K$ and $k=(1+\mu)R_K/(r_I + R_L)$ Factor k is a constant and is automatically set by the designer's choice of tube (r_I and μ) and its anode and cathode resistances R_L and R_K.

The amplitude characteristic has one zero (ω_0) and one pole $(1+k)\omega_0$. You should recognize it as the limited phase lead network recently discussed! This is a first order system or filter, so at the frequency drops below the frequency of the pole (ωP) of the pole, the gain starts falling 6 dB/octave (or 20 dB/decade) until it reaches the angular frequency of the zero, ω_0, and then it stays constant (horizontal) with the falling frequency. The asymptotes are illustrated as well as the actual amplitude curve.

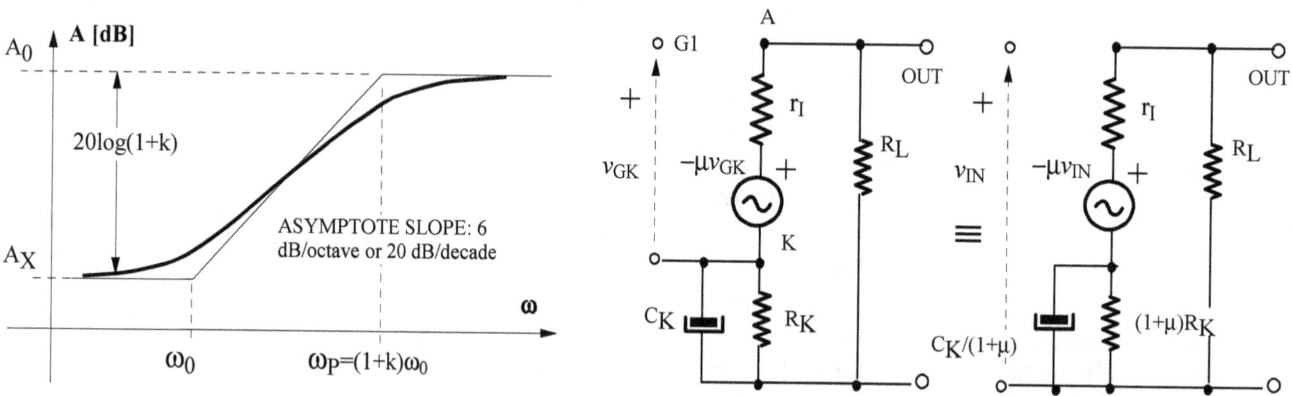

Logarithmic amplitude characteristic of cathode-biased common cathode triode stage at low frequencies

The low frequency equivalent circuit

DESIGN PROBLEM

PROBLEM: For the ECC40 stage illustrated, find the value of C_K so that the lower -3dB frequency f_L is below 5 Hz.

R_L=47kΩ, r_I=11kΩ and μ=32, so k=(1+μ)R_K/(r_I+R_L)=33*1.8/(11+47)=1.02

The drop in amplification factor at ω_0 will be 20log(1+k) = 20log(2.02) = 6.1dB

Since the drop happens at 6 dB per octave (2:1 ratio of frequencies) the -3dBpoint will be approximately at twice the ω_0 frequency, so ω_0=2πf_0=2*π*2.5 Hz= 15.71 rad/s! Since ω_0=1/$R_K C_K$, the minimum capacitance required is C_K=1/$R_K\omega_0$ = 1/(1,800*15.71) = 35.4 μF

RC-coupled stages

Since it is the simplest and cheapest to implement, AC-coupling is by far the most common type of coupling. The capacitor between the anode of the previous stage and the grid of the following stage prevents DC currents from flowing, so stages can be individually designed, DC conditions of each are independent.

We have already seen the midrange model for an RC-coupled common cathode stage. However, this model is a simplified or special case of a true or universal model that applies for all audio frequencies and includes inter-electrode capacitances as well. Instead of analyzing the full model, which would involve awkward equations and complex maths, we divide the frequency spectrum in three bands and make prudent assumptions and simplifications for each of them, which makes the analysis of the simplified circuits much easier.

The three bands are not clearly defined, but rules-of-thumb generally say that the low frequency range in audio circuit analysis is under 100Hz and the high frequency band is above 10kHz. Obviously, the midrange model will apply to the frequencies in between those two extremes.

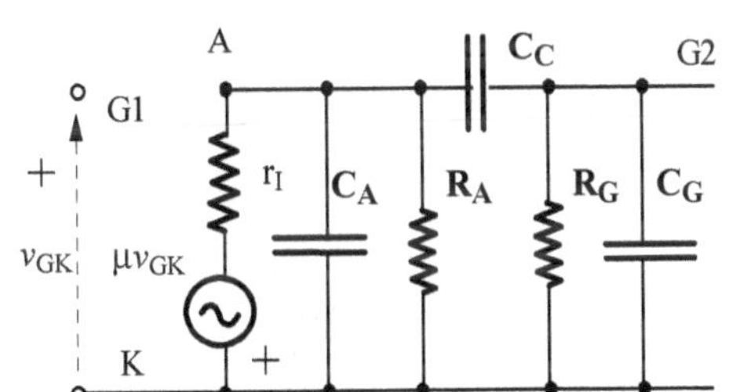

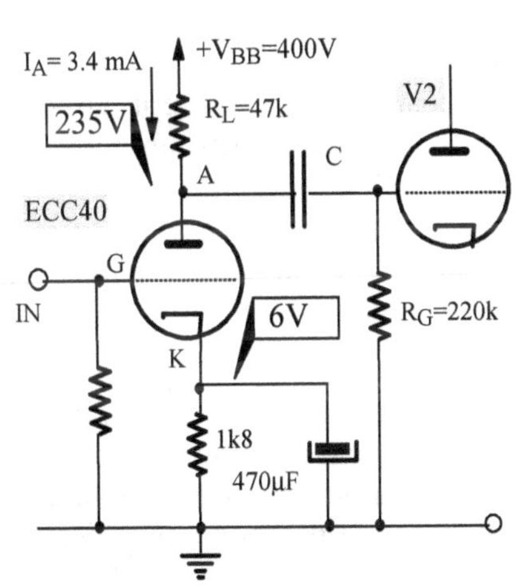

Two common cathode stages with RC-coupling (RIGHT) and their universal model (LEFT)

RC-coupled stages at midrange frequencies

In the midband of frequencies all capacitances can be neglected. The reactances of shunt capacitances C_A and C_G are very high and can be considered an open circuit and the reactance of the coupling capacitance is very low and can be considered a short circuit. The amplification factor is A_M = -μ*R_P/(r_I + R_P) where R_P=R_A||R_G

Using the alternative current model the three resistances are in parallel, so the equivalent resistance is R_E=r_I ||R_A||R_G and the midrange voltage gain is simply A_M= -gmR_E

$$R_E = \frac{r_I R_A R_G}{r_I R_A + R_G R_A + r_I R_G}$$

Midrange voltage and current models

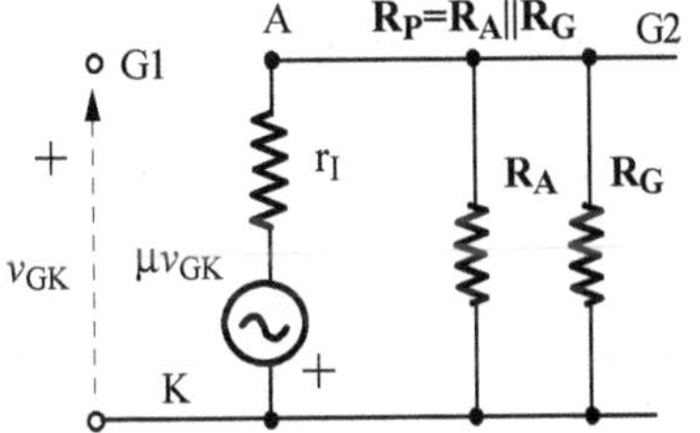

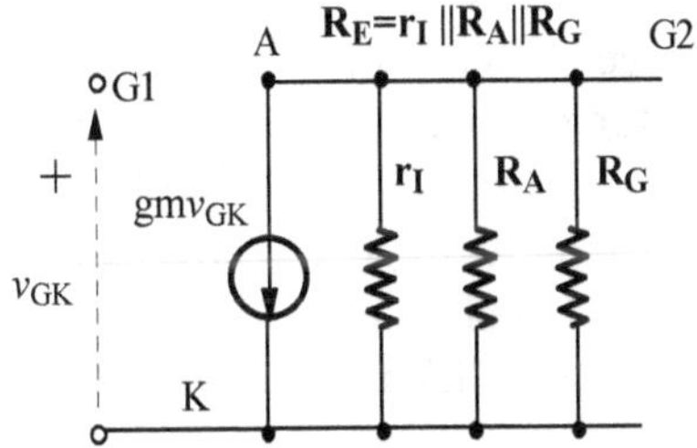

RC-coupled stages at low frequencies

At low frequencies the reactance of the coupling capacitance becomes significant, and forms an RC filter with the grid resistance of the following stage R_G. The coupling capacitor becomes part of the load and a considerable part of the signal on the anode is dropped across it, forming a voltage divider circuit.

To make matters even worse, the load is now a complex reactance and the load line is not straight any more, but an ellipse. The phase shift of the amplification stage is not 180 degrees any more!

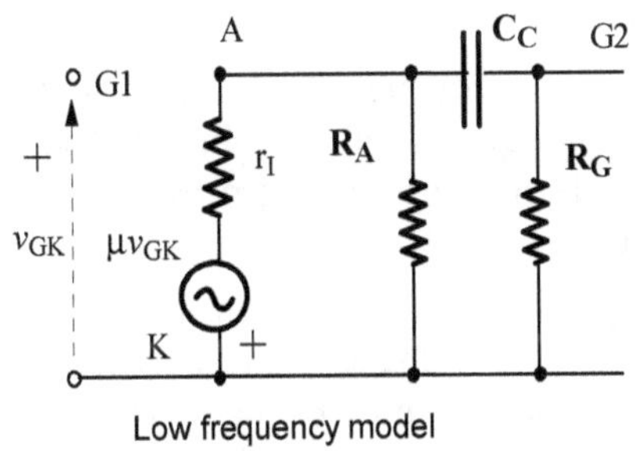

Low frequency model

$$A_L = V_{G2}/V_{G1} = \frac{-gmR_E}{\sqrt{[1+X_C^2/R_X^2]}} = \frac{A_M}{\sqrt{[1+X_C^2/R_X^2]}}$$ R_E is the same as for midrange: $$R_E = \frac{r_I R_A R_G}{r_I R_A + R_G R_A + r_I R_G}$$

$R_X = R_G + r_I||R_A = R_G + r_I R_A/(r_I + R_A)$

The expression under the radical (square root) sign is called the *low-frequency reduction factor*. When $X_C = R_X$ that factor is $\sqrt{2}$ and the frequency at which that happens is called the lower -3dB frequency f_L or the lower half-power frequency. Since $1/\sqrt{2} \approx 0.71$, at f_L $A_{-3dB} = 0.71 A_M$!

By substituting $R_X = R_G + r_I R_A/(r_I + R_A)$ and $X_C = 1/(2\pi f C_C)$ into the equation $X_C = R_X$ and by tidying it up we finally get

$$f_L = \frac{r_I + R_A}{(r_I R_A + R_G R_A + r_I R_G)(2\pi C_C)}$$

The low frequency response of RC-coupled stages is primarily determined by the size of the coupling capacitor C_C: the larger C_C, the lower f_L!

THE COMMON CATHODE AMPLIFIER STAGE AT HIGH FREQUENCIES

The Miller effect in a triode

Lets look at another dame of voltage amplification, 6SL7 duo-triode. Its main asset is a very high amplification factor (70), more than twice that of ECC40 (32), but that also means its internal resistance will be very high, 44kΩ!

Since there are three active electrodes in a triode, data sheets specify three types of capacitance between electrodes. The figures for 6SL7 and ECC40 are illustrated in the diagram. The input capacitance between grid and cathode C_{GK} is of most interest because it will divert high frequency signal down to ground and therefore limit the upper -3dB frequency f_U and bandwidth! However, at 3 pF it would take the f_U into the MHz range, so it doesn't seem to be an issue in audio.

What about the anode-grid capacitance? If we draw the model of a common cathode stage with the three capacitances included, it becomes obvious that C_{AG} provides unwanted feedback from anode back to the grid!

We will not show the circuit analysis here, but after some arithmetic it can be shown that the negative capacitive feedback from the output (anode) to the input (grid) increases the input capacitance of the amplifying stage to $C_{IN} = C_{GK} + C_{AG}(1+A)$ This increase is called the *Miller effect.*

Let's compare the input capacitances of 6SL7 and ECC40 triodes at their typical operating points. A common-cathode stage with 6SL7 can achieve A=50, so the input capacitance is $C_{IN} = C_{GK} + C_{AG}(1+A) = 3+2.8(1+50) = 3+143 = 146$ pF

Notice that C_{GK} is much smaller than $C_{AG}(1+A)$ and can be neglected for high amplification factor tubes such as 6SL7!

We have already calculated the midband amplification factor A = 25 for our ECC40 circuit , so $C_{IN} = C_{GK} + C_{AG}(1+A) = 2.8+2.7(1+25) = 3+143 = 73$ pF

Coincidentally, the C_{AG} capacitance of the two tubes is almost identical, but since the amplification of the ECC40 stage is half that of the 6SL7stage, its input capacitance is also roughly halved.

TUBE PROFILE: 6SL7GT/12SL7GT

- Indirectly-heated high μ dual triode
- 6.3V/0.3 A - 12.6V/0.15A
- Maximum anode voltage 300 volts
- Maximum V_{HK}: $90V_{DC}$
- P_{AMAX}=1W, I_{AMAX}= 10 mA
- TYPICAL OPERATION:
- V_A=250V, V_G=-2.0V, I_0=2.3mA
- gm=1.6 mA/V, μ=70, r_I = 44 kΩ

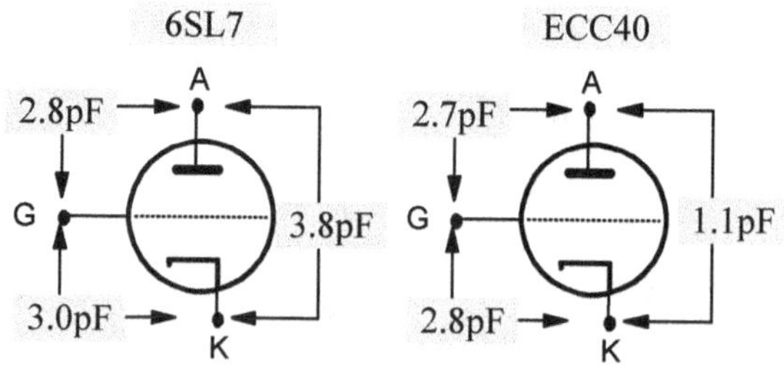

Inter-electrode capacitances of 6SL7 and ECC40 triodes

Neutralization

One consequence of the Miller effect is the reduction of gain at higher frequencies, which limits the upper -3dB frequency of triode amplifiers. The other is the possibility that at high frequencies the capacitive coupling between anode and grid can turn from a negative into a positive feedback, causing the amplifier to oscillate at ultrasonic frequencies. It is possible to neutralize both effects by a simple measure.

If you found the common saying "to fight fire with fire" a bit strange (after all, you fight fire with water, not another fire!), here it makes more sense. Since the capacitive coupling from the anode back to the grid is a negative feedback, we neutralize it by using a positive feedback. We need to provide feedback current, equal in magnitude but opposite in phase, to cancel the feedback current through the Miller capacitance.

This is achieved by capacitively coupling the anode of the offending tube to a point of the same phase as the voltage on its grid. In push-pull amplifiers (covered in Volume 2) that is easily done by capacitively cross-coupling the grid of the upper tube to the anode of the lower tube and vice versa, since their signals are always of the opposite phase.

The positive feedback increases gain at higher frequencies by coupling a anode of each tube to the grid of the other tube.

As the frequency increases, the reactances of C_{F1} and C_{F2} decrease, causing more signal voltage to be fed back, so the shunting action of parasitic capacitances C_{P1} and C_{P2} is compensated for.

If there are phase-shift problems and NFB becomes PFB, the same will happen to our neutralizing circuit, so the two capacitive currents between anode and grid will always be of opposite phase!

The schematic at the bottom of the page shows a push-pull amplifier with the second voltage amplification stage (1) neutralized in the same manner as the output stage (2). This is where small trimmer capacitors of 1-50 pF come handy. Or, by using fixed capacitors and trial-and-error approach, one can quickly get a feel for the required values.

Since this reverse transadmittance from anode to grid varies with individual tubes, circuits and transformers, the value of the neutralizing capacitor has to be experimentally determined for each amplifier. This is very expensive for mass-produced amplifiers, but is recommended for DIY projects, if for no other reason then as an educational exercise in using oscilloscopes and fine-tuning amplifiers.

While neutralization compensates for the Miller effect and extends the high frequency response, it also increases the output capacitance. This capacitance lowers the upper frequency limit, thus reducing the effect of neutralization on the amplifier's bandwidth.

Using physically smaller components (tubes, capacitors, resistors) reduces parasitic capacitances between the metal bodies/parts of these components and ground (amplifier chassis), as does increasing the distance between components and metal chassis. Another measure is to use low capacitance tubes. Generally, miniature tubes have lower capacitances than Noval tubes (12AU7 at al.), which in turn beat the Octal tubes such as 6SN7.

The inter-electrode capacitances of the 9002 miniature triode are 2-3 times lower than those of 6SL7 triode (1.4pF between G and A, 1.2pF between G and K!) The glass bulb of 9002 is 17 mm in diameter and 35 mm tall, compared to a typical 6SL7 tube which is 30 mm in diameter and 70 mm tall, so 1/2 the size overall.

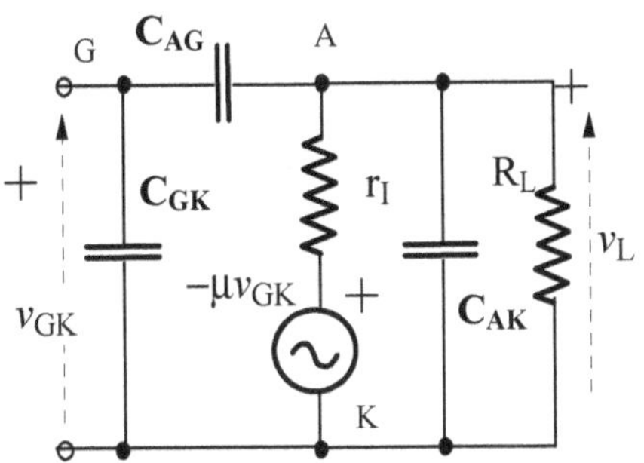

The high frequency model of the common-cathode stage, including inter-electrode capacitances

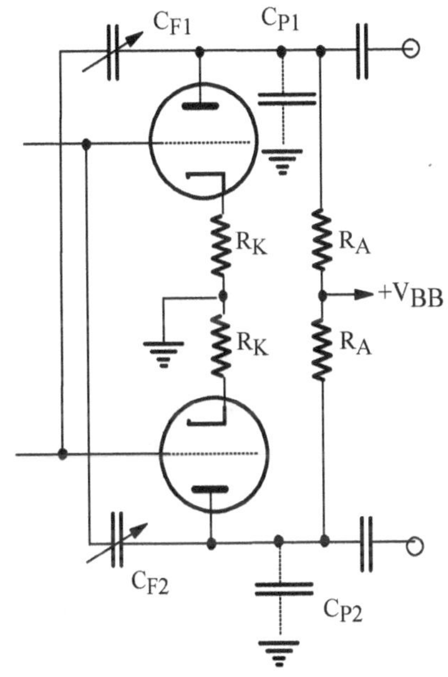

High frequency compensation by capacitive cross-coupling

BELOW: Practical neutralization in a balanced push-pull amplifier

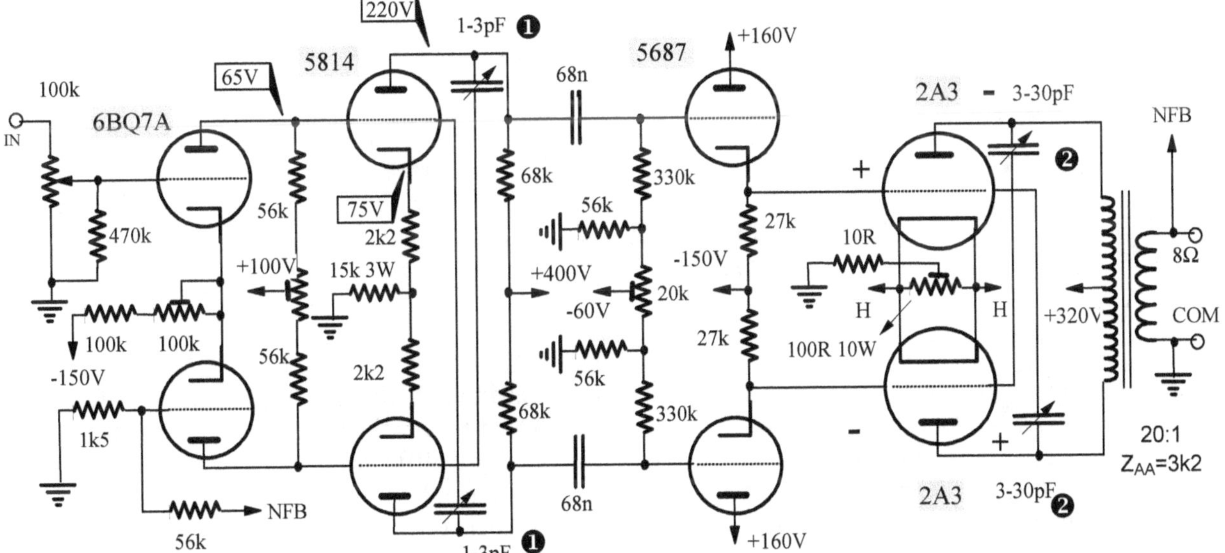

Phase correction networks

Phase correction and neutralization can be implemented in preamp and single-ended stages too. The RC network is connected between the anode (output) and ground, or in parallel with the anode load. Since the impedance of the power supply is low, and can be considered at ground potential for AC signals, that is almost identical to the previous connection. Many classic designs use this trick, for instance the Williamson amplifier in its first stage.

The frequencies of the shelf-circuit are $f_1=1/(2\pi RC)$ and $f_2=1/[2\pi(R+R_A)C]$.

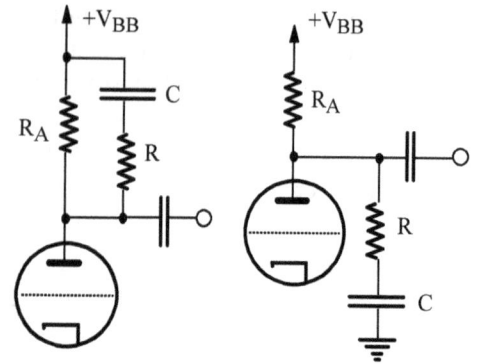

RC networks can be connected in parallel with anode load or between anode and ground

HF compensation by partial cathode bypassing

The simplest method of high frequency compensation is partial bypassing of cathode resistor. For a stage with given values of R_K, R_A and the parasitic or stray capacitance C_P, there is an optimal value of C_K, one that will provide the widest frequency range. C_K will only decouple the cathode resistor R_K at high frequencies, increasing the local negative feedback this resistor provides at lower frequencies and thus decreasing the amplification at low frequencies relative to high frequencies.

Since the gain of tube amplifiers in the bass region is at premium, due to limitations in interstage and output transformers, this method is rarely, if ever used in audio.

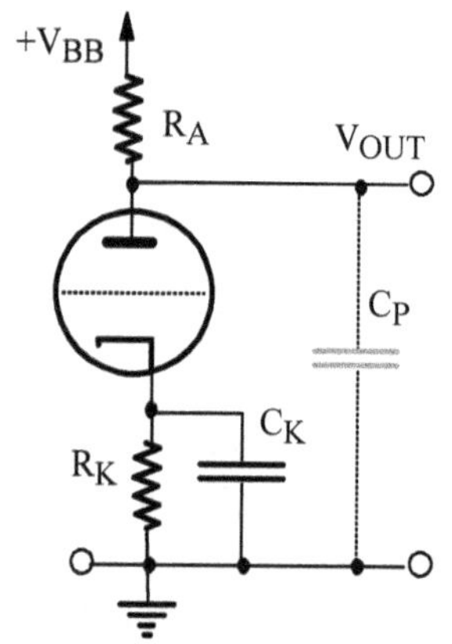

Partial cathode decoupling for high frequencies

Grid stopper resistors

A grid of a tube doesn't differentiate between the wanted audio signal and the unwanted radio-frequency and other "interference" signals. High mutual conductance tubes are especially troublesome, tending to burst into oscillation at every opportunity.

The solution is to limit the upper frequency range of any signal at the grid, and that is easily accomplished by the addition of a suitably selected resistor in series with the signal. This "grid-stopper" R_{GS} stops the HF oscillations by forming a low pass RC filter together with the input capacitance of a tube.

Grid stoppers should be as close to the grid as possible, so solder them directly onto the tube socket.

The upper angular frequency is $\omega_U=1/R_{GS}C_{IN}$ and since $\omega_0=2\pi f_U$, we get $f_U=1/(2\pi R_{GS}C_{IN})$ If we choose the cutoff frequency, the required resistor's value is $R_{GS}=1/(2\pi f_U C_K)$.

Our ECC40 stage has around 73pF of input or Miller's capacitance, but we need to add to that any stray wiring capacitance between the grid and the ground. The accuracy of the whole calculation will hinge on how accurately we estimate this stray capacitance. With careful wiring (short & tidy wire runs) we add about 1/3 of the tube's Miller capacitance, in this case 27pF, for the total C_{IN} of 100pF.

For say 120 kHz cutoff $R_{GS}=1/(2\pi f_U C_K)$ = 13,263Ω, so we would choose the standard value of either 12kΩ or 15kΩ.

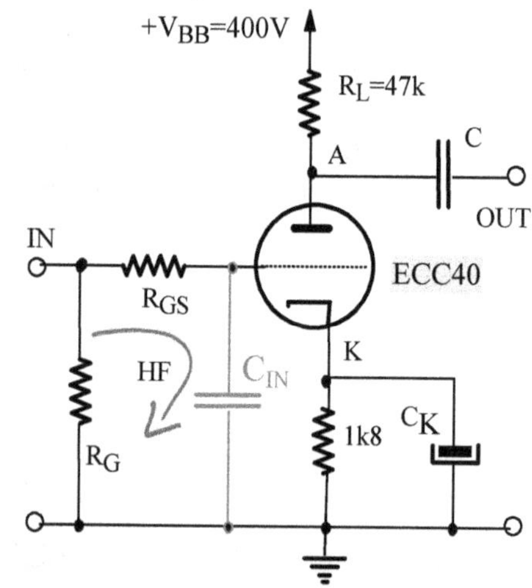

The grid stopper resistor R_{GS} and the input capacitance C_{IN} form a low pass filter

RC-coupling at higher audio frequencies

Now that we understand the unfortunate Miller effect we can proceed with high frequency behavior of RC-coupled triode stages. In the HF model C_A is the sum of the Miller capacitance between the first tube's anode and grid, and all parasitic capacitances to the left of the coupling capacitor C_C. These are difficult to estimate since they depend on many factors, primarily the lead dressing, the component layout, the tube socket size and type, the length of component leads, their distance from the metal chassis and the tube socket , and so on and so forth!

Likewise, C_G is the sum of the grid-to-cathode capacitance of the second tube, and all parasitic capacitances to the right of the coupling capacitor C_C. These depend on the same factors as C_A, the lead dressing, the component layout, etc. Again, it is easier to use the current model.

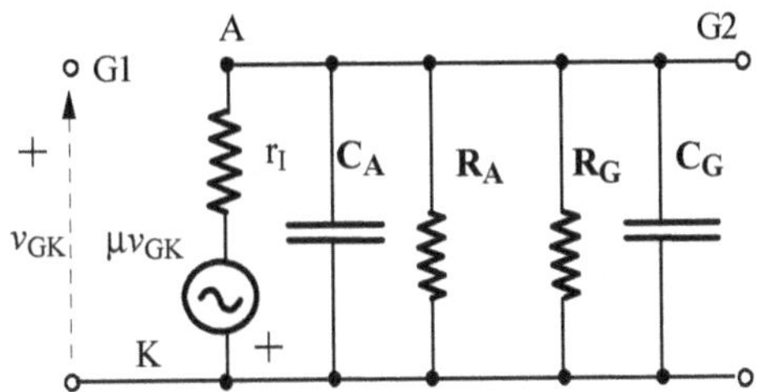

LEFT: high frequency equivalent circuit of RC-coupled stages using the voltage model

RIGHT: high frequency equivalent circuit of RC-coupled stages using the current model

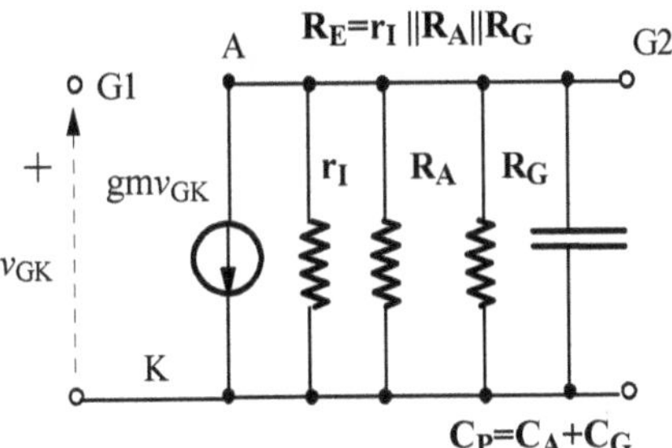

$$A_H = V_{G2}/V_{G1} = \frac{-gmR_E}{\sqrt{[1+R_E^2/C_P^2]}} = \frac{A_M}{\sqrt{[1+R_E^2/C_P^2]}}$$

R_E is the same as for midrange: $R_E = \dfrac{r_I R_A R_G}{r_I R_A + R_G R_A + r_I R_G}$ and $C_P = C_A + C_G$

The expression under the radical (square root) sign is called the *high-frequency reduction factor*. When $X_P=R_E$ that factor is $\sqrt{2}$ and the frequency at which that happens is called the upper -3dB frequency f_U or the upper half-power frequency. Since $1/\sqrt{2} \approx 0.71$, at f_U we have $A_H=0.71A_M$!

By substituting $R_E= r_IR_AR_G/(r_IR_A+R_GR_A+r_IR_G)$ and the reactance of the parasitic or parallel capacitance C_P, $X_P=1/(2\pi fC_P)$ into the equation $X_P=R_E$ we get $f_U = 1/(2\pi C_PR_E)$ or $f_U=(r_IR_A+R_GR_A+r_IR_G)/(2\pi C_PR_Ar_IR_G)$

At high frequencies the three shunt capacitances can be agglomerated into one, while the coupling capacitance is again a short circuit. We have a parallel RC circuit whose time constant $\tau=R_EC_P$. When a square wave signal is brought to the grid, this time constant is the rise time of the signal at the input of the following stage.

Ideally, this rise-time should be zero, which can only be achieved if one of the parameters is zero. Obviously, R_E cannot be zero, since it involves R_A and R_G in parallel. If R_A was zero there would be no amplification, and if R_G was zero the grid of the following tube would be short-circuited to ground.

To maximize the amplification of the stage R_E should be as high as possible. If we want a rapid response to the signal and a quick rise time, R_E should be as low as possible. These are clearly contradictory requirements, so the only way to minimize this rise time is to minimize the shunt capacitances C_A and C_G. They include the input capacitances which are increased by the Miller effect. Again, if we want a high amplification factor, the Miller capacitance is also increased and the rise time increases, so a similar contradiction is encountered.

In summary, the high frequency response of RC-coupled stages is primarily determined by the size of the parasitic capacitance between anode and cathode of the first tube and the parasitic capacitance between grid and cathode of the following tube. The smaller these parasitic capacitances, the higher f_U!

THE AMPLITUDE AND PHASE CHARACTERISTICS OF RC-COUPLED STAGES

A typical A-f characteristics of an RC stage is illustrated, with a bandwidth of 7Hz to 100kHz. Notice a linear vertical scale (amplification factor) and a logarithmic horizontal (frequency) scale. Now we can summarize the findings and illustrate them in a universal amplitude-frequency characteristic of an RC-coupled stage (next page).

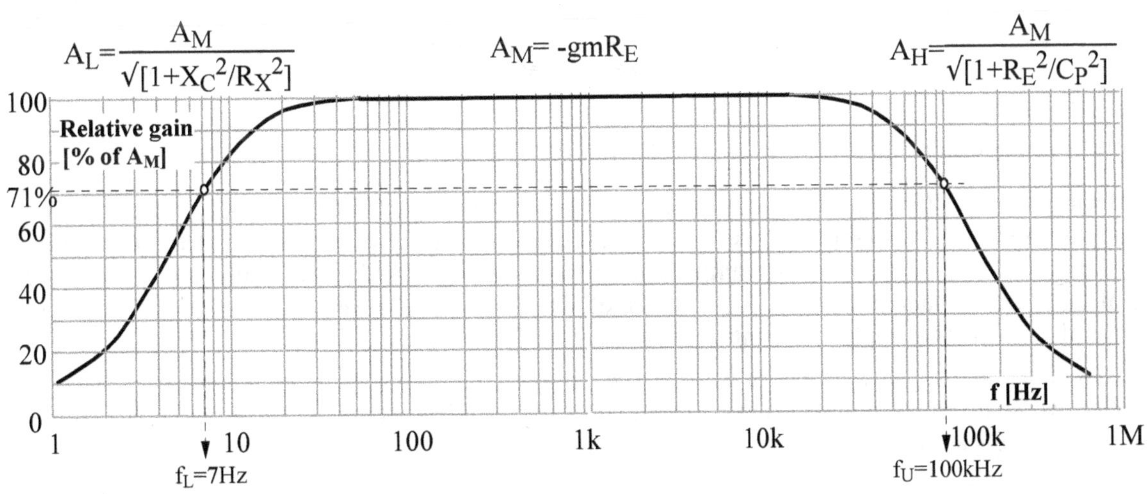

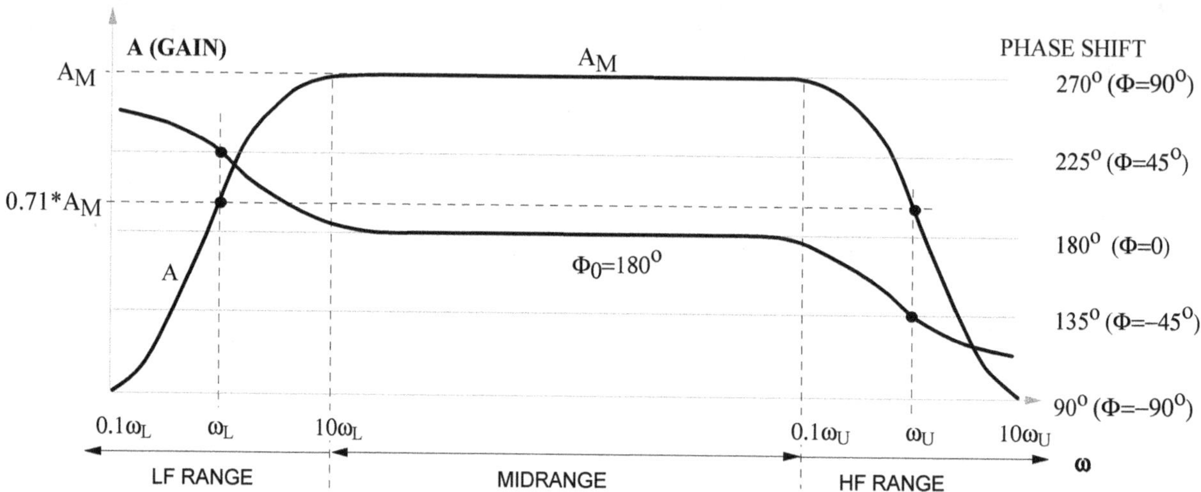

The phase shift between the output and input signal goes through a similar change, from 180° at midrange frequencies, down to 180°-90°=90° at high frequencies, and up to 180°+90° = 270° at low frequencies! The LF range is defined as the range below the frequency ten time higher than f_L, while the HF range is defined in a similar fashion, as the range above the frequency ten times lower than f_U.

Notice that at f_L the phase shift between the input and output is 180°+45°=225°, while at f_U the phase shift between the input and output is 180°-45°=135°. Looking at the pole-zero plot, the meaning of Φ is illustrated for any frequency ω_X. When $\omega=\omega_L$, Φ=45°. As ω is increased from ω_X to ω_Y, Φ reduces towards zero (midrange). As ω is decreased from ω_X towards zero, Φ increases towards 90°.

The frequency response at low frequencies can be described as $A_L=-Kj\omega/(j\omega+\omega_L)$The vertical distance between the origin (zero) and the specific value of ω (for instance ω_X or ω_Y) is the magnitude of the numerator, while the distance between the pole ω_L and ω_X (the hypotenuse) is the magnitude of the denominator.

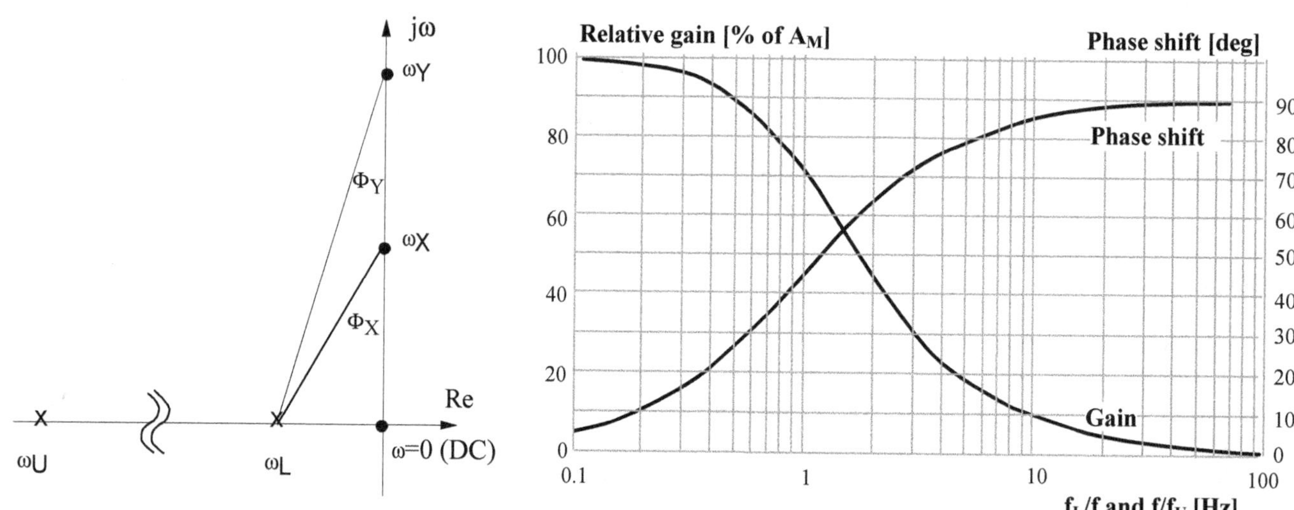

A pole-zero plot of a RC-coupled amplifier

Universal gain and phase curves for RC-coupled amplifier stage, where f_L = lower -3dB frequency, f_U = lower -3dB frequency, A_M = midrange gain

The frequency response and pulse response of cascaded stages

We have seen that each RC coupled stage behaves as a 1st order bandpass filter. At both frequency extremes the voltage gain drops with a slop of -6dB/octave (2:1 frequency ratio) or -20dB per decade (10:1 frequency ratio). What happens when two, three or more voltage amplification stages are cascaded? The graphs above illustrate the HF behavior of 1-, 2- and 3-stage amplifiers. For the sake of simplicity and the ease of illustration identical stages are assumed, but the results also apply to dissimilar stages with different upper and lower -3dB frequencies.

Likewise, only the high frequency behavior is discussed, similar cumulative attenuation happens at low frequencies as well. With two identical RC coupled stages the gain drops at double the rate, or -12dB per octave, and with three identical RC coupled stages the gain drops at triple the single-stage rate, or -18dB per octave!

This is bad news, since both half-power frequencies drop according to formula

f_{-3dB} (N stages) = f_{-3dB} (1 stage) / $\sqrt{[2^{1/N}-1]}$

The curves don't seem that bad, but when you calculate the attenuation you get the following disturbing results.

For two stages N=2 and $f_{-3dB}(2) = f_{-3dB}(1) / \sqrt{[2^{1/2}-1]} = 0.64*f_{-3dB}(2)$

For three stages N=3 and $f_{-3dB}(3) = f_{-3dB}(1) / \sqrt{[2^{1/3}-1]} = 0.51*f_{-3dB}(3)$

Assuming that one stage has the upper -3dB frequency of f_{-3dB} (1)=60 kHz, two stages will be 3dB down at $0.64*f_{-3dB}$ (1) = 0.64*60 = 38.4 kHz, while three stages will be 3dB down at 0.51*60 = 30.6 kHz!

The next time you see that old & tired circuit diagram with two cascaded common-cathode stages using 6SN7 triodes (high Miller capacitance) driving the 300B output triode without any negative feedback, you can bet that the amp will sound soft and seductive, but both the treble and bass frequencies will be heavily attenuated.

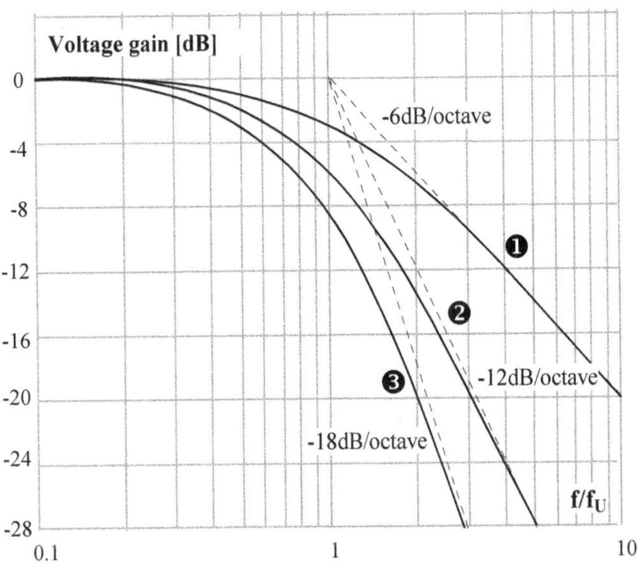

The upper frequency attenuation and asymptotes of a single stage (1), two identical RC-coupled stages (2) and three identical RC-coupled stages (3)

Defining the parameters of a pulse

Compared to the ideal square wave or a single pulse, a real pulse suffers from numerous flaws and distortions. Firstly, there is a time delay between the rise of the input pulse and the moment the output pulse reaches 10% of the amplitude A.

Secondly, the rise time is defined as time required for the leading edge to rise from 10 to 90% of the amplitude A. The fall time of the trailing edge is defined in a similar way. The rise and fall times are not necessarily equal.

The pulse width is defined as the time between the leading edge reaching half the amplitude A and the trailing edge dropping to the same level (50% points).

The sag of the ideally flat top is a percentage of the amplitude A. In tube amps it does not happen at the same time as the overshoot and ringing, which are more pronounced at higher frequencies, while the sag is dominant at low frequencies, although both effects are illustrated on the same pulse here.

Depending on the design of the amplifier and the signal frequency, there could either be a rounding of the edge or overshoot and ringing. In the later case another parameter is defined, and that is the settling time, the time required for the oscillations ("ringing") to "die down" and settle to level A. Indeed, in many inferior amplifiers the settling time is longer than the pulse width and such oscillations never settle down!

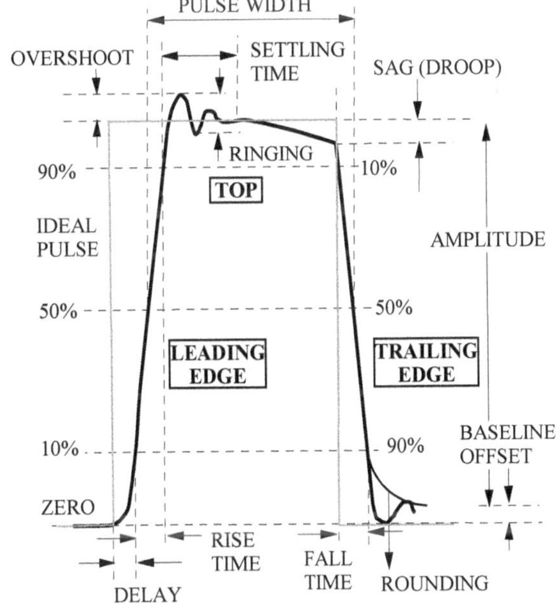

Pulse parameters and definitions

Pulse response of amplification stages

In our analysis of low pass filters we have seen that the leading and the trailing edge of amplifier's response to a square wave or pulse depend on the filter's time constant and the upper -3dB frequency f_U.

Thus, the rounding of the leading and trailing edges depends on amplifier's high frequency behavior. The higher the f_U, the less rounded the pulse's edges will be.

The 10-90% rise time is $\mathbf{T_R=2.2R_EC_P}$, where as we have already seen, $R_E=r_I \| R_A \| R_G$ and $C_P=C_A+C_G$ or the total parallel capacitance to ground, including wiring capacitances.

Likewise, from the response of high-pass filters to a square wave or pulse we have seen that the sag of the pulse's top is determined by the filter's time constant and the lower -3dB frequency f_L, i.e. its low frequency behavior. Thus, since an amplifier is a bandpass filter, its sag will also depend on its low frequency behavior: the lower the f_L, the less will the top of a square wave sag while being amplified.

The high fidelity transmission and amplification of a flat top of the pulse is determined by three circuits: the cathode bias circuit, the screen bias circuit (for tetrodes and pentodes only) and the gridleak-coupling capacitor circuit.

The sag caused by the grid leak coupling circuit is $\boldsymbol{\omega_L T = T/(R_P+R_G)C_C}$, where C_C is the coupling capacitance, R_G is grid-leak resistance, and R_P is the parallel resistance of r_I and R_A (internal tube's resistance and anode resistance, $R_P=r_I||R_A$). T is the period of the pulse.

So, to minimize the sag, R_P should be as low as possible, meaning low impedance tubes are preferred over high impedance ones. R_G and C_C should also be made as high as practically possible. However, there is an upper limit placed on grid-leak resistance R_G by tube's manufacturer, since small grid currents flowing through large grid-leak resistors can create significant positive voltage drops which can upset the tube's bias.

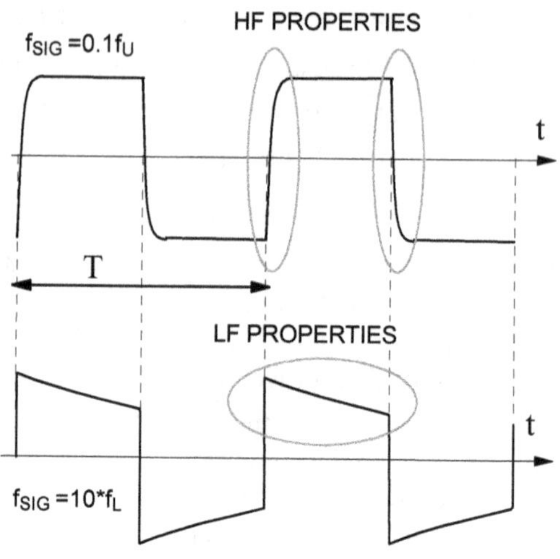

The HF (upper) and LF (lower) response of a tube amplifier to a square wave signal

IMPEDANCE-COUPLED STAGES AND INTERSTAGE TRANSFORMERS

- IMPEDANCE COUPLING
- THE FREQUENCY RESPONSE OF TRANSFORMER-COUPLED STAGES
- INTERSTAGE TRANSFORMER DESIGNS

"If it sounds good, it *is* good."
Duke Ellington

10

IMPEDANCE COUPLING

RC interstage coupling is the cheapest and easiest to implement. However, we have seen that resistors have the same impedance for DC and AC currents, and to maximize the amplification factor of the stage, R_L should be as high as possible, ideally infinite. However, then the DC voltage drop on such a resistor would be very high.

An inductor has low resistance for DC and very high impedance for AC currents and seems ideal for an anode load, Since the DC resistance of the anode choke is very low (typically 1-3 kΩ) there is very little DC voltage drop across it. In our LC coupled version of the ECC40 stage the DC voltage drop on the anode choke is $I_A R_A$ = 3mA*2k3=6.9V, while the DC voltage drop on the 47k anode resistor was 3mA*47kΩ = 155V! Instead of 400V anode supply we now only need 242V.

This means a lower voltage power supply can be used, or, using the same available power supply, the tube can work with much higher DC voltage on the anode than with anode resistor.

Since the reactance X_L of the choke is very high, the amplification A of the stage approaches that of μ of the tube. If we substitute X_L in the basic formula for the amplification of the common-cathode stage $A = -\mu R_L/(r_I + R_L)$, we get $A = -\mu X_L/(r_I + X_L)$, and if X_L is much larger than r_I, as it is, at least at higher frequencies, we get $A \approx -\mu$.

For those who like chokes and dislike resistors, there is a special treat: double-impedance coupling! This concoction involves the use of an anode choke in the driver stage and a grid choke in the driven stage.

The two chokes and the coupling capacitor form a series-tuned circuit that is resonant at a certain low frequency, so with careful calculation and proper tuning, this will increase the gain in the bass region.

However, since two chokes are in parallel for AC signals it will also double the value of the parasitic distributed capacitance of the dual-choke arrangement. This higher shunt capacitance will reduce the gain at high frequencies, so again, improving the frequency response of an amplifier at one frequency extreme (bass in this case) negatively affects its response at the opposite end (treble).

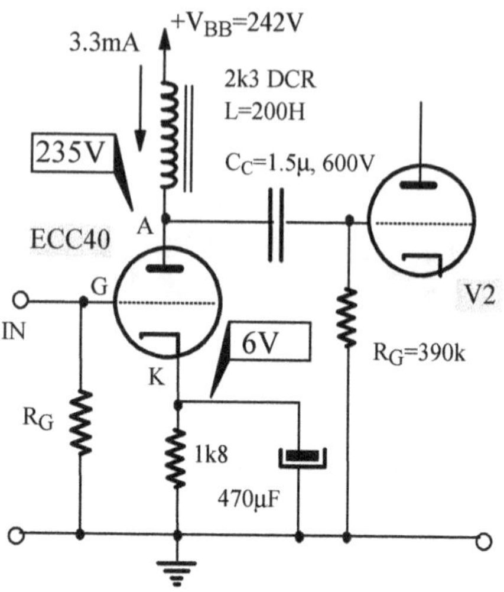

Impedance or LC-coupling

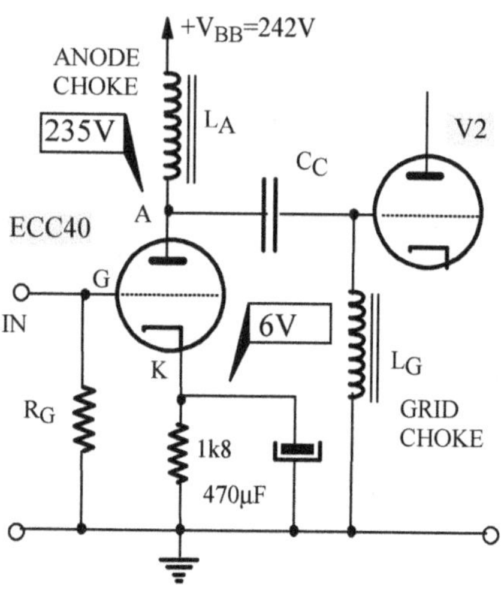

Double-impedance or LCL-coupling

Low- and high-frequency models

As with RC-coupling, we simplify the full model into three approximations for three frequency bands. C_A, C_C, C_G and R_G have the same meaning as with RC-coupling. L is the inductance of the anode choke, C_L is its distributed capacitance and R_F is choke's core loss resistance, which is mostly its iron loss. Do not confuse it with its DC resistance R_L.

The two chokes and the coupling capacitor form a series tuned circuit that is resonant at a certain low frequency. This will increase the gain in the bass region, but since two chokes are in parallel for AC signals it will also double the value of the parasitic distributed capacitance of the chokes. This increased shunt capacitance will reduce the gain at high frequencies, so again, improving the frequency response of an amplifier at one frequency extreme (bass in this case) makes its response at the opposite end (treble) worse.

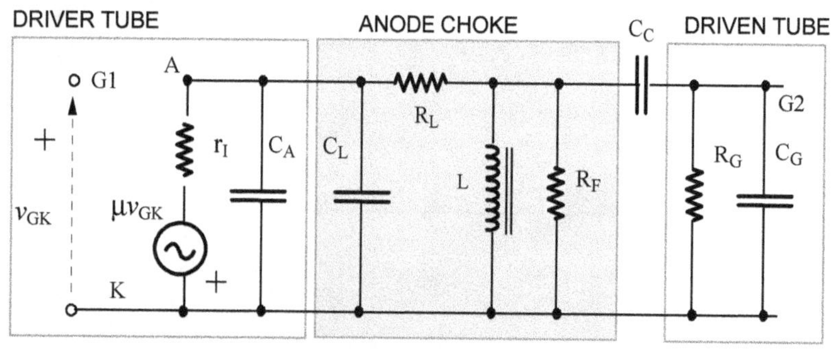

In the midband of frequencies, just as with RC-coupling, capacitances can be neglected, the shunt capacitances are open circuit and the coupling capacitance is a short circuit.

LEFT: The full model of impedance-coupled stages

At low frequencies the reactance of C_C becomes significant, but in contrast with the RC-coupling where it plays the dominant role, with LC-coupling it can be neglected without a significant error. This is because the decreasing reactance of the anode choke has a much more dominant effect on the low frequency response than the capacitor's increasing reactance.

At high frequencies the three shunt capacitances can be lumped into one and the coupling capacitance can again be considered a short circuit. We will not repeat the mathematical analysis here, because it is very similar to that of RC-coupled stages, so let's look at a practical amplifier instead.

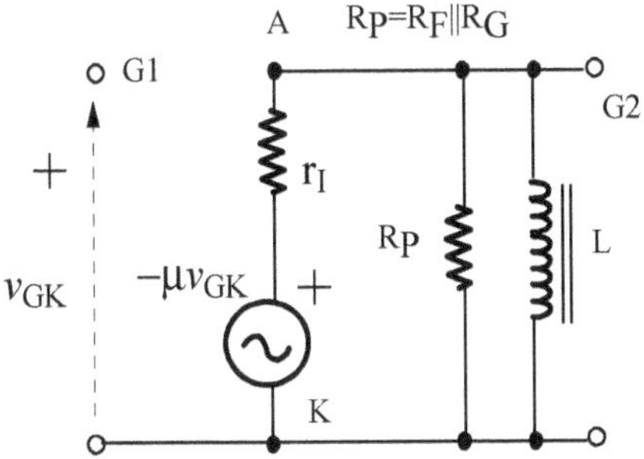

Low frequency model of impedance - coupled stages

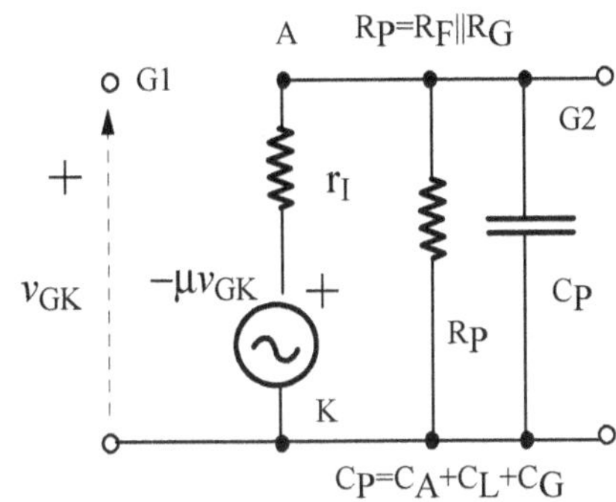

High frequency model of impedance-coupled stages

LC-coupled ECC40 stage

Since the impedance of an anode choke raises linearly with frequency, HF response of LC-coupled stages isn't of concern, but LF response is, because at low frequencies the choke's impedance may drop to a value that is too low for the required LF response of the stage.

At a midrange frequency of 1,000Hz, the 200H choke's reactance is $X_L=2\pi fL$ = 1.257MΩ! The anode load Z_L is X_L in parallel with R_G of the output stage, in this case R_G=390kΩ, so $Z_L=X_L||R_G$ = 298kΩ

The voltage amplification of the stage with one triode is $A_1=-\mu Z_L/(Z_L+r_I)$ = -32*298/(298+11) = -32*0.96 = -30.86

Since the choke's impedance is much higher (3X) than the grid resistance of the output stage, the grid resistance is the dominant factor in gain calculations. Connecting two triodes in parallel does not increase μ, but halves the internal resistance, so $A_2=-\mu Z_L/(Z_L+r_I/2)$ = -32*298/(298+5.5) = -32*0.98 = -31.42

The gain is slightly higher, but look what happens at low frequencies. At f=20Hz, the choke's reactance is only X_L=25kΩ! The anode load Z_L is X_L in parallel with R_G of the output stage, in this case R_G=390kΩ, so $Z_L=X_L||R_G$ = 23.5kΩ Using one tube, the amplification factor has dropped to $A_1= -\mu Z_L/(Z_L+r_I)$=-32*23.5/(23.5+11)=-32*0.68 =-21.8, or 21.8/30.86 = 70.6% of the midrange gain. This means that (by pure coincidence) we have stumbled upon the -3dB of the stage, f_L, which with one tube is around 20Hz.

Using two paralleled triodes, $A_2= -\mu Z_L/(Z_L+r_I)$ = -32*23.5/(23.5+5.5) = -32*0.81 = -25.9, a much better result, a drop to 25.9/31.42 = 82.5%, a significant improvement over 70.6% with one triode.

To reduce the drop of amplification at low frequencies, the triodes used in LC-coupled stages should have a low internal resistance r_I and/or should be paralleled to reduce such resistance as much as possible. At the lowest desired frequency of operation, the reactance of the anode choke should be at least twice the internal resistance of the tube.

LC-coupling compared to RC-coupling

For impedance and transformer-coupled stages, the static load line will be very steep, almost vertical, since the DC resistance of the choke or transformer's primary winding is low (in our case 2k3), while its AC impedance is very high. At 1kHz, $X_L=2\pi fL$ = 2*3.14*1,000*200 = 1.257MΩ, so the parallel with 220k grid resistor presents a load of 1.257MΩ||220kΩ = 187kΩ!

The voltage amplification factor of the LC coupled stage can be determined from the curves. V_{MIN}=80V and V_{MAX}=375V, so ΔV_A=375-80=295V. The grid swing is the same as for RC-coupling, ΔV_G=10V, so voltage amplification is $A_V=-\Delta V_A/\Delta V_G$=29.5 times or 29.4 dB.

While inductive coupling is clearly advantageous at higher frequencies, at lower frequencies it is inferior, because X_L is frequency-dependent and R_L isn't! At frequency zero (DC), X_L drops to zero and so does the amplification factor, so inductive coupling cannot be used in DC amplifiers!

Depending on the value of the R_L and L, you can find at what frequency the inductive coupling results in the same A_V as that of the resistive coupling: $R_L=X_L$ or $R_L=2\pi f_X L$ so $f_X=R_L/2\pi L$.

For our anode choke used in a real amplifier (L=200H), instead of using a resistor R_L=47kΩ we get $f_X=R_L/2\pi L$ = 47,000/400π = 37 Hz! This is bad news, you have to think carefully if the higher amplification of a certain stage is worth the price paid - a compromised low frequency response.

The midrange gain with impedance coupling is higher than with RC coupling, but it drops off more rapidly at frequency extremes so the frequency bandwidth is narrower. The graph below illustrates the low frequency response of a CC stage with a resistive load R_L and inductive load X_L.

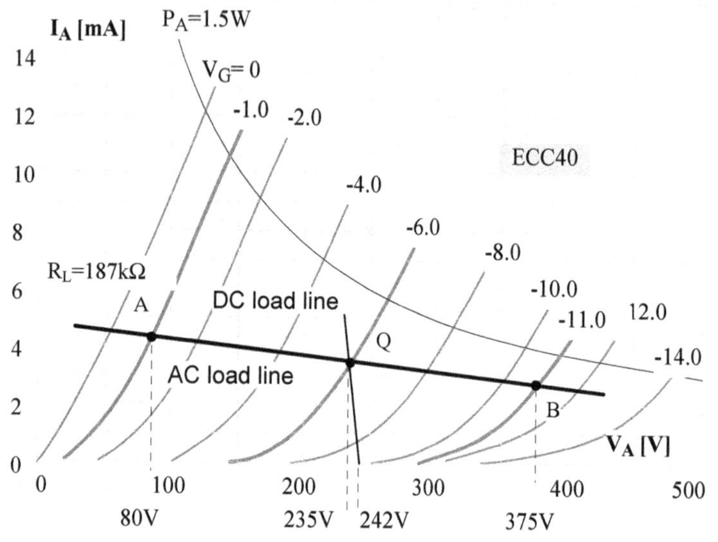

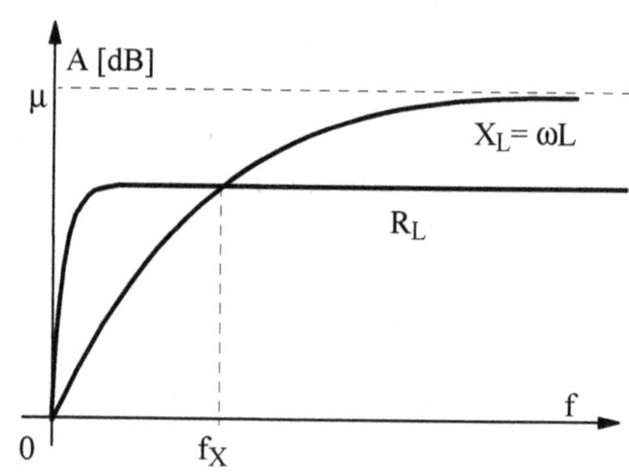

ABOVE: The frequency dependence of the amplification factors of a RC and LC-coupled common cathode stage

LEFT: The AC and DC load lines for LC-coupled ECC40 stage at 1 kHz

TUNING THE IMPEDANCE AND DOUBLE-IMPEDANCE COUPLED STAGES

CALCULATION

The anode choke in the ECC40 stage has L_A=200H. Find the value of the coupling capacitor needed to tune the resonant circuit to 20Hz, to improve the bass response. How would that value change if a grid choke L_G=5,000H is used instead of R_G (double impedance coupling)?

At resonance the reactive impedances of the capacitor and the choke are equal: $X_C=X_L$ or $1/\omega_R C=\omega_R L_A$, so we have $C=1/(\omega_R^2 L_A)$. In this case $\omega_R=2\pi f_R=2*3.14*20=125.7$ rad/s and $C=1/(\omega_R^2 L_A) = 0.317\mu F$, so we would use the nearest standard value, which is 0.33μF!

If a grid choke is also used the effective inductance is $L=L_A+L_G$=5,200 H

Since in class A_1 operation there is practically no current through them, grid chokes are wound using a very fine wire to get a high number of turns. They have ten or more times higher inductance than anode chokes, which must pass 5-20 mA of current and are wound using thicker wire, meaning less turns. The required coupling capacitance is now much lower: $C=1/(\omega^2 L) = 0.0122\ \mu F = 12.2$ nF

Since $\omega=1/\sqrt{(LC)}$ or $f=1/[2\pi\sqrt{(LC)}]$, using a larger value capacitor will shift the resonant frequency lower. For instance, if a 22nF coupling capacitor is used instead of 12.2nF just calculated, the resonant frequency would drop from 20 Hz to 14.9 Hz!

THE FREQUENCY RESPONSE OF TRANSFORMER-COUPLED STAGES

Believing that capacitors, even of PIO or film&foil kind, should be avoided at any cost, audiophiles advocate the merits of transformer coupling between stages, instead of the cheaper and simpler RC coupling. Transformer coupling, they claim, results in a more natural, less colored or stringent sound.

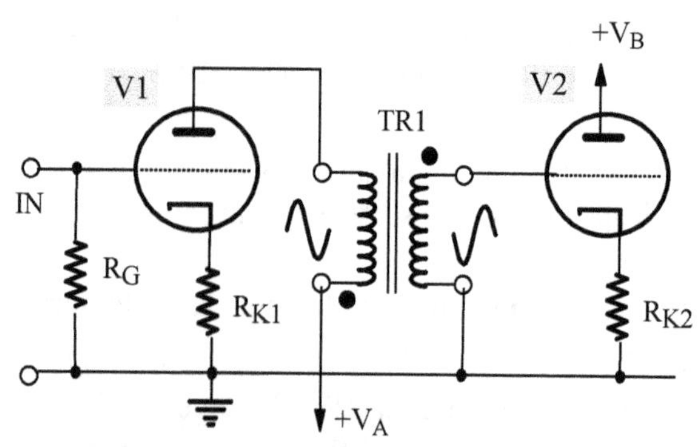

Single-ended driver to single grid, series output, DC current flowing through transformer's primary, phase inverting

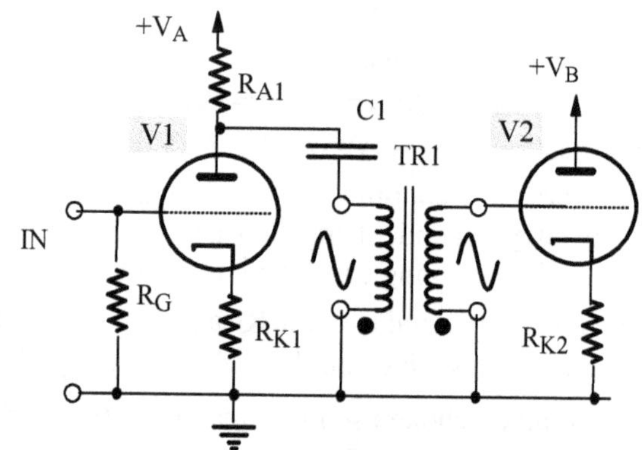

Single-ended driver to single grid, shunt-fed output, no DC current flowing through transformer's primary, no phase inversion

Interstage transformers also make precise matching of impedances between the stages possible. They can serve as phase splitters, converting a single-ended signal from the driver stage into two signals of the opposite phase to drive the output tubes in a push-pull configuration.

However, quality interstage transformers are even more difficult to design and build than output transformers, and are costly, especially if there is a primary DC current. For that reason a parafeed arrangement can be used where a coupling capacitor prevents DC current from entering the primary winding. Much wider bandwidth can be achieved without the air gap, but we are back to having a capacitor in the signal path, so the parafeed arrangement seems to be an exercise in futility.

Audio transformer models

Real transformer windings use copper wire, and such wire has resistance. R_P in the model represents the DC resistance of the primary winding and R_S of the secondary winding.

Not all flux produced by the primary current goes through the secondary coil. The same applies to the secondary coil. This "leakage" of flux into the surrounding air is modeled by a leakage inductance. L_{LP} is the leakage inductance of the primary winding and L_{LS} is the leakage inductance of the secondary.

Equivalent diagram of an audio transformer at low and midrange frequencies.

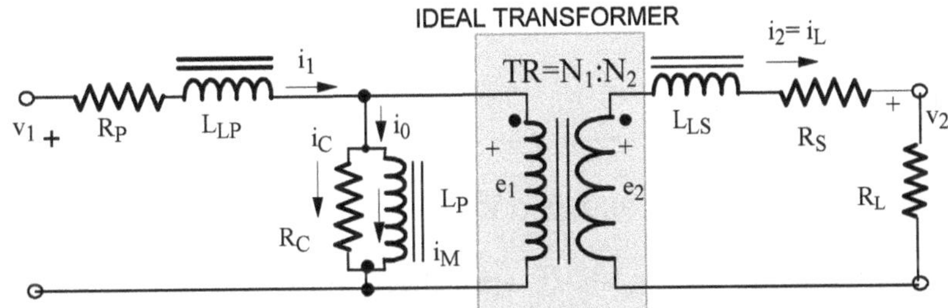

"Copper losses" is the power dissipated in R_P and R_S. While R_P and R_S can easily be intuitively understood and measured, the R_C and L_P in the vertical branch are not real components, they are only a symbolic representation of the physical phenomena in the magnetic core. The excitation or primary current i_1 has two components. i_M is the current needed to magnetize the core. L_P is the primary's inductance with open secondary, or primary inductance for short. Te other component of the primary current, i_C, is the core-loss current, which together with resistor R_C symbolizes core power losses, due to eddy currents and hysteresis.

Parasitic capacitances

The complete equivalent diagram of an audio transformer, including parasitic capacitances.

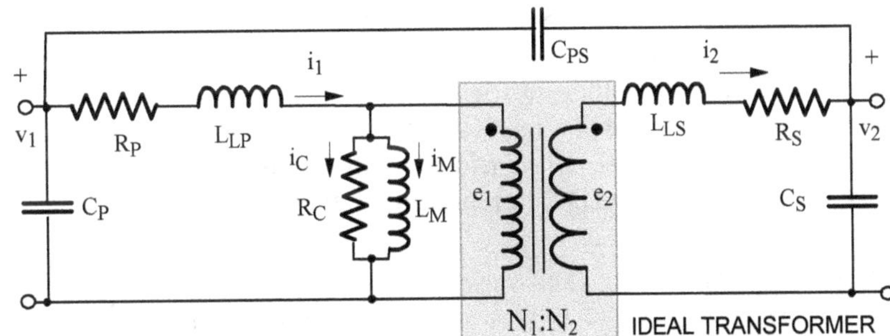

The model gets further complicated by another type of imperfection, and that is parasitic capacitance. There are capacitances between each winding and the core (or earth), distributed capacitances between adjacent turns, between layers within each winding and between windings.

Each layer and the whole winding acts as a capacitor, so by winding your transformer you have just wound a few unwanted capacitors as well! The first simplification is representing all these distributed capacitances with only three lumped parameters: C_1 (primary winding's capacitance), C_2 (capacitance of the secondary) and C_{PS} (between the primary & secondary windings). From the circuit analysis point-of-view, C_{PS} complicates things, the whole circuit becomes a messy network requiring complex mathematical modeling. Luckily, in the first approximation, it can be omitted. Likewise, the C_P and C_S are irrelevant for power transformers and have very little impact on the behavior of output transformers. They are mostly of interest to the design of input and interstage transformers.

Low frequency response of an interstage transformer

The low frequency model for interstage transformer coupling is a simple RL high pass filter. The inductive impedance is $Z_L = \omega L_P = 2\pi f L_P$ where L_P is the primary inductance. The total resistive component of the filter is equal to the series combination of tube's internal resistance r_I and the transformer's primary resistance R_1, which we will call R_S, where $R_S = r_I + R_1$.

Of most interest is the frequency at which the inductive reactance of the transformer primary inductance L_P becomes equal to the resistive component of the filter ($Z_L = R_S$):

$f_L = R_S/(2\pi L_P) = (r_I + R_1)/(2\pi L_P)$ so $L_{PMIN} = (r_I + R_1)/(2\pi f_L)$

This is why high μ tubes such as 12AT7 or 12AX7 should not be used as drivers. Due to their high internal resistance r_I their low frequency response would be poor in this situation.

Primary inductance can be increased by using larger laminations and a higher number of primary turns, but then parasitic capacitances also increase and the HF response deteriorates. Primary inductance can also be increased by using magnetic materials of increased permeability such as Permalloy and Hiperm, but they are expensive and saturate at much lower levels of B, so no luck there either.

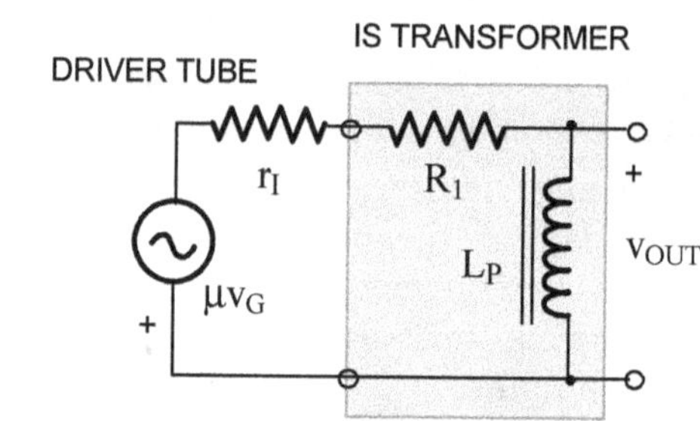

Low frequency model for Class A_1 and AB_1 interstage transformers

LF RESPONSE OF INTERSTAGE TRANSFORMERS

$f_L = (r_I + R_1)/(2\pi L_P)$

$L_{PMIN} = (r_I + R_1)/(2\pi f_L)$

Choosing the driver tube and its anode current

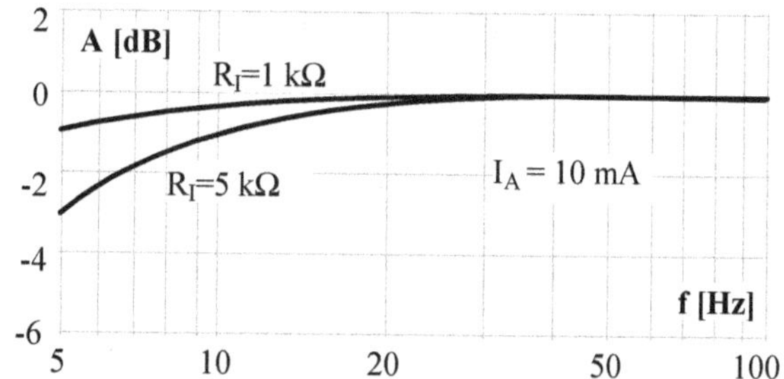

Low frequency response of Tango NC-20 interstage transformer for r_I=1 kΩ and 5 kΩ

These graphs from Tango NC-20 spec-sheet illustrate two important points. In the upper graph, if you compare the two curves for the same primary current (10 mA in this case), the higher the internal impedance, the faster the low frequency response drops and the earlier such a drop starts. At 5 Hz and 10 mA, the response is down only -1dB with 1kΩ driver tube, compared to -6dB for a 5kΩ driver tube.

Primary DC current also affects the low frequency response. The higher the current, the stronger the low frequency attenuation and the higher the f_L! Looking at the graph for a 5kΩ driver tube, at 10 mA the -3dB f_{LA} is 9 Hz, which is an acceptable result, while at 30 mA f_{LB} is around 21 Hz, too high for a hi-fi amp. Thus, choose driver triodes with low internal resistance r_I and minimize the DC current through transformer's primary.

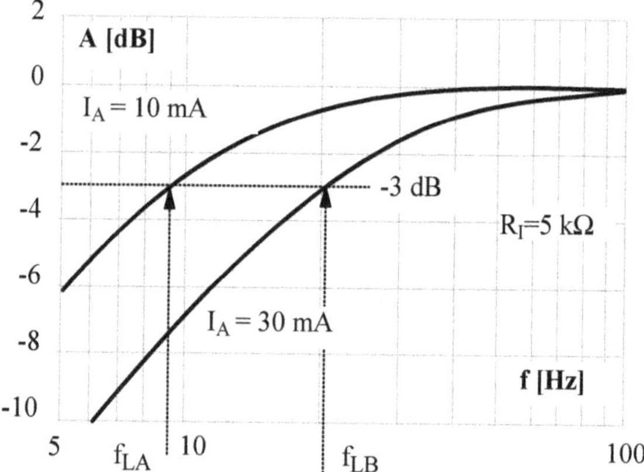

Low frequency response of Tango NC-20 interstage transformer for r_I=5 kΩ and two different values of primary DC current

The internal impedance of a tube varies inversely with plate current. The graphs for 6SN7 triodes show r_I dropping from about 35 kΩ at 1 mA to between 7 and 8 kΩ above 10 mA. This is an obvious contradiction. Minimizing plate current through transformer's primary winding will cause the driver tube's internal resistance to go up, and vice versa.

In any case you should never go bellow the current where amplification factor drops, in this case around 5 mA. Don't go too high, either, since r_I does not drop much further (in this case r_I is constant from 7 to 17 mA). Depending on the transformer used, the optimal range for 6SN7 would be 7 - 13 mA.

We could connect two or even more triodes in parallel and substantially reduce the overall internal resistance. At 9 mA one 6SN7 triode has r_I of 8 kΩ, so two in parallel would have half that, which would lower f_L and improve the LF response of the amp! But, the current through the interstage transformer primary would then double, so it would need to be wound with thicker wire, necessitating the use of a larger size EI laminations or C-core.

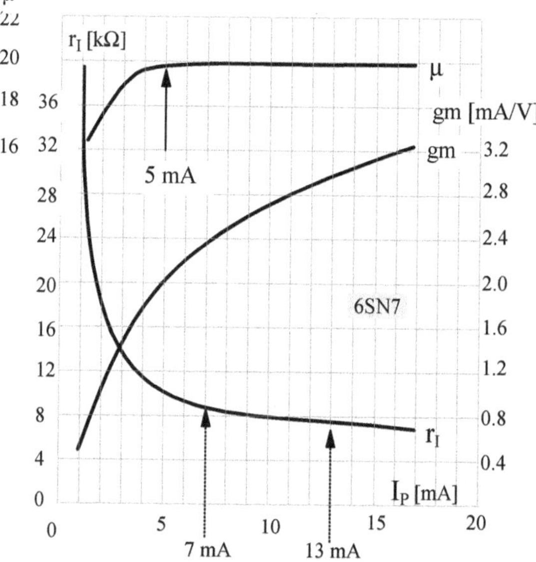

The three main parameters of 6SN7 triode as a function of anode current

Both the inter-electrode capacitances of the two paralleled tubes and the interstage transformer's parasitic capacitances would also increase, and, as a result, the upper -3dB frequency would drop significantly, up to half of the single-tube value. So, if the upper -3dB frequency with one tube was, say, 120 kHz, it would now drop to about 60 kHz! Again, we face an insurmountable obstacle our old nemesis, contradictory requirements!

High frequency response of an interstage transformer

As another practical example, let's look at Hashimoto A-105 interstage transformer with the following specs: 7kΩ primary, 25-25,000Hz +-2dB, with 5V input, r_I=5kΩ, I_A=10mA. With two primaries in series the primary inductance is L_P=60H, measured at I_A=7mA at 5V and the frequency of 50Hz.

The graphs published by the manufacturer show high frequency response of this transformer, for 1:2 ratio (secondaries in series) and 1:1 ratio (secondaries in a phase splitting arrangement). Notice how f_U increased from around 47 kHz for a 1:2 connection to around 75 kHz for the 1:1 connection. This confirms our rule-of-thumb: The higher the step-up ratio, the lower the bandwidth of an interstage transformer!

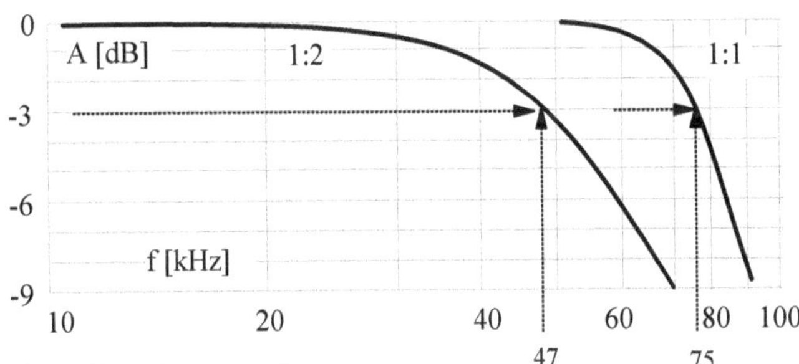

High frequency response of Hashimoto A-105 interstage transformer for two different voltage ratios

High frequency model for interstage and output transformers

At high frequencies the transformer behaves like a series resonant circuit formed by leakage inductances and parasitic capacitances. The model is simplified, of course, since we are using lumped capacitance, while in reality they are distributed parameters, but the results are very close to measured figures.

$R_{SER} = r_I + R_1 + R_2TR^2$, $L_L = L_{LP}+L_{LS}TR^2$, and $C = C_1+(C_2+C_{IN})/TR^2$

At the upper frequency limit (-3dB) $Z_C = R_{SER}$ or $1/(\omega C) = R_S$, so $\omega_U = 1/(R_SC)$

The resonant frequency of the LC circuit is $\omega_R = 1/\sqrt{(L_LC)}$ The optimal HF response is achieved when the resonant frequency f_R equals the upper -3dB frequency f_U: $\omega_R/\omega_U= 1$

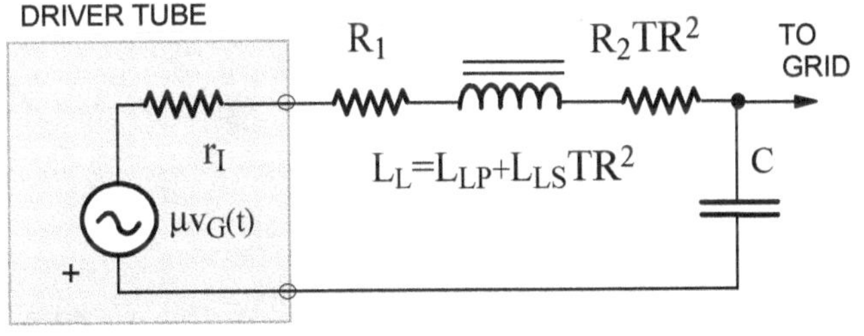

The high frequency model of an interstage transformer in class A_1 (infinite grid resistance so no grid current)

Theory versus practice

The primary inductance is $L_P=\mu_{EF}\mu_0N_1A_{EF}/(g+LMP/\mu_{EF})$, where LMP is the length of magnetic path and g is the air gap. Once we determine the L_P required for the low frequency response, we need to find N_1. We know the laminations (A_{EF} and LMP) but we don't know the size of the gap g and effective permeability μ_{EF}.

There are ways to estimate these, but basing our designs on assumptions is not just tedious, but can also be grossly inaccurate, since even a small discrepancy in one or both parameters has a huge impact on the calculated number of turns, requiring two, three or even more iterative steps. Only an anal-retentive personality would pursue this tedious approach. A practical and prudent alternative is to fit as many turns as possible, assemble the transformer and adjust the air gap, just as we do with chokes, and then measure its performance and see what we have achieved.

The step-up transformer problem

A line-level input transformer has a ratio of 1:10, the leakage inductance is 10mH, primary and secondary capacitances are roughly the same, around 40pF. It drives the grid of EF86 pentode with an input capacitance of around 15pF. What is the resonant frequency of the circuit and what upper -3dB frequency can we achieve?

$C= C_1 + (C_2+C_{IN})/TR^2 = 40 + (40+15)/(1/10)^2= 40 + 55*100 = 5,540$ pF = 5n5

When reflected to the primary, C_2 and C_{IN} are divided by the square of the TR, and since TR in this case is smaller than 1 (N_1/N_2=0.1), the capacitances are multiplied by 100! For 1:20 ratio, they would be multiplied by 400!

The resonant frequency is $\omega_R= 2\pi f_R=1/\sqrt{(L_LC)}$ so $f_R= 1/\sqrt{(L_LC)}/2\pi = 21,382$ Hz, barely enough for a hi-fi amplifier. To make things worse, this is one of the best cases using a low leakage transformer and a low input capacitance pentode.

With a triode at the input with C_{IN} of say 115 pF, we would get C= 40+155*100 = 15,540 pF = 15nF and f_R= 1//(0.01*15.54*10-9)/ 2*π = 4,037 Hz Assuming $f_R = f_U$, an upper frequency around 4 kHz would not be enough even for a guitar amp, let alone hi-fi. If you've ever wondered why quality wideband IS transformers never go above 1:2 ratio, now you understand.

The impact of Q-factor on the high frequency response

The high frequency resonant peak is an issue with interstage transformers, where the load impedance is very high (the grid impedance of the following tube). For step-down interstage and output transformers the quality factor Q is usually below 1 and there is no peak.

The optimum Q s about 0.7-0.8, resulting in the flattest and widest response curve. In the first approximation we can assume that the upper -3dB frequency f_U is the same as the resonant frequency f_R. The curves are not to scale, the differences are smaller than illustrated!

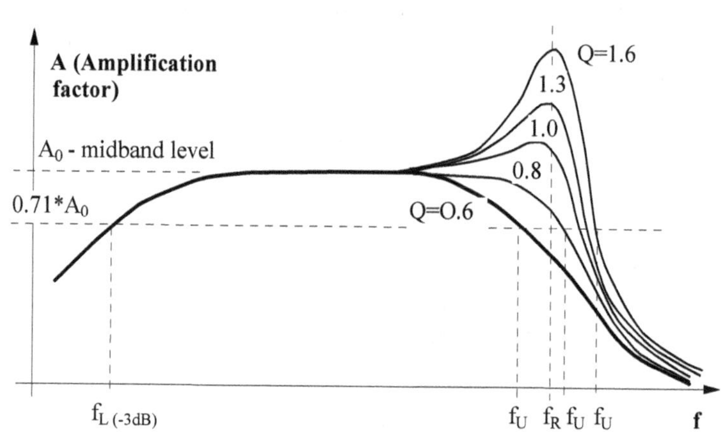

Frequency response of an interstage transformer depends heavily on the Q-factor

RC- versus transformer coupling

In conclusion, it is impossible to obtain the same bandwidth with transformer coupling as can be achieved with capacitive coupling. The flatness of the frequency response is usually also inferior. Technically, transformer coupling cannot be justified.

Listening tests show that transformer coupling sounds different from capacitive coupling, but does it sound better or just different is the ultimate question, one that divide audiophiles and cause endless debates and disagreements.

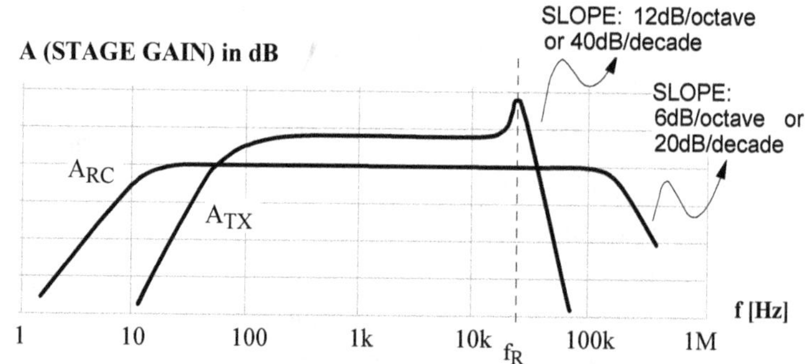

Stage gain (log scale) versus frequency of a typical RC-coupled and transformer-coupled triode stage

INTERSTAGE TRANSFORMER DESIGNS

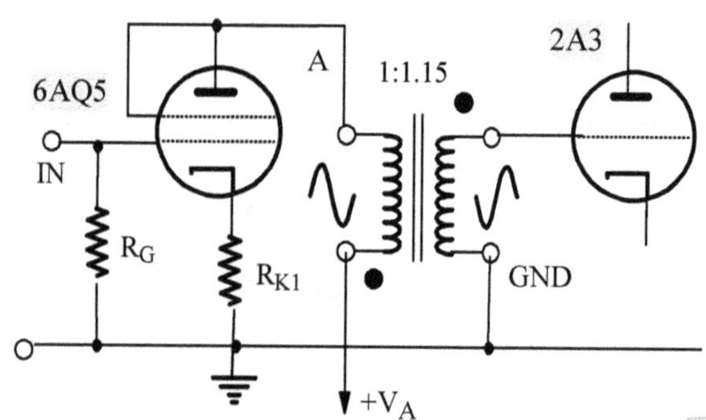

TUBE PROFILE: 6AQ5 (EL90)

- Miniature Noval beam power tube
- Heater: 6.3V, 0.45 A
- V_{AMAX}= 250V_{DC}, V_{HKMAX}=100V_{DC}
- Maximum plate dissipation: P_A=12W
- TYPICAL OPERATION (triode):
- V_A=250V, V_G=-12.5V, P_A=9W, I_0=50 mA
- gm=4.8mA/V, μ=10, r_I =2kΩ

Single-ended driver for 2A3 and 300B tubes

Now that we understand at least some of the issues and problems arising from the use of interstage transformers, let's look here at a few practical designs.

We needed a 1:1.15 interstage transformer between triode-connected 6AQ5 driver at 20 mA DC anode current and 2A3 power tube in single-ended configuration. 6AQ5 has plate dissipation of 12 Watts as a beam tube or 9 Watts when triode connected.

Its performance is similar to the octal 6V6-GT. Connected as a triode, it has an amplification factor of around 10 and low internal impedance of under 2 kΩ (1,970Ω), 3-4 times lower than 6SN7.

EI66 laminations from 70V line transformers were used, 3:2 sectionalizing, 3 primaries in series and 2 secondaries in series.

P_1=P_2=P_3=1,100 turns, wire 0.13mm dia, 7 layers in each section

S_1 =S_2=1,900 turns, wire d=0.13mm, 11 layers per section

The turns ratio is TR=$V_1/V_2 = N_1/N_2$ =3,300/3,800 = 0.8684 so $V_2 = 1.15V_1$

Measured results: L_1=19.5 H R_1=750 Ω, I_0=20 mA, air gap g= 0.1mm

The lower -3dB frequency can be estimated by the formula $f_L = (r_I+R_1)/(2\pi L_1)$ where r_I is driver tube's internal resistance, R_1 is the transformer's primary DC resistance and L_1 is the transformer's primary inductance. In this case we get f_L = (1970+750)/(2*π*19.5) = 2,720/122 = 22.3 Hz, very close to the measured f_L of 24 Hz. To lower the f_L we would need larger laminations and a thicker stack.

+V_A GRID 1 2 3 4 5 A GND 3x1,100 turns in 7 layers per section, wire d=0.13 mm 2x1,900 turns in 11 layers per section, wire d=0.13 mm

Connect the primary start lead to the anode supply (+V_A) and its end to the anode (A). The secondary winding's start point should be connected to the output triode's grid. This leaves adjacent turns in both windings at zero audio potential, and the effective primary-secondary capacitance is zero.

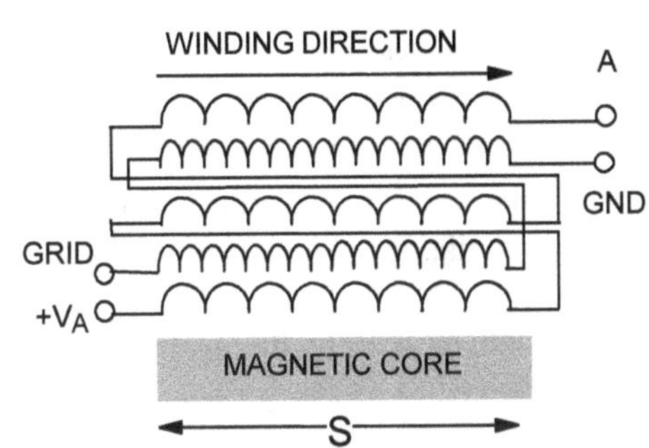

The way it used to be: Triad HS-35 interstage transformer

Triad HS-35 interstage transformer appears in their 1961 catalog, and is specified as a single plate to single or push-pull grids. The primary impedance is 15 kΩ and the total secondary impedance (both secondaries in series) is 111 kΩ, resulting in a 1:2.72 voltage ratio. The frequency range at 26 mW power level is given as 20Hz - 20kHz.

The DC resistance of the primary is 1.67 kΩ and its inductance is above 200H. The secondary DC resistance is 3,940Ω.

49% nickel alloy laminations were used, size 26-27 EE. The center leg is 3/8" wide (9.525mm) and the stack is 13/32" thick (10.32mm). Simple 1-2 sectionalizing was used.

Window Length is WL=0.6875" = 17.5 mm, allowing for 2 mm bobbin the Coil Length is CL=WL-4 mm = 13.5 mm

The 0.04mm is the dimension of the bare wire, with heavy Formvar insulation used at the time, its diameter is around 0.055mm

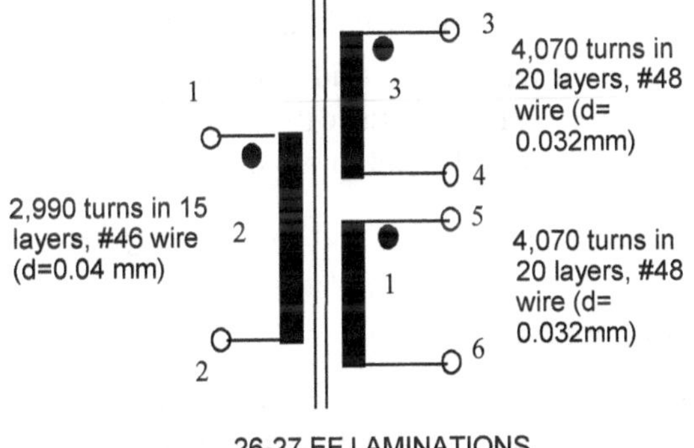

Triad HS-35 interstage transformer winding diagram

Primary Turns-Per-Layer: TPL_{MAX}=CL/d= 13.5/0.055 = 245 T

With 0.8 horizontal fill factor (meaning we can comfortably fit 80% of the turns) we get TPL=245*0.8 = 196

Indeed, the dissection of this transformer by one keen enthusiast who posted his results online showed 15 layers of 196-204 turns per layer (their turns-counting was a bit lax!), so our theoretical estimate is spot-on. So, we should aim at 200 TPL (2,990/15 =200)!

The window height is 0.25"=6.35mm, -2mm bobbin thickness, gives us the 4.35mm maximum winding height,

Let's check the vertical fit. Primary Height: PH= 15*0.055 = 0.825 mm, Secondary Height: SH= 40*0.047 = 1.88 mm, total Winding Height WH = PH+SH= 2.7 mm

4.35 - 2.7 = 1.65 mm for insulation and bulging factor!

1:1.5 SE to PP driver/inverter transformer for 2A3 or 300B output stage

We bought a Chinese-made 300B amp on ebay ("Soundtrack" brand), attracted by the large transformers (or rather large transformer covers!) and the engraved tops on the two smaller cases, indicating interstage transformers. Despite its good looks, the amp turned out to be the biggest lemon we have ever seen or heard. The output transformers were tiny, less than 1/2 of the height of the mostly empty covers and the two "interstage transformers" flanking the center-mounted power transformer were actually filtering chokes! The amp sounded so bad that we took it apart and reused some of its components.

There was one piece of good news, the said chokes were actually made of 0.35mm GOSS laminations which we made into true interstage trafos. The laminations were EI76.2, with 1" center leg (25.4mm). The stack was 15mm thick, resulting in the cross-sectional area of 2.54*1.5=3.75 cm^2

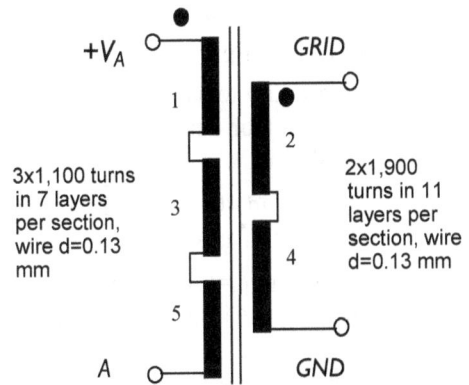

We wound it with 3 primaries in series, each with 800T of d=0.11mm wire, and 4 secondaries in CT configuration, each with 1,800T of d=0.11mm wire.

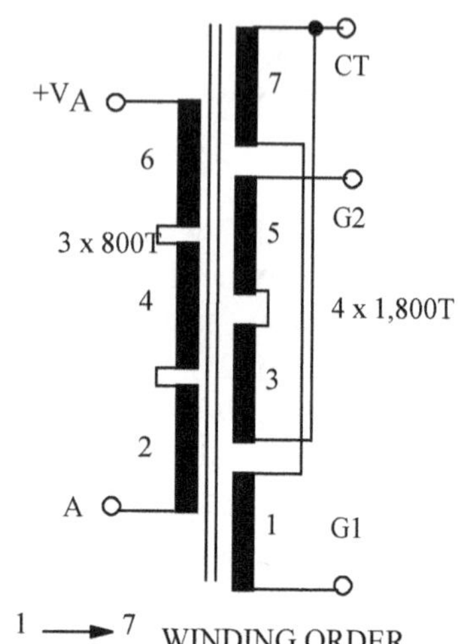

The turns ratio primary to half f the secondary is N_1/N_2=2,400/3,600 = 0.67 = V_1/V_2 so V_2 = 1.5V_1 (for each grid). Grid-to-grid voltage is twice that, or $V_{GG}=2*V_2=3*V_1$

L_1=20H at 120 Hz and 40H at 1 kHz, DC resistance of the primary is R_1=700 Ω For I_0=15 mA the air gap needs to be g=0.1mm!

For EL86 tube with r_I =1.1 kΩ in triode connection as a driver, the lower -3dB frequency is f_L = (1,100+700)/(2*π*20) = 1,800/125= 14.4 Hz, a very good result!

Interstage transformers for output stages operating in Class A_2

Instead of step-up action, no matter how counterintuitive it sounds, the best choice for IS transformers driving output tube's grid in class A_2 is a step-down interstage transformer. The grid driving voltage is reduced, but two advantages are achieved.

The secondary winding will have a much lower DC resistance, eliminating the danger of bias change due to grid current flowing. Secondly, the step-down ratio reduces the output impedance of the driver stage, as seen by the grid of the power tube.

With one 6SN7 triode and a 2:1 transformer, the source resistance reflected onto the secondary side is cut from 6k8 to 1k7 or 4 times (the impedance ratio IR=TR^2= 2^2 = 4) With two paralleled triodes the source resistance would drop from 3k4 to 850Ω.

Despite halving of the grid voltage by the transformer, the same gain can be achieved, since the voltage drop due to internal source impedance r_I is also cut to 1/4 of its previous value!

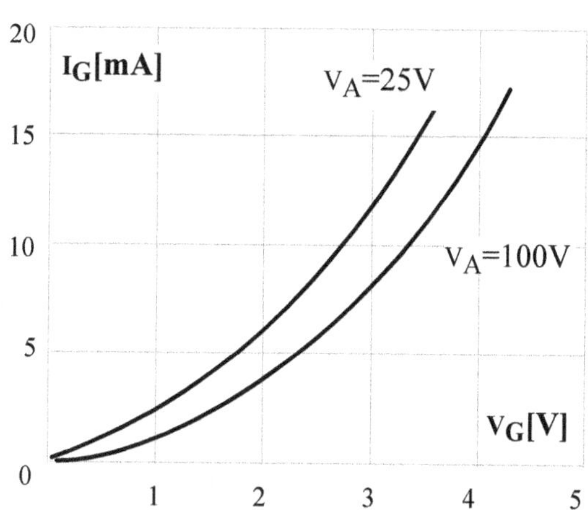

Control grid current as a function of positive grid voltage

Grid chokes

Our personal experience confirms the claims of many audiophiles: tube amps using grid-chokes instead of grid resistors sound better, more dynamic and musical.

The arrangement on the right shows a simple yet very good sounding amplifier topology with a SRPP input stage and a choke-driven grid of a single-ended output stage.

The end of the choke connected to ground can instead be taken to a negative source of DC voltage, -70 to -90 V_{DC}, for an even larger swing of the driving voltage on the output tube's grid and higher output power.

If a grid choke is wound with a center tap (or any other % tap), it can be used as a step-up auto-transformer, which increases gain, or in a step-down mode. In this case the step-up ratio is $V_{OUT}/V_{IN}= (N_1+N_2)/N_1$

One example of commercial grid-choke is made by SILK in Thailand. They claim the inductance of more than 7,000H at 12Hz, DCR of 1.3 kΩ, AC impedance at 1kHz of more than 5 MΩ (meaning that $L_{1,000}$=796H) and parasitic capacitance of less than 20pF.

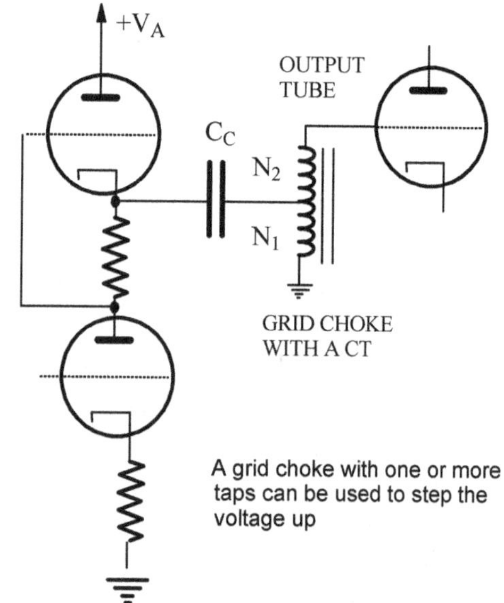

A grid choke with one or more taps can be used to step the voltage up

Chokes have a center tap for phase splitting applications and use Supermalloy material. Permalloy is a term for a nickel iron magnetic alloy with about 20% iron and 80% nickel content. Supermalloy has 5% molybdenum, 79% Ni and 15% Fe. It saturates very early, at 0.7 to 0.8 Teslas, the usable induction (the flux level at which the incremental permeability has substantially decreased) is 0.65 to 0.75 Teslas, much lower than silicon steel at around 1.6 T. The lower end of range is for Supermalloy, the higher end for Permalloy.

NEGATIVE FEEDBACK

- HOW NEGATIVE FEEDBACK WORKS
- TYPES OF NEGATIVE FEEDBACK
- THE BENEFITS AND DRAWBACKS OF NEGATIVE FEEDBACK
- PRACTICAL NEGATIVE FEEDBACK TECHNIQUES
- REDUCING UNWANTED POSITIVE & NEGATIVE FEEDBACK
- THE STABILITY ASPECT OF NEGATIVE FEEDBACK
- FREQUENCY COMPENSATION METHODS

"There are two things to be considered with regard to any scheme. In the first place, 'Is it good in itself?' In the second, 'Can it be easily put into practice?' "
Jean-Jacques Rousseau, *Émile*

HOW NEGATIVE FEEDBACK WORKS

Feedback ratio β

We have already learned a bit about NFB, so let's broaden our analysis and make some important general conclusions. Global NFB is usually taken from the secondary of the output transformer to the cathode of the first stage, as illustrated.

The feedback voltage is determined by the ratio of the feedback resistor R_1 and the cathode resistor R_2, which must be left un-bypassed. Only the un-bypassed cathode resistance counts, in this case R_K is bypassed for audio signal by C_K, so it does not enter the NFB calculations.

The circuit is a simple voltage divider. The lower that ratio, the weaker the feedback. This feedback ratio is $\beta = R_2/(R_1+R_2)$ **x100%** Usually $R_2 << R_1$, so $\beta \approx R_2/R_1$ **x100%**

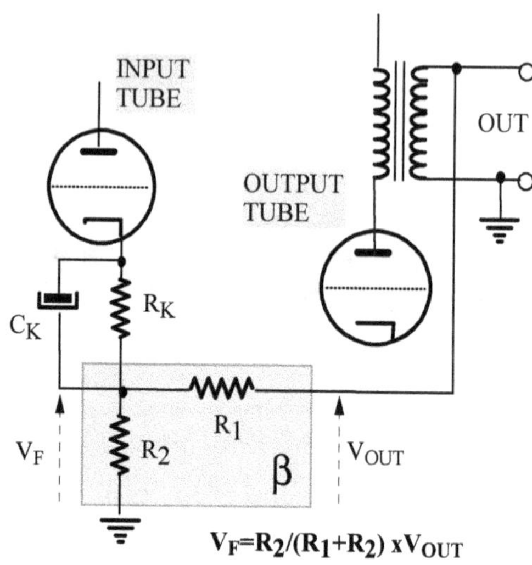

The most common way of applying global negative voltage feedback in a single-ended amplifier, in series with the input voltage (across R2).

Feedback factor FF

If we take an amplifier with amplification factor A as a starting point, to get amplitude A at its output we need signal amplitude of 1 at its input. The NFB network has its own amplification (or rather attenuation) factor β, it takes the output voltage A and multiplies it by β, so the signal amplitude at its output is βA.

This signal is added in series to the original input (1), so now the new input voltage into the amplifier with feedback, 1+βA is needed to get the output voltage A! 1+βA is called Feedback Factor (FF).

The new amplification factor (with feedback) is $A_F=V_{OUT}/V_{IN}=A/1+\beta A=A/FF$, so $FF=A/A_F$.

The factor βA is called *open loop gain*, where the "loop" is the path from the input point X to point Y, which includes the two blocks (A and β), but is *not* closed.

Negative feedback does NOT change the amplification factor A of the amplifier, it reduces its input voltage instead, thus reducing the output voltage. Perhaps that is where this confusion originates. This seems obvious once this block diagram is drawn and firmly committed to your memory.

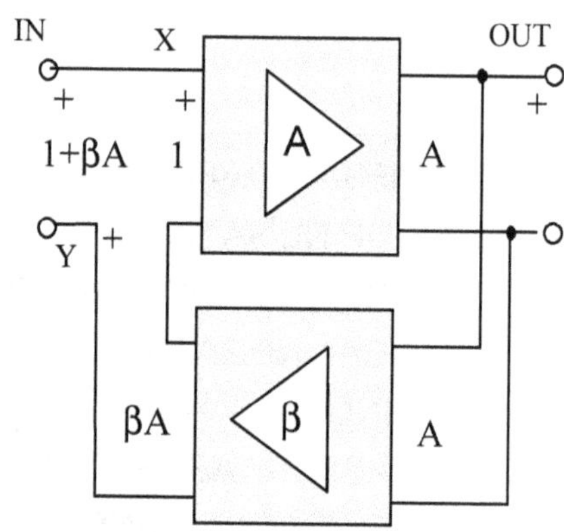

The general block diagram of a feedback amplifier

Negative feedback in dB

When you read that a certain amplifier has a certain number of dB of negative feedback, what does that really mean? Let's use an easy and convenient example, say 20dB of feedback is used in a certain design. We have $20\log(A/A_F) = 20\log(1+\beta A) = 20dB$, so $\log(1+\beta A)=1$ which means the $FF=(1+\beta A)=10$ (because log10=1!) Finally we have the open loop gain $\beta A=10-1=9$

Assuming the global NFB arrangement illustrated above, back to the cathode of the input stage, the feedback ratio is $\beta = R_1/(R_1+R_2)$. Let's say that amplification factor of our amplifier without feedback is A=18 (18V out for 1V in, about 40.5 Watts into an 8Ω load). Applying 20dB of NFB would reduce the gain to $A_F=A/10 = 18/10 = 1.8$, which would produce an output of 1.8V for 1V input signal. This is obviously far too much, so let's see what a mere 3dB of NFB would do: $20\log(A/A_F) = 3$ so $A/A_F = 10^{0.15} = 1.41$, or $A_F=A/1.41 = 18/1.41 = 12.74$

On an 8Ω load this is equivalent to $P=V^2/R = 20.25$Watts We see that 3dB of NFB reduced the voltage to 71% of its previous value and halved the output power.

TYPES OF NEGATIVE FEEDBACK

As we have seen, negative feedback can be local (un-bypassed cathode resistor in one stage) or global (two or more stages inside the feedback loop). Global feedback can include the preamp stages (without the output stage), or, if taken from these secondary of the output transformer), it will include ALL stages.

Four types of NFB based on the way the feedback is derived and applied

There are four types of negative feedback, classified as per how the feedback signal is derived (is it proportional to the output voltage or output current) and how it is applied in the input circuit (in series with the signal or in parallel with it). The four arrangements are illustrated below in a block-diagram form.

Since vacuum tubes are voltage-controlled devices, even if current feedback is used, what is fed back is still a voltage, but it is a voltage drop on a series resistor (in series with the load), thus the feedback voltage is ultimately proportional to the output current.

The voltage feedback lowers the output impedance, while the current feedback increases it. The series-applied feedback increases the input impedance, while the parallel-applied feedback reduces the input impedance.

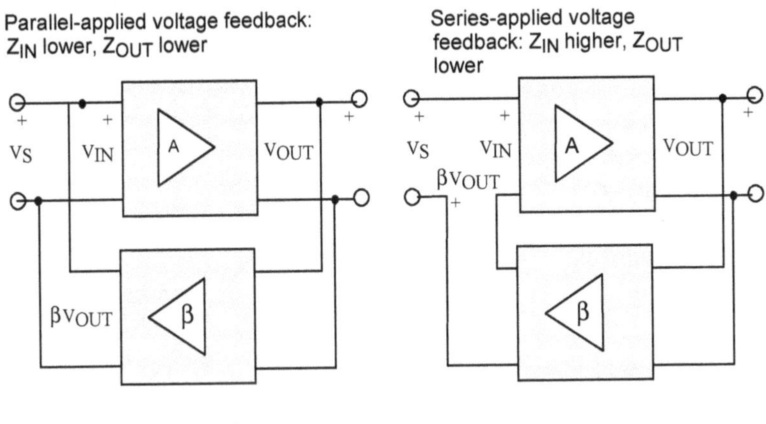

Parallel-applied voltage feedback: Z_{IN} lower, Z_{OUT} lower

Series-applied voltage feedback: Z_{IN} higher, Z_{OUT} lower

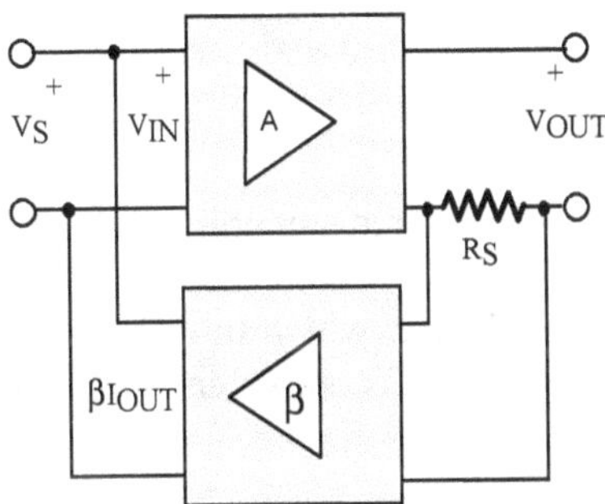

Parallel-applied current feedback: Z_{IN} lower, Z_{OUT} higher

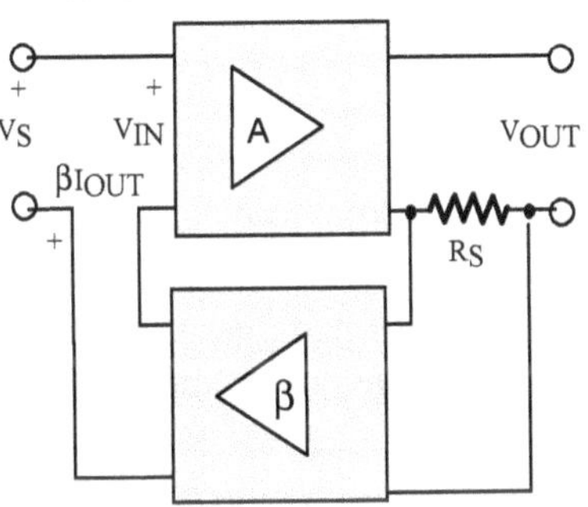

Series-applied current feedback: Z_{IN} higher, Z_{OUT} higher

The simple NFB circuits we have already studied

We have already analyzed simple examples of NFB, such as the common cathode stage with un-bypassed cathode resistor, its the special case (without the anode resistor), called cathode follower or common anode stage, and the resistive feedback from the anode to the grid of the CC stage, often called anode follower.

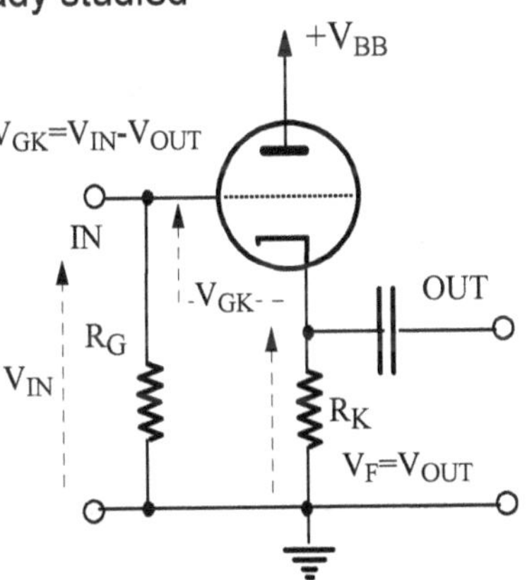

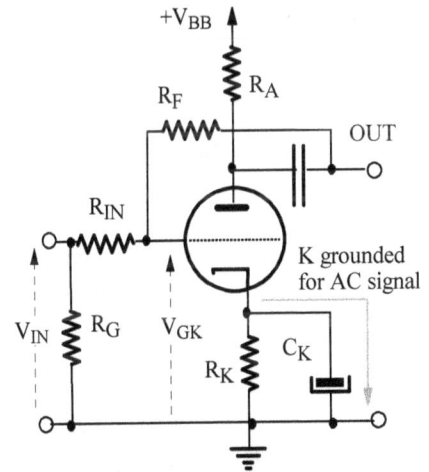

Series-applied voltage feedback: Cathode follower. The whole output voltage signal is fed back into the cathode circuit and added in series with the input voltage.

Parallel-applied voltage feedback: Anode follower. The whole output voltage signal is fed back into the grid circuit.

THE BENEFITS AND DRAWBACKS OF NEGATIVE FEEDBACK

NFB has many advantages:

- widens the frequency range (lowers the lower -3dB frequency and increases the upper -3dB frequency)
- reduces distortion and noise
- increases input impedance
- reduces output impedance, improves damping factor
- provides phase correction and reduces phase distortion
- stabilizes the amplification factor and reduces changes due to drift, component aging and mismatching

Negative feedback widens the frequency range but reduces gain

As an example, let's take a simple case of an amplifier with a 1st-order amplitude characteristics (low pass filter).

$A = V_{OUT}/V_{IN} = A_0(1+jf/f_U)$ Assuming resistive feedback network resulting in real feedback ratio β (independent of frequency), the transfer function after the application of NFB becomes:

$$A_F = A/(1+\beta A) = \frac{\dfrac{A_0}{1 + jf/f_U}}{1+\dfrac{A_0}{1 + jf/f_U}} = \frac{A_{F0}}{1 + jf/f_{UF}}$$

NFB REDUCES GAIN BUT WIDENS THE FREQUENCY RANGE

$f_{UF} = f_U(1 + \beta A_0)$ $\qquad A_{F0} = \dfrac{A_0}{1 + \beta A_0}$

Three conclusions can be made from this analysis. Firstly, when applied to 1st-order systems, NFB does not change the shape or the character of the system or its transfer function. Secondly, the midrange gain A_0 is reduced to A_{F0}, in proportion to the feedback factor $1+\beta A_0$. And finally, the upper -3dB (half-power) frequency is increased by the same feedback factor.

Notice that f_C, the frequency at which the amplification drops to 1 or 0dB (log1=0), has not changed. $f_C = A_0 f_U = A_{0F} f_{UF}$ The gain has decreased but the upper corner frequency has increased by the same factor, so we can conclude that the application of NFB does not change the gain-bandwith product of an amplifier!

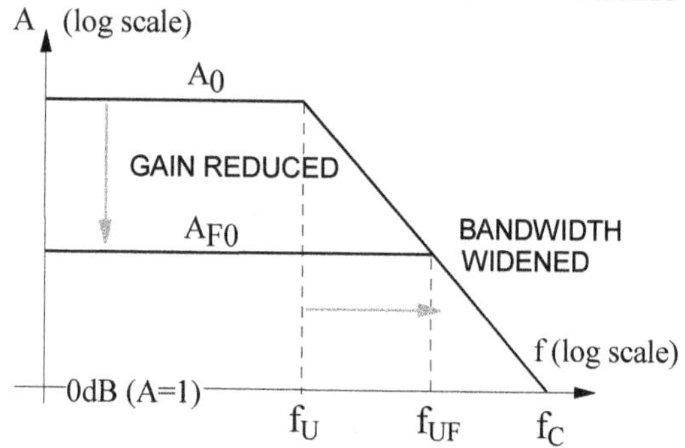

NFB reduces the midrange gain but widens the bandwidth, keeping the GBW product constant!

Negative feedback reduces distortion and noise

Distortion, interference and noise (DIN) can enter the amplifier or be generated by the amplifier at any point, from the input to the output. The illustration shows the simplest amplifier with one voltage amplification stage and the power or output stage. The test circuit demonstrates the impact of negative feedback on interference, which in this case in poorly filtered DC voltage from the amplifier's high voltage power supply. Apart from the DC component $+V_{BB}$, it also has the superimposed AC ripple component V_{RP}.

The ripple component has been attenuated to practically zero by the decoupling circuit supplying power to the input stage, so we are assuming that its power is perfect DC.

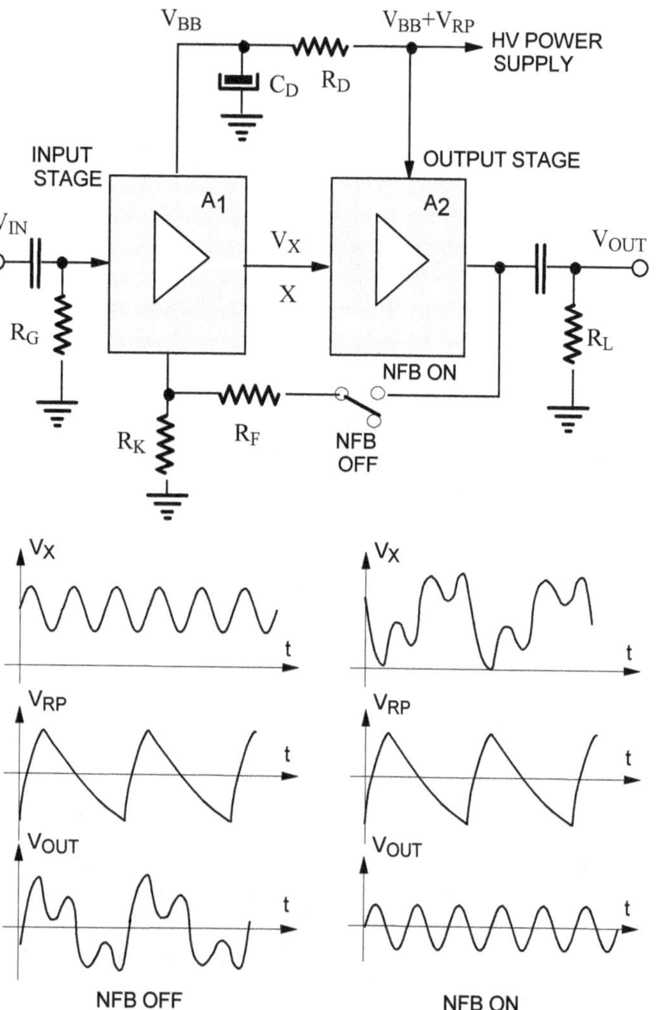

Without any NFB (switch in the bottom position, as illustrated) the signal V_X at the output of the first stage is undistorted. However, the signal at the output of the second stage is distorted (since its supply voltage is "dirty", not filtered enough), superimposed on the ripple voltage V_{RP}. The frequency of the signal voltage is shown about three times higher than that of the ripple, which is either 100 or 120 Hz.

With the NFB switch closed, the negative feedback takes the distorted output voltage and brings it back to the first stage, where it is mixed with the undistorted input signal V_{IN}. Notice that the signal V_X is now distorted, but of inverted phase (the meaning of "negative" in negative feedback), a mirror image if you like of the previous output voltage V_{OUT} without feedback!

Since we only have two stages, one of them must be non-inverting in order for this feedback to be negative. With two inverting stages the output voltage would be in phase with the input voltage and the feedback would be positive, causing the whole circuit to become an oscillator!

So, if DIN appears at the point X (in this case between the 1st and 2nd stage) where the downstream amplification (from that point to output) is A_2, or the upstream amplification factor (form the input to that point) is A1 we have

$V_{OUT} = A/(1+\beta A)V_{IN} + A_2/(1+\beta A)V_{DIN}$ or $V_{OUT} = A/(1+\beta A)V_{IN} + (A/A_1)/(1+\beta A)V_{DIN}$

Since $A=A_1A_2$, and A_2 is smaller than A, the second term tells us that DIN signals are amplified less than the input signals if DIN enters the circuit within the amplifier or is generated by the amplifier itself. This improves the signal to noise ratio. However, in case when $A_2=A$, meaning the DIN signal has entered together with the signal at the amplifier's input, both the useful signal and the DIN signal will be reduced equally by negative feedback, so in that case NFB does not improve the S/N ratio!

The amplitude of the 2nd harmonic in an amplifier with NFB is the amplitude of the 2nd harmonic of the same amplifier without the NFB, reduced by the portion of the 2nd harmonic returned back to the input through the feedback loop βA: $D_{2F} = D_2 - \beta A D_{2F}$ Once we express D_{2F} we get $D_{2F} = D_2/(1 + \beta A)$ NFB reduces harmonic distortion by the same factor as the midrange gain. The open loop gain βA in this formula must be calculated or measured at the frequency of the 2nd harmonic. The same formula applies to all other harmonics.

The linearizing effect of negative feedback

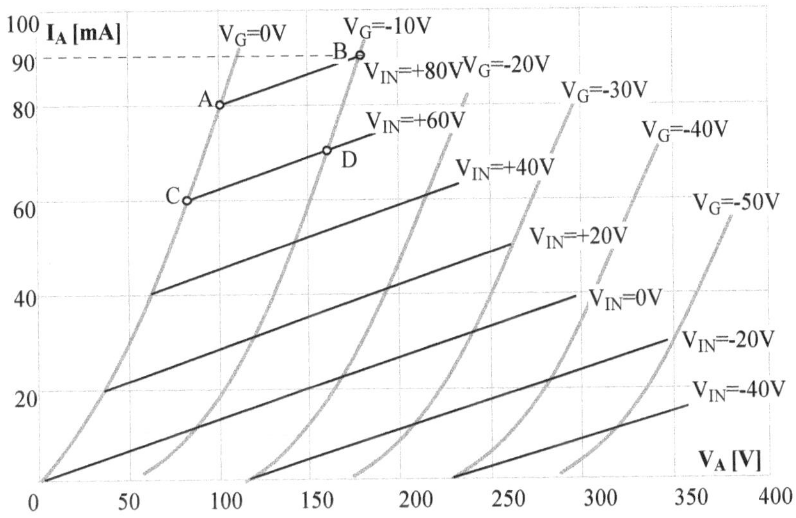

Anode characteristics of a power triode, original and after the application of negative current feedback via un-bypassed 1kΩ cathode resistor: higher internal resistance, lower sensitivity and lower amplification factor but more linear curves (lower harmonic distortion)

The equation for the input circuit of a triode with un-bypassed cathode resistor is $V_{IN}=V_{GK}+I_AR_K$ For the tube illustrated and R_K=1kΩ, we can start with point A, for which V_{GK}=0V, V_A=100V and I_A=80mA, thus V_{IN}=0+1,000*0.08 = 0+80V=80V

To position a second point B on that curve, lets choose V_{GK}=-10V and since V_{IN}=80V, V_{IN}= -10+1,000*I_A=80V, so I_A=90/1,000 = 90mA

Now we have two points and can draw a line through them. These new anode curves are not perfectly straight lines, but in the first approximation can be considered as such, depending on the strength of the NFB.

The same approach and calculations apply to the line CD (for V_{IN}=+60V), and all the others. Notice that the new characteristics have the same slope.

Series-applied feedback increases input impedance

Since this topology is the most common in tube audio, we will study the effect of NFB on the input resistance using an amplifier with a series-applied voltage feedback. The results will make it clear why is this usually the most desirable arrangement.

The input resistance is the ratio of the input voltage V_{IN} and input current I_{IN}: $R_{IN}=V_{IN}/I_{IN}$. We have just seen that NFB reduces the input voltage V_{IN} by the feedback factor 1+βA, which means that the input current I_{IN} which flows through R_{IN} is also reduced by the same factor and that the input resistance which the signal generator "sees" is increased by the same factor: $R_{INF}=(1+\beta A)R_{IN}$!

Higher input resistance (or impedance in general) would not be desirable in low impedance solid state circuits with bipolar transistors, but is beneficial in high impedance tube circuits, so that the next stage does not load the previous voltage amplification stage!

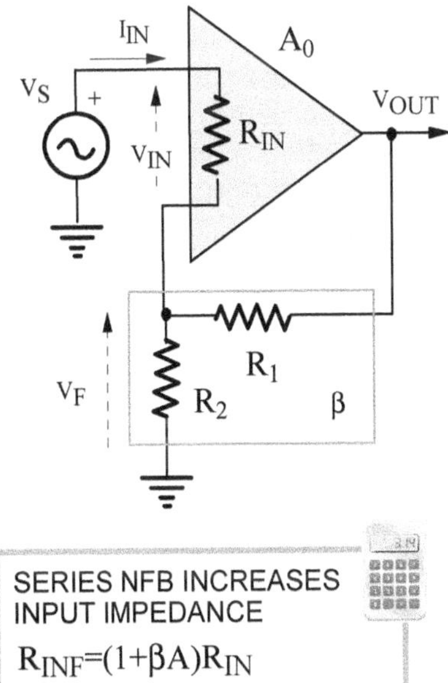

SERIES NFB INCREASES INPUT IMPEDANCE
$R_{INF}=(1+\beta A)R_{IN}$

Voltage feedback reduces output impedance

There are two way to determine the output resistance or, in general, the modulus of its output impedance. In the first step, with input test signal V_S and *without NFB*, we measure the output voltage without any load connected (unloaded) and then with the load R_L. In the second step we repeat the same two measurements but this time *with NFB* applied. We get: $V_{UL}= A_FV_SR_L/(R_{OUTF}+R_L)$ and $V_L= AV_SR_L/(R_{OUT}+R_L)$

A_FV_S and AV_S are output voltages in unloaded cases (with and without NFB). Likewise, R_{OUTF} and R_{OUT} are output resistances with and without feedback.

The output voltage in the unloaded case is higher than when R_L is connected. With NFB the connection of the load reduces the output voltage, but it also reduces the voltage fed back (βV_L) thus increasing the input voltage $V_{IN}=Vs-\beta V_L$.

This means that with NFB the output voltage is less dependent on the load resistance, meaning the voltage drop on the output resistance of the amplifier is lower, and lower voltage drop means that the output resistance with NFB is lower.

With NFB the unloaded output voltage is

$V_{UL}=-AV_{IN}R_L/(R_{OUT}+R_L)$ and if we substitute $Vs-\beta V_L$ for V_{IN} and rearrange, we get

$$V_{UL}=\frac{-A}{1+\beta A}\,\frac{R_L}{R_L+\mathbf{\dfrac{R_{OUT}}{1+\beta A}}}$$

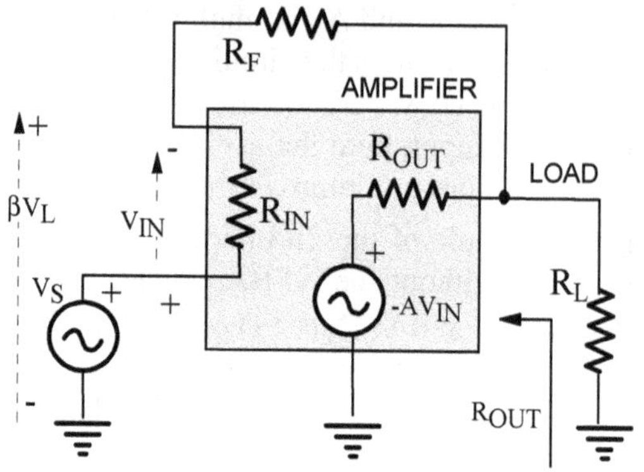

Determining output resistance with NFB applied R_{OUTF} - method #1

The factor in bold is the new, lower output resistance with NFB, $R_{OUTF}=R_{OUT}/(1+\beta A)$!

Using the second method we remove the signal source from the amplifier's input (short-circuit the input terminals), connect a test generator to the output terminals and measure the test voltage V_T. The only voltage at the amp's input is the feedback voltage $V_{IN}=-\beta V_T$ so the electromotive force in the output circuit is $-\beta AV_T$. The test current is $I_T=(V_T-\beta AV_T)/R_{OUT}$ so $R_{OUTF}=V_T/I_T = R_{OUT}/(1+\beta A)$, the same result as that obtained by the first method.

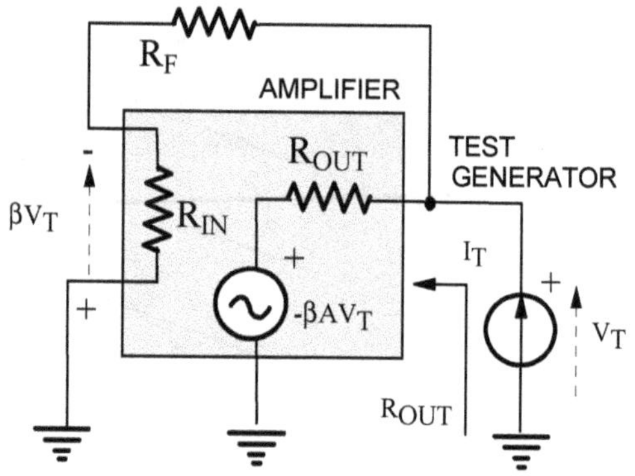

Determining output resistance with NFB applied R_{OUTF} - method #2

NFB provides phase correction and reduces phase distortion

The impact of NFB on the variations in phase is best illustrated with a phasor diagram. In a real amplifier without NFB, at a certain frequency, the phase shift between the input signal V_{IN} and the output signal V_{OUT} is the angle marked Θ, which is less than ideal 180°. We are assuming here an inverting amplification stage or the overall amplifier.

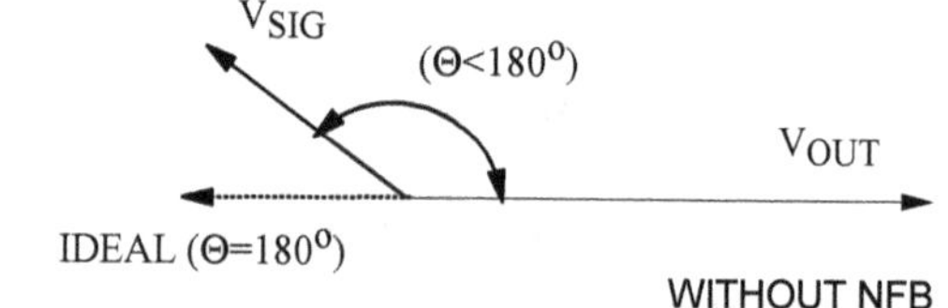

Using the output voltage V_{OUT} as a referent phase angle, once the NFB is applied, the feedback voltage V_F is in phase with the output voltage and the length of its phasor is determined by the feedback factor β. This feedback voltage brought to the input combines with the actual signal input V_{SIG} into a new net input V_{IN}. The angle between the input and output voltage has been increased, it is now much closer to the ideal case, which is 180°.

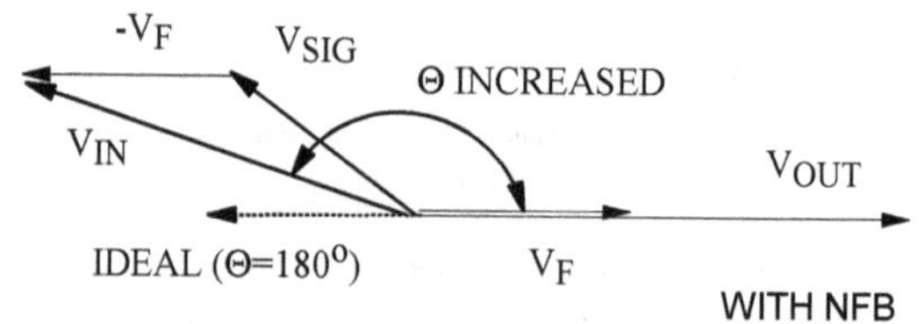

NFB provides phase correction and reduces phase distortion

Negative feedback stabilizes the amplification factor

Since $A_F= A/(1+\beta A)$, if we could make A very high, then $A_F\approx 1/\beta$! In other words, the amplification of such amplifier would be determined only by the passive feedback network (two resistors), and not at all dependent on the amplification factor and the components used (tubes, transistors, capacitors, resistors, etc.) Even with finite gain we can see another benefit of NFB. If we differentiate the equation $A_F= A/(1+\beta A)$ with A_F as independent variable X (dX/X), we get $dA_F/A_F= dA/[A(1+\beta A)]$

This means that relative changes of amplification with NFB are $(1+\beta A)$ times smaller than changes without feedback! The feedback factor $(1+\beta A)$ stabilized the gain, which is now much less dependent on factors such as aging of tube and other components (changes in their values), changes due to temperature drift of component parameters and changes in operating points and load lines. NFB reduced all those unwanted effects due to temporal, temperature-related and manufacturing tolerances-related causes.

In 1950s and 1960s, manufacturers, of course, loved the idea. Products did not require manual tuning and adjustment, needed less servicing, and stayed within specifications longer. The uniformity between batches of amplifiers improved greatly! As gain was getting cheaper (pentodes, transistors, integrated circuits), NFB was getting stronger. As is always the case in life in general and in engineering in particular, NFB also has some drawbacks. It reduces gain and output power, and if there is too much phase shift within the feedback loop, negative feedback can turn into a positive feedback at frequency extremes and cause instability and oscillation, effectively turning an amplifier into an oscillator. Many audiophiles claim that too much NFB has a detrimental effect on the sound quality. Some claim that even low levels of NFB adversely affect the sound!

The undesirable gain increase at frequency extremes

The second issue mentioned above is very important and will be covered in more detail very soon, when we discuss the (in)stability aspects of NFB. For now, let's illustrate it with an illuminating comparison of amplifiers with various amounts of NFB applied. The trend is unmistakable - the stronger the NFB, the higher the peaks at frequency extremes. These peaks in amplification factor mean that negative feedback at midrange is turning into a positive feedback at high and low frequencies due to the phase shift in those regions.

The positive feedback increases gain, but also increases potential instability and threatens to turn a stable amplifier into an unstable oscillator.

While this change from negative into positive feedback is generally undesirable, there are some designs that use positive feedback to compensate for inadequacies in other aspects.

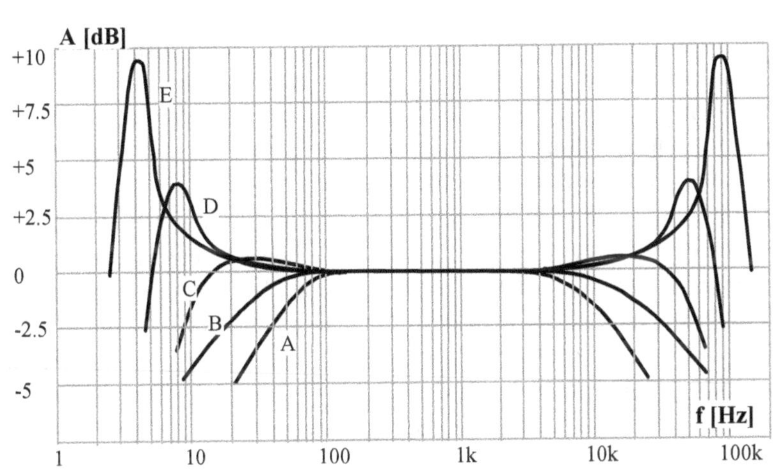

Relative gain (referenced to midrange gain) versus frequency of a triode amplifier A) without any feedback B) with mild negative feedback C) with moderate NFB D) with strong NFB E) with excessive NFB

NFB and other types of distortion

When the use of NFB is debated, reduced harmonic distortion is usually one of the factors considered. However, it is possible that in some case negative feedback may reduce the benign sounding 2nd and higher even harmonics, but increase the discordant and much more unpleasant IM (intermodulation) distortion.

In their promotional article, Electra-Print (a transformer-making company in USA) claims "Driver and voltage amp tubes with un-bypassed self bias resistors used for decreasing THD and/or used for local feedback for lowering gain, will increase IMD due to the many different motions at their cathodes."

The awkward sentence construction and vague terms (what is meant by "different motions"?) leave us none wiser. We wanted to release this book as soon as possible (it is already 5-10 years late!), but perhaps in one of the future revisions we could look into this issue deeper and evaluate these claims by conducting our own experiments. Our listening tests indicate that un-bypassed cathode resistors do seem to negatively impact sound quality.

PRACTICAL NEGATIVE FEEDBACK TECHNIQUES

Approaches to amplifier design from the NFB perspective

It is interesting to note the historical link between developments in tube technology and the approach to amplifier design. Of course, audiophile preferences are the intervening variable. In the 1930s, low powered directly heated triodes such as AD1 in Europe, 45, 2A3 and 300B in USA, ruled the field. Coupled to efficient speakers, realistic sound levels were achieved. Negative feedback or "degeneration" was not used.

As pentodes and beam tetrodes were developed and push-pull amplifiers became dominant, negative feedback also increased in popularity. Unfortunately, in terms of sensitivity, speaker design went backwards, especially in USA. Low efficiency loudspeakers were easier to design and produce, so the increased power levels gained by the use of pentodes and push-pull circuits were gobbled up by power-hungry speakers. The advancement (or should I say the retardation) was "cleaner" but more irritating and fatiguing sound.

Luckily, the trend reversed again in the early nineties of the last century, when the tube revival started. A high proportion of audiophiles (perhaps not the most, since many were limited by their low efficiency speakers, which still dominated the field at the time) went back to low powered single-ended triode amplification.

Four principal approaches to amplifier design:

a) No NFB of any kind

b) Stages linearized as much as possible by various means, very weak global NFB

c) Stages linearized as much as possible by various means, including mild local NFB (stage gains sacrificed), but no global NFB

d) Stage gains maximized, high open loop gain but also high distortion, both reduced by the strong global NFB

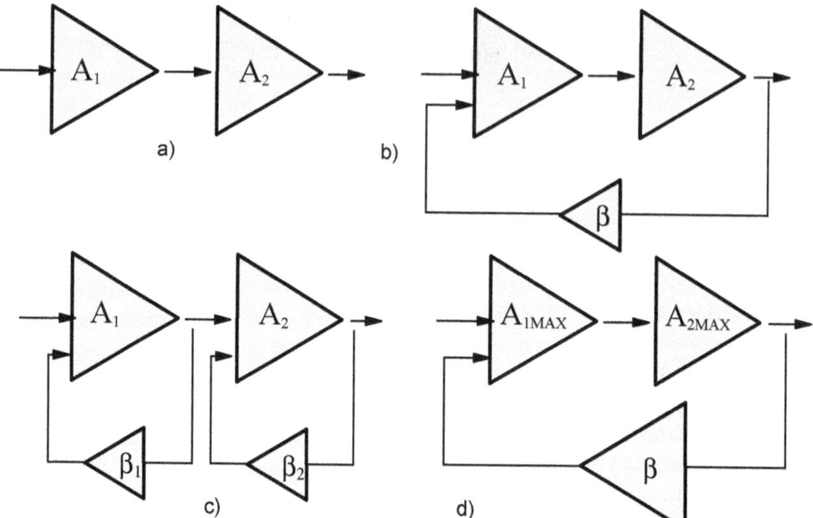

The dominant design approach of the 1950s and 60s, illustrated in d) above, was to maximize the gain of every stage, which would also raise distortion levels "through the roof" and then to apply copious amounts of NFB to bring distortion to an acceptable level. This discredited approach is far less popular today, but still persists in higher powered tube push-pull amplifiers produced by manufacturers without enough knowledge or motivation to design their own circuits, those who copy circuits from the 60s but claim to be at the forefront of tube technology.

On the other side of the barricades is the "no NFB at all" brigade, illustrated in a). They abhor pentodes and use triodes exclusively, mostly of the directly-heated kind. These amplifiers have high 2nd harmonic distortion, low damping factors and limited frequency range, but sound sweet and seductive. They have to be carefully paired with suitable speakers, since their performance depends heavily on the speaker's impedance curve.

The two "middle-of-the road" design approaches are illustrated in b) and c). Each stage is designed to be as linear as possible, including a mild local or global negative feedback. "Mild" means 2-5 dB, compared to strong global NFB levels of 10 - 15dB! Of course, the price needs to be paid, in this case as lower amplification factors of each stage, and increased circuit complexity.

Simultaneous use of positive and negative feedback

Some tube designs compensate the gain reduction of NFB by careful and judicial use of positive feedback (PFB).

The RIAA circuit illustrated below was used in many vintage integrated amplifiers and preamplifiers. It needs a much higher gain at lower frequencies than its two-stage circuit can provide, so apart from the NFB with the standard RIAA filtering (1) it also uses positive feedback (2). Notice the 68k feedback resistor between pins 3 and 8. Such positive feedback boosts the amplification of lower frequencies.

The positive feedback may solve one problem, but it creates at least three bigger ones, increasing distortion and reducing stability in the bass region. It also makes the accuracy of the RIAA amplification highly dependent on the parameters of individual tubes.

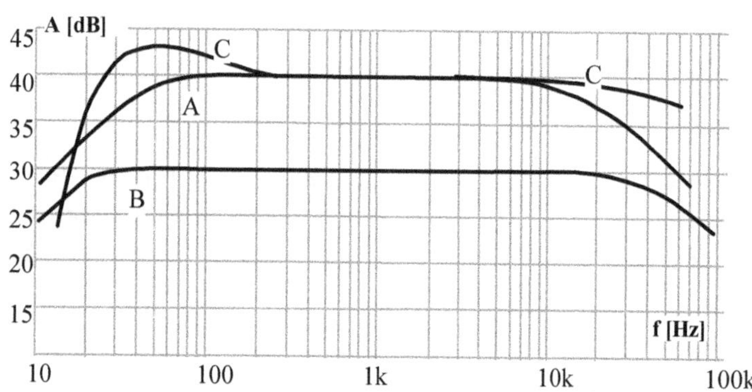

ABOVE: The frequency range of a triode amplifier A) without feedback B) with negative feedback C) with both positive and negative feedback

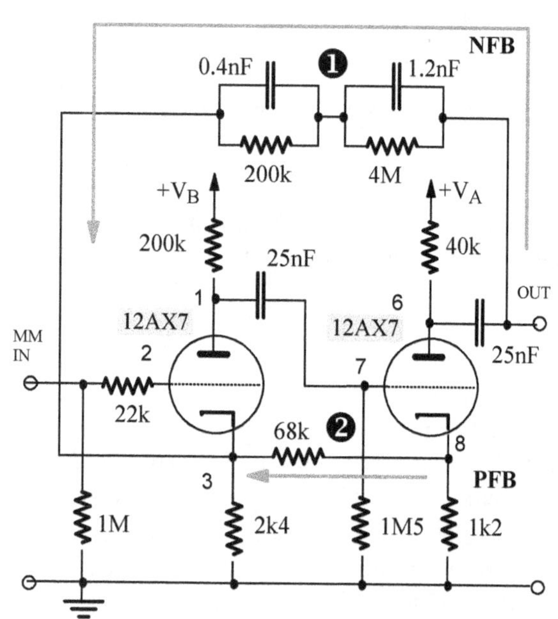

RIGHT: Phono preamplifier with positive feedback bass boost

To find the output impedance of an amplifier using both voltage and current feedback, we apply a signal E to its unterminated output (no load), with no external signal at its input. If we measure or calculate the current I, the output impedance is then $Z_{OUT} = E/I$. Even without any external signal at its input (the test signal E is now applied to its output), the feedback voltages βE and $-\alpha I$ will appear at the input.

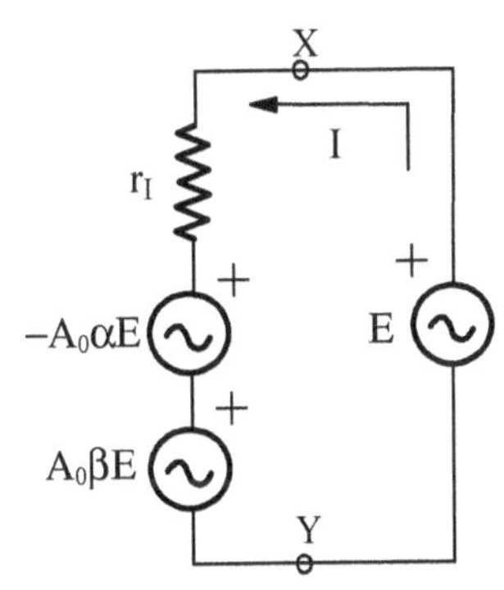

The equivalent output circuit of an amplifier using both voltage and current feedback, with a test signal E applied to the output.

The voltage caused by current feedback has a negative sign since it flows in the opposite direction from the amplified current. A_0 is the voltage gain of the whole amplifier, from the control grid of the input tube to the output terminals X and Y, with unloaded output (open circuit).

We have $I = [E-A_0(\beta E-\alpha I)]/r_I$, which, after some rearranging, gives us $Z_{OUT} = E/I = (r_I-\alpha A_0)/(1-\beta A_0)$, where α and β are negative for negative feedback and positive for positive feedback.

If only voltage feedback is used (α=0) $Z_{OUT} = (r_I)/(1-\beta A_0)$. Since β is negative for negative voltage feedback, that means that negative voltage feedback lowers the output impedance (the denominator is then always larger than one) and that positive voltage feedback increases it ($1-\beta A_0$ is then smaller than 1).

If only current feedback is used (β=0) $Z_{OUT} = (r_I-\alpha A_0)$. Since α is negative, the negative current feedback increases the output impedance and positive current feedback lowers it, the opposite of the voltage feedback!

Damping factor control

Damping factor control is one of the features of vintage amplifiers that fell out of favor even before it became widely accepted. A few manufacturers tried it on some of their models and then even they abandoned it. Just as loudness control, it seems to be a great and useful idea on paper, but somehow it does not deliver in practice.

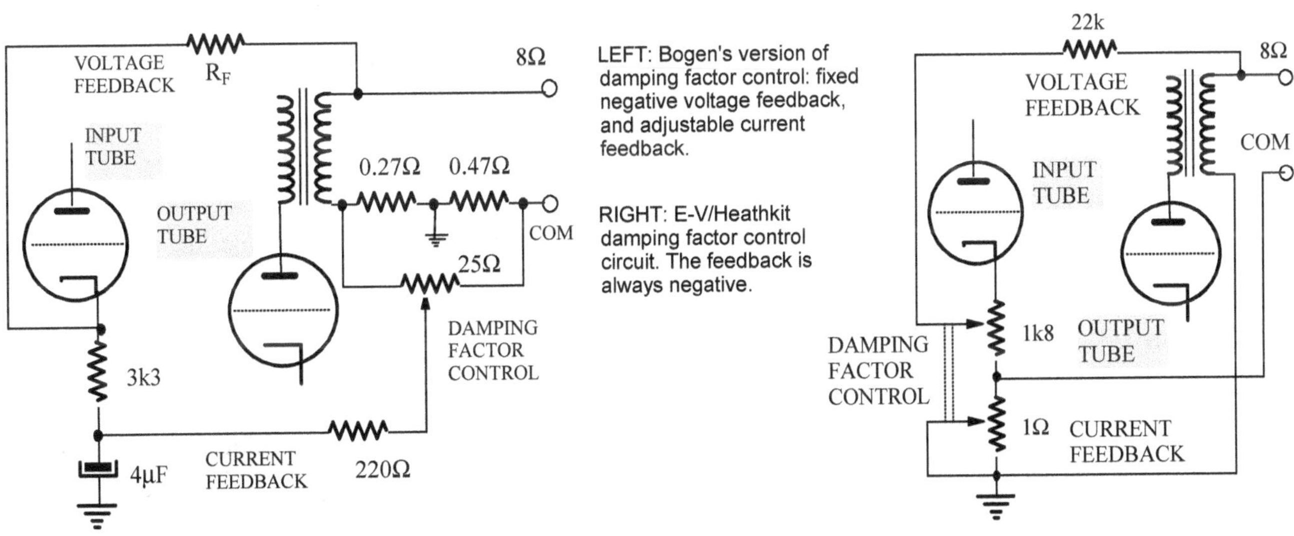

LEFT: Bogen's version of damping factor control: fixed negative voltage feedback, and adjustable current feedback.

RIGHT: E-V/Heathkit damping factor control circuit. The feedback is always negative.

There are two principal approaches to the issue. Both use a combination of voltage and current feedback. With Bogen's arrangement, the amount of negative voltage feedback remains constant, but the current feedback can be either positive or negative. At some point of the potentiometer's rotation the positive and negative feedback cancel each other and no current feedback is applied.

When the slider moves towards the transformer side of the ground, (left on the diagram) the feedback becomes positive. Towards the speaker's side of the ground (right), the feedback is negative. Notice the low-pass filter (220Ω resistor and 4μF capacitor), which makes current feedback operational only at low frequencies (in this case below approx. 600 Hz.)

Electro-Voice and Heathkit used only negative feedback, in a much smaller range, resulting in damping factors of roughly 0.125 to 10. By using carefully matched dual potentiometers, the amount of negative voltage feedback is reduced by the same amount as the current feedback is increased, making total NFB constant.

When the slider or wiper of the DF dual control potentiometer is moved upwards (towards the cathode side of the 1k8 and 1Ω pots, the voltage feedback is increased and the current feedback is weakened, since it the bottom resistance is reduced towards zero. This results in lowered output impedance and increased damping factor DF.

The opposite happens by moving the slider downwards (on the drawing). The negative voltage feedback is reduced, the current feedback is increased, the output impedance rises and the DF is reduced. Note that the two sliders are mechanically locked together, as on most dual potentiometers.

Making NFB variable

The jury is still out on the issue of negative feedback. Audio purists of the single-ended triode persuasion usually say the best feedback is no feedback. I wish things were that simple. Some amps will sound better without any feedback, but most will sound better with some mild feedback. As for the vintage amplifiers, by modern standards, most have way too much feedback.

Since the sound with and without feedback depends on so many factors (how well the amp was designed, how good are the output transformers and so on), a prudent approach is to make NFB variable and then experiment with your amp and your speakers. There are two ways to do this.

The first option involves the use of a linear potentiometer of 10 or 22 kΩ. Wire it in series with a smaller feedback resistor of say 2.2-8.2kΩ. When the slider is in the far left position, the feedback is at its maximum. At far right, the feedback is at its minimum (maximum feedback resistance).

Option B involves the use of a selector switch with 3 or more positions. Just as with the potentiometer, for a mono amp, a single switch is fine, but with a stereo amp you have to change both channels simultaneously, so a dual-ganged pot or a double-pole switch is needed. In our example there are 3 positions: "NFB off", "Low NFB" (large resistance) and "High NFB" (low resistance).

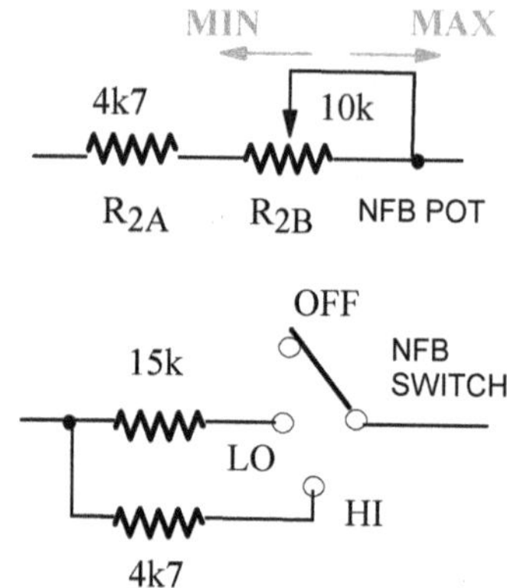

Two ways to make NFB adjustable, a potentiometer provides a continuous adjustment, while switched resistors provide discrete levels of feedback.

Repositioning NFB from one output tap to another

Most vintage amps have global negative feedback taken from the 16 ohm tap of the output transformer, yet most modern speakers are of nominal 8 or 4 ohm impedance. Ideally, the NFB should be taken from the tap to which the speaker is connected to. Here are two formulas to help you calculate the value of NFB resistor and its bypass capacitor (if used) when you reposition NFB point from the 16 ohm tap to another tap:

$$R_{NEW}= R_{16}/\sqrt{(16/R_{NEW})} \quad \text{and} \quad C_{NEW}= C_{16}\sqrt{(16/R_{NEW})}$$

EXAMPLE: Dynaco ST70 has a 1kΩ NFB resistor taken from the 16Ω tap. To move it to the 8Ω tap, the new resistor's value should be $R_8 = R_{16}/\sqrt{(16/8)} = R_{16}/\sqrt{(2)} = R_{16}/1.41 = 707$ ohms, so we would use a standard value of 720 ohms. To move NFB to the 4Ω tap, R_4 needs to be exactly half of the 16Ω tap resistor or 500 ohms.

If there is a phase lead compensation capacitor in parallel with the feedback resistor, the value of the repositioned capacitor should be $C_8 = C_{16}\sqrt{(16/8)} = C_{16}\sqrt{2} = 1.41C_{16}$

REDUCING UNWANTED POSITIVE & NEGATIVE FEEDBACK

Decoupling between amplification stages

Audio signal from all stages passes through the power supply, so the DC anode supply voltage for all stages is "modulated" by the signal AC voltage, which is superimposed on top of it. This unwanted signal appears on the anode resistors or directly on the anodes of cathode followers. Depending on the amplifier design, this unwanted feedback signal can be in phase (positive feedback) or out-of-phase with the amplified ("wanted") signa (negative feedback), as indicated by the pulse waveforms indicated on the block diagram in A, B and C.

The block diagram illustrates a three-stage amplifier and its power supply with a finite internal impedance Z_{OUT}. The highest anode voltage is usually in point A, the output stage supply, and then it gets progressively smaller for the driver stage (point B) and the preamp stage (point A). This is achieved by voltage dropping resistors or chokes.

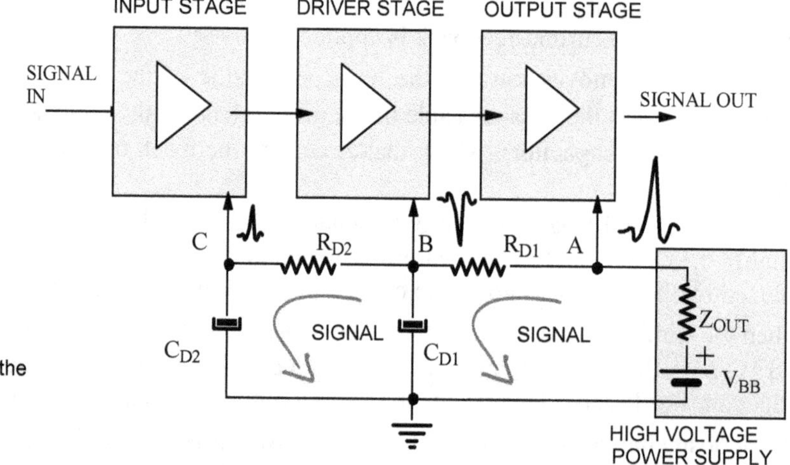

Signal coupling through the common power supply and the operating principle of decoupling filters

If out-of-phase, the power-supply-propagated signal acts as a negative feedback in that particular stage, which may be undesirable but is not detrimental to the operation of the amplifier. However, if the two signals are in-phase, positive feedback occurs and the amplifier may start oscillating. "Motorboating" is a popular name for low frequency oscillations (a few Hertz), since the sound the amplifier makes is similar to the sound of the outboard boat engine (put-put-put).

High frequency oscillations above 20 kHz are even more troublesome. Since they cannot be heard, they may go undetected by the inexperienced amplifier builder. Only careful examination with an oscilloscope will reveal the presence of a HF signal at the output.

In one amplifier we noticed that the output transformers were getting warm. Unless a very thin wire was used for the windings and unless the amp was played at very high volume levels for hours, this was not normal. After further investigation it was discovered that high frequency (ultrasonic) oscillations were present, increasing thermal losses in transformers. After the HF instability was fixed, the output transformers were cool as cucumbers.

A decoupling circuit is an RC filter which acts as a voltage divider and reduces the feedback due to the common power supply impedance in the ratio of V_{OUT}/V_{IN}. It also reduces the ripple voltage (AC component superimposed on the DC voltage bus).

This voltage divider effect is only effective if the capacitive reactance X_C is small in comparison with R_D at low frequencies. A typical frequency at which such effectiveness is evaluated is the lower -3dB frequency f_L.

Of course, the DC anode current(s) of the previous stage(s) must pass through R_D so the resistance of R_D cannot be too high, otherwise too much of the supply voltage V+ would be lost on such a series resistor.

Circuits are usually drawn from left (input) to right (output), but because decoupling filters are drawn in the opposite direction, their inputs are on the right and outputs on the left side (the power flow is right-to-left).

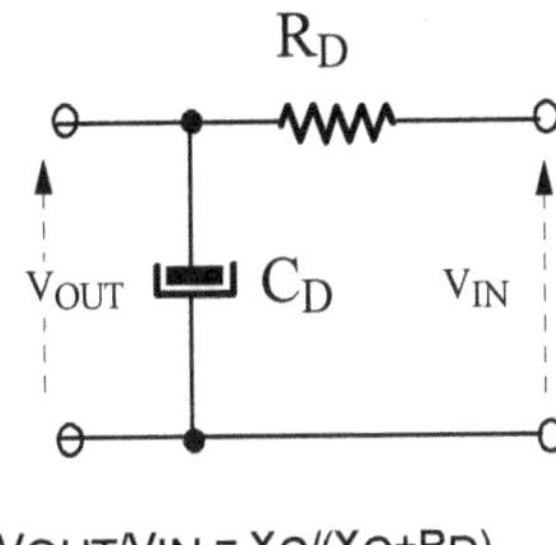

$V_{OUT}/V_{IN} = X_C/(X_C+R_D)$

HOW EFFECTIVE ARE DECOUPLING CIRCUITS AT REDUCING RIPPLE?

CALCULATION

Determine ripple voltage V_{ROUT} at the output of the typical decoupling circuit, if the ripple voltage at its input is $V_{RIN}=1V_{PP}$. Assume 50Hz mains frequency.

Since $X_C=1/(\omega C)$, at 50Hz mains frequency the ripple frequency is double that or 100 Hz, so we have $X_C=1/(2*\pi*100*22*10^{-6}) = 72\Omega$

The attenuation of ripple is $A=V_{OUT}/V_{IN}=X_C/(X_C+R_D) = 72/(72+10{,}000) = 0.00715$

Or, in dB: A= 20log0.00715 = 20*-2.146 = 42.9dB

Finally, $V_{ROUT}= V_{RIN}A = 1*0.00715 = 7.15\ mV_{PP}$

The peak-to-peak value of the ripple will be reduced from 1 Volt to 7.15 mV!

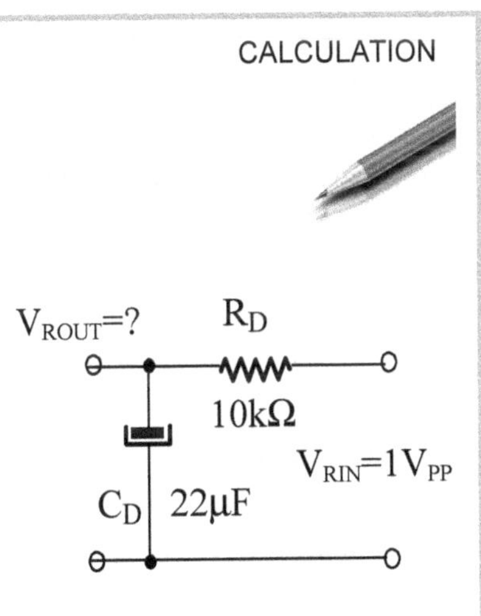

The voltage divider and the bleeder resistor

This partial diagram shows the decoupling circuits from one of our 300B amplifiers. The 340V supply goes to the output stage and then supplies the second voltage amplification stage, illustrated here, which draws 2.8mA of current, and the input SRPP stage which draws 1mA. The DC voltages in important points are marked.

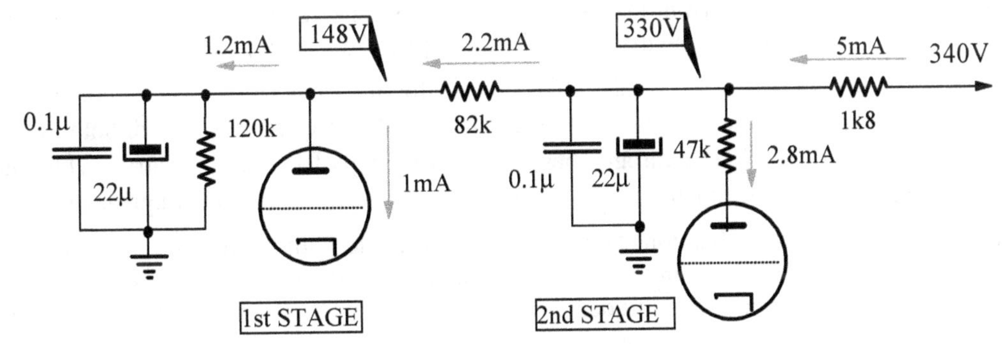

A typical dual-decoupling circuit with a bleeder resistor

The 120k resistor is called a bleeder. Its function is twofold. When the amplifier is switched off, good quality filtering capacitors in the power supply (not shown) and the decoupling capacitors shown in this diagram (22µF) stay charged for hours and even days and present a high voltage hazard during servicing. The bleeder resistor provides a discharge path for all power supply capacitors, since they are all interconnected through decoupling resistors (82k and 1k8).

The bleeder resistor also serves as a part of the voltage divider circuit formed by the three resistors shown. It acts as a simple shunt voltage regulator and stabilizes the DC voltages in the important points (148V and 330V).

Active decoupling circuits

The effectiveness of the passive RC decoupling filter $V_{OUT}/V_{IN} = X_C/(X_C+R_D)$ relies on very low X_C, meaning a high value of shunt capacitor's capacitance, and a high value of the series resistor R_D.

In this active decoupling filter ("active" because a an active element such as bipolar transistor is used) the transistor works in the common-collector (emitter follower) configuration, since the "load" comprising of the anode resistor R_A and the internal tube resistance r_I are in the collector's circuit.

V_S is the unwanted audio signal or the AC ripple riding on top of the DC voltage. The unwanted signal feedback voltage V_F is essentially equal to the signal voltage V_B on the base, which is the voltage across the decoupling capacitor C_D.

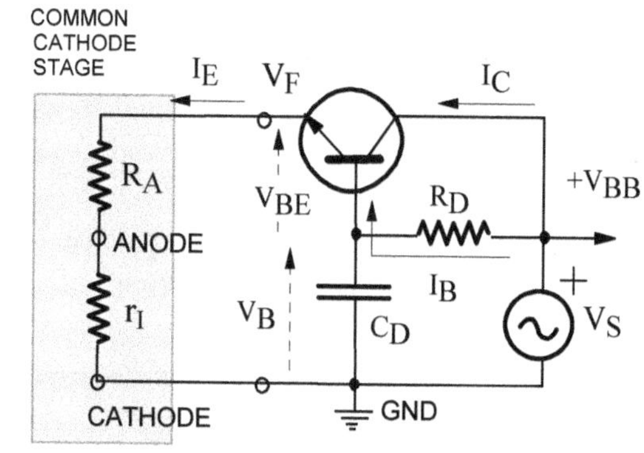

Active decoupling circuit with a bipolar transistor

The first advantage of this circuit is that series resistor R_D can now be much larger than before without causing a significant DC voltage drop. This is because now only a small base current I_B flows through it, not the much larger collector current I_C! The second advantage is that now capacitance C_D can be much smaller. The advantage is not financial, since 22-47µF 450V elcos are very cheap these days, but results in sonic benefits, because now this 1-10µF capacitor can be a good quality film&foil or PIO type, not an electrolytic.

DESIGNING AN ACTIVE DECOUPLING CIRCUIT

CALCULATION

The load current (also emitter current I_E in the model above) to an input stage of a tube amplifier is 4 mA. The transistor used has current amplification factor β_0=100. If the maximum allowable voltage drop across the series transistor is 10V and attenuation of 100Hz ripple signal of at least 100 (-40dB) is required, determine the values for R_D and C_D.

Since I_E=4mA and $I_E= \beta_0 IB$, $IB=0.004/100 = 40\mu A$! Assuming a base-emitter voltage drop V_{BE} in the "on" state as 0.6V, the voltage across R_D is the maximum voltage drop minus V_{BE} or 10-0.6 = 9.4V. Thus, R_D must be R_D=9.4V/40mA= 235 kΩ.

$X_C=1/(\omega C)$, at 50Hz mains frequency the ripple frequency is double that or 2100 Hz, so we have $X_C=1/(2*\pi*100*22*10^{-6}) = 72\Omega$

The attenuation of ripple is $A=V_{OUT}/V_{IN} = X_C/(X_C+R_D) = 1/100$ so $100X_C=X_C+R_D$, so $99X_C=R_D$ or $X_C=R_D/99$, and finally we have $C=99/(R_D*2*\pi*100) = 0.67\mu F$

A 10µF capacitor would increase the attenuation to more than 0.001 or -60dB!

THE STABILITY ASPECT OF NEGATIVE FEEDBACK

Nyquist loops

The transient behavior of an amplifier is determined by the poles of its transfer function. The criterion for stability is that all poles must lie in the negative half of the complex frequency plane. The same applies to zeroes, too.

The Nyquist criterion is based on the polar plot of the open loop gain Aβ, which, as a complex number, may be represented as point in the complex frequency plan. As the frequency is varied, both the real (σ) and the imaginary (jω) component of the loop gain are changing, and their plot is a closed curve. The curve starts at the (0,0) for frequency f=0 (DC), and finishes in the same point for the infinitely high frequency (f = ∝).

For an RC-coupled amplification stage the curve is a circle with a diameter $A_0\beta$, the open loop gain at midrange. Since the gain and the phase are constant throughout the midrange band of frequencies, the whole band is in one point f_M. Only with increasing and decreasing frequency, when the phase starts changing, is the curve continued. Notice the -45° and +45° angles between the vector Aβ and the horizontal axis at f_U and f_L!

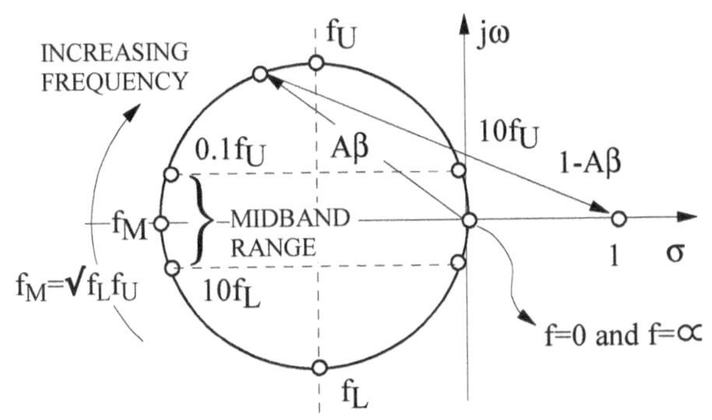

The locus of the (1-Aβ) vector for an RC-coupled stage is a circle, which is unconditionally stable.

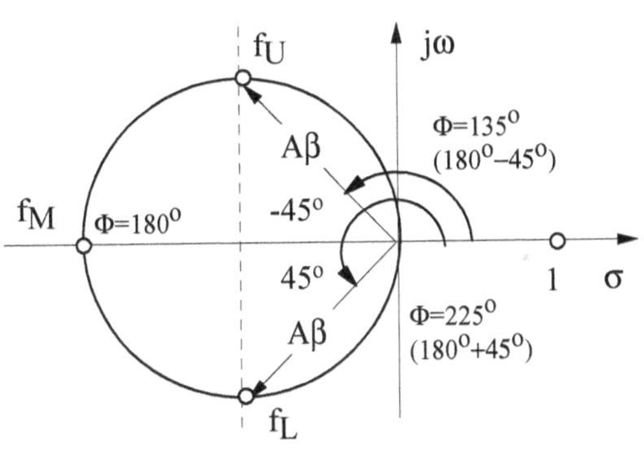

The location of significant frequencies and their phase angles on a Nyquist loop

Nyquist loops for transformer-coupled amplifiers

Since the gain of such amplifiers is not constant (the amplitude characteristic has one or more high frequency peaks), the polar plot of transformer-coupled amplifiers is not a circle any more, but resembles a warped, kidney-shaped curve. The decrease in amplitude at very low frequencies is due to the saturation of the magnetic core, while the increase in amplitude (gain) at high frequencies is, as we have seen, a result of the resonant effects. Due to the transformer's phase shift at high frequencies, the polar plot crosses over to the positive half of the complex plane, which, depending on the loop gain, may or may not result in oscillations and instability.

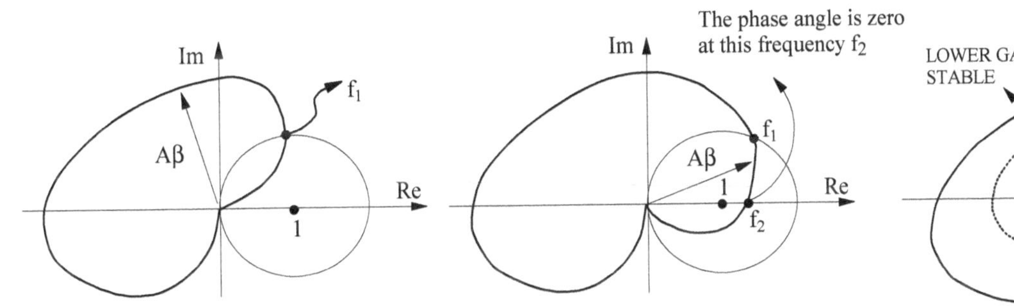

Negative feedback becomes positive at frequency f_1, but there are no oscillations since point (1.0) is not encircled

This amplifier's Nyquist loop encloses the critical point (1,0) or 1+0j and oscillation will result

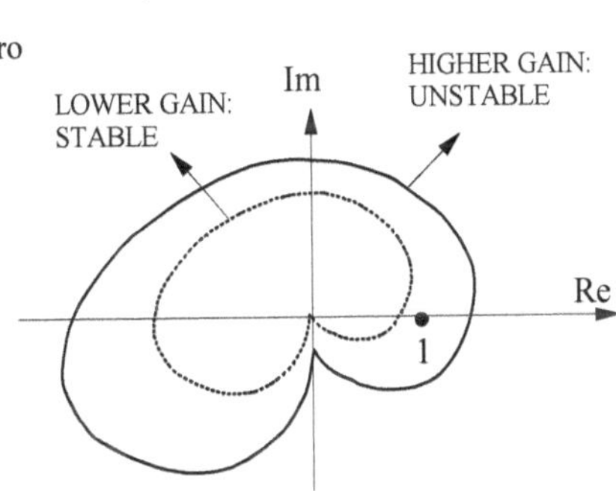

A conditionally stable amplifier where increasing gain Aβ causes instability

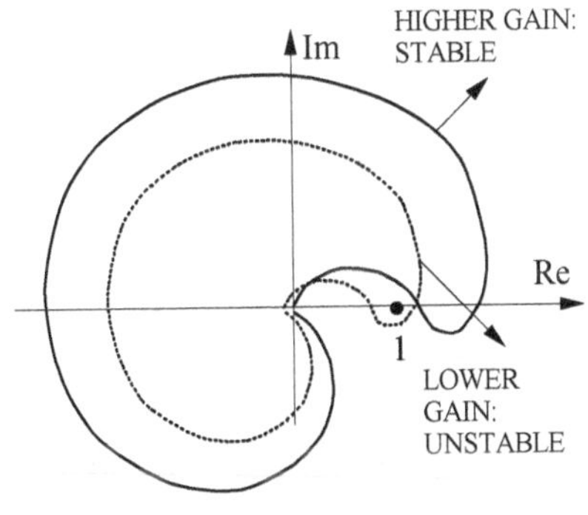

A conditionally stable amplifier where decreasing gain Aβ causes instability

Of particular interest is the loop gain Aβ's trajectory on the positive side of the real axis. When, at some frequency f_1, the locus of loop gain Aβ crosses the locus of the (1-Aβ) vector, the circle with radius r=1, centered in the point 1+0j, the negative feedback turns into a positive one. That is a necessary condition for oscillations to occur, but it is not a sufficient factor. Nyquist established that sustained oscillations will only occur if polar curve encircles the point 1+0j.

It is possible that an amplifier is stable for some magnitudes of open loop gain but unstable for others. Usually increasing the open loop gain causes instability, but there are some cases where an amplifier may be stable for higher gains, but reducing gain "shrinks" the loop so it starts encircling point (1) and thus starts the instability!

Phase- and Gain-margins

So that designers can get a better idea of how close an amplifier is to instability, two numerical factors have been devised. One deals with the phase angle "left over" after the gain Aβ drops to 1 (or 0dB). That is point B on the illustration, where the loop crosses the unity gain circle! This angle θ is called the phase margin. The larger the phase margin the more the open loop gain Aβ can be increased until the system becomes unstable.

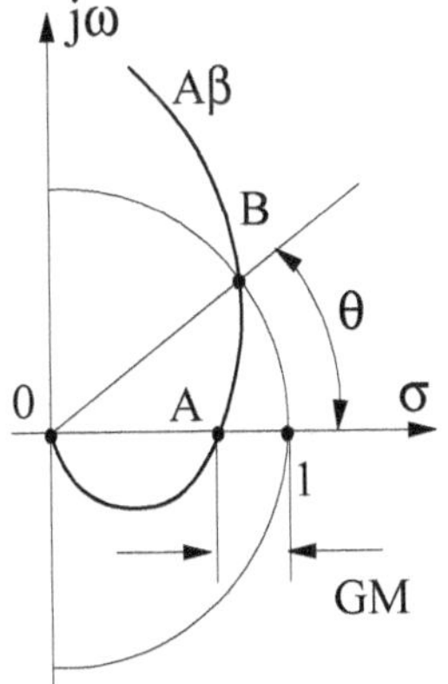

The phase margin θ and gain margin GM on a Nyquist diagram

Since two factors, amplitude and phase of the gain are involved, there is another stability factor, called the gain margin. It looks at the open loop gain of a stable amplifier when the phase margin is reduced to zero, meaning when the polar plot crosses the horizontal axis (point A). A stable amplifier will have a gain βA of less than 1, so the polar plot does not encircle point (1,0). The gain margin tells us how far point A is from the critical point (1,0).

Stability evaluation using Bode plots

Although the Nyquist criterion looks cool and is relatively simple to comprehend, it isn't easy to apply in practice. It can be expressed as a rule to be remembered, though, that oscillations cannot occur if the magnitude of open loop gain Aβ is below unity when the phase angle is zero.

Bode diagrams, on the other hand, make direct reading of the phase- and gain-margin possible. Again, similar deliberations apply to LF and HF situation, so let's just look at a high frequency response of an amplifier.

To calculate the Gain Margin (GM) we start at the phase shift of 180° and move horizontally until we hit the phase curve (point X). Then we read the gain at that frequency f_X, which in this case is around -17dB. That is the Gain Margin.

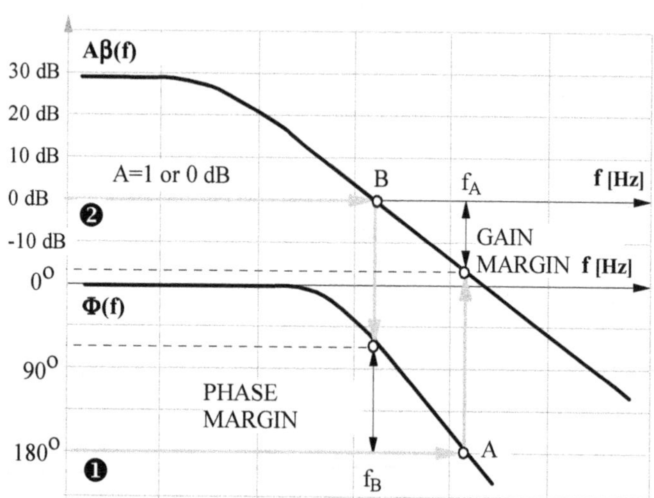

The phase- and gain-margins can be read directly from Bode's diagrams

It means we can increase the loop gain Aβ by up to 17 dB and the system will remain stable. A negative GM means a stable system. A positive loop gain at that point would mean the gain is larger than 1 when the phase shift is 180°, and the feedback would be positive and cause oscillation!

The GM readout started with the phase characteristics, so the Phase Margin (PM) readout will start with the gain curve at 0dB (A=1). We go across until we hit the gain curve (point B). Then we read the phase shift at that frequency f_B, which in this case is around 67.5°. The Phase Margin is the number of degrees left until the phase shift reaches 180°, when oscillations would start. In this case PM=180-67.5 = 112.5°. In this example both our margins are large, meaning this amplifier is far away from oscillation!

Points A and B on this graph have the same meaning as A and B on the Nyquist diagram on the previous page.

FREQUENCY COMPENSATION METHODS

The frequency compensation principle applies to attenuators, oscilloscope probes and many other systems where a uniform response over a wide range of frequencies is required, which includes tube amplifiers as well.

Compensated oscilloscope probes

Perhaps the most common or best known application of the frequency compensation principle is in the test & measurement field, and that is the compensation of test probes for oscilloscopes and vacuum tube voltmeters.

The capacitance of the coaxial cable used to connect the tip of the probe to the oscilloscope and the input capacitance of the oscilloscope itself, together with the signal source impedance R_S, form a low-pass filter. At higher frequencies the reactance of the two capacitances in parallel becomes smaller and smaller and thus high frequencies are shorted to ground. This severely limits the upper frequency limit f_U of oscilloscopes and other test equipment.

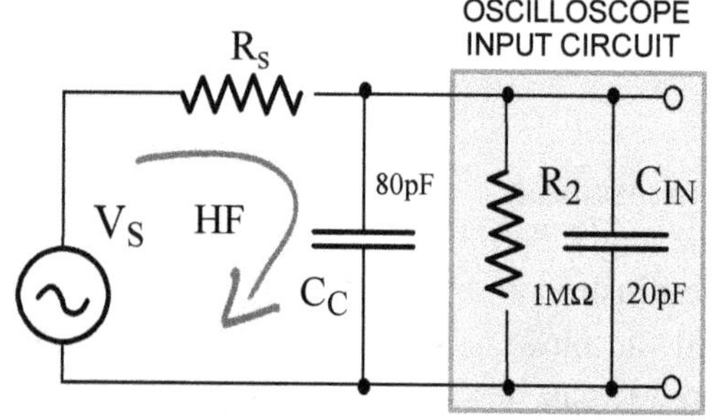

Problem: the capacitance C_C of the shielded test cable, together with the input capacitance C_{IN} of the oscilloscope, attenuates high frequencies

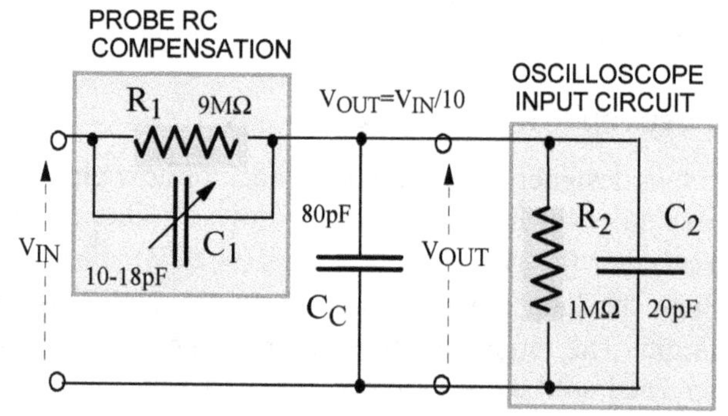

Solution: a compensated probe

Since they are in parallel, we can group the cable's and the instrument's capacitances together $C_2 = C_C + C_{IN}$

$V_{OUT}/V_{IN} = A(j\omega) = R_2||XC_2/(R_2||XC_2 + R_1||XC_1)$

$A(j\omega) = [(R_2/j\omega C_2)/(R_2+j\omega C_2)]/[(R_2/j\omega C_2)/(R_2+j\omega C_2) + (R_1/j\omega C_1)/(R_1+j\omega C_1)]$

$$A(j\omega) = \frac{\dfrac{R_2}{j\omega R_2C_2+1}}{\dfrac{R_2}{j\omega R_2C_2+1} + \dfrac{R_1}{j\omega R_1C_1+1}}$$

For the compensated voltage divider $R_1C_1=R_2C_2$, so the "$j\omega R_2C_2+1$" and "$j\omega R_1C_1+1$" factors cancel out and the transfer function becomes independent of the frequency, which was the aim from the very start:

$A(j\omega) = A = R_2/(R_1+R_2)$

Compensation made the whole input circuit behave as a simple resistive voltage divider, and also increased the input impedance of the oscilloscope. However, there is no free lunch in electronics. This has been achieved at the expense of the signal voltage reduction in the ratio of $R_2/(R_1+R_2)$ which is usually the attenuation of 10:1! In the typical case (illustrated in the drawings) $R_1=9M\Omega$ and $C_1=100/9 = 11.1$ pF!

Frequency compensation is a typical example of the gain-bandwith compromise. Just as for vacuum tubes, the gain-bandwith product is constant. We widen the useful bandwidth of a test instrument but we pay the price in 10X reduced input signals! More on oscilloscopes in Volume 2 of this book.

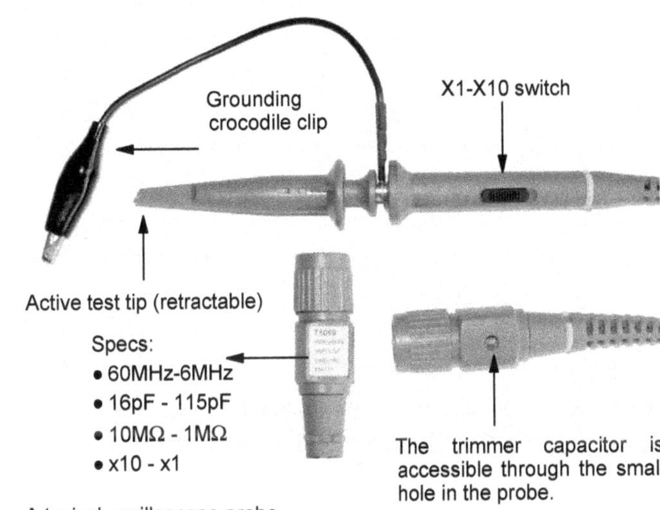

A typical oscilloscope probe

Phase-lead compensated NFB circuit

No matter how important it may be to have fully-functioning, calibrated and accurate instruments and test probes, the previous discussion had an ulterior motive behind it. The identical principle is used in a very common tube amp topology, where the negative global voltage feedback signal is fed into the cathode circuit of the amplifier's first stage, with a feedback resistor R_1 bypassed by a low value compensating capacitor C_1 (a few pF). We don't want any additional capacitance in parallel with R_2, so that resistor is *always* left un-bypassed!

This capacitive compensation is only required if the uncompensated response is underdamped, as illustrated below. If the response is overdamped or critically damped (or even slightly underdamped), there is no need for this measure!

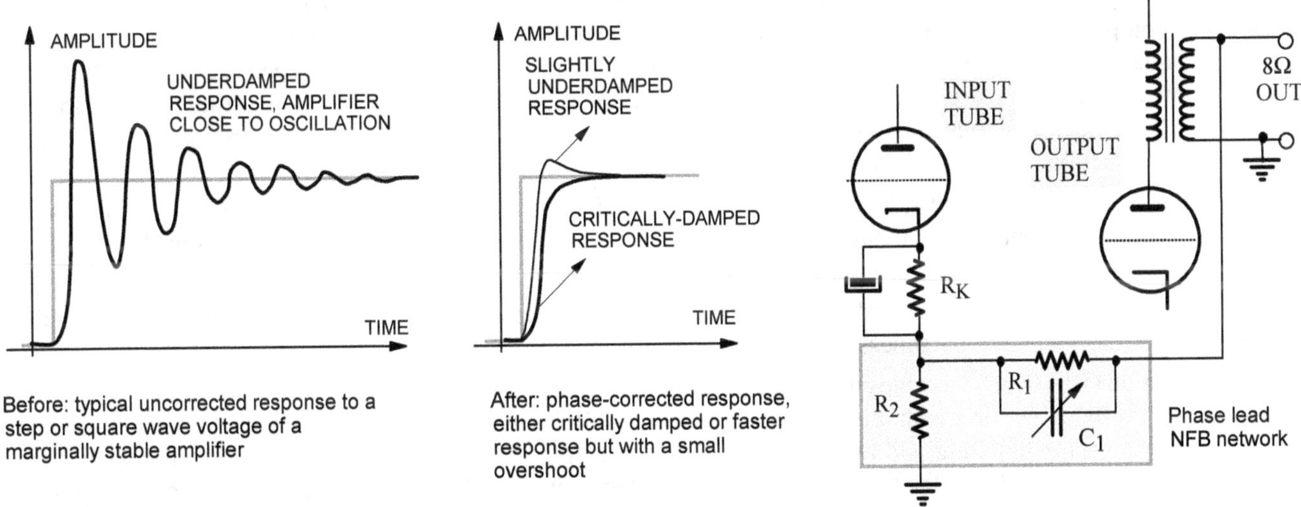

Before: typical uncorrected response to a step or square wave voltage of a marginally stable amplifier

After: phase-corrected response, either critically damped or faster response but with a small overshoot

Root-locus method of analyzing feedback circuits

The root locus is a plot of poles (roots) of the transfer function of an amplifier or linear system in general, as one parameter is varied. Usually that parameter is the open loop gain βA of an amplifier using NFB. Root locus method is an advanced topic that even many electronic engineers don't study and/or understand, and as such it is beyond the scope of this book. A few vintage electronics books listed in the reference section cover it, albeit superficially. The best coverage is in "Electronic Engineering" by Alley and Atwood.

Without phase lead compensation, with increasing loop gain the pole at ω_6 moves to the left, wile the two poles at ω_5 and ω_4 converge, until they become one double pole, after which they separate again and become a pair of complex conjugated poles, diverging from the real axis. The two complex poles closer to the imaginary axis are of more interest, since they reach and cross the $j\omega$ axis after a relatively small increase in gain and cause the amplifier to become unstable. It can be said that the pole ω_4 by moving leftwards pushes the two critical poles faster into the right half of the s-plane.

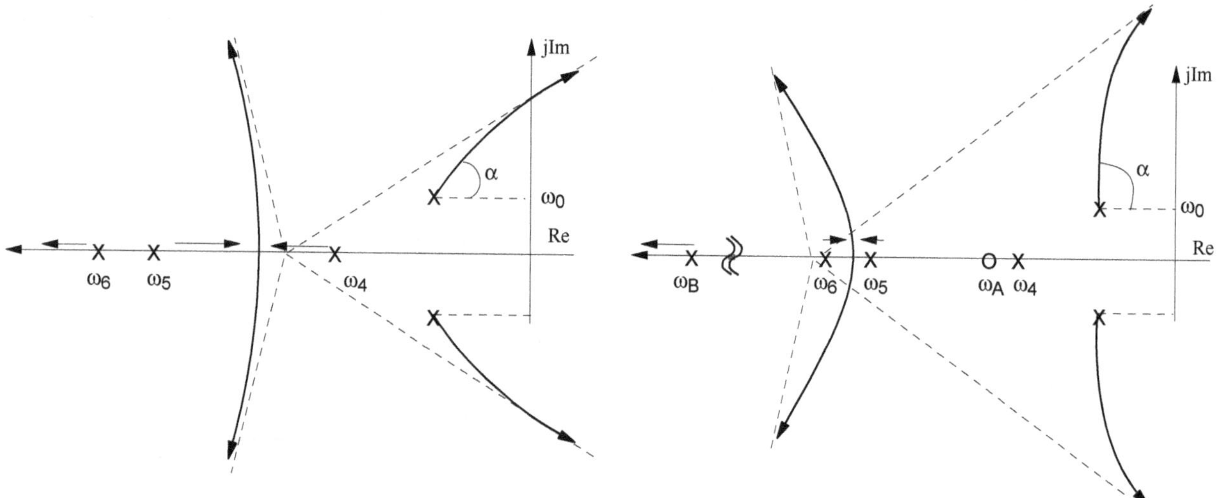

The high frequency root locus plot of an amplifier with two RC-coupled stages, a transformer-coupled output stage and resistive global negative feedback.

The HF root locus plot of the same amplifier with phase lead compensation

Phase-lead compensation introduces an extra zero at ω_A which "cancels" the pole at ω_4, or, by being very close to it, holds it "captive" and prevents it from converging towards pole ω_5. Now ω_5 and ω_6 interact, which is of no interest, since they move away from the unstable region. Some authors call this principle "slugging the dominant pole".

The main difference is the angle at which the two conjugate poles diverge with reference to the real axis (α) and the angle at which they approach the $j\omega$ axis. Notice that with compensation they diverge at almost 90° to the horizontal axis and instead of moving rapidly towards the $j\omega$ axis they now travel almost parallel to it and don't reach it at all, or reach it at extremely high values of loop gain. They have thus being effectively prevented from crossing over into the unstable territory!

Adjusting the feedback capacitor for flattest response (no ringing)

Since most vintage push-pull amplifiers used large amounts of negative feedback, feedback bypass capacitors were necessary to compensate for all sorts of high frequency misbehaviors, to stabilize the amp at high frequencies and to smooth the square-wave response. Many of these amps have a high input sensitivity, not needed these days when CD players produce $2V_{RMS}$ or more. If you are modifying these amps by reducing gain, by converting them to triode operation or by replacing output transformers, you may or may not need these capacitors. The same applies to the design and construction of a modern amplifier.

There are two ways to go about this, and, don't worry, using calculations from the previous pages or the root-locus method are not among them. As often happens, the main purpose of these theoretical deliberations is to shed light on the issue, not necessarily to encourage their use in practice.

If you don't have an oscilloscope, but have a superior hearing, you can learn to tune amplifiers by ear. Ringing in tube amplifiers sounds as increased harshness in the treble region.

Install a low-value (10-100pF) variable capacitor (can be salvaged from old radio receivers or tube oscilloscopes) but don't connect it fully, listen first without any capacitance in parallel with R_1. Then start with the lowest setting and listen again, progressing to higher and higher values of the capacitance. Finally, settle on the setting the sounds the best to you, in your system. Once the sound mellows down, stop changing the capacitance. It may not be the optimal setting according to our frequency - compensation formula, but amplifiers are about sound, not mathematics!

The other way, usually preferred by obsessive-compulsive personalities, is to feed a square wave of 10 or so kHz frequency into the amp, hook up an oscilloscope at the output and tune the variable capacitor until the ringing (if any) is eliminated and the shortest rise time is achieved.

TONE CONTROLS, ACTIVE CROSSOVERS AND HEADPHONE AMPLIFIERS

- TONE CONTROLS
- ACTIVE CROSSOVERS WITH TUBES
- HEADPHONES AMPLIFIERS
- CATHODE-RAY TUBES (TUNING INDICATORS OR "MAGIC EYES")

"Though the parameters R, L and C are considered 'linear' in this book, nonlinear parameters are common. Another book might be titled 'Nonlinear Electric Circuits.' The author has no intention of writing it."

Wallace L. Cassell, "Linear Electric Circuits"

TONE CONTROLS

The "classic" tone controls

This most common of all tone control circuits features independent bass and treble controls. The filter cannot really "boost" anything, so even when both controls are in their maximum or "boost" positions, there is still an overall loss. In this context "boost" really means lower attenuation, while "cut" means higher attenuation.

Notice that 0dB in the graphs usually published, such as the one bellow, is for the "flat" position of both controls, where potentiometers R_3 and R_4 are in the middle of their range. That position is usually indented so there is a positive feel when the middle is reached.

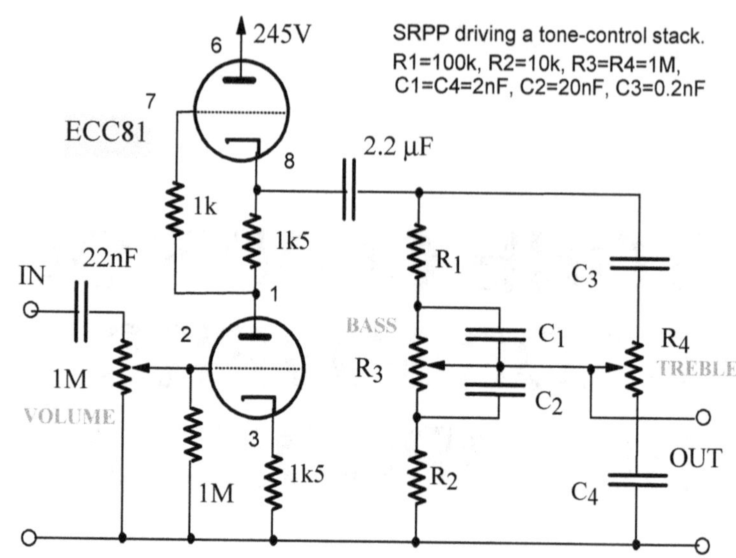

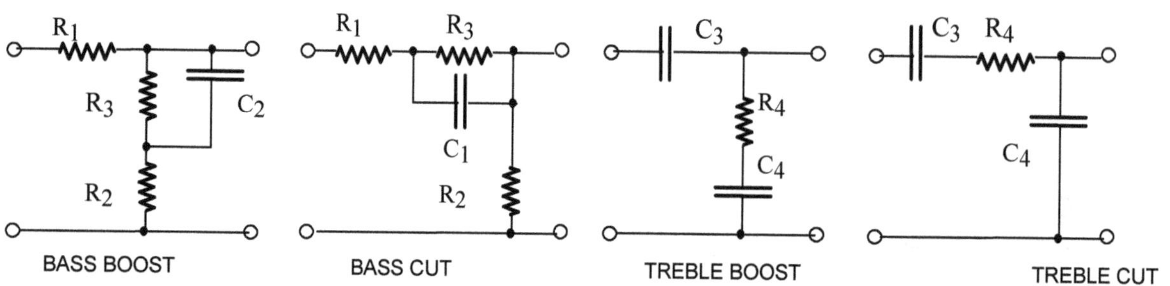

The attenuation for any frequency and any position of controls is $A=20\log(V_{OUT}/V_{IN})$

When R_3 is in max. boost position (upper end), C_1 is short-circuited and C_2 is in parallel with R_3. At low frequencies the reactance of C_2 is much larger than R_3 and can be considered open circuit. The output impedance is R_2+R_3, which is much greater than R_1, so the output impedance is a high percentage of the total input impedance. At high frequencies the reactance of C_2 is much smaller than R_3 and the output impedance is approximately $\sqrt{(R_2^2+X_{C2}^2)}$. Thus attenuation is $A=\sqrt{(R_O^2+X_S^2)}/\sqrt{(R_I^2+X_S^2)}$, where $R_O=R_2+R_S$, $R_I=R_1+R_2+R_S$, $R_S=R_3X_{C2}^2/(R_3^2+X_{C2})$ and $X_S=R_3^2X_{C2}/(R_3^2+X_{C2})$

When R_3 is the max. cut position, C_2 is short-circuited and C_1 is across R_3. At low frequencies the output resistance is approx. equal to $\sqrt{[(R_1+R_2)^2+X_{C1}^2]}$. The attenuation is $A=R_2/\sqrt{(R_I^2+X_S^2)}$.

When treble control R_4 is in max. boost position the output impedance is R_4 and C_4 in series. At high frequencies the reactances of C_3 and C_4 are low and Z_{OUT} is approx. equal to R4. At low frequencies the output impedance is approximately $\sqrt{(R_4^2+X_{C4}^2)}$, since the reactances of C_3 and C_4 are high.

Attenuation: $A=\sqrt{(R_4^2+X_{C4}^2)}/\sqrt{(R_4^2+(X_{C3}+X_{C4})^2)}$

When R_4 is the max. cut position, the output impedance is only C_4. The attenuation is

$A=X_{C4}/\sqrt{(R_4^2+(X_{C3}+X_{C4})^2)}$.

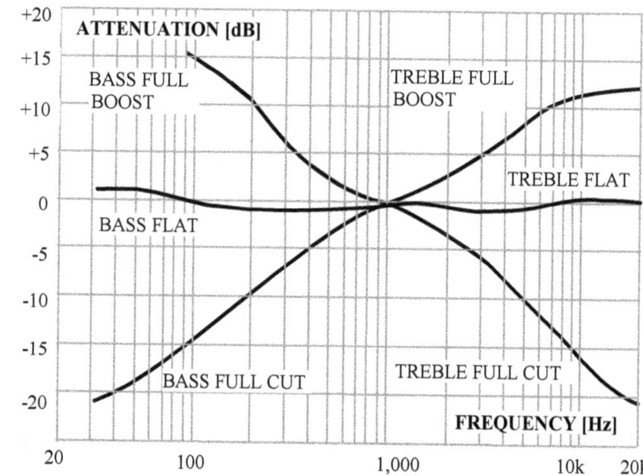

Attenuation curves for the classic tone control circuit

Baxandal tone controls

First published in 1952 in his Wireless World (now Electronics World) article titled "Negative Feedback Tone Control - Independent Variation of Bass and Treble Without Switches", this tone control method has been named after Peter James Baxandall. It offers quite a few advantages over the passive circuit. It is fully symmetrical, there is no interaction between the bass and treble controls, and, with the controls in the mid-position, the frequency response is flat and, most importantly, there is no loss.

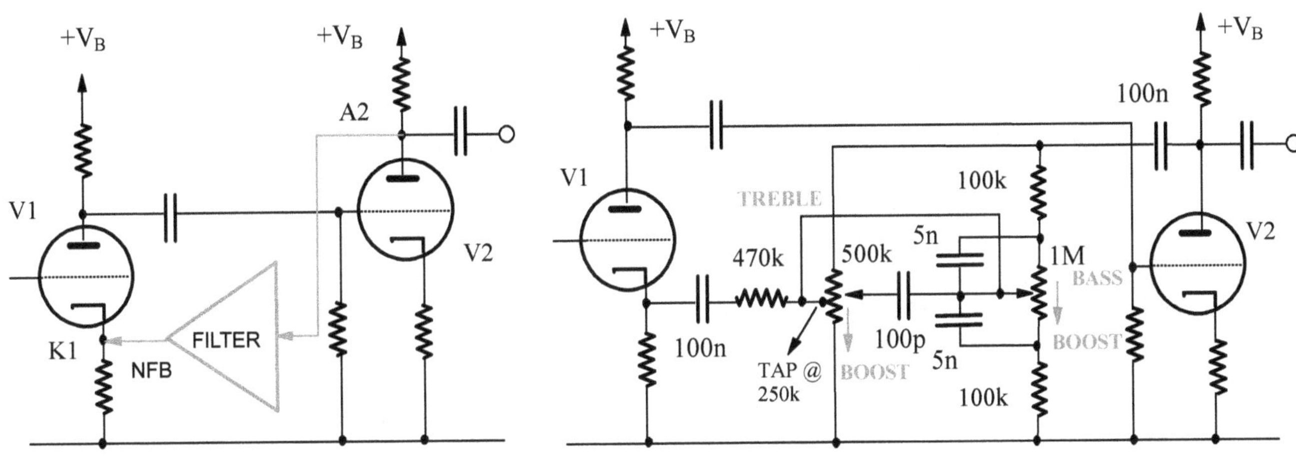

The operational principle behind the Baxandall circuit: the filter is in the NFB branch from A2 back to K1

The Baxandall tone control circuit

Low-loss tone controls

Tone controls attenuate signals a lot and require at least one stage of amplification to bring the signal back to the original level. The midband attenuation of this simple circuit is only around 9dB, compared to 15 or more dB for conventional tone controls.

A detailed explanation of the circuit's operation would take too much space here, you can easily download the whole document, search online for US patent #2,680,231!

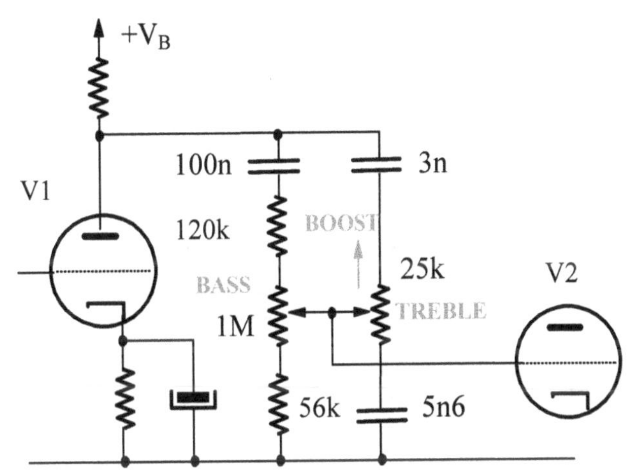

The low-loss tone controls as described in US patent #2,680,231

ACTIVE CROSSOVERS WITH TUBES

An active crossover is a frequency-selective preamplifier usually connected between the signal source (CD player, phono stage or preamplifier) and the power amplifier.

A two-way crossover has two filters, a low-pass filter, whose output feeds the low frequency or woofer amplifier, and a high-pass filter, whose output is connected to the input of a HF or tweeter amplifier. These amplifiers can be but don't have to be identical, and that is one of the benefits of this kind of bi-amping.

A LF amplifier needs to drive the power hungry voice-coil of a bass driver, or "woofer" in popular jargon, and thus requires significant power (current) capability.

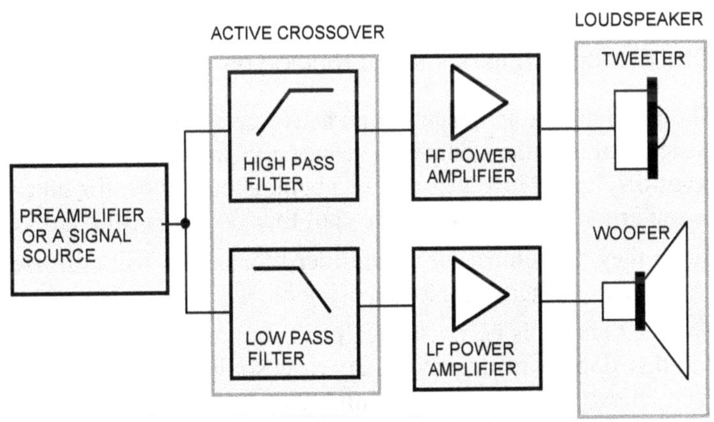

The block diagram of a two-way active crossover setup

Main benefits of active crossovers

Since low powered single-ended amplifiers often struggle in this department, and since the spectral beauty and soundstage nuances of SET amplification are all but lost on a woofer, high powered but cheaper push-pull or even (blasphemy?) solid-state amplifiers could be used for this purpose. Tweeters require significantly less power, so can be driven by (very) low power triode amps. This whole rationale is based on the most commonly used speaker type, using dynamic (moving coil) drivers. With full-range horn and electrostatic speakers (except hybrid designs such as Martin Logan) no crossovers are needed.

The LCR components in passive crossover networks must handle large signals and currents from the output of power amplifiers. As such, they introduce power losses which reduce speaker sensitivity and SPL levels of the whole system.

Active crossovers do not introduce power losses, so speaker efficiency is increased significantly. Amplifiers, especially low powered SET ones, will work at reduced power levels and produce lower distortion, have more dynamic headroom and sound better, with improved microdynamics and faster and cleaner transients.

Many passive crossovers use bipolar electrolytic capacitors and iron-cored inductors, which are nonlinear and cause significant distortion. This problem is eliminated in active crossover-based systems, which don't use inductors or elcos, resulting in further lowering of distortion.

Since the bandwidth of signals to each driver (woofer, midrange, if any, and tweeter) is strictly controlled and limited, the cross-modulation is if not eliminated then significantly reduced. For instance, a distorting amplifier will pass a 2kHz tone and a 2.5kHz tone together, but will also produce a 500Hz intermodulation distortion tone and a 5.5kHz tone. In a three-way active crossover the two original tones will pass through the midrange crossover. The 500Hz and 5.5kHz distortion products will be significantly attenuated by such filter, the 500Hz tone will not propagate through the LF fitter and the 5.5kHz tone will not even appear at the input of the HF filter since the filters are independent. The end result will be a significant reduction in IM (Intermodulation) distortion.

Assuming they feature volume or "level" controls for each frequency band, which are very easy to implement, active crossovers bring us the flexibility to adjust power levels of the high and low frequency bands. With inbuilt, passive crossovers that decision was made by the speaker manufacturer and no user adjustment is usually possible.

Active crossovers do not interact with the frequency-dependent (read "fluctuating") impedance of speaker drivers as passive ones do, so a more precise and stable spectral division is achieved. The damping factor is also improved.

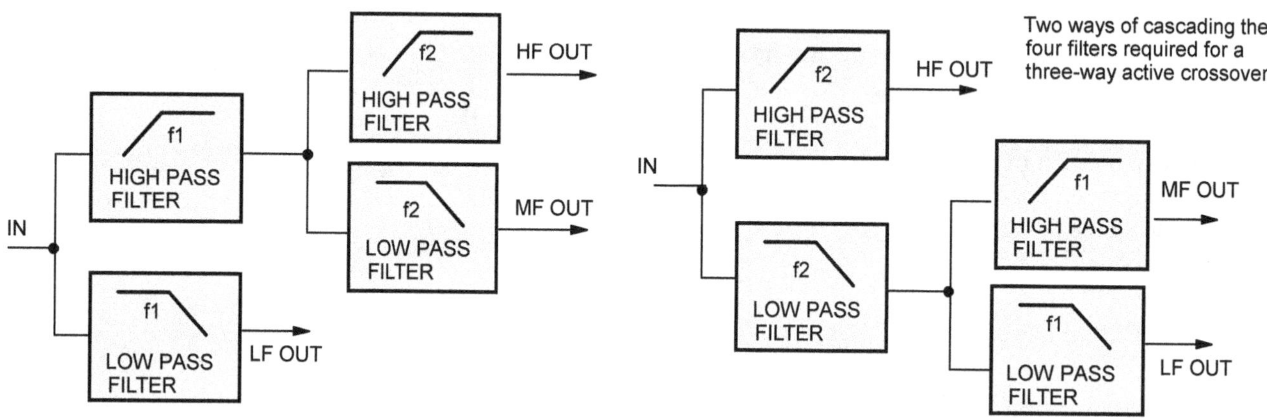

Two ways of cascading the four filters required for a three-way active crossover

Choices of technologies and topologies

The main choice in designing an active crossover is between operational-amplifier and tube technology. Op-amp designs are smaller, simpler, cheaper and easier to design and build! They also perform better (from a filtration point-of-view). However, this is a book about tubes, for tube aficionados, and if you want to build things using a dozen or more duo-triodes instead of four op-amps, who am I to try to dissuade you?

Since they are almost ideal amplifier blocks, we will still use op-amps in our discussions and analysis, and then implement the same topologies using tubes.

The next choice is between 1st, 2nd, 3rd or higher order filters. First order filters attenuate at only -6dB per octave, and this slope is inadequate in any case, so they are out! 2nd order filters are simpler than the higher order ones and can achieve a maximally flat amplitude response. They produce a 180° phase shift, but that can be dealt with by reversing the polarity of the tweeter. However, their 12dB/octave slope does not minimize the intermodulation distortion of speaker drivers enough.

An almost flat amplitude response can also be achieved by third-order filters, and their 18 dB/octave slope reduces IM distortion further. On the negative side, the complexity and component-count is increased. They produce a 270° phase shift between the input and the outputs, which can result in irregularities and anomalies in the speaker's coverage pattern. A 4-th order filter would overcome this issue (360° phase shift) but at even higher complexity and cost!

The final choice is between two most common topologies, the Sallen-Key and the Multiple Feedback (MFB), for which component values can be so chosen to result in different transfer functions and attenuation curves, named after the mathematicians who first described them, such as Butterworth, Bessel, Chebyshev and Linkwitz-Riley.

Multiple FeedBack topology is less sensitive to component variations and has better high-frequency behavior. The advantage of Sallen-Key topology is that its unity-gain does not depend on the values of resistances and capacitances that determine crossover frequencies.

Negative feedback filter topologies

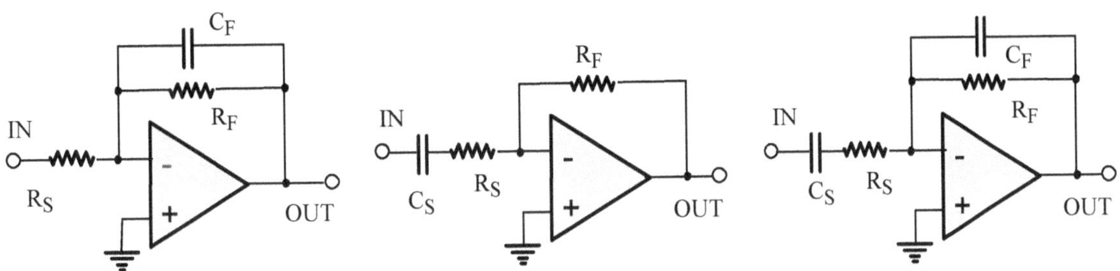

1st order negative feedback active filter (L-R): Low-pass, high-pass, bandpass

The closed-loop gain of an inverting amplifier is $A=V_{OUT}/V_{IN} = -Z_F/Z_S$ An infinite amplification factor is assumed, applicable to operational amplifiers, but in the first approximation the equation can be applied to tube amplifying stages, to illustrate the principle behind the topology.

For the low-pass (LP) filter $Z_F=R_F||X_C = R_F||1/j\omega C = R_F/(1+j\omega C_FR_F)$ and $Z_S=R_S$ so $A(j\omega) = (R_F/R_S)/(1+j\omega C_FR_F)$

The first factor, R_F/R_S is the frequency-independent amplification factor of an inverting amplifier, the second factor is a transfer function of a low-pass filter with time constant $\tau=R_FC_F$ For the high-pass (HP) filter $Z_S=R_S+X_S = R_S+1/j\omega C$ and $Z_F=R_F$ so we have $A(j\omega) = -Z_F/Z_S = j\omega C_SR_F/(1+j\omega C_SR_S)$

Both filters are 1-st order type, with the attenuation slope of 20dB/decade! The two feedback RC networks can easily be combined to form a bandpass (BP) filter: $Z_F= R_F/(1+j\omega C_FR_F)$ and $Z_S=(1+j\omega C_SR_S)/j\omega C_S$ so we have $A(j\omega) = j\omega C_SR_F/[(1+j\omega C_FR_F)(1+j\omega C_SR_S)]$

First order filters don't attenuate "fast" enough for audio crossovers, so 2nd and 3rd-order topologies are needed.

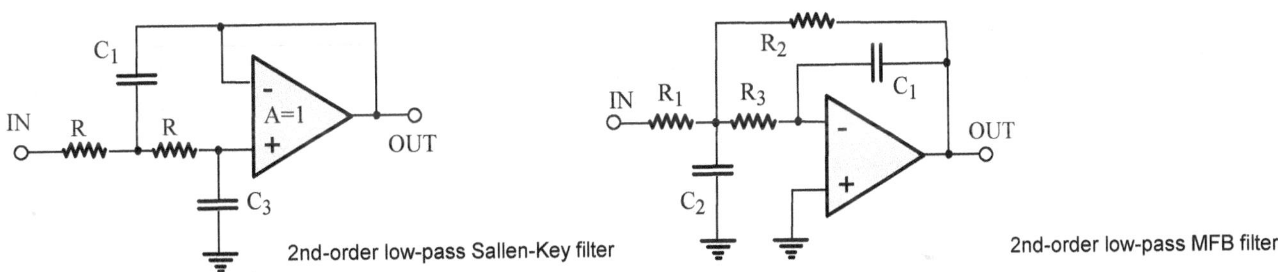

2nd-order low-pass Sallen-Key filter

2nd-order low-pass MFB filter

Butterworth response

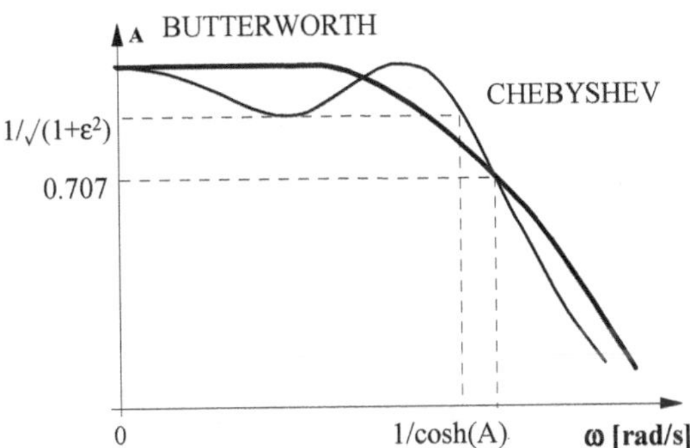

A comparison of normalized (unscaled) attenuation curves of a 3rd order Chebyshev and Butterworth low-pass filters

Even order filters can be realized using feedback topology, but to realize 3rd and higher odd order filters, a 1st order filtering stage has to be added in cascade at the input of an even order filter (next page). An active three pole low-pass filter has a transfer function $T(s) = 1/(As^3+Bs^2+Cs+1)$ where $A=C_1C_2C_3$, $B= 2C_3(C_1+C_2)$ and $C=C_2+3C_3$. It is assumed that R=1.

The Butterworth response is characterized by the flattest amplitude response and is thus optimal for electronic crossovers.

Its transfer function is $T(s) = 1/(s^3+2s^2+2s+1)$, and by equating the coefficients of the two transfer functions we get A=1, B=2 and C=2. After substituting these into the three equations for A, B and C and solving them for C_1, C_2 and C_3 we get C_1= 3.546F, C_2= 1.392F and C_3= 0.2024F!

Frequency and impedance scaling

The filter values and its response curve have been normalized for angular frequency ω=1 rad/s, which is the -3dB frequency of the filter. To shift the filter's response curve to a different frequency range, the values of capacitances are divided by the Frequency Scaling Factor (FSF): FSF= 2*π*[desired -3dB frequency]/[reference frequency]

Let's say we want the -3dB frequency of our low-pass filter to be at 400Hz. The FSF is 2*π*400/1 = 2,513, so we have to divide all capacitance values by 2,513: C_1= 3.546/2,513 = 1.41 mF, C_2= 0.55 mF and C_3= 80.4 μF!

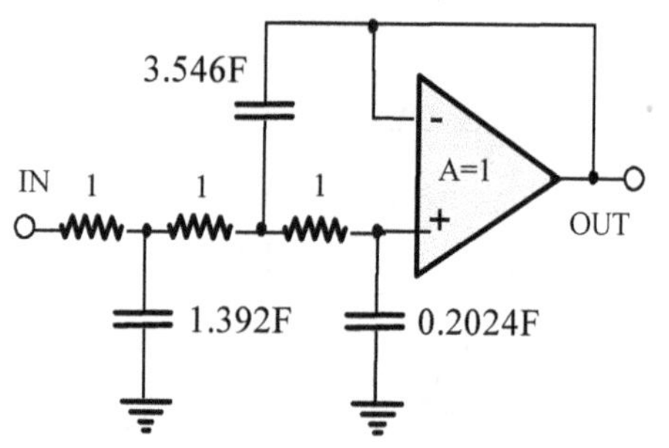

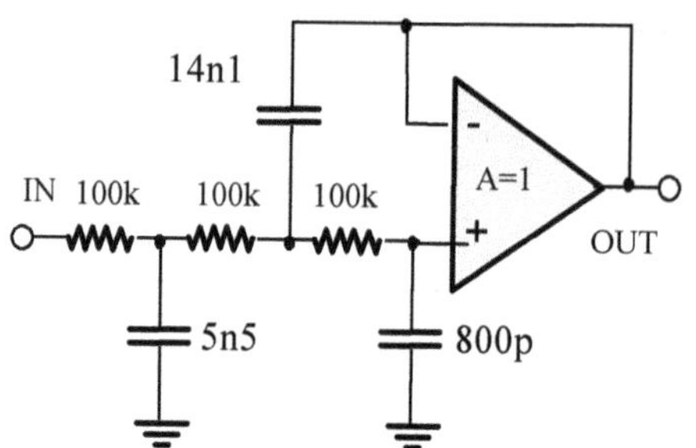

FAR LEFT: Unscaled (normalized) 3rd-order Butterworth low-pass filter

LEFT: 3rd-order Butterworth low-pass filter scaled to 400Hz and 100k

Resistance values of 1Ω and very high capacitances of 1,410 μF are not practical. Luckily, since this is a linear system, their values can be "scaled", just as we did with frequency scaling. The roots and transfer function of linear network do not change if resistor and inductor values are multiplied and capacitance values are divided by an Impedance Scaling Factor (ISF)!

Let's say we want the resistors to be R=100 kΩ instead of 1Ω. ISF is therefore 100,000! Once we divide all capacitance values by 100,000 we get C_1= 14.1 nF, C_2= 5.5 nF and C_3= 0.8 nF!

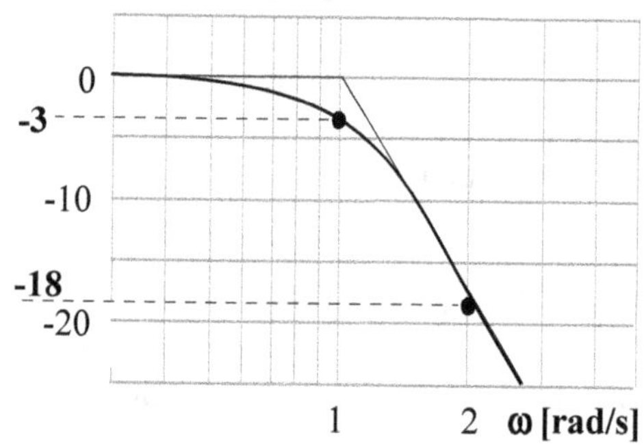

Frequency response curve of unscaled 3rd-order Butterworth low-pass filter

Low-pass to high-pass transformation

Once you have the design of the low pass (LF) filter done, its high pass (HF) "companion" filter can be derived by swapping resistors and capacitors around. Replace each capacitor by a resistor having the reciprocal value of its capacitance ($1/C_{LF}$) and each resistor by a capacitor with value $1/R_{LF}$!

After the LF to HF transformation, the frequency and impedance scaling is performed in the same manner as with LF filters. Resistors are multiplied by ISF (Impedance Scaling Factor) and capacitors are divided by FSF*ISF! In our 400Hz case FSF=2,513 and ISF=100,000 so resistors are multiplied by 100,000 and capacitors are divided by 251,300,000!

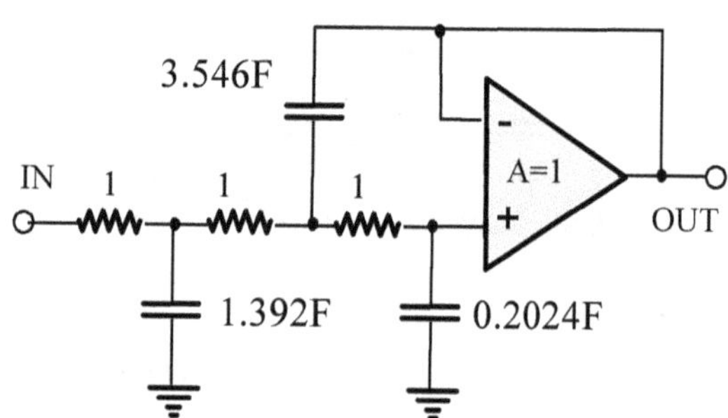

START: unscaled 3rd-order Butterworth low-pass filter

From op-amps to vacuum tubes

Now that we know a little about building active filters with operational amplifiers, how would we implement such designs in vacuum tube technology? Firstly, notice that the op-amp is working as voltage follower with a gain of 1. The output is directly connected back to the inverting (- or negative) input, and that is 100% voltage feedback! However, simply replacing the op-amp with a cathode follower will not work. A cathode follower does not have two inputs as an op-amp does. We need four cathode followers in a cascade arrangement, as per the block diagram(next page).

The cathode followers can be identical, although we bootstrapped the input grid leak resistor to achieve an even higher input resistance. There is no need to go through that trouble with the other three stages. The values and the scaling techniques stay the same as before. The 100Ω resistors in anodes are not mandatory and can be omitted should you wish to reduce the component count.

The tube choice is left to you, almost any duo-triode will work, although the use of high μ triodes minimizes overall attenuation. For instance if A=0.92 is achieved with 12AX7, the overall attenuation of four CF stages (without the filter attenuation) is $0.92^4 = 0.716$!

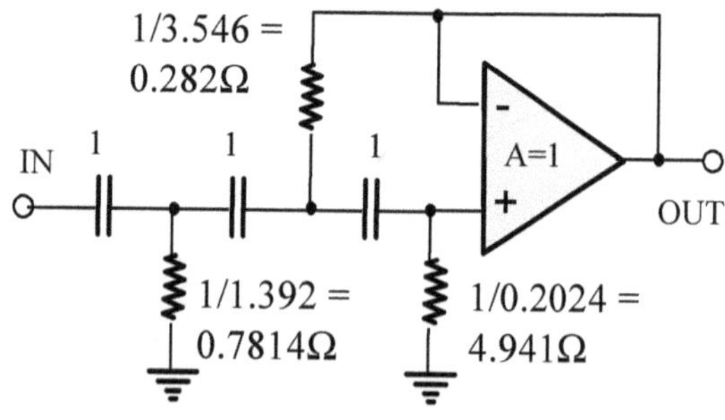

STEP 1: Transformation into an unscaled high pass filter

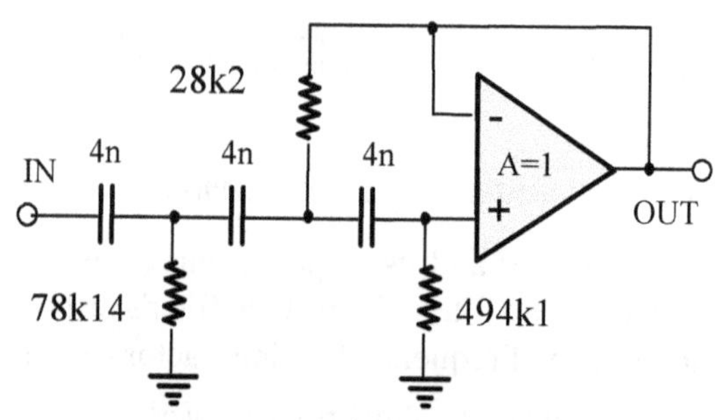

STEP 2: 3rd-order Butterworth high-pass filter frequency & impedance scaled to 400Hz and 100k

With 12AU7 the attenuation of one stage would be around A=0.84, so four CF stages would result in $A=0.84^4=0.5$!

Eight duo-triodes are needed for a stereo two-way crossover (for per channel). For a three-way system, the complexity and component count increase significantly. The midrange filter is essentially the LF and HF filters in one, so the component count doubles. Sixteen duo-triodes are needed, and that clearly violates the "Simplicity Rule"!

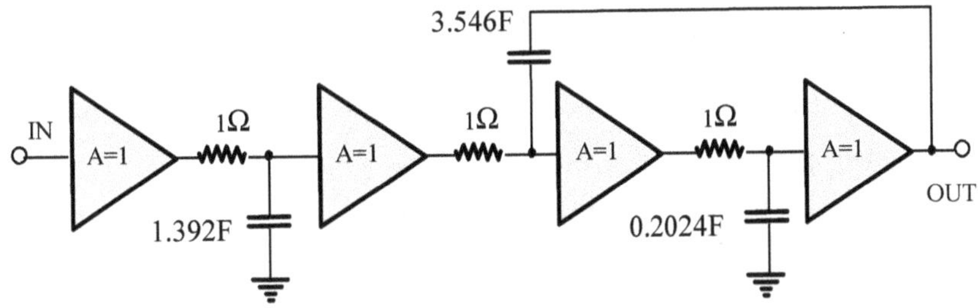

Block diagram of an unscaled 3rd-order Butterworth filter with discrete cathode followers ABOVE: low-pass and BELOW: high-pass

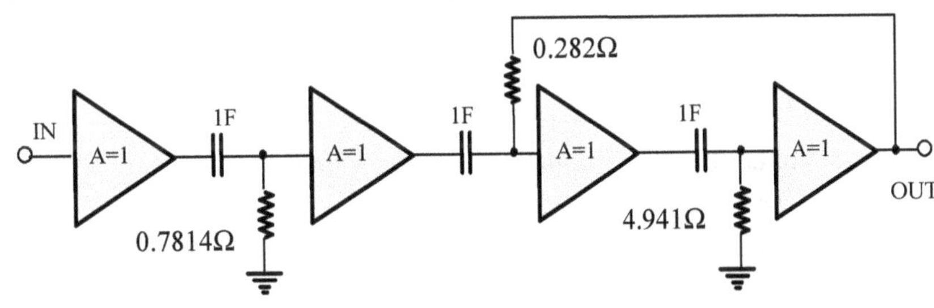

Unscaled low-pass 3rd-order Butterworth filter with discrete cathode followers

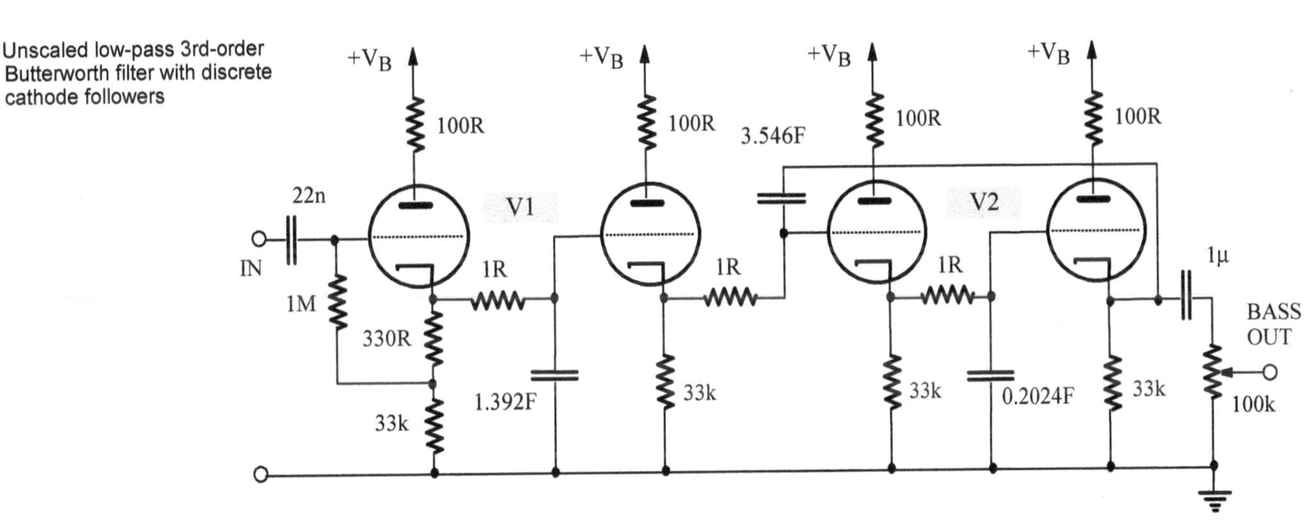

Unscaled high-pass 3rd-order Butterworth filter with discrete cathode followers

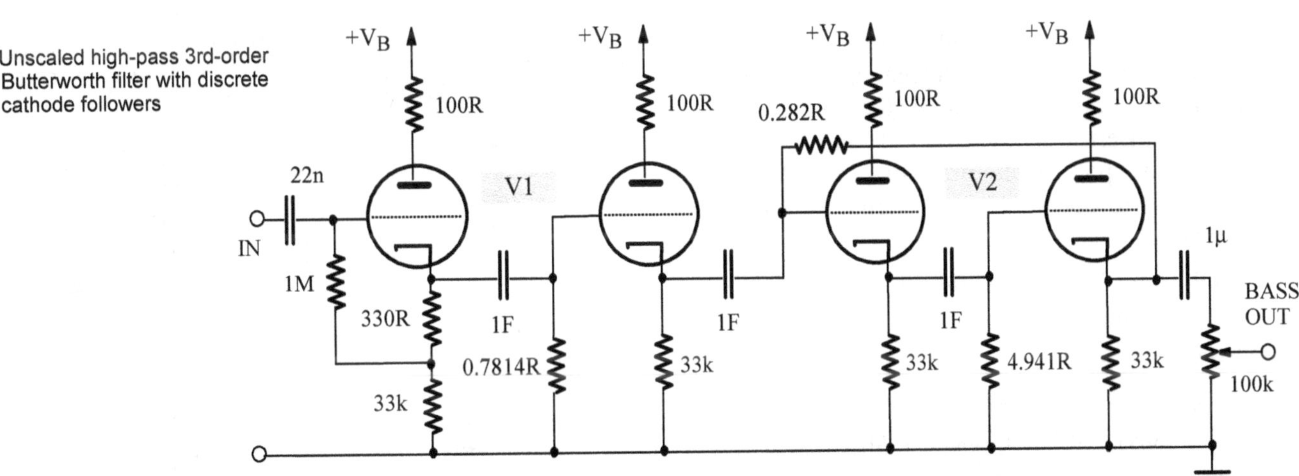

Staggered RC network designs

The design of 2 and 3-stage RC filters is based on the following requirements: 1) the attenuation at the crossover frequency of each stage is -1dB, meaning their time constants are equal, 2) no stage should appreciably load the previous stage at any frequency and 3) the overall network should not load the driving cathode follower (CF) appreciably at any frequency.

Let's start with the low-pass filter design. The attenuation of each RC filter is $A=20\log V_{IN}/V_{OUT}$. In the first step, this figure is chosen. Let's say we want A=1dB at the crossover frequency f_C. Thus, $V_{IN}/V_{OUT}=10^{1/20}=1.12$! Since $V_{IN}/V_{OUT}=(R_1+1/\omega C_1)/(1/\omega C_1)$ at f_C we have $\sqrt{(\omega_C{}^2R_1{}^2C_1{}^2+1)}=1.12$ and $\omega_C R_1 C_1=1/2$, so $C_1=1/4\pi f_C R_1$

To satisfy the loading criteria R_1 should be at least 10k, and $R_3=10R_2=100R_1$. All three time constants must be the same ($R_1C_1 = R_2C_2 = R_3C_3$) so $C_2= (R_1/R_2)C_1$ and $C_3= (R_2/R_3)C_2$! Let's say we again want a 400Hz filter and choose R_1=10k. Thus, $R_2=10R_1$=100k and $R_3=10R_2$=1M. The capacitors required?

$C_1 = 1/4\pi f_C R_1$ = 19.89nF, so we would use 22nF. Now $C_2= (R_1/R_2)C_1 = 1/10C_1$ = 2n2 and $C_3= (R_2/R_3)C_2 = 1/10C_2$ = 0.22n

High-pass filter design: We must chose the same attenuation of each RC filter as in the LF step, in this case A=1dB at the crossover frequency f_C. Thus, $V_{IN}/V_{OUT} = 10^{1/20} = 1.12$

Since $V_{IN}/V_{OUT} = R_1/(R_1+1/\omega C_1)$ at f_C we have $\sqrt{}(\omega_C^2R_1^2C_1^2+1)/(\omega_C R_1 C_1)$=1.12 and $C_1 = 1/\pi f_C R_1$! Notice that this is a 4 times higher value than in the case of the LF circuit.

Again, to satisfy the loading criteria R_1 should be at least 10k, and $R_3=10R_2=100R_1$. All three time constants must be the same ($R_1C_1 = R_2C_2 = R_3C_3$) so $C_2= (R_1/R_2)C_1$ and $C_3= (R_2/R_3)C_2$!

This means we can simply invert the position of resistors and capacitors, use the same resistor values and 4x higher capacitor values, so no calculations are necessary!

Thus, $R_1^*=R_1$=10k, $R_2^*=R_2$=100k and $R_3^*=R_3$=1M. $C_1^* = 4C_1$ = 88n, $C_2^* = 4C_2$ = 8n8 and $C_3^* = 4C_3$ = 0.88n!

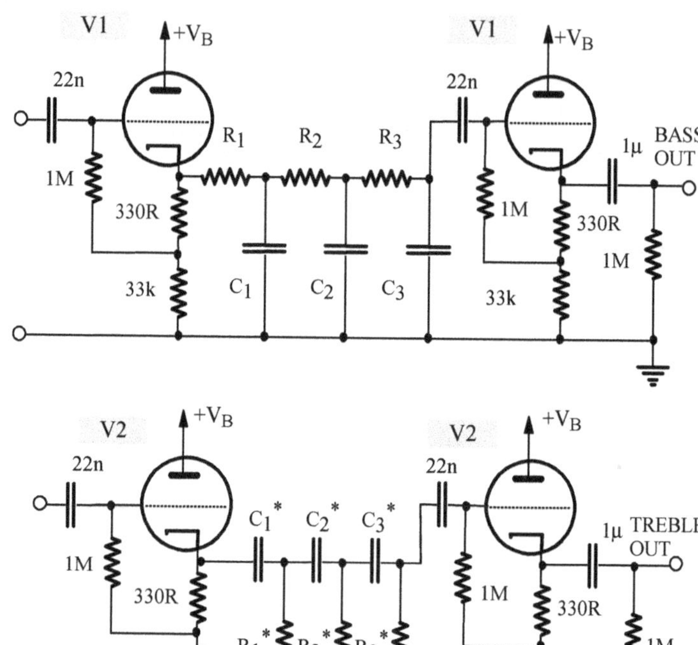

2-way 18dB/octave electronic crossover

The crossover frequencies can be made continually adjustable, but that requires tripple-ganged potentiometers of unequal values, which are impossible to find. A two-way crossover needs dual-gang potentiometers of unequal (10:1) values, which are also rare but may be salvaged from vintage equipment.

Alternatively, get a few different brands of 10k and 100k linear potentiometers, open them up and swap one wafer around, i.e. take a 10k wafer from a 10k dual pot and insert it instead of one 100k wafer of the dual 100k pot. Some brands/types will be relatively easy to open and modify, others next to impossible.

The formulas for the two-section filters are:

LF: $C=1/(9.7f_CR)$

HF: $C=1/(4.06f_CR)$

The same criteria for resistor values apply, so choose $R_2=10R_1$ and R_1 or 10kΩ or larger.

Of course, you can change potentiometers R_1/R_2 and R_1^*/R_2^* into fixed resistors if you don't require variable crossover frequency. Alternatively, for some flexibility, once you determine the ballpark of your required f_C (again say around 400Hz) install a switch with a few values around that frequency and do listening tests to see which one sounds the best. For instance, a 4-position selector switch and resistor values for 350Hz, 400, 450 Hz and 500Hz.

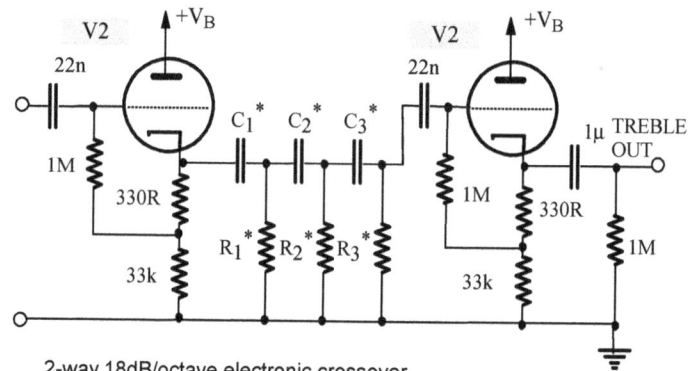

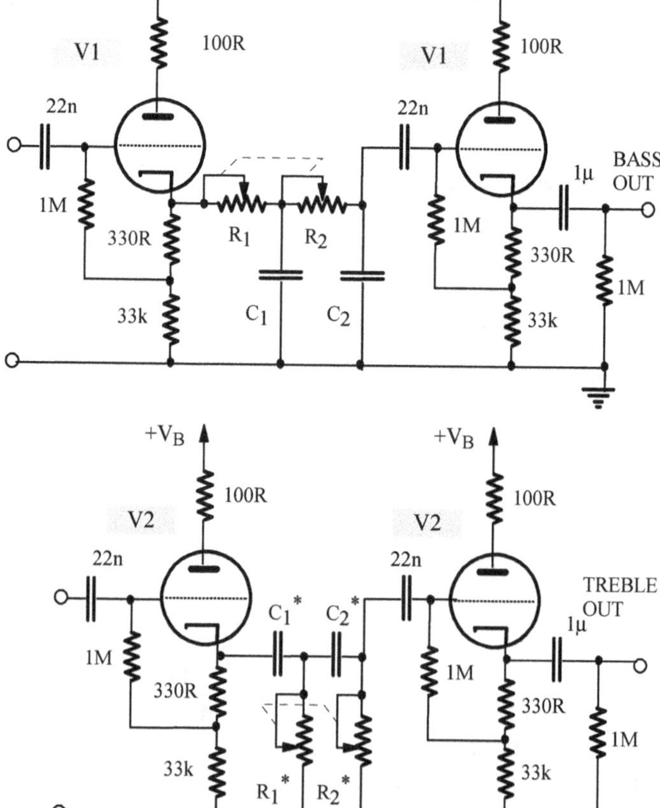

2-way 12dB/octave electronic crossover with continually adjustable crossover frequencies

The KISS version

This single CF stage 2-way crossover is the simplest active crossover one can build. Two-, three- or even 4-stage RC filters can be used. The values of R and C in the two circuits don't have to be the same. Most preamp tubes can be used, 6SN7, ECC40, ECC88, etc. The $+V_{BB}$ supply is not critical, any voltage 150-250V would work fine.

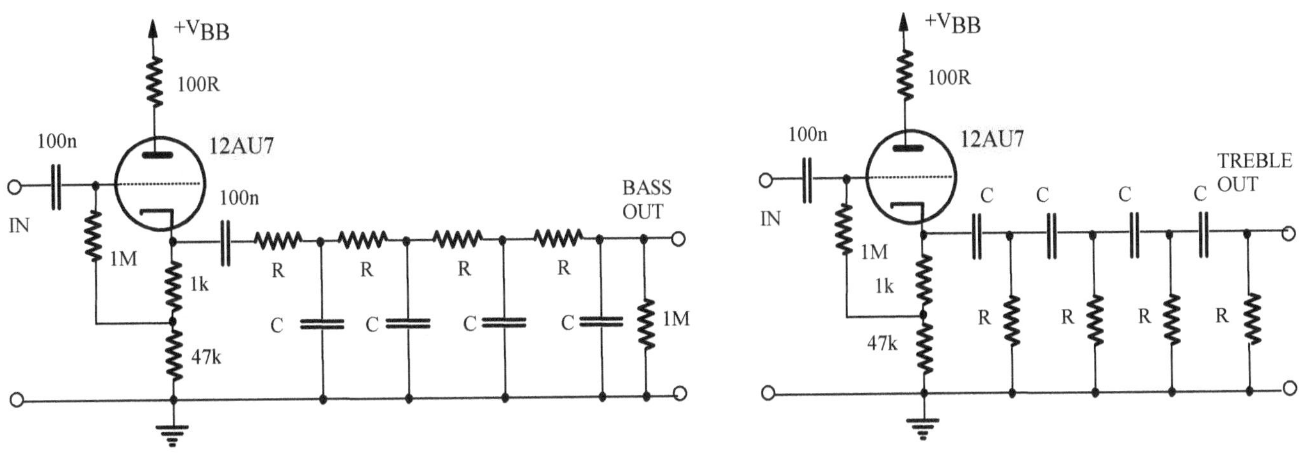

A simple single-stage active crossover: Low pass (left) and high pass (right)

LF formula: After choosing the value for R and f_C, calculate C from $C = 1/16\pi f_C R$. Again, let's say we want a 400Hz filter and choose R=47k. Thus, $C = 1/16\pi f_C R_1 = 1.06nF$, so we would use 1nF.

HF formula: After choosing the value for R and f_C, calculate C from $C = 4/\pi f_C R$. For a 400Hz filter and R=470k we need C =6.77 nF, so we would use 6.8nF.

The way it used to be: Audio Research EC-4

This vintage electronic crossover features three topologically identical circuits, a common cathode stage directly-coupled to a cathode follower, each with an adjustable level control at the input. There is NFB from the output of the cathode follower to the grid of the first stage. The filtering components are shaded in gray.

The HF filter uses simple CR network at the input (passive filtering), followed by the amplifier, meaning it is a 1st order filter. This is clearly a design mistake, since high energy low frequencies will not be attenuated enough before they reach a sensitive tweeter!

The midrange circuit uses a passive CR filter at the input and a CR network in the NFB circuit, so it's a 2nd order design.

The bass circuit also uses a passive RC filter at the input and a CR network in the NFB circuit, so its also a 2nd order design.

The midrange and bass filters are textbook cases, if you replace the two-stage tube amplifier with the general operational amplifier symbol, you will get topologies from the previous page.

Another weakness is that a change in the level will also change the set frequencies! The level potentiometer is in parallel with 158k resistors which determine the corner frequencies, and as the volume is changed so are those frequencies.

EC-4 audio section, © Audio Research Corporation

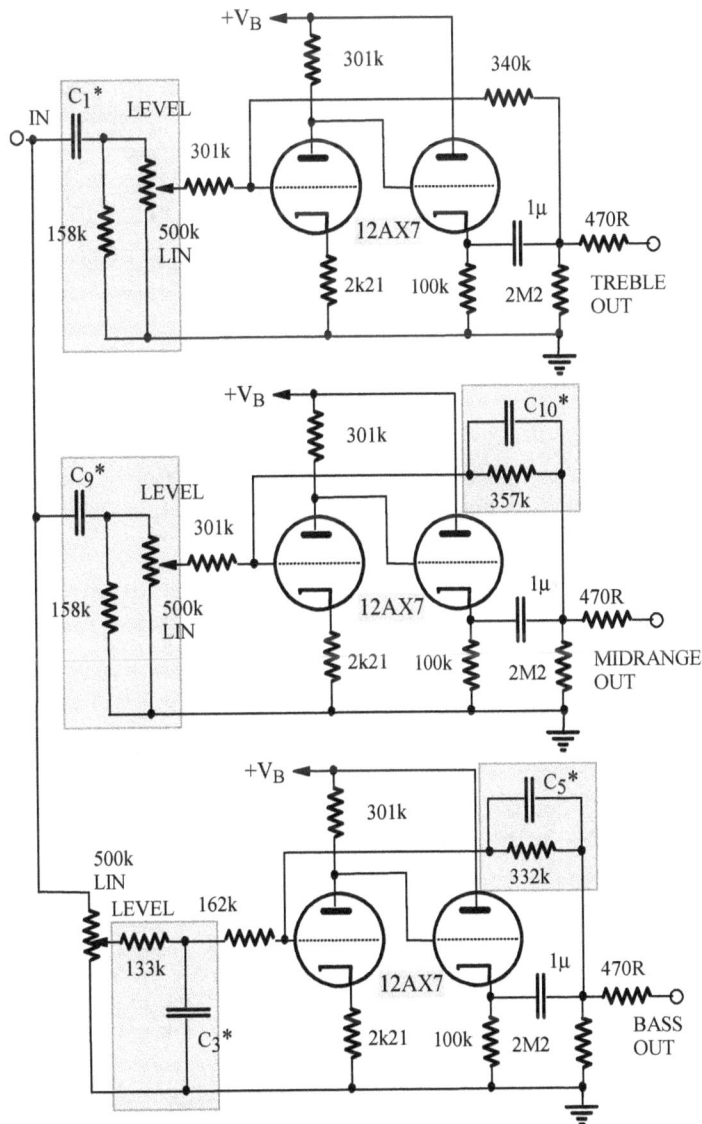

The way it used to be: Marantz Electronic Crossover

Marantz Model 3 active crossover features 12 switch-selectable crossover points at ½ octave intervals, from 100Hz to 7 kHz. The slope is 12 dB per octave, with -3dB (non-ringing) "knees". It is estimated that only 500 units were ever made and as such it is very rarely offered for sale.

The input and output stages are cathode followers, with one amplification stage in between. All three stages use 12AX7 triodes. With level potentiometers (Marantz calls it "Balance") at maximum, 10 dB of gain (about 3x) is available.

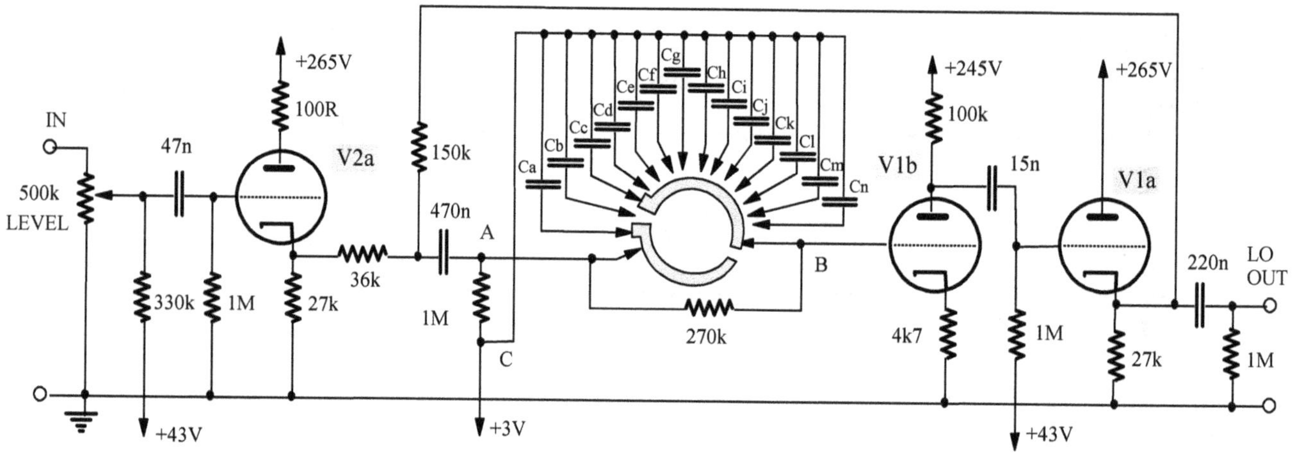

Ca	Cb	Cc	Cd	Ce	Cf	Cg	Ch	Ci	Cj	Ck	Cl	Cm	Cn
47n	33n	22n	15n	10n	6n8	4n7	3n3	2n2	1n5	1n	630p	430p	300p

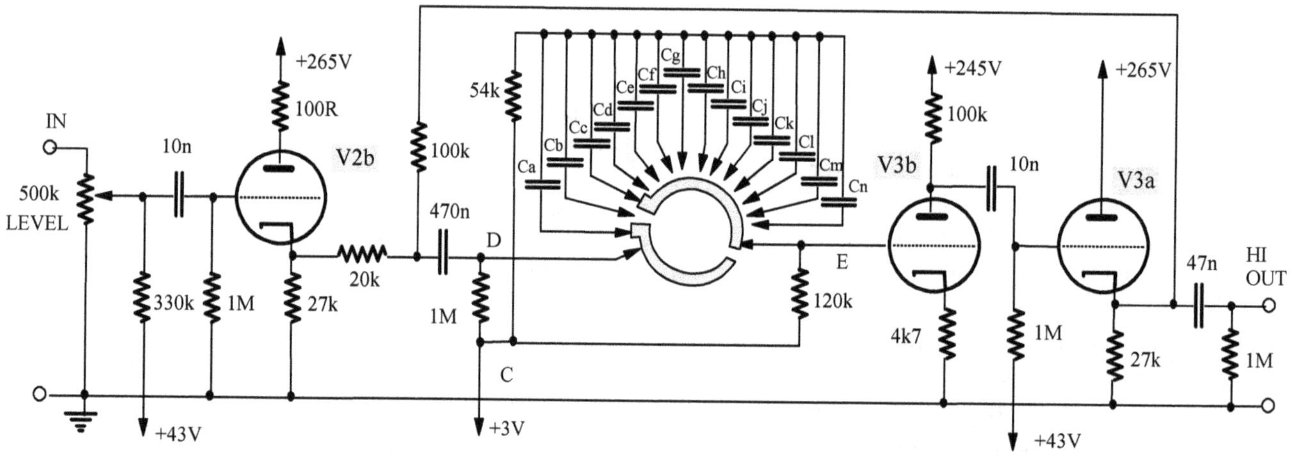

Ca	Cb	Cc	Cd	Ce	Cf	Cg	Ch	Ci	Cj	Ck	Cl	Cm	Cn
13n	10n	6n8	4n7	3n3	2n2	1n5	1n	630p	470p	330p	220p	150p	91p

The audio section of Marantz Model 3 electronic crossover

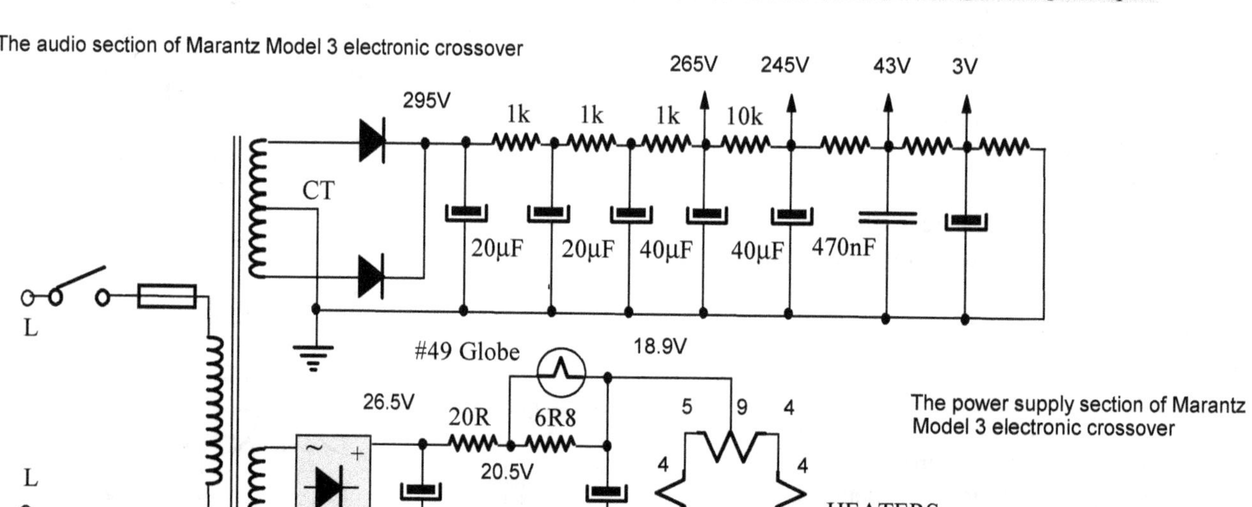

The power supply section of Marantz Model 3 electronic crossover

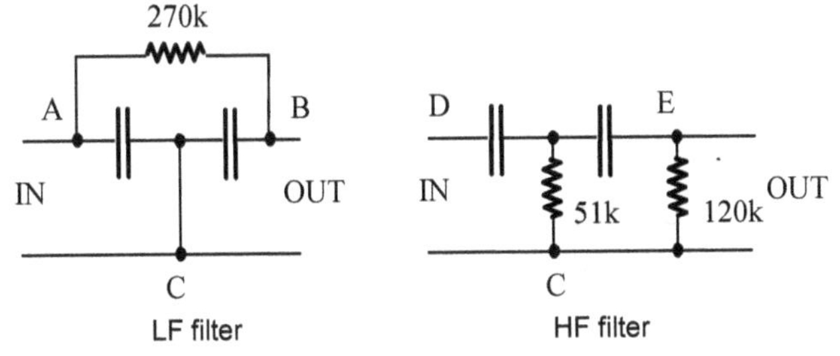

The switching design is very clever. In both LF and HF filters the frequency selector switches two capacitors simultaneously. These are connected in series, with their midpoint going straight to +3V point (ground for AC signal) in the case of the LF filter (point C), and through 51k resistor to ground in HF filter.

The LF filter is a bridged-T filter, while the HF filter is a cascade of two CR filters, as illustrated.

The way it used to be: Heathkit XO-1

Heathkit XO-1 is a 2-way mono tube crossover with 7 switch-selectable frequencies, 100, 200, 400, 700, 1,200, 2,000 and 3,500 Hz. The design uses two RC filters, a common cathode amplification stage between them, a cathode follower at the output and a global NFB around both filters.

In the HF crossover both RC filters are in the feedback loop, but in the LF crossover only one filter is in the feedback loop. Once the switches are removed (single setting illustrated) the circuit diagrams looks like this:

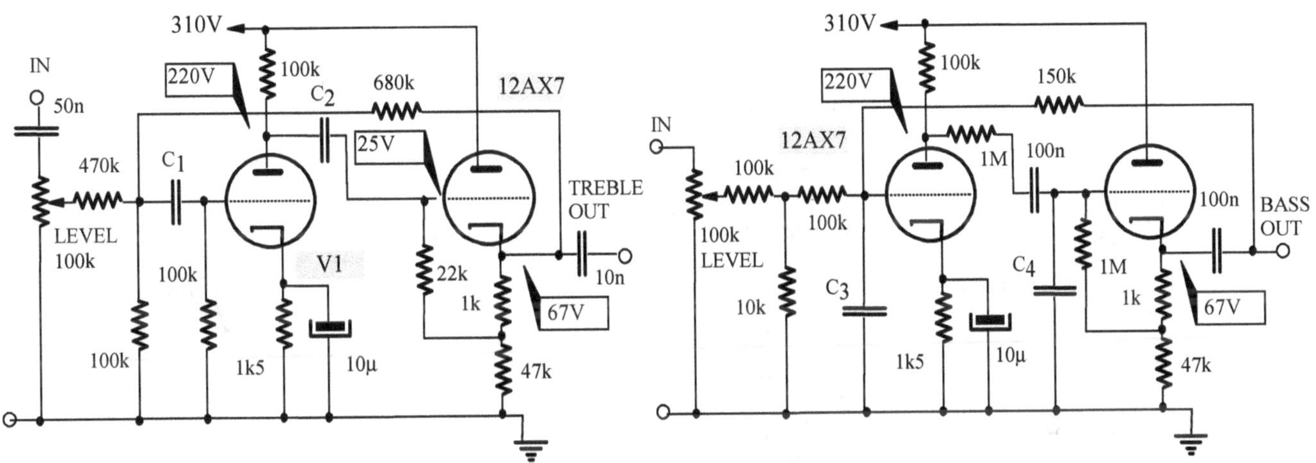

The audio section of Heathkit XO-1 crossover (selector switches and the rest of the four capacitor banks not shown for clarity)

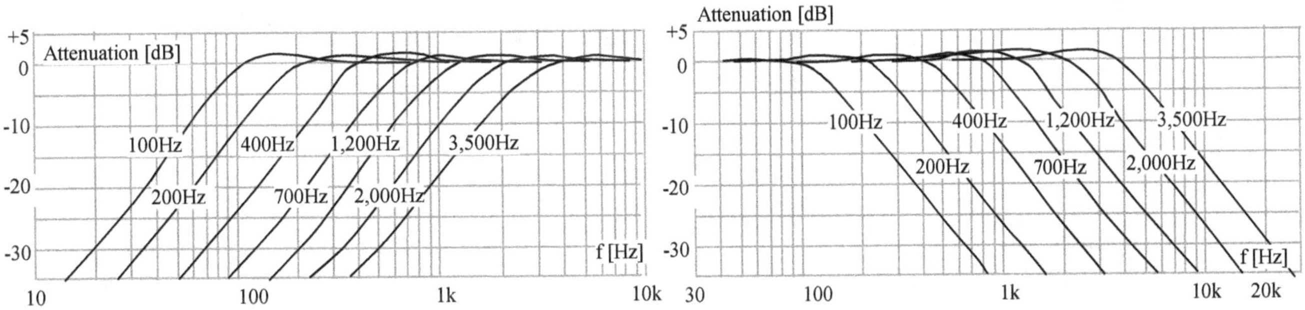

Attenuation curves for Heathkit XO-1 crossover: High pass (left) and low pass (right)

Heathkit XO-1 is not as rare as the Marantz unit and occasionally shows up on ebay, however, unless you get it for a song at an auction, sellers usually price it at $300 or more. You need to replace all of its capacitors and many or most of its carbon composition resistors whose values have drifted. Plus, it's a mono unit, so you may as well build a new stereo version cheaper and faster.

HEADPHONE AMPLIFIERS

There are two main types of amplifiers designed to drive headphones, with and without output transformers. It may be argued that the use of output transformers negates many of the aims and benefits of the headphone use. There is no fundamental difference in the design of a tube amplifier for loudspeaker or headphone loads when output transformers are used, and this approach is discussed in detail in this book and its companion volume.

Among the two main approaches to designing an OTL headphone amplifiers, one uses the standard series topology common in OTL tube amplifiers driving low impedance loudspeakers (a series push-pull circuit), while the other is based on the cathode follower output stage. We will focus on the later approach here.

PRACTICAL DESIGN: Triode-strapped EL84 headphones amplifier

The top triode is an "active resistor" for the common cathode stage of the lower triode. Without determining the exact parameters in the operating point from ECC82 graphs, in the first approximation, gm=2.2 mA/V, μ=17 and r_I = 7.7 kΩ Now the active anode resistor has the value of $R_{A1}=r_{I2} + (\mu 2+1)R_2$ = 7.7 + (17+1)0.82 = 2.6+23= 22.5kΩ, thus the amplification of the whole quazi-SRPP stage is $A= -\mu R_{A1}/(R_{A1}+r_{I1})$ =-17*22.5/(22.5+7.7) =-12.6

We have ignored the un-bypassed cathode resistor (220Ω), due to its negative feedback effect the actual A will be lower than the one just estimated. The measured value was A=-10!

The current through the EL84 cathode follower output stage is 30mA, so from the graph below we find that in that Q point r_I=1.8kΩ and μ=19.5

TUBE PROFILE: EL84 (6BQ5)

- Indirectly-heated power pentode
- Heater: 6.3V, 0.75 A
- Maximum plate voltage: 300 V
- P_{AMAX}=12 W P_{SMAX}=2 W
- SE OUTPUT TRIODE OPERATION:
- $V_A= V_S$= 250V
- I_{A0} = 34 mA, 49.5 mA with max. signal
- Bias voltage -9.2V, Load: 3.5 kΩ

With a 220Ω cathode resistor now we can calculate the amplification factor of the cathode follower:

$A_K=\mu R_K/[(1+\mu)R_K+r_I]$= 19.5*220/(20.5*220+1,800)= 4,290/6,310 = 0.68

The overall voltage gain of the amplifier is A=-10*0.68=-6.8

The output impedance is

$Z_O=[r_I/(1+\mu)]||R_K$ = 1,800/20.5|| 220 = 62.8Ω

The measured -3dB frequency range without load was 5Hz - 32kHz.

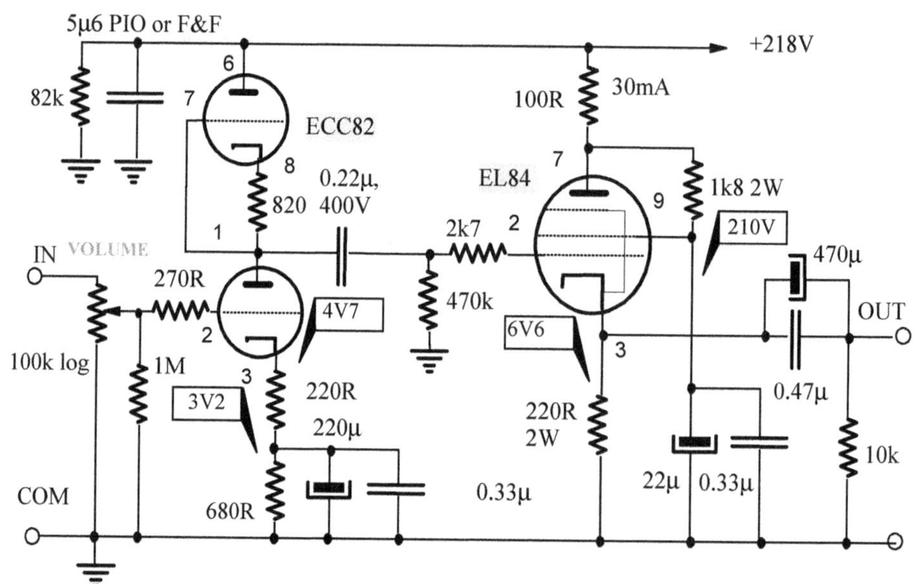

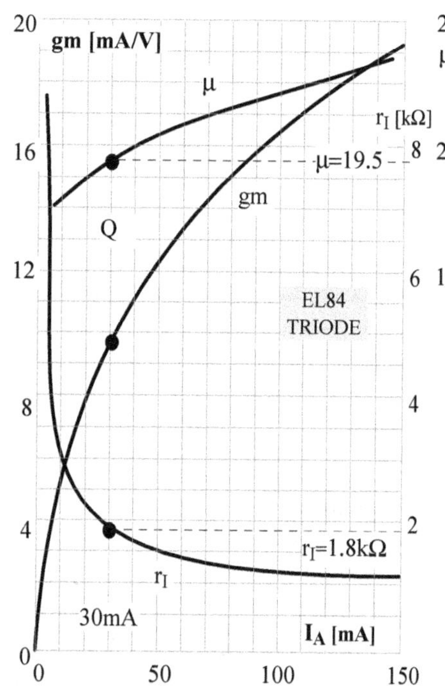

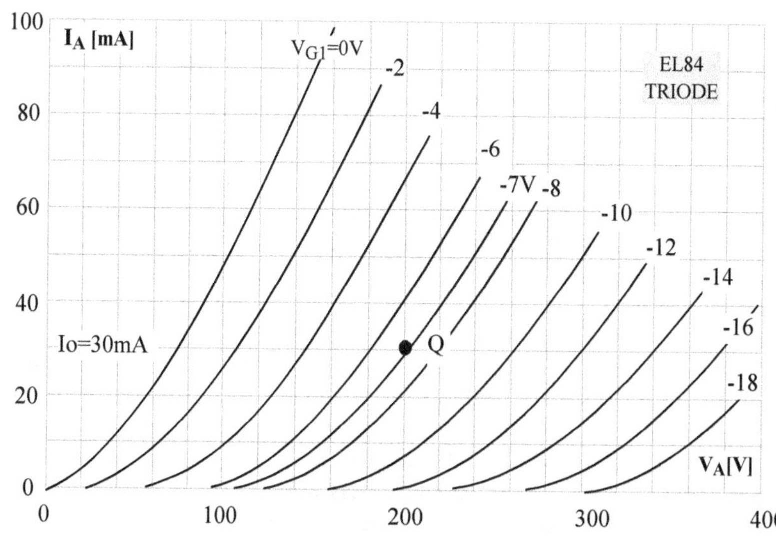

How the three tube parameters vary for triode-connected EL84 (ABOVE LEFT) and anode characteristics of triode-strapped EL84 with the quiescent operating point Q (ABOVE RIGHT)

PRACTICAL DESIGN: ECL82 (6BM8) headphones amplifier

ECL82 (6BM8) is a triode-pentode developed for audio use in radios, tape recorders/players and low cost stereo systems. The 300mA heater version, PCL82, was used in the audio stages of domestic TV sets. The successor, ECL86 (6BW8), was released in 1962.

The anode current of the first stage is 2mA and its amplification factor is $A= -\mu R_L/(r_I+R_L) = -70*27/(28+27) =-34$ The current through the cathode follower is 28 mA.

The amplification factor of the cathode follower is $A_K= \mu R_K/[(1+\mu)R_K+r_I]=6.7*470/(7.7*470+1{,}600) = 4{,}290/5{,}220 = 0.82$ The overall gain is $A=-34*0.82 = -28$.

TUBE PROFILE: ECL82 (6BM8)

- Indirectly-heated audio triode-pentode
- Noval socket, heater: 6.3V/780 mA
- P_{AMAX} (T): 1W, P_{AMAX} (P): 7W
- I_{AMAX} (T): 15mA, I_{AMAX} (P): 40mA
- Max. plate/screen voltage: 300V
- TYPICAL OPERATION TRIODE:
- V_A=300V, V_G=-1.3V, I_0=1.1 mA
- gm = 2.5 mA/V, r_I=28kΩ, μ=70
- Pentode section triode-connected:
- gm=4.2 mA/V, r_I=1.6kΩ, μ=6.7

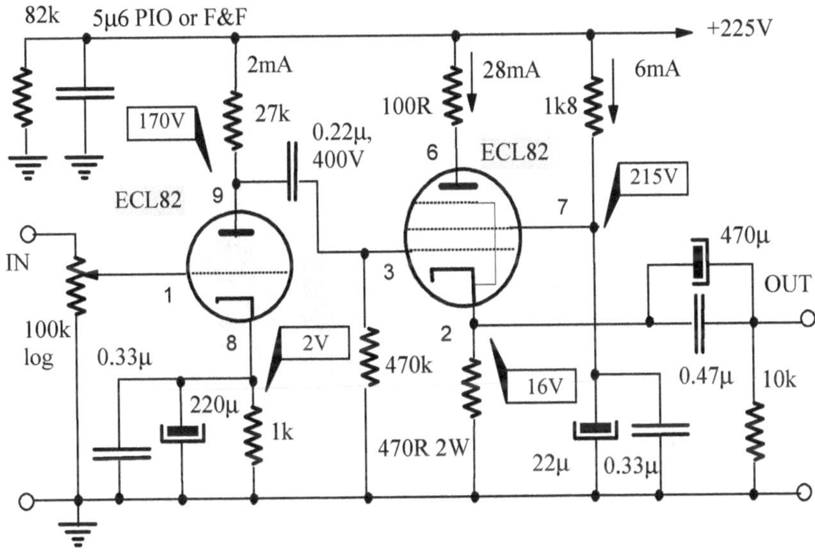

The output impedance is $Z_O = r_I/(1+\mu) \parallel R_K = 1{,}600/7.7 \parallel 470 = 208\parallel 470 = 144\Omega$

The screen current is 5.5-6mA making the cathode current $I_K=I_A+I_S$ = 34mA

The measured -3dB frequency range was 2Hz - 64kHz without load.

Should you wish to eliminate the coupling capacitor between the stages, reduce the DC anode voltage in the first stage by increasing the load resistor, say to 47 or 56kΩ.

Assuming we keep the anode current at 2mA that would increase gain, and since we already have more gain than we need, we can remove the bypassing capacitors in the cathode circuit to reduce gain by local NFB.

Using a 56k in the anode, the DC voltage drop is now 2*56 = 112V, bringing the anode voltage down to 113V. Since the output tube needs a bias of -16V, its cathode needs to be at the potential of 113V+16V = 129V!

The new value of the cathode resistor is R_K=129V/0.034A=3k8. The power dissipated on it would be almost 5W ($P=I^2R_K$=4.39W), so a 10W rated resistor must be used.

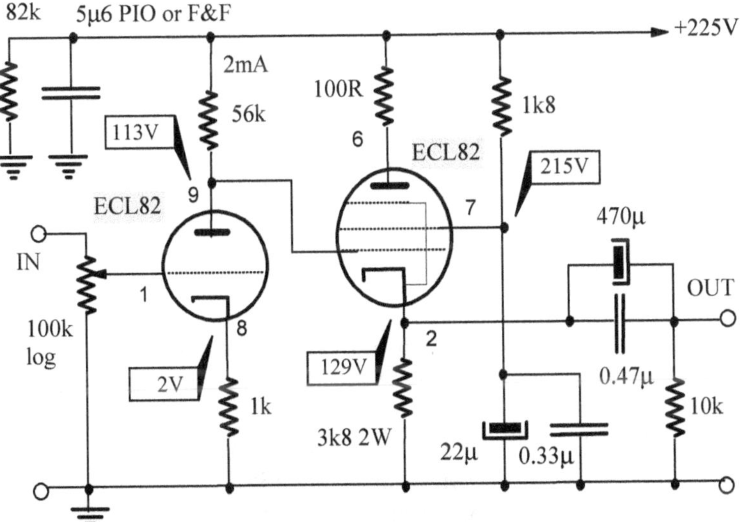

Directly-coupled version

The output impedance is now $Z_O = r_I/(1+\mu) \parallel R_K = 1{,}600/7.7\parallel 3{,}800 = 208\parallel 3{,}800 = 197\Omega$, a rise of 37% compared to 144Ω with capacitive coupling. Plus, we still have an electrolytic coupling capacitor at the output, so the sonic benefits of a direct interstage coupling are questionable to say the least.

Also, should the output capacitor fail (become short-circuited), a high DC current would flow through the headphones which would be instantly destroyed! So, for these reasons, this option is *not* recommended, it was presented here as an educational exercise only.

ELECTRON-RAY TUBES (TUNING INDICATORS OR "MAGIC EYES")

Electron-ray tubes, better known as "magic eyes", are voltage indicators that were used in the golden tube era as cheap and sturdy replacements for analog moving coil meters. They can be found in vintage test equipment such as LCR bridges where they indicated that the bridge was balanced and as recording level indicators in tape recorders. Their most common use was as tuning indicators in radio receivers, so they became known as "tuning eyes".

The circular anode dish is coated with a fluorescent material and is known as the "target. When hit by the electrons from the cathode, the target glows green. The center area is shielded by a round'disk, so any light from the cathode is blocked. The round target tubes need to be suitably mounted so they are viewed from the top.

There are two triodes in one glass envelope. The first triode is a voltage amplifier, the output from its anode drives the deflector electrode of the second assembly.

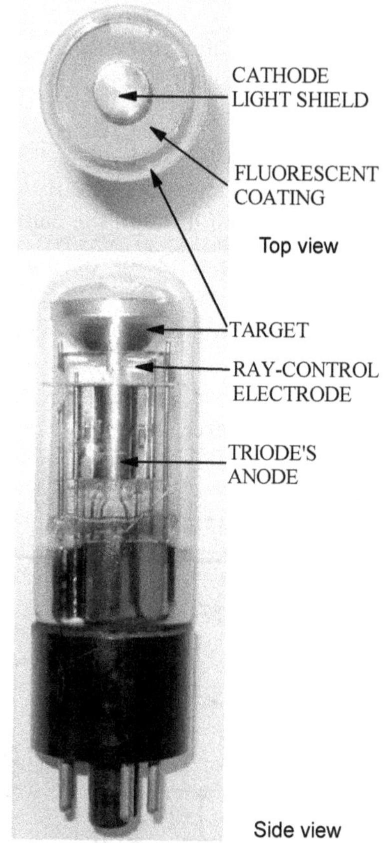

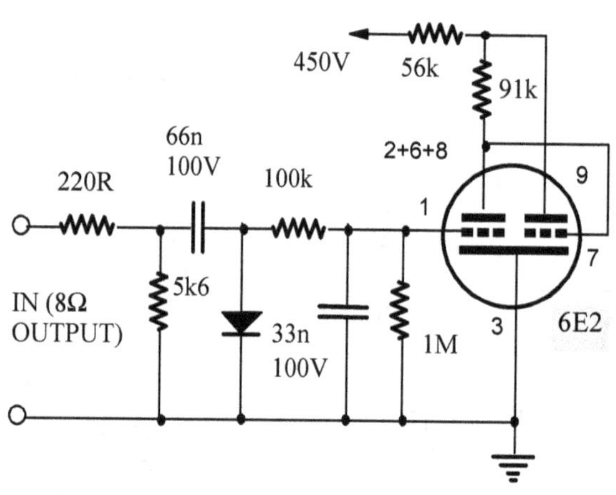

VU-meter with a "magic eye" tube

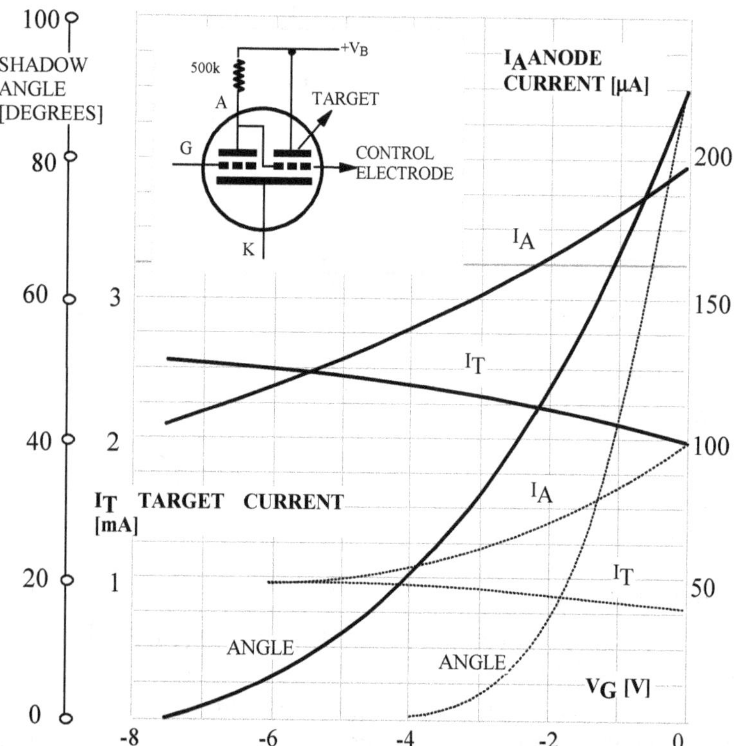

Characteristics of a typical electron ray tube (I_A and I_T vs. V_G)

This "ray-control electrode" is shaped as vertical wire parallel to the cathode. It is typically 150-200V negative with respect to the target, and as such it repels some of the electrons and thus alters the path of the electron beam.

This results in an unlit conical or trapezoidal segment on the target, called a shadow. The more negative the ray-control electrode is made with respect to the target, the wider this shadow becomes.

Not all ER tubes have circular targets. EM84, introduced in 1959, is an example of a more modern dual linear-bar type that uses a Noval socket. These are viewed from the side and are useful as VU-meters (indicators). One bar stretches from top downwards, the other from the bottom upwards.

PRACTICAL LINE-LEVEL PREAMPLIFIER DESIGNS

- ECC40 LINE LEVEL PREAMP
- CASE STUDY: CONRAD JOHNSON'S "ART" LINE-LEVEL PREAMPLIFIER
- E86C LINE-LEVEL TRANSFORMER-OUTPUT PREAMPLIFIER
- PROMITHEUS AUDIO TVC LINE STAGE
- TRIODE - CONNECTED C3g DRIVER STAGE
- ECC82 LINE-LEVEL PREAMP WITH CATHODE FOLLOWER OUTPUT

"The secret is that there is no secret."
Sheldon B. Kopp, in his book *If You Meet the Buddha on the Road, Kill Him!*

ECC40 LINE LEVEL PREAMP

One ECC40 duo-triode would be enough for a stereo line-level preamplifier, but we will use this case also to study the behavior and analysis of triode stages with paralleled tubes, so lets use one duo-triode per channel.

With a bypassed cathode resistor the amplification would be too high, so we decided to introduce a mild local negative feedback in the cathode circuit by leaving it un-bypassed.

The two paralleled triodes must behave like one, meaning the tubes must be matched to the highest degree possible. Matching them in one quiescent point is absolutely necessary, but to ensure they are also matched dynamically, across the operating range, their anode characteristics should be identical.

We will start with the circuit and actual measured results and work backwards, to check if graphical estimation form the curves would give us the same results.

Firstly, since there are two triodes in parallel, we change the current values on the vertical axis of the anode curves to double their values for one tube. Secondly, since we have an un-bypassed cathode resistor, we need to draw a load line for $R_L+R_K = 48k\Omega$

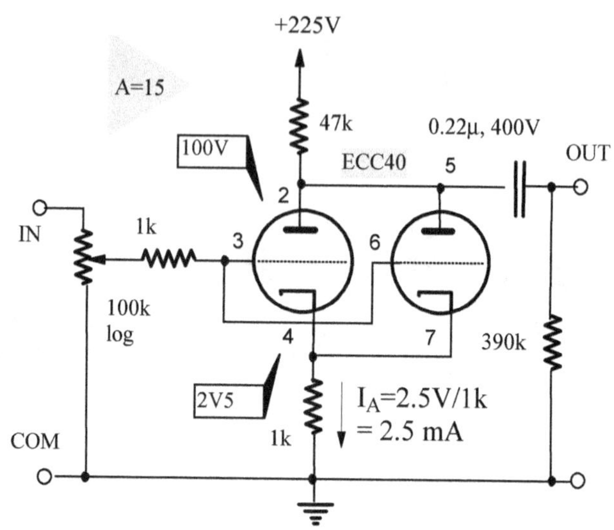

PARALLELED TUBES

REDUCED INTERNAL IMPEDANCE: r_I/N
INCREASED TRANSCONDUCTANCE: gm*N
AMPLIFICATION FACTOR DOES NOT CHANGE
r_I = internal resistance of one tube
gm = transconductance of one tube
μ = amplification factor
N= number of paralleled identical tubes

Audio circuit analysis

Point Y is our supply voltage of 225V, and point X is calculated as $I_X = E_{BB}/(R_L+R_K) = 225/48{,}000 = 4.7mA$

Once we draw the loadline we can position our point Q at its intersection with I_A=2.5mA or with vertical line V_A=100V, which, according to our measurements, must be the same point on the graph, and it is.

Next we need to change the values of bias voltages as well, by using the formula $V_G' = V_G+I_AR_K$, as we have seen before in the section dealing with cathode followers. For V_G=0 we get $V_G' = 0+4mA*1{,}000\Omega = 0+4=4V$ Instead of V_G of -2V as on the original graph, the new V_G' is $V_G' = -2+3*1 = 1V$!

Since Q is on the new V_G=0V, with a grid signal of +/-3.5V around that bias $\Delta V_G = (3.5V-(-3.5V)) = 7\ V_{PP}$.

Now we draw vertical lines down to the voltage axis and read $\Delta V_A = V_{MAX}-V_{MIN} = 155 - 45 = 110\ V_{PP}$, so the voltage amplification factor is $A_V = -\Delta V_A/\Delta V_G = -110/7 = -15.7$

The measured amplification factor was -15, so very close results and a very good approximation from the graphs.

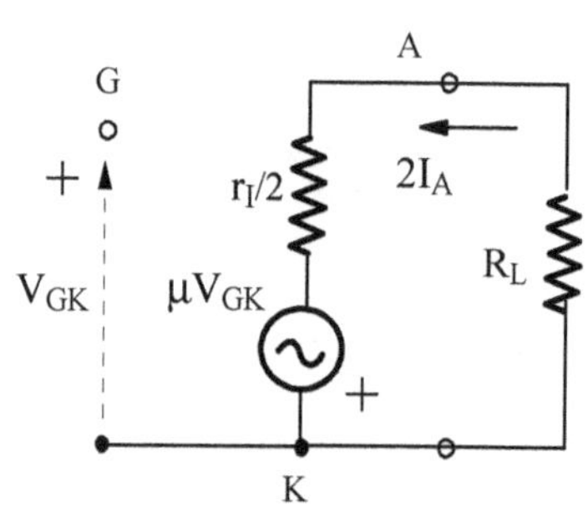

ABOVE: The voltage source model of a common cathode stage with two paralleled triodes

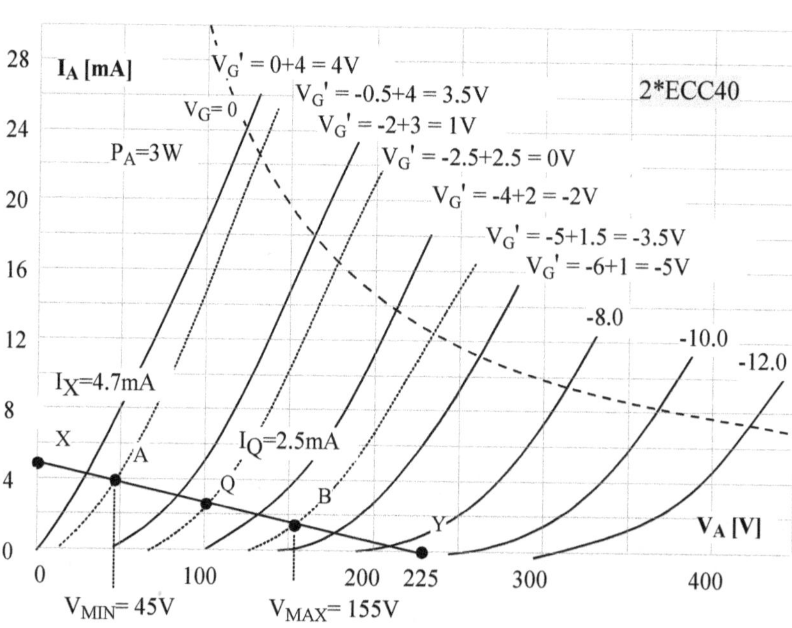

RIGHT: Quiescent point Q, 47k load line and voltage swings for the chosen design

Power supply (with ECC82 used as a duo diode rectifier!)

A few unusual design features deserve to be discussed here. Firstly, the mains transformer TR1 was from a solid-state application and only had two 12 Volt secondaries. To get a high voltage for anode supply, we connected a second 240V to 12 V mains transformer (TR2) with CT in a back-to-back, balanced arrangement. The CT of the second transformer was grounded, which also grounded the heater of the rectifying tube V3. Neither of the power transformers had an electrostatic shield, but due to this clever arrangement and a balanced connection, such a shield was not necessary!

The balanced transformer connection eliminates common-mode signals such as hum and RF interference. More on that ingenious arrangement in Volume 2 of this book, in the chapter on power supplies.

Secondly, you can see how a tube rectifier can be used together with two solid state diodes in a full bridge configuration. A four-diode bridge is necessary when the HV secondary winding does not have a CT (center tap).

Next, notice that 12AU7 (ECC82), an audio amplifying duo-triode, was used as a rectifier tube! The two triodes were connected as diodes, their grids strapped to anodes. 6CG7, 5687, 6SN7 and many other beefier duo-triodes also make great rectifiers for preamps.

The heater rectifiers are 50V Schottky diodes pulled out of the PC switch-mode power supplies and the choke was a small mains transformer with open secondary, a trick also covered in Volume 2.

Finally, the last filtering capacitor is not an elco, but a good quality 5.6µF film&foil capacitor, either polypropylene or PIO. The value is not critical, anywhere from 5.6 to 22µF would do.

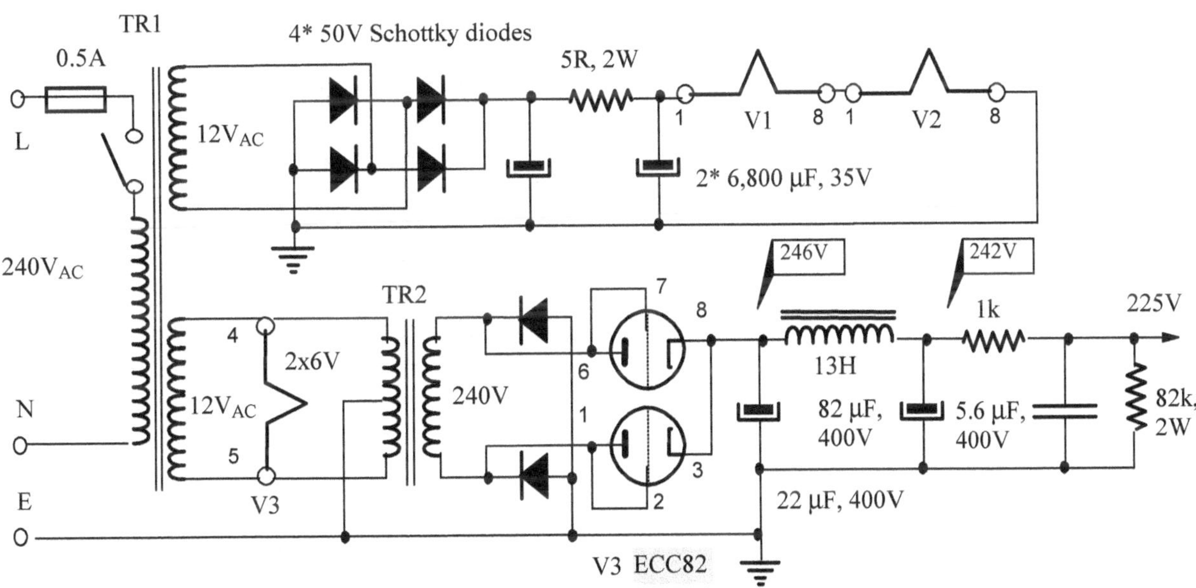

Construction details

The prototype was built on a pair of smallish timber photo frames, finished in black lacquer. The power supply was on one and the audio section on the other "chassis", with a 4-core umbilical cord connecting them. The top and bottom covers were cut out of a thick solid aluminium slab.

Fitting the power supply into its frame was a challenge, especially the two power transformers and a choke, which all had to be low-profile units. Due to well filtered heater and anode voltages and totally separate power supply section, the preamp was very quiet.

The final result was esthetically pleasing, the natural anodized aluminium finish and solid aluminium control knobs contrasting with black bases, with a hint of gold from RCA sockets and the front logo. The petite preamp looked very cute.

CASE STUDY: CONRAD JOHNSON'S "ART" LINE-LEVEL PREAMPLIFIER

This case study is neither a critique nor a recommendation of this particular model, but an exercise to show you how even a few pieces of information about a particular design can serve as a base for reverse-engineering of the audio circuit, without ever having it in front of you on a test bench.

The preamplifier uses no less than ten ECC88 triodes in parallel. Its diagram was not available online, but according to a magazine review, the measured gain was 25.4dB and the output impedance was 475Ω.

The output impedance is $Z_O=R_Lr_{IP}/(R_L+r_{IP})$ and the amplification factor is $A=\mu R_L/(r_{IPI}+R_L)$. The internal resistance r_{IP} is for 10 tubes in parallel. We have two equations with two unknowns. After expressing R_L from the second equation we get $R_L=Ar_{IP}/(\mu-A)$, which we substitute into the first equation and get $Z_O=Ar_{IP}/\mu$, from which we calculate $r_{IP}=766\Omega$ and $R_L=1{,}250\Omega$

If $r_{IP}=766\Omega$ for 10 tubes in parallel, then $r_I=7.66$ kΩ for one tube. On the anode characteristics we draw a horizontal line through 7.66kΩ, find the Q-point and read $I_0=3$mA By drawing a vertical line through Q we get $\mu=28$!

$A_{dB}=20\log A$, so $A=\text{invlog}(A_{dB}/20)= 18.6$ The amplification or voltage gain of the stage is

$A=\mu R_L/(r_{IP}+R_L)$

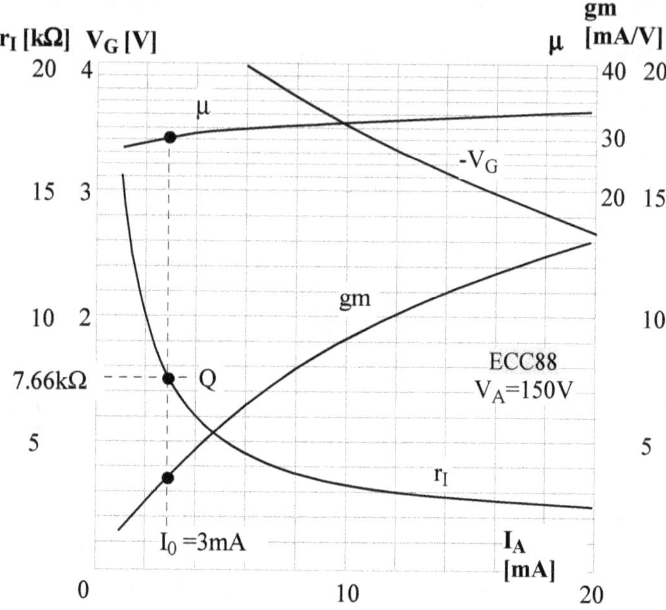

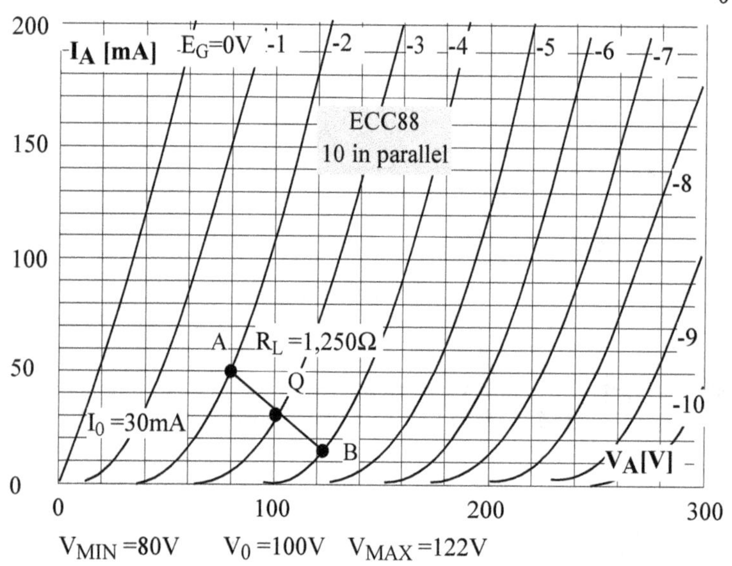

Now we can calculate the load resistance likely to be used:

$R_L= Ar_{IP}/(\mu-A) = 18.6*766/(30-18.6) = 1{,}250\ \Omega$

The current through ten paralleled triodes is ten times the current through one tube or $I_A=10*3\text{mA} = 30\text{mA}$. The ordinary anode characteristics (for one tube) are valid, but we must multiply the Y-axis values by 10 (the anode curves on the left).

The grid bias is around -3V. To estimate voltage gain, from $V_G=-3V$ we go 1V to the left, to $V_G=-2V$ and then 1V to the right, to $V_G=-4V$ so $\Delta V_G=2V_{PP}$ The anode voltage swing is $\Delta V_A=122-80= 38V_{PP}$, so $A=\Delta V_A/\Delta V_G= 38/2 = 19$ The measured gain in the review was 18.6 so it seems we are on the right track! To get 3V on the cathode with 30mA cathode current, we need $R_K=V_K/I_K = 3/0.030 = 100\Omega$

Of course, our analysis assumes that a common anode and a common cathode resistor is used. This may or may not be a correct assumption. We have also assumed that the common cathode resistor is bypassed by a capacitor.

In the next version of the ART preamplifier CJ changed over to the Russian tube known as 6N30P. A few other manufacturers jumped on the bandwagon, BAT (Balanced Audio Technology) and Audio Research among others. Not to be outdone, if not in the audio sphere, then at least in the marketing hype, CJ's competitor BAT even touted it as "The Super Triode".

From its tube profile we immediately notice a very low internal resistance of 800-900Ω and a high transconductance. Alas, that means the amplification factor is low, only 15! This triode operates at low anode voltages and high anode currents. V_{HKMAX} is a very high 400V, compared to only 90V in many other triodes (150V for ECC88). This tube would be great in totem pole arrangements such as SRPP or μ-follower.

Three times lower r_I means that ten ECC88 can be replaced by only three 6N30P, albeit at halved amplification factor of the preamplifier .

TUBE PROFILE: 6N30P

- Indirectly -heated duo-triode
- Noval socket, 6.3V/825 mA heater
- $V_{HKMAX}=400V_{DC}$, $P_{AMAX}=4W$
- $V_{AMAX}= 250V$, $I_{KMAX}=100$ mA
- $R_{GMAX}=300k\Omega$
- TYPICAL OPERATION:
- $V_A=80V$, $V_G=-2V$, $I_0=40$ mA
- gm = 18 mA/V, μ=15, $r_I = 840\Omega$

E86C LINE-LEVEL TRANSFORMER-OUTPUT PREAMPLIFIER

E86C (EC86)

E86C is a special quality (gold-plated pins) version of EC86 or American 6CM4. It's a long life single triode with medium-to-low internal resistance, very high amplification factor and high mutual conductance. Low inter-electrode capacitances result in a wide frequency bandwidth, so EC86 has a very high figure of merit compared to other triodes.

It is a very linear triode, which means low distortion, it operates at relatively low anode voltages of 150-200V at high anode currents of up to 20 mA. Not microphonic at all, it's relatively cheap to buy (mainstream manufacturers have not discovered it yet), and has a pleasing sonic character. Considering all factors, despite its higher internal resistance, E86C is more deserving of the "Super Triode" title than 6N30P.

Please note that despite its similar name, ECC86 is a totally different tube, you should recognize by now that "CC" in the name means it's a duo-triode, while EC86 is a single triode!

TUBE PROFILE: E86C

- Indirectly-heated SQ triode
- Noval socket, gold pins
- Heater: 6.3V, 165 mA
- Life: 10,000 hours minimum
- V_{AMAX}=250V, V_{HKMAX}=100V_{DC}
- P_{AMAX}=2.4W, I_{AMAX}=20 mA
- TYPICAL OPERATION:
- V_A=185V, V_G=-1.5V, I_0=12 mA
- gm=14.0 mA/V, μ= 68, r_I = 4.8 kΩ

Capacitorless line stage

In this design, there are no capacitors in signal path. The output transformers are STANCOR WF-35, an equivalent of UTC A-25, with 15kΩ primary impedance, and various output impedances, from 600Ω down to 50Ω, depending on which secondary terminals are used. The DC resistance of the primary is R_P=1.6kΩ

Voltage swings from the graph below: ΔV_G = -3.4V, ΔV_A=250-85 = 165V, so $A_V=\Delta V_A/\Delta V_G$=165/-3.4 = -48.5 The measured figure is A_V=-50, remarkably close to the predicted -48.5V!

The positive anode voltage swing is 170-85V = 85V, while the negative swing is 250-170 = 80V. The 2nd harmonic distortion is D_2=(85-80)/2*(250-85)=5/330=1.5%! Considering that distortion figure is at the maximum swing, it's a great result.

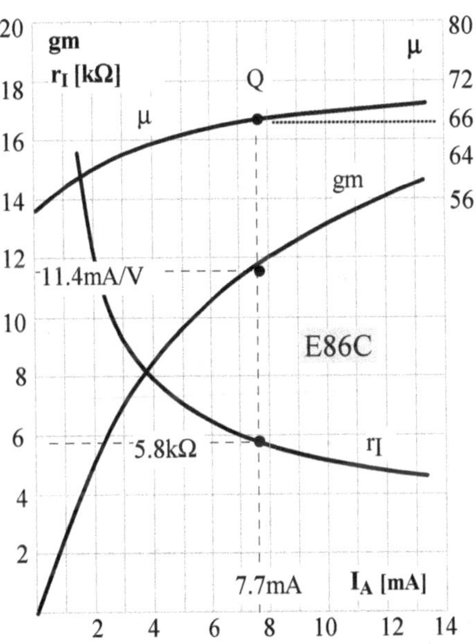

The chosen quiescent operating point Q on the μ-gm-r_I diagram of E86C is illustrated above.

Notice the very high amplification factor of 66 and low internal resistance (6k8), a rare combination. This means the mutual conductance is high (11.4 mA/V), resulting in a dynamic sound.

Since the anode load is a transformer with low DC resistance of its primary winding, high anode current does not demand a very high anode supply voltage, as with resistive loads. In this case, V_{BB} is only 170V!

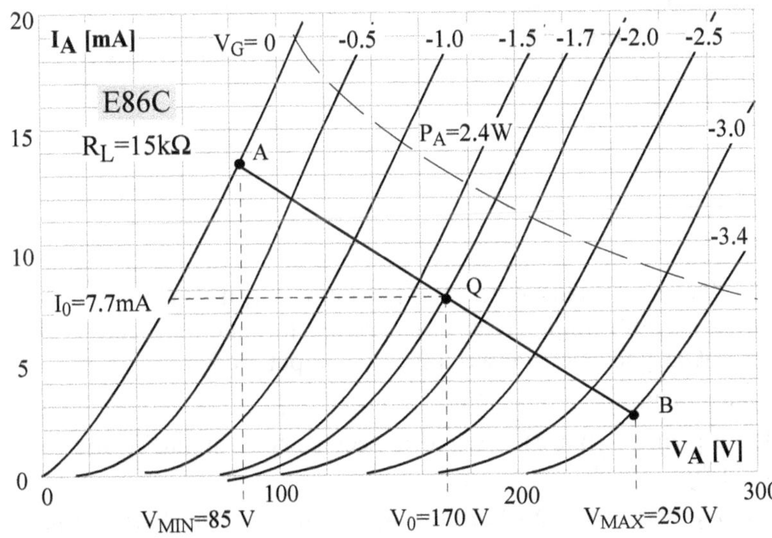

Why negative feedback?

The maximum allowed DC current through the WF-35 primary is 8 mA, we have 7.7 mA. The un-bypassed 220Ω cathode resistor provides local negative feedback and also enables a mild global negative feedback to be brought in from the output transformer's secondary via the 5k6 resistor.

NFB improves the bass response. If removed, the lower -3dB frequency would increase significantly and you would lose the bass frequencies bellow 40-50Hz!

15kΩ to 600Ω means the impedance ratio of the output transformer is IR=25 and the voltage ratio is 5. The triode circuit amplifies around 50 times, and the overall amplification factor is 12, so with 1V input up to 12V are available at the output.

The power supply is simple, a solid-state bridge, followed by a CLC and CRC filters. Of course, another choke could be used instead of the 4k7 resistor for better filtration if required.

The 2μF decoupling capacitor should be the best quality you can find or afford, either film&foil (copper, silver, etc.) or paper-in-oil.

82k is a bleeder resistor that not only discharges the three elcos once the preamp is switched off, but also improves voltage regulation during operation.

The heater voltage is not regulated, it's not even DC but AC. Firstly, we wanted this to be a simple design, and secondly, there was no sonic advantage in DC heating.

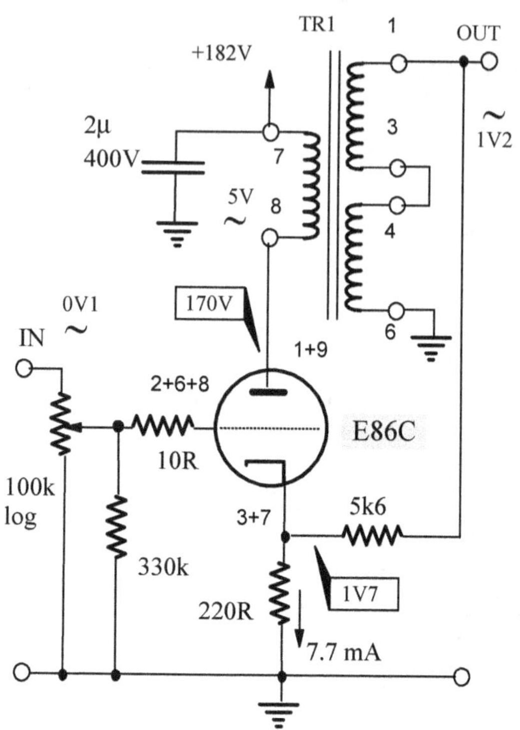

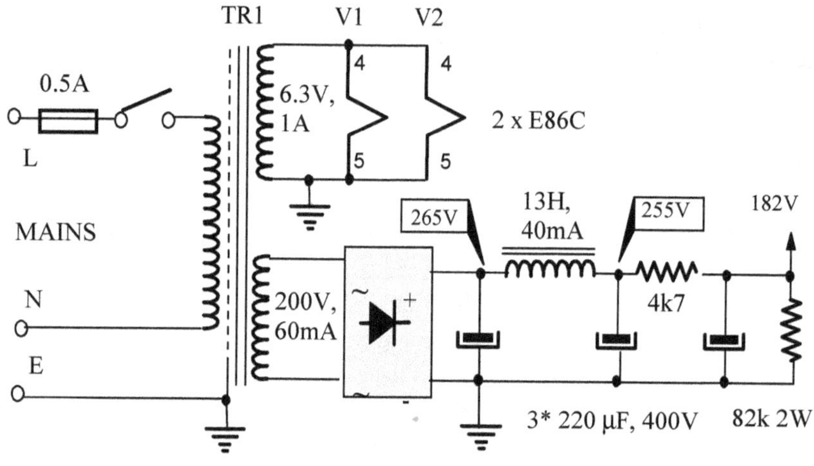

How does it sound? Detailed, clean and transparent, without artificial tube sweetness, so typical of many other designs. It is also free from the frizziness and other artifacts associated with even the best of coupling capacitors.

The output transformers do introduce their own subtle coloration. Luckily, their sonic signature is pleasant, balancing well the relatively neutral (but never sterile) character of E86C triodes.

Balanced outputs? No worries!

Dual secondaries on output transformers lend themselves to a balanced preamplifier output or, if used in a power amplifier, as a phase splitter for driving push-pull grids! If both single-ended (RCA connector) and balanced (XLR connector) outputs are needed, a simple switch makes it easy to select between the two output modes.

The circuit diagram on the next page shows this arrangement.

Pin 1 of the balanced output is always grounded, the positive (pin 2) is wired to secondary terminal #1 and the negative output (pin 3) is wired to secondary terminal #6.

When the switch is in the "RCA" position, the center tap (terminals 3&4) of the output transformer is disconnected from ground and a full secondary (between terminals #1 and #6) is connected across the unbalanced output.

RIGHT: The split secondary winding of the output transformer makes the implementation of a balanced output easy.

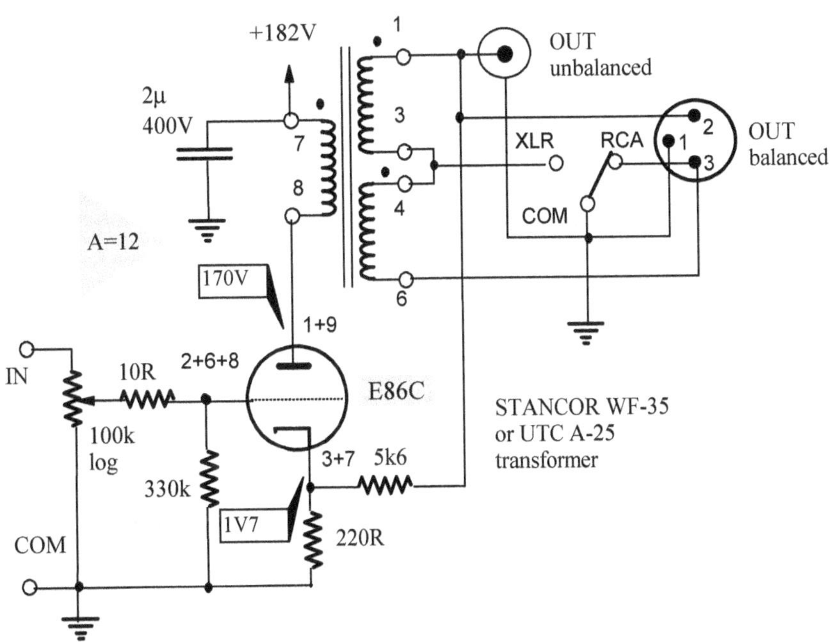

The loaded versus unloaded transformers

While output transformers of power amplifiers are always loaded with the speaker impedance, interstage and preamplifier output transformers can have various loads on their secondaries. One option is to leave the secondaries unloaded, the situation when an interstage transformer drives a grid circuit of the next stage. With no grid current flowing, the grid impedance is very high, compared to the transformer's own primary impedance it's practically infinite. With no impedance reflected back to the primary, primary impedance at low and midrange frequencies is the reactance of the primary winding, determined by its inductance L_P: $Z_P=X_L=\omega L_P$.

The drawbacks of unloaded secondary are best identified by comparing its A-f curve to those of loaded transformers. Unloaded secondary yields the highest midrange amplification, but low frequencies are rolled-off very early, so the lower -3dB frequency f_L is relatively poor (high). Also, due to low damping, there is a pronounced peak at the transformer's resonant frequency, which for lesser quality units may fall into the audible range (at 20kHz as illustrated). This is a general discussion and the curves illustrated are NOT for the amplifier described above!

As the secondary load is increased, the midrange gain of the stage drops (from A=40 to A=30 in the illustrated example), but f_L is lowered, f_U is raised slightly and the resonant peak flattened.

Larger loads (R_S>$R_{NOMINAL}$) completely remove the resonant peak and widen the frequency range further, but at the expense of significantly reduced amplification factor (halved in this case).

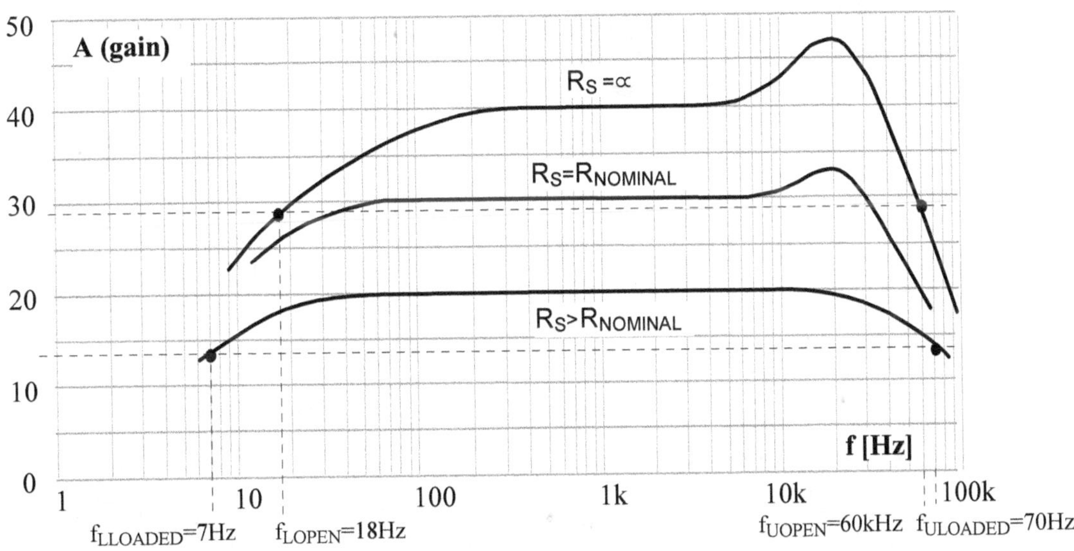

Amplitude vs. frequency characteristics for a transformer-coupled preamplifier or a driver-stage with three different secondary loading: open secondary, nominal secondary load impedance R_S and load impedance much higher than nominal.

PROMITHEUS AUDIO TVC LINE STAGE

Made in Malaysia, this unorthodox line-level preamp is two preamps in one. It can be used as a passive TVC, transformer volume control preamp, or an active preamp, a paralleled triode stage with a transformer output. The TVC is in circuit in both cases. There are four RCA inputs and two paralleled RCA outputs.

There is no name plate, no model number, not even a manufacturer's logo, which is weird to say the least. The knobs on the photo are not original, the original ones were made of a lighter-colored timber and two kinds of timber just didn't look right together. The preamp did not have any feet, so we installed those as well.

This case study was included in the book for two reasons. Firstly, this was the only preamp with transformer volume control that we came across. Secondly, to illustrate how a preamp that measures very poorly (see oscillograms on the next page) can sound OK despite its technical imperfections.

RIGHT: Block-diagram of one channel

BELOW: Circuit diagram of the power supply and one audio channel. The input switching circuitry and transformer volume control are not shown.

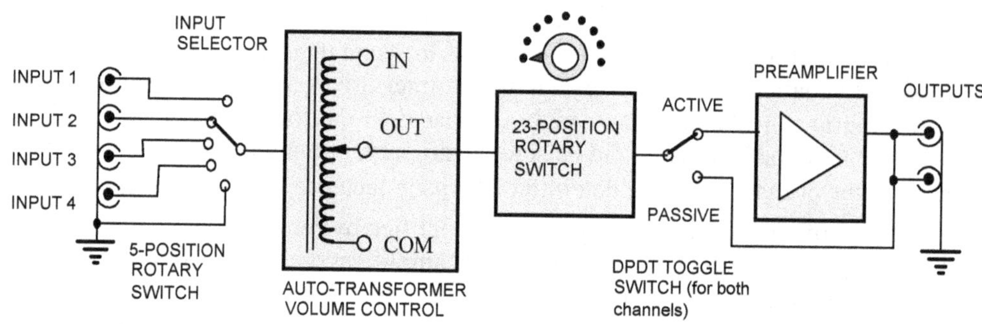

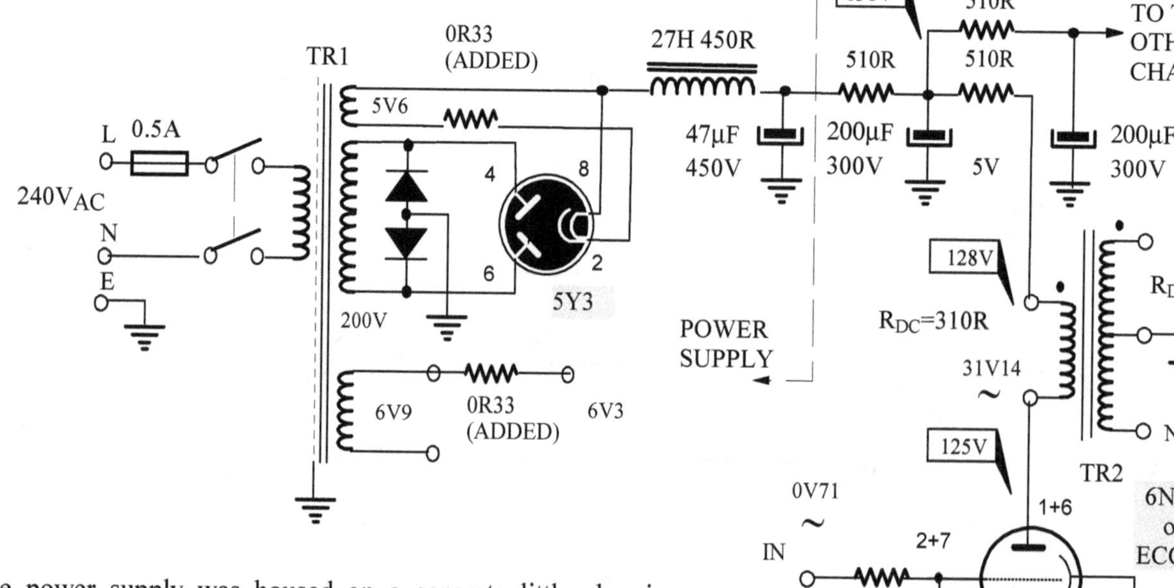

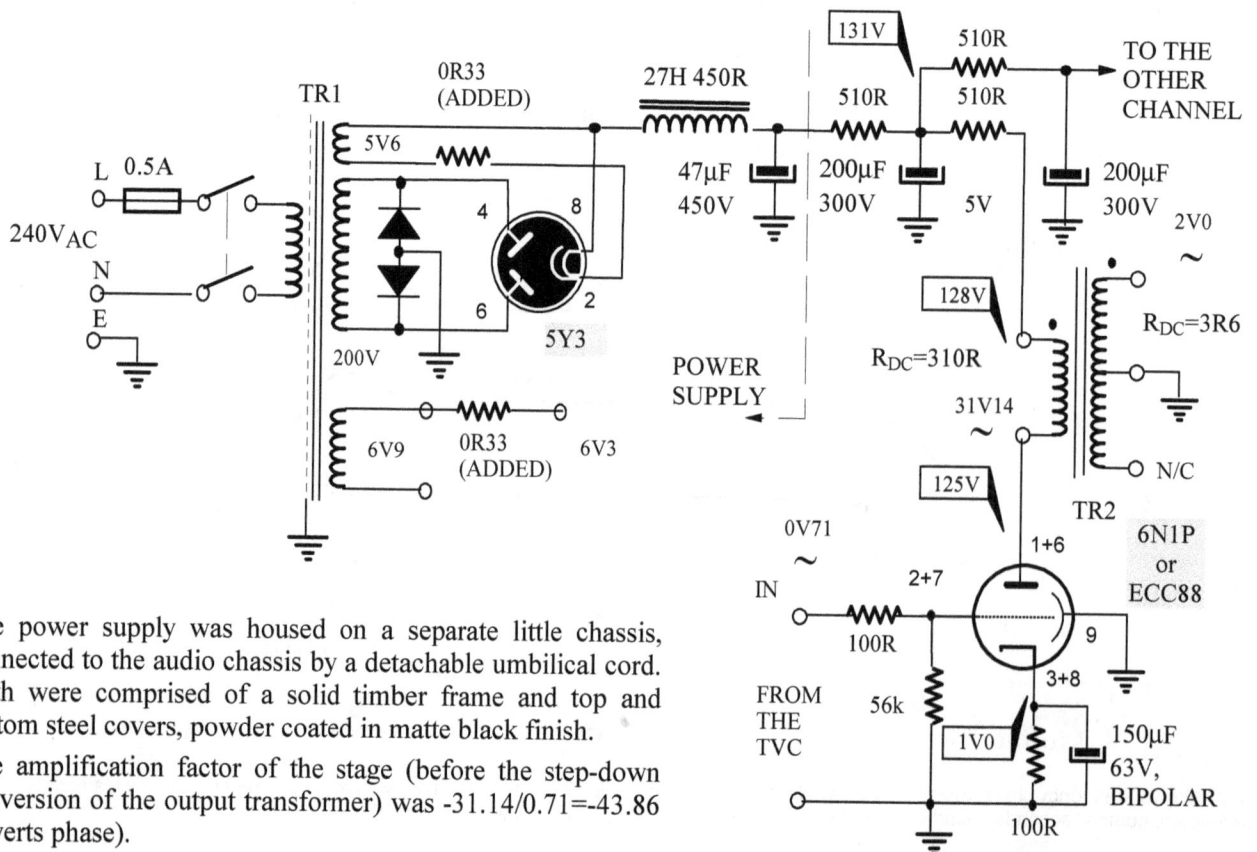

The power supply was housed on a separate little chassis, connected to the audio chassis by a detachable umbilical cord. Both were comprised of a solid timber frame and top and bottom steel covers, powder coated in matte black finish.

The amplification factor of the stage (before the step-down conversion of the output transformer) was -31.14/0.71=-43.86 (inverts phase).

Since μ for 6N1P is declared as 35+/-7, it seems that these tubes had their μ at the high end. The output signal was $2.0V_{RMS}$, meaning the output transformer's voltage/turn ratio was 31.14/2=15.6 (step-down). The impedance ratio was thus $IR=TR^2=15.6^2=244$! The overall (output-to-input) amplification factor at 1 kHz was 2/0.71=2.82.

It seems the preamp was designed for 220V, since all heating voltages were too high. The rectifier heater voltage was 5.6V and the output tubes' heaters had 6.9V on them. Low value resistors were added in series with heaters to bring voltages down to the right levels. The anode voltage was not critical and needed not to be lowered.

Due to a low secondary high voltage ($200V_{RMS}$), a hybrid bridge rectification needed to be used, a tube rectifier plus two solid state diodes. There was only one choke (1) shared between channels. The separate volume controls used a 23-position rotary switch, connected to its auto transformer with 23 taps (3). EI66 laminations were used with S=3.6cm stack thickness.

The two halves of ECC88 were paralleled, each operating at Q illustrated on the graph, passing 5mA of anode current. The internal resistance of each triode was around 5kΩ or 2k5 for both together, with estimated μ=30.

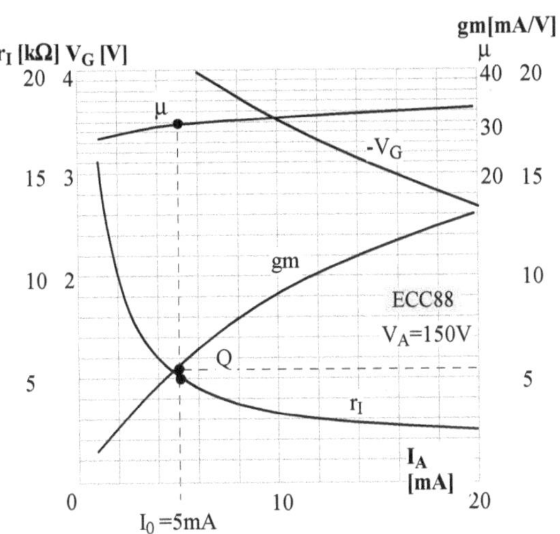

MEASURED RESULTS:

- A= 2.82 @1kHz
- BW: 14Hz - 20 kHz (-3dB)
- Z_{OUT} = 215Ω

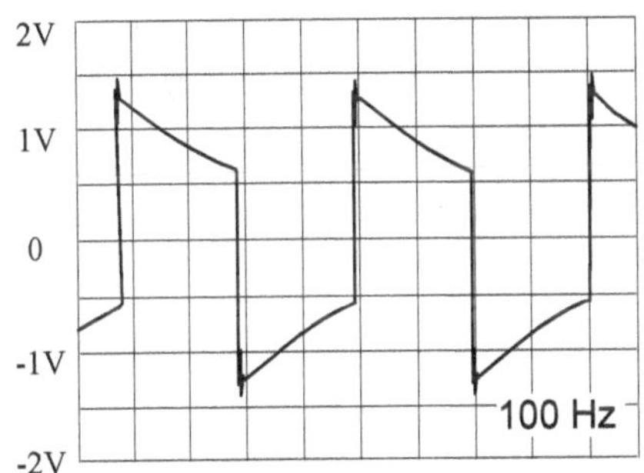

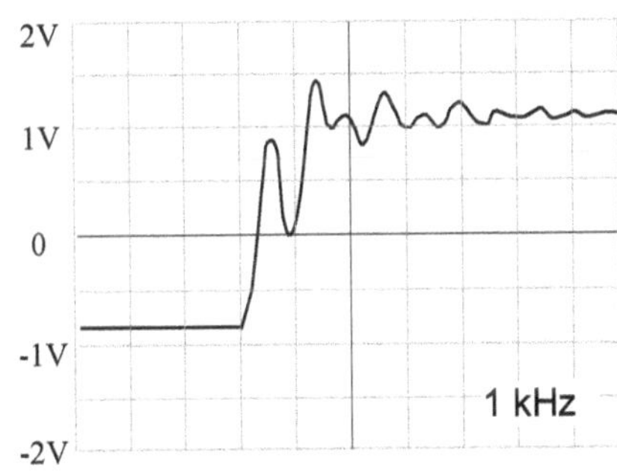

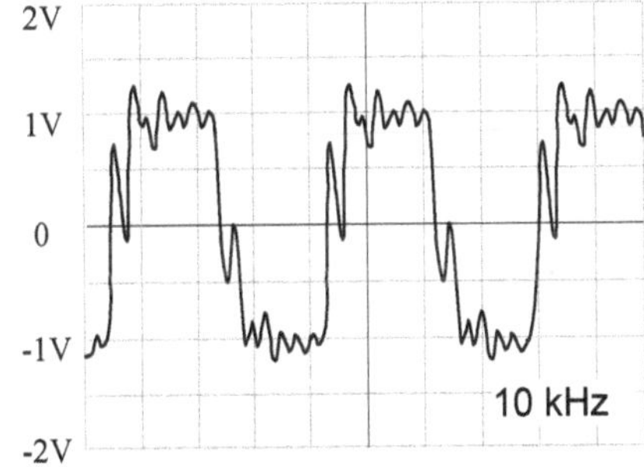

ABOVE: Square wave tests at three common frequencies reveal a severe distortion, slew-rate limitation and narrow bandwidth.

The frequency range was very limited, only up to 20kHz (-3dB). The square wave was severely distorted at all frequencies, with pronounced oscillation.

This was due to the resonances in input and output transformers. All tests and measurements were performed with unloaded outputs, but loading them with a 100k power amplifier did not change the shape or the quality of waveforms, so the loading was not an issue.

The input selector switch was dual, common for both channels (4), but the switches selecting the operating mode, passive TVC or active preamplification, were separate for each channel (5).

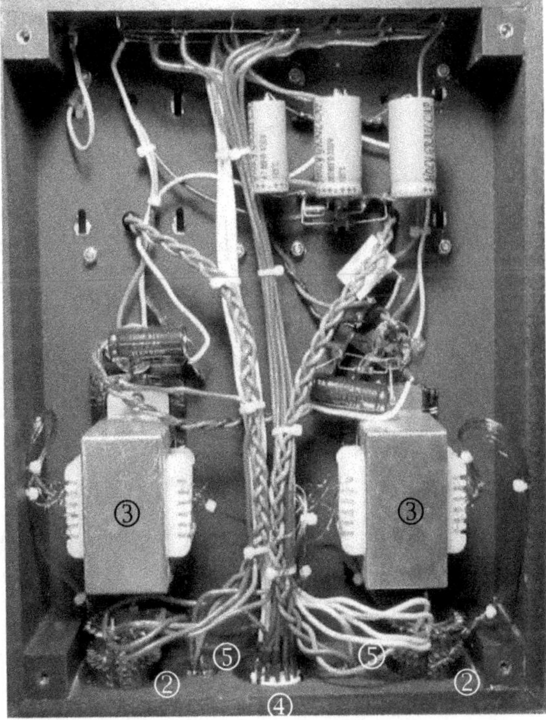

RIGHT: The internal view of the preamplifier

TRIODE - CONNECTED C3g DRIVER STAGE

C3g and a slightly more powerful C3m were developed for German Post and used in repeater amplifiers in telephone applications. Characterized by high reliability, long life, low noise and low microphonics, these pentodes make extremely linear triodes, with a decent amplification factor (40), low internal impedance of around 2kΩ and very high mutual conductance of 17 mA/V.

Despite their looks, these are not metal but glass tubes, the black metal housing is just an external shield and can be easily removed. Even the loktal base can be removed and pins soldered onto an octal base should you have the time and patience to do so.

This tube is sensitive and powerful enough to be used in single-stage headphone amplifiers. Its low output impedance as a triode makes it a great driver for Class A_2 triode output stages. This suggestion for such a driver operates at high anode currents of around 15 mA, so it sounds good, and has enough amplification to drive 2A3 and 300B tubes to full power with input voltages of 1.5V or more.

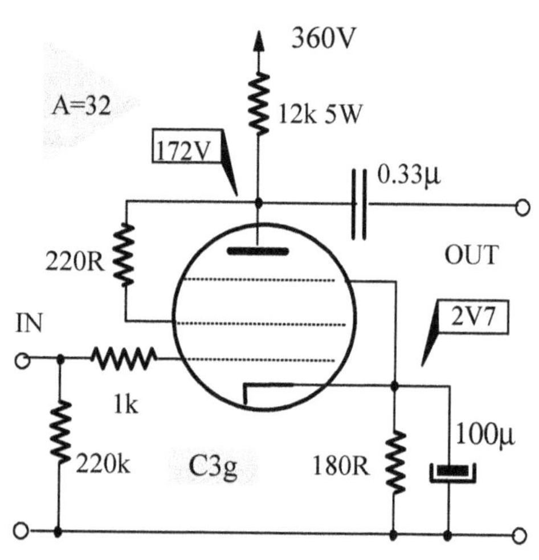

TUBE PROFILE: C3g

- Pentode for wideband amplifiers
- Loktal socket, heater: 6.3V, 370 mA
- V_{AMAX}=250V_{DC}, V_{SMAX}=220V_{DC}
- P_{AMAX}=3.5W, P_{SMAX}=0.7 W
- I_{KMAX}= 30 mA, V_{HKMAX}=120V max.
- TYPICAL OPERATION (pentode):
- V_A=220V, V_S=150V, R_K=115Ω
- I_0 =13mA, I_S=3.3 mA, r_I=15kΩ, P_{OUT}=1.2W
- TYPICAL OPERATION (triode):
- V_A=V_S=200V, R_K=180Ω, I_0=17mA
- gm=17 mA/V, μ=40, r_I = 2.3 kΩ

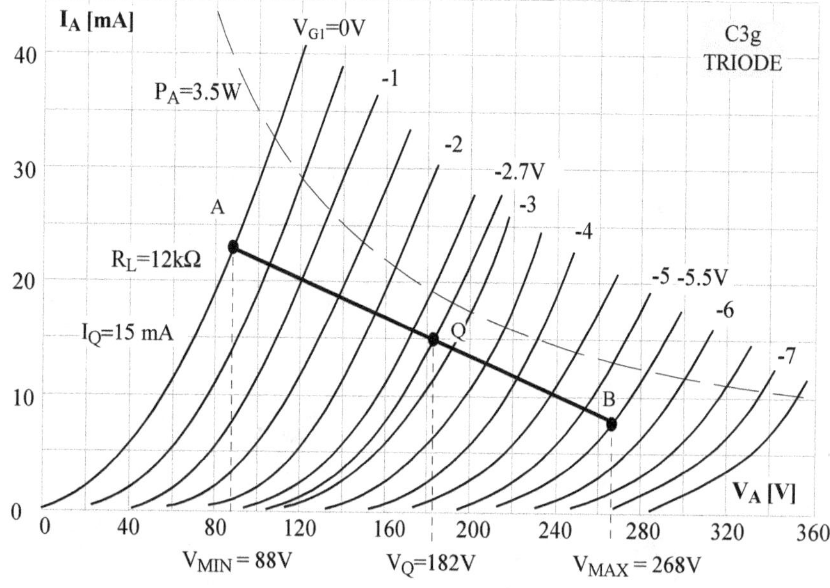

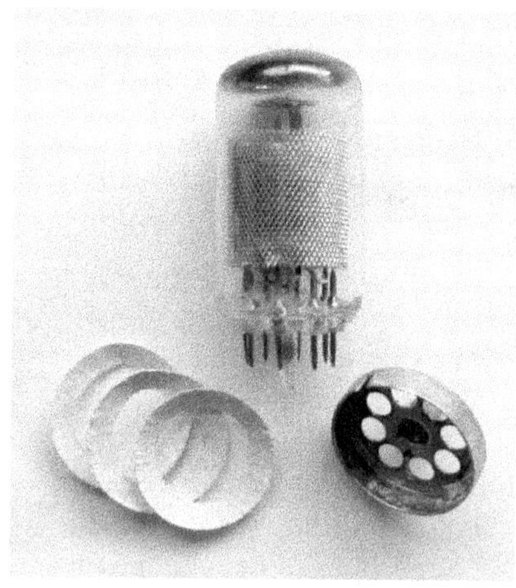

LEFT: The naked C3m pentode after the removal of its metal shield, Loctal base and padding rings.

Voltage amplification and 2nd harmonic distortion estimation from the anode characteristics:

ΔV_G = -5.5V, ΔV_A=268-88 = 180V, A_V=$\Delta V_A/\Delta V_G$=180/-5.5= -32.7

$\Delta V+$ = 94V, $\Delta V-$ = 86V, D_2= $(\Delta V_+ - \Delta V_-)/2\Delta V_A$=8/360= 2.2%

This is a remarkably low distortion for such a large voltage swing, so the tube looks very promising. With an anode choke of 200H or higher inductance, the amplification would be around 40!

ECC82 LINE-LEVEL PREAMP WITH CATHODE FOLLOWER OUTPUT

Why ECC82 and not 6SN7?

Let me guess what some of you are thinking right now: "ECC82? 12AU7? No way, wrong tube, man, these Noval tubes don't sound right!" Over the years a few audiophiles criticized ECC82 and their alleged high distortion, probably after reading "Valve Amplifiers" by M. Jones, in which his tests different preamp tubes for distortion. Most would rather see a 6SN7-based design. Others, all alarmed, will get out their vampire stakes: "What? A cathode follower? Are you serious? Didn't you know that they are bad for the sound??" and so on and so forth ...

Well, firstly, just because Morgan Jones says something does not mean that it should be taken as gospel or that it's correct. Measurements are one thing and listening is another.

Sure, 6SN7 preamps do sound velvety and seductive but nothing beats the transparency, resolution and speed of this design! We tried paralleled tubes, differential amplifier, SRPP and a myriad of other designs, but none sounded as good as this baby.

Finally, the myth that cathode followers are "bad for the sound" is just that - a myth. One of the best amplifiers of all times, the venerable Ongaku, uses a cathode follower to drive the output 211 tubes. If that doesn't put your mind at ease, nothing will.

Simply, this line stage sounded better than all 6SN7 preamps we tried.

The audio circuit

The anode current of the first stage is I_1=2.6V/1k2 = 2.2 mA and its amplification factor is $A= -\mu R_L/(r_I+R_L)$ = -17*60/(7.7+60) = -17*0.886 = -15

The current through the cathode follower is I_2=72V/10k = 7.2 mA.

Its amplification factor is

$A_K= \mu R_K/[(1+\mu)R_K+r_I]$ =

= 17*10/(18*10+7.7) = 0.91

The overall amplification without NFB is A=-15*0.906 = -13.6

There is a mild negative feedback through the 180k resistor (1), which brings the overall amplification factor from 13.6 down to around 8 (-4.6dB). For a NFB-free design disconnect this resistor from the output and connect that end to ground, so it's connected between the grid of V1 and ground.

The measured output impedance is very low, around 215Ω. No need for fancy and complex cathode followers, the simpler the better!

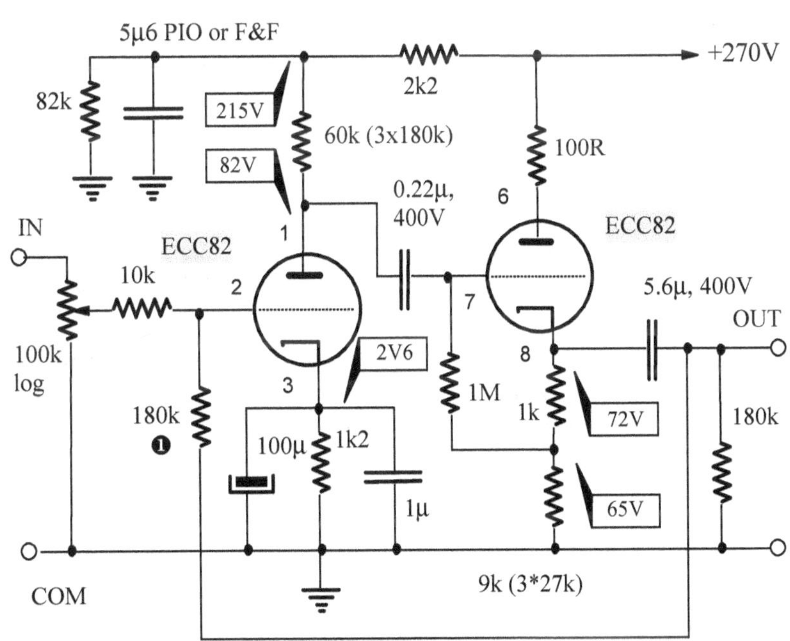

The power supply

You can use this universal power supply for any phono stage or line preamp (next page). If you are allergic to solid state devices simply replace each diode bridge with a dual-rectifier tube such as EZ90, EZ81, 6X4, 6V4 and similar. Of course, the mains transformer must have CT (Center-Tapped) secondaries in that case.

Also, if you don't want to use a choke in the heater supply (2), just replace it with a suitably sized resistor, or use a solid state voltage regulator if you wish. Notice how the choke brings the ripple in the heater power supply from 2 V_{PP} down to 0.1V_{PP}!

The use of two CLC filters in the high voltage section reduces the AC ripple from 2.5 V_{PP} to only 1 mV_{PP} (3)! Such result would be impossible to get with CRC filtering, which would also significantly reduce the available DC voltage. With two chokes in series, the voltage drop is only 10 volts!

Instead of ordinary silicon diodes use Schottky diodes (4) for the heater circuits and soft and fast recovery diodes for the high voltage circuit. A good source of duo-diodes (two in one 3-legged case) are old PC power supplies.

They use these Schottky diodes in their 5V supplies because of their low forward voltage drop (0.3V instead of the usual 0.8V)! Unfortunately, Schottky diodes are only available up to 100V.

The last filtering capacitor is directly in the signal path, it should be a quality film&foil or PIO unit (5)!

The mains transformer MUST have an E/S (electrostatic) shield, which should be connected to the main star grounding point at the first electrolytic capacitor. All components strapped to the points with the ground symbol (the two heater balancing resistors and the three filtering capacitors) must be connected to that point, which is then grounded to the metal chassis.

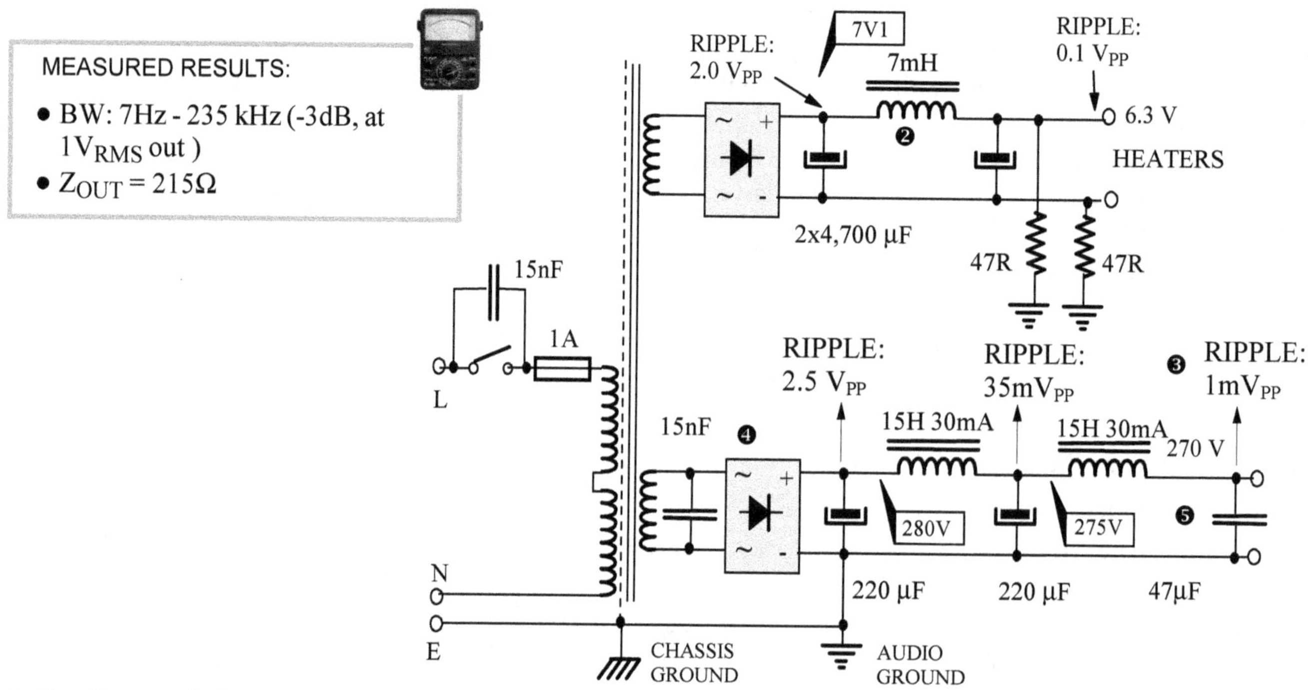

A directly-coupled version

Notice how close the anode DC voltage of the first stage (82V) is to the grid bias voltage of the cathode follower (65V). By making them equal, we can eliminate the coupling capacitor. This can be done in various ways, the simplest is to increase the resistance of the anode resistor in stage 1.

We need to drop 82V-65V=17V, and with 2.2mA anode current flowing the resistance increase needs to be 17/0.0022 = 7,727Ω! Since the existing anode resistance is 60kΩ, a standard value 68kΩ resistor would be perfect for the job, giving us the needed 8kΩ increase!

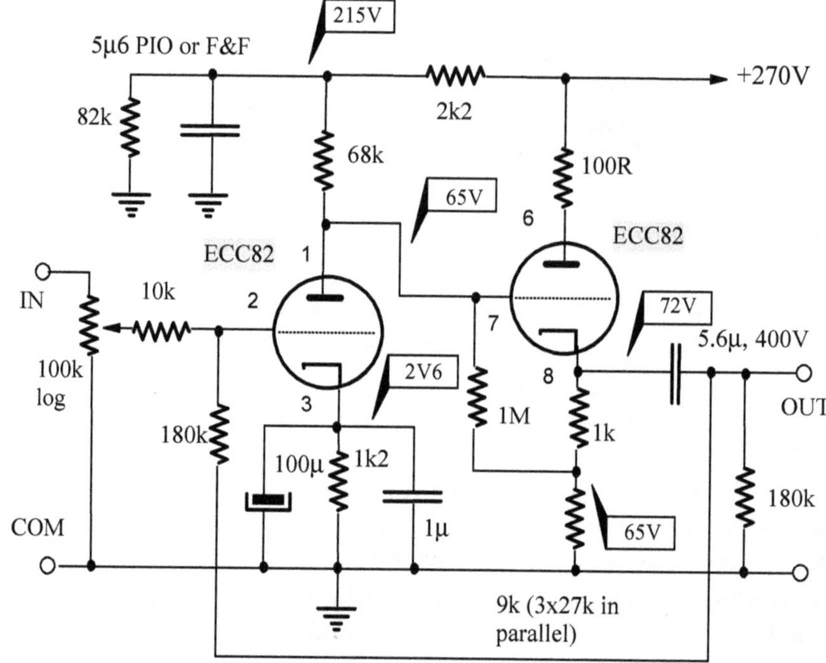

PHONO PREAMPLIFIERS

- DESIGN CRITERIA, TUBE & TOPOLOGY CHOICES FOR PHONO STAGES
- RIAA DE-EMPHASIS CURVE
- PASSIVE DE-EMPHASIS FILTERS
- FEEDBACK OR ACTIVE DE-EMPHASIS FILTERS
- LOW VOLTAGE AND BATTERY POWERED PHONO STAGES
- BATTERY-POWERED SUPPLIES FOR PHONO STAGES
- CARTRIDGE-CABLE-PHONO STAGE INTERFACE
- BALANCED PHONO STAGES

"My mother says I was vaccinated with a phonograph needle.
I love to talk. I just love to talk."
Jerry Hunt, composer

DESIGN CRITERIA, TUBE & TOPOLOGY CHOICES FOR PHONO STAGES

After the music or speech is recorded on a magnetic tape, it goes through an "emphasis" filter which attenuates low frequencies, then a power amplifier drives a cutting lathe which produces the master recording. The master goes through a few stages during which metal stampings are produced, which in the presses produce LPs.

During the reproduction process, a "de-emphasis" filter is required to boost the low frequencies. This filter is a mirror image of the emphasis filter used during the mastering process.

This RIAA filter, and its almost identical relative, the NAB filter, is part of a phono preamplifier or phono stage. In other words, a phono stage is a special type of frequency-selective audio preamplifier, since its amplitude response varies with frequency according to the RIAA de-emphasis curve.

Design & build quality criteria for phono stages

- Low hum and noise, resulting in high signal-to-noise ratio
- Accurate frequency response curve (correctly designed RIAA de-emphasis filter)
- Adequate gain (especially at frequency extremes)
- Low distortion (high linearity)
- Optimized input impedance (cartridge matching)
- Low output impedance
- Performance must not be seriously affected by the aging of tubes and the drift of component values

Design choices

- MC or MM input
- Passive or active equalization (or a combination of both)
- CR, LR or LCR RIAA filtering
- Discrete components (tubes, transistors) or IC (integrated circuits)
- Bipolar, FET transistors or tubes (or a combination of two or all three)
- Mains or battery power supply
- Circuit topology: common cathode (or emitter or source), cascode, SRPP, m-follower, differential amplifier
- RC-, direct- or transformer-coupling
- High impedance- or low impedance - output (cathode follower or output transformer)
- Regulated or unregulated power supply

Equivalent noise resistance of a tube

You may have noticed a parameter called R_{EQ}, specified in some tube data sheets. That resistance has nothing to do with internal resistance of a tube, but is a way to specify the level of noise generated by the tube. You are probably puzzled now, asking how and why is noise level expressed in ohms, shouldn't that be in dB or Volts?

Indeed, the noise can be expressed in volts, the formula is $V_{N\ RMS}=\sqrt{(4kT_IR_{EQ}B)}$, where we again have R_{EQ} plus a few other constants, such as Boltzman's constant k. So, we are none wiser. The idea behind this resistance figure is that a noisy tube can be modeled as a noise-free tube, with a noisy resistor R_{EQ} at its input (grid).

To measure R_{EQ} connect a resistor to tube's input terminals (grid and cathode) and using a sensitive AC μA-meter measure the output voltage v_A. Then short-circuit the input terminals and repeat the measurement (v_B). Finally, combine the two measurements in one single formula: $R_{EQ} = R_X(v_B/v_A)^2$

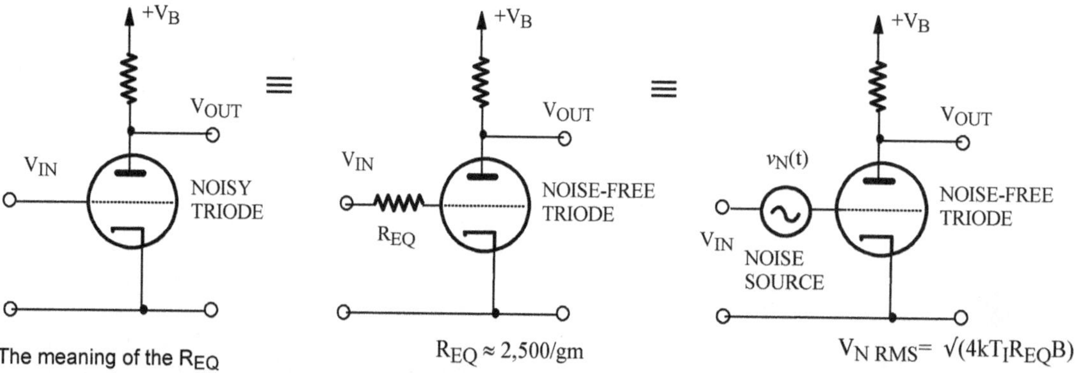

The meaning of the R_{EQ} $\qquad R_{EQ} \approx 2{,}500/g_m \qquad V_{N\ RMS}= \sqrt{(4kT_IR_{EQ}B)}$

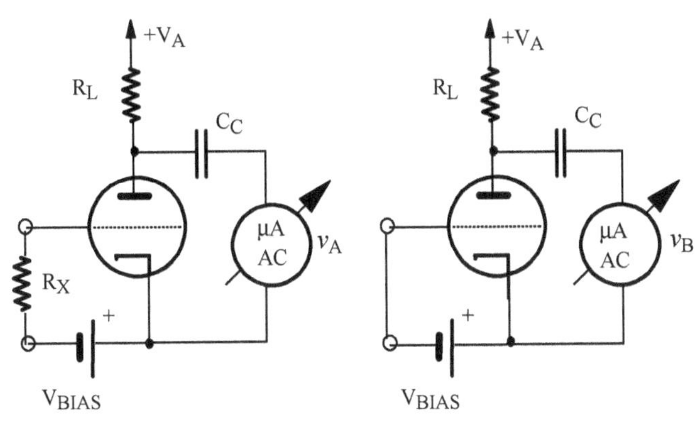

How to measure the equivalent noise resistance of a triode

Using a 100k resistor and ECC88 tube, we got v_A=30 mV, v_B=560mV, so $R_{EQ}=100K*(30/560)^2 = 287\Omega$

Equivalent noise resistances for a few triodes and triode-connected pentodes are given in the table below. Low figures are in the 220-300 Ω range. Notice how ECC85 and C3m are twice as noisy as that average, while d3A has the lowest noise of all tubes here.

From the practical audio perspective, the only application where a designer should consider this issue is in phono stages, which deal with low (MM cartridges with average 1-5mV) and very low input voltages (MC cartridges that produce typically 0.1mV signals).

TUBE	REQ [Ω]
d3A	150
6SL7	1,560
6SN7	960
E188CC	250
E86C	250
6J4	210
E88CC	300
ECC85	500
E280F (triode con.)	220
C3m (triode con.)	650

The highly esteemed octal duo-triodes (6SL7 and 6SN7) are very noisy, and thus unsuitable for phono stage use!

An approximate formula for estimating the equivalent noise resistance of a triode is $R_{EQ}\approx 2,500/gm$, where gm is mA/V.

The cathode current in pentodes divides randomly or irregularly between two or three positive electrodes (anode, screen grid and suppressor grid) and this "partition noise" adds to the shot noise and makes pentodes noisier than triodes. For pentodes, R_{EQ} depends on anode and screen currents as well, and is $R_{EQ}\approx I_A/(I_A+I_S)[2,500/gm + 20,000I_S/gm^2]$. The currents must be in mA and gm in mA/V.

Let's estimate R_{EQ} for 6AC7 when used as a pentode and when triode-connected. In a typical operating point as a pentode I_A=10mA, I_S=2.5mA and gm=9mA/V So, $R_{EQ}\approx I_A/(I_A+I_S)[2,500/gm+20,000I_S/gm^2] = 10/(10+2.5)*[2,500/9+20,000*2.5/81] = 0.8(278+618) = 720\Omega$

As a triode, gm=11mA/V so $R_{EQ}\approx 2,500/11 = 225\Omega$, very close to the specified value of 220Ω.

RIAA DE-EMPHASIS CURVE

The limited phase lag network

We have already studied the limited phase lag network and that knowledge will come very useful now, since the RIAA filter is nothing but two of these networks in a cascade (in series)! The final output level A_X is equal to the ratio $\omega_P/\omega_Z = R_2/(R_1+R_2)$ or, if logarithmic scale is used, $A_X = 20\log[R_2/(R_1+R_2)]$ To achieve proper de-emphasis in accordance with the prescribed RIAA curve, at midband central frequency f_C = 1,000 Hz, this level needs to be 20dB below the initial level at frequency zero, A_0 = 1! Since -20 dB = 0.1, we have $R_2/(R_1+R_2) = 0.1$ or $(R_1+R_2)/R_2 = 10$, which finally gives us the required ratio of resistances used in this filter of $\mathbf{R_1=9R_2}$

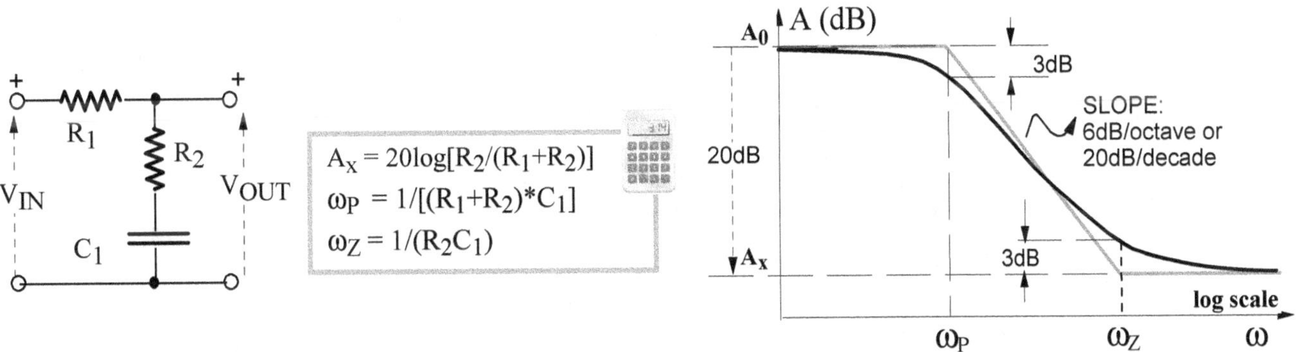

The RIAA de-emphasis curve essentially boosts bass frequencies from 500 Hz down to 50 Hz by 20dB and attenuates treble (high frequencies) by the same amount, 20dB, from 2,120 Hz to 21,200 Hz. The referent or 0dB level is exactly at the central frequency (f_C) of 1 kHz!

The turnover frequencies and associated time constants are named accordingly to the seminal article by Stanley Lipshitz, “On RIAA Equalization Networks”, JAES 1979.

f (Hz)	2	4	8	16	20	30	40	50	80	100	150	200	300
dB	-0.2	+5.7	+11.2	+15.4	+16.3	+17.0	+16.8	+16.3	+14.2	+12.9	+10.3	+8.2	+5.5

f (Hz)	400	500	800	1k	2k	3k	4k	5k	6k	8k	10k	15k	20k	21.2k
dB	+3.8	+2.6	+0.7	0.0	-2.6	-4.8	-6.6	-8.2	-9.6	-11.9	-13.7	-17.2	-19.0	-20.0

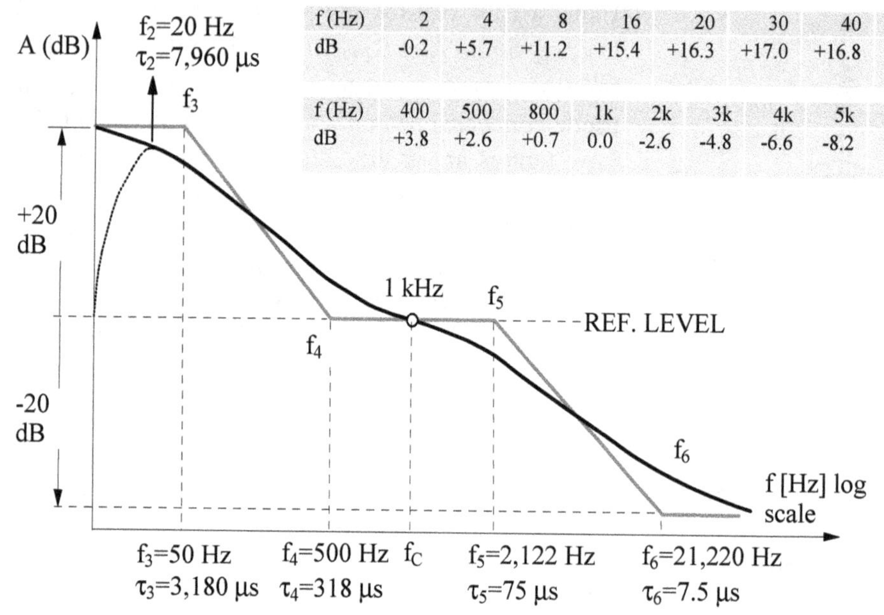

RIAA de-emphasis A-f curve, with asymptotes, turnover frequencies and associated time constants

There are three original time constants, corresponding to turnover frequencies f_3, f_4 and f_5: 3,180 μs (50 Hz), 318 μs (500 Hz), and 75μs (2,120 Hz). The newly introduced f_2=20 Hz gives as the time constant of T_2 =7,950*10^{-6} s. Some designers argue that frequencies over f_6=21,200 Hz should not be continued to be attenuated indefinitely, so they introduce another τ at that frequency, τ_6= 7.5*10^{-6} s!

Another point of disagreement is the optional corner frequency f_3. Sometimes called "New Orthophonic" standard, it is based on recommendation of 1953 by NARTB, of 1955 by IEC No.98, and B.S. No. 128. The original RIAA de-emphasis curve is shown in full line and the later (1976) low frequency modification to the curve is in dotted line. This modification doesn't allow for the amplification of sub 20 Hz frequencies, so not to amplify the turntable rumble and to avoid the tone-arm resonance range of frequencies.

PASSIVE DE-EMPHASIS FILTERS

Separate filters (Type E)

τ_3 and τ_4 are usually done together in one filter, as are τ_5 and τ_6. Since the signal coming off a disc has its highest amplitudes at high frequencies, they should be attenuated first to avoid high frequency overload in the first stage.

Also, it is better to defer filtering the low frequencies until they have been amplified to a reasonably high level. In this arrangement we have two amplification stages and an impedance buffer (cathode follower) at the output.

There are four types of active filters, types A to D (more on them soon), as per the nomenclature devised by Lipshitz, that is why I call this arrangement type E.

The simple formulas are fine with op-amps which have practically infinite input impedance. However, discrete amplification stages with tubes don't have infinite input impedance, primarily because they usually feature resistors between the grid and ground. Unless the values of these resistors are much higher than the values for R_1 and R_3 (they are almost always much higher than R_2 and R_4, so these are not an issue), we have to include them in the amended formulas. This is because the AC signal effectively sees these grid resistors in parallel with R_1 and R_3.

R_1' is a parallel combination of R_1 and R_Y: $R_1' = R_1R_Y/(R_1+R_Y)$ and R_3' is a parallel combination of R_3 and R_X: $R_3' = R_3R_X/(R_3+R_X)$

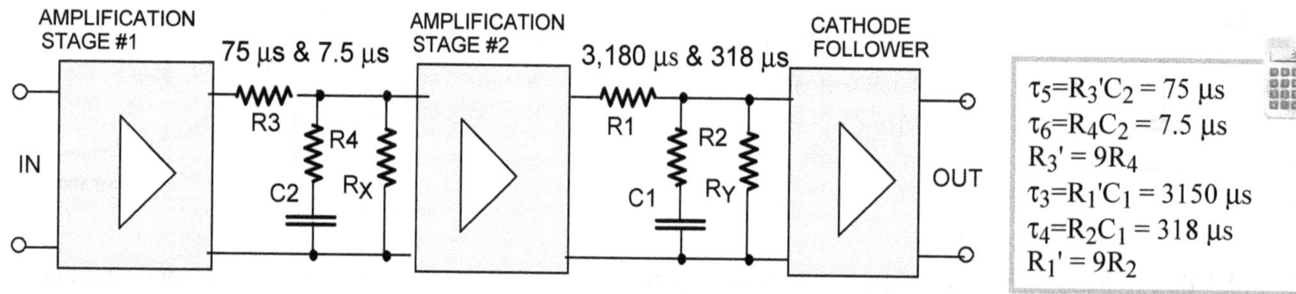

All-in-one passive filter (Type F)

The calculations for this circuit are complex (refer to Lipshitz paper, page 468), but these are the final formulas: τ_4 = 318 μs = $R_2(C_1+C_2)$ and τ_5 = 75 μs

Normally, R_2C_2 would give us 75µs , but with this network this formula the accurate value needs to be $R_2C_2 = 81.2$ µs

$\tau_3 = 3{,}180$ µs is NOT R_1C_1 but R_1C_1 can be calculated form the following equation: $R_1C_1 = \tau_3 - \tau_4 + \tau_5 = 2{,}937$

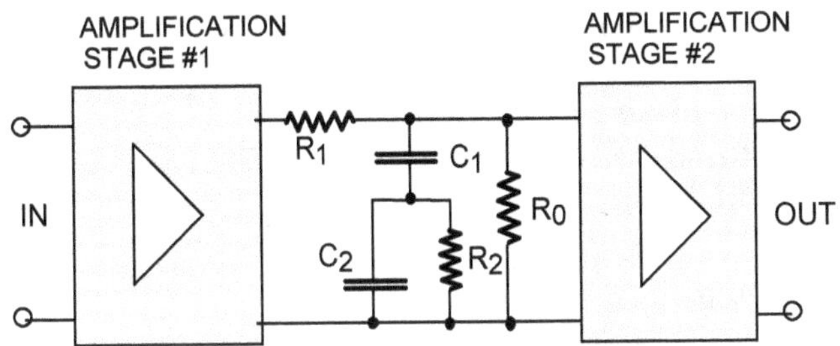

Accurus phono stage

This moving magnet design is based on type-F RIAA filter. It sounds very good and has enough gain do drive power amplifiers directly. With a quality MC-stepup transformer it forms a formidable pair that only highly complex topologies and ten times more expensive phono stages can beat sonically. Only one 12AX7 duo-triode per channel simplifies construction.

Time constants. $\tau_4 = R_2(C_1+C_2)$ We have 24*13.3 = 319µs, R_2C_2 should be 81.2 µs, we have 3.3*24 = 79µs, a very good result. R_1C_1 = 300*10 = 3,000µs instead of 3,150µs, an error of 2.15% Good enough!

If you want to make the bass faster & tighter, replace the 300k resistor with 470 k. Notice two decoupling circuits in the HV power supply, plus each elco is bypassed by a good quality film cap.

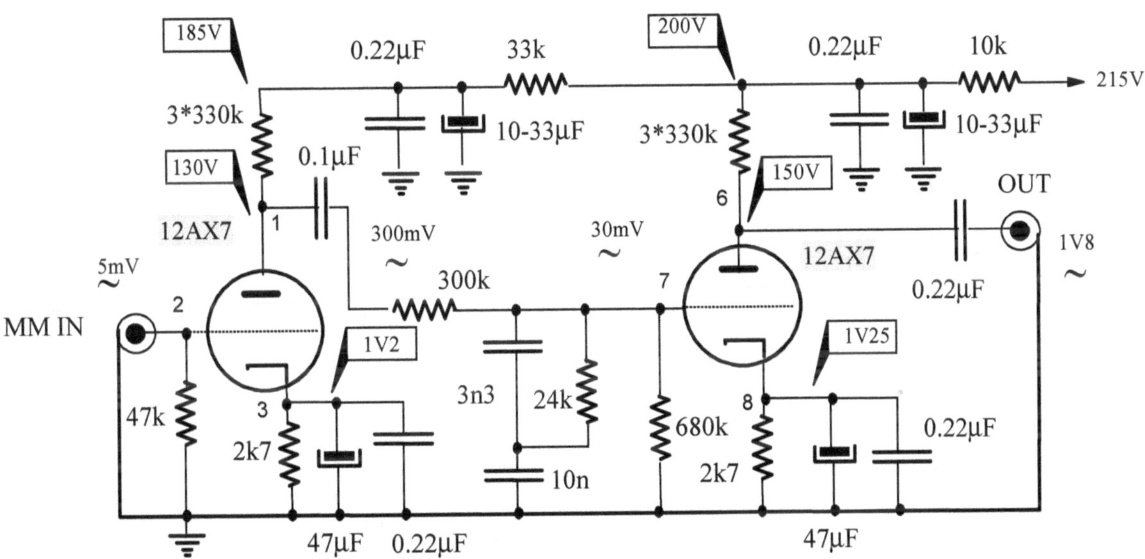

Instead of using one 110k anode resistor, we discovered that a parallel combination of three or four resistors sounds better! Why? I don't know, it does not make technical sense. Sure, resistors in parallel have a lower thermal noise, if they were in the grid circuit, perhaps you would notice the difference, but in the plate circuit?

AC voltages in significant points are indicated. Theoretically, the gain of both stages is identical: $A_1 = A_2 = -\mu R_L/[r_I + R_L] = -100*110/(65+110) = -62.9$ so the overall gain without RIAA feedback is $A = A_1A_2 = 3{,}951$ or 20log[3,951] = 72 dB After 20 dB of filter attenuation at 1 kHz, this leaves us with 72-20 = 52dB (398 times) of gain.

Commercial example:
Audio Note M2 Phono

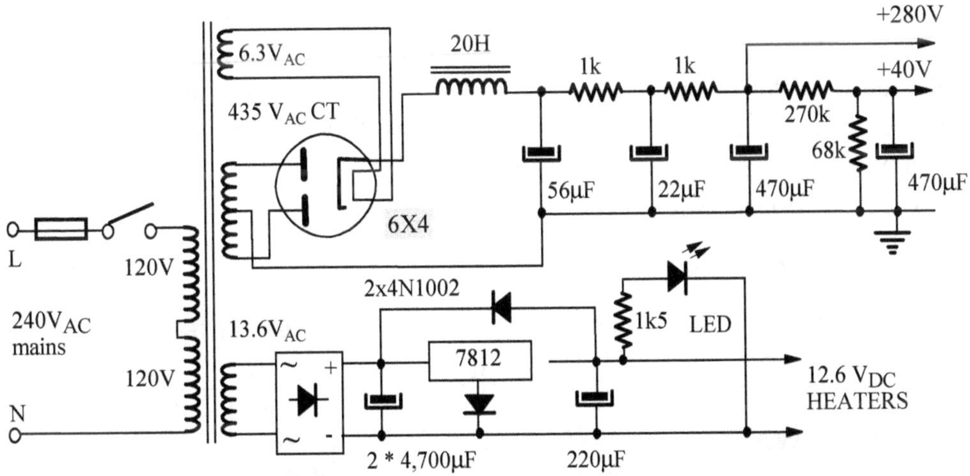

A relatively simple kit by Audio Note UK. A decent design, two SRPP stages and an all-in-one passive RIAA filter in between. $R_P=R_1R_0/(R_1+R_0)$ = 620k||470k = 267k The formulas are:

a. R_PC_1= 2,187 µs We have 267*8.2 = 2,189 Perfect.

b. R_PC_2= 750µs We have 267*2.76 = 734 µs Close enough.

c. R_2C_1= 318 µs We have 39*8.2 = 320 µs OK

d. C_1/C_2=2.916 (we have 8.2/2.76=2.97)

e. R_P/R_2=6.88, we have 267/39=6.84, which is OK.

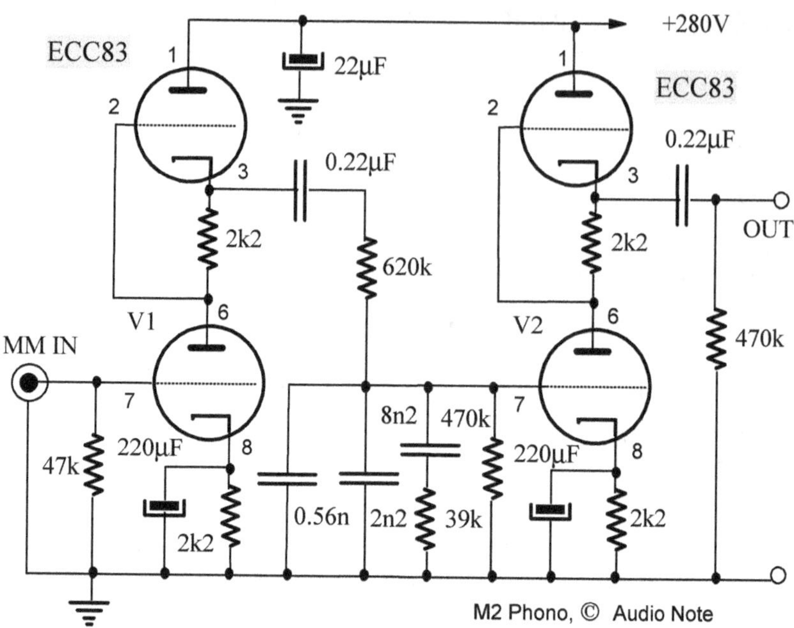

The power supply features tube rectification and choke-input (LC) filtering, with regulated 12.6V_{DC} heater supply. The +40V_{DC} voltage is only used to elevate (bias) the heaters for ECC82 tubes in the line stage (not shown).

Although the upper tubes' cathodes are at approximately half of V_{BB} voltage (140V), ECC83 tubes have V_{HKMAX}=200V, so the maximum heater-cathode voltage is not exceeded, and the heaters of these tubes don't have to be referenced to an elevated voltage.

All-in-one passive filter (Type G)

Since the signal sees R_0 as effectively in parallel with R_1, we need to define R_P as a parallel combination of R_1 and R_0: $R_P=R_1R_0/(R_1+R_0)$, where R_0 is the grid resistor of the following stage. The formulas are: R_PC_1= 2,187 µs, R_PC_2= 750µs, R_2C_1= 318 µs andR_P/R_2= 6.88 (C_1/C_2 = 2.916)

To determine component values:

1. Choose C_1 and from equation a) calculate R_P
2. From equation b) calculate C_2
3. From equation c) calculate R_2
4. from equation d) calculate the required value for R_1.

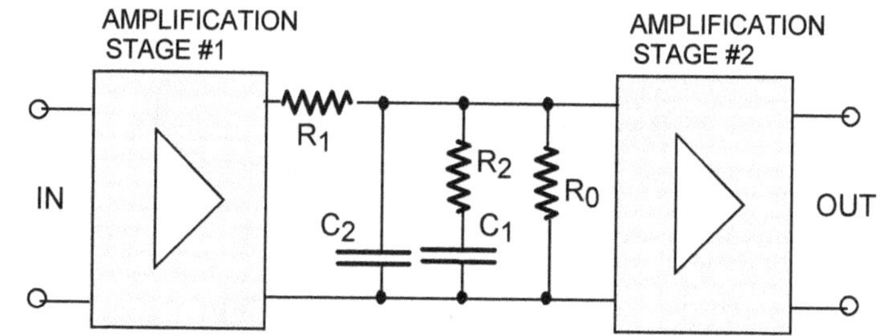

FEEDBACK OR ACTIVE DE-EMPHASIS FILTERS

This filter type filters has its filtering network in the feedback branch (the β block). While there are four types of equalization networks used in feedback or active filter arrangements, types A and B are most commonly used by far, and among the two, Type A is more prevalent. Types C and D are seldom used in commercial preamps.

The often-used "simplified" equations for Type A network: τ_3=3,180 µs =R_1C_1, τ_4= 318 µs = R_2C_1, τ_5=75 µs = R_2C_2 While equations for τ_3 and τ_5 are correct, the one for τ_4 is only approximate. The correct equation is τ_4= $(R_1||R_2)(C_1||C_2) = (C_1+C_2)R_1R_2/(R_1+R_2)$ The DC resistance is $R_{DC}=R_1+R_2$

The formulas for Type B network are: C_1R_1 = 2,937 , C_2R_2 =81.2, C_1R_2 = 236.8 and C_1/C_2 = 236.8/81.2 = 2.92 The DC resistance is $R_{DC}=R_1$

Type A, B, C and D filter networks

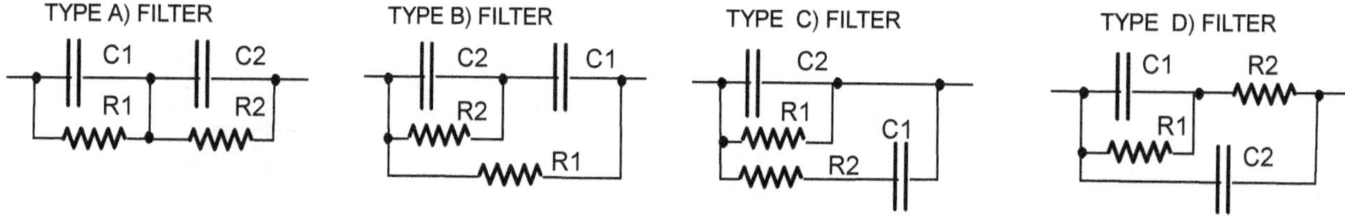

Improving a one-tube NFB phono stage

This RIAA circuit (exact values) was used in Eico ST40 and ST70 integrated amplifiers' phono stages, and with slightly different values it was also used in budget equipment such as Dynaco PAS2/3 preamp, Eico ST84 preamp and many others of that era. We have already mentioned it in the chapter on negative feedback.

It needs a much higher gain at lower frequencies than its two-stage circuit can provide, so apart from the NFB with the standard RIAA filtering it also uses positive feedback (the 68k resistor between pins 3 and 8), which boosts the amplification of lower frequencies.

The positive feedback may solve one problem, but it creates at least three bigger problems, increasing distortion and reducing stability in the bass region. It also makes the accuracy of the RIAA amplification highly dependent on the parameters of individual tubes.

The improvement is simple, yet profound: 1. Remove the positive feedback resistor 2. To increase the gain, bypass the cathode resistor of the second stage and increase its anode resistor 3. Load the input properly (47k resistor) 4. Increase the value of the coupling capacitor 5. Change the feedback component values. 6. If possible increase the power supply voltages $+V_A$ and $+V_B$ to 300V or higher, but make sure you adjust the values of cathode and anode resistors in that case. Normally these voltages are at a much lower value (150-200V).

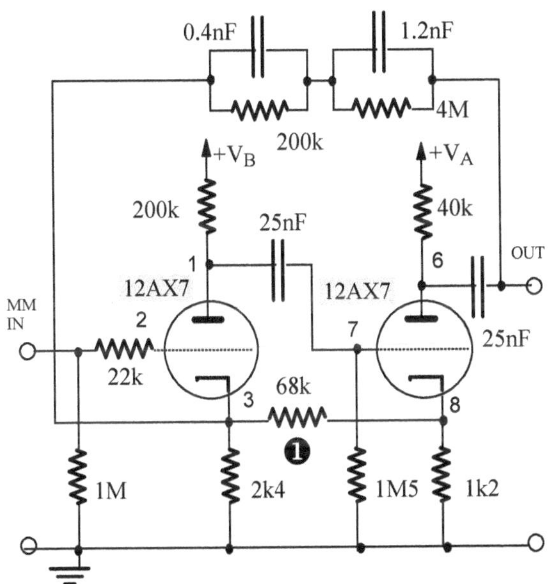

BEFORE: The standard two-stage feedback RIAA stage

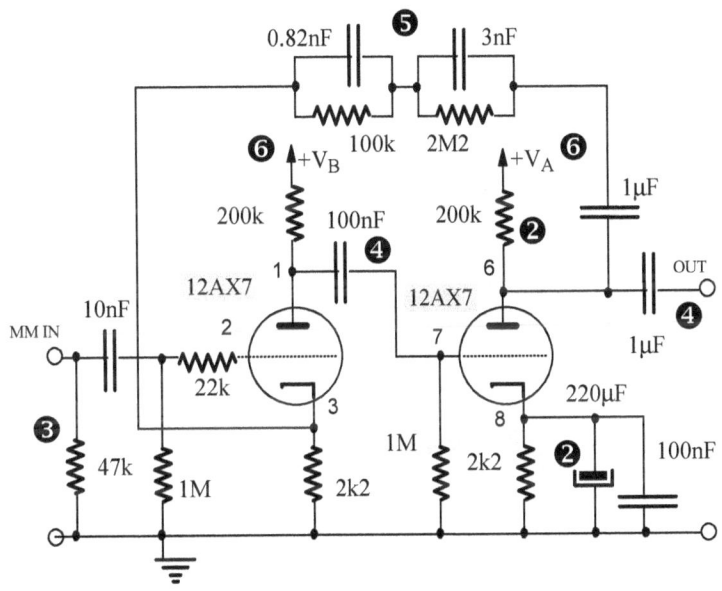

AFTER: Improved two-stage feedback RIAA stage

The Cascode Baby

Instead of using two common cathode stages in a cascade why not use a cascode at the input and a cathode follower at the output? The voltage gain of the first stage is $A=g_mR_A= 1.5\ mA/V*270k\Omega = 405$

Now you realize why the relatively high voltage of $400V_{DC}$ was used. With the usual 275-300V power supply, the plate resistor would have been around 100k and the gain would be "only" around 150, although even that would be a huge increase over the basic two-stage RIAA circuit. The gain of the second stage is

$A_F=-\mu R_L/[r_I+(\mu+1)R_K+R_L]=$
$-100*300/[65+(100+1)*2.2+300]$ or $A_F = -100*300/587 = -51$

The total gain without NFB is thus A=405*51*0.97 = 20,035 or 86dB!

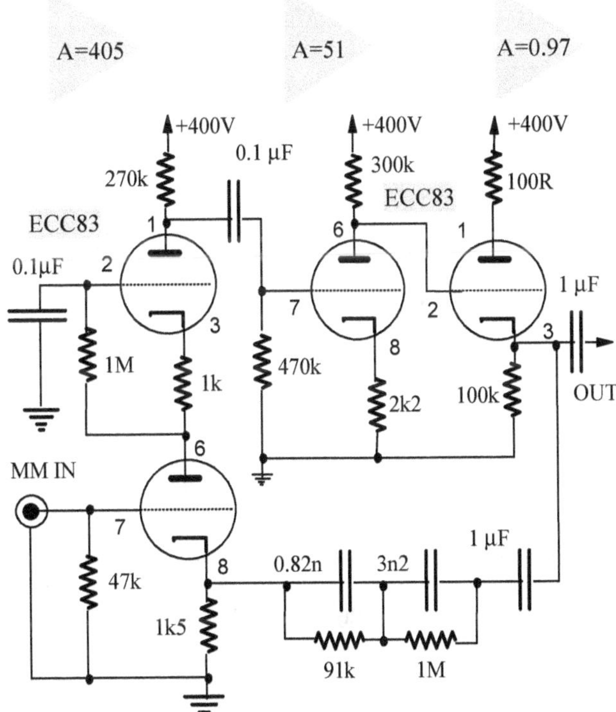

RIGHT: A high gain cascode, followed by a common cathode stage and a cathode follower

Commercial example: VTL Ultimate + SDL phono stage

Since MC step-up transformers are expensive ($500 - $5,000+), some manufacturers use an additional amplification stage for MC cartridges. In the VTL design MC input goes through a pair of coupling capacitors to the cathode of the first stage, which is connected as a common-grid stage. As we have seen, CG stage has a similar gain to CC stage, but it does not invert the phase and it has a very low input impedance. MC cartridges are low impedance devices so the low Zin of the grounded-grid stage matches them well.

As further benefits go, the grounded-grid stage has a very wide bandwidth (frequency range) and low noise, which is important when dealing with 0.1-0.5 mV moving coil signals!

The MM signal goes straight into the second stage, which is a common cathode amplifier with un-bypassed 4k7 cathode resistor. The rest is standard stuff, the output of the second stage capacitively coupled to the grid of the third amplification stage, which is in turn again capacitively coupled to the cathode-follower at the output.

The MM magnet signal passes through 4 coupling capacitors, and MC signal through six. Plus, there are two high-frequency feedback capacitors (marked *), most likely needed to improve the poor shape of square wave signals. R_1-R_6 are input resistors for the adjustment of the input impedance for cartridge matching purposes.

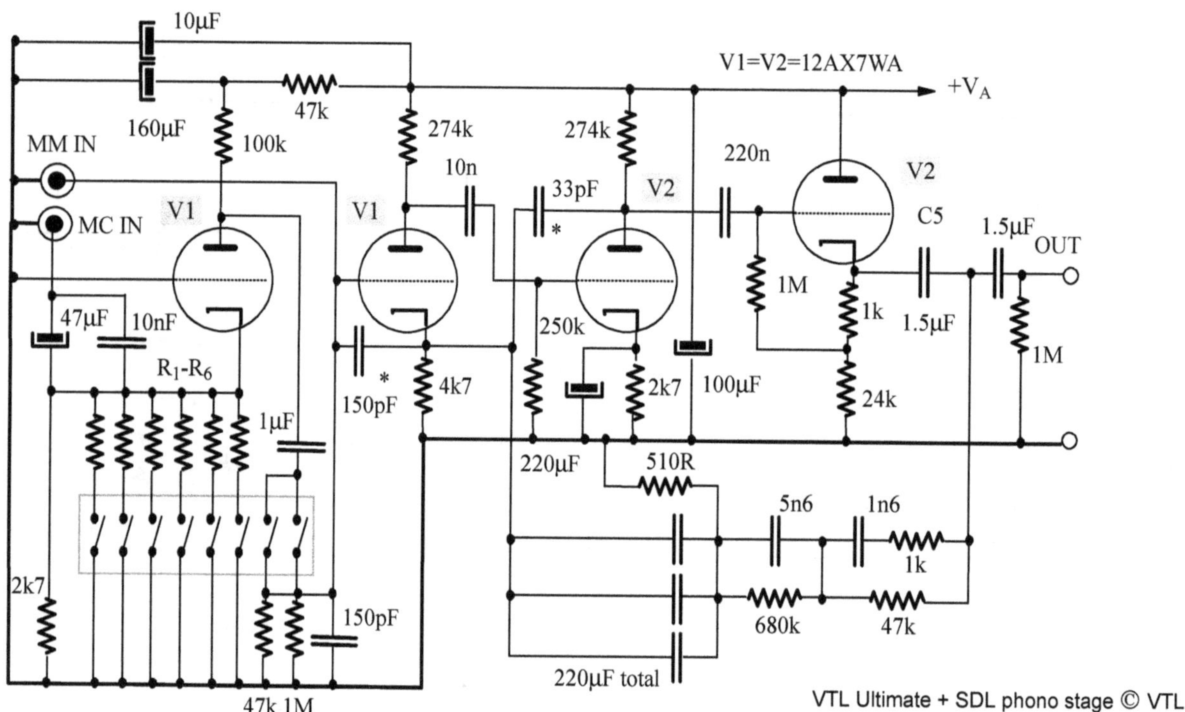

VTL Ultimate + SDL phono stage © VTL

CG stage analysis:

$A_o = [(\mu+1)R_L]/[r_I+(\mu+1)R_S+R_L]$

Our Ortofon MC-1 cartridge has R_S=5Ω, so the gain would be A_o = [(100+1)*100,000] / [65,000+(100+1)*5 + 100,000] = 101*100,000/165,505 = 101*0.6 = 60

This seems way too high! For the 0.5mV output of MC-10 we would get 30 mV at the output, while we only need about 5mV, which an amplification factor of A=10 would give us.

The output impedance is $R_o=r_I+(\mu+1)R_S$, while the input impedance between terminals H and G is $R_{IN}=(V_G/I_P)$ or $R_{IN}=(rI+R_L)/(\mu+1)$ For 12AX7 μ=100, r_I = 65kΩ, and with a 100k load, the input resistance is 1,634 Ω.

MM circuit analysis:

The gain of the second stage is $A = -\mu R_L/[r_I + R_L]$ = -100*274/(65+274) = -80.8 For 12AX7 r_I is 65 kohm, R_L is the anode resistor (274 kΩ here), and μ is the amplification factor of the tube, which for 12AX7 is 100.

The un-bypassed cathode resistor reduces the gain of the first stage $A_F = A/(1-A\beta)$ The feedback factor is quite high (strong NFB): $\beta = R_K/R_L$ = 4k7/274k = 0.01715 Without feedback the gain would be $A= -\mu R_L/[r_P + R_L]$ = -100*274/(65+274) = -80.8 and with feedback A_F = -80.8/(1+80.8*0.01715) = -80.8/(2.386) = -33.8

Without RIAA NFB the open loop gain for the MM preamp would be $A=A_1A_2$ = -80.8 * -33.8 = 2,736 or 68.7 dB, which would give us 20dB less at 1 kHz, or 48.7dB.

Commercial example: E.A.R. 834P

The internal view of Tim de Paravicini's E.A.R.834P phono stage does not impress. Tiny chassis, a single PCB and a simple power supply.

The first thing one notices is the breaking of two rules that purists swear by: don't use a voltage doubler in the power supply and don't share the valves between channels. We have found that these rules are not technically relevant and have used voltage doublers very often.

Actually, arranging the valves in such a fashion gives you more freedom in modification. You can have one type as V1, another type as V2 and yet a third type, if you really wish, as the output valve in the cathode follower (V3). The designer has not taken advantage of this opportunity, though, all three tubes are ECC83!

The power supply is very simple, unregulated DC on series- connected heaters, and for high voltage supply a voltage doubler followed by two RC filters in cascade.

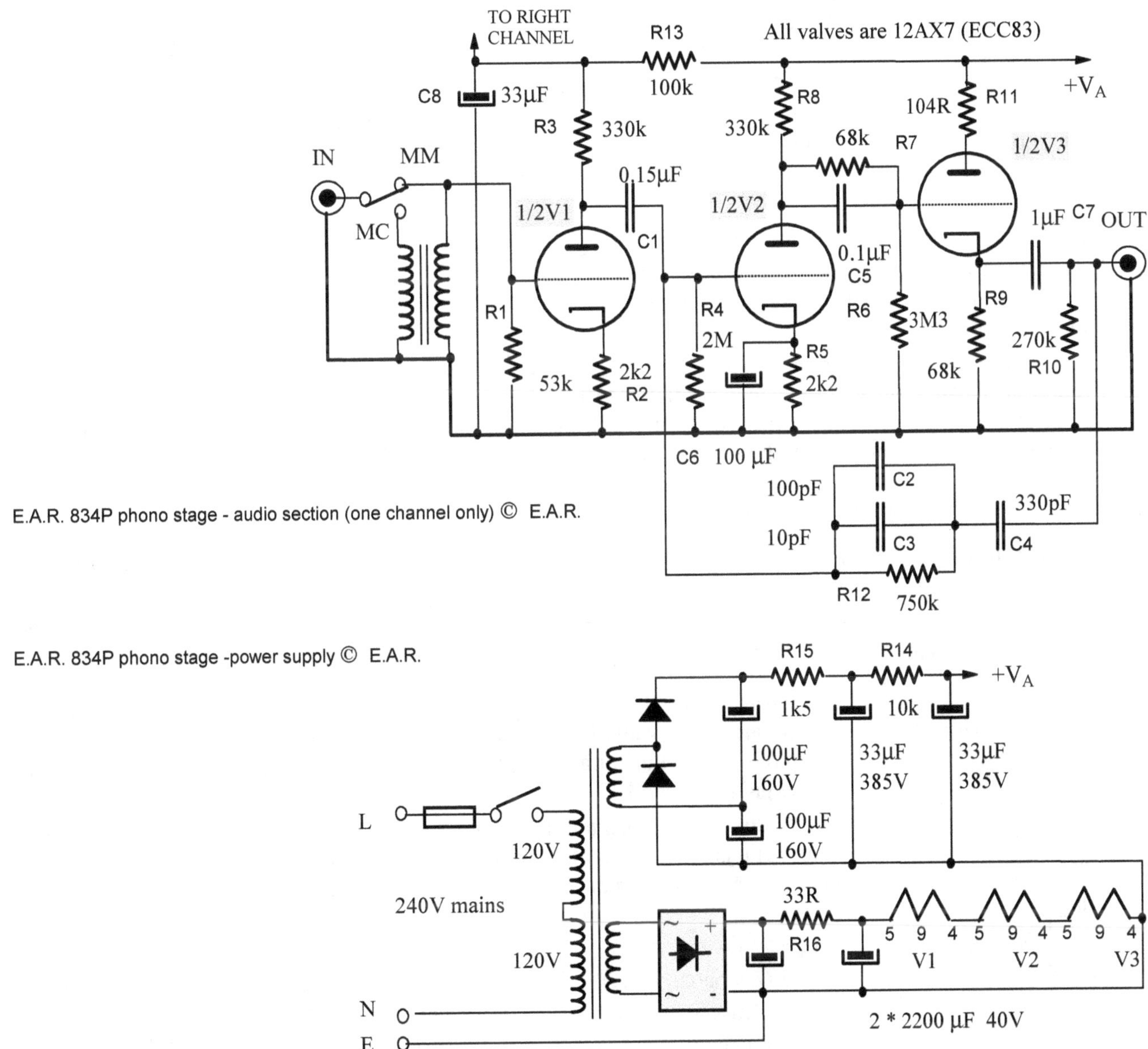

E.A.R. 834P phono stage - audio section (one channel only) © E.A.R.

E.A.R. 834P phono stage -power supply © E.A.R.

LOW VOLTAGE AND BATTERY POWERED PHONO STAGES

Since signal levels involved in phono stages are very low, the highest being the output line-level signal of around $1V_{RMS}$, powering a tube phono stage from a battery or a low voltage DC source is a viable option.

You don't even have to design and build your own power supply. Just get an external DC power supply of $24\text{-}48V_{DC}$ output, such as those used to power up printers, scanners, X-boxes and myriad of other IT and general electronic devices. Most (but not all!) have regulated DC outputs. Newer ones are almost exclusively switch-mode designs, without a bulky mains transformer, older ones may be the conventional linear types.

The designs that follow can be powered up by such power supplies or directly from two 12V sealed lead-acid batteries, the type used in alarm systems and UPS (uninterruptible power supplies). Afterwards, we will return to have another look at the battery-based power supplies that can be used.

The main issue is the fact that tubes are naturally high voltage devices, and when anode (plate) voltages are low, they operate at very nonlinear portions of their transfer curves, which results in distortion. There are tubes that were specially developed for the first car and truck radios, designed to operate at voltages of 12 volts or even lower (2 volts!), but they are rather limited in supply and performance and hard to get.

Contact bias (Grid-leak bias)

Although we always say that grid current starts flowing once the grid voltage becomes positive, that simplification is not entirely true. A small (a few μA) positive grid current starts flowing a bit earlier, while the DC grid voltage V_G is still negative. Voltage V_{CB} is called the contact potential of a tube and the region to its right, towards the zero grid voltage, is called the contact-current region.

Even such a small current of a few microamperes can cause a sufficient voltage drop on a grid-leak resistor of high value, so this arrangement can be used to provide a small bias voltage, from a few mV up to 1V, depending on the tube used and its grid leakage current. R_G is typically in the order of 2-12MΩ.

Since this grid-leak potential is small, and the signal voltage must always be smaller than the bias voltage, the signal handling capacity of this circuit is low, so it can only be used in phono stages, otherwise signal rectification will occur.

The value of the anode resistor has little effect on the gain of the amplification stage using contact bias, which is solely dependent on tube's m. Since there are no frequency-dependent elements such as the cathode capacitor, this high gain is constant all the way down to DC.

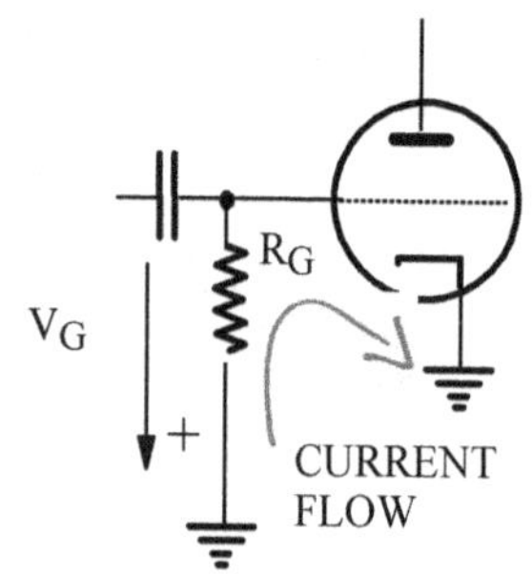

Electrons flowing from the grid down the grid-leak resistor to ground bias the grid negatively

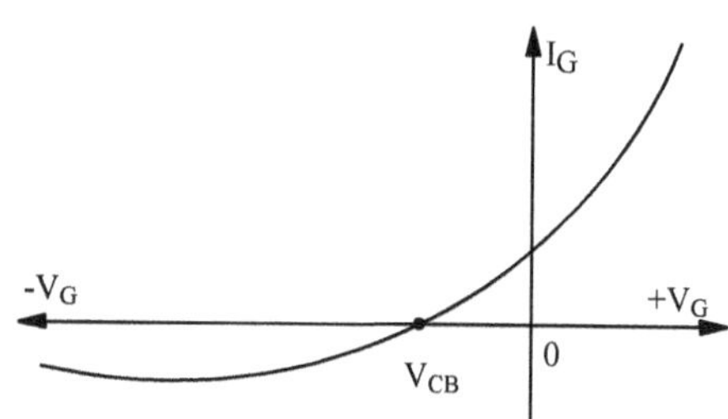

The grid current versus grid voltage. A small positive grid current starts flowing while V_G is still negative. V_{CB} is the contact bias voltage.

Experiment: Gain comparison of triodes at low anode voltages

ECC88 triode is usually the first to be thought of when low voltage operation is mentioned. Indeed, it was designed to operate at relatively low (for tube circuits) anode voltages of around 90V. How well would it work at 24V?

A typical phono preamplifier input stage (below) was used for this experiment with $5mV_{AC}$ signal (1 kHz) at the input. The tabulated results are the average values. Samples comprised of at least half-a-dozen tubes of each kind.

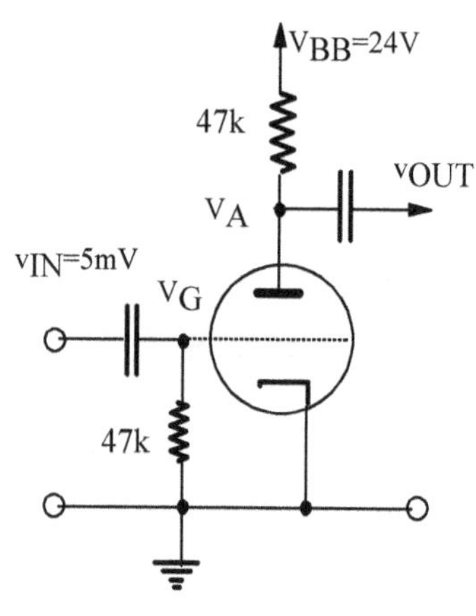

6N1P was supposed to be the Russian equivalent of ECC88, but it achieved the gain of only around 6, while USA-made Amperex tubes had a gain of 11. Zaerix, UK brand of unknown manufacture (probably of eastern-European origin), had a much higher gain of around 13-14.

The E280F pentode is a totally different beast, not an equivalent of ECC88 in any way (but it is pin-compatible). Triode-connected, it achieved by far the highest gain in this circuit, 19 times or 25.6 dB.

v_{IN} = 5 mV	V_G	V_A	v_{OUT}	A_V
6DJ8 (Amperex, USA)	-0.75V	3.6V	55 mV	11.0
6N1P (Russia)	-0.65V	15.0V	32 mV	6.4
ECC88 (Zaerix, UK)	-0.13V	5.7V	68 mV	13.6
E280F (Siemens, Germany)	-0.17V	6.5V	95 mV	19.0

ECC88 high-gain passive RIAA phono stage for MC cartridges

This design uses a passive RIAA filtering, but instead of one input stage with a high gain ECC83 triode, two stages are needed, since the gain of ECC88 is 3 times lower. DC voltages are marked next to significant nodes, signal (AC voltages) are framed, measured using a 5mV, 1 kHz sine wave at the input. The time constants are exactly the same as for the Accurus phono stage, since the RIAA filters are identical.

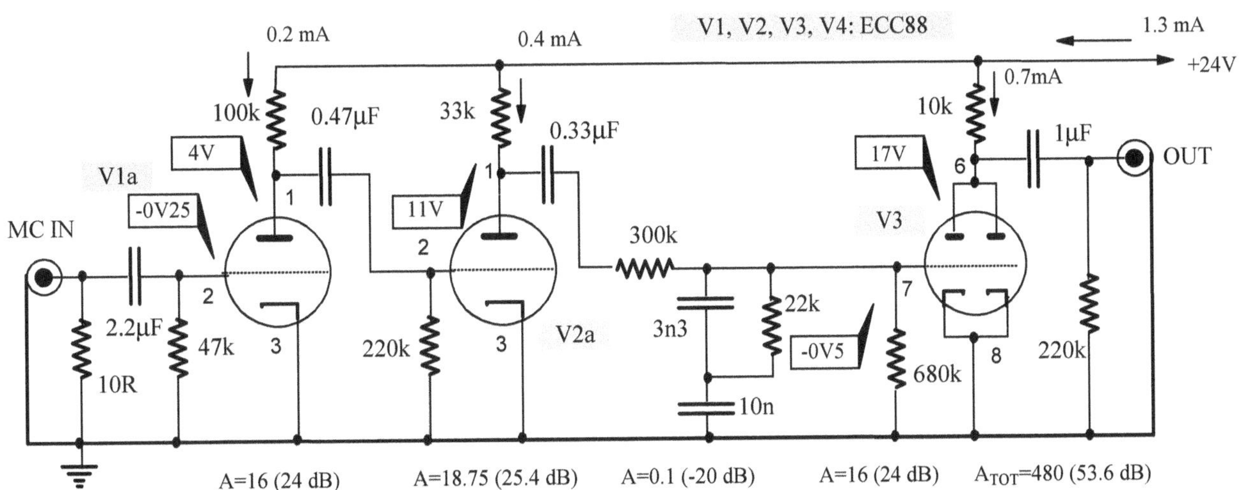

The total amplification factor is 16x18.7x0.1x16 = 480 After 20 dB of attenuation at 1 kHz, this leaves us with midrange gain of 53.6 dB, which is quite high. For a 0.5 mV input, the output should be around 0.24V. If followed by a line stage preamp with a gain of 7 or higher, it should be enough for MC cartridges. Simply add a 10 - 47 ohm resistor between the input and ground for proper loading of the low impedance MC cartridge and you are in business!

The input capacitor, together with the 47k input resistance forms a high pass filter and determines the low cutoff frequency. 2.2 µF gives us 1.53Hz, while 0.22 µF would result in the -3dB f_L of 15.3 Hz, which would be too high and would affect the bass reproduction.

The same applies to the coupling capacitor between the 1st and 2nd stages, 0.47 µF and 220k grid resistor, which result in f_L of 1.8 Hz. Finally, the 1 µF output capacitor with 220k loading resistor forms another high pass filter with f_L= 0.72 Hz. The input resistance of the following preamp or power amp will be added in parallel to this resistance and will lower it, unless there is a series capacitor at its input. Say, for instance the input resistor on the following amplifier is 100 kΩ. That would lower the parallel resistance to 68.75 kΩ and raise the f_L to 2.3 Hz. So far so good. However, if a solid state preamp with a 10 kΩ input impedance is used, the f_L would shoot up to 16.6 Hz. This would be akin to implementing the $\tau 6$ constant!

Triode-connected E280F passive RIAA design

E280F pentode is SQ (Special Quality) tube developed for use as a wide band amplifier. It is shock and vibration resistant and guaranteed to last at least 10,000 hours!

The published anode curves (next page) are equidistant, indicating that this pentode makes a very linear, low distortion triode with very high µ=60 and very low internal resistance (1k8). Most importantly, it sounds great.

TUBE PROFILE: E280F

- Indirectly-heated SQ pentode, Noval socket
- Heater: 6.3V/315 mA
- V_{HKMAX}=100V_{DC}, P_{AMAX}=4 W
- V_{AMAX}= 220V, V_{SMAX}=/80V
- I_{KMAX}=30 mA
- TYPICAL OPERATION TRIODE:
- V_A= 160V, V_G= -9.6V, I_0= 24 mA
- gm = 33 mA/V, µ= 60

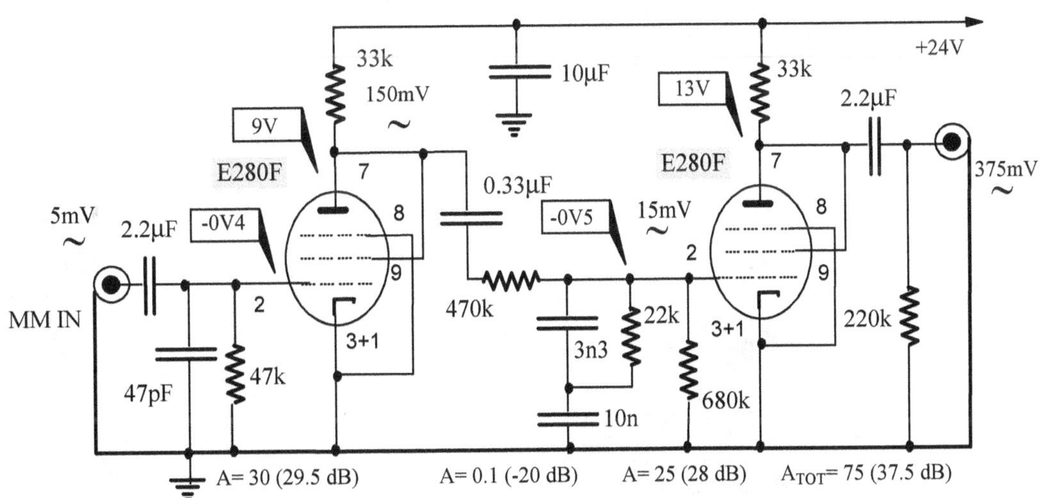

7586 nuvistor + 5670 passive RIAA design

5670 was developed for industrial and communications applications. It uses a Noval socket, but is not pin-compatible with the 12A*7 series. It is a smallish but robust tube, vibration and shock resistant, long lasting and very low in microphony. It has a decent μ and sounds very good indeed.

Type 7586 was the first Nuvistor tube, introduced in 1959. At that time, Nuvistors were superior to transistors, especially at higher frequencies. The extended skirt of the metal case protects the fragile pins and also serves as a guide into the socket. The metal case is 10 mm in diameter and 20 mm tall.

The main issue with this Nuvistor design is their low amplification factor. Even with two stages in a cascade, we managed to get amplification of only around 40 (6 x 6.6) before the passive filter. The output stage compensates for the 0.1 attenuation of the passive filter, so the overall midrange gain is only 40 times or 32 dB. A preamplifier will be needed to amplify the 0.2V output voltage to a 1V line level.

Since two 5670 triodes in series draw 0.35A and four 7586 tubes in 2x2 connection draw 2x0.135=0.27A, we must add a shunt resistor in parallel with 7586 heaters that will draw the difference, or 0.35-0.27 = 80mA. Its resistance should be R_P= 12.6V/0.08A = 158Ω and minimum power rating P=12.6*0.08 = 1W, so 3 or 5W resistor is needed.

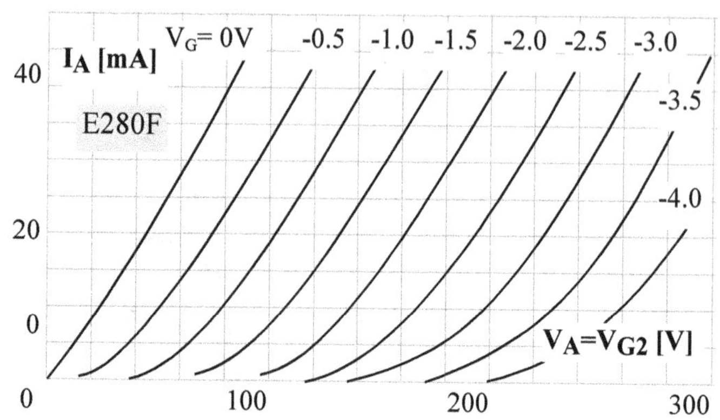

TUBE PROFILE: 5670

- Indirectly -heated duo-triode
- Noval socket, heater: 6.3V/350 mA
- V_{HKMAX}: 100V_{DC}
- Max. anode dissipation: 1.65 W
- V_{AMAX}: 330V, I_{AMAX}:18 mA
- TYPICAL OPERATION:
- V_A=150V, V_G=-2.0V, I_0=8.2 mA
- gm = 5.5 mA/V, μ= 35

TUBE PROFILE: 7586

- Indirectly-heated nuvistor triode
- Heater: 6.3V, 135 mA
- TYPICAL OPERATION:
- V_A=75V, V_G= 0V, I_0=10.5 mA
- gm=11.5 mA/V, μ=35, r_I=3.0 kΩ

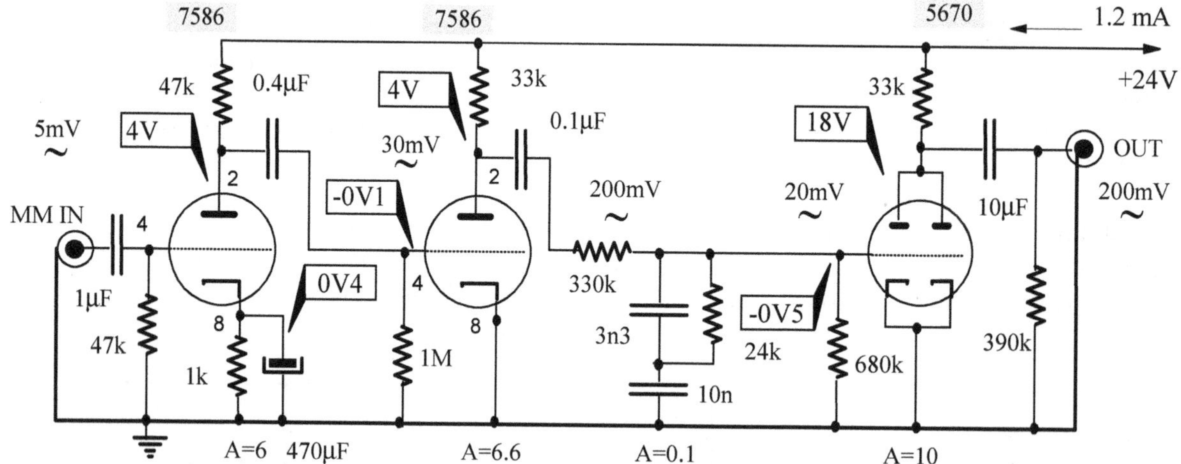

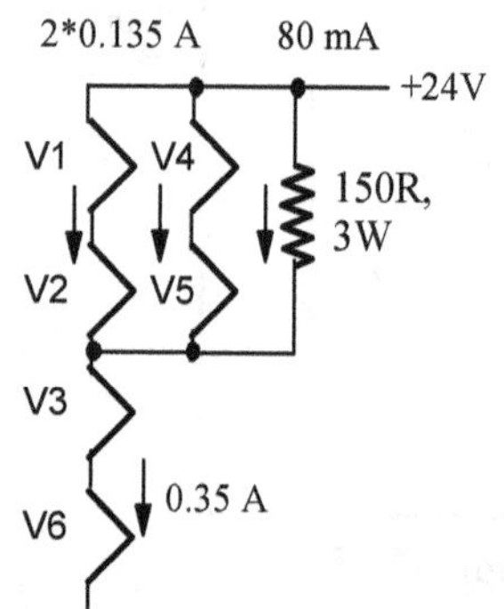

LEFT: Combining heaters of different current ratings

RIGHT: The prototype was made on a PCB, with each component leg and tube pin connected to a circular "island", drilled out together with its hole. The islands are interconnected using wires in a point-to-point fashion. So while the substrate is a blank copper circuit board, the construction is still point-to-point!

BATTERY-POWERED SUPPLIES FOR PHONO STAGES

The advantages of battery-based power supplies

Linear power supplies have a limited capacity to deliver power when needed by sudden increases of demand by the audio circuitry. Although quality amplifiers have large capacitor banks for energy storage, the energy stored in a large lead-acid battery is many times greater. The ultimate power source in hi-fi is a battery bank of a suitable voltage and current capability. When battery powered, the audio circuitry is disconnected from the mains, meaning that any distortions and fluctuations in the mains voltage or frequency become irrelevant.

It also means that mains-propagated spikes, surges and noise cannot reach the audio circuitry. This benefit on its own is enough for us audio fanatics to stop dreaming about breasts and buttocks and start dreaming about batteries of all shapes, sizes and technologies.

Ordinary linear power supplies have a limited capacity to deliver power when needed by sudden increases of demand by the audio circuitry. Although quality amplifiers have large capacitor banks for energy storage, the energy stored in a large lead-acid battery for instance is many, many times greater. Look at it this way: If such a battery can crank-over a car or truck engine, it can certainly supply enough power even during the loudest crescendos in music reproduction!

Transformers radiate electromagnetic waves. This radiation is picked up by the audio circuitry and amplified as a signal, since it is of an audio frequency (50/60 Hz for the mains sine voltage or 100/120Hz for a full wave rectified voltage pulses). This is what we call hum. Transformers can also vibrate mechanically and produce their own sound, called a buzz. Hum and buzz are nasty twins, the biggest enemies of high fidelity. With battery power, there is nothing to hum or buzz and nothing to generate RFI (Radio Frequency Interference).

The more technically savvy amongst you will now scream "Hey, wait a minute Igor, you've forgotten one of the biggest benefits of battery supplies, their low internal impedance!" Dead right. An ideal voltage source has a zero internal impedance, so all the power is transferred to the load, nothing is wasted on its internal impedance. Batteries have a very low internal impedance, and arguably, voltage regulation in battery power supplies may not even be required. The internal resistance of a 12Volt 7.0 Ah rechargeable sealed lead-acid battery is typically 20-30 mΩ, 50 times lower than the 1Ω Z_{OUT} of the best regulated power supply!

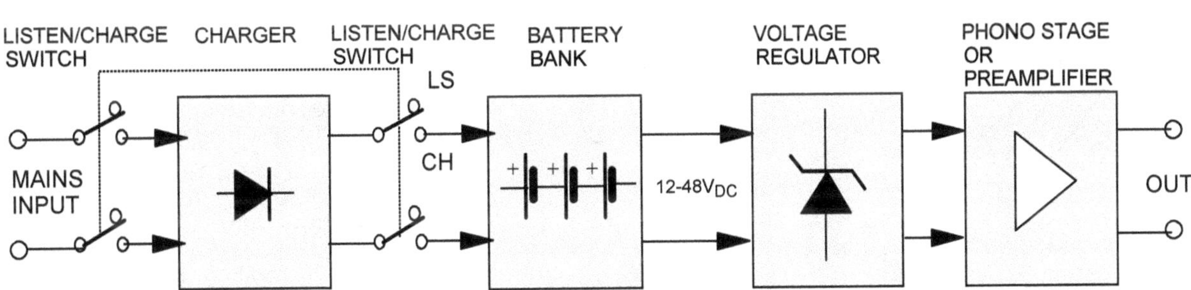

Block diagram of a battery-powered low voltage phono stage or line preamplifier

Designing a battery power supply

How would we go about designing and building a battery-based power supply? If you are of the solid state persuasion, you can stop reading right now and start stuffing your black boxes with batteries.

While using batteries to provide heater power of power amplifiers and preamplifiers is a viable and reasonable proposition, connecting thirty 12 Volt batteries in series to get a typical anode voltage of 360V_{DC} would be sheer madness. Plus, there serious safety issues, such as the flammable vapors that develop during the battery charging process and the high voltage hazards present at all times, not just during charging.

Battery-powered solid state phono stages and line preamplifiers are easy to design and build - semiconductors work on low voltages anyway. "Low voltage" in audio power supplies usually means DC voltages under 48V. Battery powered tube phono stages and preamps are a bit more difficult.

Firstly, a typical duo-triode of the 12AX7-12AT7-12AU7 variety, needs 300 mA (milliAmperes) of heater current at 6.3V or 150 mA at 12.6 Volts. Their heaters can be configured in two different ways. Most other preamp tubes don't have that capability. Let's say you have a preamp with two duo-triodes in the phono stage and two in the line stage (that would be a minimum required).

At 12.6V, their heaters will draw 0.6 A of current. We should not discharge the battery below say 50% of it's Ah capacity. Ah is short for "Ampere hours", a unit describing capacity of batteries.

Theoretically, a 7 Ah battery can supply 7 A of current for 1 hour, or 1 A for 7 hours, and so on. The number of discharge-charge cycles of a battery is limited. The deeper the discharge, the less cycles are available. So, to discharge a 7 Ah battery to 50% of its capacity at a heater current of 0.6A would take roughly 5 hours (3 Ah/0.6A = 6 hours). Very few audiophiles would listen to their systems for more than 5 hours in one "sitting" anyway.

The use of voltage regulators will lower the output voltage (there is always some voltage drop on the regulator itself) and will affect the sonics of the phono stage since the power supply is directly in the signal path. Luckily, their use is entirely optional. If you limit your listening time to an hour or two, the output voltage of a high capacity battery bank will drop from say 4*12.7V = 50.8V to 4*11.5V = 46V, and that will have no impact on the sonics, so powered up your phono stage directly from the battery bank!

Power supplies, including tube and solid-state voltage regulators will not be covered here, we will analyze them in detail in Volume 2 of this book.

Sealed lead-acid batteries

PS-1270 is a 12Volt 7.0 Ah rechargeable sealed lead-acid battery by PowerSonic, typical of this technology, Absorbent Glass Mat (AGM) . Its internal resistance is 23 mΩ, so even connecting two in series for $24V_{DC}$ operation would only bring R_I to half a milliohm, which is as close to an ideal voltage source as we can get.

The three graphs redrawn from data sheet give us a deeper insight into its behavior. The first is the impact of "standing period" on capacity retention ratio with ambient temperature as a parameter. The longer a battery sits on a shelf the more capacity it loses. Above 80% retention (A) an initial charge before use is not required.

Below 80% and above 60% (B), initial charging is necessary to recover the full capacity, while below 60% the manufacturer warns "Charge may fail to restore full capacity. Do not let batteries reach this state."

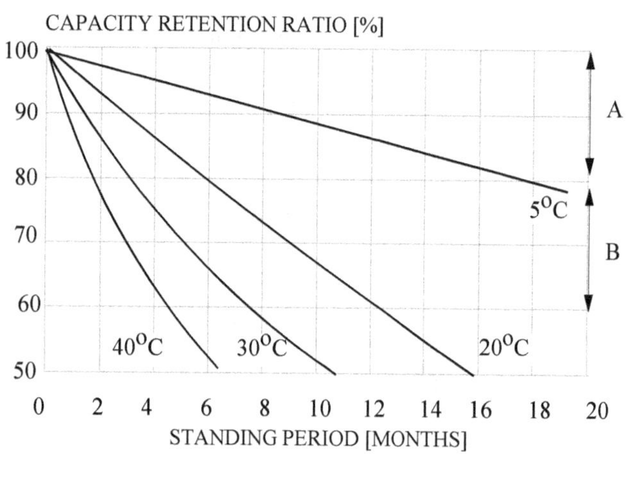

Shelf life

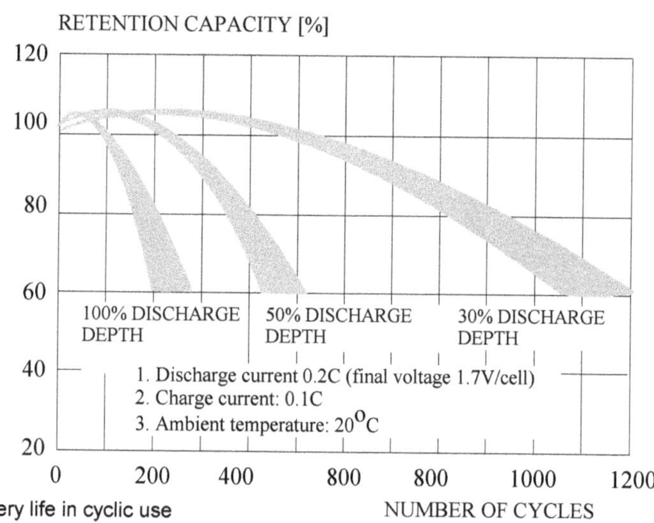

Battery life in cyclic use

Notice how a relatively mild temperature rise from 20 to 40°C decreases the critical 60% shelf life limit from 12 months down to just about 4! The lesson is clear - never buy batteries more than 6 months old, buy them from a reputable supplier with a large turnover and based in a cold country or city! Plus, recharge and test the batteries immediately after purchase and ask for money back guarantee, in case the full capacity cannot be restored!

The discharge depth will affect the number of cycles you will get out of a battery. With 30% discharge more than a thousand cycles are possible, while 50% discharge depth halves that to less than 500 cycles.

A 100% discharge is a no-no, it would reduce the battery life to only 200-250 cycles! Looking at discharge curves, at 0.35A current it takes about 20 hours to discharge the battery from 12.7V down to 10.7V, which is in accordance with it s capacity: 7Ah/0.35A = 20 hours.

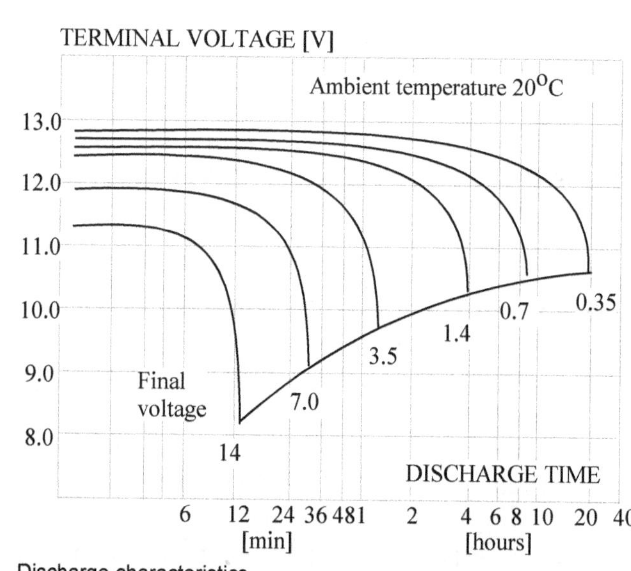

Discharge characteristics

In our application, assuming a typical draw of a stereo phono stage at less than 4mA, or thousand times smaller, it would take 20,000 hours to reach that level! Clearly, the anode power draw can be neglected in comparison with the heater power. Assuming four ECC88 tubes with their 6.3V heaters in series (for 24V operation), the heater current of 0.365A will discharge two batteries in about 18 hours. Thus, it would be prudent to limit the listening time to about 30% of 18 hours or about 5 hours. If that isn't enough for you, use three batteries in series for 36V or four for 48V operation, or use higher Ah (capacity) batteries.

CARTRIDGE-CABLE-PHONO STAGE INTERFACE

A moving magnet cartridge can be modeled by a signal source in series with its internal resistance and inductance. The shielded cable has its distributed resistance R_C and capacitance C_C, and the input stage of the phono preamp is represented by the parallel RC combination of its input resistance and capacitance. R_G (pickup's resistance) and cable's resistance can be lumped together into one equivalent resistance R, as can the cable's lumped capacitance C_C and the preamps input capacitance C_{IN}.

You should recognize a series LRC resonant circuit, the same circuits that describes the behavior of an audio transformer at high frequencies. It's a 2nd order low pass filter, and, as we have studied recently, its response to a square wave excitation depends on the three parameters, R, L and C and the frequency of the signal!

Moving coil cartridge interfacing

Moving coil cartridges have a very low inductance, so their equivalent circuit is simpler, just an AC signal source in series with cartridge's internal resistance R_G, typically 10-50Ω. The disappearance of L for MC cartridges makes the whole setup a 1st order filter.

However, the necessary inclusion of a step-up transformer means we are dealing with the 2nd order system here as well! The step-up transformer shown has two primary windings, so two impedance and voltage ratios are possible, as with many commercial models.

C_{IN} is around 100 pF for a typical input stage with 12AX7 triode in common-cathode arrangement, R_{IN} is the input loading resistor, usually 47 kΩ.

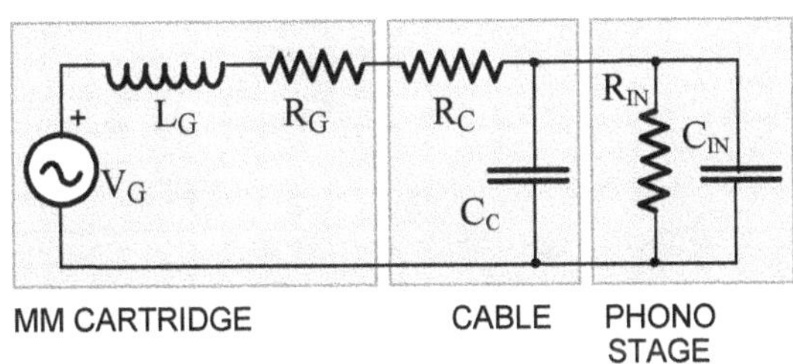

The simplified model of the cartridge-cable-phono stage chain

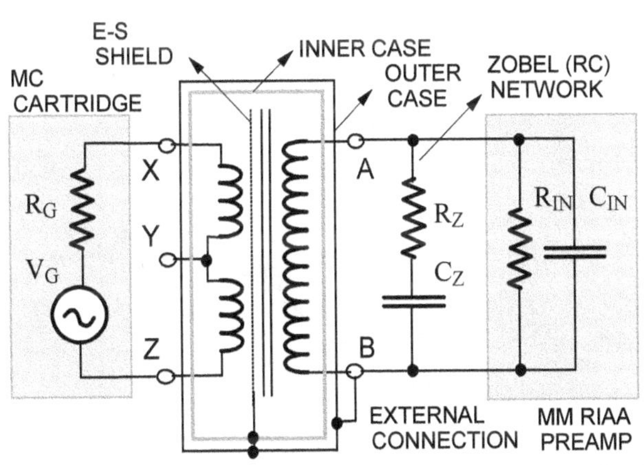

E-S SHIELD, INNER & OUTER CASE INTERNALLY CONNECTED

The square wave response of a phono cartridge

The RC network often used across the transformer's secondary is called Zobel network. Its aim is to optimize or correct the square wave response, from underdamped to critically damped curve, or, if preferred, to mildly underdamped response.

The same optimizing methods as for NFB compensation apply here, use of an oscilloscope or tune it by ear.

Signal-to-noise ratio and comparison between MC and MM circuits

A stray magnetic field will introduce the same amount of noise or interference voltage in both moving magnet (MM) and moving coil (MC) circuits. However, if the shielding is inadequate, the signal-to-noise ratio of the two circuits will be very different.

With a 10:1 voltage step-up transformer for MC signals, the signal to noise ratio at the source (cartridge) is $S/N = V_S/V_N = 0.1*10^{-3}/1*10^{-6} = 100$ or S/N [dB] = 20log(S/N) = 20log100 = 20*2=40 dB At the input of the phono stage, or the output of the step-up transformer, the S/N ratio is 1/0.01 = 100

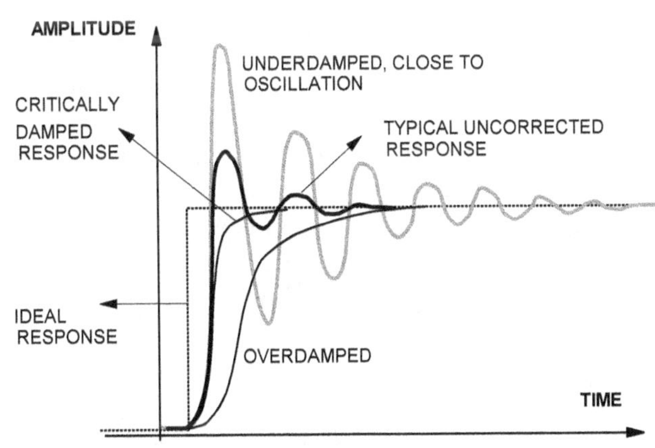

Square wave response of a phono cartridge (2nd order system)

The S/N ratio didn't change since the step-up transformer lifts both voltages by the same ratio.

With the high impedance MM circuit, we would still get the same 1μV of noise, but the signal is already 1mV, so our S/N ratio is S/N = 20log 1/0.001 = 20log1,000 = 60dB, or 20dB more than for the MC cartridge!

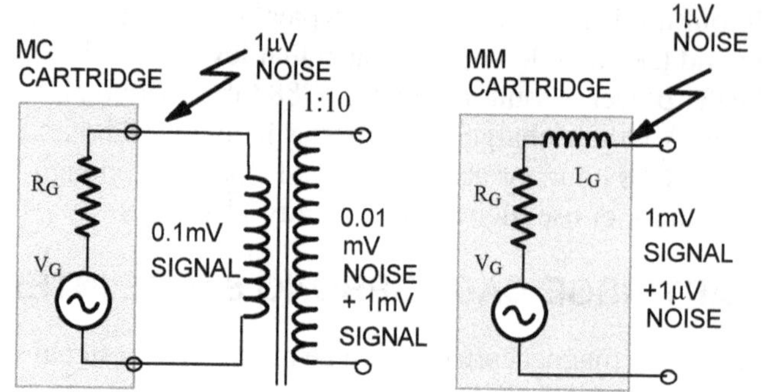

BALANCED PHONO STAGES

As we have seen, a differential amplifier has two inputs and two outputs. It amplifies only the difference in the voltage levels between the two inputs. Any signal common to both inputs is NOT amplified. Obviously differential amps lend themselves to balanced topologies, and in power amps they act as perfect phase inverters as well since the output signals are identical and out-of-phase.

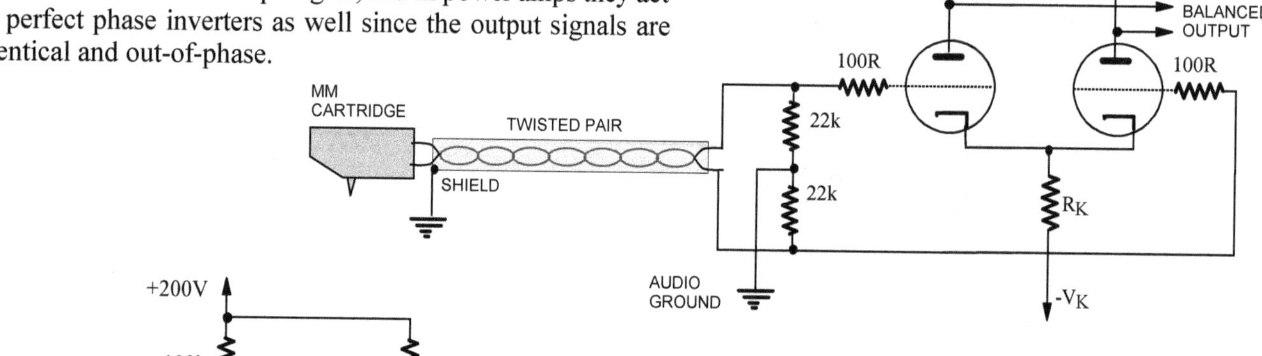

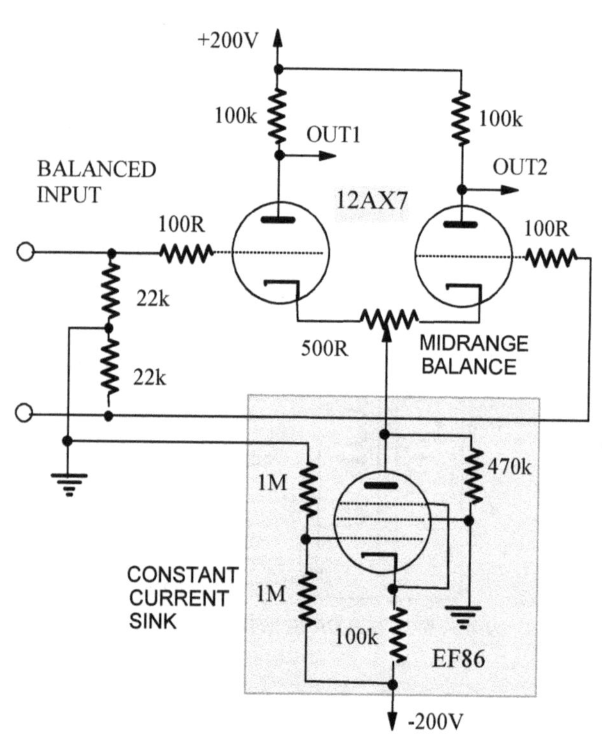

Instead of 47k resistor as the cartridge loading resistor, we now need a symmetrical arrangement with reference to ground, with two 22k resistors, each between one end of the signal input and ground.

A practical balanced input stage is illustrated on the left. ECC83 triodes are used for high gain, with an EF86 pentode as a constant current sink for the differential amplifier. A +/-200V power supply is easily built if a voltage doubler is used, with the center point of the doubler grounded. Its schematics is shown in the chapter on single-ended output stages, see the fixed bias 300B SE design.

Since tubes are seldom completely balanced, the 500Ω trimpot balances the anode currents and gain in the midrange.

Unbalanced (RCA) to balanced (XLR) signal conversion

A transformer with a center-tapped secondary is a natural phase splitter. Grounding the CT (point B) makes it the "reference" terminal, so the signals in points A and C will be identical in amplitude, but of the opposite phase, a true differential signal.

eve if the step-up transformer doesn't have split secondary, all is not lost. Simply create an artificial CT or reference (ground) by connecting two high value (100k-470k) resistors in series, as illustrated. The transformer in question can be an MC step-up kind, or, in general, a transformer of any impedance ratio, including 1:1 isolation transformers such as 600Ω:600Ω!

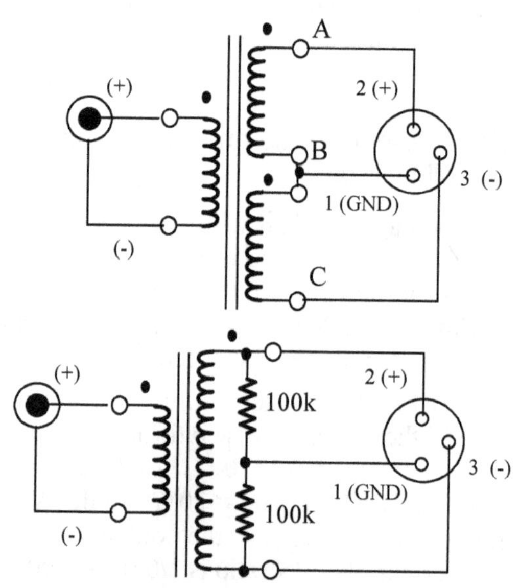

SINGLE-ENDED TRIODE OUTPUT STAGE

- LOAD LINES AND CLASSES OF OPERATION
- OPTIMAL ANODE LOAD
- EFFICIENCY OF TRANSFORMER- AND IMPEDANCE-COUPLED AMPLIFIERS
- ESTIMATING HARMONIC DISTORTION
- 2A3 OUTPUT STAGE
- 300B SE OUTPUT STAGE
- ESTIMATING HARMONIC DISTORTION
- REACTIVE LOADS AND THEIR LOAD "LINE"

"The 24watts that Unison Research claim for the Simply 845 turned out to be rather closer to 18watts on measurement (will single-ended valve amp manufacturers please stop quoting such optimistic output figures!)"
Jon Marks in his review of Unison Research "Simply 845" amplifier, Hi-Fi World, March 1998

LOAD LINES AND CLASSES OF OPERATION

Series- and shunt-fed output stages

The most common single-ended stage has the primary of the output transformer in series with the output tube. The DC anode current I_A flows through the primary winding and magnetizes the magnetic core, reducing the incremental inductance significantly. That is why SE transformers must use larger cores and more windings than push-pull transformers, through which the net primary DC current is zero (ideally, with perfectly-matched tubes).

The shunt-fed output stage has the output transformer in parallel with the output tube ("shunting" it to ground). The DC anode current I_A does not flow through the transformer's primary winding, so smaller cores can be used and higher primary inductances can be achieved compared to series-fed stages.

However, the price to pay is twofold. Firstly, there is large-value film coupling capacitor C in the signal path, to prevent the DC current from flowing through the primary. Secondly, an additional choke is needed as series anode load. This choke must be designed to accommodate high anode DC current through its winding. It must have an air gap and be of large physical size so it does not saturate even at the lowest frequencies of interest, 10-20Hz!

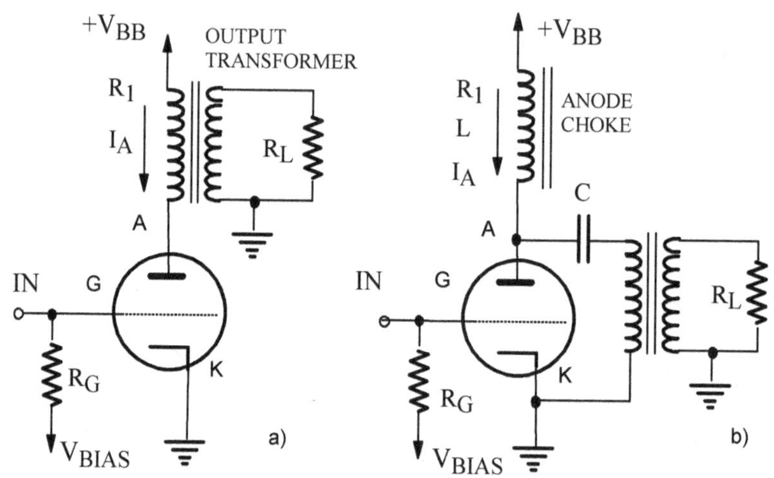

Series-fed (a) and shunt-fed (b) single-ended output stage with a fixed (external) bias

Static and dynamic load lines

The primary winding of output transformers has a small DC resistance R_1, 70-300Ω. Anode current flowing through the primary winding crates a voltage drop on such resistance so the quiescent voltage on the anode V_{AQ} is lower than the anode supply voltage V_{BB}. There is a difference between the static and dynamic load lines.

Output transformer with a primary resistance higher than 200Ω (wound with a thin wire on small size laminations) should be avoided due to the significant voltage loss on their primary windings.

The static load line has slope $\tan\beta = V_{R1}/I_0$. It passes through the operating point Q_A and the point V_{B+} on the V_A axis. Usually we neglect this small voltage drop V_{R1} and analyze the output stage as V_{AQ} would be equal to V_{BB}. For instance, typically, $V_{BB} = 400V$ and $V_{R1}=10V$, so $V_{AQ}=390V$, introducing the error of (400-390)/390 =2.5%

In our graphical estimations we will neglect this DC voltage drop , but you need to keep in mind that it's just another simplification we are making for the sake of convenience.

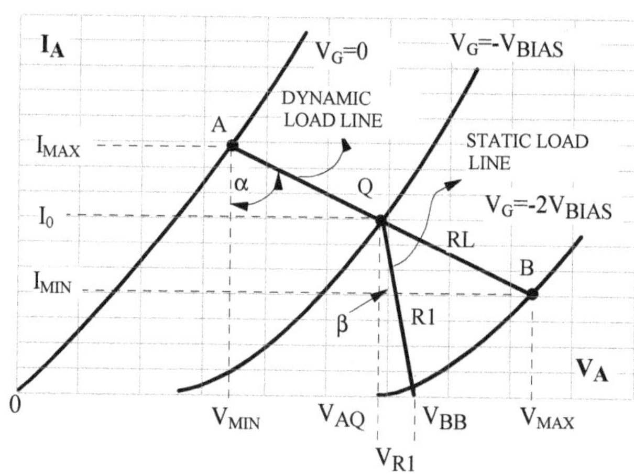

The static and dynamic load lines for a transformer-loaded or impedance-loaded single-ended stage

HOW CAN THE ANODE VOLTAGE EXCEED THE POWER SUPPLY VOLTAGE V_{BB}?

Theoretically, for ideal triodes and no losses of any kind, the anode voltage can swing between zero and $2V_{BB}$. Even in reality V_{MAX} is much higher than the power supply voltage V_{BB}. This is possible due to the transformer action, more precisely, the self-induction in the primary winding of the output transformer.

With the grid signal falling from $V_G=0$ (point A) towards point Q, the voltage in transformer's primary is induced to prevent or oppose such a drop (to keep I_A constant). That induced voltage keeps rising for as long as I_A keeps falling (towards point B). After I_A starts rising (operating point swinging back from point B up towards Q) the induced voltage changes, now trying to oppose the rise in I_A and reduces back towards lower anode voltages! Chokes operate on the same induction principle, opposing any change in current.

SE transformer-coupled output stage

The model of the output stage for a series-fed single-ended output stage is quite simple. V_{BB} is the anode DC power source, supplying current I_A, which is also the DC current I_1 flowing through the transformer's primary. The AC (signal) voltage that the output valve as an AC generator provides is $v_s(t)=-\mu v_G(t)$, where $v_G(t)$ is the signal on the tube's grid and μ is tube's amplification factor.

Points A and B are the primary transformer terminals. For AC signal point B is at ground potential since the ideal DC source V_{BB} has a zero internal impedance. In practice that is not the case, the power supply always has some internal resistance. The signal $i_1(t)$ flows through the DC power supply, which is why the quality of a single-ended tube amp's power supply directly impacts its sound! The same applies to interstage transformers, the driver stage's plate current flows through the primary, but the secondary (grid) impedance is very high so there is no secondary current.

Unbalanced DC current I_1 flows in transformer's primary. A transformer does not pass or "transform" DC currents or voltages, so there is no secondary DC current, only the AC signal current $i_2(t)$. The load is reflected onto the primary side as $(N_1/N_2)^2Z_L$, the ratio $(N_1/N_2)^2$ is called an impedance ratio of the output transformer.

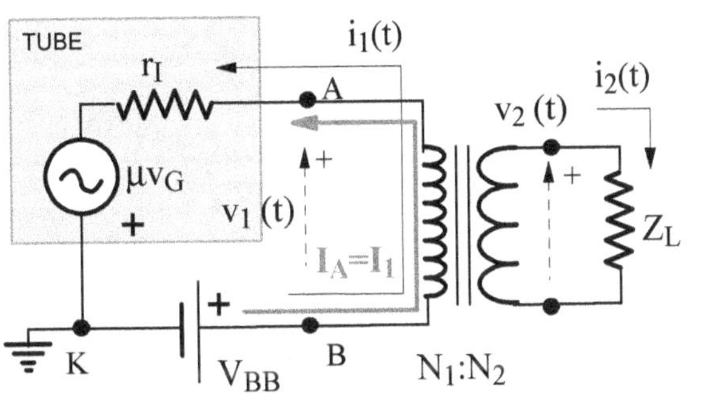

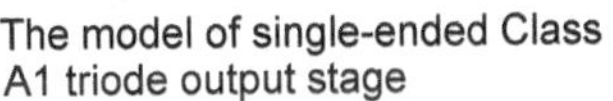
The model of single-ended Class A1 triode output stage

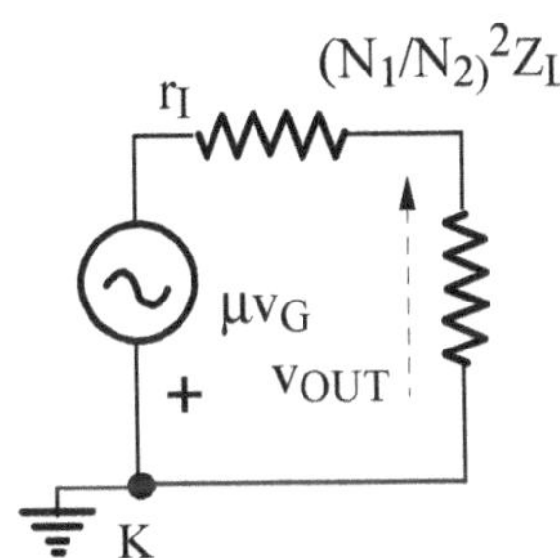

The small signal model of Class A1 SET stage

In the small signal AC model the DC power supply is a short circuit and does not impact AC conditions, so the model is ultimately a simple resistive voltage divider between the tubes internal resistance and the reflected secondary resistive load. This assumes an ideal transformer and a purely resistive load, both of which are sheer fantasies, but make it easier to understand the fundamental behavior of the output stage.

Class A_1 and class A_2 operation

In class A_1 operation, the grid is biased negatively with respect to cathode and no grid current flows, meaning the input resistance of the grid circuit is infinite.

However, if the grid of a power tube is driven into the positive region $+V_G$ (grid positive with respect to cathode), as in Classes A_2 and AB_2, the grid will start attracting electrons from the cathode and acting as a mini-anode. The grid current will start flowing, and the input impedance of the output stage (both single-ended and push-pull) will drop to a very low value, typically around 1 kΩ.

That will cause the loading of the driver stage (increased driver tube's anode current) and a large distortion, unless the driver stage was properly designed for such operation.

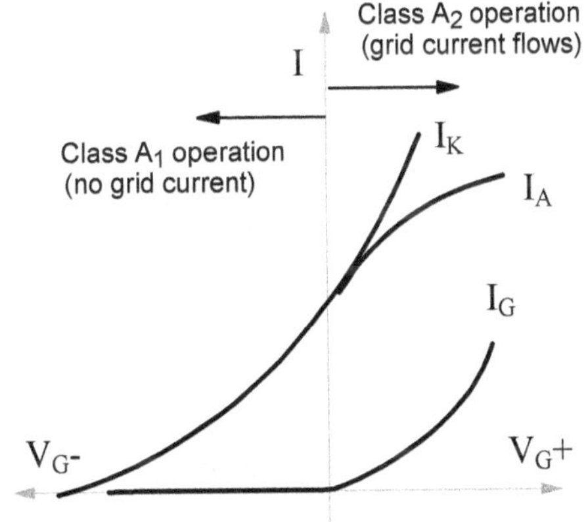

Cathode, anode and grid currents in a triode with fixed anode voltage, as a function of grid voltage: $I_K = I_G + I_A$

This situation is illustrated in the equivalent diagram where the switch models the presence or absence of grid current. The driver tube is a voltage source with its internal impedance r_I, R_A is its anode resistor, R_G grid resistor of the output stage. With the switch open (no grid current), the voltage divider r_I - R_A ratio is constant and there is no distortion.

The flow of grid current is equivalent to switching a very low resistance R_{GK} between the grid and the cathode of the power tube, in parallel with R_A and R_G.

RC coupling should never be used for class 2 output stages. One option is to use an interstage transformer specially designed to operate with secondary DC current (grid current). Commercial transformers of that kind are rare and expensive, so a cathode follower driver stage, directly coupled to the output tube's grid is a cheaper and easier alternative.

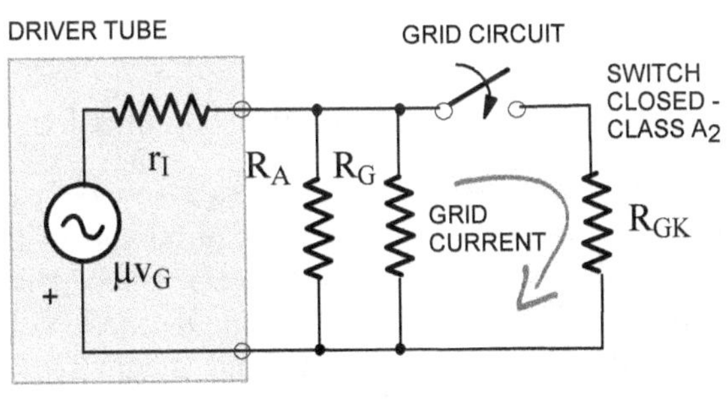

The model of single-ended Class A_2 triode driver stage: the closed switch simulates grid current flowing through the low grid-cathode resistance R_{GK} of the power tube

The driver triode should have a high anode dissipation and low internal resistance. A single 6SN7 triode section has r_I=6,800 Ω. You can parallel two sections together and get half that, r_{IP}=3,400Ω A cathode follower would bring that impedance down to a couple hundred ohms, which will drive higher grid impedance very well. Lower r_I driver tubes such as E86C and power triodes such as 45, 2A3 or 300B would be even better (r_I around 700-800Ω).

Nonlinear operation and the shift in quiescent point

In single-ended amplifiers, as signal amplitude increases, so does the distortion: the negative dips become larger than positive ones, and in the extreme case, for the maximum voltage swing as shown, they end up being 130V and 110V respectively. In that case the quiescent point Q will not stay where it was, but will move down, since there is now a DC component of the distorted signal, due to the fact that rectification of the signal has taken place! The undistorted signal would have no DC component.

This effect is sometimes noticeable on class A power amplifiers that have anode current DC meters. As the volume control is increased, once the output power reaches a certain point, the DC anode current actually starts falling, which is counterintuitive, but is the result of the already-mentioned rectification effect.

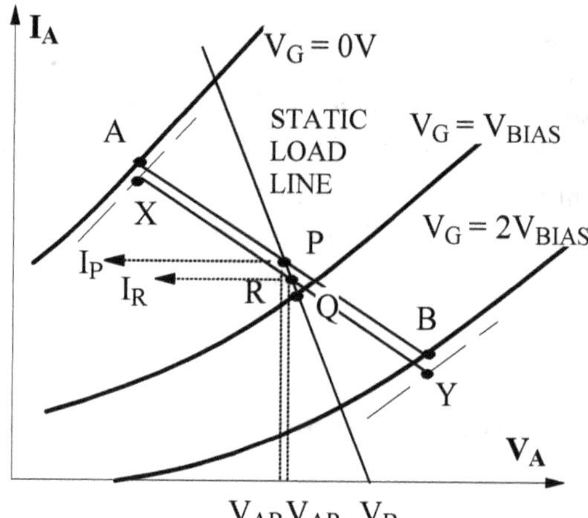

ABOVE RIGHT: The rectification of the signal shifts both the operating point and the grid swing range down, from A-P-B to X-Q-Y.

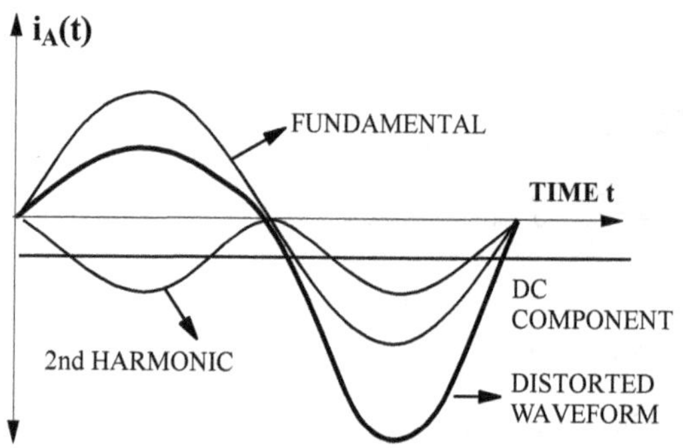

RIGHT: The distorted waveform with a significant DC component and the 2nd harmonic.

OPTIMAL ANODE LOAD

The choice of the anode load for output tubes is a compromise between maximum power and minimum distortion, especially so for triodes.

For triodes the load impedance should be 3-4 times the internal impedance of the tube. For 300B triode, with r_I=800Ω, Z_A should be between 2k4 and 3k2, while 3k5 is commonly used. These are maximum power rules. For lower distortion levels, higher load impedances should be used.

Notice that at the load impedance producing maximum power Z_{AP} the distortion is higher compared to larger load impedances, such as Z_{AD}, and that power and distortion drop off at approximately the same rate with increased load.

As an example, for 300B with 350V anode-to-cathode voltage, $V_{G0} = 0.75V_A/\mu = 0.75*350/3.85 = -68V$, and $Z_{AOPT} \approx 3r_I - 4r_I$ or from 3*790= 2k4 to 4*790 = 3k2!

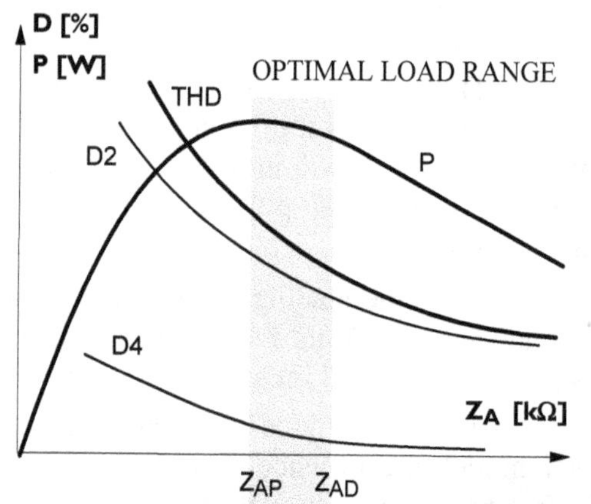

Typical triode's power and distortion curves as a function of anode resistance. D2 and D4 are the 2nd and 4th harmonics

OPTIMAL ANODE LOAD AND GRID BIAS FOR SE TRIODES

$\mathbf{Z_{AOPT} \approx 3r_I - 4r_I}$, r_I = internal resistance of a tube

$\mathbf{V_{G0} = 0.75V_A/\mu}$ **[V]**, V_A = DC anode voltage

μ = amplification factor of the output triode

The graphical interpretation of "optimal load"

The slope of the load line $\Delta I_A/\Delta V_A$ is inversely proportional to the load impedance Z_A and since the output power is a product of I_A and V_A, the issue of maximizing the output power can be graphically interpreted as maximizing the area of the triangle whose hypotenuse is the load line and whose base is the I_0 horizontal line. Three examples of load lines are illustrated. The quiescent DC anode voltage is the same in all three cases (V_{AQ}), while the bias voltages E_G vary, resulting in very different quiescent DC currents I_{01}, I_{02} and I_{03}.

Z_{L1} has the smallest slope (it's the most "horizontal" of all) and therefore the highest load impedance. The voltage swing, the difference between the V_{Amin} and V_{AQ}, is very large, but its anode current swing (current in point A minus quiescent current I_{01}) is very small. Z_{L3} goes to the other extreme.

The load impedance is low, resulting in a large current swing but a minuscule voltage swing, again resulting in low output power.

The intermediate impedance Z_{L2} results in the optimal balance between maximizing current and voltage swings. Tube manufacturers have determined this impedance for each of their tube types, these are the recommended load impedances listed in tube data sheets and application notes.

For tubes not originally developed for audio use, such as TV sweep tubes, screen-driven tubes, series voltage regulator tubes and pentodes & beam tubes strapped as triodes, optimal load impedances were not published and have to be determined by experiment.

A good starting point for choosing the load impedance and positioning the quiescent point Q is at anode current I_0 at half of the maximum current I_{MAX} and then fine tuning it to get the best performance.

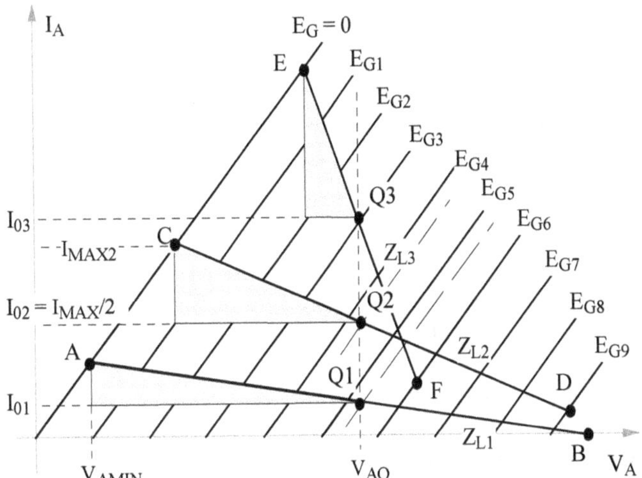

Load line choices for SE triodes: maximizing the area of the power triangle

EFFICIENCY OF TRANSFORMER- AND IMPEDANCE-COUPLED AMPLIFIERS

Class A transformer-coupled load

You may recall a dismally low efficiency (a maximum of 25%) of an amplifying stage where the load was DC-coupled or in series with the anode. However, the efficiency of the preamplifier stages where such arrangements are used is generally of no concern to designers, since our aim is to amplify the voltage signal and not power.

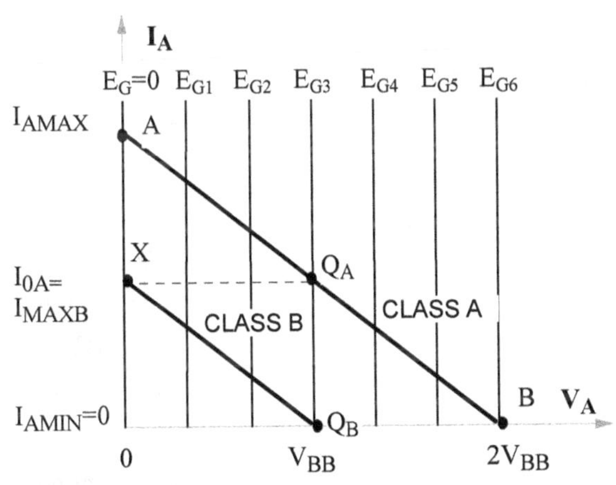

Anode curves and load lines for ideal triode with transformer-coupled resistive load in class A and class B

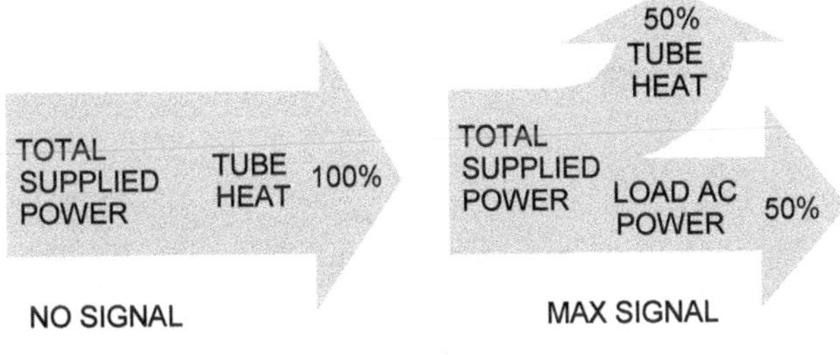

Power tubes in Class A run cooler with AC signal than when idle. At amplifier's maximum output power the output tubes run coolest.

Because output stages amplify power (both voltage and current), to reduce the size of the power transformers and other components in power supplies, and to minimize the heat that amplifiers dissipate into listening rooms, transformer- and impedance-coupled output stages should be made as efficient as possible.

Let's see if the situation is any better with transformer-coupled resistive load in Class A. The transformer prevents the DC anode current from flowing through the load on its secondary. We will assume an ideal transformer, without any losses.

The quiescent point is Q_A, and the load line is AB. $I_{AMAX} = 2I_0$, the anode voltage in the quiescent point is the voltage of the high voltage battery V_{BB}.

η= (AC output power)/(DC input power)*100%

$\eta = (2V_{BB}-0)(I_{AMAX}-0)/8V_{BB}I_0*100\%$

Since $I_{AMAX} = 2I_0$ the efficiency is

$\eta = 2V_{BB}2I_0/8V_{BB}I_0)*100\% = 1/2*100\% = 50\%$

Well, that is certainly a big improvement. By eliminating the DC current through the load and by capitalizing on the transformer action, where due to induction the anode voltage can swing to double the DC voltage of the power supply (V_{BB}), the efficiency is doubled to 50%.

With no signal all the power is dissipated on the output tube. With a maximum signal, half of the power supplied by the power supply is dissipated as heat in the tube, the other half is converted into AC or useful signal power.

Class B transformer-coupled load

In class B operation, the quiescent point Q_B is at the same anode voltage V_{BB}, but the quiescent current is zero. When no signal is present, the current is zero and the dissipated power is also zero! Again, we will assume a maximum voltage swing without any distortion.

The mathematics is a bit more involved, to get the average current we need to integrate the sinusoidal waveform from 0 to π radians (half-period), since the current is zero for the other half-cycle. This integration gives us the average current as $I_{AV}=I_{MAXB}/\pi$, so the input power is $P_{IN}=I_{AV}V_{BB} = V_{BB}I_{MAXB}/\pi$. The efficiency is η= (AC output power)/(DC input power)*100% = $(V_{BB}-0)*(I_{MAXB}-0)/4V_{BB}I_{MAXB}/\pi*100\% = \pi/4*100\% = 78.5\%$

Hi-fi amplifiers of single-ended kind cannot operate in class B, since half of the input waveform would be cut off and not reproduced at all. For class B operation we need a push-pull arrangement with two tubes, where each tube would amplify half of the signal. However, the analysis above is equally valid for class B push-pull stages.

The efficiency of real amplifiers, regardless of the type of loading and the class of operation, are significantly lower, typically 50-70% of these ideal figures. So, a single ended transformer-coupled triode stage will achieve the efficiency of 25-35%, instead of the ideal 50% maximum!

2A3 SE OUTPUT STAGE

In 1929 RCA engineers developed type 45 triode. Very quickly the need for a higher power tube became obvious and being prudent and practical Americans, instead of developing a new tube from scratch, they paralleled two 45 tube structures inside a larger glass bulb and in 1932 the 2A3 triode was born.

To this day there are many audiophiles who claim that 45 triodes sound better than 2A3s, however their minuscule output power requires very sensitive speakers, if not horns.

TUBE PROFILE: 2A3

- Directly-heated power triode
- Heater: 2.5V, 2.5 A (6.25W)
- V_{AMAX}=300V, P_{AMAX}=15W
- TYPICAL OPERATION:
- V_A=250V, V_G=-45V, I_0=60 mA
- Z_L=2.5 kΩ
- gm=5.25 mA/V, μ=4.2, r_I=800Ω
- P_{OUT}= 3.5 W @ 5% THD

Optimal conditions for 2A3 triode

RCA data sheet for 2A3 recommends the anode voltage V_0=250V. The grid bias in the operating point is calculated from the empirical equation for triodes $V_{GQ} = 0.75V_0/\mu$ = 0.75*250/4.2 = 44.6 V

For directly heated triodes, half of the AC heating voltage needs to be subtracted from this figure, or 1.25V in the case of 2A3 tube, so we get the bias of 43.35V, which is almost identical to the point specified by RCA engineers, namely V_{GQ} = -43.5V!

Using a 3.5kΩ load instead of the usual 2.5kΩ loses a bit of power but the distortion drops 2.7%, from 6.7% to 4%! That is especially important in low- or no-feedback designs.

The graph below left shows dynamic transfer curves of 2A3 SE stage for various load impedances. The larger the load impedance, the straighter the dynamic transfer characteristics, meaning lower distortion. However, it also means smaller anode current swing and lower output power!

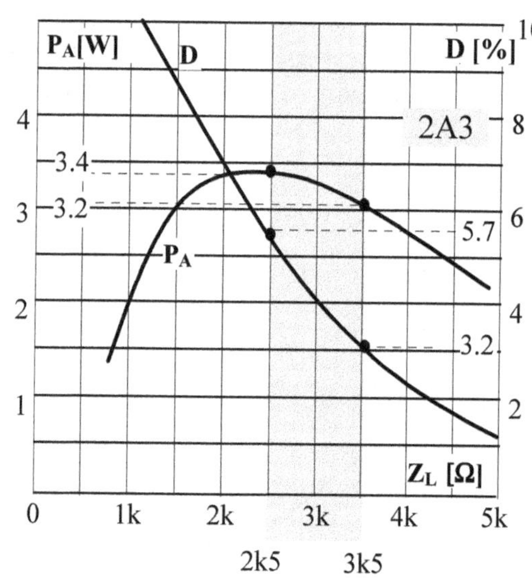

The power and distortion curves for SE 2A3 triode.

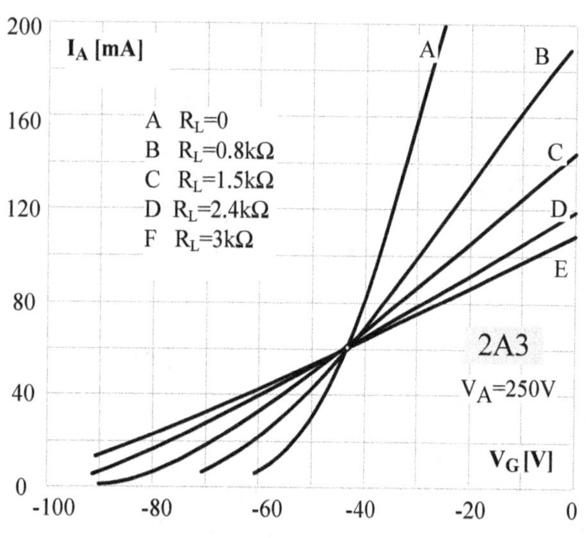

Dynamic transfer curves of 2A3 SE stage for various load impedances

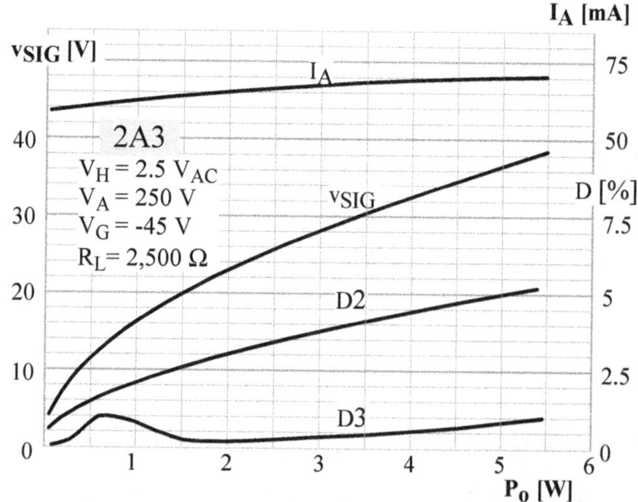

Anode current, distortion and required drive signal as a function of output power for SE stage with a 2A3 triode

The load line and efficiency of a 2A3 SE stage

2k5 load:
V_{MIN} =105V, V_{MAX}=370V
ΔV = 265V
I_{MAX} =120 mA I_{MIN} =15 mA
ΔI = 120-15 = 105 mA
$P_{IN}= I_A V_0$ = 0.06*250 = 15
$P_{OUTMAX}= V_0 I_0/2$ =
= 250*0.06/2 = 7.5 W
$P_{OUT}=\Delta V \Delta I/8$ = 265*0.105/8
= 3.5W
$\eta = P_{OUT}/P_{IN}$ =3.5/15=23.3%

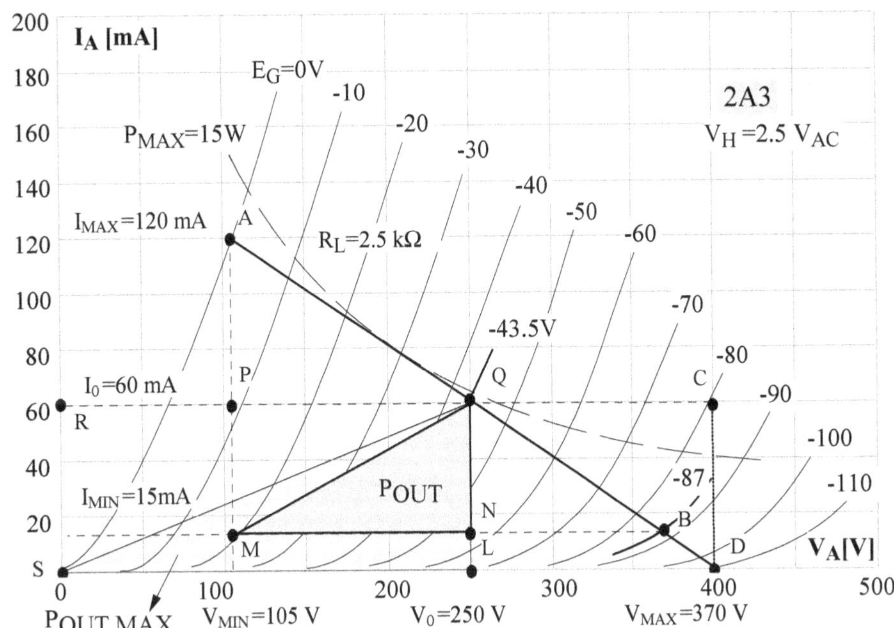

Power diagram of a 2A3 SE output stage

300B SE OUTPUT STAGE

TUBE PROFILE: 300B

- Directly-heated power triode
- Heater: 5V, 1.2 A
- Maximum anode voltage: 450 volts
- Maximum anode dissipation: 36W
- TYPICAL OPERATION:
- V_A=350V, V_G=-74V, V_{SIG}=74V_P = 52V_{RMS}
- I_0=60 mA, I_{MAX}=77 mA, R_L=4kΩ
- gm=5.0 mA/V, μ=3.85, r_I =790Ω
- P_{OUT} = 7 W @ 5% THD

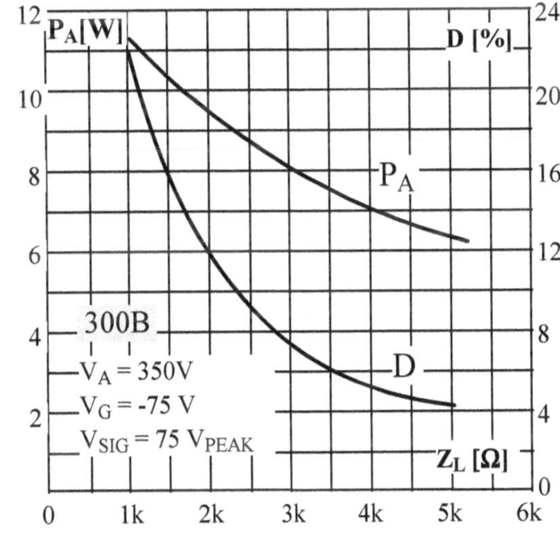

Operating conditions from WE catalogue

Let's start with the operating point listed by Western Electric in 1936. They positioned the operating point Q at 60 mA and 350V, meaning the power dissipation on the anode is 21 Watt, way below the maximum plate dissipation curve of 36 watts.

Obviously, they didn't want their tubes to fail after a few months, as would happen if biased at the maximum ratings, say 350V and 100mA anode current.

Next we draw the AC load line for 4 kΩ.

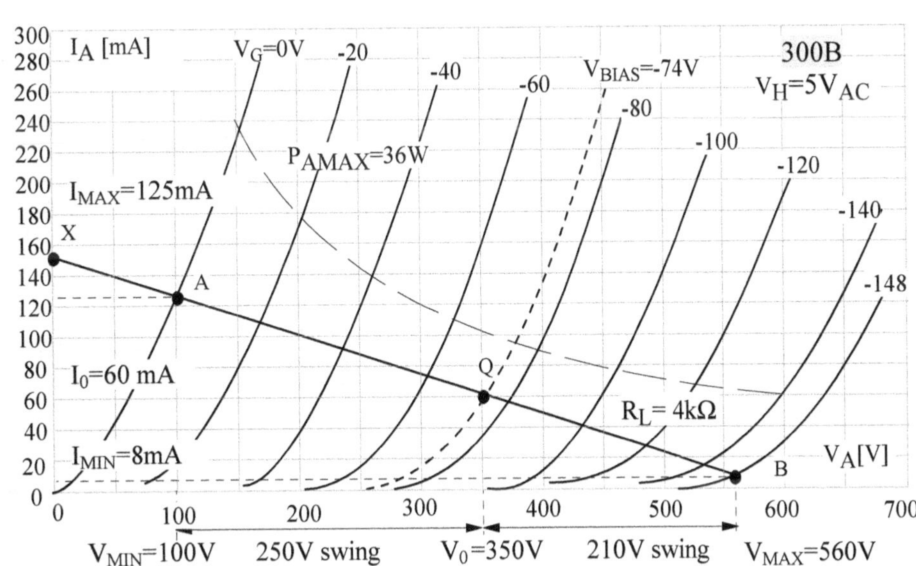

To find the intersection with Y-axis we divide the voltage in operating point Q with load impedance 350V/4,000Ω = 87.5 mA and add it to I_0 = 60 mA to get 147.5 mA. Let's call that point X. Now we have two points, Q and X, enough to draw the AC load line for R_L=4kΩ.

Western Electric engineers specified signal voltage of 74V peak, exactly the same as the bias voltage, which brings us exactly to V_G=0 and its intersection with the load line. That is point A. Since the bias is -74V, we need the anode curve for 74+74= -148V on the grid. Since such a curve is not given in data sheets and graphs, we have to estimate it (draw it provisionally). The intersection of the load line with that curve is our point B.

For points A and B we draw vertical lines down to the V_A axis and estimate (read) the minimum and maximum anode voltage V_{MIN} and V_{MAX}. As the grid signal swings from Q (V_G=-74V) to A (V_G=0V and AC signal v_G=74V peak) and than down towards B, the operating point Q moves up the load line to point A, then down to point B, and so on. The anode voltage swings 490V, between V_{MIN}=100V and V_{MAX}=560V. Since the grid voltage swings from 0V through 74V and then to -74V, the peak-to-peak grid driving signal must be around 148V_{PP} or around 52.5V_{RMS}!

$P_{OUT} = (\Delta V_A \Delta I_A)/8 = (V_A-V_B)(I_A-I_B)/8$ = (560-100)*(0.125-0.008)/8 = 53.8/8 = 6.7 Watts

Total power input is $P_{IN}=V_0I_0$ = 350*0.06 = 21 Watts. Since the anode is rated at 36 Watts, this design is considered very conservative, since it is only 58.3% of the rated dissipation.

The efficiency is $\eta = P_{OUT}/P_{IN}$*100 [%] = 6.7/21*100 [%] = 31.9 % Remember, the maximum efficiency of this topology is 50% (assuming an ideal triode), so our efficiency here is only 31.9/50 = 64% of the maximum.

Pushing tubes to their limits

Anode dissipation of less than 60% of its rated power is low by any standard, so let's look at the other extreme, biasing the output tube at its very limit. Q is positioned at 400V between anode and cathode, with 80V on the cathode (or -80 on the grid), so the V_{BB} voltage of 480V is needed. In this operating point the anode current is 100mA and $P_{IN}=V_0*I_0$ = 400*0.1 = 40 Watts According to manufacturers' data, modern 300B are rated at 40Watts, slightly higher than the original WE 300B, which was rated at 36Watts.

We will also reduce the load impedance slightly, from Western Electric's choice of 4kΩ down to the most commonly used value of 3.5kΩ! In class A_1 operation no grid current flows, so the grid voltage input will swing between -80B and 0V! With a 3k5 load we have:

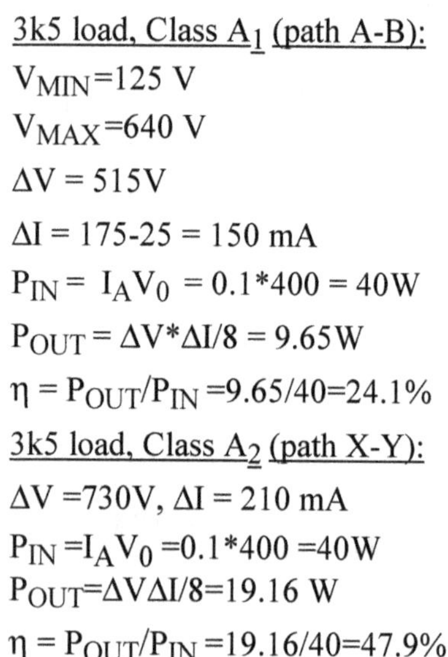

3k5 load, Class A_1 (path A-B):

V_{MIN}=125 V

V_{MAX}=640 V

ΔV = 515V

ΔI = 175-25 = 150 mA

$P_{IN} = I_AV_0$ = 0.1*400 = 40W

P_{OUT} = ΔV*ΔI/8 = 9.65W

$\eta = P_{OUT}/P_{IN}$ =9.65/40=24.1%

3k5 load, Class A_2 (path X-Y):

ΔV =730V, ΔI = 210 mA

$P_{IN} = I_AV_0$ =0.1*400 =40W

P_{OUT}=ΔVΔI/8=19.16 W

$\eta = P_{OUT}/P_{IN}$ =19.16/40=47.9%

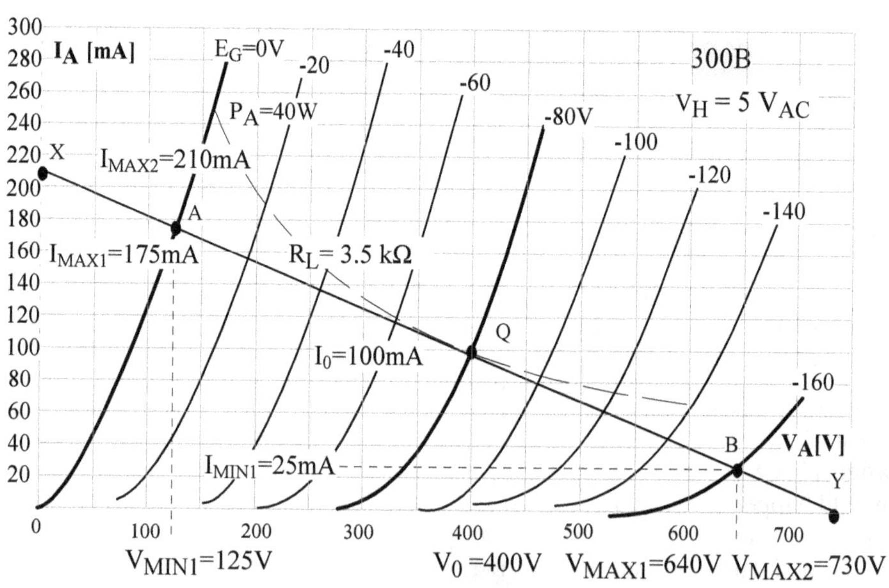

So, it is possible to get 10 Watts from 300B in class A_1, and almost double that in class A_2! However, the tube would be operating at the maximum of its rated power and would not last long. Unless you have money to burn on extortionary- priced 300B tubes, reducing the anode current down to a maximum of 80 mA is mandatory! For long & stress-free life of the precious 300B princesses, even lower quiescent currents of 60-65mA are recommended.

The 2nd harmonic distortion of this design in Class A_1 is D_2= [(I_{AMAX} +I_{AMIN}) - 2I_0]/2(I_{AMAX} -I_{AMIN})*100% = [(175+25) - 200]/2(175 -25)*100% = 0/300 = 0%

This is not a trick or a magically-positioned operating point, but simply a coincidence, helped by imprecise readout of figures from a smallish drawing and a simplistic formula. Instead of quietly tiptoeing around it we decided to mentioned it in order to illustrate such a possibility.

Using the voltage swing version of the D_2 formula, $D_2 = [(V_{AMAX} + V_{AMIN}) - 2V_0]/2(V_{AMAX} - V_{AMIN})*100\% = [(640+125) - 800]/2(640-125)*100\% = 35/1{,}030 = 3.4\%$

Although some books claim that the two formulas are interchangeable, they clearly are not, and if you use both to estimate the distortion of the same design, you will usually get different results.

ESTIMATING HARMONIC DISTORTION

The 5-point and 3-point harmonic distortion estimation method

We have already illustrated the distortion of the anode current waveform due to triodes' parabolic dynamic transfer curve. In this case the second harmonic is added to the fundamental during the positive half-wave, making it taller and slightly narrower than the fundamental, and is subtracted from the fundamental during the negative half-wave, making it slightly wider and flatter than the undistorted signal. The 5-point method for harmonic distortion determination uses the five significant current levels in the points illustrated in the graphs.

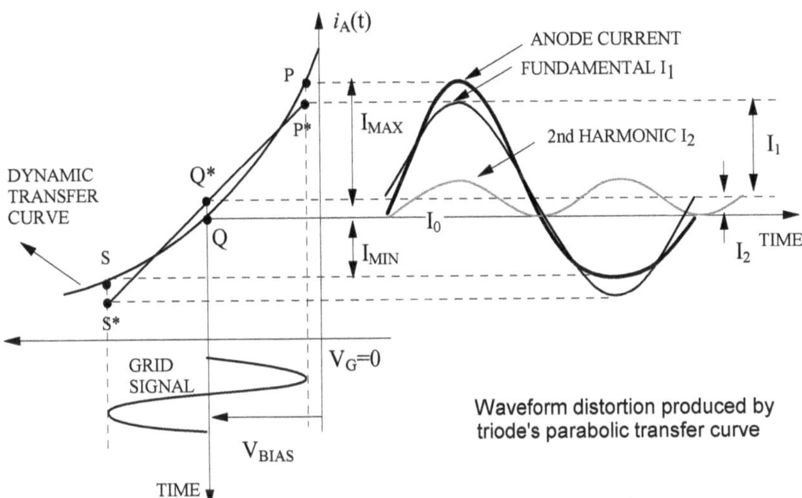

Waveform distortion produced by triode's parabolic transfer curve

The 3-point method uses only I_{MAX}, I_0 and I_{MIN}, for $E_G=0V$, $E_G=V_{BIAS}$ and $E_G=2V_{BIAS}$ respectively, to which the 5-point method adds current levels that are obtained by the intersection of the load line with anode characteristics for grid voltages $E_G=0.5V_{BIAS}$ (I_X) and $E_G=1.5V_{BIAS}$ (I_Y).

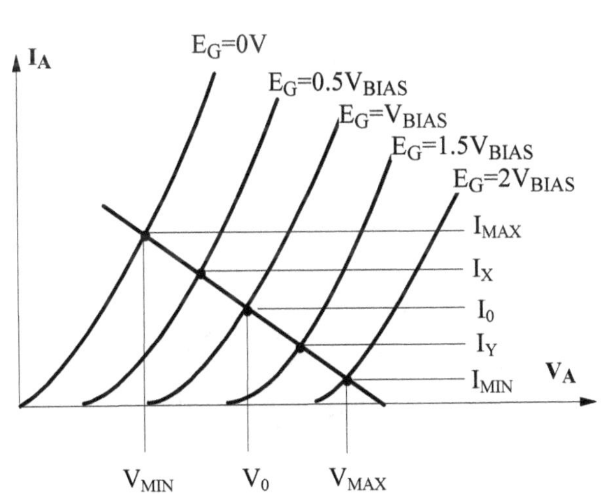

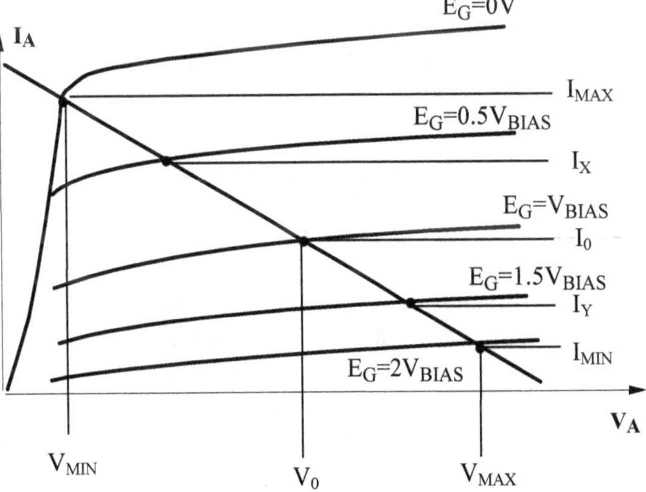

The 5-point method for triode (LEFT) and pentode (RIGHT)

The values of the rectified (DC) component of the signal and the first four harmonics are given in the table below. The individual harmonic distortion coefficients can be calculated by $D_2=I_2/I_1*100$ [%], $D_3=I_3/I_1*100$ [%] and $D_4=I_4/I_1*100$ [%] The total harmonic distortion factor is THD = $(D_2^2+D_3^2+D_4^2+D_5^2+ ...)$

The final formula for the 2nd harmonic distortion is $D_2= [(I_{AMAX} + I_{AMIN}) - 2I_0]/2(I_{AMAX} - I_{AMIN})*100\%$

If higher level harmonics are negligible (as with triodes), the use of the 3-point formulas is faster and easier. However, higher harmonics cannot be neglected in the case of tetrodes and pentodes, so the 5-point formulas must be used.

SIGNAL COMPONENT	5-POINT METHOD	3-POINT METHOD
Rectified (DC) component	$(I_{MAX}+2I_X+2I_Y+I_{MIN})/6 - I_0$	$(I_{MAX}+I_{MIN}-2I_0)/4$
Fundamental I_1	$(I_{MAX}+I_X-I_Y-I_{MIN})/3$	$(I_{MAX}-I_{MIN})/2$
2nd harmonic I_2	$(I_{MAX}+I_{MIN}-2I_0)/4$	$(I_{MAX}+I_{MIN}-2I_0)/4$
3rd harmonic I_3	$(I_{MAX}-2I_X+2I_Y-I_{MIN})/6$	
4th harmonic I_4	$(I_{MAX}-4I_X+6I_0-4I_Y-I_{MIN})/12$	

Harmonic distortion of WE 300B stage

Referring to the figures from the operating conditions chosen by Western Electric for the 300B output stage we have I_0=60 mA, I_{MAX}=125mA and I_{MIN}=8mA,so D_2= [(I_{AMAX} +I_{AMIN})-2I_0]/2(I_{AMAX} -I_{AMIN})*100% = [(125 +8) - 120]/2(125 -8)*100% = 13/234 = 5.55%

REACTIVE LOADS AND THE LOAD "LINE"

Real loads such as dynamic and electrostatic loudspeakers aren't purely resistive. Dynamic speakers are inductive and in the first approximation they can be modeled by a series RL circuit, the impedance being $Z_L=R_L+X_L$.

Electrostatic speakers are predominately a capacitive load, modeled by a resistor R and capacitor C in parallel. In that case the current source model makes calculations easier.

In both cases, instead of the operating point moving up and down along the straight load line as with a purely resistive load, now the operating point follows an elliptical trajectory. For capacitive loads it rotates counter- clockwise, for inductive loads it moves in the clockwise direction.

If you draw a horizontal line through any value of anode current between I_{MAX} and I_{MIN}, it will intersect the ellipse in two points. There are two solutions to the loadline equation, for each current there are two voltages that will produce such a current. The same applies to anode voltages, if you draw a vertical line for any anode voltage between V_{MIN} and V_{MAX} you will get two different plate currents.

Intuitively, without going into complex maths, the larger the reactive component of the load (or in this case inductive reactance), the wider or "fatter" the ellipse. The smaller the reactance, the flatter the ellipse, and the more such an ellipse will approach a straight line.

However, the reactive component of the load will change with frequency. The higher the frequency of the signal, the larger $X_L=2\pi fL$ will be and the fatter the ellipse. So, with a complex signal comprising of many different components of various frequencies, one ellipse will turn into dozens. If you observe that on an oscilloscope, the load "line" will be a messy, fuzzy, filled "ellipse" of ever changing size and shape.

I put even ellipse in quotation marks, because the tangled mess cannot be called an ellipse any more, unless a single sinewave signal of fixed frequency is being reproduced, which is never the case with music, only during testing.

We will not analyze these models for a few reasons. Firstly, these are small signal models which do not fit well the large signal output (power) stages. Secondly, the mathematics would be quite advanced and tedious. Thirdly, and most importantly, nothing would be gained by such tedious analysis, it is neither practical nor pleasurable!

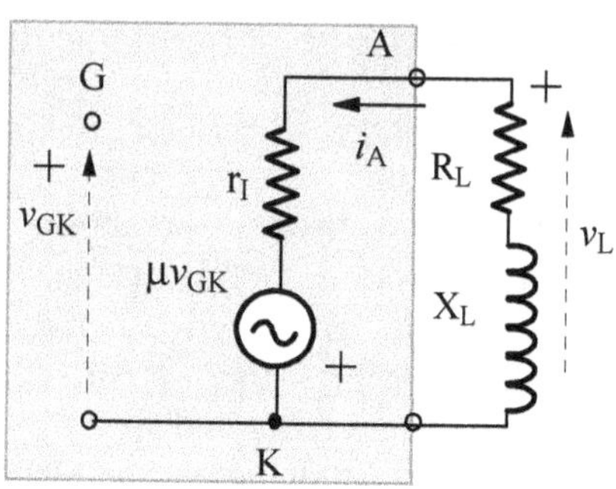

Small signal model of a triode with inductive-resistive load, such as a dynamic (moving coil) loudspeaker

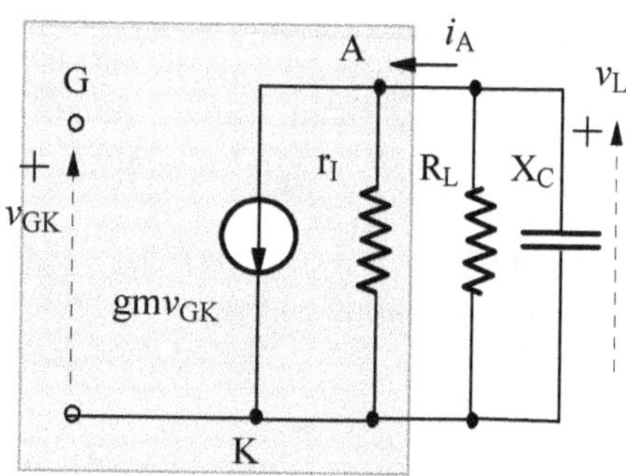

Small signal model of a triode with capacitive-resistive load, such as an electrostatic loudspeaker

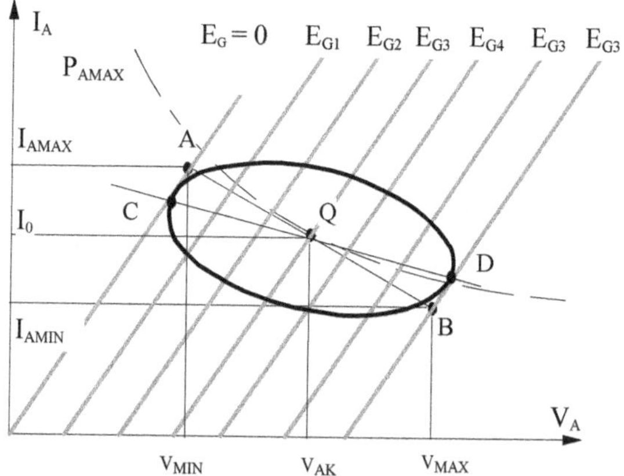

The load "line" of a tube with a complex load (resistive + reactive component) is an ellipse.

PRACTICAL SINGLE-ENDED TRIODE AMPLIFIER DESIGNS

- FIXED BIAS 4-STAGE 300B SE DESIGN
- MINIMALIST 2-STAGE 300B SE DESIGN
- CASE STUDY: AUDIO NOTE "CONQUEST"
- CASE STUDY: AUDION "GOLDEN NIGHT"
- SE AMPLIFIERS WITH LOW IMPEDANCE TRIODES
- 6C33C-B SET MONOBLOCKS
- SV572-10 SINGLE-ENDED AMPLIFIER
- OUTPUT STAGES WITH OTHER HIGH VOLTAGE TRANSMITTING TUBES
- CASE STUDY: "ONGAKU" BY AUDIO NOTE
- DESIGN PROJECT: 211 - 845 - GM70 AMPLIFIER
- CASE STUDY: CAZ-TECH "SE845" MONOBLOCKS
- SET AMPLIFIERS - DESIGN AND CONSTRUCTION ISSUES AND CHOICES

"I made a number of 211-S amplifiers, but was not thoroughly satisfied with their tone quality although they showed excellent characteristic. They lacked 'tenderness' of 2A3 and 'depth' of 300B."

Mr. Hiroyasu Kondo, the late founder of Audio Note

16

FIXED BIAS 4-STAGE CLASS A2-CAPABLE 300B SE DESIGN

We have already analyzed the cathode-follower driver stage of this circuit. Since it is directly coupled to the grid of 300B, it can supply significant grid current making operation in class A_2 possible with low distortion! Driver tubes such as 12BH7, 6CG7, 6SN7, 5687 and many others can be used.

The 10R resistor is for measuring of the cathode current and, indirectly, the bias voltage while adjusting it to the desired value. 0.8V means the DC current is 80mA. If you don't like its sound, you can replace the quazi-SRPP input stage with a common cathode stage or a differential amplifier. Likewise, for a NFB-free design remove the 100k and 100R resistors and connect point X to GND.

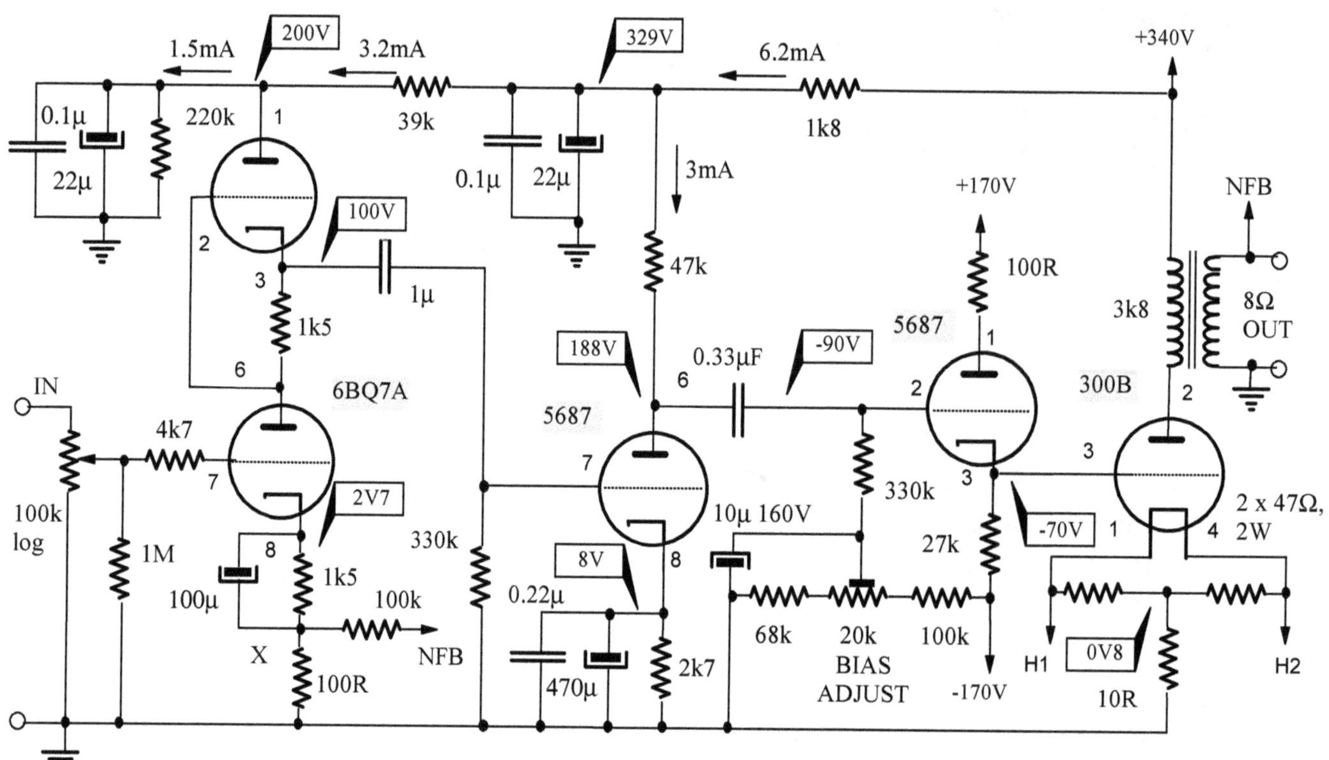

6BQ7 (6BZ7) and 6AQ8 (ECC85) duo-triodes

TUBE PROFILE: 6BQ7 (6BZ7)

- Indirectly-heated Noval duo-triode
- Heater: 6.3V/400 mA
- V_{HKMAX}: $200V_{DC}$, P_{AMAX}: 2W
- Max. anode voltage & current: 250V/20 mA
- TYPICAL OPERATION:
- V_A=150V, V_G=-2.0V, I_0=9 mA
- gm = 6.0 mA/V, μ= 35, r_I=5.8kΩ

TUBE PROFILE: 6AQ8 (ECC85)

- Indirectly-heated Noval duo-triode
- Heater: 6.3V/400 mA
- V_{HKMAX}: $90V_{DC}$, P_{AMAX}: 2.5 W
- Max. anode voltage & current: 300V/15 mA
- TYPICAL OPERATION:
- V_A=250V, V_G=-2.3V, I_0=10 mA
- gm = 5.9 mA/V, μ= 57, r_I=9.7kΩ

Introduced in 1954, ECC85 was developed for front end stages of FM radio receivers. The two boxed anodes are divided by an electrostatic screen. UCC85 (26V/0.1A) and PCC85 (9.5V/0.3A) have very similar parameters. They were used in television tuners as RF amplifiers and oscillators.

ECC85 has a high μ, almost 60, and a high anode current and dissipation. In our designs it can be replaced with 6BQ7A if less gain is required.

6BQ7 is a low noise triode developed for cascode RF amplifiers. The 6BQ7A and its different heater voltage versions 4BQ7A and 5BQ7A have controlled heater warm-up characteristics and were used in TV sets with series heater strings. Notice a high maximum heater-cathode voltage of 200V, so when used in SRPP, cascode and other totem-pole circuits 6BQ7 heaters don't have to be at a raised DC level, which simplifies the power supply.

Although it would otherwise work without any modifications, in those cases ECC85 (due to its low V_{HKMAX} of only 90V) cannot be plugged in instead of 6BQ7 to get more gain.

The input stage in a SRPP configuration using 6BQ7A duo-triode. The upper triode is an active anode resistor for the lower one, whose value is $R_{A1}=r_{I2}+(\mu_2+1)R_2 = 5.8+(33+1)1.5 = 57k\Omega$. The estimated voltage amplification of the first stage is $A = -\mu R_{A1}/(r_I + R_{A1}) = -35*57/(6+57) = -0.9*35 = 31.6$ The measured value was A=30!

Voltage amplification of the 2nd stage is $A = -\mu R_L/(r_I + R_L) = -16*47/(3+47) = 15$, which agrees with the measured value. Should you not need that second amplification stage in your design, parallel the two 5687 triodes in the cathode follower for an even lower output impedance and doubled current capability.

A very mild global NFB is used (β=100k/100R= 0.001), from the secondary of the output transformer back to the split cathode resistor of the first stage. It does not affect the sonics of the amplifier as high levels of NFB do, but it improves the waveforms and lowers the output impedance, thus increasing the damping factor.

Graphical analysis using anode characteristics

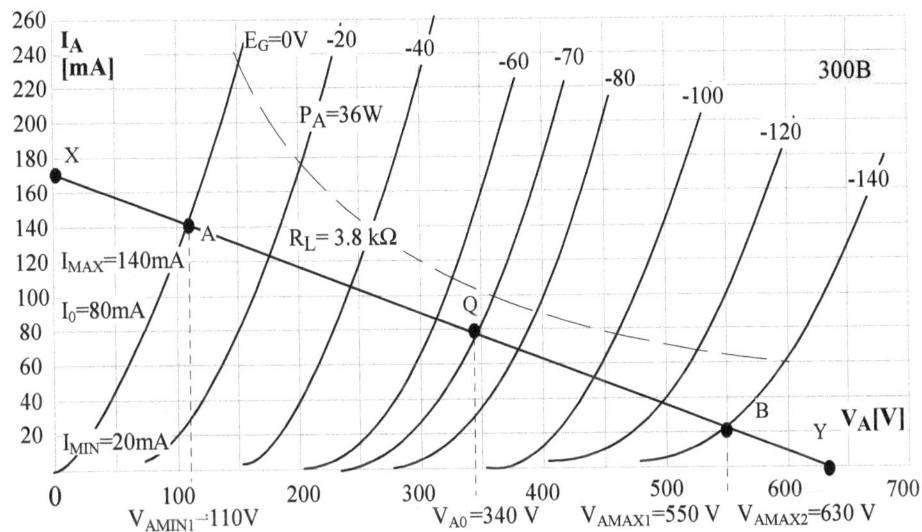

CLASS A_1 (A-Q-B):

$P_{IN} = 340V*0.08 = 27W$

$P_{OUT} = (\Delta V \Delta IA)/8$

$P_{OUT} = 440*0.12/8 = 6.6$ W

$\eta = P_{OUT}/P_{IN}*100[\%$

$\eta = 6.6/27 = 24.4$ %

CLASS A_2 (X-Q-Y):

$P_{IN} = 340V*0.08A = 27W$

$P_{OUT} = (\Delta V \Delta IA)/8$

$P_{OUT} = 630*0.17/8 = 13.4$ W

$\eta = P_{OUT}/P_{IN}*100$ [%]

$\eta = 13.4/27 = 49.6$ %

Power supply

The high voltage power supply uses a voltage doubler, providing $+V_A$ (anode voltage) of 390 V_{DC}. Why voltage doubling? Because the number of secondary turns (and layers) of the power transformer is halved, ideal if you think your window area will be too small for a standard winding, as ours was (we wind our own transformers). The NTC resistor limits the inrush current upon switch-on. R_1 and R_2 discharge the capacitors once the amp is turned off.

Each channel has its own filtering circuit. This improves channel separation and results in better imaging, so the sonic benefits of such investment (an additional choke and a few capacitors) are immense.

Notice a relatively low voltage of 390V! However, since there is no voltage drop on the cathode resistor, this whole voltage will appear between anode and cathode. 70V would be lost with cathode bias, so to get the same 390V on the anodes the power supply would need to produce 460V had cathode bias been used.

The regulation of a voltage doubler is not as good as that of a ordinary full-wave rectifiers, meaning there would be a larger voltage sag under heavy load, but this applies to Class AB amplifiers only, not Class A where the power draw is constant. World-class designers such as Tim de Paravicini and many others use voltage doublers, and if it's good for them it will be good for you and me too.

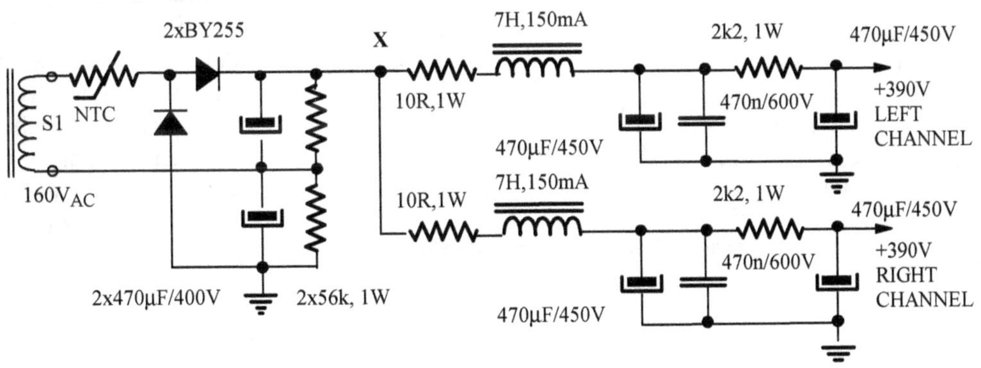

The biasing circuit illustrates another very useful application of the voltage doubler. Instead of grounding the anode of diode D4 and the filtering cap as in the top application, we can ground the middle point where caps are joined and therefore get two symmetrical (+/-) voltages instead of one doubled voltage!

The -200V will provide negative bias for the cathode follower driving 300B grids, while +200V will be its anode voltage.

The circuits powered up by secondaries S3, S4 and S5 are identical in topology. CT full wave rectification and filtering, providing 12.6V_{DC} heater supply for duo triodes in the preamp and driver stages and two separate 5V_{DC} supplies for 300B heaters. DB3-004 are duo-diodes.

To get the exact heating voltages, the series resistors R26, R27 and R28 need to be adjusted in situ once the DC voltages are measured under load, with all tubes plugged in! This assumes that the voltages will be slightly higher than required, so any excess voltage is dropped on these power resistors. In our case they are 1R5 - 2R7, 5W.

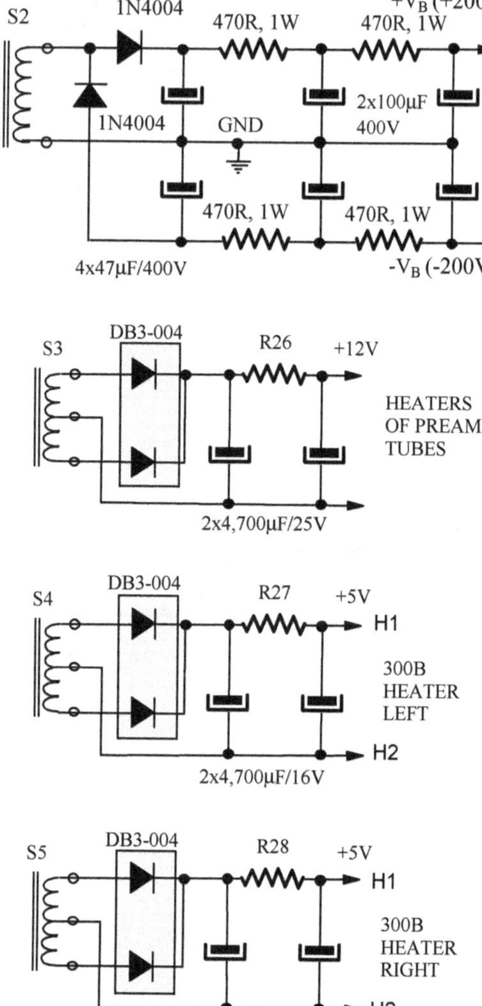

Comparison with other commercial 300B amplifiers

In this side-by-side comparison are four single-ended 300B amplifiers tested at 10 kHz (waveforms below), three from the Stereophile magazine reviews and one of our designs.

The top two have quite similar square-wave responses, Allnic with a higher overshoot, Cayin with a smoother but also slower response. USA-made Sophia Electric's has the narrowest bandwidth of all four, so even at a relatively low frequency of 10kHz the square wave response has degenerated into a wobbly mess.

Now, we are not talking about sonic merits of these amplifiers, only about one aspect of their "fidelity".

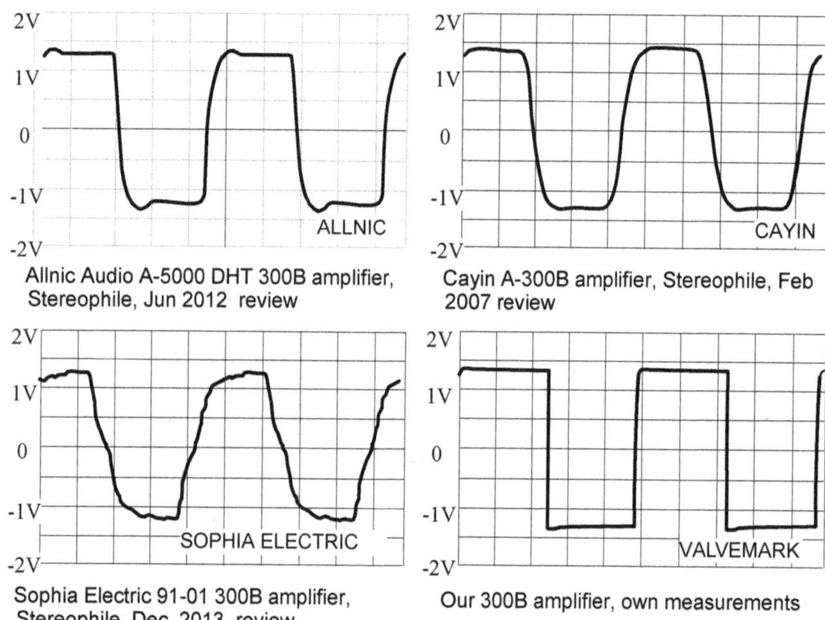

Allnic Audio A-5000 DHT 300B amplifier, Stereophile, Jun 2012 review

Cayin A-300B amplifier, Stereophile, Feb 2007 review

Sophia Electric 91-01 300B amplifier, Stereophile, Dec. 2013 review

Our 300B amplifier, own measurements

It is possible that the worst measuring amplifier (in this case Sophia Electric) in some aspects could sound better than the one with best test results (in this case our own design).

I still remember one of our later projects, a SET amplifier using RS1003 power tubes (featured on p262). As soon as the prototype was finished I had a quick listen and loved the sound. It had presence, it conveyed emotion, it was dynamic, punchy, and above all, "natural", in a sense that it sounded like performers were right in front of me in a live concert.

ABOVE: 10 kHz square wave reproduction by four 300B SET amplifiers. Reprinted with permission from Stereophile (www.stereophile.com)

When I returned it to the workshop I got a rude awakening - it was the worst measuring amplifier I had ever seen. The square wave was warped, the frequency range was limited, the distortion was high. After some fine tuning it measured very well indeed, and sounded great, but its sound was not as magical or palpable as it was the first time, in its rough state.

KEY FEATURES:

- Capable of low-distortion Class A_2 operation
- Cathode follower driver stage with fixed bias
- Only two coupling capacitors in the signal path

MEASURED RESULTS:

- BW: 5Hz - 57 kHz (-3dB, 1W into 8Ω)
- BW: 10Hz - 49 kHz (-3dB, 10W into 8Ω)
- Class A_1: V_{OUTMAX}: $11V_{RMS}$, $P_{MAX} = 15W$
- Class A_2: $V_{OUTMAX}=14V_{RMS}$, $P_{MAX} = 24W$
- $Z_{OUT}= 2.1\Omega$ (DF=3.8)

SINGLE DRIVER-STAGE 300B SE DESIGN WITH ANODE CHOKE

The fixed bias design discussed so far gives you a higher output power, but the price to pay is increased complexity. Also, fixed bias amplifiers sound different from self- or cathode biased designs. Some audiophiles feel that the midrange magic (read "pleasant-sounding distortion") is lost in fixed-biased stages.

On the positive side, this simple design is much easier to build than the previous one. It uses LC-fed driver stage and cathode- biased output stage. There is only one driver stage and the +/- 200V biasing circuit is gone. However, instead of more than 20 Watts in Class A_2 you will only get 7-8 Watts in Class A_1.

Two interesting design features can be found here. The first is paralleled triodes in the preamp section. ECC40 is a duo-triode, just like 12AU7, so one would be enough for both channels. Here we paralleled the two sections for halved internal resistance, doubled transconductance (results in a faster and more dynamic sound!) and a slightly higher amplification of the stage.

The other departure from the ordinary designs is the use of anode or plate chokes. A resistor has the same DC and AC impedance, equal to its resistance. So, a 100kohm resistor has a 100k DC resistance and a 100k AC resistance.

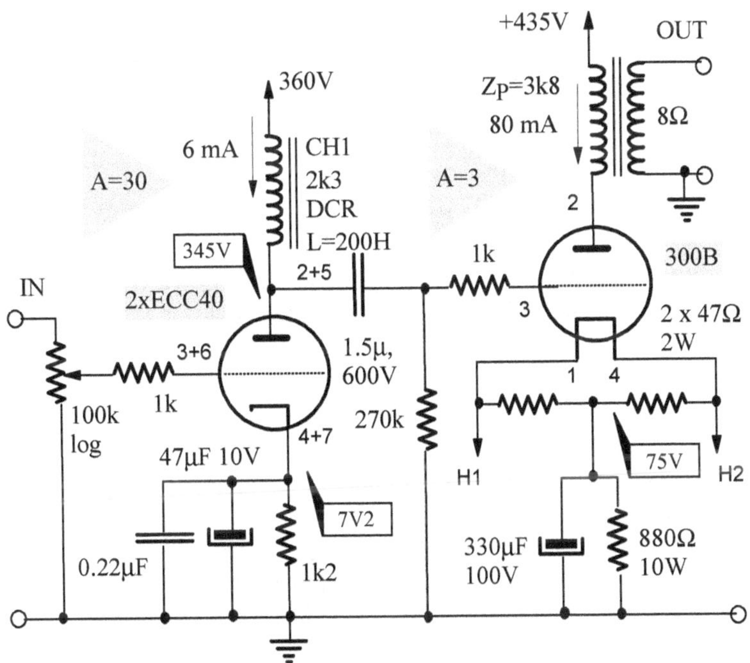

A choke has a low DC resistance (in this case 2.3 kΩ), but a very high AC impedance. With 200H inductance at 1kHz, its impedance at that frequency is $Z=2\omega L = 2\pi fL = 2*3.14*1{,}000*200 = 1{,}257$ kΩ!

This brings us two benefits. Firstly, low DC resistance means low DC voltage drop on the anode choke, so if V_B is 360 Volts, the anode current is 6 mA, the voltage lost on the plate resistor is only $V = IR = 0.006*2{,}300 = 13.8$ Volts That means the anodes of ECC40 work on 360-13.8 = 346.2 V (345V measured).

With a typical 100 kohm plate resistor the voltage drop would be 600V, meaning we would need to raise V_B to at least 800 Volts, so the anodes can be at 200V, which is about minimum for this design. Preamp and driver stages sound much better if tubes used draw high current, as in this case, rather than stages with 12AX7 that draw 0.5-1 mA only.

The second benefit is even more important. The higher the plate impedance, the closer the amplification factor of the stage approaches μ (amplification factor) of the tube used, in this case around 32. With a 100kΩ plate resistor the first stage would only give us the amplification of around 22, which would not drive the 300B output stage.

The bias is at -75V, so the maximum peak grid signal is 75V or 53V_{RMS}. This way we get the amplification factor of around 32, which means we need 53V/32 = 1.66 V_{RMS} at the input for the full output, so a typical CD player will give us enough drive. With plate resistors used instead of chokes, we would either need two amplification stages or retain the anode choke and change the preamp tube to a tube with a higher μ.

Power supply

Indirectly-heated vacuum rectifier 5V4G is used for its low voltage drop. With directly heated rectifiers such as 5U4 instead of 450V on the first filtering elco, you'd get only 415V!

Also, indirectly-heated rectifiers provide a soft start protection, 300B triodes will heat up before the high voltage appears on their anode. Finally, I believe indirectly-heated rectifiers sound better and cleaner.

This first capacitor should not be larger than 47 μF, otherwise rectifier tube could get damaged due to a high inrush current upon power-up.

Preamp tubes are heated from the same secondary windings of the power transformer, using AC, so twisted wires and careful layout is needed.

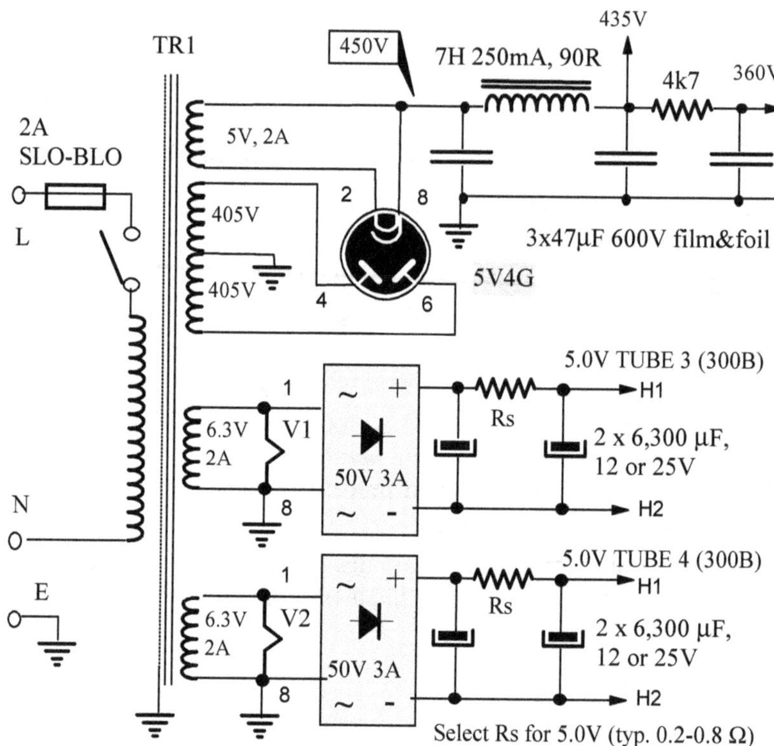

Only one filtering choke is used, supplying both channels, but if you prefer, you can use two chokes, one for each channel, or you can replace the series resistor in the second filter (1 kΩ) with a second choke. That will improve filtration of the AC ripple on the high voltage line even further.

You may have noticed that the amplifier in the photo has two 0-100mA DC analog meters not shown on the circuit diagram. They are connected in series with the cathode resistor of 300B tubes, so their ohmic resistance needs to be subtracted from the total needed cathode resistance.

VALAB ANODE CHOKE - MEASURED RESULTS:

- L_P (120Hz): 220H, L_P (1kHz): 558H
- R_{DC}: 2,302 Ω
- Z(120Hz): 3.98MΩ, Z(1kHz): 9.36MΩ

KEY FEATURES:

- Single driver stage with choke-loaded parallel triodes
- Only one coupling capacitor in the signal path
- Cathode bias of the output stage
- No elcos in high voltage power supply
- No negative feedback of any kind

CASE STUDY: AUDIO NOTE "CONQUEST"

Going through my tube amp "scrapbooks", I found two interesting debates, both from the UK magazine "Hi-fi News & Record Review". In the first one, dated Nov. 1996, the late Riccardo Kron, then with "Vaic Valve Productions" in Italy, in his letter to the Editor, claimed that it was impossible for Audio Note's Conquest amplifier to produce 18 Watts from two 300B tubes in parallel SE circuit. Peter Qvortrup of Audio Note in UK replied with a calculation. We will analyze the circuit of this amplifier, which is in public domain, and see which of the two was right.

The other article, which was probably a paid advertisement, is by David Chessel from Alema UK, better known for their "Audion" brand of valve amps. The article was written in a pseudo-discussion format, referring to an earlier article by Tim de Paravicini, the UK designer of E.A.R. fame. The author uses about half-a-dozen statements by Mr. de Paravicini to agree (the title of the article is "We agree with Tim") with those statements initially, but to then try to prove the opposite.

Mr. Paravicini said about 300B amplifiers "Driven hard in single ended mode these tubes give 10 watts but with poor reliability.

At the same time he was espousing the strengths of EL509 power tubes which he was using in E.A.R. 859 amplifier and other designs, namely their low price, high reliability and long life.

Mr. Chessel claimed in this ad: "A new SINGLE TUBE 300B design we worked on recently, operating within long life conditions measured fully, returned the following results without feedback into 8 ohms 20hz to 20khz (sic). 7 Watts < 1%, 13 Watts < 5%, 21 Watts <15% maximum output."

He didn't specify if those figures were in class A_1 or A_2, but I can bet that those 21 watts weren't in class A_1! I wonder what Mr. Kron would have said about that when he doubted 18W from two tubes in Audio Note monoblocks?

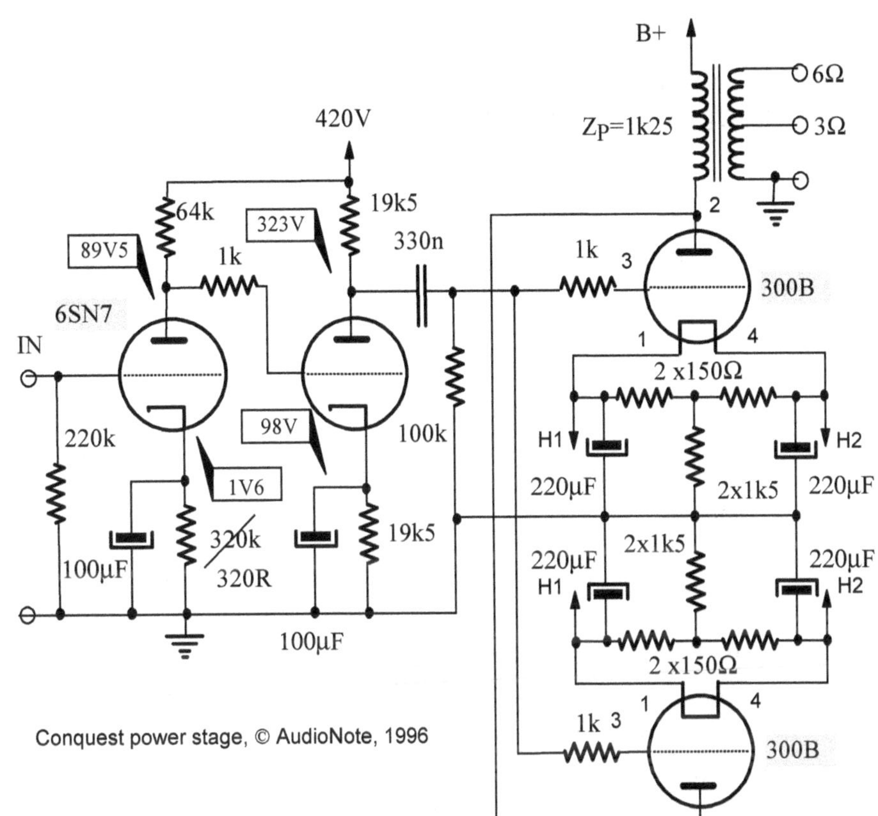

Conquest power stage, © AudioNote, 1996

How to figure out DC voltages and operating points

The design is plain vanilla. Two common cathode amplification stages, each using one half of the 6SN7 duo-triode. They are directly-coupled, so there is only one capacitor in the signal path, coupling the 300B grids to the second stage. Therefore, the amp can only operate in class A1.

The output tubes use self- or automatic cathode bias. Each 300B has its own cathode resistor, 750Ω (two 1k5 resistors in parallel), bypassed by two 220µF elcos. The same heater power supply is used for both tubes.

Transformers of at least 3k5 primary impedance are used for 300B tubes. The output transformer's primary impedance in this design seems low, 1k25 for two tubes (equivalent to 2k5 for one tube). This would maximize the output power but also increase distortion!

The original power supply diagram did not specify DC voltages, so we have to do a bit of detective work. We know that the second stage anode current is I_{A2}=5mA and the anode-to-cathode voltage is 225V (disclosed in the article), so from the anode curves we get the bias of -8.5V. Thus, if the cathode is at +98V, the grid must be at 98-8.5 = 89.5V.

The voltage drop on the anode resistor is equal to the drop on the cathode resistor (98V), so the $+V_{BB}$ is around 420V. The first stage also uses the same $+V_{BB}$, so the voltage drop on its 64k anode resistor is 420-89.5V = 330.5V, which gives us the first stage anode current as I_{A1} = 330.5/64,000 = 5.2 mA, almost identical to I_{A2}.

R_{K1} is specified as 320kΩ, but that is impossible, 5mA current flowing through it would create a voltage drop of 1,600V! The anode is only at 89.5V, so to get 5mA anode current, the graphs say that the voltage on the cathode should be slightly less than 2V. Therefore R_{K1} should be 1.8/0.0052 = 346Ω So, perhaps its value was supposed to be 320R, not 320k! With 320R we get V_K=320*0.0052 = 1.67V which agrees with the graph.

Let's see what the curves tell us about two parallel 300B tubes. We don't have the DC voltages or currents on the Conquest circuit diagram, but in his rebuttal of Kron's accusations, Peter Qvortrup gave us all we need: grids at -65V, anodes at 350V, 85mA anode current per tube, 30% efficiency, anode voltage swing between 550V maximum and 125V minimum, current swing between 170 mA and zero, so let's put these figures onto the graph (next page).

Assuming the AB operating line, $\Delta V = 550\text{-}125 = 425V$, $\Delta I = 340$ mA, $R_L = 425/0.34 = 1k25$

$P_{IN} = 340*0.17 = 58W$, $P_{OUT} = \Delta V \Delta I/8 = 18W$, $\eta = P_{IN}/P_{OUT} = 18/58 = 31\%$

It seems possible to get the output power of 18 Watts under the specified assumptions. However, notice that grid signal is not symmetrical. The AQ swing is 65V, while QB is anywhere from 85 to 95V (hard to tell from the graphs), making the anode current dip all the way down to zero.

This is always a bad idea, since that low anode current region increases distortion immensely and should be avoided in all designs, let alone those without any negative feedback!

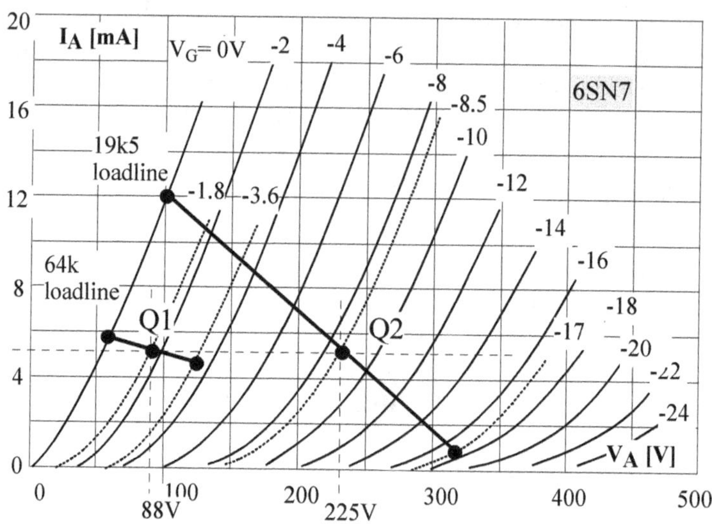

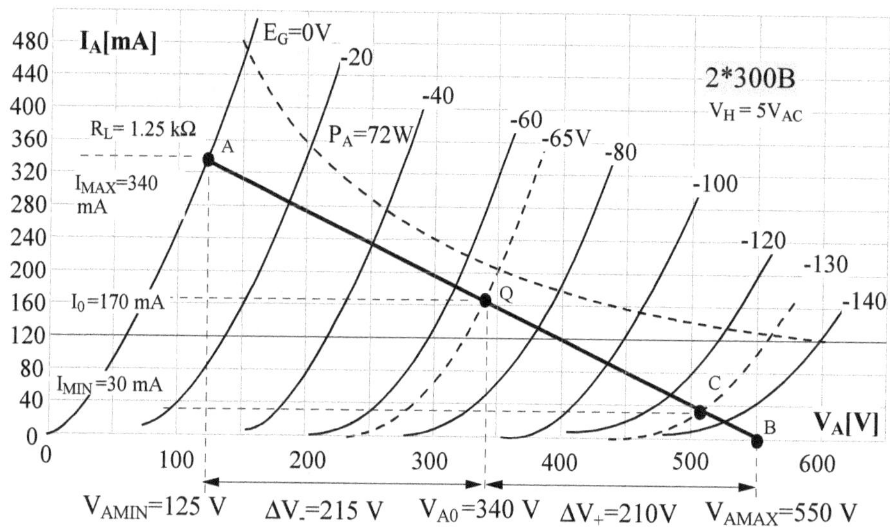

For a more realistic symmetrical grid drive, assuming the AC operating line (V_G=0-65V-130V): $\Delta V = 510\text{-}125 = 385V$, $\Delta I = 340\text{-}30 = 310mA$, $R_L = 425/0.34 = 1k25$

$P_{IN} = 340*0.17 = 58W$, $P_{OUT} = \Delta V \Delta I/8 = 385*0.31/8 = 15W$, $\eta = P_{IN}/P_{OUT} = 15/58 = 25.9\%$

So, while Mr. Qvortrup figures stack up, based on the curves alone, without actually testing this amp on a test bench, it seems that a more realistic power rating for the Conquest would be 15 Watts, rather than 18Watts.

CASE STUDY: AUDION "GOLDEN NIGHT"

This amplifier is interesting from an educational viewpoint for three reasons: the marketing claims, the attempt to hide and hype the preamp tubes used and the interesting output transformer. We did not have this amplifier on our test bench and can only base our superficial assessments on the information and diagrams available on the web.

The Audion website claims "This triode based monoblock amp uses no (zero) feedback in the design." This is not true. Without NFB it would be impossible to achieve the claimed levels of distortion: "Distortion @ 1Watt: <0.1% No Feedback". The primary windings of the output transformer reveal the truth. The amp uses two kinds of negative feedback. The first is partial cathode loading of the output stage. The second is voltage feedback from another dedicated winding back to the anode circuit of the driver stage, applied in series.

One weakness of this approach is that the input stage is not included in the feedback loop, and neither is the output transformer, which is usually the most distorting component in any power amp. Only the driver and the anode circuit of the output stages are affected by the feedback. A specially designed and wound output transformer is needed, standard commercial designs cannot be used, so such a design is out of bounds for a typical DIY amateur.

As for the preamp tubes, the manufacturer claims "They use JJ 300B and Audion's own CVX100 & CVX120 as driver and input tube." Does that mean Audion is actually manufacturing tubes? No way!

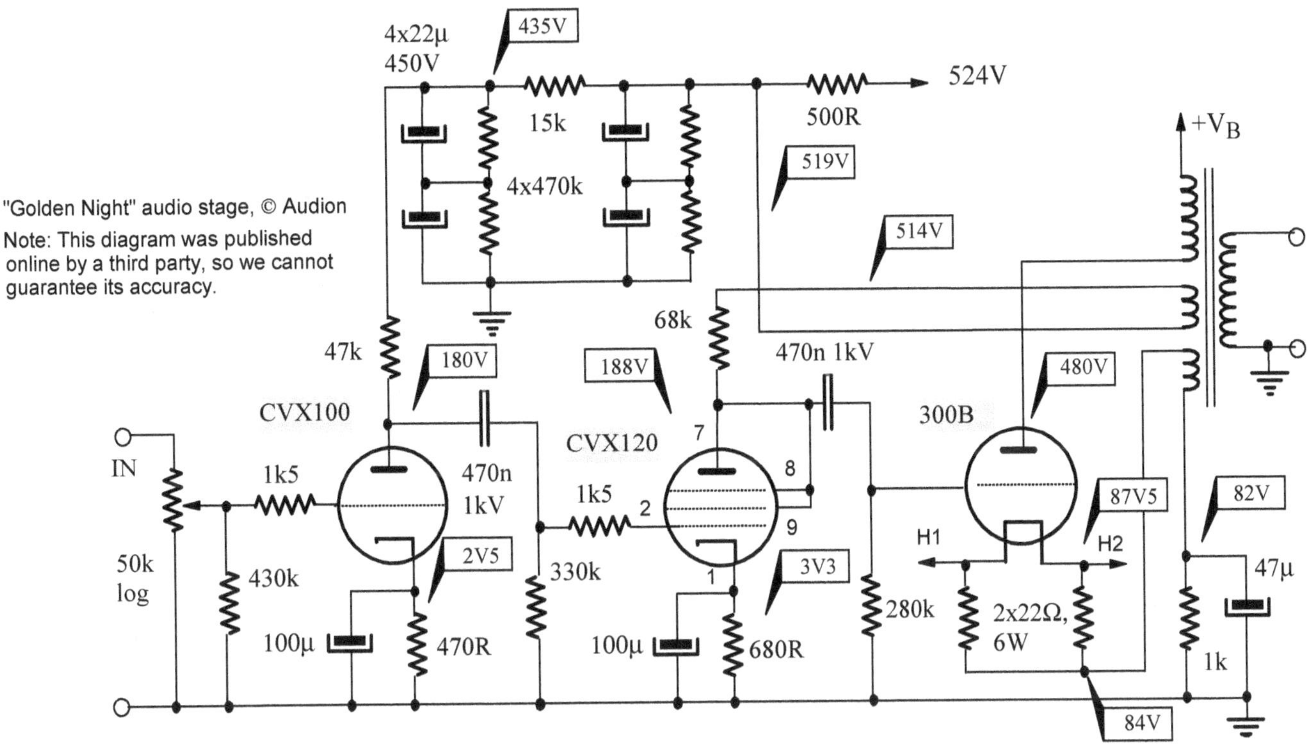

"Golden Night" audio stage, © Audion
Note: This diagram was published online by a third party, so we cannot guarantee its accuracy.

Finding the triode used in the first stage

As an educational exercise, let's analyze the circuit and try to identify which of the commercial tubes could hide behind the numbers. CVX120 should be easier to find since the person who published this circuit diagram on the web identified the pin numbers. We have no pin numbers for CVX100, but it is a single triode with a Noval (9-pin) socket, and there aren't too many of those!

The DC cathode current of the first stage is I_{K1}=2.5V/470Ω = 5.3 mA The anode-cathode voltage is 180-2.5 = 177.5V Our task now is to find a single Noval triode that when biased at -2.5V will pass around 5mA of current with anode voltage of around 180V. The loadline is for 47k||330k = 41k.

6C4 is half of 12AU7, but that tube use a miniature 7-pin socket. There aren't many single triodes using a Noval socket: 12B4 is a power triode and 5842 and 6S4/A do not match the V-I data.

How about our esteemed friend, E86C? The intersection of the vertical line through V_A=180V and the horizontal line through I_A=5.3mA is at approx. -1.9V grid bias, which is reasonably close to the 2.5V assumed.

The person measuring the voltages in this amp could have misread the value of the cathode voltage or resistor, or perhaps the actual E86C tube in his amplifier was not that close to the graph. Whatever the case, CVX100 could be E86C.

The maximum peak-to-peak grid signal is ΔV_{G1}= 2*1.9V = 3.8V, while the anode voltage swing is ΔV_{A1}= 280-54 = 226V, so the amplification factor of the first stage is A_1= 226/3.8 = 59

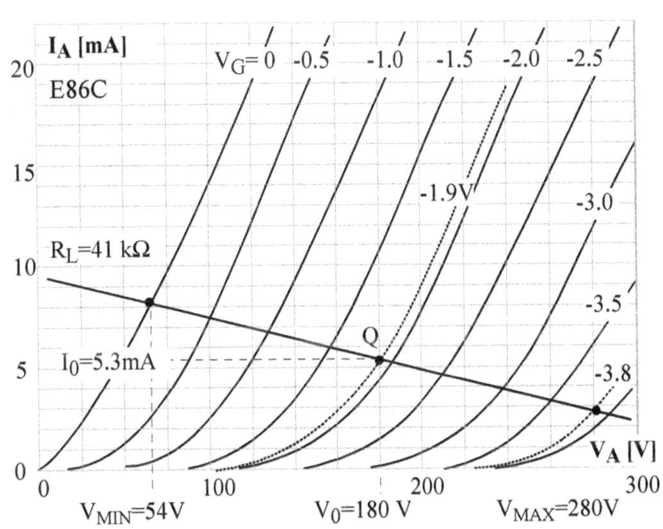

Finding the triode used in the second stage

2nd stage: If the cathode voltage is 3.3V, the anode current is I_{A2}=3.3/680 = 4.85mA The anode- cathode voltage is 188-3.3 = 185V. After other candidates such as 6JC6 and EF86 (different pinout) are eliminated, E280F, E810F (7788) and D3a are the most likely candidates. If only we could get hold of the actual tubes used, even just a visual examination would be enough to identify them, without a need for testing and curve tracing.

The maximum peak grid signal swing is ΔV_{G2}= 3.3V, while the anode voltage swing is ΔV_A = 185-18 = 167V, so the amplification factor of the second stage is A_2= 167/3.3 = 50

If our estimates are correct, the combined amplification of the two stages is 60*50 = 3,000! Since the cathode of 300B is at approx. 88V, an amplification factor of around 80-90 is needed for 1V input sensitivity.

However, negative feedback from the output transformer to the plate of the second stage reduces the gain of that stage, but we have no data on that winding so cannot calculate the strength of the NFB used. All we know is that the DC voltage drop on the NFB winding is 519-514V, or 5V. Since the DC current through that winding is 4.85mA, its resistance is R_{NFB} = 5/0.005 = 1kΩ. That is a high resistance, meaning that either this winding was wound using a small diameter wire, or that the number of turns is high (or both)!

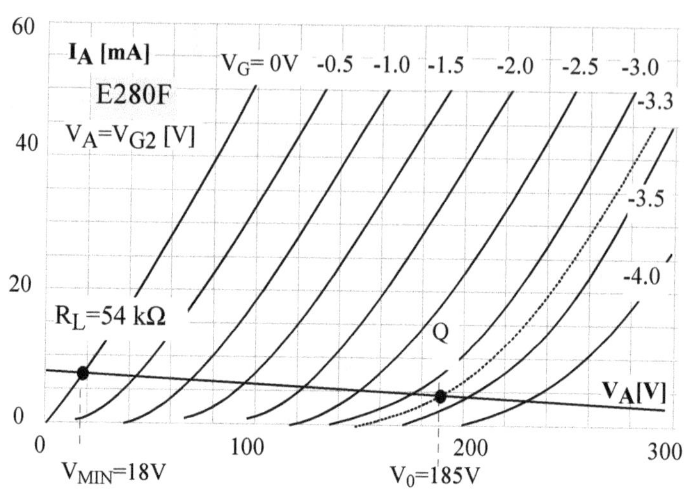

The higher the number of turns in that winding, the stronger the negative feedback would be. Since the claimed distortion of this amplifier is below 0.1% (although it is not specified at which output level, rendering it meaningless), this feedback seems to be *very* strong.

Since the DC voltage drop on 300B's cathode resistor of 1kΩ is 82V, the quiescent current of the output stage is I_0=83mA, which is on a high side and could shorten the life of expensive 300B tubes. This current creates a DC voltage drop of 1.6V on the cathode feedback winding, so the resistance of this winding is 1.6/0.082 = 19.5Ω. Compared to 1kΩ of the other feedback winding, this seems a milder NFB, although it had to be wound with a much thicker wire, since the full 300B's anode current passes through it.

TWO STAGE 300B SE DESIGN WITH A GRID CHOKE

A similar approach to the single ECC40 driver stage design, but his time the input stage is a quazi-SRPP amplifier with ECC85 (6BQ8) duo-triode, amplifying about 55 times, capacitively coupled to a self-biased 300B stage with a grid choke, with a measured amplification factor of around 2.7. This is *truly* a feedback-free design.

Since output tubes are AC-heated, the use of hum-minimizing potentiometer across 300B's heater terminals is mandatory. This is usually done by ear, but can be also done by connecting a multimeter (set on Volts AC) across speaker terminals.

The pot is adjusted until AC ripple (hum) is at its minimum value.

The output transformer's primary impedance is a lowish 2.7kΩ, anything up to 5kΩ can be used. Higher load impedance means lower distortion but also slightly lower maximum obtainable power.

If you don't have an exact value and/or power rating of the cathode resistor, feel free to connect two or more higher resistance/lower power rating resistors in parallel. They don't even have to be of the same value. Use the third or the fourth one to precisely "trim" the value down to the require value. Here we use two 2k7 and one 3k3 5W resistors.

Notice that the cathode of the upper ECC85 triode is at DC potential of 140V, which exceeds its heater-to-cathode insulation rating.

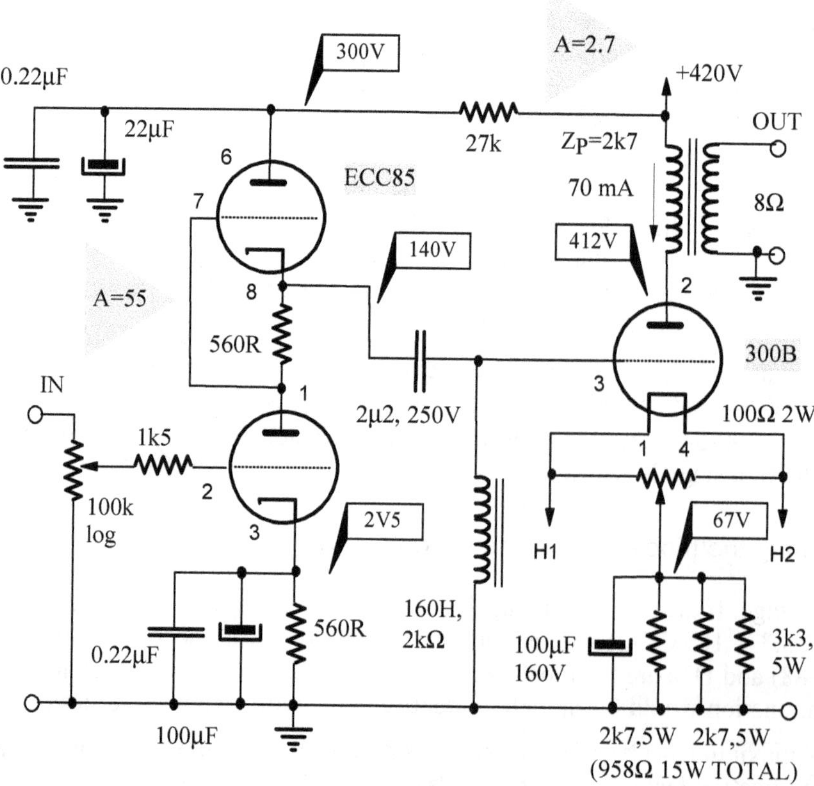

These two input tubes are DC-heated, but such voltage is not referenced to ground, but to a higher DC voltage of +78V, roughly half of the upper cathode's DC voltage (140V). This is done by a voltage divider (220k and 51k) since 51/(51+220)=0.19 and 410V*0.19 = 78V!

With reference to the heater, the upper cathodes of ECC85 tubes are now at 140-78= +62V potential, while the bottom triodes' cathodes are at 2.5-78= -75.5V, both within the specified V_{HKMAX}=100V!

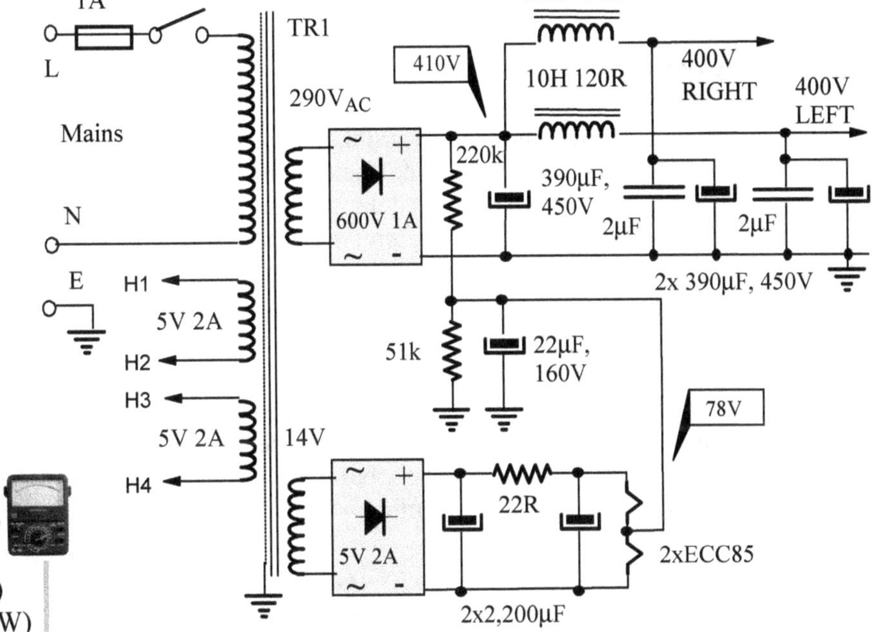

MEASURED RESULTS:

- BW: 7Hz - 35 kHz (-3dB @ 1W)
- BW: 15Hz - 27 kHz (-3dB @ 10W)
- V_{OUTMAX}: 12V_{RMS}, P_{MAX} = 18W

SE AMPLIFIERS WITH LOW IMPEDANCE TRIODES

The 6080 "family" of triodes was designed for use as series-pass tubes in voltage regulators. They can pass high currents and have a very low internal impedance (below 300Ω). However, their amplification factors are low, typically around 2-3, so high grid voltage signals are needed. The notable exception is 6528, with µ of 9, which makes it an easier tube to drive.

Another headache these triodes cause to amplifier builders is a high variation in their parameters, making them notoriously difficult to match. So, to end up with a matched quad, you will need to buy 10-20 tubes and try matching them yourself.

Notice very high parasitic capacitances of 6528 and 6336 triodes, 17-24pF. These will cause HF attenuation and reduction in bandwidth!

SIDE-BY-SIDE	6080	6528	6336
Heater voltage	6.3V	6.3V	6.3V
Heater current	2.5A	5A	5A
P_A max. [W]	13	30	30
V_A max. [V]	250	400	400
I_K max. [mA]	125	300	400
µ	2	9	2.7
r_I [Ω]	280	245	200
gm [mA/V]	7	37	13.5
C_{AG} [pF]	8.6	23.8	21.8
C_{GK} [pF]	5.5	17.8	16.7

6528 CLASS A_1 FROM THE GRAPH:

V_{MIN}=65V, V_{MAX}=440 V, ΔV = 375V

ΔI = 185 - 45 = 140 mA, R_L=2k6

P_{IN} = $I_A V_0$ = 0.1*280 = 28 W

P_{OUT} = ΔVΔI/8 = 375*0.14/8 = 6.5W

DISTORTION (2nd harmonic):

ΔV_+=215V, ΔV_-=160V, D_2 =(ΔV_+ - ΔV_-)/2ΔV *100 [%] = (215-160)/2*375*100 = 7.3 %

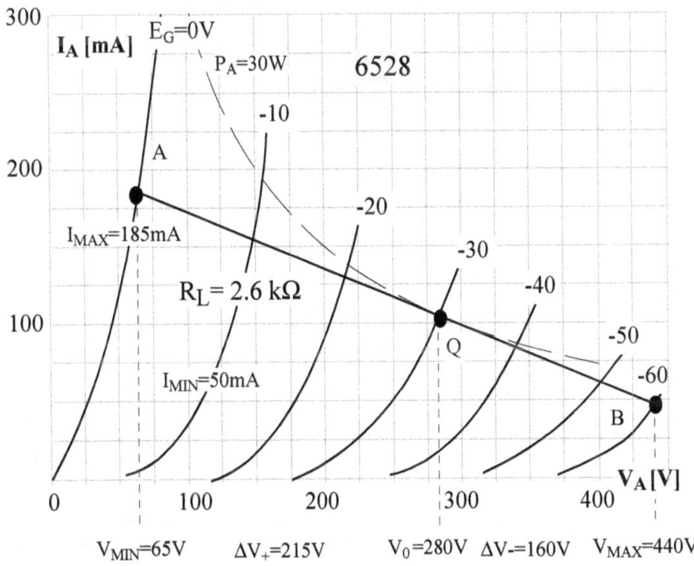

Under almost identical conditions, the performance of 6528 and 6336 is similar. However, 6336 require 160V peak-to-peak or 57 V_{RMS} grid drive, while 6528 needs only 60V_{PP} or 21V_{RMS}. In other words, 6528 is a higher-merit tube. Assuming 1V input sensitivity, it can be driven by a simple single-stage driver of around A=20!

However, less than 6 Watts of audio for a heater power of 32 Watts is depressing, with triode-connected gems such as E130L, EL12 and EL153 around, why would you choose one of these?

6C33C-B SET MONOBLOCKS

6C33C-B power triode

6C33C-B is the original tube number in Cyrillic, which would be 6S33S-V in Latin alphabet. Only one factory in Soviet Union (in Ulyanov, today's Russia) used to produce these power tubes for the Soviet military, before the production stopped in 1980.

Due to their extremely low internal resistance and high current capability, the tube was used as series pass tube in voltage regulators. There are actually two identical triodes internally connected, and there is a version with only one system, called 6C41C. Normally it would be silly to even discuss a tube whose production stopped more than 35 years ago, but there were so many produced that NOS stocks are plentiful.

TUBE PROFILE: 6C33C -B

- Indirectly-heated duo-triode
- B7A socket
- Heater: 6.3 or 12.6, 6.6/3.3A
- Maximum anode voltage 250 volts
- Maximum V_{HK}: +/-300V_{DC}
- P_{AMAX}: 60 W (both triodes together)
- gm: 40 nom. (28-50) mA/V
- μ=2.7 nom. (2.5-4), r_I =80-120Ω

One brief look at the anode curves and we realize we are dealing with a low plate voltage high current beast here, so the first limitation will be the output transformer, more precisely the maximum DC current it can take. The higher the primary DC current, the lower the inductance we can get out of the output transformer, which is bad news. However, since this triode is of extremely low internal impedance, a very low primary inductance L_P is required for good bass reproduction, so luckily, the two requirements compensate.

Our output transformer could easily take 200 mA in the idle mode, so we chose this operating point. At 180V the tube can take up to 350 mA, so we are way below the maximum dissipation curve, meaning the tube would not be stressed at all and would have a long and uneventful life in this amplifier. The transformer's turns and voltage ratio was TR=1,000 : 80 = 12.5, meaning the impedance ratio was $IR=TR^2=12.5^2$ = 156.25, so that an 8Ω speaker would reflect onto the primary as 1,250Ω. Now we can draw the loadline.

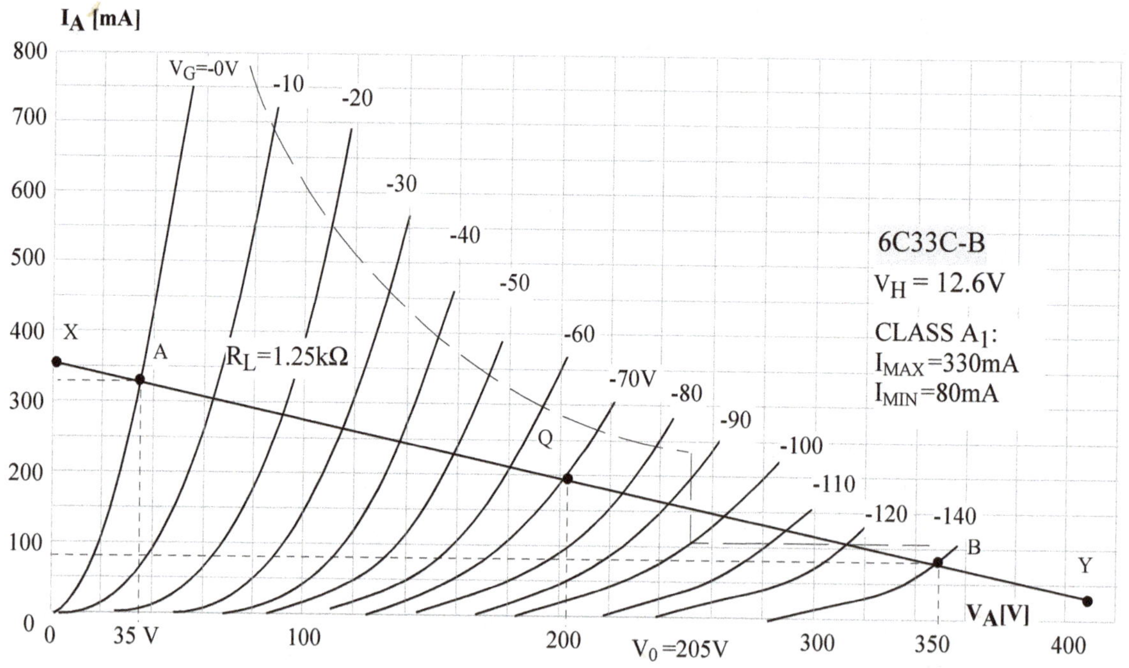

Anode curves for 6C33C-B heated with 12.6V. Beware, the curves for 6.3V heating show a much reduced power rating, due to the fact that at 6.3V one of the two 6.3V heaters is not operational, so if you decide to use 6.3V heating you will not utilize this tube fully!

<u>1k25 LOAD, CLASS A1 (AB):</u>

V_{MIN}=35 V, V_{MAX}=350 V, ΔV = 315V   ΔI = 330 - 80 = 250 mA R_L=ΔV/ΔI = 315/0.25 = 1,260Ω

P_{IN}= I_0V_0 = 0.2*205 = 41 W P_{OUT} = ΔVΔI/8 = 315*0.25/8 = 9.8 W

D_2= [(I_{MAX} +I_{MIN}) - 2I_0]/2(I_{MAX} -I_{MIN})*100% = [(330+80) - 400]/2(330-80)*100% = 2.0%

CLASS A_2 (XY):

$\Delta V=410V$

$\Delta I=35\text{-}30=325mA$

$P_{IN}=I_AV_0 = 41$ W

$P_{OUT}= \Delta V\Delta I/8 =$

$410*0.325/8 = 16.6W$

$\eta=P_{OUT}/P_{IN}=16.6/41=$ 40.6%

A pretty good result, almost 10 Watts in Class A_1 with only 2% distortion. In Class A_2, a whopping 70% increase in output power. Let's lower the load impedance to 1kΩ and see what happens.

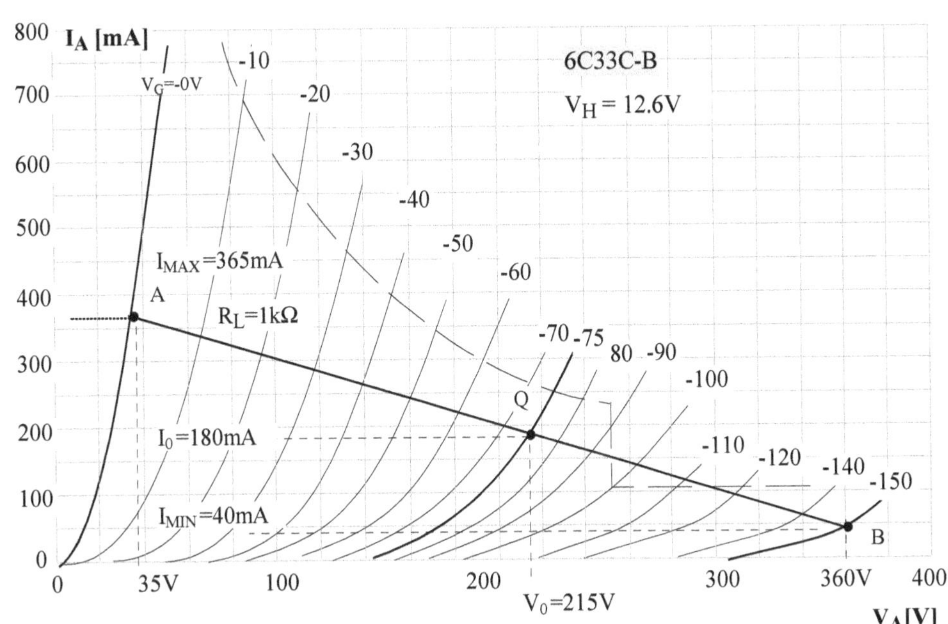

V_{MIN}=35V, V_{MAX}=360V, ΔV = 325V, ΔI =365-40=325 mA

$R_L=\Delta V/\Delta I$ = 1,000Ω$P_{IN} = I_AV_0$ =0.18*215=38.7 W, $P_{OUT}= \Delta V\Delta I/8$ = 325*0.325/8 = 13.2 W

D_2= $[(I_{MAX} +I_{MIN}) - 2I_0]/2(I_{MAX} -I_{MIN})$*100% = [(365+40) - 360]/2(365-40)*100% = 6.9%

35% more power but much higher distortion, so let's stick to 1k25 load!

Our favorite topology, quazi-SRPP input stage, driving a common cathode second stage, with a cathode follower driver stage DC coupled to the grid of the output triode. The fixed bias of the output stage is provided by the cathode follower's symmetrical power supply (+/-180V_{DC}).

Test point TP1 for measuring the cathode current of the output tubes. A very mild global negative feedback from the 8Ω output to the un-bypassed lower cathode resistor of the first stage.

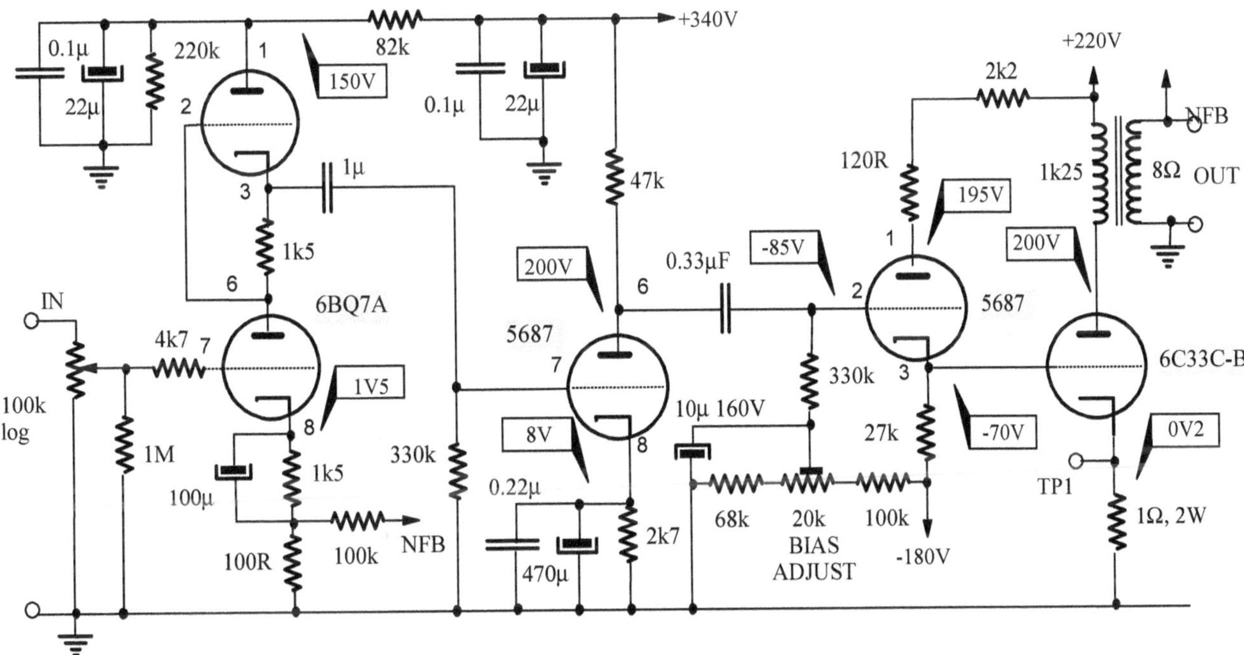

The measured results exceeded all expectations, a very wide frequency bandwidth, both at 1W and at 10 Watts, 18 Watts in class A_1 and a whopping 32 Watts in class A_2!

Due to the output tube's low internal impedance, the output transformer didn't have many primary turns, but the winding job wasn't easy due to a large diameter primary wire, to allow for the primary current of up to 400mA!

MEASURED RESULTS:

- BW: 5Hz - 57 kHz (-3dB, 1W into 8W)
- BW: 10Hz - 49 kHz (-3dB, 10W into 8W)
- Class A_1: $V_{OUTMAX}=12V_{RMS}$, P_{MAX} = 18W
- Class A_2: $V_{OUTMAX}=16V_{RMS}$, P_{MAX} = 32W

SV572-10 SINGLE-ENDED AMPLIFIER

Svetlana SV572-10 is a modern variant of the American 572 tube. The Russians removed the annoying top cap and improved the tube overall, hoping that it will challenge the supremacy of 300B triode. It didn't.

SV572-10 can produce up to 50 Watts in a single-ended mode, compared to 8-12 Watts for 300B.

Released in 1996, SV572-10 was produced for only a few years, but stocks are still available.

The filament is made of thoriated tungsten. If you study the tube carefully, there are no dark or silver gettering areas on the glass. How is that possible? The getter is titanium, which is bonded with the anode's material (graphite). Titanium is far superior to the commonly used barium, because its coefficient of gas absorption is ten times greater.

TUBE PROFILE: SV572-10

- Directly-heated power triode
- Socket: 4-pin, heater: 6.3V, 4 A
- Anode dissipation: 125W
- Maximum anode/grid current: 250/50 mA
- gm=4.5 mA/V, μ=9.5, r_I = 2,100 Ω
- Typical operation:
- V_A=750V, V_G=-40V, I_0=60 mA
- I_{MAX}=150 mA
- P_{OUT}= 25 W, Class A_2 @ 5% THD

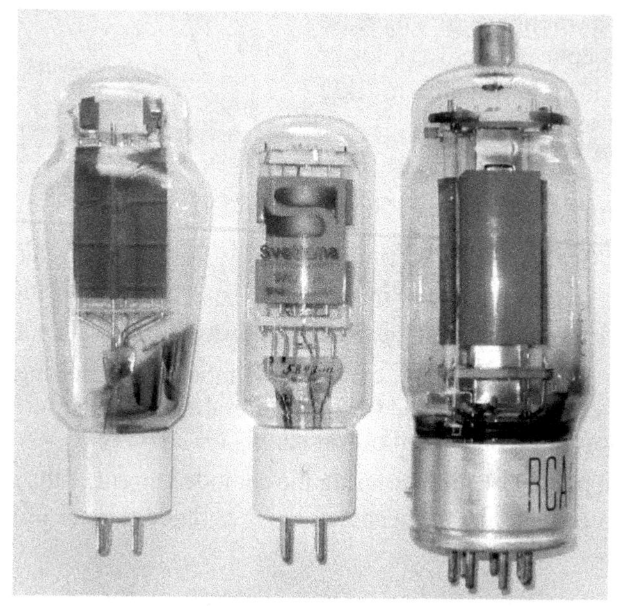

ABOVE: Both the 300B and 813 are physically bigger than SV572-10 (middle). However, SV572-10 has the same anode dissipation (125 Watts) as 813! Its smaller anode has more than 3x higher power rating than 300B's.

Design estimate from anode curves

CLASS A_1 TRIODE (A-Q-B):

V_{MIN} =380V, V_{MAX}=940V, ΔV = 560V

ΔI = 120 - 25 = 95 mA

$P_{IN} = I_0 V_0$ = 0.075*670 = 50 W

$P_{OUT} = \Delta V \Delta I/8$ = 560*0.095/8 = 6.7 W

CLASS A_2 TRIODE (X-Q-Y):

V_{MIN}=170V, V_{MAX}=1,100V, ΔV=930V

ΔI = 155mA

$P_{IN} = I_0 V_0$ = 0.075*670 = 50W

$P_{OUT} = \Delta V \Delta I/8$ = 930*0.155/8 = 18W

The power figures seem low, in the 300B territory, but let's see the actual circuit and the measured results.

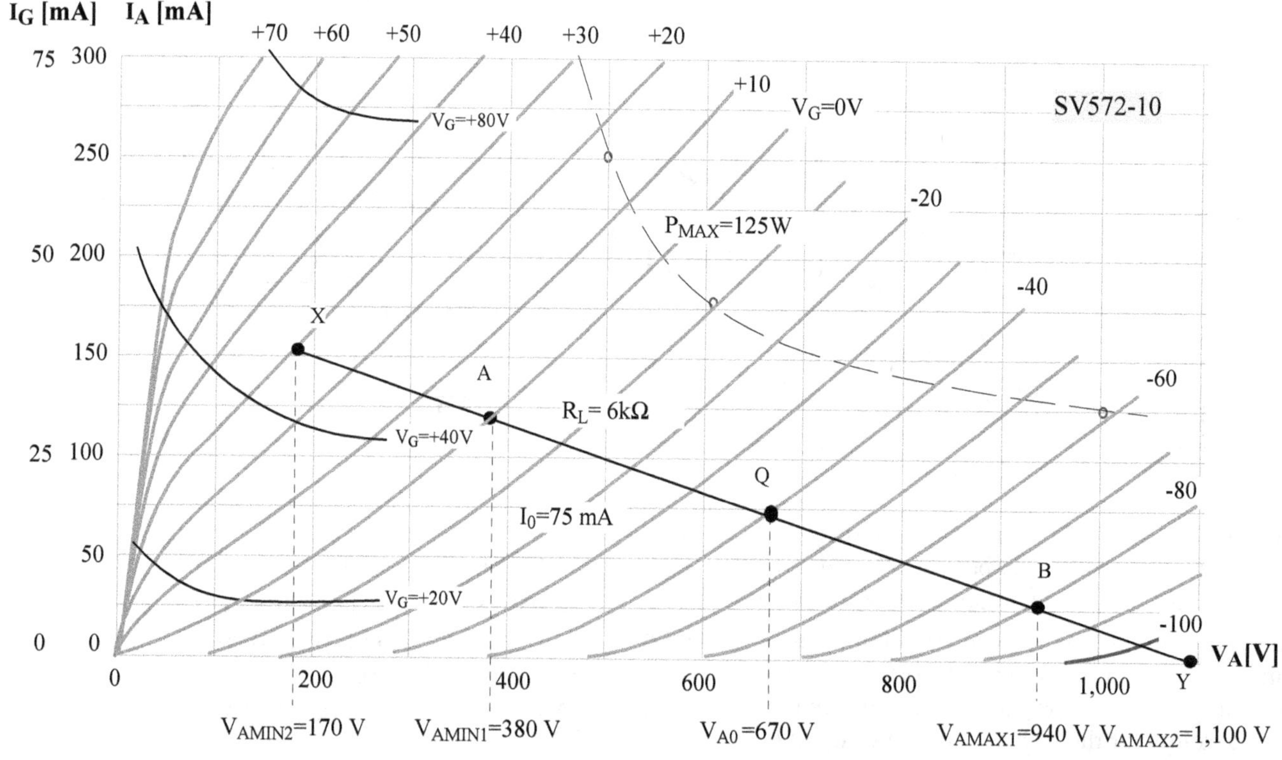

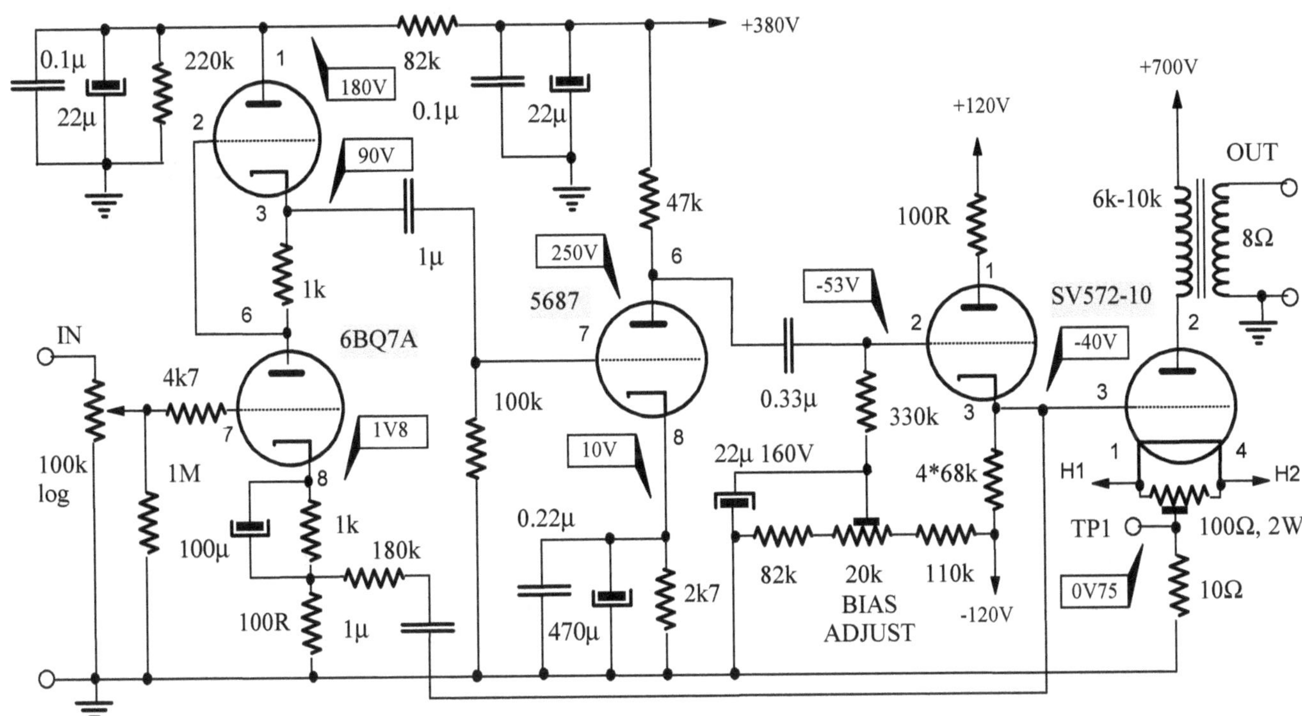

The trusted universal 3-stage circuit works well with pretty much any output tube. SRPP input stage, 5687 or 6N6P second stage with a cathode follower directly coupled to the grid of the power triodes, capable of driving the output triodes deep into class A2 without any significant increase in distortion. Instead from the speaker output, the mild negative feedback is taken from the output of the cathode follower.

6kΩ SE output transformer

The 6kΩ output transformer is a relatively simple, yet very well performing design with 4/3 sectionalizing on EI96 laminations: a=32mm, S=6.25cm, $A=a*S=20cm^2$

We haven't covered transformer designs yet (Volume 2), for now let just say that "a" is the width of its center leg, "S" is lamination stack thickness and "A" is the area of the center leg's cross section.

As a rule-of-thumb, the power rating of this transformer's core is $P \approx A^2 = 20^2 = 400W$, yet it works on audio power levels of only up to 40 Watts, so it is 10X oversized!

MEASURED RESULTS:

- BW: 10Hz - 73 kHz (-3dB, 1W into 8Ω)
- BW: 13Hz - 63 kHz (-3dB, 10W into 8Ω)
- V_{OUTMAX}: $17V_{RMS}$, $P_{MAX} = 36W$

KEY FEATURES:

- Capable of low-distortion Class A2 operation
- Cathode follower driver stage with fixed bias
- Only two coupling capacitors in the signal path
- SV811-10 power triodes can also be used

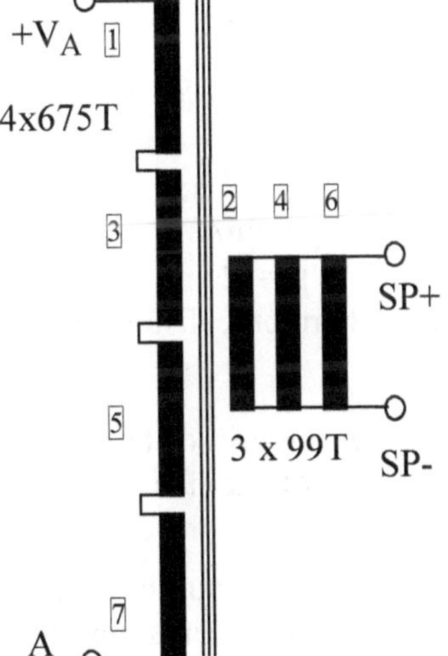

Remember this figure well, use it for evaluating commercial SE output transformers and estimating their oversize factors. If it's under five, the transformer is considered undersized!

PRIMARY: 4 x 675 turns = 2,700 turns, wire d = 0.25 mm, SECONDARY: 3 x 99 turns in parallel, wire d = 1 mm, g = 0.15 mm (air gap)

$TR = 2,700/99 = 27.27$ $ZR = TR^2 = 744$ Z_P (8Ω load) = 744 x 8 = 5,952 Ω

For multiple secondary taps, you need to change the sectionalizing to either 3/4 or 5/4, so you can have 4 secondary sections. Then you can connect them all in parallel, all in series or two and two in parallel, and then in series.

Should you want to increase the anode voltage to 1,000V to get higher output levels, increase the impedance of the output transformer to 8kΩ or even higher. A design of a 10kΩ SET output transformer is detailed in Volume 2 of this book, in the chapter on practical output transformer designs.

ABOVE: The winding diagram of 6kΩ SET output transformer. Framed numbers indicate the winding order of the sections.

OUTPUT STAGES WITH OTHER HIGH VOLTAGE TRANSMITTING TUBES

Load and quiescent point choices

TUBE PROFILE: GM70

- Directly-heated triode
- Heater: 20V, 3 A (60W)
- Plate dissipation: 125 W
- Amplification factor: 9.5
- r_I= 2,300 Ω

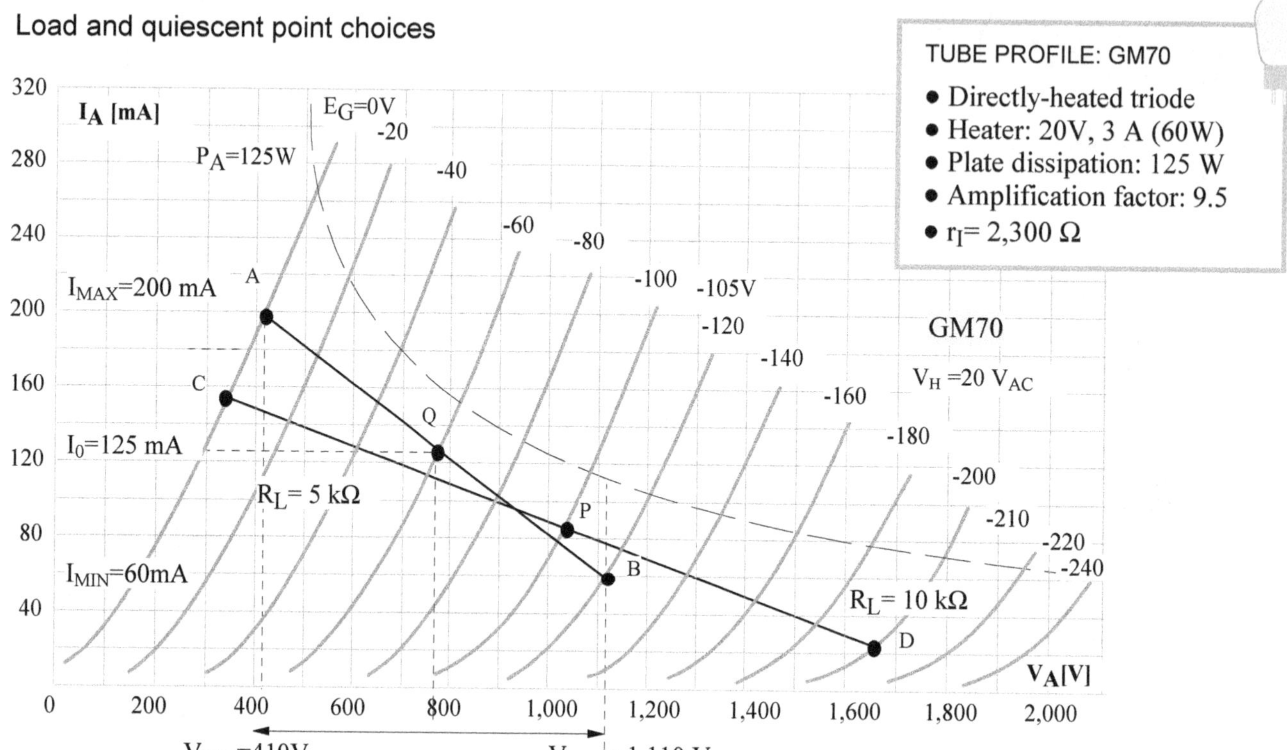

CLASS A_1 TRIODE WITH 5K LOAD (A-B):
V_{MIN}=410V, V_{MAX}=1,110V, ΔV = 700V
ΔI = 200 - 60 = 140 mA
$P_{IN} = I_0V_0 = 0.125*760 = 95W$
$P_{OUT} = \Delta V\Delta I/8 = 700*0.125/8 = 11$ W

CLASS A_1 TRIODE WITH 10K LOAD (C-D):
V_{MIN}=340V, V_{MAX}=1,660V, ΔV = 1,320V
ΔI = 154 - 22 = 132 mA
$P_{IN} = I_0V_0 = 0.086*1{,}040 = 90W$
$P_{OUT} = \Delta V\Delta I/8 = 1{,}320*0.132/8 = 22$ W

The output power with 5k load and low anode voltage (760V) is only 11W, the power that can be achieved much easier using low voltage tubes. With 10kΩ load, the operating point must shift towards much higher voltages, between 1,000 and 1,100V. The current swing remains roughly the same, around 130-140mA, but the voltage swing is now almost doubled, and so is the output power, 22W instead of 11W!

Case study: Ongaku by Audio Note

TUBE PROFILE: 211 (VT-4C)

- Directly-heated thoriated tungsten triode
- Heater: 10V, 3.25 A
- Anode dissipation: 75 W continuous, 100W maximum
- TYPICAL OPERATION:
- V_A=1,250V, V_G=-80V, I_A=60mA
- r_I=3,600Ω, gm=3.3mA/V, μ =12
- P_{OUT}=19W @ 5% 2nd harmonic distortion

Compared to triode-strapped 813 beam tetrode, 211 has higher internal resistance, lower gm and a lower anode power rating. On the plus side, 211's bias is typically around -45 to -50V compared to -100V for 813 triode, thus 211 is easier to drive.

A brainchild of the late Mr Kondo of Audio Note Japan, Ongaku was for quite a few years hailed as the best (and most expensive) tube amplifier in the world.

A pseudo-SRPP stage with unequal cathode resistors is directly coupled to the common cathode second stage, which eliminates one coupling capacitor.

The second stage is capacitively coupled to a cathode follower driver, which directly drives the grid of the output tube, meaning class A_2 operation is possible. A fixed bias is used, with a relatively low anode voltage (for 211 tube) of around 950V. Negative feedback is not used.

The output transformer's primary impedance of 16kΩ seems too high (7 or 10k would yield higher output power). The current model uses a 12k transformer. The graphical estimation of the output power (next page) seems low, only 9W in class A_1 and only 18W in class A_2, way below the specified Ongaku's output of 27 Watts! The specified frequency range 8Hz - 21kHz (+0dB, -3dB @1W) is very limited at the high frequency end, this is probably the best Kondo-san could achieve without any negative feedback.

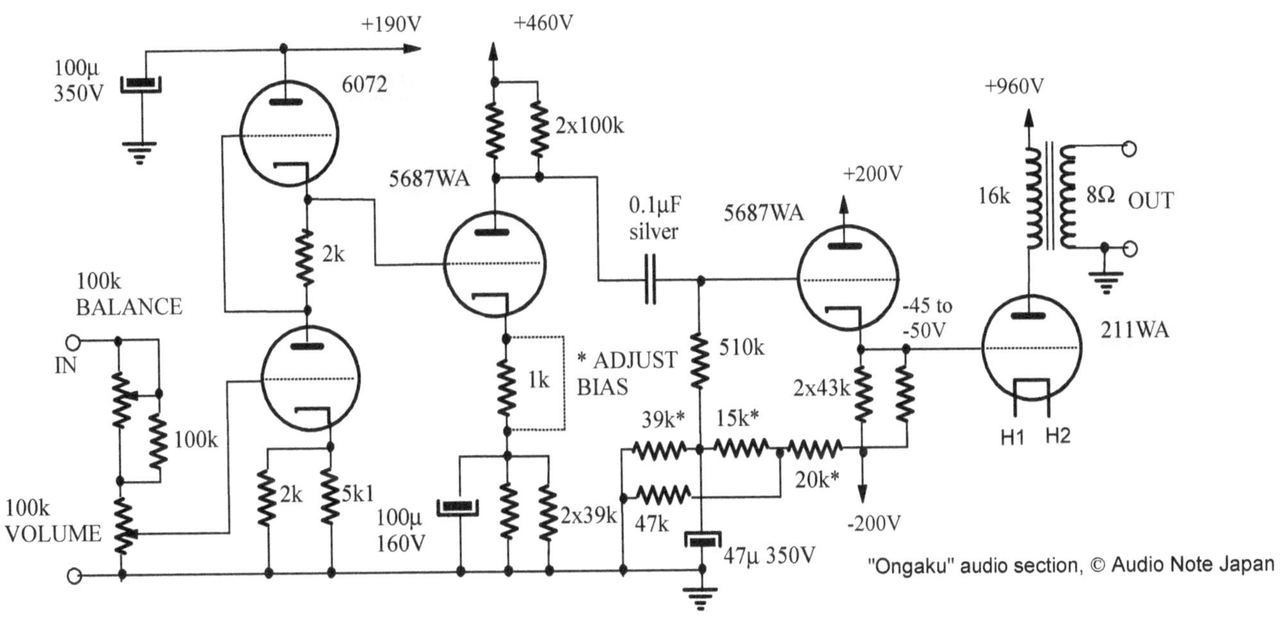

"Ongaku" audio section, © Audio Note Japan

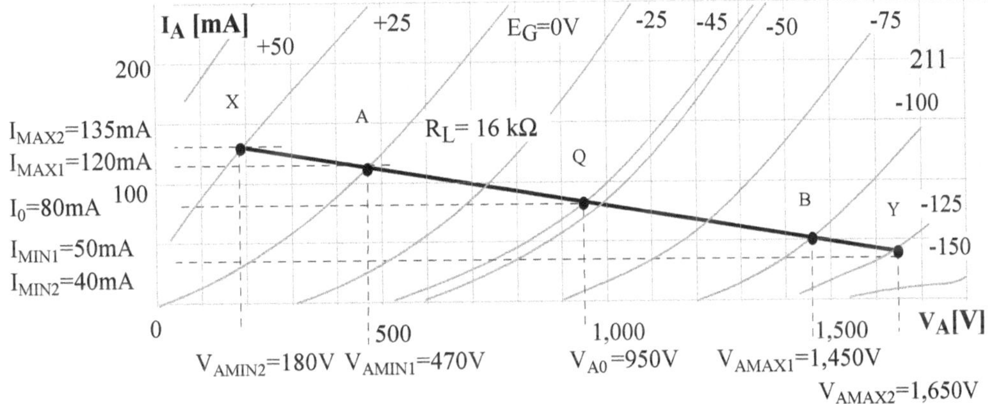

<u>CLASS A_1 TRIODE FROM THE GRAPH (A-B):</u>

V_{MIN}=470V, V_{MAX}=1,450V, ΔV = 980V ΔI = 120 - 50 = 70mA

$P_{IN} = I_0V_0$ = 0.08*950 = 76W $P_{OUT} = \Delta V \Delta I/8$ = 980*0.07/8 = 8.6W, $\eta = P_{OUT}/P_{IN}$ = 8.6/76 = 11.3%

D_2= [(V_{AMAX} +V_{AMIN}) - 2V_0]/2(V_{AMAX} -V_{AMIN})*100% = [(1,450+470) - 1,900]/2(1,450-470)*100% = 20/1,030 = 1.0%

<u>CLASS A_2 TRIODE FROM THE GRAPH (X-Y):</u>

V_{MIN}=180V, V_{MAX}=1,650V, ΔV = 1,470V ΔI = 135 - 40 = 95mA

$P_{IN} = I_0V_0$ = 0.08*950 = 76W $P_{OUT} = \Delta V \Delta I/8$ = 1,470*0.095/8 = 17.5W, $\eta = P_{OUT}/P_{IN}$ = 17.5/76 = 23%

A variation on the Ongaku theme

This amp, very loosely-based on Ongaku design, was brought in for a checkup after 15 years of faithful service. There was no brand or manufacturer marked. Since the power supply was extremely complex, with regulated voltages for preamp stages, we had no inclination to waste hours producing its circuit diagram. The output tubes were heated with fully-rectified but unfiltered voltage, a pulsating rectified sine waveform.

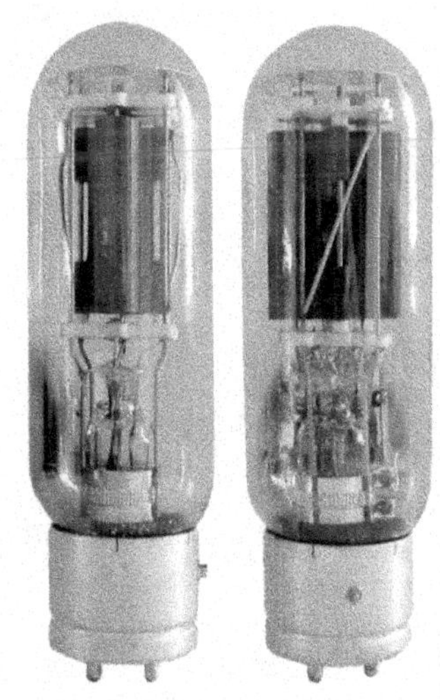
211 (VT4-c) triodes

In the anode power supply two 470µF/400V capacitors were series-connected, and the anode voltage was a low 760V_{DC}. Instead of tube rectification, the high voltage for the output stage was provided by four silicon diodes in a bridge.

Instead of the Ongaku's SRPP, the first stage was triode-connected E280F pentode and instead of 5687 duo-triode, Russian 6N6P was used. We have already met E280F pentode in the phono preamplifier section.

The cathode resistor of the first stage was bypassed by only a 10nF film capacitor (Wima), while the cathode resistor of the second stage was in two parts, the 2k2 resistor was not bypassed.

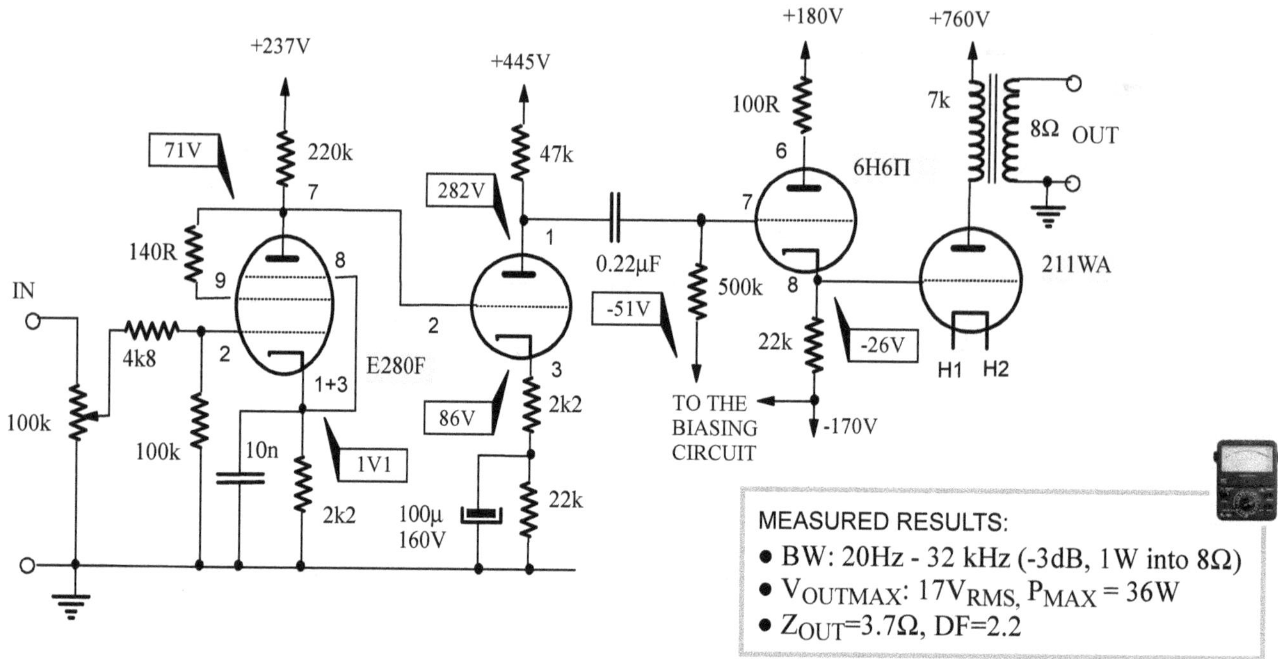

This amp did not sound as good as our 211 amp (next page). Also, in comparison with the EL153 SET design (detailed in the next chapter), the EL153 amp had better microdynamics, more refined top-end and superior transparency. Even its midrange was slightly smoother.

The biasing circuit (not shown), instead of being fixed as in Ongaku's case, used 10-turn trimmer potentiometers, one for each channel's bias adjustment.

Due to its lower anode voltage and different positioning of the operating point, the bias was much lower than Ongaku's, only -26.3V. While global NFB was not used, negative feedback was implemented as cathode degeneration in the first two stages (un-bypassed 2k2 cathode resistors). The 10n capacitor does not bypass the cathode resistor of the 1st stage for audio frequencies.

Despite its huge chassis, the insides were overcrowded. The positioning of major connections was wrong. Look how far the front on-off switch (1) is from the mains inlet and the fuse (2). They should be in the corner (3) and the switch should be on the side, right next to it (4).

The RCA inputs (5) are also far from the preamp tubes, necessitating the use of high frequency sucking shielded cables (6). The preamp/driver section is cramped, very difficult to access the resistors and capacitors with a soldering iron without damaging something else (7). There are five Soviet PIO capacitors used for bypass purposes, but their location in the opposite corner from the elcos (8) makes the wiring long and messy. Only one type and color (white) of hookup wire was used, making wire tracing and troubleshooting difficult and time consuming.

Despite their modest capacitance, the elcos are physically huge. This led to space problems down the line.

The constructor could not fit all four in a row, so one was orphaned in the bottom corner (6), again, lengthening the wire runs!

There is only one low inductance filtering choke for both channels (9), a small and sad affair, a serious oversight by the designer. This kind of amplifier requires at least two, or ideally three large chokes (one common, one for each channel)!

The messy wiring and long wire runs stem from topological problems

DESIGN PROJECT: 211 - 845 - GM70 AMPLIFIER

In this design, due to a low anode voltage on 211 triodes, their grids are biased at only -24V. This means that low driving signals are required. The output tubes idle at a very low DC current, meaning the quiescent power dissipation on their anodes is also very low, P_0=660V*55mA= 37W! Since the anode power dissipation of 211 tubes is 75Watts continuous, they will have a long and easy life in this amplifier.

If 211 triode is the heart of this amp, the input tube, the tiny but mighty E86C is its brain. Everything depends on its performance, and what a performance it is - holographic imaging and stunning dynamics.

As for the driver tube, 6CG7 has the same pinout as 6N6P, and can be substituted in this design, giving tweakers some tube rolling freedom.

6N6P duo-triode is very similar to ECC99 and E182CC, the main difference being that 6N6P operates only on 6.3V heater voltage. Its pin 9 is the shield between the two triode systems, which should be grounded. Pin 9 on ECC99 and E182CC is the heater center tap, which is left unused on 12.6V heater operation or is one end of the heater supply if 6.3V is used (the other heater end is then pins 4 and 5 strapped together). Due to these heater wiring differences, the two western tubes are not a drop-in replacement for 6N6P.

For the harder-to-drive triodes such as 845 or GM70 the amplification of the first stage is increased to around 15 by increasing the anode resistance to 12k, while the second stage amplification is increased by bypassing the cathode resistor with a 220µF elco.

The power transformer needs to be very large and have a good regulation. Ours produced 800V_{DC} at idle, but it dropped to only 700V_{DC} under full load. Although the diagram shows one transformer supplying all voltages, we used a separate auxiliary transformer for the two 211 heater voltages.

KEY FEATURES:

- Capable of low-distortion Class A_2 operation
- Cathode follower driver stage with fixed bias
- Regulated power supply for the input stages
- Choice of 6CG7 or 6N6P driver tube, 12BH7, ECC99, 5687 and E182CC with a simple wiring change

SIDE-BY-SIDE	ECC99	6N6P	E182CC (7119)
Heater	6.3V/0.8A or 12.6V/0.4A	6.3V/0.75A	6.3V/0.64A or 12.6V/0.32A
P_A max. [W]	5	4	4.5
I_A max. [mA]	60	45	60
V_A max. [V]	400	300	300
V_{HK} max. [V]	200	100	200
gm [mA/V]	9.5	11 (+/-3)	15
µ	22	20 (+/-4)	24
r_I [kΩ]	2.3	1.8	1.6
Pin 9	Heater CT	Shield	Heater CT

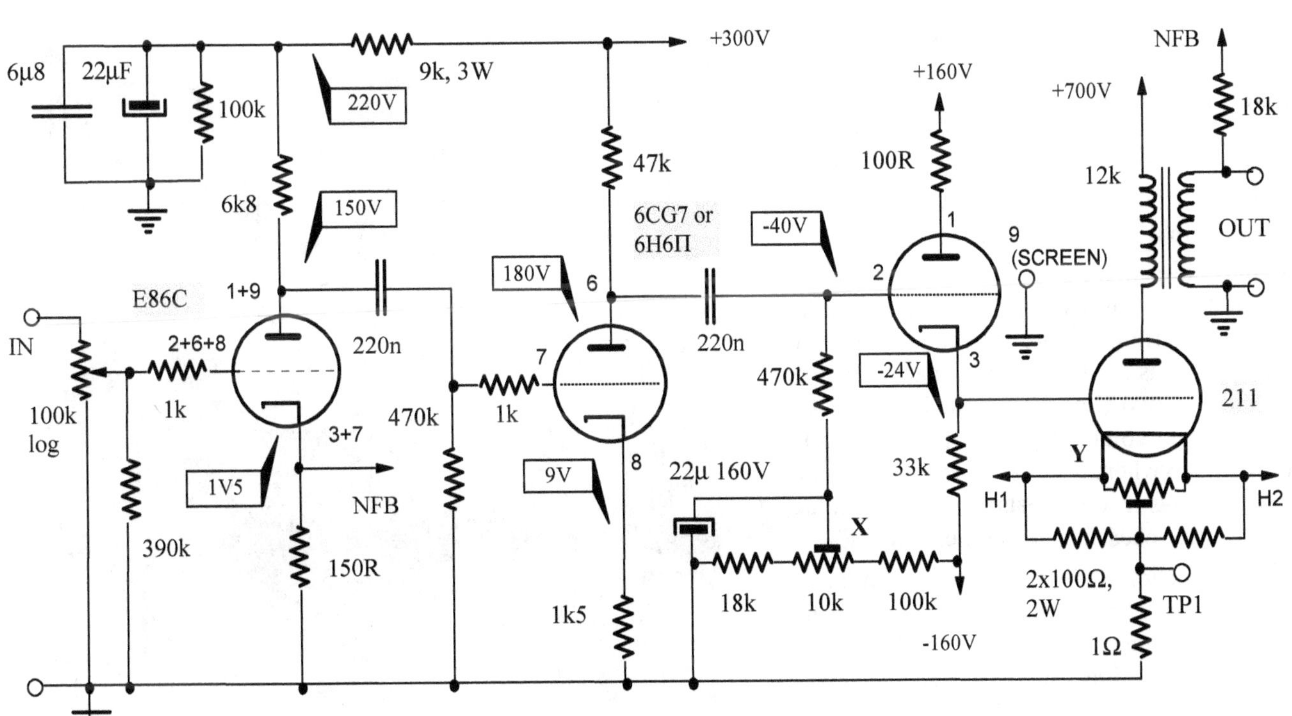

Since the middle point of the voltage doubler is used to supply DC voltage to the input stage, it needs to be fully regulated due to its high ripple factor. Alternatively, use a separate winding on the mains transformer to get 250-300V_{DC} after filtration.

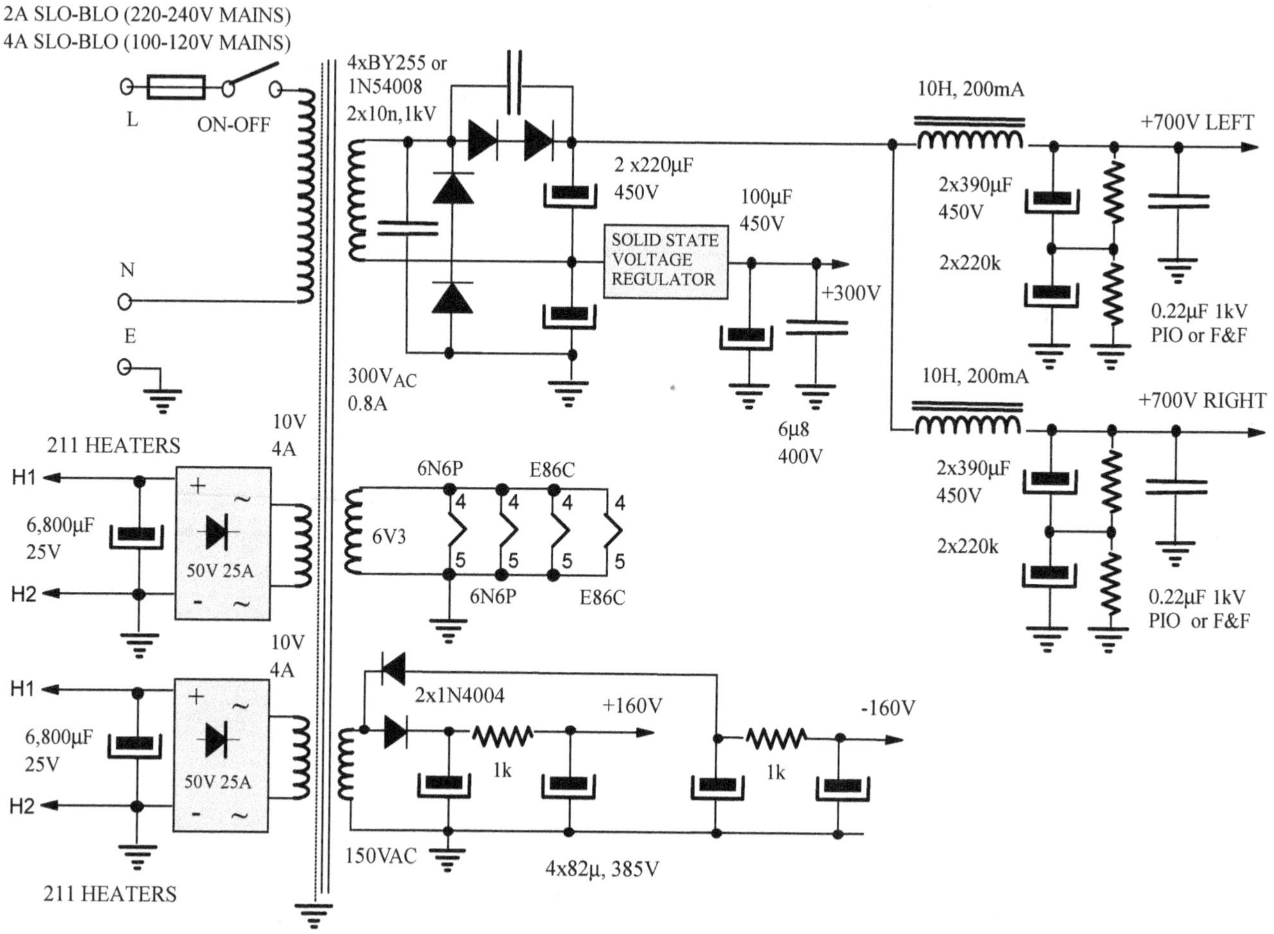

ABOVE: This amplifier demands a very complex power supply, and since 211 tubes and even the power transformer get very hot, a large chassis with good air flow is absolutely necessary!

The internal view of the prototype amplifier. The RCA inputs are very close to the volume control pot, minimizing the length and the capacitance of the shielded cables. Since the front fascia is made of Perspex, the metal case of the volume control potentiometer had to be grounded, tied to the chassis (1).

The bias adjustment potentiometers are quality 10-turn Bourns units (2), but ordinary 270° single turn trimmers are perfectly fine. The mains switch is at the front of the amplifier, so twisting the two-core mains-rated cables is a must (3). The voltage regulator module can be mounted anywhere (4), it does get hot, and although it has its own heatsink, a good contact with the steel chassis is recommended.

The two chokes were initially mounted symmetrically in the back corners (5), but the mains transformer was getting very hot due to the two 10V windings supplying 211 heaters.

We fixed it by adding an auxiliary transformer (6) to provide those voltages and moved one choke from its original location adjacent to the other channel's choke.

Although rated at 35A (!) the rectifiers for 211 heaters (7) get very hot supplying only 10A each, even when bolted onto a heatsink and then to the chassis. The whole power supply was pre-wired on one huge terminal board and mounted on the rail that also locks the 211 sockets into place.

The two hum trimmers and the two power resistors were mounted on a small terminal strip (one for each output tube) and bolted onto the same rail (9). Such rheostats with integral plastic knobs (salvaged from old TVs) would be very difficult to find these days.

Power tubes don't have to be sunk under the chassis that much, this was done so that they don't protrude above the mains transformer's enclosure. That way when the amplifier is turned upside down for testing or servicing, the weight of the amplifier is fully on the transformer's enclosure (mild steel) and the top of 211 tubes is clear of the work bench.

Unfortunately, due to the weird design of the 4-pin jumbo sockets, the sockets always have to be sunken, since their exposed metal lugs are on the top side, facing the chassis. There are nicer looking and much more expensive current production sockets that have their lugs facing downwards, these can be flush-mounted on top of the chassis.

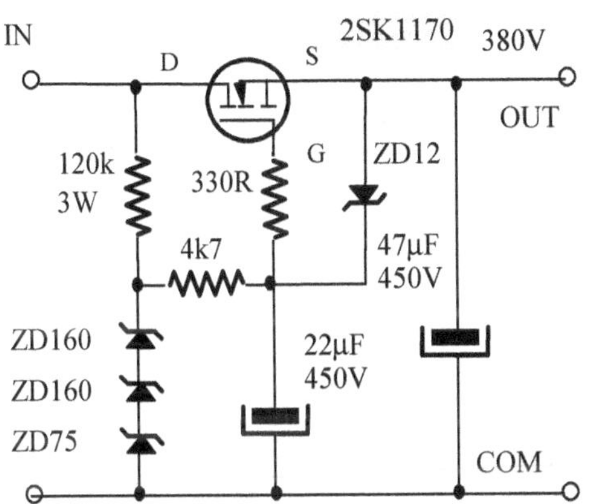

MEASURED RESULTS:

- BW without NFB: 18Hz - 13 kHz (-3dB, 1W into 8Ω)
- BW with NFB: 13Hz - 43 kHz (-3dB, 1W into 8Ω)
- CLASS A_1: $V_{OUTMAX}=11V_{RMS}$, $P_{MAX}=15W$
- CLASS A_2: $V_{OUTMAX}=16V_{RMS}$, $P_{MAX}=32W$
- $DF = R_L/Z_{OUT} = V_L/(V_O-V_L) = 2.0/(2.8-2.0) = 2.5$
- $Z_{OUT}=3.2\Omega$

LEFT: The MOSFET voltage regulator for the Bad Ass power supply. Similar regulator was used in the 211 amplifier to provide $300V_{DC}$.

Bad Ass totem-pole regulated power supply

If you want to use LC filtering voltage doubling is out of question, since voltage doublers inherently include the first filtering capacitor (a pair of them actually), so the filtering is unavoidably of the CLC kind.

Likewise, if you don't want to use voltage doubling, finding a mains transformer with a 700-1,000V, 300mA+ secondary is not easy, and even if you do, they are very expensive.

This concept enables you to use any number of secondary windings, even windings from separate transformers, rectify their outputs individually and run each through a voltage regulator. Finally, you pile all the regulated outputs up on top of one another until you get the final DC voltage you desire.

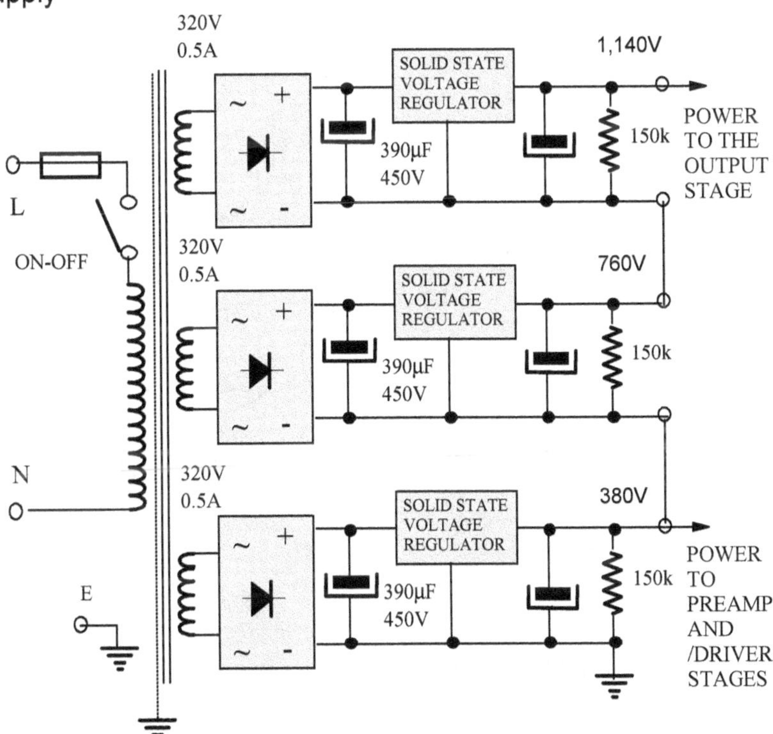

No chokes are used, which will save you another small fortune and make the amplifier a few kilograms lighter. The Bad Ass power supply is cheap, flexible, saves space and weight, and yet fully regulated, with lower ripple and better regulation than any choke can achieve! Yes, I'm talking solid state regulators here. If you belong to the puritan tube school of denial, you can always go back to backbreaking chokes and custom-wound power transformers.

CASE STUDY: CAZ-TECH SE845 MONOBLOCKS

The 845 triode has a lower internal resistance than the triode-strapped 813 or 211 true-triode, so it is much more difficult to drive. A typical bias is between -150 and -200V, meaning the grid drive signal must be up to 150-200V peak or 106-140V_{RMS}!

This design is from a defunct Canadian firm Caz-tech. There were two variants, with 22 and 28 Watts output power rating, depending on the bias circuit. The schematics below is for the 22Watt version, with -150V bias.

Interestingly, the biasing circuit was not on the original drawing of the power supply section, so it is not included here either.

TUBE PROFILE: 845

- Directly-heated power triode
- Socket: 4-pin jumbo
- Anode power rating: 100W
- Thoriated tungsten filament/cathode
- Heater: 10V, 3.25 A
- Max. anode voltage: 1,250V
- gm=3.1 mA/V, μ=5.3, r_I = 1,700 Ω

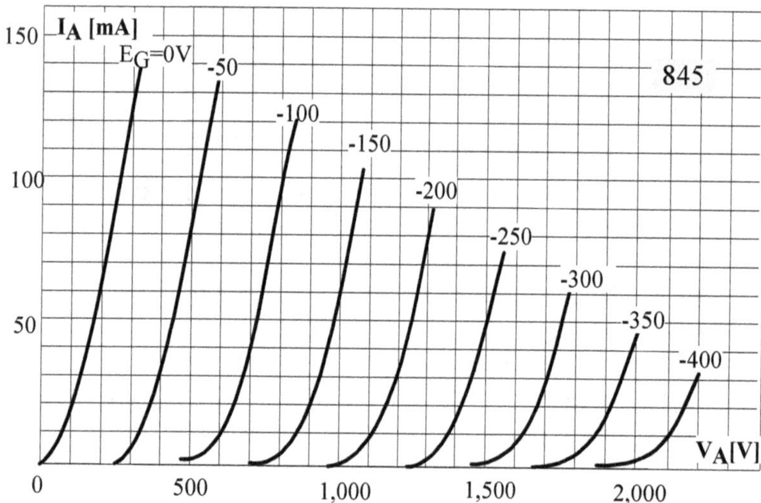

Two 6SN7 duo-triodes are used. The first works as a pseudo-SRPP stage, the second as a cathode follower, directly-coupled to the first stage, with both triodes in parallel.

The output of the cathode follower is capacitively coupled to the input of the SRPP driver stage, for which 6BX7 duo-triode was selected. The driver is then capacitively coupled to the 845 grid, meaning the output stage can only work in class A_1.

We haven't tried this circuit in its exact form, but we did use the 6BX7 SRPP driving GM70 triodes, with our favorite E86C input stage, and it sounded sublime.

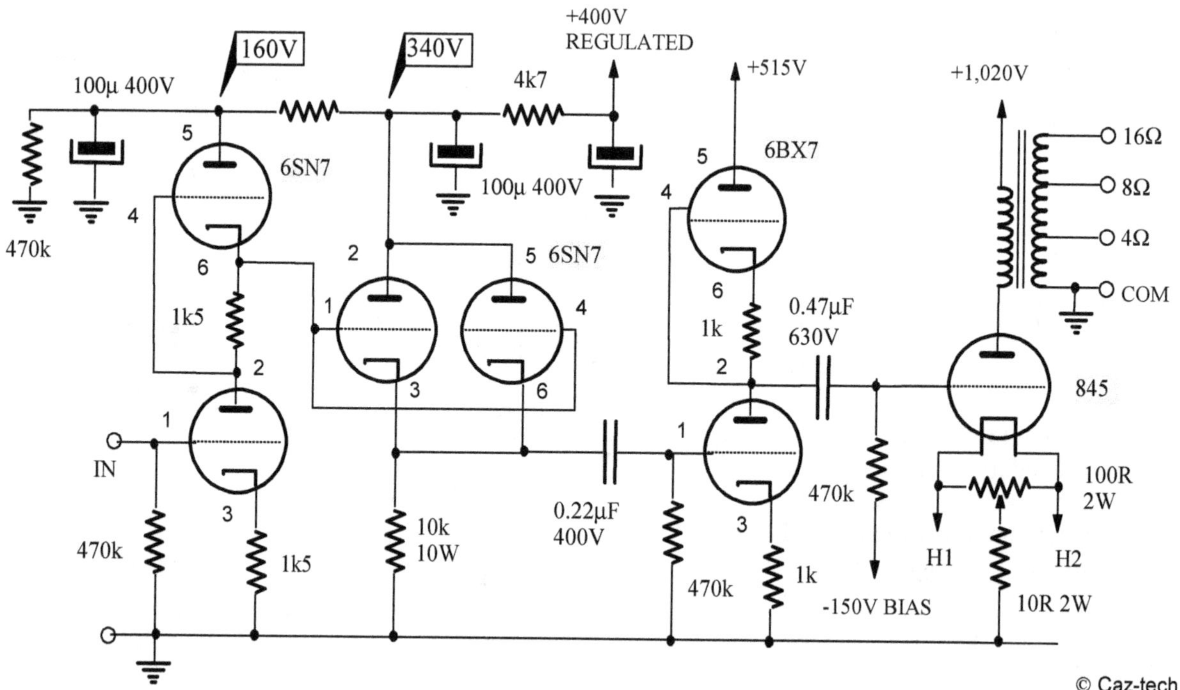

6BX7 as a driver

The 10 Watt rating of each 6BX7 triode is promising for a low-power push-pull amplifier, but due to its small physical size (limited heat dissipation capability), when paralleled the two triodes can only dissipate 12 Watts. 6BX7 exhibits excellent linearity, not to mention that it can easily supply 40-50 mA of current (each triode)!

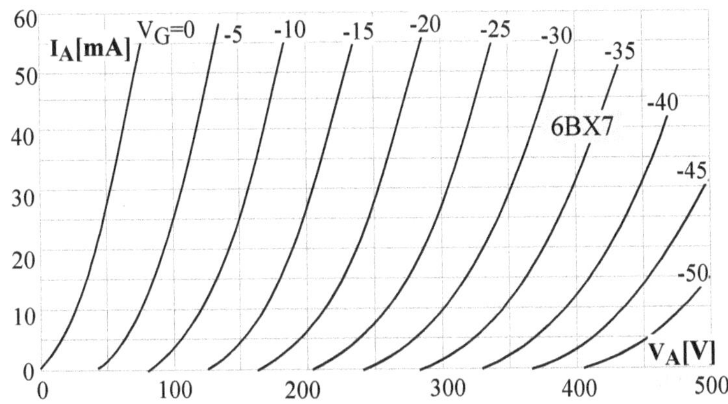

TUBE PROFILE: 6BX7GT

- Indirectly-heated twin triode
- Socket: octal, heater: 6.3V, 1.5A
- Anode dissipation: 10W each, 12W both
- V_{HKMAX}= 200V
- Typical operation:
- V_A= 250V, V_G=-16.4V, I_0=42 mA,
- gm=7.6 mA/V, μ=10, r_I = 1,300 Ω

Still, this tube makes a great output tube for headphone amplifiers, or driver for power-hungry class A_2 or class B power triodes' grids. It has a very low internal impedance, and, when doubled up, its impedance is lower than that of 300B and 2A3, around 650Ω. Plus, it doesn't require separate heating as directly heated tubes such as 2A3 and 300B do, so it can heated by the same voltage as the input tubes.

Finally, 6BX7 is easier to drive, requiring a bias of only -15V, compared to -45V for 2A3 or -75V for 300B! However, it is an obsolete tube, not in current production, so the supplies are limited.

The first time we used 6BX7A in a SRPP driver stage (driving the grids of GM70 triodes), the upper frequency limit of the whole amplifier was only 23 kHz.

6BX7 is a good-looking and very linear octal duo-triode but serious questions remain about its high frequency performance.

The power supply

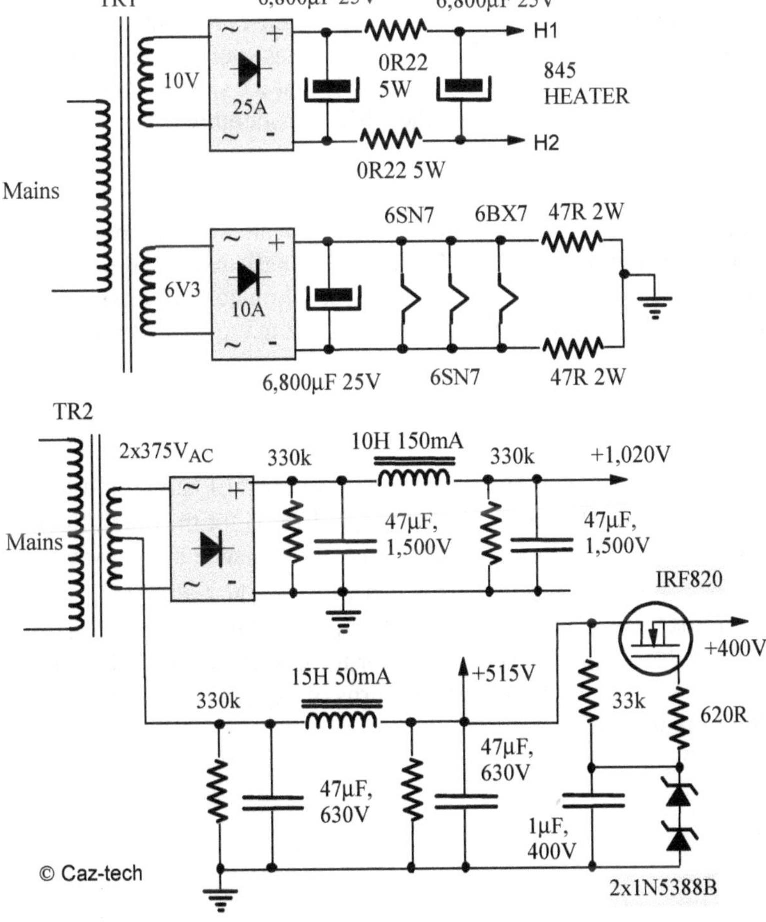

The limit was not in the output but in the driver stage. As soon as we replaced 6BX7A with Russian 6N6P triodes, the upper frequency limit almost doubled to 40+ kHz!

We thought the culprit was 6BX7A's large input capacitance, but both tubes have an identical input capacitance of 4.4pF so the reason for such a sub-par performance will remain a mystery. We have never used 6BX7A since.

Going back to Caz-tech design, the 845 triode is DC-heated, the heater voltages for all tubes are provided by a separate transformer.

The high voltage transformer has a $2x375V_{AC}$ center-tapped secondary. One rectifier bridge is connected between the secondaries' ends, so there are $750V_{AC}$ at its input. The filtering is extremely simple, only a single CLC filter, supplying 1,020-$1{,}040V_{DC}$.

The center tap supplies another solid state bridge rectifier, again with a single CLC filter, providing the +515V unregulated DC voltage for the 6BX7 driver stage.

The same voltage is then fed into a simple MOSFET voltage regulator. The regulated +400V supplies the anodes of the first two stages. Notice that elcos are not used at all, obviating the need to string two or three in series. High voltage film capacitors are used instead, a great idea. However, their capacity is only 47μF, so the energy storage of this amplifier is on the low side.

SET AMPLIFIERS - DESIGN AND CONSTRUCTION ISSUES AND CHOICES

Self- or cathode-bias as a dynamic compressor

The choice between a fixed or external and cathode- or self-bias is a major design decision. The cathode bias voltage is created by the flow of DC current through a resistor connected between the cathode and ground (or reference point), making cathode positive, which is equivalent to making the grid negative with reference to the same point.

Assuming we want +70V on the cathode and a DC current of 80mA, we need R_K=70/0.08 = 875Ω, so we would use a standard value of 880Ω! The power dissipated on that resistor (wasted into heat) would be $P=I_K^2R_K$ = 0.082*880 = 5.6W That is quite a lot compared to a useful audio output of 8-10W!

However, we have just seen that in class A_1 the anode current swings up to 140mA for 300B tube and even higher for others, so in that point A the dissipation would be 17.3W! The current also drops down to 20mA for 0.4W of heat, thus we only concern ourselves with the average dissipation over the whole period (cycle).

Our simplified analysis is always based on the premise that the bias voltage is fixed at the value it has in the quiescent point Q. If an external fixed bias is used, that is a reasonable assumption, providing the negative bias power supply is properly designed. The grid current is negligible in class A_1, but since some designs such as this one can crossover into class A_2, the bias supply must be capable of delivering a significant current into the grid, 10 or even 20mA!

The situation is different with cathode bias. The main problem is that as the anode current swings up and down between 20 and 140mA, the cathode voltage changes between 0.02*880 = 17.6V and 0.14*880= 123V, and so does the grid bias voltage! In other words, the bias is not stable at all.

Firstly, with increasing music signal on the grid and increasing anode current, the bias becomes more negative and thus opposes the dynamic swing of the anode current. This is a classic example of negative feedback, resulting in dynamic limiting or compression of the output signal. A direct consequence of this phenomenon is that the same amplifier provides a much higher output power when fixed bias is used instead of the cathode bias.

Secondly, the varying bias and the resulting dynamic compression cause a significant increase in harmonic and IM (Intermodulation) distortion. This could be one explanation of why amplifiers with fixed bias sound cleaner and less colored than their cathode-bias counterparts. The main excuse for using cathode bias is that it's simple, cheaper and foolproof - the user cannot run the expensive 300B tubes too hot and shorten their life considerably.

DC or AC heating?

300B tubes are heated using dedicated DC supplies. General consensus is that directly-heated or "filament-type" tubes sound better when AC-heated and there is some truth in it. However, AC heating usually results in increased mains-frequency hum, which hum-neutralizing rheostats can never completely remove! Ultimately, if you speakers are very efficient (say above 93 dB/W), use DC heating, with lower efficiency speakers 1-3 mV of AC hum that you will get using AC heating will be tolerable.

Grid chokes

The grid chokes for our projects were wound in-house, using 0.1mm diameter wire on a A=4cm^2 EI lamination stack. Their DC resistance was around 2kΩ and the inductance at 120Hz was 155H for one and 163H for the other. Such a variation in the final inductance is normal and must be expected. It does not affect the performance in any way.

Valab GC640-5 grid chokes from an ebay seller in Taiwan cost $60 a pair+$10 postage (in Jan 2015), a tiny investment compared to outrageously priced "audiophile" capacitors, but a significant sound improvement. These have inductance of 640H at 120Hz and over 2,500H at 20Hz, so they seem superior to ours, most likely due to the better quality of their magnetic laminations. They use Z11 laminations and 20,000 turns of oxygen free copper wires. At such a reasonable price there is no point wasting time winding your own!

PRACTICAL SINGLE-ENDED PSEUDO-TRIODE DESIGNS

- 829B PSET AMPLIFIER
- SINGLE-ENDED DESIGNS WITH TRIODE-CONNECTED PENTODES & BEAM TUBES
- F2a and F2a11 BEAM TETRODES
- RS1003 SET AMPLIFIER
- HORIZONTAL AND VERTICAL TV DEFLECTION TUBES IN AUDIO SERVICE
- SINGLE-ENDED TRIODE AMPLIFIER WITH EL519/PL519 POWER TUBES
- GIANT KILLER: EL153 SET AMPLIFIER
- CATHODE-LOADED 6V6 TRIODE AMPLIFIER

"Perfection is not when there is no more to add, but no more to take away."
Antoine de Saint-Exupery, aviator and author of "The Little Prince"

829B PSET AMPLIFIER

829B in its many western, Chinese and Russian versions is one of the prime candidates for a first DIY project. It is available as NOS from a variety of manufacturers and it's cheap because very few commercial amplifiers use it (apart from a few Chinese push-pull designs). Furthermore, it's a robust tube, sensitive (easy to drive), requires low-impedance (easy to wind) output transformer, the two anodes together are rated at 60 Watts, and, most importantly, it sounds clean and transparent, yet musical.

It uses the same 7-pin socket as 6C33C-B, with which it also shares dual heater configuration. The two heaters in series need 12.6V at 1.3A, or 6.3V at 2.6A in parallel connection, giving the constructor flexibility. Triode-connected, a single tube in SE topology needs an output transformer with 2.5-3.5kΩ primary, two triodes in parallel (one physical tube) need half that, so 1.5-2kΩ is in the ballpark.

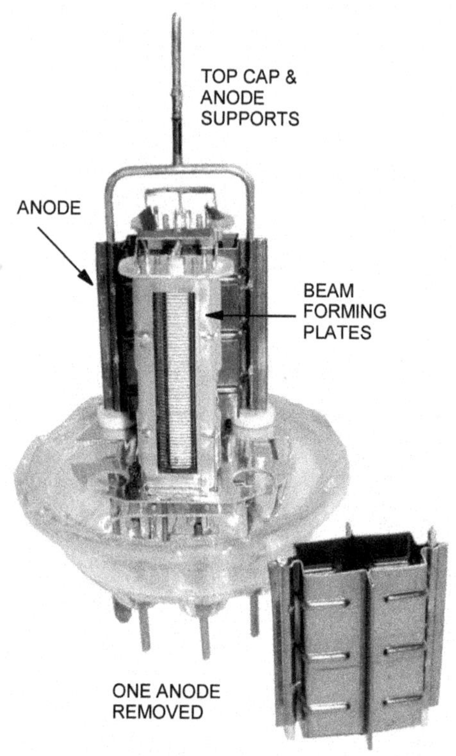

ABOVE: The rather unusual and thus even more fascinating internal construction of 829 duo-beam power tube

There are weaknesses as well. While anode voltage can go up to 750V, the maximum allowed screen grid voltage is only 225V, so, unless you use 829B as a pentode, the output power will be limited by the low screen voltage.

Despite a robust construction, one tube developed an air leak through the glass so we opened it up and removed one anode. The anode is one piece and is held in place by the glass around the top pins. You can see the beam-forming plates, and through their opening the screen grid, control grid and the white oxide-coated cathode behind them.

In triode connection a reasonable operating point at V_A=280V and V_G=-25V would result in I_0=80mA, gm=12mA/V, μ= 9, r_I=750Ω These are very friendly figures. Low internal impedance, an par with the 300B, and low bias, so a single driver stage with a voltage gain between 10 and 20 is sufficient.

TUBE PROFILE: 829B

- Indirectly-heated twin beam power tube
- Heater: 12.6V,1.3A (series) or 6.3V, 2.6A (parallel)
- Maximum anode/screen voltage 750/225V
- Maximum anode dissipation: 30W+30W
- Maximum screen grid dissipation: 7W

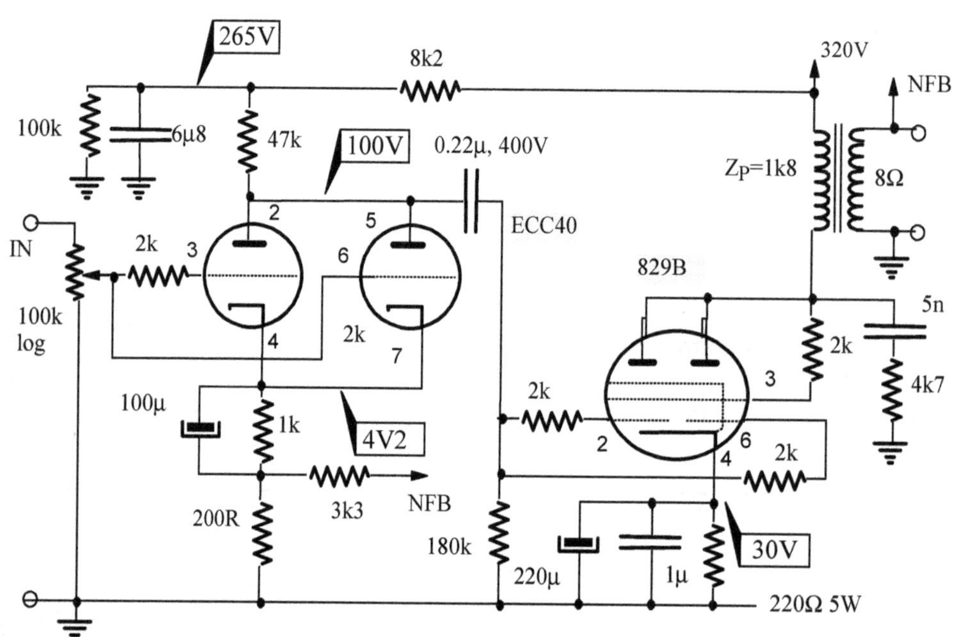

A simple design, paralleled ECC40 input stage with cathode biasing, capacitively coupled to the grid of the output stage, two paralleled triode-strapped 829B tubes, also cathode biased.

Notice that the screen voltage is around 320-30 = 290V, although V_{SMAX} is specified as only 225V! In triode connection, the 70V higher screen voltage seems to be of no consequence to these robust tubes.

829B is a transmitting tube and those have a tendency to oscillate at the first opportunity, so grid and screen stopper resistors are mandatory. Even then, in our case, one channel oscillated at 130kHz. A snubber between the anodes and ground suppressed the oscillation (5n+4k7).

Although the frequency bandwidth was limited, most likely due to lower grade output transformer laminations used, the prototype amplifier sounded great.

For two paralleled triodes with a 1k8 load: V_{MIN}=75V, V_{MAX}=510V, ΔV=435V, ΔI=270-25=45mA

$P_{IN}=I_AV_0=0.13*320=41.6W$

$P_{OUT} = \Delta V \Delta I/8=435*0.245/8 = 13.3$ W

MEASURED RESULTS:

- BW: 15Hz- 25 kHz (-3dB, 10V_{RMS} into 8Ω)
- V_{MAX}: 11V_{RMS}, P_{MAX} = 15W

The predicted output power from two paralleled triodes (13.3 W) is quite close to the measured 15W. That was with with 30V lower anode voltage at 290V.

It should be possible to get 17W with a higher anode voltage, but the question is how high can we push the screen voltage, which has already exceeded its 225V limit. However, even if the output tubes don't last too long working at such a high screen voltage, 829B lovelies are cheap and cheerful.

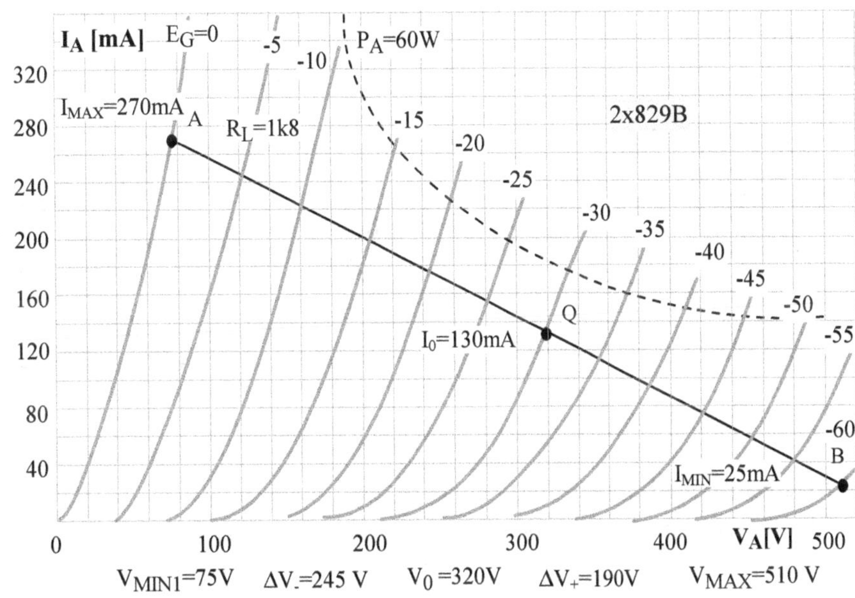

the directly-coupled version

To eliminate the coupling capacitor we need to raise the output's tube cathode from 30V to 130V. At 130mA cathode current a 1kΩ cathode resistor is needed. The power dissipation on such a resistor is high ($P=I^2R_K = 0.13^2*1{,}000 = 16.9W$), so a 25W resistor is needed, bolted to the metal chassis for better heat dissipation!

To keep the anode-cathode voltage at the same 290V level, the power supply needs to be 290+130 =420V, which requires a different mains transformer. Electrolytic capacitors rated at 450V should be used in the anode power supply, although 500V units would be safer and would last longer. For best sonics, if you have space under the chassis, use multiple 47μF 630V film capacitors.

To reduce the driver stage supply voltage to 265V, the decoupling resistor must be increased to (420-265)/0.0035 = 44kΩ.

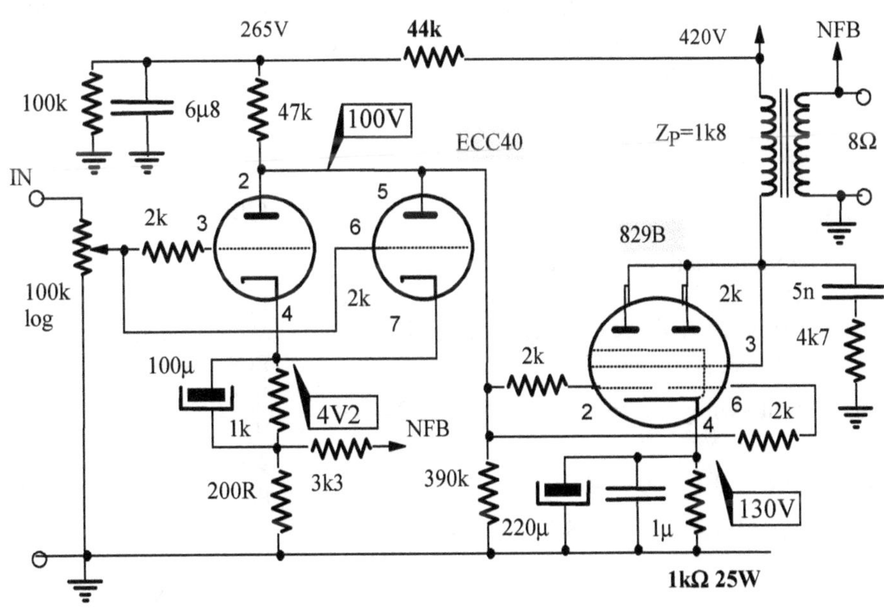

SINGLE-ENDED DESIGNS WITH TRIODE-CONNECTED PENTODES & BEAM TUBES

Triode-strapped KT90

The KT90, or "Kinkless Tetrode 90", is a beam power tetrode developed by Elektronska Industrija Niš (Ei) in former Yugoslavia (today Serbia). It may be used as a substitute for KT88, with which it shares the octal socket, given appropriate re-biasing when used in push-pull configuration.

Electro-Harmonix in Russia also started manufacturing a tube called KT90, but this is a totally different tube, more similar to KT88. The use of the same name was probably an attempt to capitalize on the original tube's good reputation and its unavailability due to the bombing and destruction of Ei factory in NATO's bombing of Yugoslavia in 1999.

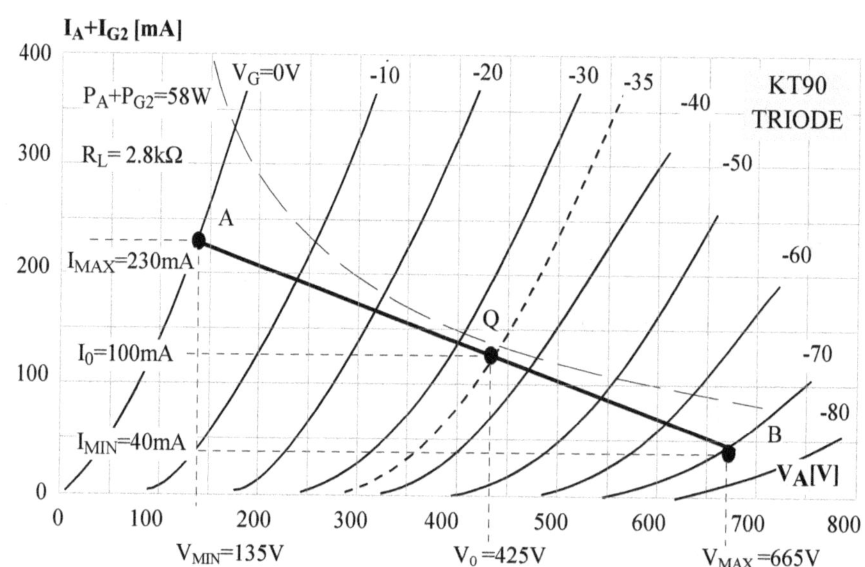

CLASS A_1 TRIODE:

Maximum voltage swing: V_{MIN}=135V, V_{MAX}=665 V, ΔV = 530V

Maximum current swing: ΔI = 230 - 40 = 190 mA

Load impedance: R_L=2k8

The input power is $P_{IN}=I_AV_0$=0.1*425= 42.5 W $P_{OUT}=\Delta V\Delta I/8$=530*0.19/8=12.6 W and the efficiency of the output stage is $\eta=P_{OUT}/P_{IN}$=12.6/42.5=29.6%

The 2nd harmonic distortion coefficient is D_2= (290-240)/2*530 = 4.7%

Almost 13 Watts in Class A_1 is some serious single-ended power, in our designs with Ei KT90 we regularly exceeded that figure, up to 16W maximum. The second harmonic distortion at the highest power level is acceptable, under 5%!

Triode-strapped KT88

CLASS A_1 TRIODE:

For a similar load (R_L=4kΩ), referring to the graphical analysis on the next page, we get a slightly smaller voltage swing: V_{MIN}=130V, V_{MAX}=580 V, ΔV = 450V

The current swing is also smaller: ΔI = 160-42 = 118 mA,

$P_{IN} = I_0V_0$ = 0.1*360 = 36 W

$P_{OUT} = \Delta V\Delta I/8$ = 450*0.118/8 = 6.6 W

SIDE-BY-SIDE	KT90 EI	KT88	EL156
	Beam tetrode	Beam tetrode	Power pentode
Heater	6.3 V/1.6 A	6.3 V, 1.6 A	6.3 V, 1.9 A
P_A max.	50 W	42 W	50 W
V_A & V_{G2} max.	750/550 V	800/600 V	800/450 V
PG2 max.	8 W	8 W	12 W
I_K max.	230 mA	230 mA	180 mA
r_I	25 kΩ	12 kΩ	25 kΩ
gm	8.8 mA/V	11.5 mA/V	10 mA/V
μ / rI as a triode	9/650Ω	8/670Ω	13.6/640Ω

$\eta = P_{OUT}/P_{IN} = 6.6/36 = 18.4\%$

$D_2 = (230\text{-}220)/2*450 = 1.1\%$

The distortion is much lower, but the price to pay is very low output power, half of the KT90 case.

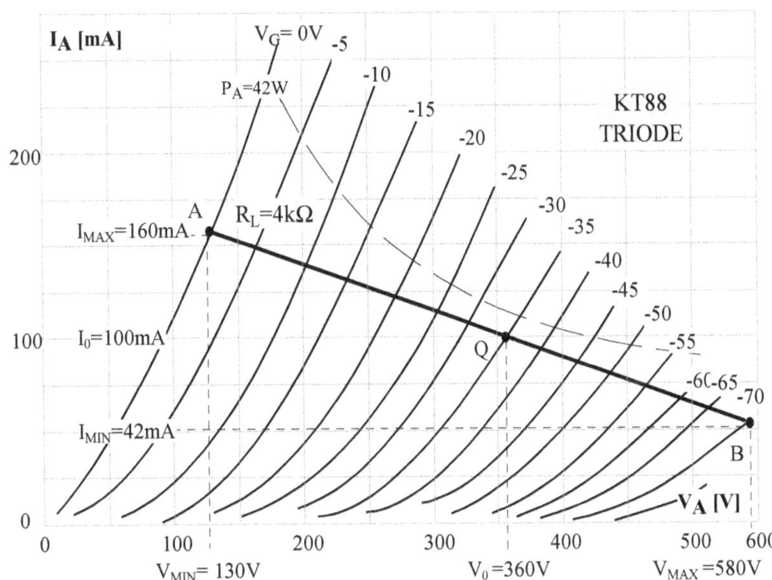

I will not show you designs with these two tubes, you can substitute the output tubes in other designs from this section with any tube from the KT88-KT150 family, whatever lights your fire.

The bias is -35V, so assuming 1.5-2V_{RMS} signal at the input, many preamp tubes in common cathode arrangement will amplify enough to get the maximum allowed 35V peak drive signal (which is only 24.8 Volts effective or RMS value). This means a single preamp triode stage is sufficient, providing it has a decent current capability as well.

German power tubes: EL12N, E235L, EL152/153

I never truly enjoyed the sound of EL34 and KT88/90/100 family. We've heard many Chinese and Western amplifiers with KT88, and made a few of our own, and I cannot recall even one that I really loved. I simply don't enjoy their sonic signature, either in push-pull or single-ended configuration.

I must admit, I do have a weak spot for less common European tubes, such as F2a, F2a11, EL12, E130L, LS50, E235L, RS1003, EL153 and EL152. They sound open, detailed, refined, but never sterile.

Three lesser known but superb-sounding German power tubes (L-R): E235L, EL12N (East German manufacture) and EL153 (Telefunken). On the far left is EL84 power pentode, included here for size comparison purposes.

Relatively cheap (except F2a and F2a11, which are priced at 300B levels), reliable, long lasting, these tubes have a great bass, transparent and detailed treble, superior soundstaging and well balanced sonic presentation. Most importantly, they convey emotion.

SIDE-BY-SIDE	E235L	EL12N	EL153
Heater	6.3 V, 1.2 A	6.3 V, 1.2 A	6.3 V, 1.55 A
P_A max. [W]	20	18	40
V_A max. [V]	400	425	650
P_{G2} max. [W]	2	2.8	5
V_{G2} max. [V]	300	425	300
I_K max. [mA]	220	90	230
r_I [kΩ]	5	30	50
gm [mA/V]	14	15	4
SE class A1	4W	10W@5% THD	18W@10% THD
PP class B	30W@6% THD		120W@10% THD
PP AB1 self bias		35W@5% THD	50W@5% THD

EL153 (beam tetrode) and EL152 may look identical, but EL152 is a pentode. Both are also available in 12.6V heater versions as well (FL153 and FL152). Pretty-looking and relatively small in size, they pack a lot of punch and sound very good when triode-connected.

To our knowledge, only Telefunken made those tubes in their Ulm factory in Germany, so the prestige is insured for you Telefunken aficionados.

Triode-connected E235L

Triode-connected, E235 has a decent amplification factor (μ=6.5) and very low internal impedance of 350Ω, meaning output transformers don't need a large primary inductance to get a good bass! However, the output power levels are low. Connecting two tubes in parallel is a viable option, the only issue being that the idle current would be around 200mA, raising to 350mA with signal peaks. Thus, SE output transformers wound with a large-diameter primary wire on a large magnetic core (A=20cm^2 or more) must be used, such as those designed for 6C33C-B.

Since it sounds so good in Class AB_1 push-pull and produces a decent power output (25W), we haven't tried it in a SE triode mode, but if you are contemplating building a 6080 SE amplifier, it is an option worth investigating. Since both tubes use the same octal socket, if you aren't happy with the 6080 sound (not bad, but not great either), you can always convert your amp and use a pair of these per channel. Obviously, leave some room for two more sockets.

6080 needs 6.3V 2.5A heating, E235L requires 6.3V, 1.2A (7.6W), so no issues there. 6080 (two triodes in parallel - one physical tube) has anode power rating of 26W, compared to E235L anode+screen power rating of 22W, so no big change there either. Notice E235L's superiority in the efficiency stakes, its heater efficiency is 22W/7.6W = 2.9, while 6080 is only 26W/15.8W = 1.65. This means for each watt of the heater consumption E235L will produce 2.9W of audio power, while 6080 will produce only 1.65W!

The anode is rated at 20W and the screen grid at 5.5W (very high!), but together they are derated in Telefunken data sheet to 22 Watts, probably due to limited heat dissipation ability of such a physically small tube!

Notice that the anode characteristics for triode connection, also by Telefunken, place the dissipation parabola at only 16 Watts. We added the 22W curve for comparison.

The estimations below are very rough. We only have curves up to 250V and grid bias only up to -35V, so the whole signal swing cannot be determined, the maximum voltage and minimum current levels must be estimated.

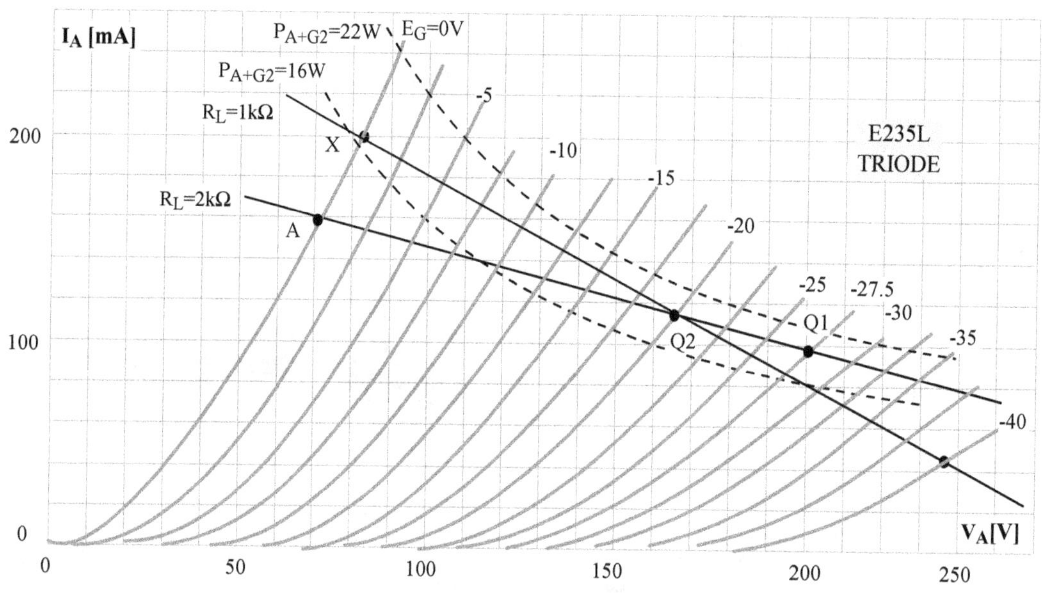

CLASS A_1 TRIODE:	CLASS A_1 TRIODE:
R_L=2k, Q1, V_{G0}=-27.5V:	R_L=1k, Q2, V_{G0}=-20V:
V_0=200V, V_{MIN}=70V, V_{MAX}=330V (est.), ΔV=260V	V_0=165V, V_{MIN}=85V, V_{MAX}=245V (est.), ΔV=160V
I_0=100mA, ΔI = 160 - 40 (est.) = 120 mA	I_0=115mA, ΔI = 200 - 45 = 155mA (est.)
$P_{IN}=I_AV_0$ =0.1*200= 20 W	$P_{IN}=I_AV_0$ =0.115*165= 19W
P_{OUT}=ΔVΔI/8=260*0.12/8=3.9 W	P_{OUT}=ΔVΔI/8=160*0.155/8=3.1W

The power output levels for loads between 1 and 2 kΩ are in the 3-4W class, the 2A3 territory. The power sensitivity of triode-strapped E235L is double that of 2A3, so the grid driving signals can be half that needed for 2A3, making it much easier to design the preamp/driver stage (only one stage is required).

Heater precautions are unnecessary, AC heating is fine, since E235L is indirectly heated and much quieter than filament type tubes such as 2A3. Heater current is half that of 2A3, meaning the radiated magnetic field is also weaker. Finally, E235L is at least 10 times cheaper than 2A3 and will outlast it many, many times. Tube rolling is possible, there are Valvo, Telefunken, Siemens and RTC branded tubes. So, in all aspects except sonics, E235L as a pseudo-triode is superior to 2A3. As for the most important aspect, the sonics, I leave that judgment to you.

F2a and F2a11 BEAM TETRODES

F2a11 and sonically slightly inferior F2a are special quality power beam tetrodes. The p-factor is 0.15%, meaning that 0.15 tubes out of every 100 will fail within each 1,000 operating hours period, or 15 tubes per 10,000! F2a has lower voltage ratings and doesn't sounds as good as F2a11. It also uses a different socket.

While being far from rare, the ebay sellers keep prices high despite large stocks of NOS F2a Siemens tubes. The prices rose significantly in the last few years due to their use in Shindo amplifiers.

While being a well constructed tube with a pleasing sonic signature, there is nothing that these tubes can do that EL12, EL153 or E130L cannot do equally well (perhaps even better!) and much, much cheaper.

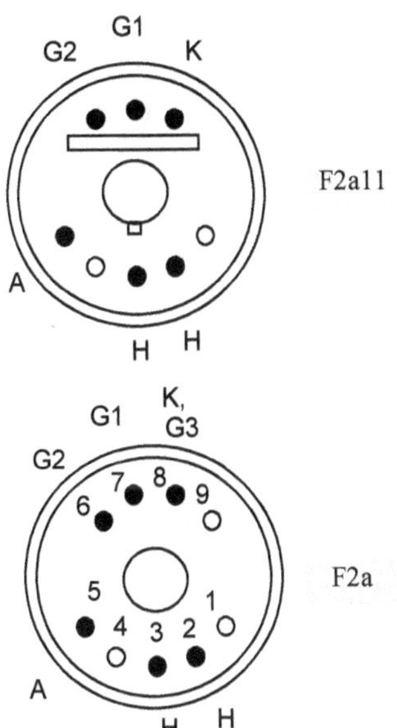

TUBE PROFILE: F2a & F2a11

- Indirectly-heated beam tetrode, 6.3V, 2.0A heater
- Max. anode/screen voltage 425V/425V (600V/425V for F2a11)
- Max. H-K voltage: 80V (120V for F2a11)
- Max. anode / G2 dissipation: 30W / 5W, max. cathode current: 140 mA
- STATIC DATA (PENTODE):
- gm=18 mA/V, r_I=23kΩ, μ=414, μ_{G1G2}=17.5, C_{GK}=18pF, C_{AK}=128pF
- STATIC DATA (TRIODE): gm=21 mA/V, r_I=800Ω, μ=17

Triode operation according to Siemens data sheet

V_G= -13.16V, V_0=330V

I_0 =90mA, R_L=1.5kΩ

V_{MIN}=180V, V_{MAX}=470V, ΔV=290V

ΔI=190-18 = 172 mA

P_{IN}= I_0V_0 = 0.09*330 = 29.7W

P_{OUT}= ΔVΔI/8 = 290*0.172/8 = 6.2W

The Siemens data sheet claims 5.5W output at 10% distortion. This load line seems totally wrong. Firstly, the 1k5 load is way too low, 3-4k would be much more appropriate. Secondly, half of the load line is above the maximum power dissipation. Thirdly, the operation enters the low anode current range where the characteristics are very nonlinear and the distortion is horrendous!

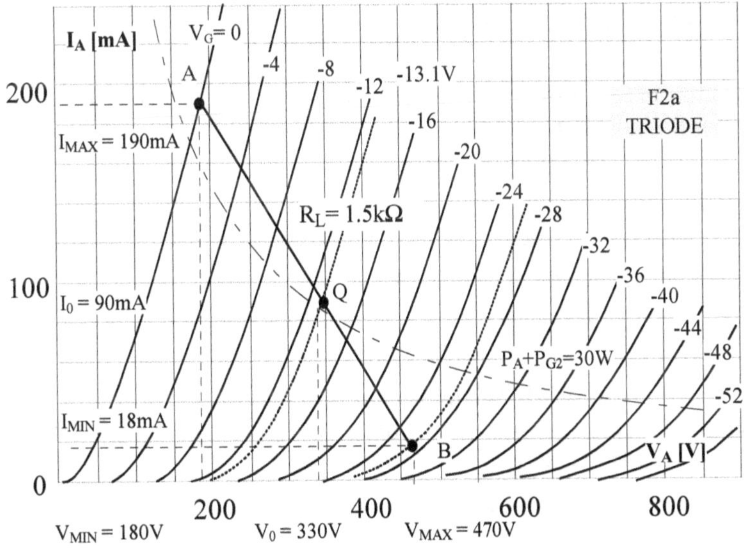

Pentode operation according to Siemens data sheet

V_G= -6.7V, V_0=250V

I_0=97mA, I_{S0}=14mA, R_L=2.2kΩ

V_{MIN}=50V, V_{MAX}=450V, ΔV=400V

ΔI = 190-10 = 180 mA

P_{IN} = I_0V_0 = 0.097*250 = 24.2W

P_{OUT} = ΔVΔI = 9W

Not quite the claimed 10W, but close enough!

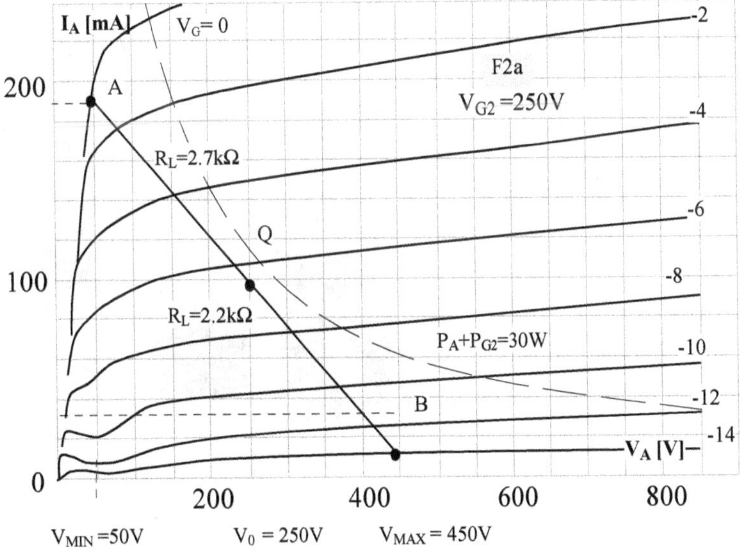

RS1003 SET AMPLIFIER

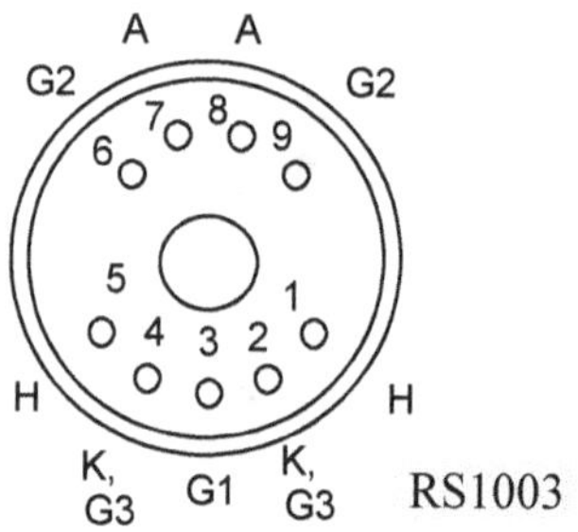

TUBE PROFILE: RS1003 (SRS551)

- Indirectly-heated transmitting pentode, Heater: 6.3V, 2.1A
- V_{AMAX}=1,000V, V_{SMAX}=600V, max. V_{HKMAX}200V_{DC}
- Max. anode / G2 dissipation: 60W / 10W, I_{KMAX}=260 mA
- STATIC DATA (PENTODE):
- $V_A=V_S$=400V, V_{G1}=-12V, I_0=100 mA
- I_{G2}=10mA, gm=18mA/V, μ_{G1G2}=20

Since RS1003 and its East German version SRS551 were not meant to be used in audio, there are only basic static figures given for pentode operation (gm=18mA/V and μ_{G1G2}=20) and no data for triode connection at all. However, we now know that the μ_{G1G2} approximately equals μ in a triode connection. We also know that pentode's and triode's transconductance are almost identical, so in the first approximation we can assume triode gm=18mA/V! Thus, the estimated internal impedance in triode mode is $r_I = \mu$/gm = 20/18 = 1.1kΩ! Thus, according to another rule-of-thumb prescribing 3-5 times higher load, output transformers with primary impedance of 3.5-5kΩ seem appropriate.

<u>CLASS A_1 TRIODE (Q1):</u>

V_G= -14V, V_0= 400V, I_0=70mA

R_L=5kΩ, V_{MIN}=150V, V_{MAX}=615V

ΔV=465V, ΔI=125-30 = 95mA

P_{OUT}=ΔVΔI/8=465*0.095/8 = 5.5W

Such a low output from a 60W tube seems an exercise in futility, so let's push the tube harder (Q2):

V_G=-152V, R_L=5kΩ, V_0=450V

I_0=105mA

V_{MIN}=170V, V_{MAX}=700V,

ΔV=530V, ΔI=160-50 =110mA

P_{OUT}=ΔVΔI/8= 7.3W

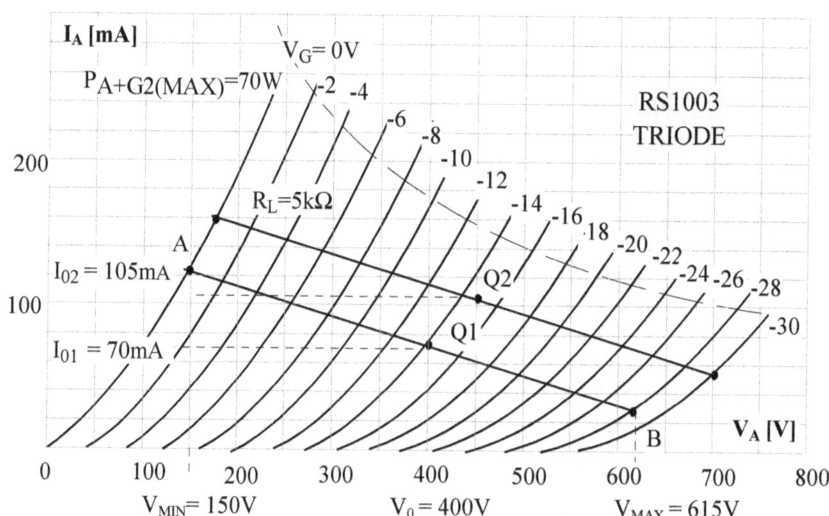

RS1003 is physically smaller than KT88 or EL56 power tubes, not just in terms of the glass bulb, but its anode is smaller as well. However, its combined anode+screen power rating is 70 Watts, higher than the other two (50 Watts and 62Watts respectively).

The amplification factor of the first stage with ECC82 triode (next page) is $A = -\mu R_L/(r_I+R_L = -17*52/(7.7+52) = -14.8$

The big guns: (L-R): 6L6GC (RCA, USA), EL34 (Mullard, UK), KT88 (EH-Russia). EL156 (Shuguang), RS1003 (Siemens, Germany)

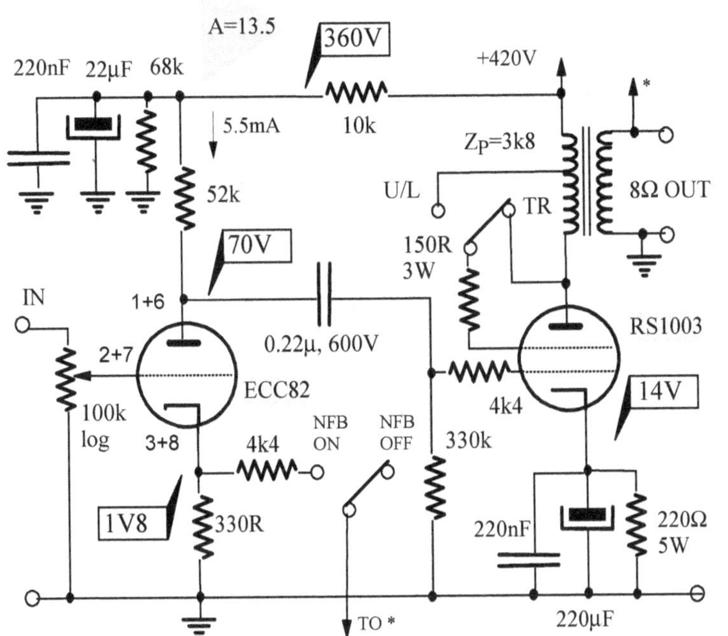

The load R_L=52k, r_I=7.7kΩ and µ=17

With two triodes in parallel and a bypassed cathode resistor, R_L=52k, r_I=3.8kΩ and µ=17, so A = -17*52/(3.8+52)= -17*0.93 = -15.85

The µ stayed the same, but the amplification of the stage is slightly higher due to less signal loss on the smaller internal resistance of the two paralleled triodes!

With both paralleled triodes and an un-bypassed R_K we have

$A=-\mu R_L/[r_I+R_L+(1+\mu)R_K]=$
$-17*52/[3.8+52+(18)0.33] = -14.3$

The measured result was A=13.5 The drop in amplification factor (from 15.85 to 14.3) was small because the R_K was also small, only 330Ω, resulting in very mild NFB.

The NFB ON-OFF and TRIODE-U/L switches give the listener four possible combinations. My favorite is U/L mode with a mild negative feedback, although in most other cases a triode with a mild negative feedback sounds better.

This tube is another example of discrepancy between graphs' prediction of output power (5-7 Watts) and reality (up to 18 Watts)!

MEASURED RESULTS:

- BW: 19Hz-27kHz (-3dB, 10W into 8Ω)
- $V_{OUTMAX}=12V_{RMS}$
- P_{MAX} = 18W

SRPP version using 6CG7 or 12BH7 duo-triodes

TUBE PROFILE: 12BH7

- Indirectly-heated Noval duo-triode
- Heater: 6.3V/600 mA or 12.6V/300mA
- V_{HKMAX}: $100V_{DC}$
- P_{AMAX}: 3.5W each, 5W both
- Max. anode voltage & current: 300V/20 mA
- TYPICAL OPERATION:
- V_A=250V, V_G=-10.5V, I_0=11.5mA
- gm = 3.1 mA/V, µ= 16.5, r_I=5.3kΩ

TUBE PROFILE: 6CG7

- Indirectly-heated Noval duo-triode
- Heater: 6.3V/600 mA
- V_{HKMAX}: $100V_{DC}$
- P_{AMAX}: 3.5 W each, 5W both
- Max. anode voltage & current: 300V/20 mA
- TYPICAL OPERATION:
- V_A=250V, V_G=-8V, I_0=9 mA
- gm = 2.6 mA/V, µ= 20 r_I=7.7kΩ

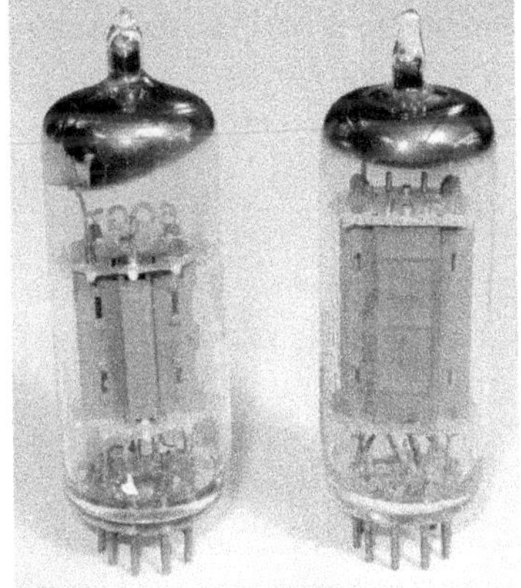

6CG7 is electrically identical to 6SN7-GTB, but it is a more modern tube with a Noval socket.

12BH7 has an advantage that the heater can be configured for both 6.3 and 12.6V, while 6CG7 only works on 6.3V. The amplification factor of 6CG7 is slightly higher, but the internal resistance of 12BH7 is lower.

The inter-electrode capacitances of 6CG7 are higher than those of 12BH7. Grid-to-anode capacitance of 6CG7 is 4pF, compared to 2.6pF for 12BH7.

ABOVE LEFT: 6CG7 printed Penta Labs USA, but of unknown manufacture and Australian-made 6CG7 by Radiotron. Notice a much larger anode of the Australian tube. The Penta tube does not seem to have the power dissipation of 6CG7 at all, looks more like 12AU7!

ABOVE RIGHT: The 12BH7 tube with larger and darker anode is by Radiotron Australia, the one with smaller anode is by Matsushita Japan.

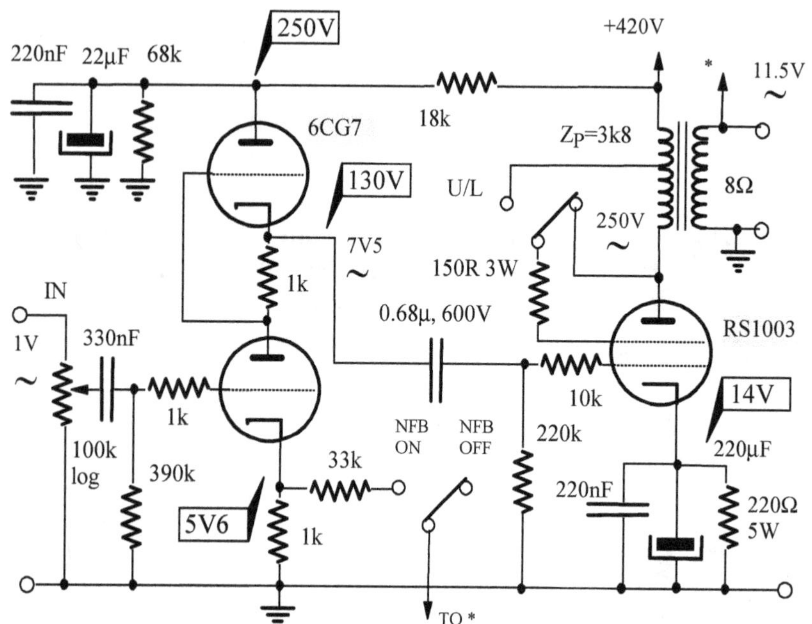

The 1st stage of the SRPP version has A=7.5, so almost 2V at the input are needed for full power. 6CG7, 6N6P and 6BQ7A can be plugged-in without any changes.

6CG7 sounded the softest while 6N6P sounded detailed and crisp but not as emotional or musical as 6CG7.

Since it drives the high input impedance of the Class A_1 output tube's grid, the SRPP stage here does not work in a true SRPP fashion, there is no output current (apart from small current charging the input capacitance). The upper triode is simply an active load for the bottom triode, and not even a constant current source as many claim online.

The power supply is simple. Individual LC filters are used for each channel, resulting in lower crosstalk, better imaging and superior overall performance. Notice a hybrid rectification scheme, with two solid state diodes and one dual vacuum tube rectifier, 5V4G, in a bridge configuration.

The heaters of both tubes in each channel are powered from the same AC secondary winding.

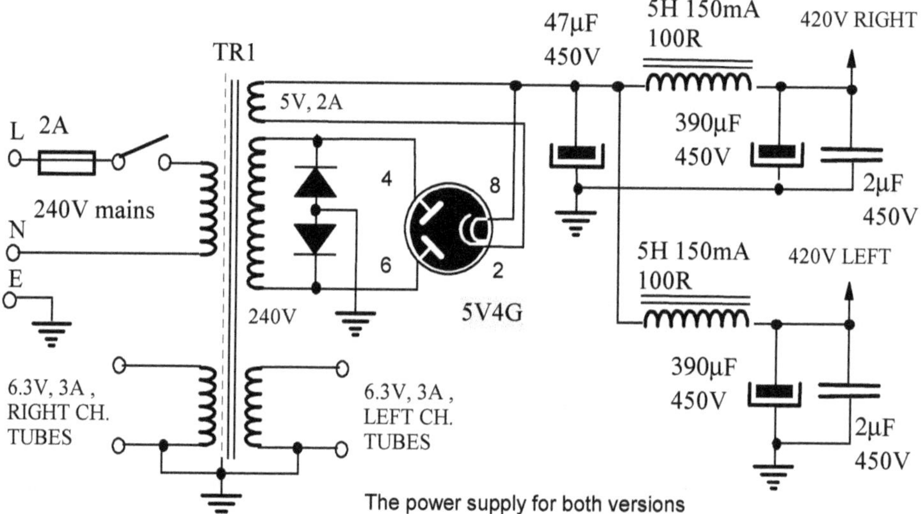

The power supply for both versions

HORIZONTAL AND VERTICAL TV DEFLECTION TUBES IN AUDIO SERVICE

PL509 was released in 1969, and PL519 followed a year later. Both tubes use a B9D or "Magnoval" socket with 9 pins. Both were designed to serve as sweep tubes, since they "sweep" the electron beam in large color TV screens horizontally. For that reason they are also known as horizontal deflection tubes. Radio amateurs also used similar American-designed and made sweep tubes in short-wave transmitters. PL519 is rated at 35W, while the anode dissipation of PL509 is slightly lower, 30Watts.

In PAL-standard countries the tubes worked on a line frequency of 15,625 Hz (625 lines × 50 Hz ÷ 2), which is an audio frequency. Likewise, PL508, their smaller brother, was a vertical deflection driver in European TV sets, it operated on the mains frequency of 50 Hz, which is obviously also an audio frequency.

Both tubes perform extremely well in audio applications. The PL519 has an obvious advantage of three times higher anode dissipation, but it uses an unfortunate top cap. PL508 is much cheaper and in less demand (since no commercial audio amplifier uses that tube, apart from a few of our models), but in my opinion it sounds even better than PL519. While PL519 as a triode sounds extremely open, detailed and transparent, PL508 has a better bass, and a better balanced tonal balance, with emotionally more engaging sound.

Everything we say about PLXXX tubes applies equally to their ELXXX counterparts. The EL versions are generally more in demand and more expensive. DIY builders don't seem to like dealing with nonstandard heater voltages, probably because most of them simply buy available mains transformers off the shelf (instead of winding their own) and very few of those have a 40V or 17V winding needed for PL519 and PL508 respectively.

Tbbes with higher heater voltages pull less heater current, which in turn reduces the magnetic field around heater wiring and hum pickup, so there is a real and audible benefit of higher heater voltages! There are surplus power transformers designed for solid state equipment with voltages suitable for the heaters of these TV tubes (2x9V = 18V for 17V heaters, 2x20 or 2x24V for 40V heaters).

The Russian "equivalent", 6P45S (6П45C in Cyrillic) and usually labeled EL509 (as in the photo below) is an even beefier tube, not just by its dimensions (see how much wider it is than PL519) but its glass is thicker and therefore heavier. Its maximum plate dissipation is 50W!

There is also the oddball EL509 by JJ (Slovakia), which does not have a top cap and uses octal socket. However, there is only one manufacturer of these tubes in the world so tube rolling is not an option should you choose to go that way, which totally negates the benefit of the octal socket, so the only benefit you get is the absence of the top cap. We tried a quad of JJ EL509 and were not impressed by the results so decided against their further use.

Many other sweep tubes of various power ratings are available as NOS, most notable being EL36, PL500, PL504, 6BQ6 and 6AV5.

SIDE-BY-SIDE	PL500 - EL500 Line output pentode	PL504 Line output pentode	PL508 - EL508 Frame output pentode	PL519 - EL519 Line output pentode
Heater	27V/300mA - 6.3V/1.4A	27V 300mA	17V/ 300mA - 6.3V/ 0.825A	40V/300mA - 6.3V/2A
P_A max.	12W	16W	12W	35W
V_A max.	250 V	250V	400 V	700V
P_{G2} max.	5 W	6W	4 W	9 W
V_{G2} max.	250 V	250V	275 V	275V
I_K max.	250 mA	250 mA	100 mA	500 mA
V_{HK} max.	220V	250V	220V	250V
gm	11.5 mA/V		9 mA/V	18 mA/V
As a triode			1.1kΩ, 7mA/V, μ=7.5	280Ω, 12mA/V, μ=3.3
SE class A1			7W@ 5% THD	14W @ 5%THD

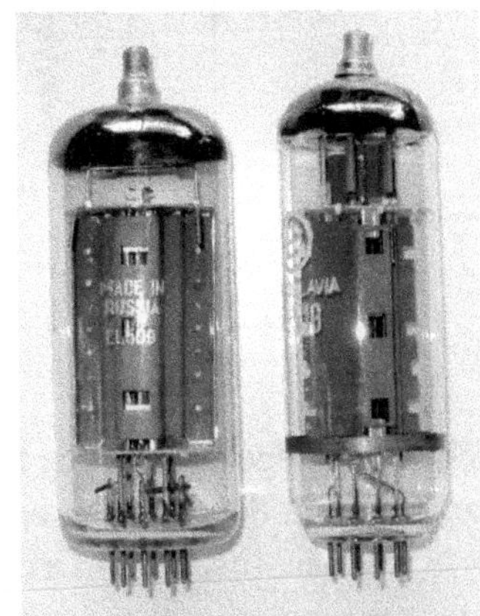

SINGLE-ENDED TRIODE AMPLIFIER WITH EL519/PL519 POWER TUBES

Any of the PL519 family of tubes (PL509, PL519, EL509, EL519 and the similar Russian tube 6P45S) can be used, depending on what secondary heater voltages you have available on your mains transformer.

Despite the graph's prediction of only around 8Watts of class A_1 power, the actual amplifiers produced almost double! Don't get discouraged by the estimated power output or distortion figures from these graphical methods. They are only a tentative starting point. Build the amplifier and ascertain reality.

The primary impedance can be anywhere from 1k5 to 3k, the tube is quite flexible in that regard. The higher the load impedance the lower the distortion, but also slightly lower maximum power. Connected as a triode, EL519 has a low amplification factor (μ=3), so it requires a large driving signal and two preamp amplification stages.

Triode parameters in the Q-point:

- $r_I = \Delta V_A/\Delta I_A = (230-200)/(0.21-0.1) = 272\ \Omega$
- $\mu = \Delta V_A/\Delta V_G = 30/10 = 3$
- gm=3/272=11 mA/V

<u>CLASS A_1 TRIODE:</u>

V_{MIN}=45V, V_{MAX}=470 V

ΔV=425V, ΔI =170-20=150mA

R_L=2k75

$P_{IN}=I_0V_0=0.085*290=25W$

$P_{OUT}=\Delta V\Delta I/8=425*0.15/8=8.0W$

$\eta = P_{OUT}/P_{IN} = 8/25 = 32\%$

$D_2=(\Delta V_{OUT+}-\Delta V_{OUT-})/2\Delta V_{OUT}$

$D_2=(245-180)/2*425 = 7.6\ \%$

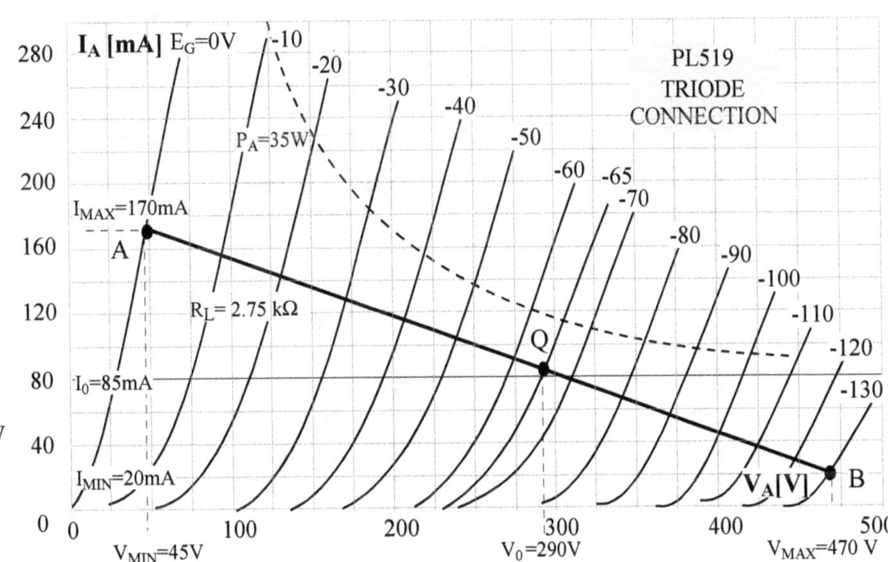

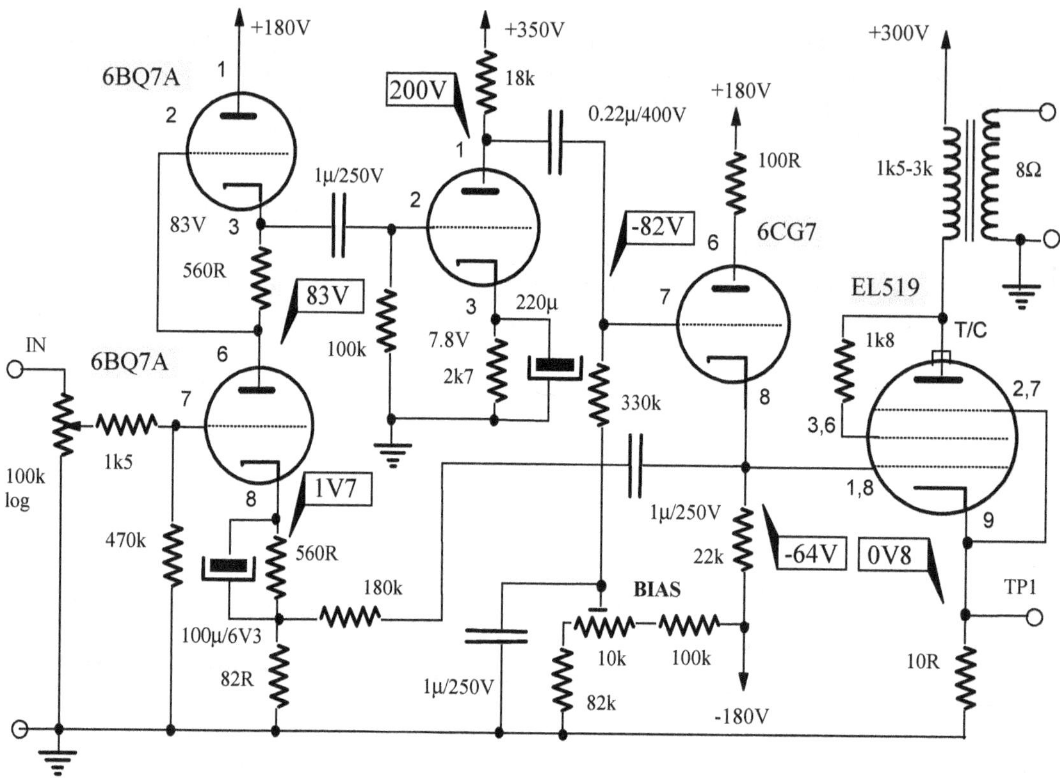

The third stage is a cathode follower, directly coupled to the output stage, enabling operation in Class A_2. The output tube idles at 80mA (0.8V on a 10Ω cathode resistor in test point TP1), with maximum signal the current jumps to 175mA. The first stage is a common cathode triode stage with an active anode load $R_L=r_{I2}+(\mu_2+1)R_2$! For the 6BQ7 stage this anode resistance is $R_L=5.8 + (35+1)0.56 = 26k\Omega$.

Power supply

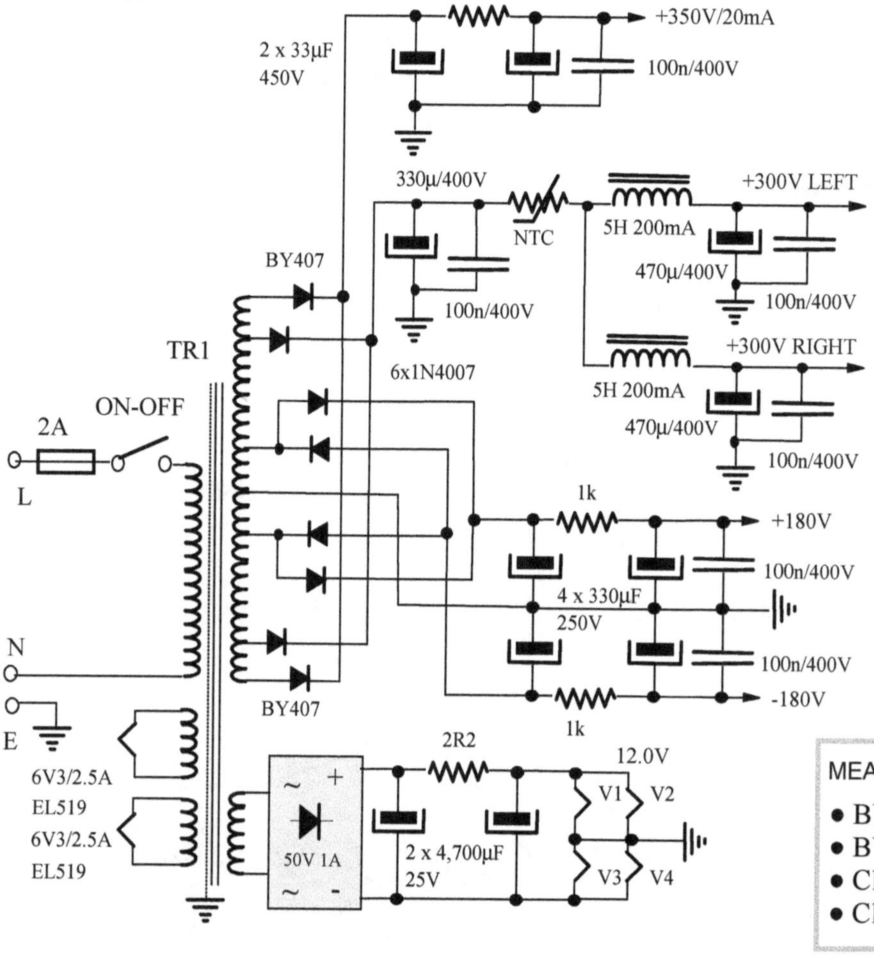

The calculated amplification factor is $A=\mu R_L/(r_I+R_L)= -35*26/(26+5.8) = 28.6$, which is slightly reduced to A=25 due to the fact that one of the two cathode resistors is un-bypassed (82R) so it can take the negative feedback signal.

The second stage amplifies around 14 times. There is a very mild NFB from the output of the cathode follower back to the first stage, but there is no global feedback from the output transformer's secondary.

MEASURED RESULTS:

- BW: 5Hz-57 kHz (-3dB@1W)
- BW: 10Hz-49 kHz (-3dB@ 10W)
- Class A1: P_{MAX} = 15W
- Class A2: P_{MAX} = 28W

GIANT KILLER: EL153 SET AMPLIFIER

This is if not the best, certainly one of the three best sounding SET amplifiers we have ever made. Smooth and seductive midrange, refined microdynamics, transparent and detailed treble, and a 3D soundstage that will leave most 300B, F2a and similar amps in the dust! 211 and similar transmitting triodes cannot match its tenderness and subtlety.

If you want to put to shame many of your friends' "famous brand" amplifiers, build this baby. For even better sonics, instead of SS diodes use mercury vapor rectifiers. In that case you don't need the 5V4G rectifier in series with the DC output.

A single driver stage is all that is needed due to EL153's high sensitivity, especially if you use it in pentode connection, so use any preamp tube of your liking.

The 5V4G rectifier tube does not work as such here, it is simply a series-connected diode (actually two diodes inparallel) that decouples the solid state rectifier from the rest of the power supply and the audio circuit, since it conducts only away from the mains transformer, so it cannot feed anything back the other way. Also, since it is indirectly heated, just as the audio tubes are in this design, it provides a soft-start feature. The high DC voltage at voltage doubler's output passes through to the LC-filters and appears on the audio tubes' anodes after the 5V4G rectifier and the audio tubes have warmed up.

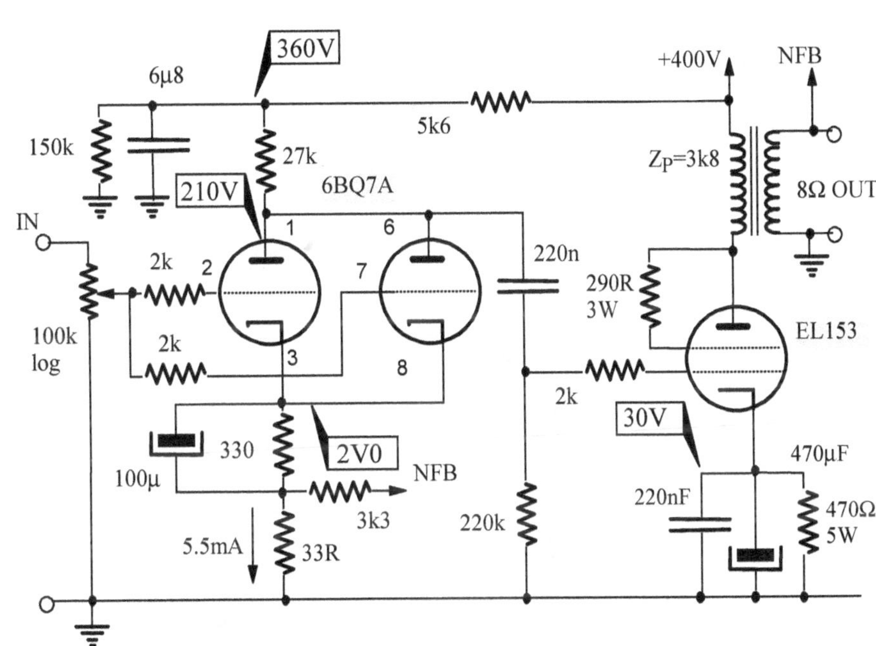

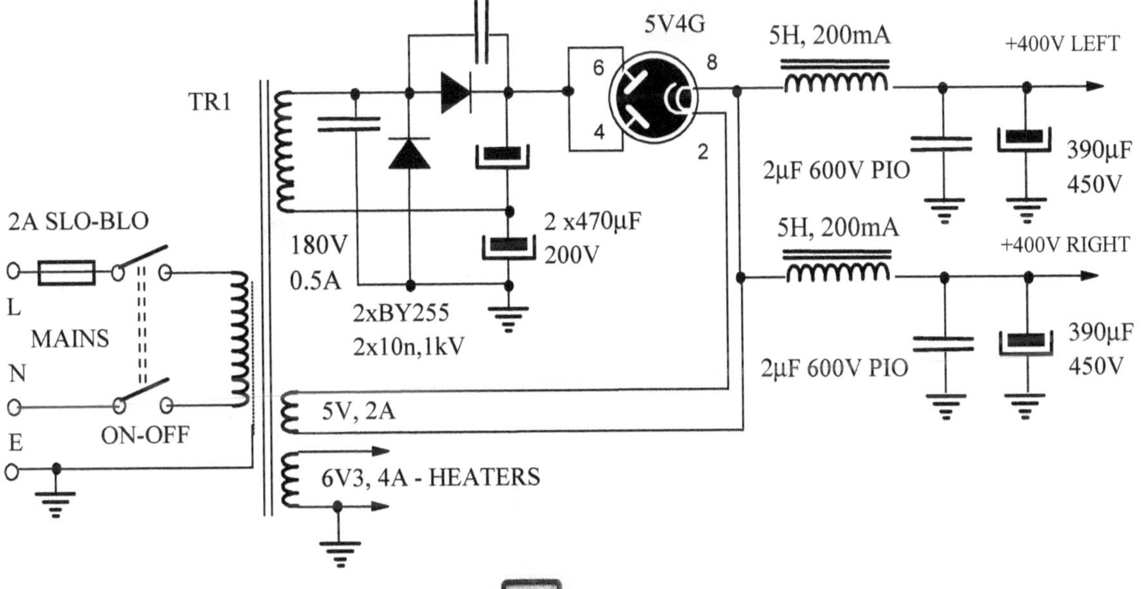

MEASURED RESULTS:

- BW: 12Hz - 69 kHz (-3dB, 1W into 8Ω)
- BW: 17Hz - 44 kHz (-3dB, 10W into 8Ω)
- TRIODE: V_{OUTMAX}=9V_{RMS}, P_{MAX}=10W

CATHODE-LOADED 6V6 TRIODE AMPLIFIER

If cathode followers are so good in linearizing preamplifying stages and as drivers of output tubes, why don't we use the output stage in a CF mode, i.e. connect the load between the output tubes' cathodes instead of the anodes?

This kind of output stage should yield many benefits. Firstly, the output impedance of the cathode drive is very low. Secondly, there is 100% NFB, meaning distortion in the output stage should be practically eliminated. Thirdly, the frequency range should be widened and the impedance peaks in both the output transformer and the load (speaker) should be dampened (flattened). Also, cathode followers are immune to the interference and hum (ripple) on top of the DC power supply, which should also contribute to the quietness of this topology.

6V6 seems the best choice for this circuit. The graph below is for the anode load, the graph for cathode load is identical, except the V_G bias values, different (much higher) grid voltages apply, meaning the output stage needs a much larger grid drive, identical to anode voltage AC swings in the anode-loaded stage!

6K6 and 6F6 pentodes can be substituted for 6V6 for slightly different sonics and slightly lower maximum power. Notice that 6F6 requires higher heater current, so if your heater supply is operating close to its maximum current capability with 6V6, do not insert 6F6!

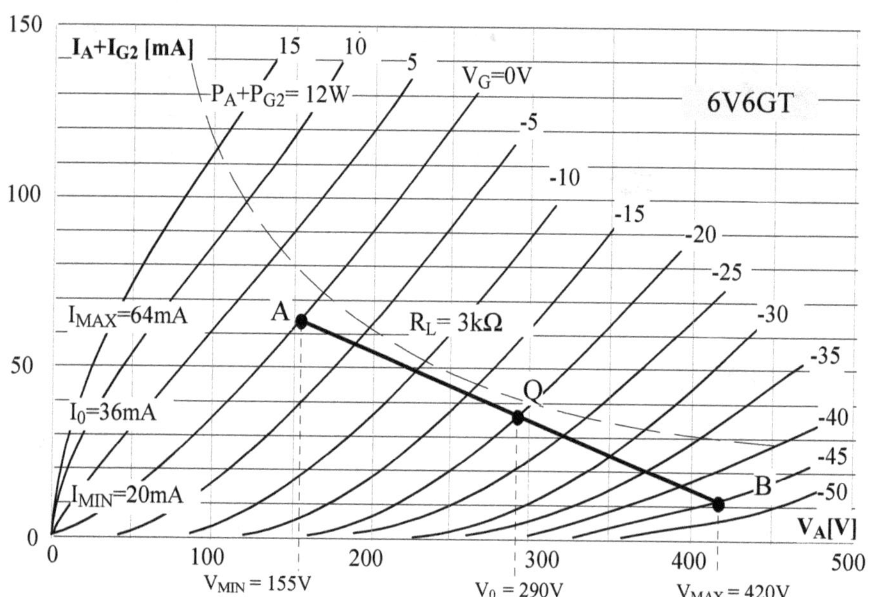

To analyze the performance of cathode-loaded output stages, cathode-follower anode characteristics should be used. However, to draw them requires hours of measurements, so standard anode characteristics can be used for quick estimation purposes.

CLASS A_1 TRIODE:
V_G=-20V V_A=290V
V_{MIN}=155V, V_{MAX}=420V
ΔV = 265V, ΔI=64-20=44 mA
$R_L=\Delta V/\Delta I$ = 270/0.18 = 3kΩ
$P_{IN}=I_0V_0$= 0.036*290 = 10.5W
$P_{OUT} = \Delta V\Delta I/8$ = 1.45W

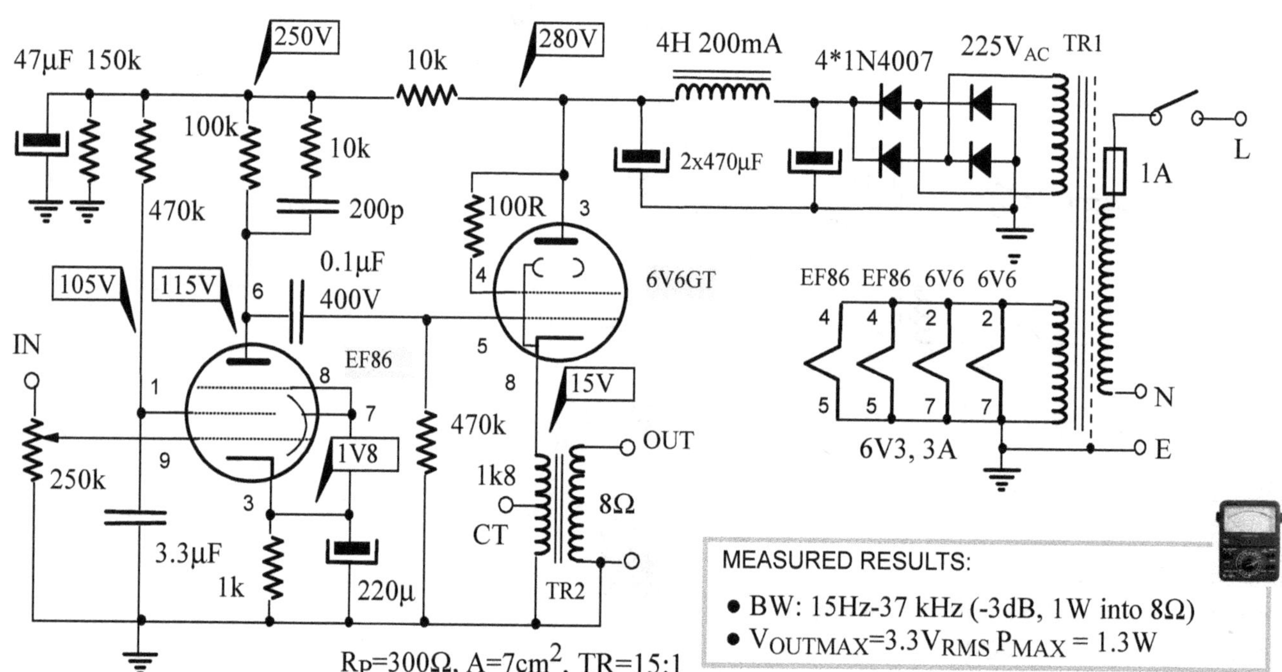

The first stage with EF86 pentode has an amplification factor of around 90. The output transformers were smallish (A=7cm^2) push-pull units salvaged from a vintage amplifier, so no air gap, but they did not saturate at all in this application. Respectable measured results, although the lower -3dB frequency could be improved by using output transformers with a larger cross-section and an air-gap.

SIDE-BY-SIDE	6V6	6K6	6F6
Heater	6.3V/0.45A	6.3V/0.4A	6.3V/ 0.7A
P_A / P_{G2} max.	12W / 2W	8.5W / 2.8W	11W / 3.75W
V_A / V_{G2} max.	315 / 285 V	315 / 285 V	375 / 285 V
SE class A1 pentode P_{OUT}	5.5W	4.5W	4.8W
V_A, V_S, V_G	315V, 225V, -13V	315V, 250V, -21V	285V, 285V, -20V
Load R_L	8.5kΩ	9kΩ	7kΩ
I_{A0}, I_{S0}	34mA, 2.2mA	26mA, 4mA	38mA, 7mA
Parameters as a triode	gm= 4mA/V, r_I=2400Ω, μ=9.6	gm= 2.7mA/V, r_I=2,500Ω, μ=6.8	gm= 2.6mA/V, r_I=2600Ω, μ=6.8
V_A max. triode	300V	315V	350V
$P_A + P_{G2}$ max. total triode	12.5W	7W	10W
V_A, V_G, I_0	250V, -20V, 39mA	250V, -18V, 38mA	350V, -38V, 48mA
Load R_L	4.8kΩ	6kΩ	6kΩ

Before finalizing the design with output transformers, we tested the output stage using a resistive load, in this case 390Ω. With 30V_{AC} signal at the cathode, we reduced the value of the 1 kΩ test potentiometer until the voltage measured with a high input impedance electronic voltmeter dropped to half this value, or 15V_{AC}. The pot was then disconnected and its resistance between points A and B was measured at 180Ω, meaning that the output impedance of the stage at the test frequency of 1kHz was 180Ω!

Interestingly, the -3dB frequency range of the resistively-loaded stage was 3Hz-64kHz, much better than when transformers were used, so the claim that the quality of output transformers does not matter at all in this topology is obviously not true!

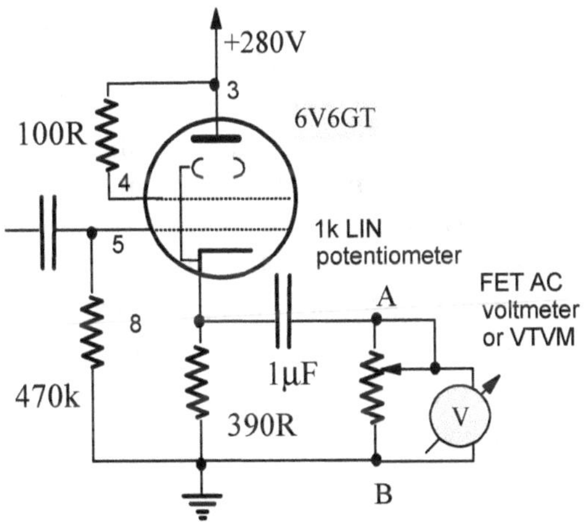

Listening impressions

The listening evaluations were positive. The 1.3 Watts maximum power sounded very loud indeed and no clipping could be heard. Even on an oscilloscope there was no harsh clipping, only a smooth rounding and "fattening" of the sine wave. The sound was very clean, transparent and detailed, but not flat or sterile.

Initially, it seemed that the bass was a bit understated, but after prolonged listening the conclusion was that the bass was all there, clean and tight, without that "trademark" artificial boominess in the mid-to-upper bass rang of lesser tube amps, and thus seemed less pronounced. This was a situation similar to hearing horn speakers for the first time, where audiophiles expect a loud and boomy bass, but get a total opposite, clean, controlled and composed bass.

Even more impressive was the treble, very detailed and refined, without a trace of harshness or sibilant distortion.

Overall, if you have very efficient speakers and like the sound of type 45 triode, you will love the sound of this amplifier. It is simple, easy to make and the parts are very cheap! An abundance of various NOS 6V6, 6F6 and 6K6 pentodes makes this baby a tube-roller's dream!

Parallel version with a mains transformer as an output transformer

Encouraged by the success of the single-tube design, we investigated two issues: how would a mains transformer perform as an output transformer with this design, and secondly, what would be gained by paralleling two or more output tubes.

Instead of 3.3V output voltage across 8Ω with a single tube, two tubes increased that to 4V, or, in terms of Watts, an increase from 1.3W to 2W of output power.

The output transformer was 240V to 24V mains transformer using EI66 laminations, again without any air gap. The measured primary inductance was L_P= 2.26H at 120Hz and 2.0H at 1kHz. As mentioned elsewhere, the smaller the difference in measured inductance values at these two frequencies, the more suitable the transformer laminations are for use in audio transformers!

Since the primary and secondary windings are not sectionalized at all, the measured leakage inductance was on a high side, L_L= 126mH @ 1kHz.

A typical SE output transformer has L_L of around 10-30mH, so not a bad result for this non-sectionalized transformer.

The frequency range of the amplifier was respectable 15Hz-22kHz, so it seems that mains transformers can be used successfully in this topology, saving you a fortune on a pair of "audiophile" output transformers. However, the -3dB frequency range of the first stage was quite wide, 1Hz to 85 kHz, meaning that the limiting factor was still the output stage, just as in the anode-loaded designs!

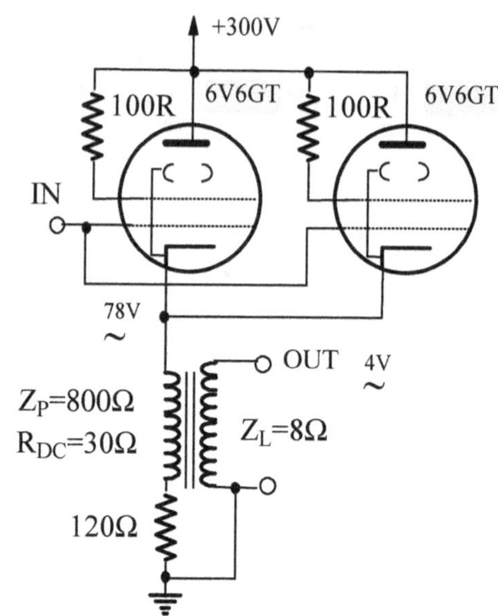

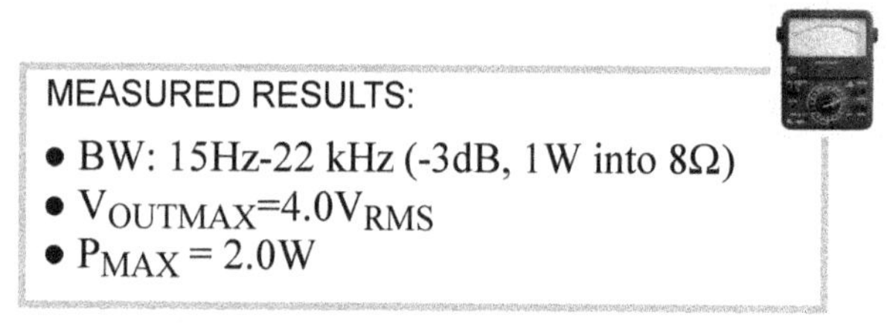

MEASURED RESULTS:

- BW: 15Hz-22 kHz (-3dB, 1W into 8Ω)
- V_{OUTMAX}=4.0V_{RMS}
- P_{MAX} = 2.0W

The final version

The previous experiment using a mains transformer as an output transformer did not yield a significant increase in the output power with two tubes in parallel, indicating that such a transformer may be the "bottleneck". We finally wound a pair of proper output transformers and tested the amplifier again. The bandwidth increased from 22kHz to 30kHz, so the prediction that the quality of output transformers should not matter in this topology was not correct.

However, the new transformer had the primary DC resistance of only 55 ohms so we ended up using a hybrid biasing scheme. The 100mA current through two tubes produces 5.5V voltage drop on the primary, while the grid is externally biased at -13V, for a total of -18.5V bias. The external bias voltage (-17V_{DC} maximum) was obtained by a voltage doubler powered by the 6.3V heater supply, a very neat trick!

The output transformer was made using 3% silicon steel (non GOSS) EI66 laminations and simple 3/2 sectionalizing: a=22 mm, stack thickness S= 4.8cm, so A= aS=2.2*4.8=10.5cm^2, R_P=55Ω, TR=600:140 = 4.3, IR=18.4 The damping factor was very low, around 1.5, meaning the output impedance of the amplifier was very high 5.3Ω! With 8Ω load Z_P=18.4*8 = 147Ω, L_P=1.6H, L_L=22mH, g = 0.2 mm (air gap)

PRIMARY: 3*200 turns = 600T, wire d = 0.27mm, SECONDARY: 2*70 turns = 140T, wire d = 0.5mm

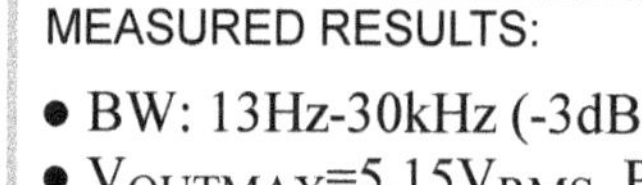

MEASURED RESULTS:

- BW: 13Hz-30kHz (-3dB, 1W into 8Ω)
- V_{OUTMAX}=5.15V_{RMS}, P_{MAX} = 3.3W
- DF=1.5

Listening impressions

The increase in the maximum output power was noticeable, the sound was faster and more dynamic, with more headroom and reserve power. It sounds strange saying that a 3.3 Watt amplifier can have "reserve" power but it does. Driving it directly from a CD player, we had to keep the volume control below the 11 o'clock level with 87 dB/W speakers in a 6x5m listening room.

The amp lost some of its former warmth, it sounded more neutral, which was probably due to the use of the predominately fixed bias scheme.

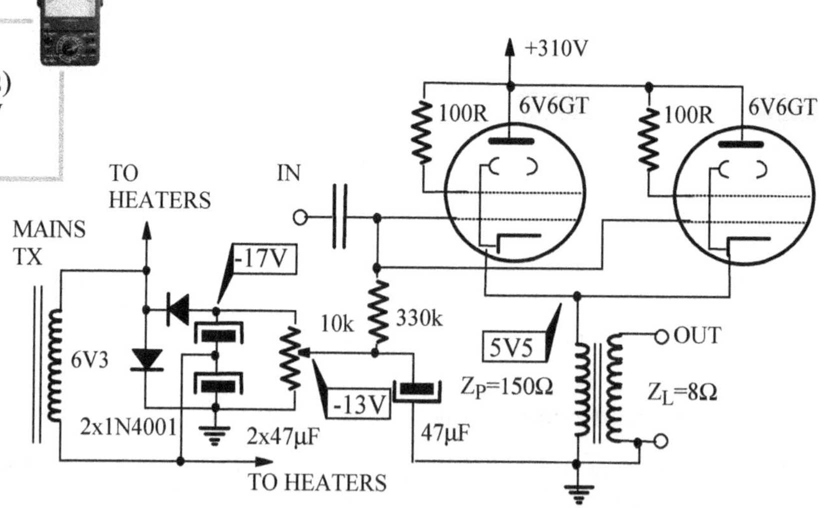

The output stage and biasing circuit of the final design. The input stage is the same as in the single-tube version.

KEY FEATURES:

- Cathode-loaded output stage
- No global negative feedback
- Mixed bias of the output stage (cathode + fixed bias)
- Only one coupling capacitor in the signal path
- Various power tubes can be used: 6L6, 6V6, 6K6, 6F6

SINGLE-ENDED PENTODE AND ULTRALINEAR OUTPUT STAGES

- SINGLE-ENDED OUTPUT STAGES WITH PENTODES AND BEAM POWER TUBES
- DESIGN OF EL84 SE PENTODE OUTPUT STAGE
- LOW-POWER OCTAL PENTODES AND BEAM TETRODES: 6V6, 6K6, 6F6, 6Y6
- DESIGN OF 6F6 SE PENTODE OUTPUT STAGE
- ULTRALINEAR CONNECTION
- LINEARIZING THE OUTPUT STAGE
- 6L6 SINGLE-ENDED STAGE: TRIODE, ULTRALINEAR AND PENTODE COMPARISON

"Tube technology enables a return to DIY as a creative act, and access to a totally different world of sound, of a totally subjective kind."

Nikola Vukušic, the late doyen of Serbian tube school, in his seminal book *Hi-Fi Lampaška Pojacala*

18

SINGLE-ENDED OUTPUT STAGES WITH PENTODES AND BEAM POWER TUBES

Three most common connections of multigrid output tubes

The existence of the screen grid leads to various ways of connecting it. In a pentode connection of the power tubes the screen is usually tied to the same high voltage point that supplies the output transformer. However, many pentodes have a much lower maximum screen voltage than anode voltage. In that case a lower source of DC voltage V_S is needed for the screen!

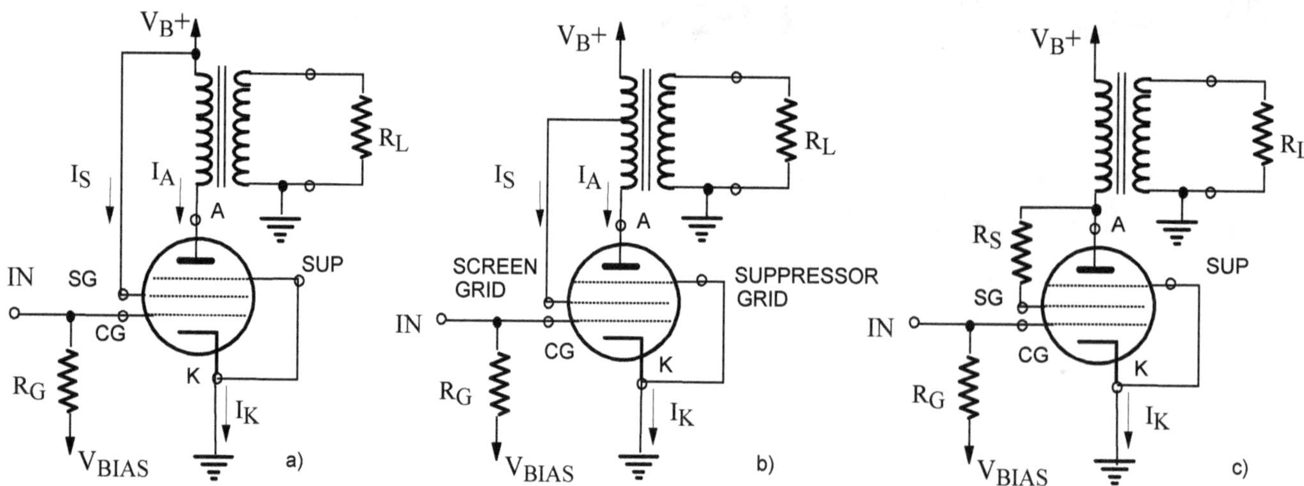

Pentode (a), ultralinear (b), and triode-strapped single-ended output stage (c). All are series-fed and fixed-biased.

For instance, E130L has the maximum anode voltage of 900V continuous and up to 2,000V in peaks (that is the reason for a top cap anode, so it is physically as far as possible from other pins), but its screen grid is rated at a maximum of only 250V.

Connecting the screen to the anode (usually through a low value resistor, to limit the screen current) turns a pentode or tetrode into a pseudo-triode. Again, if V_{SGMAX} is much smaller than V_{AMAX}, this will limit the anode voltage in triode mode to the lower value and reduce the available output power.

However, this not a rule that must be strictly adhered to. We have designed and built many amplifiers with triode-strapped pentodes that happily took screen voltages 50-100V higher than the maxima prescribed in their data sheets! Again, experiment for yourself.

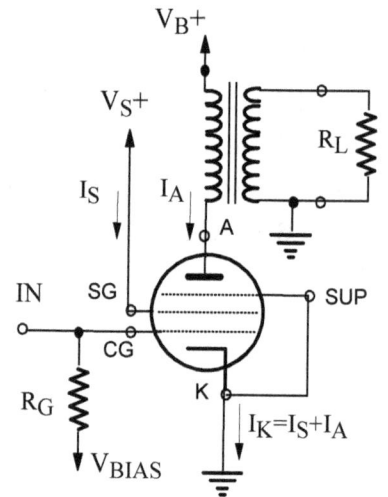

RIGHT: Many pentodes have a much lower screen voltage rating than anode voltage, necessitating a separate source of DC voltage V_S for the screen, well filtered and preferably regulated!

Optimal anode load and distortion

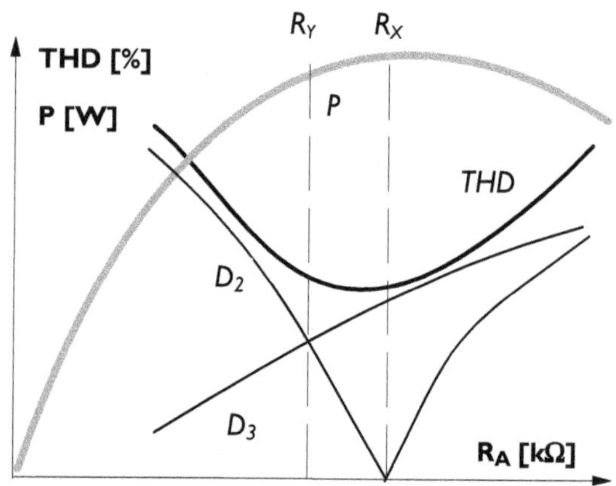

ABOVE: Typical pentode's power and distortion curves as a function of anode load. D_2 and D_3 are the 2nd and 3rd harmonics.

For common audio tubes such as EL84, 6L6 or EL34 there are numerous graphs and tables available, so guesswork and rules-of-thumb are not necessary. With tubes not originally designed for audio use, there is no data as for the optimal load impedance or parameters when triode connected.

One way to calculate the optimal load Z_A for pentode is to divide the DC plate voltage with the idle DC current. For instance, for EL84 with V_A=250V and I_0=48mA, the optimal primary load is $Z_{AOPT} \approx V_A/I_A = 250/0.048 = 5{,}208\ \Omega$, close to the commonly used value of 5 kΩ.

OPTIMAL ANODE LOAD & GRID RESISTANCE FOR SE PENTODES

$R_{AOPT} \approx r_I/5$

$R_G \approx 2r_I$

r_I = internal resistance of a pentode

Notice that in comparison with triodes, the second harmonic curve D_2 for pentodes dips to zero at a certain load impedance R_X, at which the output power is at its maximum so that point is declared the optimal load (maximum output power). Unfortunately, the discordant 3rd harmonic has no such minimum and keeps rising with increasing load impedance.

Using a lower load impedance, such as R_Y, reduces the harsh-sounding 3rd harmonic which is now also "masked" by the pleasant sounding 2nd harmonic. The output power is slightly lower but the sonic benefits are indisputable. Thus, paradoxically, designing pentode output stages for minimum 2nd harmonic distortion is not a good idea!

SCALING FACTORS

Say you want to design an amplification stage using a different anode voltage from the one specified in the tube's data sheet. Let's use a SE output stage with 6F6 pentode as an example. The data sheet specifies V_A=285V, V_G=-20V, I_A=38mA, I_S=7mA, r_I=78kΩ, R_L= 7kΩ, gm=2.55mA/V, P_{OUT}=4.8W

We want to use 250V on the anode and the screen. First, we calculate the anode voltage conversion factor K_1: K_1=250/285 = 0.893 The control grid bias will be V_G=-20*K_1= -20*0.893= -17.86V

To estimate anode and screen current we need K_2: $K_2=K_1^{3/2}$ In this case $K_2=0.893^{3/2}$ = 0.844, so I_A= 0.844*38=32mA and I_S=0.844*7 = 5.9mA

To get the new load resistance we need K_3 which is $K_3=1/\sqrt{K_1}$ = 1.06, so the load resistance should be R_L=7*1.06= 7.4kΩ! The internal resistance can also be estimated using K_3 as r_I=K_3*78k= 82.7kΩ

The transconductance in the new operating point can be estimated as gm=$2.55\sqrt{K_1}$=0.945*2.55=2.4 mA/V and output power is estimated by using factor $K_4=K_1K_2=K_1^{5/2}$ so $P_{OUT}=K_1^{5/2}$*4.8= 0.754*4.8=3.6W

Signal rectification

Output stages with pentodes and beam power tubes also suffer from the signal rectification phenomenon, just as triodes do. The DC component of the distorted AC (signal) current is added to quiescent level I_0 and decreases it to I_{AR}. The whole dynamic loadline LL_{DYN} shifts downward, due to the convex curvature of the anode characteristics. With triodes, due to the concave nature of their anode curves it usually shifts upward and the rectified component signal current increases the I_0!

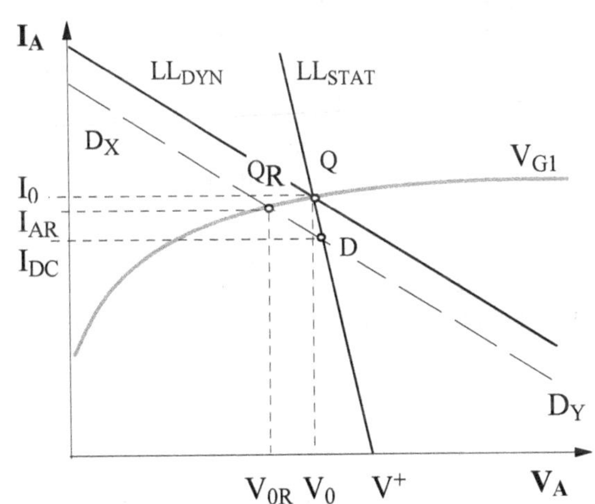

RIGHT: Pentodes and beam tubes also suffer from the distortion effect of large AC signal rectification. The quiescent point shifts from Q to Q_R and I_A drops from I_0 to I_{AR}.

DESIGN OF EL84 SE PENTODE STAGE

A European answer to the American 6V6 beam tetrode, EL84 (6BQ5), a miniature power pentode, was developed by Philips in the mid-1950s. The tube quickly become popular, not just amongst hi-fi manufacturers, even more so in guitar amps. Vox AC30 combo used four of them in a push-pull arrangement to produce 30 Watts of sweet juicy power.

EL84 also sounds very sweet in SE stages, and, just as many audiophiles prefer the tonal signature of 2A3 to that of 300B, many claim that EL84 sounds better than EL34, and I tend to agree. I have never heard an EL84 amplifier I didn't like!

The estimation formula for pentodes says we should get

$P_{OUT} \approx 0.35P_{IN}=0.35V_AI_A$=0.35*250*0.048=4.2W of single-ended power!

TUBE PROFILE: EL84 (6BQ5)

- Indirectly-heated power pentode
- Heater: 6.3V, 0.75 A
- P_{AMAX}=12W, P_{SMAX}=2W
- TYPICAL SE PENTODE OPERATION:
- $V_A=V_S$= 250V, V_G= -7.3V
- Signal (grid voltage): 6.1V_P, 4.3 V_{RMS}
- I_{A0} = 48 mA, I_{S0} = 5.5 mA
- Load resistance: 4.5 kΩ
- gm=11.3mA/V, μ=430, r_I = 38kΩ
- gm_{G2}=1.8 mA/V, μ_{G2G1}=19 (μ triode)
- P_{OUT} = 5.7 W @ THD = 10%

Choosing load impedance

For a single-ended EL84 (6BQ5) pentode, the minimum distortion Z_L=4,500–5,000Ω, but the maximum power is at Z_L=6,600 Ω. However, notice that the difference in power between the two points is only 0.2 Watt, while the increase in THD is 2%. Our rule-of-thumb says $Z_{AOPT} \approx V_A/I_A$ = 5.3kΩ.

In this example we will use a slightly lower load impedance of 4k5. The same rationale applies to push-pull operation, where we simply double the impedance and get the optimal value of Z_{PP}=2*4,500 = 9,000 Ω. Let's choose the operating point given in manufacturer's data sheets, anode and screen voltage of 250V. We will assume an ideal output transformer with no losses and zero DC primary resistance.

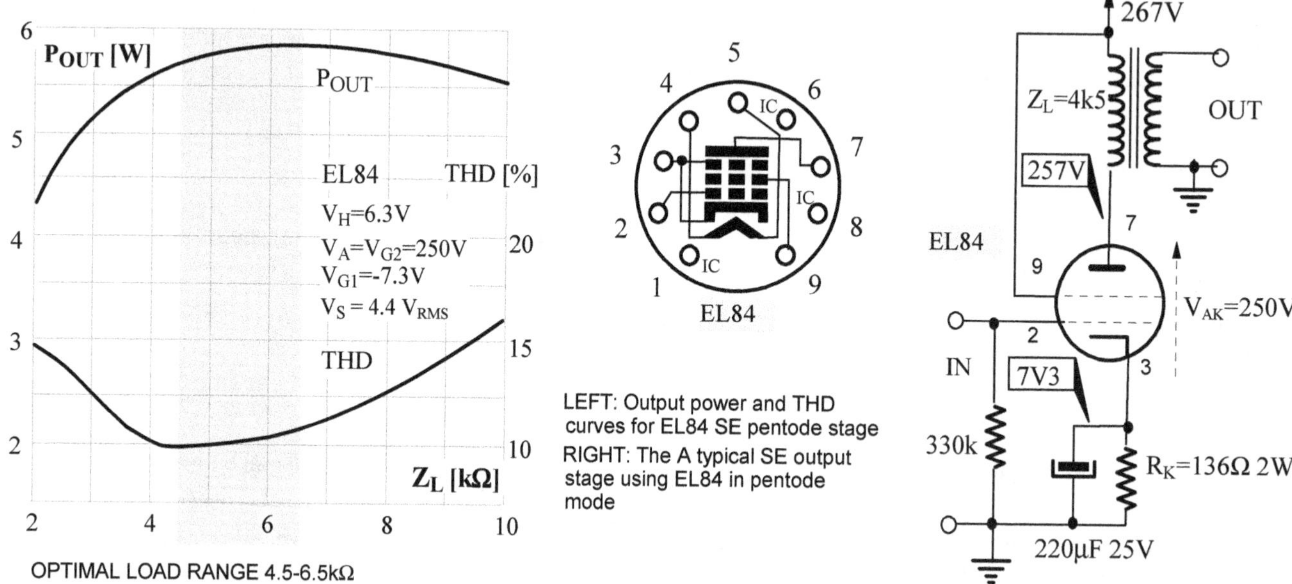

LEFT: Output power and THD curves for EL84 SE pentode stage
RIGHT: The A typical SE output stage using EL84 in pentode mode

Since the maximum anode dissipation is 12 watts, the maximum anode current in the operating point is $I_A=P_{MAX}/V_0$ = 12/250 = 48mA. From the data sheets or the anode curves we determine that grid bias voltage in that point is -7.3V and that the screen current is 5.5 mA (not shown). Cathode current is the sum of anode and screen current, so I_K=48+5.5=53.5 mA. To get 7.3V voltage drop on the cathode resistor, with the cathode current of 53.5 mA flowing through it, its value must be $R_K=V_K/I_K$ = 7.3/0.0535 = 136Ω. Power dissipation on this resistor is $P_{RK}=V_KI_K$ = 0.4W, so a 1 Watt resistor is the smallest we should use, a 2 Watt resistor would be better for long-term reliability.

Assuming we decide to limit the grid voltage to -1 Volt, half of the signal swing will be 7.3 - 1 = 6.3 Volts.

We add an identical signal swing in the negative direction to the bias of -7.3 Volts and get -7.3-6.3 = -13.6V So, our maximum peak-to-peak grid voltage swing is 12.6V, or ΔV_{GEFF} = 4.5 V_{RMS}

We now have endpoints A and B, and can draw vertical lines to get V_{MIN} and V_{MAX} of 45V and 455V respectively. That is our anode peak-to-peak voltage swing, ΔV_A=455-45=410V_{PP} or ΔV_{AEFF} = 410/2.82 = 145.4V_{RMS} The voltage amplification factor of the output stage is $A=-\Delta V_{AEFF}/\Delta V_{GEFF}$ = -145.4/4.5 = -32.5

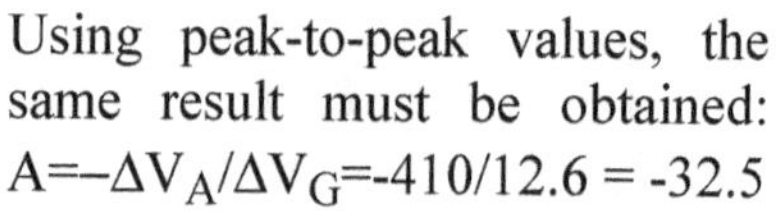

Using peak-to-peak values, the same result must be obtained: $A=-\Delta V_A/\Delta V_G$=-410/12.6 = -32.5

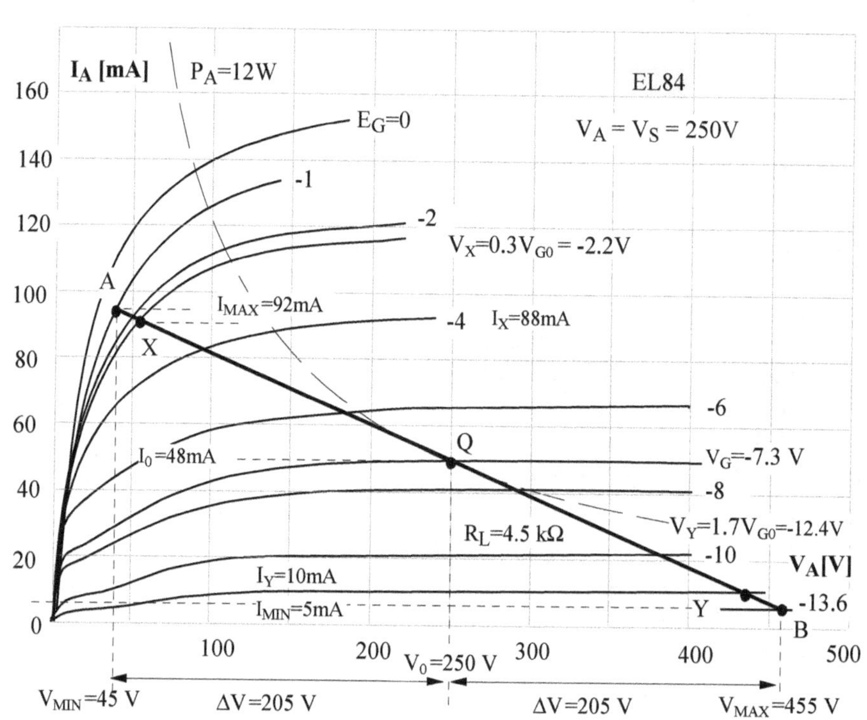

Estimating output power of SE pentode stages

RULE-OF-THUMB METHOD #1: The anode efficiency of a pentode is approximately 35%. The anode power input is $P_{IN}=V_{A0}I_{A0}$ = 250*0.048 = 12W, so P_{OUT}= 0.35P_{IN} = 0.35*12 = 4.2W

This formula estimates the "clean" power, meaning the total output power levels will be higher.

RULE-OF-THUMB METHOD #2: The output power can also be estimated using the RMS value of the output swing (205/1.41=145.4V), instead of its peak value (205V here).

$P_{OUT} = \Delta V_A{}^2/R_L = 145.4^2/4{,}500 = 21{,}138/4{,}500 = 4.7W$ This is closer to the output power declared in data sheets (5.7W), which includes significant distortion.

SE PENTODE OUTPUT POWER

$\mathbf{P_{OUT} \approx 0.35 P_{IN} = 0.35 V_A I_A}$
V_A = DC anode voltage
I_A = DC anode current

Calculating distortion and output power using the 5-point method

The second and third harmonic distortion in pentode stages can be estimated by the "5-point method". The bias voltages of interest are $V_X=0.3V_{G0}$ and $V_Y=1.7V_{G0}$. In our case they are V_X=-2.2V and V_Y= -12.4V We can read anode currents in those two points as I_X=88mA and I_Y=10mA.

The formulas for the second and third harmonic distortion coefficient are

$$\mathbf{D_2= [(I_{AMAX} + I_{AMIN}) - 2I_0]/[I_{AMAX} - I_{AMIN} + 1.41(I_X - I_Y)]*100\%}$$

$$\mathbf{D_3= [I_{AMAX} - I_{AMIN} - 1.41(I_X - I_Y)]/(I_{AMAX} - I_{AMIN} + 1.41(I_X - I_Y))*100\%}$$

After substituting our values we get

D_2= [92+5 - 2*48]/[92-5+1.41(88-10)]*100% = 1/197*100% = 0.51%

D_3= [92-5 - 1.41(88-10)]/(92-5+1.41(88-10))*100% = -23/197=11.67%

The second harmonic has been minimized, but the third harmonic distortion is large, almost 12%. The 5-point method includes distortion in the power figures, that is the third method of estimating output power:

$\mathbf{P_{OUT}=R_L[(I_{AMAX}-I_{AMIN})+1.41(I_X-I_Y)]^2/32}$ =4,500*[(0.092-0.005)+1.41(0.088-0.01)]²/32 = 5.45W

Reactive loads

An audio output stage with purely resistive load is designer's dream that never comes true. The path of operation for a partially inductive load such as a typical dynamic loudspeaker is an ellipse.

The illustration shows the worst-case-scenario, for maximum power output, with grid voltage swing between -1V and -13.5V peak-to-peak. It is obvious that for the part of the operating cycle the tube is in a cutoff, the anode current is zero. This will result in significant distortion.

Luckily, at lower power levels the ellipse stays undistorted. We can also conclude that the "fatter" the ellipse (the more inductive or capacitive the load), the earlier such cutoff situation will happen!

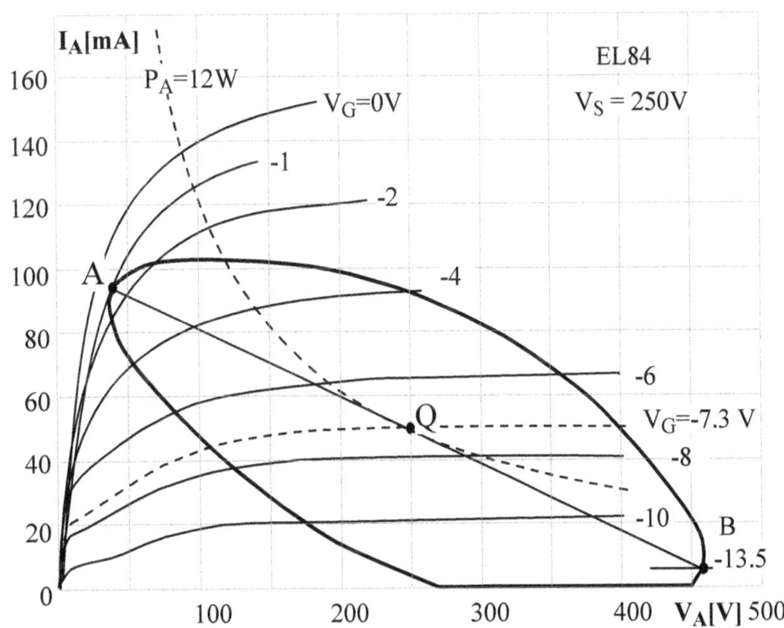

Amplification factor and output impedance using the equivalent circuit

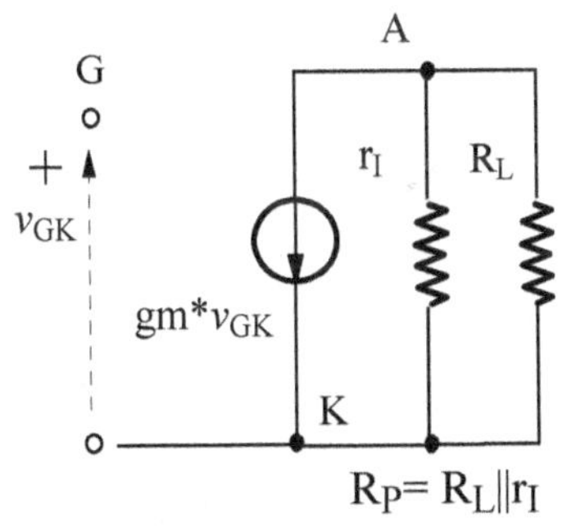

We know the anode load impedance, it is R_L=4,500Ω, so we can directly calculate the voltage gain of the stage as A=-gmR_P where $R_P = R_L || r_I$ = 4k5||38k = 4.02 kΩ, so A=-gmR_P = -11.3*4.02 = -45 This is much higher than the result (-32.5) obtained by graphical means.

The output impedance of the stage (reflected onto the secondary side, as a loudspeaker would see) is $Z_{OUT}=r_I/TR^2$, where r_I=38kΩ (the internal resistance of EL84 pentode) and TR is the turns ratio (TR=N_1/N_2). Since the impedance ratio is IR=TR^2=4,500/8 = 562.5 we get $Z_{OUT}=r_I/TR^2$= 38kΩ/562.5= 68Ω! This very high Z_{OUT} would result in very poor damping factor, one reason pentode amplifiers are not usually used without negative feedback.

ABOVE: The equivalent circuit is 1. simplified and 2. linearized, so it applies to small signals only and should not be used to estimate the voltage gain and output power of power stages, where signal swings are large! It is discussed here for comparison purposes only.

LOW-POWER OCTAL PENTODES AND BEAM TETRODES: 6V6, 6K6, 6F6, 6Y6

The 6F6 audio output pentode dates from the mid-1930s. As you can see from the Side-By-Side comparison on the next page, it is similar to the ubiquitous 6V6, but is more sensitive and requires even lower grid driving voltages. However, it needs 55% more heater power than the 6V6, meaning it is less efficient.

As pentodes, a pair of 6F6 in class A_1 push-pull can produce 10.5 Watts with only 3% distortion.

6Y6 is a low voltage-high current tube, its max. V_A is almost half that of the other three, but its anode current is doubled! Since it is not interchangeable with the other three types it is included here for comparison purposes only.

All four tube-types sound very good, especially when triode connected, and since millions were made in the golden era, they are still plentiful and relatively cheap to buy. NOS are still available as pairs or even quads.

SIDE-BY-SIDE	6V6	6K6	6F6	6Y6
	Beam tetrode	Power pentode	Power pentode	Beam tetrode
Heater	6.3V/0.45A	6.3V/0.4A	6.3V/ 0.7A	6.3V/ 125A
P_A / P_{G2} max.	12W / 2W	8.5W / 2.8W	11W / 3.75W	12.5W / 1.75W
V_A / V_{G2} max.	315 / 285 V	315 / 285 V	375 / 285 V	200 / 200 V
SE class A_1 pentode P_{OUT}	5.5W	4.5W	4.8W	6.0W
V_A, V_S, V_G	315V, 225V, -13V	315V, 250V, -21V	285V, 285V, -20V	200V, 135V, -14
R_L	8.5kΩ	9kΩ	7kΩ	2.6kΩ
I_{A0}, I_{S0}	34mA, 2.2mA	26mA, 4mA	38mA, 7mA	61mA, 2.2mA
Parameters as a triode	gm= 4mA/V, r_I=2400Ω, μ=9.6	gm= 2.7mA/V, r_I=2,500Ω, μ=6.8	gm= 2.6mA/V, r_I=2600Ω, μ=6.8	gm= 7.3mA/V, r_I=750Ω, μ=5.5
V_A max. triode	300V	315V	350V	250V
$P_A + P_{G2}$ max. total triode	12.5W	7W	10W	14W
V_A, V_G, I_0, R_L	250V, -20V, 39mA, 4.8kΩ	250V, -18V, 38mA, 6kΩ	350V, -38V, 48mA, 6kΩ	250V, -42V, 50mA, 5kΩ
SE class A_1 triode P_{OUT}	1.25W		0.9W	2.3W

DESIGN OF 6F6 SE PENTODE OUTPUT STAGE

Since 6F6 was specifically developed for audio applications, there are recommended operating conditions, the most common being V_A =250V and V_S = 250V, with 7kΩ load, so let's stick to that.

Looking at the dynamic transfer curves, the curve for 5k load is the most linear in the low V_G range (from 0 to -10V on the grid). However, when positioned on the anode characteristics, the 5k load line crosses the V_G=0 line way above the knee, meaning the voltage swing will be reduced, and the output power lower. The 7k load seems to be a better compromise, slightly lower current swing but an additional 70V of anode swing (V_{AMIN} around 30V compared to 100V for the 5k load).

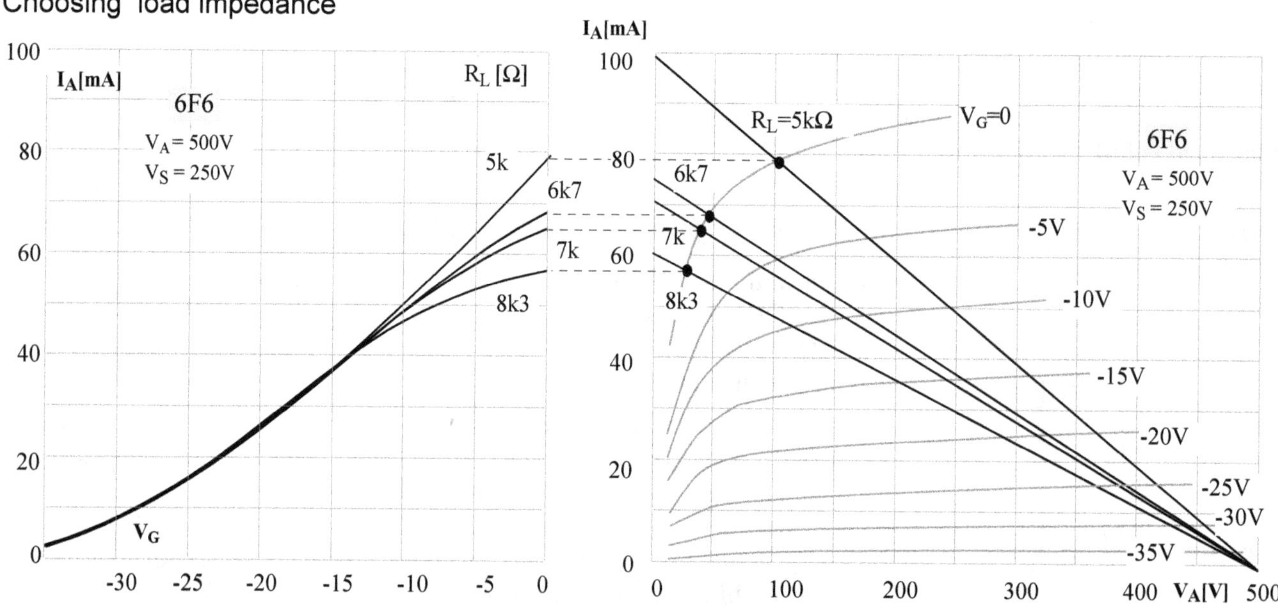

ABOVE: 6F6 pentode's dynamic transfer characteristics and anode characteristics for various load impedances.

With anode current of 35mA and screen current of 6.5mA (specified in data sheets) at the quiescent point Q, the cathode current is 41.5mA. To get the bias voltage of 16.5V the cathode resistor must be R_K=16.5/0.0415= 398Ω, so a standard value of 390Ω will do just fine.

L-R: RCA 6F6, Sylvania 6F6, Sylvania 6V6, RCA 6K6 (metal)

Estimating distortion

Using the "5-point method", the bias voltages of interest are $V_X=0.3V_{G0}$ and $V_Y=1.7V_{G0}$. In our case they are V_X=-4.8V and V_Y= -28V After marking the intersection of those two curves with the load line, we can read anode currents in those two points as I_X=61mA and I_Y=10.5mA.

The formulas for the second and third harmonic distortion coefficient are

$D_2= [(I_{AMAX} + I_{AMIN})-2I_0]/[I_{AMAX}-I_{AMIN}+1.41(I_X-I_Y)]*100\%$

$D_3= [I_{AMAX} - I_{AMIN}-1.41(I_X-I_Y)]/(I_{AMAX}-I_{AMIN}+1.41(I_X-I_Y))*100\%$

After substituting our values we get D_2= [67+5- 2*35]/[67-5+1.41(61-10.5)]*100% = 2/131.8*100% = 1.52% and D_3= [67-5 - 1.41(61-10.5)]/(67-5+1.41(61-10.5))*100% = -9.205/133.2=6.91%

Estimating the output power

The distortion of pentodes and beam power tubes is usually included in the power output figures. $P_{OUT}=R_L[(I_{AMAX}-I_{AMIN})+1.41(I_X-I_Y)]^2/32$= 7,000*[(0.067-0.005)+1.41(0.061-0.0105)]^2/32 = 3.88W

Let's see what we get using the rule-of-thumb approximation, assuming the anode efficiency of a pentode of 35%. For 6F6 output pentode working in a single-ended stage with I_A=35mA and V_A=250V, the input power is $P_{IN}=V_AI_A$ = 250*0.035 = 8.75W so P_{OUT}= 0.35P_{IN} = 3.06W

SE PENTODE OUTPUT POWER

$$P_{OUT}=R_L[(I_{AMAX}-I_{AMIN})+1.41(I_X-I_Y)]^2/32$$

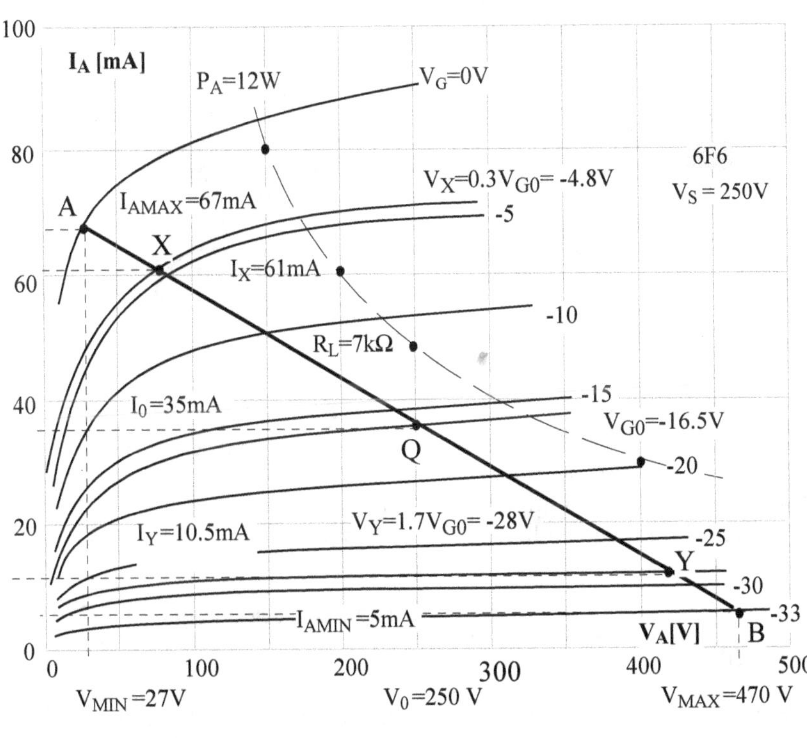

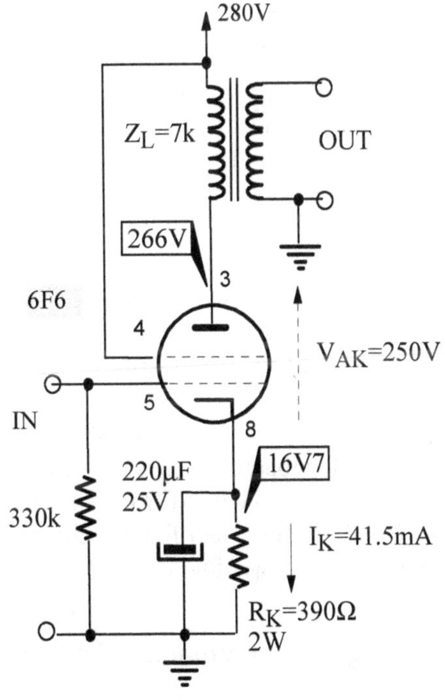

ABOVE: 6F6 SE pentode stage (right) and operating conditions (quiescent point, load line, voltage and current swings) on the left

ULTRALINEAR CONNECTION

The name was popularized by David Hafler and Herbert Keroes in Nov. 1951 issue of "Audio Engineering" magazine, in their article "An Ultralinear Amplifier".

U/L circuit requires a tap on the primary winding of the output transformer to which output tube's screen grid is connected. Instead of the screen grid being held at some stable DC voltage as in the pentode's case, now the screen's voltage varies and follows the signal, which means that a local negative feedback is applied to the screen. This results in a host of negative feedback benefits, as discussed in earlier chapters:

- output impedance is reduced to about half of the triode's circuit
- linearity is improved, resulting in reduced distortion
- the power output is higher than in triode connection, approaching that delivered by a pentode
- the power output is more constant because the output stage behaves somewhere in between the voltage (triode) and current (pentode) source.
- the circuit combines the advantages of both the triode and the pentode stage, without suffering from their respective setbacks.

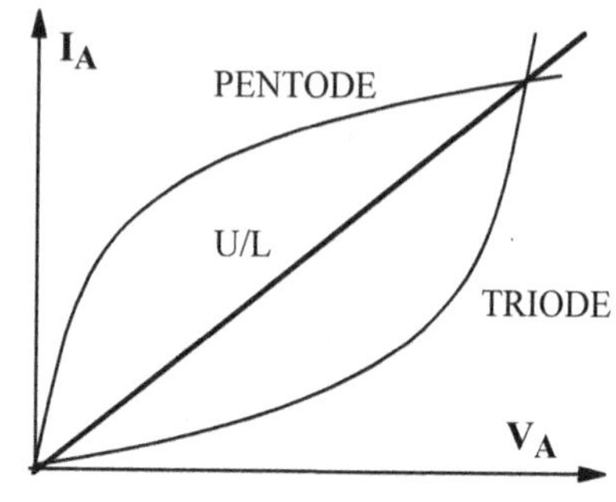

The ultralinear (UL) connection is one example of the "Linearizing Principle": Convex + Concave = Straight!

U/L stage with EL84

As in our example with EL84 in SE pentode connection, we have the same output transformer with impedance ratio $IR=(N_1/N_2)^2=4{,}500/8 = 562.5$, or turn ratio $TR=N_1/N_2=23.7$ The only addition is a primary tap at 23% of the primary turns. This means that 23% of the primary or anode AC (signal) voltage is fed-back into the screen grid as negative feedback.

The screen tap at 23% of the primary winding turns is the tap that results in minimum distortion. Thus, $N_3=0.23N_1$ and our X-factor is 0.23! We will not go into the tedious mathematics of the ultralinear stage, the final formulas for gain and output impedance are:

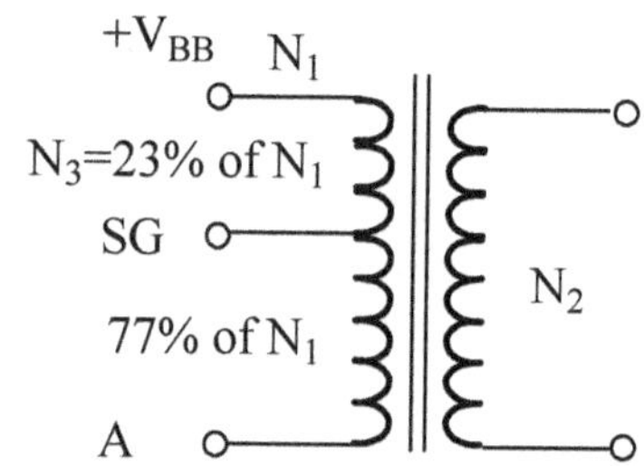

SE EL84 U/L output transformer windings and turns ratios

$$A = -\frac{(gm+Xgm_{G2})Z_A}{1+(X/\mu_{G2G1}+1/\mu)(gm+Xgm_{G2})Z_A}$$

$$Z_{OUT} = \frac{(N_2/N_1)^2}{(gm+Xgm_{G2})(X/\mu_{G2G1}+1/\mu)}$$

$$A = -\frac{(11.3+0.23*1.8)*4.5}{1+(0.23/19+1/430)*(11.3+0.23*1.8)*4.5} = -\frac{52.713}{1+0.01443*52.713}$$

$$= -52.713/1.76 = -52.713*0.568 = -29.94 \approx -30$$

$$Z_{OUT} = \frac{1/562.5}{(11.3+0.23*1.8)(0.23/19+1/430)}$$

$$= (1{,}000/562.5)/11.714*0.01443 = 10.5\Omega$$

In pentode connection we had A=-45 and Z_{OUT}= 68Ω! With a relatively mild screen feedback (X=23%) A=-30 and Z_{OUT}= 10.5Ω! The reduction in amplification due to U/L NFB is $A_F/A=30/45=0.667$, or in dB, 20*log0.667 = -3.52 dB! To get a feel for the impact of increased ultralinear feedback, calculate A and Z_{OUT} for X=33%!

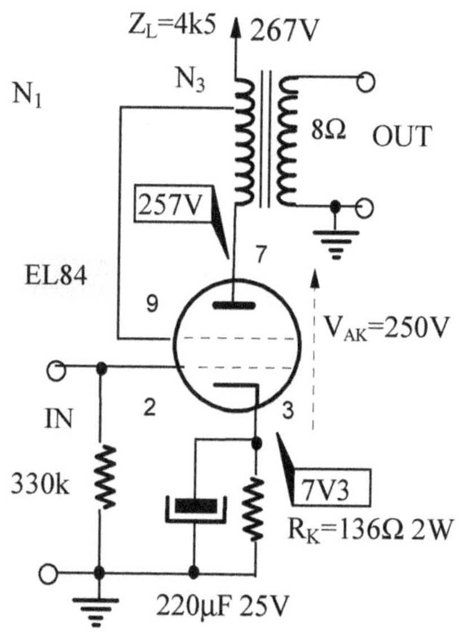

SE EL84 U/L output stage

Optimizing the performance of the suppressor grid

Most pentodes have their suppressor grids internally tied to the cathode, so amplifier designers have no choice regarding its connection. However, some power pentodes, most notably EL34, have the suppressor grid wired out to its own pin, in this case pin 1. Most amplifier builders simply follow the tradition and strap pins 1 and 8 together.

Some designers such as Peter Traynor of Traynor guitar amps, concluded that connecting the suppressor to a source of negative voltage improves the efficiency and linearity of the output stage, but also the reliability of the tubes. The most convenient negative voltage source is the bias circuit, typically with around -30V available for suppressor connection.

Another benefit of a negatively-biased suppressor is the inherent protection it provides in case of a bias failure. -30V on the suppressor will repel many electrons from the anode and limit the zero-bias anode current to approximately half of its value in the case of suppressor at zero volts, for instance 200mA instead of 400mA. This will not prevent the ultimate overheating, leading to damage or destruction of the power tube(s), should the loss of bias not be noticed and the amp not shut down quickly, but it may protect the output transformer from burning out.

Many triode puritans refer to triode-strapped pentodes in a derogatory manner as "pseudo-triodes". One reason is that with most pentodes, although the screen is tied to the anode, the suppressor is still at the cathode potential. EL34 makes it possible to strap the suppressor to the anode as well.

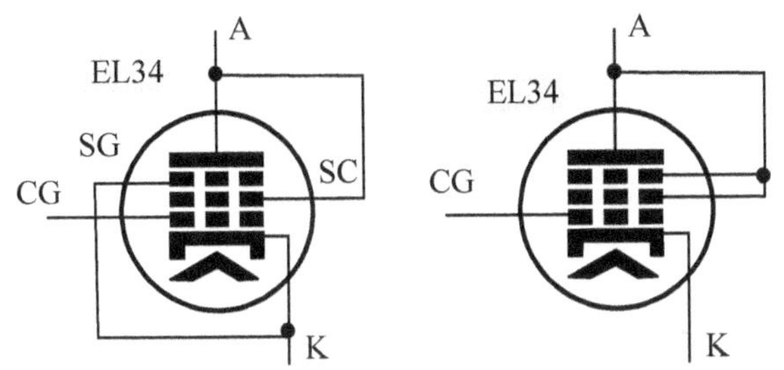

LEFT: The usual suppressor grid (SG) connection to the cathode
RIGHT: Suppressor connected to the anode

While this doesn't markedly improve the performance, it does result in lower internal resistance of such pseudo-triode, which is always welcome.

Very few other tubes have a separate suppressor pin. Interestingly, while the original Telefunken EL156 had the suppressor and cathode tied together, the Chinese octal version has the suppressor gird brought out to its separate pin. PL509/519 tubes also have a separate suppressor pin, as do PL83, LS50 (GU50), EL(FL)152 (but not EL153 which is not a pentode but a beam tetrode!), E55L, all of European origin, and the American 837 pentode.

LINEARIZING THE OUTPUT STAGE

The ultralinear circuit is a special case of "distributed loading", a technique patented by Alan Blumlein in 1937. The load winding can be placed in the anode circuit, the cathode circuit, or distributed in any proportion between them. The ultralinear circuit is fully in the anode, while its opposite would be a load placed fully in the cathode, with the output stage in cathode follower mode. Distributed loading can be applied to both single-ended and push-pull output stages.

In "unity coupling" equal loads are placed in the anode and cathode circuits, so half of the total load is in each. There seems to be a degree of confusion what unity coupling actually means, is it the 1:1 load distribution, or the fact that the output transformers used for such designs have bifilar primary windings which have no leakage inductance and so the coefficient of magnetic coupling between them is practically unity.

Driving a pentode or beam tube through its screen grid instead of the control grid is another linearizing technique. The control grid is held at zero AC potential, but it can be at any DC potential. The anode characteristics for screen-driven connection were not provided by tube manufacturers of the old, only a few current tube makers provide such information, such as the Russian tube maker Svetlana for their EL509 tubes, but only for zero grid bias. Any other bias, and you have to do your own measurements and plot your own characteristics. The slope of anode characteristics immediately tells us if the amplifying device is closer to an ideal voltage or current source.

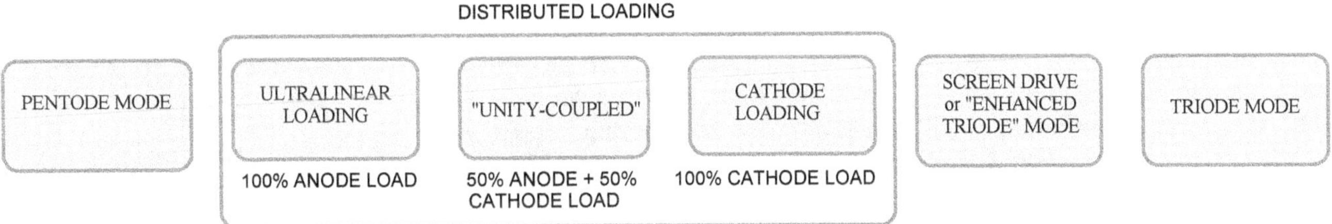

ABOVE: The pentode and triode output stages are at the ends of a "continuum", with "linearized" connections or "in-between" arrangements, the main types of which are distributed loading and screen drive or "enhanced triode" mode!

A visual comparison of the three major operational modes

The almost horizontal curves of a pentode (next page) indicate very little anode current change with an increased anode voltage, thus a very high internal resistance, typical of current sources. On the other end of the spectrum, a triode's anode current jumps rapidly with increased anode voltage, due to its low internal resistance.

Since transconductance does not differ much between triodes and pentodes, the amplification factor and sensitivity track the internal resistance, so pentodes with high μ need very little signal to drive them (high sensitivity), while low μ triodes are lazy and need large grid voltages! The same applies to grid bias voltages, which are very small for pentodes (typ. around -10V) and much higher for triodes (typ. -50 to -70V)!

The distributed load and screen-driven connections are "in-between" arrangements. Tubes so connected exhibit much lower internal resistance than pentodes but higher than triodes. Thus they are easier to drive then when triode strapped, but require higher negative bias voltages and larger grid signals for full power then when used as pentodes.

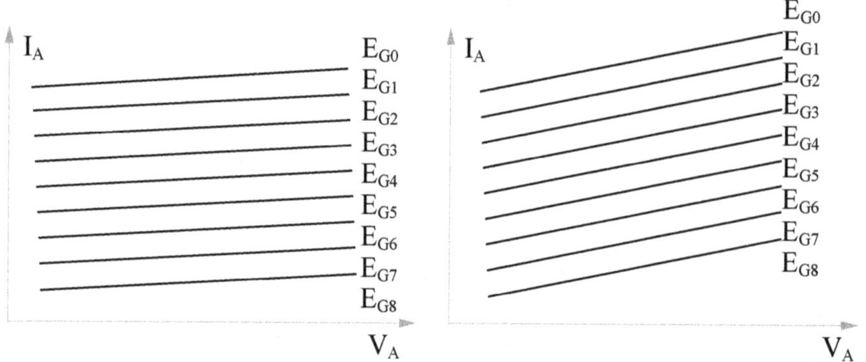

PENTODE: $\mu = g_m r_i$, high r_i, high μ, low grid drive voltages (high sensitivity)

ULTRALINEAR and SCREEN-DRIVEN TRIODE CONNECTION : $\mu = g_m r_i$, medium r_i, medium μ, medium grid drive voltages (medium sensitivity)

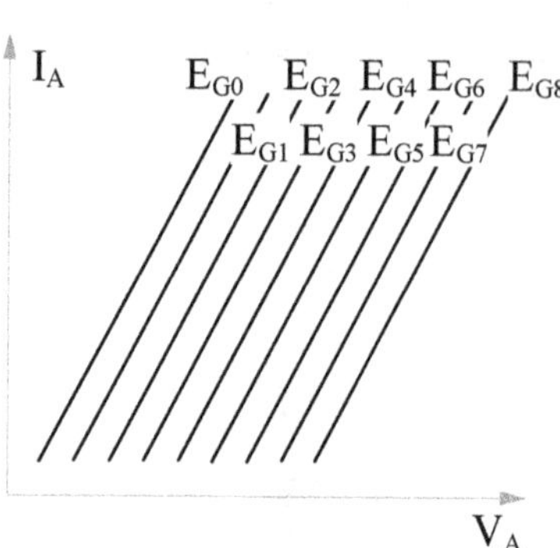

TRIODE: $\mu = g_m r_i$, low r_i, low μ, high grid drive voltages (low sensitivity)

Which tubes are the best candidates for screen-drive?

The lower the maximum allowed screen voltage of a tube, the more control the screen will have over the anode current. Sweep tubes have a low screen grid voltage rating, typically in the 175-250 V range, making them particularly suitable for screen-driven designs. They need lower drive voltage when screen-driven and achieve better linearity and lower distortion.

In comparison, tubes aimed for audio service, such as EL34, 6L6 and KT88 for instance, were designed so their screen grids can take higher voltages, 350 to 500 volts. This means they require a much larger change in screen voltage to control anode current. Such lower sensitivity of the screen makes them unsuitable for screen-driven applications.

A screen-driven amplifier using beam tetrode power tubes (EL519) is analyzed in Volume 2 of this book, de Paravicini's E.A.R. 859 SET amplifier, operating in what he calls "enhanced triode mode".

SEP stage with cathode feedback from output transformer's tertiary winding (partial cathode coupling)

The output transformer has a tertiary winding that provides a negative feedback voltage into the cathode circuit. Thus, the feedback is voltage-type, parallel-derived (the output voltage and tertiary winding voltage are in "parallel" or "in phase") and the feedback is applied in series with the input voltage, just like in case of the un-bypassed cathode resistor, but that NFB was a current type, since the input voltage was proportional to the output current.

Notice the polarities of the transformer voltages, the anode is at the primary end without a dot, while the cathode is connected to the tertiary winding's end with a dot, so these two voltages will be out of phase, as they should be to get negative or "degenerative" feedback. If they were in phase, the feedback would be positive (regenerative) and the whole stage would turn into an oscillator!

Without the NFB cathode winding with N_3 turns, the secondary load was reflected to the primary side as $Z_A = (N_1/N_2)^2 R_L = 562.5*8 = 4{,}500\Omega$

Now, N_3 must be added to N_1 in such a calculation, so to make things easier to compute let's select a value for N_3 relative to N_2, and let's make them equal, so $N_3 = N_2$ (the reason behind this choice will be clarified later).

Since $N_1 = 23.72*N_2$ and $N_3 = N_2$, now $Z_{AF} = [(N_1+N_3)/N_2]^2 R_L = (23.72+1)^2*8 = 4{,}888\Omega \approx 4k9$

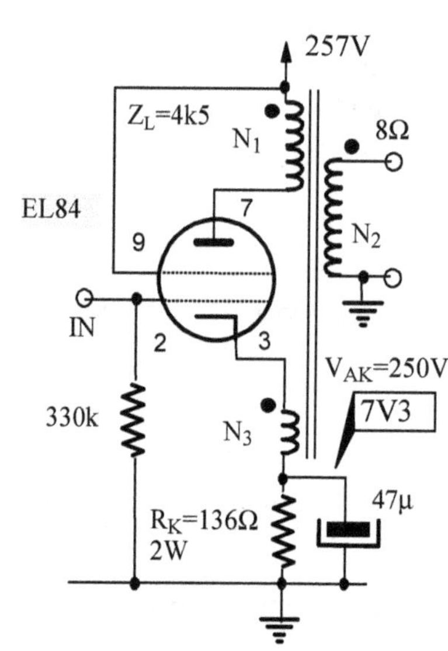

ABOVE: EL84 SEP stage with cathode feedback from output transformer's tertiary winding

The parallel combination of the internal tube resistance r_I and Z_{AF} is now R_{PF}= 38k||4k9 = 4k34 ≈ 4k3 and the voltage amplification of the output stage is

$$A_{KF} = -N_1/N_2 \frac{N_2/(N_1+N_3)gmR_{PF}}{1+(gmR_{PF})N_3/(N_1+N_3)}$$

In our case $N_2/(N_1+N_3) = N_3/(N_1+N_3) = 1/24.72 = 0.0405$

The amplification factor with cathode feedback is: A_{KF} = -23.72*(0.0405*11.3*4.3)/(1+11.3*4.3*0.0405) = -23.72*1.97*0.337 =-23.72*0.664 = -15.75

In pentode connection we had A=-45 so the reduction in amplification is A_{KF}/A= 15.75/45 = 0.35, or in dB, 20log0.35 = -9.1dB!

Since $\mu=gmr_I$=11.3*38=430, the output impedance with cathode feedback is $Z_{OKF}= r_I/[1+\mu N_3/(N_1+N_3)]$ = 38/[1+430*0.0405] = 38/18.415 = 2,064Ω

Remember that without the cathode feedback the output impedance was Z_{OUT}= 68Ω, the cathode feedback increased Z_{OUT} thirty times! In hi-fi amps we aim for as low output impedance as possible, so this type of NFB may reduce distortion but will also lower the damping factor, which is bad news.

The same benefit without the tertiary winding

Although most commercial output transformers don't have a tertiary winding, all is not lost, the cathode feedback can still be implemented by using the secondary winding as a feedback winding, as illustrated. It is important that neither of the secondary (output) terminals is grounded, the speaker must be fully floating.

The reason we previously chose $N_3=N_2$ is so that our results and conclusions would be directly applicable to this case.

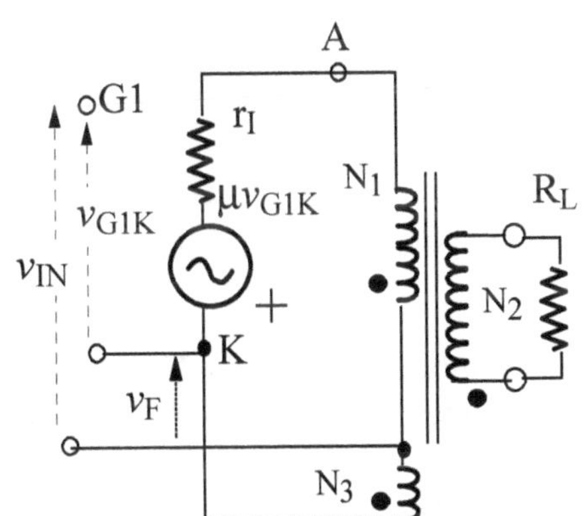

LEFT: Small signal model of SEP stage with cathode feedback from output transformer's tertiary winding

RIGHT: Using output transformer's secondary winding to provide cathode feedback

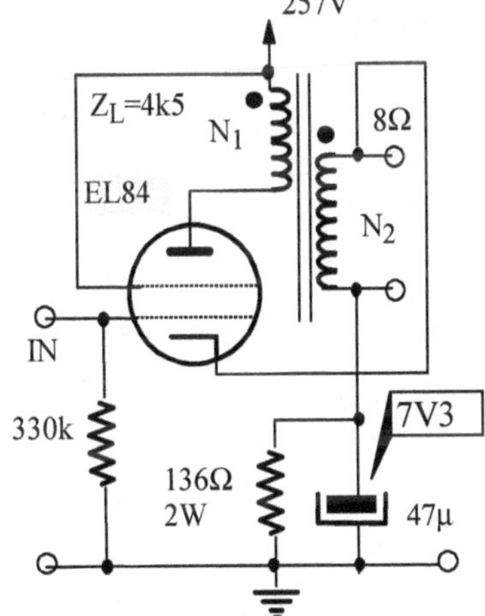

6L6 SINGLE-ENDED STAGE: TRIODE, ULTRALINEAR AND PENTODE COMPARISON

6L6 beam tetrode is a strange tube. Its sonic merits in guitar amplifiers are unquestionable. In hi-fi, however, the jury is still out after all these decades. The graphical estimations of distortion on the next page paint a bleak picture, with THD levels ranging from 9.7% in a triode connection to 9.25% second and 14.85% third harmonic distortion when working as a beam tetrode. Even in the supposedly most linear of all, so called "ultralinear" mode, the anode curves predict that THD will still be very high, 8.3%! Calling that ultralinear is delusional.

6L6 SE pentode stage

To determine the output power and distortion we also need points X and Y: V_{G0}=-13V, V_X=0.3*13=-3.9V, V_Y=1.7*13= -22V, from the graph on the next page we estimate I_X=145mA and I_Y=38mA.

D_2= [(I_{AMAX} +I_{AMIN}) - 2I_0] / [I_{AMAX} -I_{AMIN}+1.41(I_X-I_Y)]*100% =28/303=9.25%

D_3= [I_{AMAX} -I_{AMIN} - 1.41(I_X-I_Y)] / (I_{AMAX} -I_{AMIN}+1.41(I_X-I_Y))*100%= 45/303=14.85%

$P_{OUT}=R_L[(I_{AMAX}-I_{AMIN})+1.41(I_X-I_Y)]^2/32$

P_{OUT}= 2,500*[(0.175-0.023)+ 1.41(0.145- 0.038)]^2 /32= 2,500*0.3^2/32= 229/32=7.17W

Input power is $P_{IN}=I_AV_0$=0.085*320= 27.2W, so efficiency is η = P_{OUT}/P_{IN} = 7.17/27.2 = 26.3%

RIGHT: SE pentode stage using 6L6 and 2k5 load impedance

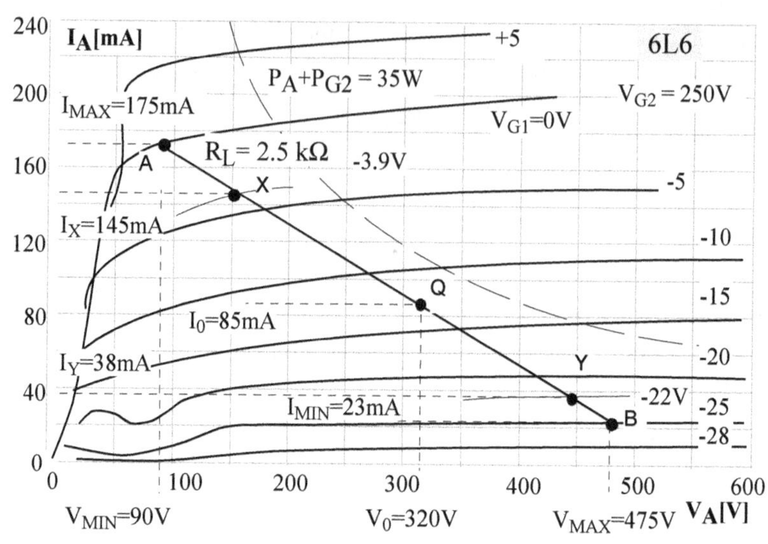

6L6 SE triode stage

$\Delta V = 490\text{-}175 = 315V$

$\Delta I = 145\text{-}25 = 120$ mA

$P_{IN} = I_A V_0 = 0.8{*}350 = 28$ W

$P_{OUT} = \Delta V \Delta I/8 = 315{*}0.12/8 = 4.7W$

$\eta = P_{OUT}/P_{IN} = 4.7/28 = 16.8\%$

For triode stages, distortion is predominately 2nd harmonic and can be estimated as:

$D_2 = (P_{OUT+} - P_{OUT-})/2P_{OUT} {*}100$ [%] = $(175{*}0.065/4\text{-}140{*}0.055/4)/2{*}4.7 =$ $(2.84\text{-}1.93)/9.4 = 9.7\%$

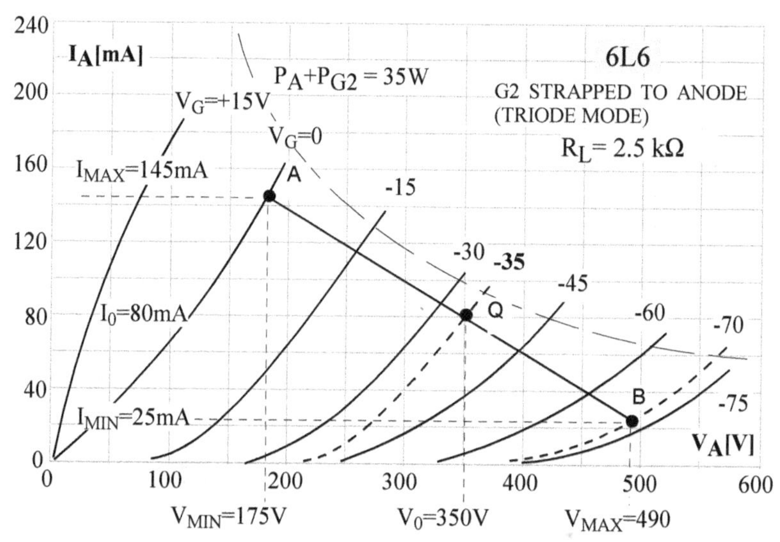

6L6 SE ultralinear stage

Even a superficial glance at the curves tell you that there is nothing linear about this tube in this regime. Notice how the spacing between the curves reduces as the grid voltage becomes more and more negative, resulting in the positive anode voltage swing $\Delta V+$ of 245V and the much smaller negative swing $\Delta V-$ of only 175!

$\Delta V = 420V$, $\Delta I = 155$ mA, $P_{IN} = I_A V_0 = 0.1{*}300 = 30W$, $P_{OUT} = \Delta V \Delta I/8 = 420{*}0.155/8 = 8.1W$

$\eta = P_{OUT}/P_{IN} = 8.1/30 = 27\%$

$D_2 = (\Delta V_{OUT+} - \Delta V_{OUT-})/2\Delta V_{OUT} {*}100[\%] = (245\text{-}175)/2{*}420 = (5.82\text{-}3.06)/8.9 = 8.3$ %

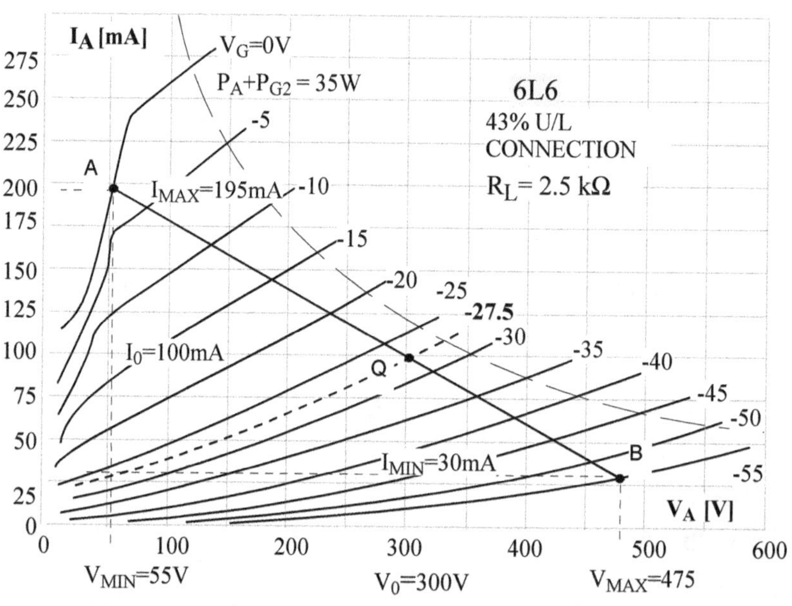

THE END MATTER OF VOLUME 1

- FURTHER READING
- INDEX
- NOTES

"Don't give up at half time. Concentrate on winning the second half."
Paul Bryant, American football college coach

FURTHER READING

IN ENGLISH

Amplifiers, H. Lewis York, 1964, 254 pages
Amplifiers (The Why and How of Good Amplification), G. A. Briggs, 1952, 216 pages
Analysis and Design of Electronic Circuits, Paul M. Chirlian, 1965, 570 pages
Applied Electronics, Truman S. Gray, 1955, 882 pages
An Approach to Audio Frequency Amplifier Design, G.E.C. Valve and Electronics Department, 1957, 126 pages
Audio Cyclopedia, Howard M. Tremaine, 1st edition, 1959, 1269 pages
Audio Design Handbook, H. A. Hartley, 1958, 224 pages
Audio Transformer Design Manual, Robert G. Wolpert, 1989, 108 pages
Basic Audio, Norman Crowhurst, 1959, Volume 1 (114 pages), Volume 2 (122 pages), Volume 3 (113 pages)
Capacitors, Magnetic Circuits, and Transformers, Leander Matsch, 1964, 350 pages
Circuit Theory of Electron Devices, E. Milton Boone, 1959, 483 pages
Designing Power Supplies for Tube Amplifiers, Merlin Blencowe, 2010, 246 pages
Designing Tube Preamps for Guitar and Bass, Merlin Blencowe, 2009, 294 pages
Electron Tubes and Semiconductors, Joseph J DeFrance, 1958, 288 pages
Electronic Amplifier Circuits, Joseph Petit and Malcolm McWhorter, 1961, 325 pages
Electronic and Radio Engineering, Frederick E. Terman, 1955, 1078 pages
Electronic Circuits, Thomas L. Martin. 1959, 708 pages
Electronic Circuits and Tubes, Harry E. & Wing, Alexander H. Clifford, 1947, 996 pages
Electronic Circuits: A Unified Treatment of Vacuum Tubes and Transistors, Ernest Angelo, 1964, 652 pages
Electronic Designers' Handbook, Landee, Davis and Albrecht, 1957
Electronic Engineering, C. Alley and K. Atwood, 1973, 838 pages
Electronic Transformers and Circuits, Reuben Lee, 1955, 349 pages
Electronics and Electron Devices, A. L. Albert, 1956
Electron-Tube Circuits, Samuel Seely, 1950, 530 pages
Engineering Electronics, George Happell and Wilfred Hesselberth, 1953, 508 pages
Essentials of Radio-Electronics, M. Slurzberg and W. Osterheld, 1961, 716 pages
Engineering Electronics With Industrial Applications, John D. Ryder, 1957, 666 pages
Fundamental Amplifier Techniques with Electron Tubes, Rudolf Moers, 2011, 834 pages
Fundamentals of Semiconductor and Tube Electronics, H. Alex Romanowitz, 1962, 620 pages
General Electronics Circuits, Joseph J. DeFrance, 1963, 526 pages
High-End Valve Amplifiers 2, Menno van der Veen, 2011, 416 pages
High-Fidelity Circuit Design, Norman Crowhurst and George Cooper, 1957, 296 pages
How to Gain Gain, A Reference Book on Triodes in Audio Pre-Amps, Vogel Burkhard, 2013
Inside the Vacuum Tube, John F. Rider, 1945, 407 pages
Instruments and Measurements For Electronics, Clyde N. Herrick, 1972, 542 pages
Maintaining Hi-fi Equipment, Joseph Marshall, 1956, 224 pages
Magnetic Circuits and Transformers, M.I.T. Electrical Engineering Staff, 1943, 718 pages
Modern High-end Valve Amplifiers Based on Toroidal Output Transformers, M. van der Veen, 1999, 250 pages
Network analysis and feedback amplifier design, Hendrik W. Bode, 1952, 551 pages
Practical Transformer Design Handbook, Eric Lowdon, 1989, 389 pages
Principles and Applications of Electron Devices, Paul D. Ankrum, 1959, 667 pages
Principles of Electron Tubes, Gewartowski and Watson, 1965, 655 pages
Principles of Electron Tubes, Herbert Reich, 1941, 398 pages
Radio Engineer's Handbook, F. Terman, 1943, 1,021 pages
Radiotron Designer's Handbook, F. Langford Smith, 4th edition, 1952, 1,482 pages

IN ENGLISH, cont.

The Sound of Silence: Lowest-Noise RIAA Phono-Amps: Designer's Guide, Burkhard Vobel, 2011
The Thermionic Vacuum Tube and Its Applications, H. J. Van Der Bijl, 1920, 391 pages
The Tube Preamp Cookbook, Allen Wright, 1997
The Ultimate Tone, Kevin O'Connor, 1995
Thermionic Valve Circuits, Emrys Williams, 1961, 427 pages
Tu-be or Not Tu-be: Modification Manual For Vacuum Tube Electronics, H. I. Eisenson, 1977, 198 pages
Theory and Applications of Electron Tubes, Herbert Reich, 2nd edition 1941, 716 pages
Theory of Thermionic Vacuum Tubes, E. Leon Chafee, 1933, 652 pages
Vacuum Tube and Semiconductor Electronics, Jacob Millman, 1958, 644 pages
Vacuum Tube Circuits, Lawrence Baker Arguimbau, 1948, 668 pages
Vacuum Tubes, Karl A. Spangenberg, 1948, 860 pages
Valve and Transistor Audio Amplifiers, John Linsley Hood, 1997, 250 pages
Valve Amplifiers, Morgan Jones, 4th edition, 2012, 700 pages

IN FRENCH

Les Tubes Electroniques et leurs Applications. Tome 1 : Principes Généraux, H. Barkhausen, 1949, 228 pages
Les tubes à vide et leurs applications, tome 2 : Les amplificateurs, H. Barkhousen, 1942, 301 pages
Physique et technique des tubes electroniques, R. Champeix, 1958, 214 pages
Théorie et pratique des circuits de l'électronique et des amplificateurs, Tome 1-2, J. Quinet, 1962
Traité moderne des amplificateurs haute fidélité à tubes, Lallie/Fiderspi, 2008, 331 pages

IN GERMAN

Audio-Röhrenverstärker von 0,3 - 10 W erfolgreich selbst bauen, Wilfried Frohn, 2005, 128 pages
Elektronische Speisegeräte, Karl Steimel, 1957, 246 pages
High-End mit Röhren, Gerhard Haas, 2007, 319 pages
Hören mit Röhren, Friedrich Hunold, 1999,
Niederfrequenzverstärker-Praktikum, Otto Diciol, 1959, 393 pages
Röhren NF Verstärker Praktikum, Otto Diciol, 2003, 393 pages
Röhrenprojekte von 6 bis 60 Volt, Burkhard Kainka, 2003, 153 pages
Röhrentechnik ganz modern, Winfried Knobloch, 1991, 103 pages
Röhrenverstärker selber bauen, Richard Zierl, 2011, 264 pages
Telefunken Laborbuch, Band I, 1970, 404 pages
Theorie der Spulen und Übertrager, R. Feldtkeller, 1957, 186 pages
Theorie und Praxis des Röhrenverstärkers, Peter Dieleman, 2007, 253 pages

IN ITALIAN

Amplificatori valvolari tecnica e pratica di autocostruzione, Ivano Incerti, 2012, 144 pages
I piccoli trasformatori : calcolo e costruzione ad uso degli elettricisti, Mario Pierazzuoli, 1949, 146 pages
I trasformatori tipo radio e simili, Dr. Ing. Enrico Baldoni, 1955, 182 pages
La construzione e il calcolo dei piccoli trasformatori, Ernesto Carbone, 1966, 80 pages
Manuale hi-fi a valvole, Luciano Macrì and Riccardo Gardini, 1994, 348 pages
Teoria e calcolo dei piccoli trasformatori, Giacomo Giuliani, 1948, 80 pages
Valvole e dintorni Hi-Fi, Atto E. Rinaldo, 2009, 228 pages
Valvole e trasformatori per Hi-Fi, Gieffe, 2004, 280 pages

IN SERBIAN

Elektronika, Dr Slavoljub Marjanovic, 1981, 580 pages
Hi-Fi Lampaška Pojacala, N.Vukušic, 2011, 726 pages

INDEX

A

A-B comparisons, 62
Acoustic feedback, 66
Active crossovers, 181-189
Active loads, 116-121
Active decoupling circuits, 174
Amplification factor, 83
Angular frequency, 22
Anode characteristics, 81
Anode choke, 154, 222, 235
Anode follower, 108, 165
Attenuation distortion, 57
Audio attenuators, 36-37
Audio Note Conquest amplifier, 236-238
Audio Note M2 phono stage, 209-210
Audio Note Ongaku amplifier, 246-247
Audio Research EC-4 active crossover, 187
Audio transformer, 157-162
Audio triodes as rectifiers, 195
Audion Golden Night amplifier, 238-240

B

Back-to-back transformer connection, 195
Balanced outputs, 199, 220
Balanced phono stage, 220
Ballast (current regulating) tubes, 34
Barkhousen's equation, 83
Battery bias, 91
Battery power, 217-219
Baxandall tone controls, 180-181
Beam tetrodes, 127
Biasing, 87, 91
Bipolar transistors, 174,
Black Gate capacitors, 42
Bleeder resistor, 173
Bode diagram, 138-141, 176, 207-208
Bootstrap principle, 104

C

Capacitor-less line stage, 197
Capacitors
- ceramic, 41
- dielectric strength, 38
- electrolytic, 42
- film capacitors, 39
- mica, 41
- parallel-connected, 28
- PIO capacitors, 40
- series-connected, 28
- trimmer, 41

Carbon composition resistors, 32
Carbon film resistors, 32
Cascaded stages, 150
- pulse response , 151-152

Cascode, 109-111, 211
Cascode differential amplifier (Hedge circuit), 115
Cathode
- activation, 80
- bias, 92, 254
- bypass capacitors, 144
- poisoning, 80

Cathode follower, 86, 102-105, 165, 185-186, 203, 212
quarter-bridge driver stage, 105-106, 232, 243, 245, 247, 249, 266
Cathode loaded SE amplifier, 268-270
Caz-tech SE845 amplifier, 252-253
Child-Langmuir equation, 72
Classes of operation: A, AB, B
Class A2 operation, 223-224
Class B operation, 225-226
Common cathode stage, 86
- efficiency, 94, 225
- graphical analysis, 89-90
- harmonic distortion, 96
- small signal linear model , 93
- voltage transfer characteristic, 95-96

Common grid stage, 86, 108-109, 212
Common Mode Rejection Ratio (CMRR), 114
Conrad Johnson Art preamplifier, 196-197
Constant current characteristics, 81, 83-84
Constant current source and sink, 116-119, 220
Contact bias, 214
Contacts, 48
Current loop analysis method, 20

D

Damping factor, 61-62, 142
control, 171
Decibel (dB), 51-52
Decoupling filters, 144, 172
Delay distortion, 57
Dielectric theory of sound, 64
Differential amplifier, 114-115, 119, 220
Directly-coupled stages, 97-98, 191, 204, 211, 238, 247, 248, 252, 257
Directly-heated triode (300B), 71, 79
Dissipation factor, 27
Distortion, 166,
Dummy load, 35
Dynamic load line, 222
Dynamic mutual conductance, 93
Dynamic transfer characteristic
- of a diode, 75
- of a triode, 89

E

E.A.R. 834P phono stage, 213
Effective Series Resistance (ESR), 27, 43
Electrolytic capacitors
- HF behavior, 43
- life, 43

Electron-ray tubes, 192
ELNA Cerafine capacitors, 42
Equivalent noise resistance of a tube, 206

F

Feedback factor, 164
Feedback ratio, 164
Filamentary cathodes, 71
Filters
- Butterworth, 183
- Chebyshev, 183
- First-order RC type, 139-140
- Multiple Feedback (MFB)
- phase lead and lag networks, 141
- Sallen-Key, 183
- second order low pass, 141-142

Frequency domain, 23-24, 58
Frequency range, 54,
- power bandwidth, 55-56

Frequency response
- of an amplifier, 54-56, 149-150, 169-170
- of a capacitor, 43
- of a transformer, 160

Fuses, 37

G

Gain-bandwith product, 166
Gain margin, 175
Getter, 79-80
Gold plating in audio, 48
Grid chokes, 162, 240, 254
Grid stopper resistors, 148

H

Harmonic distortion, 56, 229
Harmonics, 53
Headphone amplifiers, 189-191
Heathkit XO-1 active crossover, 189
HF compensation, 147-148
Hookup wire, 45
Hybrid rectifiers, 195, 200, 264

I

Ideal amplifier
- pentode, 51
- transistor, 50
- triode, 50

Ideal capacitor, 25
Ideal inductor, 25
Ideal resistor, 24
Impedance (inductance) coupling, 154-162
Inductors in series and in parallel, 29
Input impedance/resistance, 167
Interconnect cables, 46
Intermodulation Distortion (IMD), 56-57, 169, 182
Internal resistance, 83
Interstage transformers, 156-162

INDEX

J

Junction Field Effect Transistor (JFET), 118-120

K

Kinkless tetrodes (KT), 78
Kirchoff's Laws, 19-20

L

Listening room
  optimal dimensions, 65
  speaker positioning, 65-66
Loftin-White amplifier, 989
Logarithmic scale, 51-52
Loudness control, 62
Loudspeakers,
  efficiency (sensitivity), 60-61
  interface with amplifiers, 64

M

Marantz Model # active crossover, 188
Microphony, 66
Miller effect, 109, 146
μ–follower, 135-136
Motorboating, 173
Motor-start capacitors, 43-44
Moving coil cartridges, 26, 109
  interfacing, 219
  S/N ratio, 219-220
Moving magnet cartridges, 26
Mutual conductance, 82

N

Negative feedback, 211
  anode-to-grid, 107-108
  un-bypassed cathode resistor, 100
Neutralization, 146-147
Node voltage analysis, 20-21
Norton's equivalent circuit, 29-30
Nyquist stability criterion, 174-175

O

Ohm's Law, 18
Optimal anode load
  for triodes, 224-225
  for pentodes, 272
Oscilloscope probes, 176-177
Output impedance, 61, 101, 167

P

Paralleled tubes, 194, 235, 238, 252, 256, 267,
Parasitic capacitance, 44
Partial cathode coupling, 239, 280-281
Pentodes, 125
  as a CCS, 117
  choosing the load resistance, 131

Partial cathode coupling, 239, 280-281
Pentodes, 125

  distortion, 128
  driver stage, 133, 268
  μ–follower, 136
  output stages, 272-277
  providing screen voltage, 129, 132
  screen amplification factor, 126
  small signal model, 126
  the effect of screen impedance, 132
Perveance, 81
Phase correction networks, 148, 177
Phase distortion, 168
Phase lag network, 141, 207
Phase lead network, 141
Phase margin, 175
Positive feedback, 170, 211
Potentiometers, 34-35
  tracking error, 36
Power in AC circuits, 23
Power cables, 47
Power Supply Rejection Ratio (PSRR), 122
Promitheus Audio line stage, 200-201

Q

Q-factor of a 2-nd order system, 142

R

RC-coupled stages, 145, 148-149
Reactive loads, 230, 275
Rectification of the signal and operating point shift, 224, 273
Regulated DC heater supply, 209
Resistors in parallel, 20
RIAA equalization, 170, 207-208
Richardson-Dushman equation, 70
Root-locus method, 177-178
Rogers RD Senior amplifier, 115
Rules-of-thumb, 12

S

Scaling factors for power tubes, 273
Screen grid, 124
Screen driven pentodes, 279-289
Secondary emission, 70, 124
Series resonant circuit, 26
Shindo Cortese amplifier, 112
Silicon diodes, 44
Simplicity Rule, 48, 114
Solid state amplifiers, 59
Space charge, 71
Square wave spectrum, 53
SRPP, 111-112, 180, 190, 210, 232, 240, 243, 245, 247, 252, 264, 266
Stacked HV power supply, 251
Superposition theorem, 21
Suppressor grid, 125, 278-279
Switches, 38

T

Tango NC-20 interstage transformer, 158
Tetrodes, 124
Thermionic emission, 70
Thermistors (NTC and PTC), 33
Thevenin's equivalent circuit, 29
Time constant, 139
Time domain, 23, 58
Tone controls, 180-181
Total Harmonic Distortion (THD), 64
Transconductance, see: Mutual conductance
Transfer characteristics, 81
Transfer function, 138
  zeroes and poles, 138
Transient responses, 142-143
Transformer-coupled stages, 156-162
Transformer volume control (TVC), 200
Transistor, 86, 120
Triad HS-35 interstage transformer, 159
Triode coefficients (parameters), 80-84
Triode-connected pentodes, 134, 190-191, 202, 215, 239, 248, 260-261, 265-266, 268
Tube naming conventions, 78

U

Ultralinear connection, 278-280

V

vacuum diode, 72-77
  anode dissipation, 77
  as a rectifier, 74
  static resistance, 72
  dynamic resistance, 75
  dynamic transfer characteristics, 75
  heater efficiency, 77
Varistors, 45
Vector algebra, 21-22
Voltage doubler, 233-234, 250, 267,
Voltage regulator, 120-121, 251, 253
VTL Ultimate+SDl phono stage, 212
VU-meter driver with tubes, 192

W

White cathode follower, 113-114
Wirewound resistors, 32-33

Z

Zener diode, 44-45, 91, 129
Zobel network, 219

NOTES

NOTES

www.ingramcontent.com/pod-product-compliance
Lightning Source LLC
Chambersburg PA
CBHW082305210326
41598CB00028B/4444

* 9 7 8 0 9 8 0 6 2 2 3 2 4 *